CONCEPTUAL
Physical Science
Explorations

SECOND EDITION

Paul G. Hewitt
City College of San Francisco

John A. Suchocki
Saint Michael's College

Leslie A. Hewitt
San Carlos, California

Addison-Wesley
San Francisco Boston New York
Cape Town Hong Kong London Madrid Mexico City
Montreal Munich Paris Singapore Sydney Tokyo Toronto

Publisher: Jim Smith
Director of Development: Michael Gillespie
Editorial Manager: Laura Kenney
Sr. Project Editor: Chandrika Madhavan
Media Producer: Kate Brayton
Director of Marketing: Christy Lawrence
Executive Marketing Manager: Scott Dustan
Managing Editor: Corinne Benson
Sr. Production Supervisor: Nancy Tabor
Production Service: Progressive Publishing Alternatives
Project Manager: Sylvia Rebert
Illustrations: Paul Hewitt and Dartmouth Publishing, Inc.
Text Design: Emily Friel, Elm Street Publishing Services
Cover Design: John Suchocki and Susan Riley
Manufacturing Buyer: Jeffrey Sargent
Photo Research: David Chavez
Manager, Rights and Permissions: Zina Arabia
Image Permission Coordinator: Elaine Soares
Cover Printer: Phoenix Color Corporation
Text Printer and Binder: Courier, Kendalville
Cover Image: Mono Lake: Star Circle Reflection. Photo by Ian Parker.
Photo Credits: See pages 869–871.

Library of Congress Cataloging-in-Publication Data

Hewitt, Paul G.
   Conceptual physical science—explorations / Paul G. Hewitt, John A. Suchocki, Leslie A.
Hewitt. — 2nd ed.
     p. cm.
   Includes index.
   ISBN-13: 978-0-13-135933-8
   ISBN-10: 0-13-135933-9
   1. Physical sciences.   I. Suchocki, John.   II. Hewitt, Leslie A.   III. Title.
   Q161.2.H482 2010
    `500. 2—dc22   2008052530

ISBN-13: 978-0-321-56791-8 (student copy)
ISBN-13: 978-0-321-60216-9 (professional copy)

Addison-Wesley
is an imprint of

To teachers who inspire a love of learning and who help students add science to their way of thinking.

ACKNOWLEDGMENTS

For physics contributions we are grateful to Dean Baird, Howie Brand, Ernie Brown, George Curtis, Marshall Ellenstein, John Hubisz, Dan Johnson, Juliet Layugan, Tenny Lim, Iain McInnes, Fred Meyers, Diane Reindeau, and Kenn Sherey. We are also grateful to the feedback of ninth-grade integrated science teachers Barbara Toschi and Scott Savoi, of Bear River High School in Grass Valley, California. A special thank you goes to Lillian Lee Hewitt for quite wonderful contributions. For life-long inspiration I am grateful to Jacque Fresco, Burl Grey, and Dan Johnson.

For valued chemistry feedback we thank high school chemistry teachers Scott Pennington of Essex High School, Vermont, and Jon Kliegman of HARP Academy, New Jersey. For age-appropriate readability advice we are grateful to guidance counselor Wallace Dietz of James River High School, Virginia. Special thanks to John's wife, Tracy Suchocki, and children, Ian, Evan, and Maitreya. The insightful suggestions of Ian and Evan, two beginning high school students themselves, were particularly helpful. Thanks also to brother-in-law Peter Elias, for helping to review and prepare manuscript. Special thanks are also due to the chemistry author's very inspirational high school science teachers, Linda Ford (chemistry) and Edward Soldo (biology) of Sycamore High School, Ohio. Their positive impact on JS has been life-long.

For Earth science feedback we thank Mary Brown, Ann Bykerk-Kauffman, Oswaldo Garcia, Newell Garfield, Karen Grove, Trayle Kulshan, Jan Null, Katryn Weiss, Lisa White, and Mike Young. A special thank you to Leslie's husband, Bob Abrams, for his assistance in the presentation of much of the Earth science material. Thanks also goes to Leslie's children, Megan and Emily, for their inspiration, their curiosity, and their patience.

For astronomy, the authors are grateful for permission to use many of the graphics that appear in the textbook, *The Cosmic Perspective*, 4th edition. Much appreciation to its authors, Jeffrey Bennett, Megan Donahue, Nicholas Schneider, and Mark Volt. We also thank Wallace Dietz of James River High School, Virginia, for his editing efforts on our newly crafted astronomy chapters. For work on earlier versions, we remain grateful to Richard Crowe, Bjorn Davidson, Stacy McGaugh, Michelle Mizuno-Wiedner, Neil deGrasse Tyson, Joe Wesney, Lynda Williams, and Erick Zackrisson.

ABOUT THE AUTHORS

Paul G. Hewitt

Former silver-medal boxing champion, sign painter, uranium prospector, and soldier, Paul began college at the age of 27, with the help of the GI Bill. He pioneered the conceptual approach to teaching physics at City College of San Francisco. He has taught as a guest teacher at various middle schools and high schools, the University of California at both the Berkeley and Santa Cruz campuses, and the University of Hawaii at both the Manoa and Hilo campuses. He also taught for 20 years at the Exploratorium in San Francisco, which honored him with its Outstanding Educator Award in 2000.

John A. Suchocki

John is author of *Conceptual Chemistry* as well as a co-author (with Paul and Leslie Hewitt) of *Conceptual Physical Science.* John obtained his Ph.D. in organic chemistry from Virginia Commonwealth University. He taught chemistry at the University of Hawaii at Manoa and then at Leeward Community College. In addition to authoring textbooks, John is currently an adjunct faculty member at Saint Michael's College in Colchester, Vermont. He also produces science education multimedia through his company, Conceptual Productions (www.CPro.cc), writes and illustrates science-oriented children's books (www.Styraki.com), and produces music through his recording label (www.CProMusic.com).

Leslie A. Hewitt

Leslie is also coauthor of *Conceptual Physical Science* (with Paul and John). After obtaining her geology degree at San Francisco State University, Leslie's interest in teaching broadened to include educating elementary and middle school students. She completed additional graduate work in geography and education, receiving her California State Teaching Certification, also from San Francisco State University. In addition to writing, she devotes considerable time and energy to bringing science education to young people in engaging ways. She is particularly active in her local school district where she works on curriculum development and hands-on science, organizes interactive science fairs, guest teaches, and works with students on the connections between scientific concepts and life in the everyday world.

CONTENTS IN BRIEF

CONTENTS

1 ABOUT SCIENCE   1

PART ONE: MECHANICS   15

2 NEWTON'S FIRST LAW OF MOTION—THE LAW OF INERTIA   17

3 NEWTON'S SECOND LAW OF MOTION—FORCE AND ACCELERATION   37

4 NEWTON'S THIRD LAW OF MOTION—ACTION AND REACTION   57

10 ELECTRICITY   192

11 MAGNETISM   216

12 WAVES AND SOUND   235

13 LIGHT, REFLECTION, AND COLOR   260

19 HOW CHEMICALS MIX 412

20 HOW CHEMICALS REACT 437

21 TWO TYPES OF CHEMICAL REACTIONS 460

22 ORGANIC COMPOUNDS 490

23 THE NUTRIENTS OF LIFE 521

TO THE STUDENT

Welcome to *Conceptual Physical Science—Explorations*. We're convinced that the CONCEPTUAL approach is the best way to learn science, especially the physical sciences. That means that we emphasize concepts *before* computation. Although much of physical science is mathematical, a firm grasp of its concepts is crucial to understanding. Hence our motto: concepts before computation. Comprehension of concepts underlies success in computation.

This course provides plenty of resources beyond the text as well. The interactive figures, interactive tutorials, and demonstrations videos on The Companion Website (www.ConceptualSciencePlace.com) will help you visualize physical science concepts. This is especially useful when you need to understand processes that vary over time such as the velocity of an object in free fall, the phases of the Moon, or the formation of chemical bonds. The *Practice Book* provides intriguing exercises that help you to "tie it all together." The activities in the *Laboratory Manual* will build your gut-level feeling for concepts and analytical skills. All these and other ancillaries will increase your confidence and mastery of science.

As with all things, what you get out of this class depends on what you put into it. So study hard and ask all the questions you want. Most of all, enjoy your scientific tour of the amazing natural world!

TO THE INSTRUCTOR

Conceptual Physical Science—Explorations melds physics, chemistry, Earth science, and astronomy, in a manner that captivates student interest. We have taken care to match the reading level with that of the average high school student. We have done this without watering down the content, so what your students learn here is serious science. Only the most central concepts are treated. (We're mindful of courses that dwell on difficult topics that, once understood, are hardly worth the effort. Not here.) This is solid stuff, in a very readable and student-friendly format. More than enough material is included for a one-year course, which allows for a variety of course designs to fit your taste.

    Conceptual Physical Science—Explorations introduces physical science to your students using a **conceptual approach**, which is an approach that

- emphasizes central ideas over peripheral detail.
- deemphasizes technical jargon and rote memorization.
- puts concepts ahead of computation.
- relates science to everyday life.
- is personal and direct.

The conceptual approach was defined over 30 years ago by Paul Hewitt in his groundbreaking and still very popular textbook *Conceptual Physics*, which is translated and used worldwide. The conceptual approach is the backbone of this book.

    To say a science course is conceptual is not to say it is nonmathematical. For example, the mathematical foundation of physics, in particular, is quite evident in the many equations throughout this book. The equations are guides to thinking. They show the connections between concepts, rather than only being used as recipes for plugging and chugging. Our emphasis is on qualitative analysis, helping students to get a gut feel for the science they are studying. That's why the qualitative questions within this book greatly outnumber quantitative math-based problems. The challenge to the student is in understanding concepts. Students appreciate and differentiate among major scientific ideas, rather than reduce these ideas to algebraic problem solving. That can be done in another course—not this one. This course is too valuable for that!

Key Features of this Second Edition

Flip through the pages of this book or review its table of contents and you will quickly see that this second edition of CPSE is a major upgrade from the first edition, which we produced "way back" in 2001. In addition to a smoother flow of topics and updates to scientific advances (especially in astronomy) you will find numerous new and exciting pedagogical features designed to help your students learn most efficiently.

    Each chapter begins with a photo that relates to an attention-getting paragraph. The introductory paragraphs feature interest-piquing questions that relate to the chapter and provide a brief overview of chapter content. At the bottom of the opening page to each chapter is a boxed activity, *Explore!* Most of these are quickie-type activities to promote class interest in the chapter content.

Other helpful elements in this book include *Check Your Thinking* questions. These short sets of questions, sprinkled throughout each chapter, allow students to monitor their understanding. Answers are provided below the questions to give students immediate feedback. Also featured are *Reading Check* questions, which appear within every section throughout the book. Answers are conspicuously yellow highlighted key sentences in each section. Students will love the *Insights,* short margin features of quotes from our mascots, Perky the mouse, and Sneezlee the bird. Each of these mascots provides background information, interesting facts, and study tips. The *Media Icons* alert students to relevant media resources they can turn to—interactive figures, tutorials, and videos—for alternative topic presentations.

At the end of each chapter is an extensive *Chapter Review.* Much attention has been taken in the preparation of this material. It is the cornerstone to student learning and, without doubt, the most improved feature of this second edition. It consists of a listing of *Key Terms* (with definitions), *Review Questions, Think and Do, Think and Compare, Think and Explain, Think and Solve,* and lastly, a *Readiness Assurance Test (RAT),* which is a multiple-choice practice exam.

The *Key Terms* provide a glossary of the key terms that are boldfaced in the chapter. This is followed by the easy-to-answer *Review Questions,* cited by section, which frame the important ideas of each section in the chapter. They are meant solely for review, not for challenge to student intellect, for all answers are easily looked up in the chapter.

The *Think and Do* is a set of easy-to-perform, hands-on activities design to help students experience the physical science concepts for themselves. The *Think and Compare* questions ask students to analyze trends. For example, students might need to rank the momenta and kinetic energies of three vehicles with different masses and speeds. Or, they may be asked to rank atoms in order of increasing size, based upon their understanding of the periodic table.

The *Think and Explain* questions are, by a notch or two, the more challenging questions at the end of each chapter. Many require critical thinking while others are designed to prompt the application of science to everyday situations. All students wanting to perform well on exams should be directed to the *Think and Explain* questions because these are the questions that directly assess student understanding. Accordingly, many of the *Think and Explain* questions have been adapted to a multiple-choice format and incorporated into the *CPSE* test bank. We are hopeful that this will allow you to reward those students who put time and effort into the *Think and Explain* questions.

A few chapters feature *Think and Solve,* where simple mathematics blends with concepts. Lastly, there appears the *Readiness Assurance Test (RAT),* a set of ten multiple-choice questions that the student can use as a practice exam or for an easy-to-grade homework assignment. Like the *Think and Explain* questions, you will also find very similar *RAT* questions within the *CPSE* test bank.

　　Appendix A details systems of measurement and unit conversion. *Appendix B* provides extended coverage of linear and rotational motion. *Appendix C* provides practice with vector components while *Appendix D* develops the concepts of exponential growth and decay. Finally, *Appendix E* is the *Safety Appendix*, which includes safety information pertaining to the *Explore!* activities.

Organization of the Text

This second edition is organized into five main parts based on increasing complexity and scale from physics to astronomy.

　　Part One, Mechanics, which begins with Newton's laws, is followed by a study of momentum and energy with an emphasis on conservation principles. Newton's law of gravity extends to projectile and satellite motion. Part One concludes with fluid mechanics.

　　Part Two, Forms of Energy, covers heat, electricity, magnetism, wave phenomena (both sound and light), the atom, and nuclear energy. Much of this meaty material is tasty physics. It also provides a foundation for the chapters that follow.

　　Part Three, Chemistry, builds upon the foundation of concepts developed in Part Two. In the vein of all books in the *Conceptual* series, Part Three emphasizes concepts over computation, and science in everyday life. Relating chemistry to students' familiar world—the fluorine in their toothpaste, the Teflon™ on their frying pans, and the flavors produced by various organic molecules—keeps chemistry fun and relevant. The treatment of chemistry also connects to social and environmental issues.

　　Part Four, Earth Science, encompasses the sciences of geology, hydrology, oceanography, and meteorology—with emphasis of Earth as an interconnected system. The first focus is geology, the *geosphere*, which includes the topics of rocks and minerals, plate tectonics, earthquakes, volcanoes, and the processes of erosion and deposition and their influence on landforms. The next focus is on hydrology and oceanography, the *hydrosphere*, which includes Earth's freshwater (rivers and streams, glaciers, and groundwater), and saline water (Earth's oceans). The study of Earth ends with a focus on the interrelationship of the *hydrosphere* and the *atmosphere*—meteorology, the study of weather and climate.

　　In Part Five, Astronomy, students employ important ideas learned in physics, chemistry, and the Earth sciences. It begins with a study of the solar system and some details of its origin according to the nebular theory. Factual information on the planets and other heavenly bodies is tied to this central concept. The second astronomy chapter focuses on the life cycles of stars. The third and final astronomy chapter focuses on galaxies and the cosmos. This capstone chapter provides an in-depth look at concepts such as the Big Bang, Hubble's law, and the latest discoveries of such things as dark energy and dark matter.

Conceptual Physical Science—Explorations is a wide-ranging course that offers student tools to help them understand our physical universe and the many scientific issues they will confront as citizens of the 21st century. May this be one of the most interesting, amazing, and worthwhile courses your students will ever take!

Supplements

Conceptual Physical Science—Explorations contains more than enough material for a one-year physical science course. There is also a full suite of ancillary materials for students and teachers in a variety of media developed for this course. Because there are so many resources and because most of the chapters in the text are self-contained, you have great flexibility in tailoring the *Conceptual Physical Science—Explorations* program to suit your taste and the needs of your students.

Supplements for the Instructor

The following supplements are available to qualified adopters:

The **Teacher's Guide to Text and Laboratory Manual** (0-321-60217-X) is very different from most teaching guides. Every section of every chapter has discussion pointers for conceptual teaching. There are suggested demonstrations, suggested Check-Your-Neighbor questions [with answers in brackets], and topics different from but related to the topics in the textbook. It has answers to all the *Review Questions*, solutions to all the *Exercises*, and step-by-step solutions to the *Problems*.

The **Minds-On, Hands-On Activity Book** (0-321-61557-3) goes beyond the Explore! activities in the textbook and provides teachers a variety of pedagogical approaches for shaping science skills in diverse settings. Teaching strategies include groupwork (cooperative learning), concept mapping, student-designed investigations, quick hands-on activities, skill building, activity-based assessment, and oral presentations.

Also available is the **Transparency Acetates and Teaching Guide** (0-321-60215-3) containing 100 transparency acetates as well as a guide book with discussion questions and answers for each transparency.

The **Instructor Resource DVD** (0-321-59765-6) contains a rich set of resources to use in classroom presentations and to assess student understanding of material, including:

- Art and tables from text in jpeg format.
- An **ExamView test bank** of more than 2,000 questions, written by the authors of the textbook, in multiple-choice, fill-in the blank formats, that allows you to edit questions, add questions, and create multiple test versions.

- **Next-Time Questions**, a collection of illustrated "puzzlers" designed to provoke lively discussion in class. Post them at the beginning of class to focus students' attention or at the end of class to give students a puzzle to mull over until "next time."
- **Assessment Masters** include true/false, multiple-choice, and short answer questions for every chapter in the textbook.
- **Answer keys** are provided for all supplements including the Lab Manual activities, Next-Time Questions, Assessment Masters, and the Student Practice Book.

Last, but not least, is the **Student Companion Website (www. physicsplace.com)**, which includes a rich collection of media is available to enhance classroom learning. Students can explore a variety of assets, including self-paced interactive figures and tutorials, a library of videos created by the authors and other well-known sources, chapter-specific self-study quizzes, an interactive glossary and flashcard deck, an interactive periodic table, and the eBook! Instructors can also track students' completion of select tutorials and all quizzes using the gradebook feature. Access to the Website is provided with every new book.

Supplements for the Student

The **Student Companion Website** contains a wealth of media resources to aid study and comprehension, including a complete eBook, self-paced interactive figures and tutorials, a library of videos created by the authors and other well-known sources, chapter-specific self-study quizzes, and much more!

The following supplements are available for purchase:

The Laboratory Manual for Conceptual Physical Science— Explorations (0-321-60274-9) is written by the authors and high school science teacher Dean Baird. The book includes a balance of in-depth experiments that allow students to develop laboratory skills and quick activities that use readily available materials.

The **Practice Book** (0-321-60218-8) is filled with computational exercises, misconception-busting questions, analogies, intriguing puzzlers, and straightforward practice questions and problems that help students "tie it all together." Humorous and insightful cartoons by Paul Hewitt appear on every page.

Go to it! Your conceptual physical science course really can be the most interesting, informative, and worthwhile science course available to your students.

ABOUT SCIENCE

THE MAIN IDEA

Science is the study of nature's rules.

Science has given us much. Our modern world is built on it. Nearly all forms of technology—from medicine to space travel—are applications of science. But what exactly is this amazing thing called *science*? How should science be applied? Where did science come from? And what would the world be like without it?

Science is an organized body of knowledge about nature. It is the product of observations, common sense, rational thinking, and (sometimes) brilliant insights. Science has developed from group efforts as well as individuals' discoveries. It has been built up over thousands of years and gathered from places all around the Earth. It is a huge gift to us today from the thinkers and experimenters of the past.

Yet, science is not just a body of knowledge. It is also a method, a way of exploring nature and discovering the order within it. Importantly, science is also a tool for solving problems.

Science began back before recorded history, when people first discovered repeating patterns in nature such as star patterns in the night sky, weather patterns, and animal migration patterns. From these patterns, people learned to make predictions that gave them some control over their surroundings.

Explore!

Can You Change One Thing Without Changing Another?

1. Make folds in a sheet of $8\frac{1}{2}$ by 11 inch paper to make an airplane of any design.
2. Toss it and see how it flies.
3. Change it slightly by folding one of the wings a bit.
4. Toss it again.

Analyze and Conclude

1. **Observing** Note the differences in flight before you changed the wing, and after.

2. **Predicting** If you made a greater fold in the wing, would it fly differently?
3. **Making Generalizations** Can you change the design without a corresponding change in how the airplane flies? Consider the statement: "You can never change only one thing." What does it mean?

1.1 A Brief History of Advances in Science

When a sore throat keeps you home from school, you probably ask yourself, "How did I catch this cold?" You then speculate about how you could have been exposed to germs, and how you might avoid exposure next time. When a light goes out in your room, you ask, "How did that happen?" You might check to see if the lamp is plugged in, check the bulb, or even look at your neighbors' houses to see if there has been a power outage. When you think like this, you are searching for *cause-and-effect* relationships—trying to find out what events cause what results. This type of thinking is *rational thinking*. Rational thinking is basic to science.

Today we use rational thinking so much that it's hard to imagine other ways of interpreting our experiences. But it wasn't always this way. At times, people have relied more on superstition and magic to interpret the world around them—or have simply failed to ask, "Why?"

Rational thought became very popular in Greece in the 3rd and 4th centuries BC. From there it spread throughout Rome and other parts of the Mediterranean world. When the Roman Empire fell in the 5th century AD, advancements in science came to a halt in Europe. Rational thought was replaced by religion and the Dark Ages. But during this time, science continued to advance in other parts of the world. The Chinese and Polynesians were charting the stars and the planets. Arab nations developed mathematics and learned to make glass, paper, metals, and certain chemicals. Finally, the Greek philosophy of rational thinking was brought back into Europe by Islamic people who entered Spain during the 10th to the 12th centuries. Then universities emerged. When the printing press was invented in the 15th century, science made a great leap forward. This invention did much to advance scientific thought (just as computers and the Internet are doing today).

Into the 16th century, most people thought the Earth was the center of the universe. They thought the Sun circled the stationary Earth. This thinking was challenged when the Polish astronomer Nicolaus Copernicus quietly published a book proposing that the Sun is stationary and that the Earth revolves around it. These ideas conflicted with Church teachings and were banned for 200 years.

Modern science began in the 16th century when the Italian physicist Galileo Galilei revived the Copernican view. ✓ Galileo used experiments, rather than speculation, to study nature's behavior (more about Galileo in following chapters). Galileo was arrested for popularizing the Copernican theory and for his other contributions to scientific thought. Yet a century later, his ideas and those of Copernicus were accepted by most thinking people.

Scientific discoveries are often opposed, especially if they conflict with what people want to believe. Every age has its intellectual rebels

Science is a way of knowing about the world and making sense of it.

who are persecuted, condemned, or suppressed at that time. But later they seem harmless and are often essential to the elevation of human conditions. "At every crossway on the road that leads to the future, each progressive spirit is opposed by a thousand men appointed to guard the past."*

✓ READING CHECK

How did Galileo study nature's behavior?

1.2 Mathematics and Conceptual Physical Science

Pure mathematics is different from science. Math is a study of relation-ships among numbers. When math is used as a tool of science, the results are impressive. Measurements and calculations are essential parts of the powerful science we practice today. For example, it would not be possible to send missions beyond Earth if we couldn't measure the positions of spacecraft or calculate their paths through space.

Interactive Tutorial
Metric System

You will use some math in this course, especially when you make measurements in lab, but in this book we don't make a big deal about math. Our focus is on understanding concepts in everyday language. ✓ We use equations as guides to thinking rather than as recipes for "plug-and-chug" math work. We believe that focusing on math too early, especially math-based problem solving, is a poor substitute for learning the concepts. That's why the emphasis in this book is on building concepts. Only then does solving problems make sense.

You'll see many more conceptual exercises than problems at the ends of the chapters that follow. *Conceptual Physical Science—Explorations* puts comprehension comfortably before calculation.

✓ READING CHECK

What is the main role of equations in this book?

1.3 Scientific Methods—Classic Tools

In the 16th century, the Italian physicist Galileo and the English philosopher Francis Bacon developed a formal method for doing science—the **scientific method.** Although their method is not the only one, ✓ all scientific methods are based on rational thinking and experimentation. Some or all of the following steps are likely to be found in the way most scientists perform their work:

1. **Observe** Closely observe the physical world around you. Recognize a question or a puzzle—such as an unexplained observation.
2. **Question** Make an educated guess—a **hypothesis**—that might resolve the puzzle.
3. **Predict** Predict consequences of the hypothesis.

---

* From Count Maurice Maeterlinck's "Our Social Duty."

4. **Test predictions** Do experiments to see if the consequences you predicted are present.
5. **Draw a conclusion** Formulate the simplest general rule that organizes the three main ingredients: hypothesis, predicted effects, and experimental findings.

A scientific hypothesis is an educated guess. When a hypothesis has been tested over and over again and has not been contradicted, it may become known as a **law** or *principle.* A scientific **fact,** on the other hand, is something that competent observers can observe and agree to be true. For example, it is a fact that amputated limbs of certain crabs can grow back. Anyone can watch it happen. It is not a fact—yet—that a severed limb of a human can grow back.

Scientists use the word *theory* in a way that differs from everyday speech. In everyday speech a theory is the same as a hypothesis—a statement that hasn't been tested. But in science, a **theory** is a vast combination (synthesis) of facts and well-tested hypotheses. Physicists, as we will learn, speak of the quark theory of the atomic nucleus. Chemists have the theory of metallic bonding. Geologists use the theory of plate tectonics, and astronomers speak of the theory of the Big Bang.

Theories are a foundation of science. They are not fixed, but evolve. They pass through stages of refinement. For example, since the theory of the atom was proposed 200 years ago, it has been refined many times in light of new evidence. Some people argue that scientific theories can't be taken seriously because they change. Those who understand science, however, see that theories grow stronger as they evolve to include new information.

Facts are revisable data about the world.

Theories interpret facts.

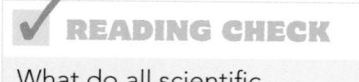

✔ **READING CHECK**

What do all scientific methods have in common?

1.4 Scientific Hypotheses Must Be Testable

In order for a hypothesis to be scientific, it must be testable. A test or series of tests can determine whether a hypothesis is valid. In scientific work, most hypotheses turn out to be wrong. Scientists have to be patient, though, and keep testing their ideas for accuracy.

A well-known scientific hypothesis that proved to be incorrect was that of the greatly respected Greek philosopher Aristotle (384–322 BC). Aristotle claimed that heavy objects naturally fall faster than light objects. This hypothesis was considered true for nearly 2000 years—mainly because of respect for this great man. Also, air resistance was not recognized as an influence on how quickly objects fall. We've all seen that

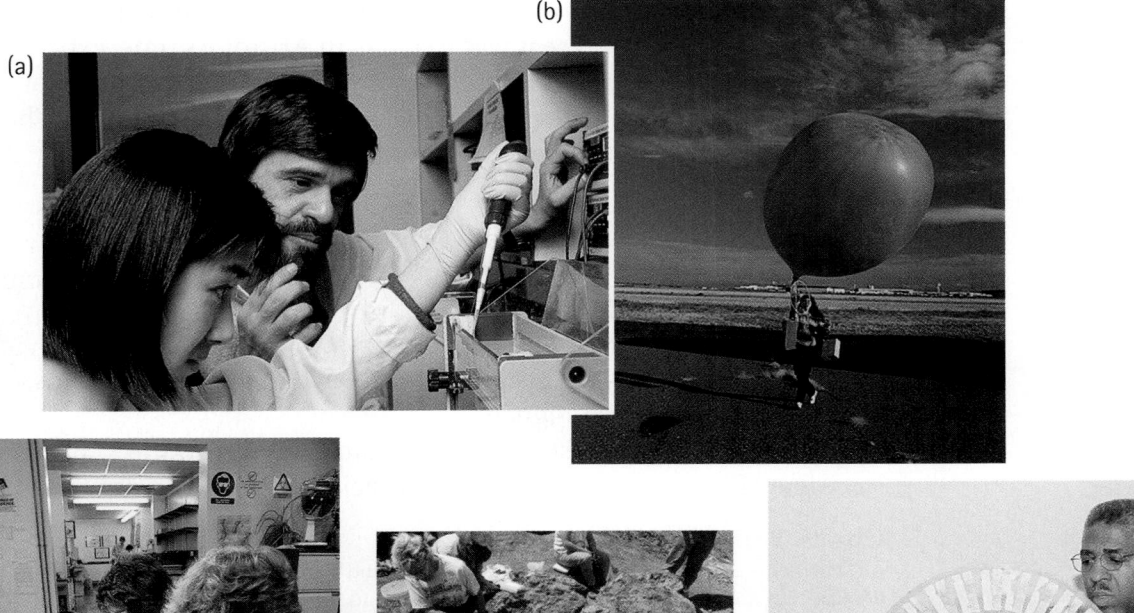

FIGURE 1.1 ▲
Physical science covers many disciplines. (a) Biochemists preparing samples.
(b) Meteorologist releasing a weather balloon to study the composition
of the upper atmosphere. (c) Technicians conducting DNA research.
(d) Paleontologist preparing fossilized dinosaur bones for carbon dating.
(e) Physicist simulating lightning strokes.

stones fall faster than leaves fluttering in the air. Without investigating
further, it is easy to accept false ideas.

Galileo very carefully examined Aristotle's hypothesis. Then he did
something that caught on and changed science forever. He *experimented*.
Galileo showed the falseness of Aristotle's claim with a single experiment—dropping heavy and light objects from the Leaning Tower of
Pisa. Legend tells us that they fell at equal speeds. In the scientific spirit,
one experiment that can be reproduced outweighs any authority, regardless of reputation or the number of advocates. Albert Einstein put it well
when he stated, "No number of experiments can prove me right; a single
experiment can prove me wrong." ✓ In science, the test of knowledge
is experiment.

Although Aristotle was wrong about falling objects, his hypothesis was scientific. That's because it was testable. In Aristotle's time, experimentation hadn't caught on. Even the claim that the Moon is made of green cheese is a scientific hypothesis, because the claim is testable. The hypothesis was tested and proved wrong with the lunar landings that began in 1969.

CONCEPT CHECK

Which statements are *scientific* hypotheses?
a. Better stock market decisions are made when the planets Venus, Earth, and Mars are aligned.
b. Atoms are the smallest particles of matter that exist.
c. Albert Einstein was the greatest physicist of the 20th century.

Check Your Answer

All statements are hypotheses, but only statements *a* and *b* are scientific hypotheses—because they are testable. Statement *a* can be tested (and proven wrong) by researching the performance of the stock market during times when these planets were aligned. Statement *b* not only can be tested but has been tested. Although it is untrue (many particles smaller than atoms have been discovered), the statement is nevertheless a scientific one. Lastly, statement *c* is an assertion or opinion that has no test. What possible test, beyond collective opinion, could prove Einstein was the greatest physicist? How could we know? Greatness is a quality that cannot be measured in an objective way.

Because the name Einstein is held in high esteem, it is a favorite of quacks (see the box on page 7). Take notice when the name of Einstein, Jesus, and other highly respected sources are cited, often by quacks who wish to bring respect to themselves and their points of view. In all fields we should be doubtful (skeptical) of people who wish to credit themselves by calling upon the authority of others.

The scientific method is effective in advancing knowledge. But many scientific advances involve trial and error, experimenting without guessing, or just plain accidental discovery. The trained eye, however, is better at noticing questions in the first place and making sense of evidence. More than a particular method, the success of science has to do with an attitude common to scientists. This attitude involves inquiry, experimentation, and being humble before the facts. A good scientist can admit to being wrong when wrong.

READING CHECK

What is the test of knowledge in science?

In your education it's not enough to be aware that other people may try to fool you. More important is being aware of your own tendency to fool yourself.

1.5 A Scientific Attitude Underlies Good Science

Scientists must accept their experimental findings even when they wish they were different. They must strive to distinguish between the results they see and those they wish to see. This is not easy. Scientists, like most people, can fool themselves. People have always tended to adopt general rules, beliefs, creeds, and ideas without thoroughly questioning their validity. And sometimes we retain these ideas long after they have been shown to be meaningless, false, or at least questionable. The most widespread assumptions are often the least questioned. Too often, when an idea is adopted great attention is given to the instances that support it. Contrary evidence is often distorted, belittled, or ignored.

None of us has the time or resources to test every idea. So most of the time we take somebody's word. How do we know whose word to accept? To reduce the likelihood of error, scientists listen to people whose findings are testable—if not in practice, then at least in principle. ✓ Ideas that cannot be tested are regarded as "unscientific."

The fact that scientific statements will be thoroughly tested helps keep science honest. Sooner or later, mistakes (or deception) are found out. A scientist exposed for cheating doesn't get a second chance in the community of scientists. Honesty, so important to the progress of science, thus becomes a matter of self-interest. There is relatively little bluffing in a game where all bets are called.

Experiment, not philosophical discussion, decides what is correct in science.

✓ **READING CHECK**

What kind of ideas are unscientific?

PSEUDOSCIENCE

Some belief systems are not scientific but pretend to be. For example, at one time people in the United States believed in the "science" of phrenology. Phrenology was the study of the surface bumps on a person's head. Phrenologists claimed to predict all sorts of things about people's health and personality based on the bumps on their heads. But phrenology was fake science—**pseudoscience.** No experimental findings backed up its claims. Can you tell real science apart from pseudoscience?

Consider astrology. Astrology tells us that human affairs are influenced by the positions and movements of planets and other celestial bodies. Yet there is no solid body of experimental evidence to back up this claim. This nonscientific view can be quite appealing. No matter how insignificant we may feel at times, astrologers assure us that we are deeply connected to the workings of the cosmos, which has been created just for humans. Astrology as ancient magic or entertainment is one thing, but astrology disguised as science is another. When it poses as a science related to astronomy, astrology is full-fledged pseudoscience.

We humans have learned much since the onset of science four centuries ago. Only by enormous effort did people along the way gain this knowledge and overthrow superstition. We have come far in comprehending nature and freeing ourselves from ignorance. We should rejoice in what we've learned. We no longer have to die whenever an infectious disease strikes. We no longer live in fear of demons. We no longer torture women accused of witchery, as was done for nearly three centuries during medieval times. Today we have no need to pretend that superstition is anything but superstition, or that junk notions are anything but junk notions—whether voiced by quacks on the Internet or by loose thinkers who write promise-heavy health books.

Yet there is reason to fear that what people of one time fight for, a following generation surrenders. The grip that belief in magic and superstition had on people took centuries to overcome. Yet today the same magic and superstition are enchanting a growing number of people. James Randi reports in his book *Flim-Flam!* that more than 20,000 practicing astrologers in the United States service millions of credulous believers. Science writer Martin Gardner reports that a greater percentage of Americans today believe in astrology and occult phenomena than did citizens of medieval Europe. Few newspapers publish a daily science column, but nearly all provide daily horoscopes. And then there are the flourishing television psychics, who gain followers daily.

Some people believe that the human condition is slipping backward because of growing technology. More likely, however, it is slipping backward because science and technology will bow to the irrationality of the past. Watch for the spokespeople of pseudoscience. It is a huge and lucrative business.

1.6 The Search for Order— Science, Art, and Religion

The search for order and meaning in the world has taken different forms: One is science, another is art, and another is religion. The domains of science, art, and religion are different, although they often overlap. ✔ The arts are concerned with personal interpretation and creative expression and are not restricted to any particular domain. The domain of science is the natural world. The domain of religion is the supernatural. The term *supernatural* literally means "above nature." Science works within nature, not above it. Science may answer the question "What is life, and how did it come to be?" But science is unable to answer philosophical questions such as "What is the purpose of life?" Although this question is valid and important, it is outside the domain of science. It is better addressed in the domains of art or religion.

Science and the arts have certain things in common. In the art of literature, we find out about what is possible in human experience. We can learn about emotions from rage to love, even if we haven't yet experienced them. The arts do not necessarily give us those experiences, but they describe them and suggest what may be possible for us. A knowledge of science similarly tells us what is possible in nature. Scientific knowledge helps us predict possibilities in nature even before these possibilities have been experienced. It provides us with a way of connecting things, of seeing relationships between and among them, and of making sense of the great variety of natural events around us. Science broadens our perspective of nature. A knowledge of both the arts and the sciences makes up a wholeness that affects the way we view the world. It helps us make decisions about the world and ourselves. A truly educated person is knowledgeable in both the arts and the sciences.

Science and religion have similarities also, but they are basically different. Science is concerned with physical things, while religion is concerned with spiritual matters. Simply put, science asks *how*; religion asks *why*. The practices of science and religion are also different. Whereas scientists experiment to find nature's secrets, many religious practitioners worship a Supreme Being and work to build human community. In these respects, science and religion are as different as apples and oranges and do not contradict each other. Science and religion are two different, yet complementary, fields of human activity.

When we study the nature of light later in this book, we will treat light first as a wave and then as a particle. To the person who knows a little bit about science, waves and particles are contradictory. Light can be only one or the other, and we have to choose between them. But to the enlightened person, waves and particles complement each other and provide a deeper understanding of light. In a similar way, it is mainly people who are either uninformed or misinformed about the deeper natures of both science and religion who feel that they must

Religion is about cosmic purpose.

Science is about cosmic order.

choose between believing in religion and believing in science. Unless one has a shallow understanding of either or both, there is no contradiction in being religious and being scientific in one's thinking. (Of course, this doesn't apply to certain religious extremists—Christian, Moslem, or otherwise—who steadfastly claim that one cannot embrace both their brand of religion and science.)

Many people are troubled over not knowing the answers to religious and philosophical questions. Some avoid uncertainty by eagerly accepting any comforting answer. An important message from science, however, is that uncertainty is acceptable. For example, in Chapter 14 you'll learn that it is not possible to know with certainty both the momentum and position of an electron in an atom. The more you know about one, the less you can know about the other. Uncertainty is a part of the scientific process. It's okay not to know the answers to fundamental questions. Why are apples gravitationally attracted to the Earth? Why do electrons repel one another? Why does energy have mass? At the deepest level, scientists don't know the answers to these questions—at least not yet. Scientists in general are comfortable about not knowing. We know a lot about where we are, but nothing really about *why* we are. Perhaps we can apply a lesson from science to our religious questions. Maybe it's okay not to know the answers to religious questions—especially if we keep exploring with an open mind and heart.

No wars are fought over science.

✓ READING CHECK

Distinguish among the domains of the arts, science, and religion.

1.7 Technology—Practical Use of the Findings of Science

Science and technology are also different from each other. Science is concerned with gathering knowledge and organizing it. **Technology** lets humans use that knowledge for practical purposes, and it provides the instruments scientists need to conduct their investigations.

Technology is a double-edged sword. It can be both helpful and harmful. We have the technology, for example, to extract fossil fuels from the ground and then burn the fossil fuels to produce energy. Energy production from fossil fuels has benefited society in countless ways. On the flip side, the burning of fossil fuels damages the environment. It is tempting to blame technology itself for problems such as pollution, resource depletion, and even overpopulation. These problems, however, are not the fault of technology any more than a stabbing is the fault of the knife. It is humans who use the technology, and humans who are responsible for how it is used.

Remarkably, we already possess the technology to solve many environmental problems. This 21st century will likely see a switch from fossil fuels to more sustainable energy sources. We recycle waste products in new and better ways. In some parts of the world, progress is being made toward limiting the human population explosion, a serious threat that worsens almost every problem faced by humans today.

We each need a knowledge filter to tell the difference between what is true and what only pretends to be true. The best knowledge filter ever invented is science.

RISK ASSESSMENT

Technology comes with risks as well as benefits. When the benefits are seen to outweigh risks, a technology can be accepted and applied. X-rays, for example, continue to be used to diagnose disease despite their potential risk for causing cancer. The benefits outweigh the risks. Of course, when the risks of a technology outweigh its benefits, it should be used sparingly or not at all.

A hard ethical problem arises when a technology benefits one group of people but poses risk to a different group of people. For example, aspirin is useful for adults, but it can cause a potentially fatal condition in children called *Reye's syndrome*. Are the benefits of aspirin to adults worth the risk that children will die by getting Reye's syndrome? Dumping raw sewage into the local river may pose little risk for a town located upstream, but for towns downstream the untreated sewage is a health hazard. Technologies involving different risks for different people raise questions that are often hotly debated. Which medications should be sold to the general public over-the-counter and how should they be labeled? Should food be irradiated to eliminate the food poisoning that kills more than 5000 Americans each year? Or are the unknown potential hazards of food irradiation sufficient cause for banning food irradiation?

People seem to have a hard time accepting the fact that zero risk is impossible. Airplanes cannot be made perfectly safe. Processed foods cannot be completely free of toxicity, for all foods are toxic to some degree. You cannot go to the beach without risking skin cancer no matter how much sunscreen you apply. You cannot avoid radioactivity, for it's in the air you breathe and the foods you eat and has been before humans first walked the Earth. Even the cleanest rain contains radioactive carbon-14; our bodies do as well. You might hide in the hills, eat the most natural foods, practice obsessive hygiene, and still die from cancer caused by radioactivity. The probability of eventual death is 100%. We have to accept that. Nobody is exempt.

Science helps to determine the most probable results. As the tools of science improve, risks can be evaluated more and more accurately. Acceptance of risk, on the other hand, is more of a social issue than a scientific one. If society were to demand zero risk from its technology, this goal would not only be impractical but selfish. Any society striving toward a policy of zero risk would consume its present and future economic resources. A society that accepts no risks receives no benefits.

READING CHECK

What is the promise of technology?

Difficulty solving today's problems results more from social inertia than failing technology. Technology is our tool. What we do with this tool is up to us. ✔ The promise of technology is a cleaner and healthier world. Wise applications of it *can* improve conditions on planet Earth.

1.8 The Physical Sciences: Physics, Chemistry, Earth Science, and Astronomy

Science is the present-day equivalent of what used to be called *natural philosophy*. Natural philosophy was the study of unanswered questions about nature. As the answers were found, they became part of what is now called *science*. The study of science today branches into the study of living things and nonliving things: the life sciences and the physical sciences. The life sciences branch into such areas as biology, zoology, and botany. The *physical sciences* branch into such areas as physics, chemistry, geology, meteorology, and astronomy—the areas addressed in this book.

Physics is the study of basic concepts such as motion, force, energy, matter, heat, sound, light, and the components of atoms. Chemistry builds on physics and tells us how matter is put together, how atoms combine to form molecules, and how the molecules combine to make the materials around us. Physics and chemistry applied to the Earth and its processes make up Earth science. When we apply physics, chemistry, and Earth science to other planets and to the stars, we are speaking about astronomy.

Biology is more complex than physical science, for it involves matter that is alive. Underneath biology is chemistry, and underneath chemistry is physics. ✔ **Physics is an essential building block for both physical science and life science.** That is why we begin with physics, then follow with chemistry and Earth science, and conclude with astronomy. All are treated conceptually, with the twin goals of enjoyment and understanding.

> Doubt and uncertainty are hallmarks of science. Most scientists feel it is far more interesting to live without knowing than to have answers that might be wrong.

CONCEPT CHECK

Which of the following activities involves the utmost human expression of passion, talent, and intelligence?

  a. painting and sculpture
  b. literature
  c. music
  d. religion
  e. science

Check Your Answer

All of them! In this book we focus on science, which is an enchanting human activity shared by a wide variety of people. With present-day tools and know-how, science types are reaching further and discovering more about themselves and their environment than people in the past were ever able to do. The more you know about science, the more passionate you feel toward your surroundings. There is physical science in everything you see, hear, smell, taste, and touch!

✔ **READING CHECK**

What field of science is basic to both the physical and the life sciences?

1.9 In Perspective

Only a few centuries ago, the most talented and skilled artists, architects, and artisans of the world directed their genius to the construction of the great cathedrals, synagogues, temples, and mosques. Some of these architectural structures took centuries to build. This meant that nobody witnessed both the beginning and the end of construction. The architects and early builders who lived to a ripe old age never saw the finished results of their labors. Entire lifetimes were spent in the shadows of construction that must have seemed without beginning or end. This enormous focus of human energy was inspired by a vision that went beyond worldly concerns—a vision of the cosmos. To the people of that time, the structures they built were their "spaceships of faith," firmly anchored but pointing to the cosmos.

Today the efforts of many skilled scientists, engineers, technicians, and artisans are directed to building the spaceships that already orbit the Earth and others that will voyage beyond. The time required to build these spaceships is extremely brief compared to the time spent building the stone and marble structures of the past. Many people working on today's spaceships were alive before the first jetliner aircraft carried passengers. Where will younger lives lead in a comparable time?

We seem to be at the dawn of a major change in human growth. For as little Evan suggests in the photo at the beginning of this book, we may be like the hatching chicken who has exhausted the resources of its inner-egg environment and is about to break through to a whole new range of possibilities. The Earth is our cradle and has served us well. But cradles, however comfortable, are one day outgrown. So with the inspiration that in many ways is similar to the inspiration of those who built the early cathedrals, synagogues, temples, and mosques, we aim for the cosmos.

We live in an exciting and a challenging time!

1 CHAPTER REVIEW

KEY TERMS

Fact A phenomenon about which competent observers can agree.

Law A general hypothesis or statement about the relationship of natural quantities that has been tested over and over again and has not been contradicted. Also known as a *principle.*

Hypothesis An educated guess; a reasonable explanation that is not fully accepted as factual until tested over and over again by experiment.

Pseudoscience Fake science that has no tests for its validity.

Science Organized common sense. Also, the collective findings of humans about nature and a process of gathering and organizing knowledge about nature.

Scientific method An orderly method for gaining, organizing, and applying new knowledge.

Technology Method and means of solving practical problems by applying the findings of science.

Theory A synthesis of a large body of information that encompasses well-tested hypotheses about certain aspects of the natural world.

REVIEW QUESTIONS

A Brief History of Advances in Science

1. What discovery in the 15th century greatly advanced progress in science?
2. Throughout the ages, how have most people responded to new ideas: with acceptance or resistance?

Mathematics and Conceptual Physical Science

3. Why is mathematical problem solving not a major feature of this book?

4. In this book, which comes first: comprehension or calculation?

Scientific Methods—Classic Tools

5. Outline the steps of the scientific method.
6. Distinguish among a scientific fact, a hypothesis, a law, and a theory.

Scientific Hypotheses Must Be Testable

7. How many experiments are necessary to invalidate a scientific hypothesis?
8. Must a statement prove correct to be a scientific hypothesis?

A Scientific Attitude Underlies Good Science

9. In science, what kind of ideas are generally accepted?
10. Why is honesty a matter of self-interest to a scientist?

The Search for Order—Science, Art, and Religion

11. Why are students of the arts encouraged to learn about science, and science students encouraged to learn about the arts?
12. How do scientists regard "not knowing" in general?

Technology—Practical Use of the Findings of Science

13. Clearly distinguish between science and technology.
14. Who uses and who is responsible for how technology is applied?

The Physical Sciences: Physics, Chemistry, Earth Science, and Astronomy

15. Name at least two fields of scientific study in the physical sciences and two from the life sciences.
16. Of physics, chemistry, and biology, which science is the least complex? The most complex? (At your school, which of these is the "easiest" and which is the "hardest" as a science course?)

THINK AND EXPLAIN

1. In daily life, people are often praised for maintaining some particular point of view, for the "courage of their convictions." A change of mind is seen as a sign of weakness. How is this different in science?

2. In daily life, we see many cases of people who are caught misrepresenting things and who soon thereafter are excused and accepted by their contemporaries. How is this different in science?

3. In answer to the question "When a plant grows, where does the material come from?" Aristotle hypothesized by logic that all material came from the soil. Do you consider his hypothesis to be correct, incorrect, or partially correct? What experiments do you propose to support your choice?

4. What is probably being misunderstood by a person who says, "But that's only a scientific theory"?

5. (a) Make an argument for bringing to a halt the advances of technology.
   (b) Make an argument that advances in technology should continue.
   (c) Contrast your two arguments.

PART ONE

MECHANICS

2 NEWTON'S FIRST LAW OF MOTION— THE LAW OF INERTIA

THE MAIN IDEA

When no net force acts on an object, no change in motion occurs—the object is in mechanical equilibrium.

If you want to move something, you apply a force to it—you either push it or pull it. To move dishes on a tablecloth, you can move them by pulling on the tablecloth. Pull slowly, and they'll move slowly. But what if you pull the tablecloth very quickly—quick enough so the force of friction between the cloth and dishes is very small and very brief? Will the dishes move? It takes a force to get things moving, but once moving, is a force needed to keep them moving? How about satellites in orbit? They are in a friction-free environment. What keeps them moving? What rules of nature guide the answers to these questions?

Explore!

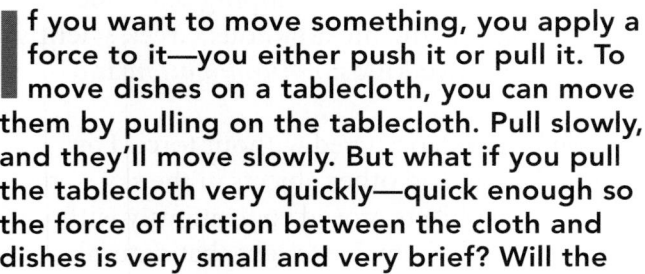

Can You Pull a Piece of Paper Out from Beneath a Coin?

1. Place a penny on top of a sheet of paper on a desk or table.
2. Pull the paper horizontally with a quick snap.
3. Repeat using a quarter.

Analyze and Conclude

1. **Observing** Did the penny remain in place when you pulled the paper? Did the speed with which you pulled

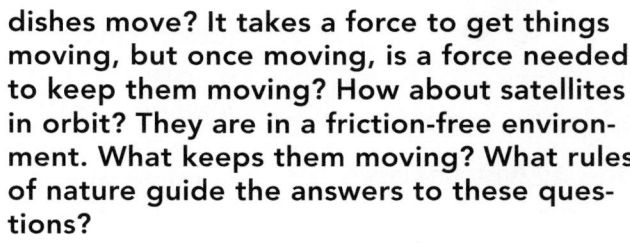

make a difference? Was the trial with the quarter more satisfactory?

2. **Predicting** Do you think this would work for a slower pull?
3. **Making Generalizations** Why were you able to snap the paper from beneath the coin with very little motion of the coin?

Aristotle was the most famous philosopher, scientist, and educator in ancient Greece. He was the son of a physician who personally served the king of Macedonia. He entered the Academy of Plato at age 17, where he worked and studied for 20 years until Plato's death. He then became the tutor of young Alexander the Great. Eight years later, he formed his own school. Aristotle's aim was to arrange existing knowledge in a system, just as Euclid had earlier done with geometry. Aristotle made careful observations, collected specimens, and gathered together and classified almost all existing knowledge of the physical world. His systematic approach became the method from which Western science later arose. After his death, his voluminous notebooks were preserved in caves near his home and were later sold to the library at Alexandria. Scholarly activity came to a stop in most of Europe during the Dark Ages, and the works of Aristotle were forgotten and lost. Various texts were reintroduced to Europe during the 11th and 12th centuries and translated into Latin. The Church, the dominant political and cultural force in Western Europe, at first prohibited the works of Aristotle. But soon thereafter the Church accepted them and incorporated them into Christian doctrine.

Does a force keep pushing a hockey puck across the ice?

✓ READING CHECK

How did Aristotle classify motion?

FIGURE 2.1 ▲
Galileo's famous demonstration.

2.1 Aristotle's Classification of Motion

The idea that motion requires a **force** (a push or a pull) goes back to the 4th century BC when the Greeks were developing scientific ideas. This was the time of Aristotle, the most famous Greek scientist. ✓ Aristotle classified motion into two kinds: *natural motion* and *unnatural motion.*

Natural motion, Aristotle believed, occurred without force. For example, motions of the Sun, Moon, and other objects in the sky were considered natural. These celestial objects moved continuously without the need for any force to push or pull them. He taught that natural motion on Earth was directed either up or down. He said, for example, that it was natural for a boulder to fall down toward Earth and a puff of smoke to rise in air. He believed that objects have resting places that they naturally seek. He went on to say that it was natural for heavy objects to fall faster than light objects.

Unnatural motion, on the other hand, required forces such as pushes or pulls by people or animals. For example, the only way to get a cart moving across the ground was to push or pull on it. Unnatural motion required forces.

2.2 Galileo's Concept of Inertia

Aristotle's ideas were taken as fact for nearly 2000 years. But in the early 1500s, the Italian scientist Galileo demolished Aristotle's belief that heavy things fall faster than light things. As mentioned in

Chapter 1, legend tells us that Galileo dropped a heavy object and a light object from the Leaning Tower of Pisa. He showed that except for the effects of air friction, objects of different weights fell to the ground at the same time.

Galileo made another huge discovery. He showed that Aristotle was wrong about forces being necessary to keep objects moving. ✓ Galileo said that a force is required to start an object moving, but once moving, no force is required to keep it moving—except for the force needed to overcome friction. (We'll study friction, a force that opposes motion, in the next chapter.) When friction is absent, a moving object does not need a force to keep it moving.

Galileo found a simple and powerful way to test his revolutionary idea. He rolled balls along flat plane surfaces tilted at different angles. He noted that a ball rolling down an inclined plane picks up speed. This is shown in Figures 2.2 and 2.3. Gravity increases the ball's speed. He also noticed the ball slows down when rolling up an inclined plane. Then gravity decreases the ball's speed. What about a ball rolling on a level surface? While a ball rolls on a level surface, it doesn't roll with nor against gravity. Galileo saw that a ball rolling on a smooth horizontal plane doesn't speed up and doesn't slow down. It has a constant speed. Galileo reasoned that a ball moving horizontally would move forever if friction were entirely absent. A ball would move of itself.

All objects show the same property of motion as the balls rolling on Galileo's planes. The tendency of things is to remain as they are. If moving, they tend to remain moving. If at rest, they tend to remain at rest. This property of objects is called **inertia.**

When Galileo changed the focus from logic to experiment, he changed the course of science.

Slope downward—
Speed increases

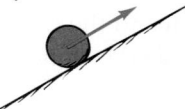

Slope upward—
Speed decreases

No slope—
Does speed change?

FIGURE 2.2 ▲
Motion of balls on various planes.

Inertia isn't a kind of force; it's a *property* of matter to resist changes in motion.

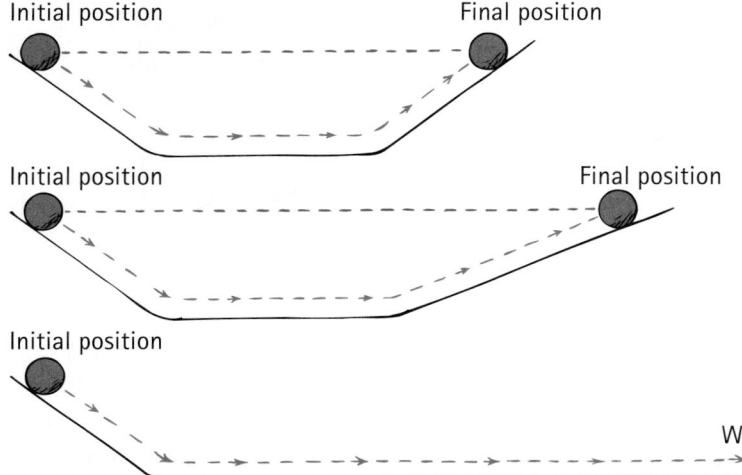

FIGURE 2.3 ▲
A ball rolling down an incline on the left tends to roll up to its initial height on the right. The ball must roll a greater distance as the angle of incline on the right is reduced.

Galileo was born in Pisa, Italy, in the same year Shakespeare was born and Michelangelo died. He studied medicine at the University of Pisa and then changed to mathematics. He developed an early interest in motion and was soon at odds with others around him, who held to Aristotelian ideas on falling bodies. He left Pisa to teach at the University of Padua and became an advocate of the new theory of the solar system advanced by the Polish astronomer Copernicus. Galileo was one of the first to build a telescope, and he was the first to direct it to the nighttime sky. He discovered mountains on the Moon and the moons of Jupiter. Because Galileo published his findings in Italian instead of the Latin typically used by scholars, and because of the recent invention of the printing press, Galileo's ideas reached many people. He soon ran into disagreements with the Church and was warned not to teach and not to hold to Copernican views. He restrained himself publicly for nearly 15 years. Then Galileo defiantly published his observations and conclusions, which didn't agree with Church doctrine. The outcome was a trial in which he was found guilty, and he was forced to renounce his discoveries. By then an old man broken in health and spirit, he was sentenced to perpetual house arrest. Nevertheless, Galileo completed his studies on motion and his writings were smuggled from Italy and published in Holland. Earlier he damaged his eyes looking at the Sun through a telescope, which led to blindness at the age of 74. He died 4 years later.

 READING CHECK

What did Galileo say about Aristotle's idea that forces are necessary to keep objects moving?

- Galileo did all his work before the advent of mechanical clocks. He timed some of his experiments with his pulse and others with the dripping of water drops. Einstein called Galileo the father of modern physics and modern science.

PhysicsPlace.com

Videos
Definition of Speed; Average Speed; Velocity; Changing Velocity

CONCEPT CHECK
A ball rolling on a pool table slowly comes to a stop. How would Aristotle explain this behavior? How would Galileo explain it? How would you explain it?

Check Your Answers
Did you think about the questions and arrive at your own answers before reading this? Please do so, and you'll find yourself learning more. Much more!
Aristotle would probably say that the ball stops because it seeks its natural state of rest. Galileo would probably say that friction overcomes the ball's natural tendency to continue rolling. Only you can answer the last question!

2.3 Galileo's Concepts of Speed and Velocity

Speed
Before the time of Galileo, people described moving things as simply "slow" or "fast." Such descriptions were vague. Galileo was the first to measure speed by considering the distance covered and the amount of time it takes. He defined **speed** as the distance covered per unit of time. The word *per* means "divided by."

$$\text{Speed} = \frac{\text{distance}}{\text{time}}$$

TABLE 2.1 Approximate Speeds in Different Units

12 mi/h	=	20 km/h	=	6 m/s (bowling ball)
25 mi/h	=	40 km/h	=	11 m/s (very good sprinter)
37 mi/h	=	60 km/h	=	17 m/s (sprinting rabbit)
50 mi/h	=	80 km/h	=	22 m/s (tsunami)
62 mi/h	=	100 km/h	=	28 m/s (sprinting cheetah)
75 mi/h	=	120 km/h	=	33 m/s (batted softball)
100 mi/h	=	160 km/h	=	44 m/s (batted baseball)

FIGURE 2.4 ▲
The greater the distance traveled each second, the faster the horse gallops.

For example, if a horse covers 20 kilometers in a time of 1 hour, its speed is 20 km/h. Of if you run 6 meters in 1 second, your speed is 6 m/s. These values are *average speeds,* for the speed at any instant, *instantaneous speed,* may be different.

Any combination of distance and time units can be used for speed—kilometers per hour (km/h), centimeters per day (the speed of a sick snail), or whatever is useful and convenient. The slash symbol (/) is read as "per." In science the preferred unit of speed is meters per second (m/s). Table 2.1 shows some comparative speeds in different units.

Velocity

When we know both the speed and direction of an object, we know its **velocity.** For example, if a car travels at 60 km/h, we know its speed. But if we say it moves at 60 km/h to the north, we specify its *velocity.*
✓ Speed is a description of how fast; velocity is how fast *and* in what direction. A quantity such as velocity that specifies direction as well as magnitude is called a **vector quantity.** Velocity is a vector quantity and is described fully in the *Conceptual Physical Science Explorations Practice Book.* Vectors are further discussed in Appendix C.

Constant speed means steady speed. Something with constant speed doesn't speed up or slow down. Constant velocity, on the other hand, means both constant speed *and* constant direction. Constant direction is a straight line—the object's path doesn't curve. So constant velocity means motion in a straight line at constant speed.

In the next chapter we will consider motion that is not constant—*acceleration.*

FIGURE 2.5 ▲
A speedometer gives readings in both miles per hour and kilometers per hour.

Velocity is directed speed.

◀ **FIGURE 2.6**
The car on the circular track may have a constant speed, but its velocity is changing every instant. Why?

Distinguish between speed and velocity.

CONCEPT CHECK

1. What is the average speed of a cheetah that sprints 100 m in 4 s? How about if it sprints 50 m in 2 s?
2. The speedometer on a bicycle moving east reads 50 km/h. It passes another bicycle moving west at 50 km/h. Do both bikes have the same speed? Do they have the same velocity?
3. "The bird flies at a constant speed in a constant direction." Say the same sentence in fewer words.

Check Your Answers

1. In both cases the answer is 25 m/s:

$$\text{Average speed} = \frac{\text{distance covered}}{\text{time interval}} = \frac{100 \text{ meters}}{4 \text{ seconds}} = \frac{50 \text{ meters}}{2 \text{ seconds}} = 25 \text{ m/s}$$

2. Both bicycles have the same speed, but they have opposite velocities because they move in opposite directions.
3. "The bird flies at constant velocity."

2.4 Motion Is Relative

Everything is always moving. Even when you think you're standing still, you're actually speeding through space. You're moving relative to the Sun and stars—though you are at rest relative to Earth. Right now your speed relative to the Sun is about 100,000 kilometers per hour. And you're moving even faster relative to the center of our galaxy.

When discussing motion, we mean ✓ relative motion—motion relative to something else. When we say a space shuttle moves at 30,000 kilometers per hour, we mean relative to Earth below. When we say a racing car reaches a speed of 300 kilometers per hour, we mean relative to the track. Unless stated otherwise, all speeds discussed in this book are relative to the surface of Earth. Motion is relative.

Aristotle used logic to establish his ideas of motion. Galileo used experiment. Galileo showed that experiments are better than logic in testing knowledge. Galileo was concerned with *how* things move rather than *why* they move. The path was clear for Isaac Newton (1642–1727) to make further connections of concepts of motion.

FIGURE 2.7 ▲
When you sit on a chair, your speed is zero relative to the Earth but 30 km/s relative to the Sun.

What is meant by "relative motion"?

PhysicsPlace.com

Videos
Newton's Law of Inertia; The Old Tablecloth Trick; Toilet Paper Roll; Inertia of a Ball; Inertia of a Cylinder; Inertia of an Anvil

2.5 Newton's First Law of Motion— The Law of Inertia

At the age of 24, Isaac Newton extended Galileo's concept of inertia and gave it the status of a fundamental law that underlies all motion. He developed two other laws of motion as well (which we'll study in the following chapters). ✓ Newton's laws were influenced by Galileo's findings. Once and for all, they put to rest mistaken Aristotelian ideas that dominated the thinking of the best minds during the previous 2000 years.

PEOPLE IN HISTORY: ISAAC NEWTON (1642–1727)

Isaac Newton was born prematurely and barely survived on Christmas Day, 1642, the same year that Galileo died. Newton's birthplace was his mother's farmhouse in Woolsthorpe, England. His father died several months before his birth, and he grew up in the care of his mother and grandmother. As a child he showed no particular signs of brilliance, and at the age of $14\frac{1}{2}$ he was taken out of school to work on his mother's farm. As a farmer he was a failure, preferring to read books he borrowed from a neighboring druggist. An uncle sensed the scholarly potential in young Isaac and encouraged him to study at the University of Cambridge, which he did for 5 years, graduating without particular distinction.

When a plague swept through London, Newton retreated to his mother's farm—this time to continue his studies. At the farm, at age 23, he laid the foundations for the work that was to make him immortal. Seeing an apple fall to the ground led him to consider the force of gravity extending to the Moon and beyond. He formulated the law of universal gravitation (which he later proved). He also invented the calculus, a very important mathematical tool in science. Newton extended Galileo's work and formulated the three fundamental laws of motion. He also formulated a theory of the nature of light and showed with prisms that white light is composed of all colors of the rainbow. It was Newton's experiments with prisms that first made him famous.

When the plague subsided, Newton returned to Cambridge and soon established a reputation as a first-rate mathematician. His mathematics teacher resigned in his favor and Newton was appointed the Lucasian professor of mathematics, a post he held for 28 years. In 1672 he was elected to the Royal Society, where he exhibited the world's first reflector telescope. Still preserved at the library of the Royal Society in London, it bears the inscription: "The first reflecting telescope, invented by Sir Isaac Newton, and made with his own hands."

It wasn't until Newton was 42 that he began to write what is generally acknowledged the greatest scientific book ever written, the *Principia Mathematica Philosophiae Naturalis.* Written in Latin, the work was completed in 18 months and appeared in print in 1687. It wasn't printed in English until 1729, two years after Newton's death. When asked how he was able to make so many discoveries, Newton replied that he solved his problems by continually thinking very long and hard about them—and not by sudden insight.

At the age of 46, Newton was elected a member of Parliament. He attended the sessions for 2 years but never gave a speech. One day he did rise and the House of Commons fell silent to hear the great man. Newton's "speech" was very brief; he simply requested that a window be closed because of a draft.

Another departure from his work in science was his appointment as warden and then master of the mint. Resigning his professorship, Newton directed his efforts toward greatly improving the workings of the mint, to the dismay of counterfeiters who flourished at that time. He maintained his membership in the Royal Society, was elected president, and then was reelected each year for the rest of his life. At the age of 62, he wrote *Opticks,* which summarized his work on light. Nine years later he wrote a second edition of his *Principia.*

Although Newton's hair turned gray at 30, it remained full, long, and wavy all his life. Unlike others of his time, he did not wear a wig. He was a modest man, very sensitive to criticism, and never married. He remained healthy in body and mind into old age. At 80, he still had all his teeth, his eyesight and hearing were sharp, and his mind was alert. In his lifetime he was regarded by his countrymen as the greatest scientist who ever lived. In 1705, he was knighted by Queen Anne. Newton died at the age of 85 and was buried in Westminster Abbey along with England's kings and heroes.

Newton showed that the universe worked according to natural laws—a knowledge that provided hope and inspiration to people of all walks of life and that ushered in the Age of Reason. The ideas and insights of Isaac Newton truly changed the world and elevated the human condition.

FIGURE 2.8 ▲
Inertia in action.

Newton's first law, usually called the **law of inertia,** is a restatement of Galileo's idea.

Every object continues in its state of rest or of uniform speed in a straight line unless a net force acts on it.

This says that things tend to keep on doing what they're already doing. Objects at rest tend to remain at rest—a force is needed to set them in motion. For example, dishes on a table are in a state of rest. The dishes tend to remain at rest even if you snap a table-cloth from beneath them.*

The law also says that when an object is moving, its tendency is to remain moving, along a straight-line path. For example, if you slide a hockey puck on the surface of slippery ice, it moves a long way until ice and air friction finally stops it. If you slide it along the surface of a city street, it is quickly brought to rest by the force of friction. It tends to keep sliding but friction acts against it. If you toss an object where there is no friction, such as in the vacuum of outer space, it will continually move in a straight-line path. An object moves by its own inertia.

While the ancients thought continual forces were needed to maintain motion, the law of inertia provides a completely different way of thinking about motion. We now know that objects continue to move by themselves. If an object is at rest, you'll have to apply force to get it moving, but once in motion no force is needed (except that to overcome any friction). The object moves in a straight line indefinitely. In the next chapter we'll see that forces are needed to change the speeds and directions of objects, but not to maintain motion if there is no friction.

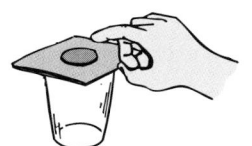

Why will the coin drop into the glass when a force accelerates the card?

Why does the downward motion and sudden stop of the hammer tighten the hammerhead?

Why is it that a slow continuous increase in the downward force breaks the string above the massive ball, but a sudden increase breaks the lower string?

FIGURE 2.9 ▲
Examples of inertia.

---

* Close inspection shows that brief friction between the dishes and fast-moving tablecloth start the dishes moving, but friction between the dishes and table stop the dishes before they slide very far. If you try this, use unbreakable dishes!

CONCEPT CHECK

When the space shuttle moves in a nearly circular orbit around Earth, is a force needed to maintain its high speed? If suddenly the force of gravity were cut off, what type of path would the shuttle follow?

Check Your Answers

No force in the direction of the shuttle's motion exists. The shuttle coasts by its own inertia. The only force acting on it is the force of gravity, which acts at right angles to its motion (toward Earth's center). We'll see later that this right-angled force holds the shuttle in a circular path. If it were cut off, the shuttle would move in a straight line at constant speed (constant velocity).

✓ READING CHECK

What influenced Newton in developing his laws of motion?

2.6 Net Force—The Combination of All Forces That Act on an Object

So we see that without force, objects don't speed up, slow down, or change direction. When we say "force," we mean the total force, or net force, acting on an object. Often more than one force acts. For example, when you throw a basketball, the force of gravity, air friction, and the pushing force you apply with your muscles all act on the ball. The **net force** on the ball is the combination of all these forces. ✓ It is the net force that changes an object's state of motion.

For example, suppose you pull on a box of chocolate bars with a force of 5 pounds. If your friend also pulls with 5 pounds in the same direction, the net force on the box is 10 pounds. If your friend pulls on the box with the same force as you but in the opposite direction, the net force on it is zero. Now if you increase your pull to 10 pounds and your friend pulls oppositely with 5 pounds, the net force is 5 pounds in the direction of your pull. We see this in Figure 2.10, where instead of pounds, the scientific unit of force is used—the **newton,** abbreviated N.

In Figure 2.10, forces are shown by arrows. Arrows are used because forces are vector quantities. As mentioned earlier, a vector quantity has both magnitude (how much) and direction (which way). When an arrow represents a vector quantity, the arrow's length represents magnitude and its direction shows the direction of the quantity. Such an arrow is called a vector. (Again, more on vectors in Appendix C.)

PhysicsPlace.com
Videos
Vector Representation; How to Add and Subtract; Geometric Addition of Vectors
Interactive Tutorial
Vectors

✓ READING CHECK

What, exactly, changes an object's state of motion?

Applied forces	Net force
 5 N / 5 N ⟹	10 N
5 N ← 5 N → ⟹	0 N
5 N ← 10 N → ⟹	5 N

◄ **FIGURE 2.10**
Net force.

2.7 Equilibrium for Objects at Rest

Suppose you tie a string around a 2-pound bag of sugar and hang it on a weighing scale (Figure 2.11) similar to the scales found in grocery stores. A spring in the scale stretches until the scale reads 2 pounds. The stretched spring experiences a "stretching force" called *tension*. The same scale in a science lab is likely in units of newtons. This scale will show the weight of the bag of sugar as 9 newtons rather than 2 pounds. Both pounds and newtons are units of weight. Units of weight in turn are units of force. The bag of sugar is attracted to Earth with a gravitational force of 2 pounds—or equivalently, 9 newtons. Hang twice as much sugar from the scale and the reading will be 18 newtons.

Note there are two forces acting on the bag of sugar—tension force acting upward and weight acting downward. The two forces on the bag are equal and opposite, and cancel to zero. Hence the bag remains at rest. We say it is in *mechanical equilibrium.*

✔️ Mechanical equilibrium is a state wherein no physical changes occur; it is a state of steadiness. When the net force on an object is zero, the object is said to be in mechanical equilibrium—this is known as the **equilibrium rule.** We express the equilibrium rule mathematically as

$$\Sigma F = 0.$$

The symbol Σ stands for "the vector sum of" and F stands for "forces." (Please don't be intimidated by the expression $\Sigma F = 0$, which is physics shorthand that says a lot in so little space—that all the forces acting on something add vectorally to zero.) For a suspended object at rest, like the bag of sugar mentioned earlier, the rule states that the forces acting upward on the object must be balanced by other forces acting downward to make the vector sum equal zero. (Vector quantities take direction into account, so if upward forces are positive, downward ones are negative, and when summed they equal zero.)

In Figure 2.13 we see the forces of interest to Burl and Paul on their sign-painting scaffold. The sum of the upward tensions is equal to the sum of their weights plus the weight of the scaffold. Note how the magnitudes of the two upward vectors equal the magnitudes of the three downward vectors. Net force on the scaffold is zero. The scaffold is in mechanical equilibrium.

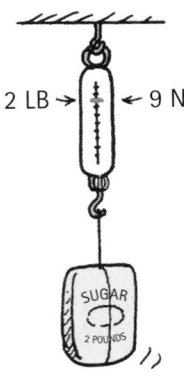

FIGURE 2.11 ▲
The upward tension in the string has the same magnitude as the weight of the bag, so the net force on the bag is zero.

FIGURE 2.12 ▲
Burl Grey, who first introduced author Paul to the concept of tension, shows a 2-lb bag producing a tension of 9 N (actually slightly more than 2 lb and 9 N).

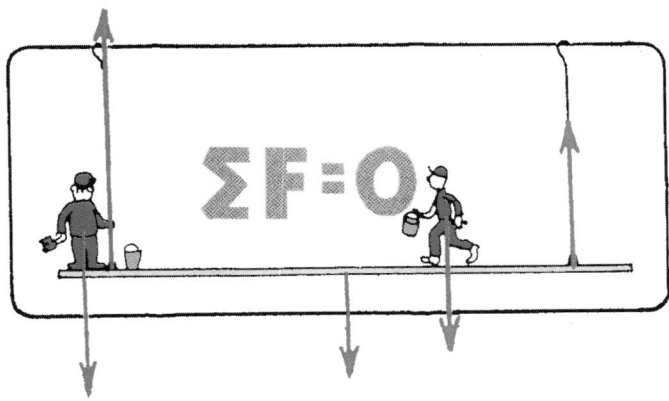

◀ **FIGURE 2.13**
The sum of the upward vectors equals the sum of the downward vectors. $\Sigma F = 0$ and the scaffold is in equilibrium.

When I was in high school, my counselor advised me not to enroll in science and math classes. Instead, I was advised to focus on what seemed to be my gift for art. I took this advice. I was then interested in drawing comic strips and in boxing, but neither of these earned me much success. After a stint in the army, I tried my luck at sign painting, and the cold Boston winters drove me south to warmer Miami, Florida. There, at age 26, I got a job painting billboards and met my greatest intellectual influence, Burl Grey. Like me, Burl had never studied physics in high school. But he was passionate about science in general. He shared his passion by discussing many fascinating questions as we painted together.

I remember Burl asked me about the tensions (stretching forces) in the ropes that held up the scaffold we were standing on. The scaffold was simply a heavy horizontal plank suspended by a pair of ropes. Burl twanged the rope nearest his end of the scaffold. He asked me to do the same with mine. He was comparing the tensions in both ropes to see which was greater. Burl was heavier than I was and he guessed the tension in his rope was greater. Like a more tightly stretched guitar string, the rope with greater tension twangs at a higher pitch. The finding that Burl's rope had a higher pitch seemed reasonable because his rope supported more of the load.

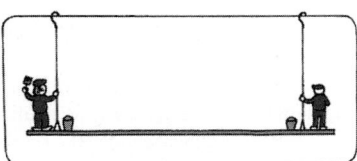

When I walked toward Burl to borrow one of his brushes, he asked if tensions in the ropes changed. Does tension in his rope increase as I get closer? We agreed that it should because even more of the load was supported by Burl's rope. How about my rope? Would rope tension there decrease? We agreed that it would, for it would be supporting less of the total load. I was unaware that I was discussing physics.

Burl and I used exaggeration to bolster our reasoning (just as physicists do). If we both stood at an extreme end of the scaffold and leaned outward, it was easy to imagine the opposite end of the scaffold rising like the end of a seesaw. And then the opposite rope would go limp. There would be no tension in that rope. We then reasoned the tension in my rope would gradually decrease as I walked toward Burl. It was fun posing such questions and seeing if we could answer them.

A question that we couldn't answer was whether or not a decrease of tension in one rope would be *exactly* compensated by an increase of tension in the other. For example, if my rope underwent a decrease of 50 newtons, would Burl's rope gain 50 newtons? (We talked pounds back then, but here we use the scientific unit of force, the newton—abbreviated N.) Would the gain be *exactly* 50 N? And if so, would this be a grand coincidence? I didn't know the answer until more than a year later. That was when Burl's stimulating questions prompted me to leave full-time painting and go to college to study science.*

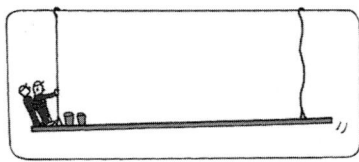

In my science classes I learned that any object at rest, such as the sign-painting scaffold I worked on with Burl, experiences no net force. It is said to be in *equilibrium*. In other words, the individual forces on the object add up to a net force with zero magnitude. So the sum of the upward forces supplied by the supporting ropes indeed do add up to the downward forces of our weights plus the weight of the scaffold. A 50-N loss in one rope would have to be accompanied by a 50-N gain in the other. Only then do the upward and downward forces cancel.

I tell this true story to make the point that one's thinking is very different when there is a rule to guide it. Now when I look at any motionless object I know right away that all the forces acting on it cancel out. We see nature differently when we know its rules. It makes nature seem simpler and easier to understand. Without the rules of physics, we tend to be superstitious and see magic where there is none. Quite wonderfully, everything is connected to everything else by a surprisingly small number of rules, and in a beautifully simple way. The rules of nature are what the study of science is about.

---

* I am forever indebted to Burl Grey for the stimulation he provided, for when I continued with formal education, it was with enthusiasm. I lost touch with Burl for 40 years. A student in my class, Jayson Wechter, doing some detective work, located him in 1998 and put us in contact. Friendship renewed, we once again continue in spirited conversations.

✔ READING CHECK

What does it mean to say that something is in mechanical equilibrium?

CONCEPT CHECK

Consider the gymnast hanging from the rings.

1. If she hangs with her weight evenly divided between the two rings, how would scale readings in both supporting ropes compare with her weight?
2. Suppose she hangs with slightly more of her weight supported by the left ring. How would a scale on the right read?

Check Your Answers

*(Again, are you reading this before you have thought about and formulated **your** reasoned answers? If so, do you also exercise your body by looking at others do push-ups? Exercise your thinking: When you encounter the many Concept Checks in this book like the one above, think before you look at the answers!)*

1. The reading on each scale will be half her weight. The sum of the readings on both scales then equals her weight.
2. When more of her weight is supported by the left ring, the reading on the right is less than half her weight. No matter how she hangs, the sum of the scale readings equals her weight. For example, if one scale reads two-thirds her weight, the other scale will read one-third her weight. Get it?

Notice that Newton's law of inertia and the equilibrium rule say the same thing: When $\Sigma F = 0$, objects don't change their states of motion.

PhysicsPlace.com

Interactive Tutorial
The Support Force—Why We Don't Fall Through the Floor

✔ READING CHECK

For an object at rest on a horizontal surface, what is the support force equal to?

2.8 The Support Force—Why We Don't Fall Through the Floor

Consider a book lying at rest on a desk. It is in equilibrium. What forces act on the book? One is the force due to gravity—the *weight* of the book. Since the book is in equilibrium, there must be another force acting on it to produce a net force of zero—an upward force opposite to the force of gravity.

Where is the upward force coming from? It is coming from the desk that supports the book. We call this the **support force**—the upward force that balances the weight of an object on a surface. A support force is often called the *normal force.*✔ For an object at rest on a horizontal surface, the support force must equal the object's weight. So in this case, the support force must equal the weight of the book. We say the upward support force is positive and the downward weight is negative. The two

---

* This force acts at right angles to the surface. Mathematically, "normal to" means "at right angles to." Hence the name *normal* force.

forces add mathematically to zero. So the net force on the book is zero. Another way to say the same thing is $\Sigma F = 0$.

To better understand that the desk pushes up on the book, think about a spring being compressed (Figure 2.14). If you push the spring down, you can feel the spring pushing up on your hand. Similarly, the book lying on the desk compresses atoms in the desk, which behave like tiny springs. The weight of the book squeezes downward on the atoms, and the atoms push upward on the book. In this way the compressed atoms produce the support force.

When you step on a bathroom scale, two forces act on the scale. One force is the downward pull of gravity, your weight, and the other is the upward support force of the floor. These forces compress a mechanism (in effect, a spring) that is calibrated to show your weight (Figure 2.15). So the scale shows the support force. When standing on a bathroom scale at rest, the amount of support force and the amount of your weight are equal. We say they have the same *magnitude.*

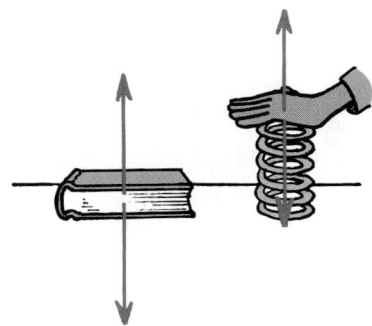

FIGURE 2.14 ▲
(Left) The desk pushes up on the book with as much force as the downward weight. (Right) The spring pushes up on your hand with as much force as you push down on the spring.

CONCEPT CHECK

1. What is the net force on a bathroom scale when a 110-pound person stands on it?
2. Suppose you stand on two bathroom scales with your weight evenly distributed between the two scales. What will each scale read? How about if you lean to put more of your weight on one scale than the other?

Check Your Answers

1. Zero, for the scale remains at rest. The scale reads *support force,* which has the same magnitude as weight—not the net force.
2. The reading on each scale is half your weight. If you put more of your weight on one scale than the other, more than half your weight will be read on that scale but less on the other. In this way they add up to your weight. Like the example of the gymnast hanging by the rings, if one scale reads two-thirds her weight, the other scale will read one-third her weight.

2.9 Equilibrium for Moving Objects

When an object isn't moving, it's in equilibrium. The forces on it add up to zero. But the state of rest is only one form of equilibrium. An object moving at constant speed in a straight-line path is also in equilibrium. We say the same thing when we say an object moving at constant velocity is in equilibrium. The forces on this object are also zero (in accord with Newton's first law).

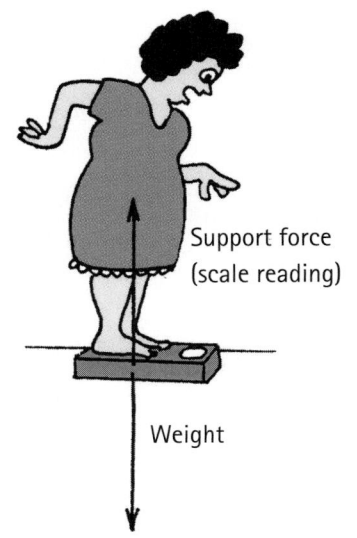

Support force (scale reading)

Weight

FIGURE 2.15 ▲
The upward support is as much as your weight.

75-N friction   75-N
force   applied force

FIGURE 2.16 ▲
When the push on the crate is as great as the force of friction between the crate and the floor, the net force on the crate is zero and it slides at an unchanging speed.

Equilibrium is a state of no change. A bowling ball rolling at constant velocity is in equilibrium—until it hits the pins. Whether at rest or steadily rolling in a straight-line path, the sum of the forces on the bowling ball is zero: $\Sigma F = 0$.

It follows from Newton's first law that an object under the influence of only one force cannot be in equilibrium. Net force couldn't be zero in such a case. Only when there is no force at all or two or more forces combining to zero can an object be in equilibrium. We can test whether or not something is in equilibrium by noting whether or not it undergoes changes in motion.

Consider a crate of canned food being pushed horizontally across a storeroom floor. If it moves at constant velocity, it is in equilibrium. This tells us that more than one force is acting on the crate—likely the other force is the friction between the crate and the floor. The fact that the net force on the crate equals zero means that the force of friction must be equal in magnitude and opposite in direction to our pushing force.

✔ Objects at rest are said to be in *static* equilibrium; objects moving at constant velocity are said to be in *dynamic* equilibrium. Both of these situations are examples of mechanical equilibrium. There are other types of equilibrium. For example, when we study heat, we'll talk about thermal equilibrium, where temperature doesn't change. For rotational motion there is rotational equilibrium.

✔ **READING CHECK**

Distinguish between static and dynamic equilibrium.

CONCEPT CHECK
An airplane flies at constant velocity. In other words, it is in dynamic equilibrium. Two horizontal forces act on the plane. One is the thrust of the propeller that pushes it forward. The other is the force of air resistance (drag) that acts in the opposite direction. Which force is greater?

Drag ←   → Thrust
      XAAPT

Check Your Answer
Both forces have the same magnitude. Call the forward force exerted by the propeller positive. Then the air resistance is negative. Since the plane is in equilibrium, can you see that the two forces combine to equal zero?

2.10 Earth Moves Around the Sun

Before the 16th century, it was believed that Earth was the center of the universe. People then believed that the Sun circled Earth. As mentioned in Chapter 1, Galileo advanced the idea that Earth moves around the Sun, instead of the other way around. There was much arguing and debate about this idea. People thought like Aristotle, and

the existence of a force big enough to keep Earth moving was beyond their imagination. One of the arguments against a moving Earth was the following.

Consider a bird sitting at rest at the top of a tall tree. On the ground below is a fat, juicy worm. The bird sees the worm and drops vertically below and catches it. It was argued that this would be impossible if Earth were moving. A moving Earth would have to travel at an enormous speed to circle the Sun in 1 year. While the bird is in the air descending from its branch to the ground below, the worm would be swept far away along with the moving Earth. It seemed that catching a worm on a moving Earth would be an impossible task. The fact that birds do catch worms from high tree branches seemed to be clear evidence that Earth must be at rest.

Can you see the mistake in this argument? You can if you use the concept of inertia. You see, not only is Earth moving at a great speed but so are the tree, the branch of the tree, the bird that sits on it, the worm below, and even the air in between. All move at the same speed. ✔ The law of inertia states that objects in motion remain in motion whenever no net force acts. So objects on Earth move with Earth as Earth moves around the Sun. When the bird drops from the branch, its initial sideways motion remains unchanged. It catches the worm quite unaffected by the motion of its total environment.

We live on a moving Earth. Stand next to a wall. Jump up so that your feet are no longer in contact with the floor. Does the moving wall slam into you? Why not? It doesn't because you are also traveling as fast as Earth, with the same speed, before, during, and after your jump. The speed of Earth relative to the Sun is not the speed of the wall relative to you.

Four hundred years ago, people had difficulty with ideas like these. One reason is that they didn't yet travel in high-speed vehicles. Rather, they took slow, bumpy rides in horse-drawn carts. People couldn't notice the effects of inertia as much. Today we flip a coin in a high-speed car, bus, or plane and catch the vertically moving coin as we would if the vehicle were at rest. We see evidence for the law of inertia when the horizontal motion of the coin before, during, and after the catch is the same. The coin keeps up with us.

Our ideas of motion today are very different from those of our distant ancestors. Aristotle did not recognize the idea of inertia, because he did not see that all moving things follow the same rules. He imagined different rules for motion in the heavens and motion on Earth. He saw horizontal motion as "unnatural," requiring a steady force. Galileo and Newton, on the other hand, saw that all moving things follow the same rules. To them, moving things require *no* force to keep moving if friction is absent. We can only wonder how differently science might have progressed if Aristotle had recognized the unity of all kinds of motion.

FIGURE 2.17 ▲
Can the bird drop down and catch the worm if Earth moves at 30 km/s?

FIGURE 2.18 ▲
When you flip a coin in a high-speed airplane, it behaves as if the airplane were at rest. The coin keeps up with you—inertia in action!

✔ **READING CHECK**

How does the law of inertia apply to objects in motion?

2 CHAPTER REVIEW

KEY TERMS

Equilibrium rule $\Sigma F = 0$.

Force A push or a pull.

Inertia The property of things remaining at rest if at rest, and in motion if in motion.

Net force The combination of all forces that act on an object.

Newton The scientific unit of force.

Newton's first law of motion Every object continues in a state of rest, or in a state of motion in a straight line at constant speed, unless a net force acts on it.

Speed The distance traveled per time.

Support force The force that supports an object against gravity.

Vector quantity A quantity that specifies direction as well as magnitude.

Velocity The speed of an object and specification of its direction of motion.

REVIEW QUESTIONS

Aristotle's Classification of Motion

1. According to Aristotle, where do moving objects tend to reach?
2. According to Aristotle, what kind of motion requires no force?

Galileo's Concept of Inertia

3. What two of Aristotle's main ideas did Galileo discredit?
4. What is the name of the property of objects to maintain their states of motion?

Galileo's Concepts of Speed and Velocity

5. Distinguish between speed and velocity.
6. Why do we say velocity is a vector quantity and speed is not?

Motion Is Relative

7. How can you be both at rest and also moving at 100,000 km/h at the same time?
8. Was it Aristotle or Galileo who relied on experiments?

Newton's First Law of Motion— The Law of Inertia

9. Who was the first to discover the concept of inertia, Galileo or Newton? Who incorporated it into a law of motion?
10. What is the tendency of a moving object when no forces act on it?

Net Force—The Combination of All Forces That Act on an Object

11. When only a pair of equal and opposite forces act on an object, what is the net force acting on it?
12. We've learned that velocity is a vector quantity. Is force also a vector quantity? Why or why not?

Equilibrium for Objects at Rest

13. Name the force that occurs in a rope when both ends are pulled in opposite directions.
14. How much tension is there in a rope that holds a 20-N bag of apples at rest?
15. What is the meaning of $\Sigma F = 0$?

The Support Force—Why We Don't Fall Through the Floor

16. Why is the support force on an object often called the *normal* force?
17. When you weigh yourself, are you actually reading the support force acting on you, or are you really reading your weight?

Equilibrium for Moving Objects

18. Give an example of something moving when a net force of zero acts on it.

19. If we push a crate at constant velocity, how do we know how much friction acts on the crate compared to our pushing force?

Earth Moves Around the Sun

20. If you're in a smooth-riding bus that is going at 50 km/h and you flip a coin vertically, what is the horizontal velocity of the coin in midair?

THINK AND DO

1. By any method you choose, determine your average walking speed. How do your results compare with those of your classmates?

2. Ask a friend to drive a small nail into a piece of wood placed on top of a pile of books on your head. Why doesn't this hurt you? (Be careful! Wear a helmet in case your partner misses and use a very small nail, safety glasses, and a wooden mallet instead of a hammer.)

3. Write a letter to grandma and explain the difference between velocity and acceleration, and how some students confuse the two concepts.

PLUG AND CHUG

These four simple problems are simply to familiarize you with the equation for average speed.

$$\text{Average speed} = \frac{\text{total distance covered}}{\text{time interval}}$$

1. Calculate Emily's average walking speed when she covers 1 meter in 0.5 second.

2. Calculate your average speed if you run 100 meters in 10 seconds.

3. Calculate the average speed of a mouse who runs across a 4-meter-long room in 0.4 seconds.

4. Calculate the speed of a bowling ball that moves 8 meters in 4 seconds.

THINK AND COMPARE

1. Jogging Johnny runs along a train flatcar that moves at the velocities shown. From greatest to least, rank the relative velocities of Jake as seen by an observer on the ground. (Call the direction to the right positive.)

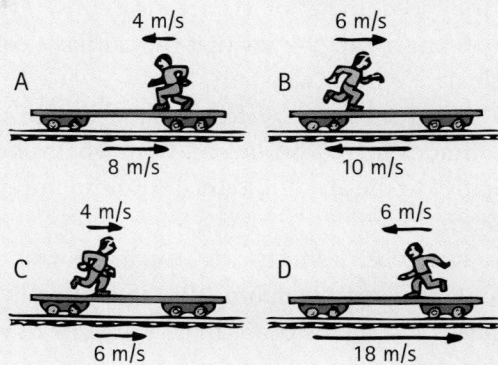

2. A track is made of a piece of channel metal bent as shown. A ball is released from rest at the left end of the track and continues past the various points. Rank the ball at points A, B, C, and D, from fastest to slowest. (Watch for tie scores.)

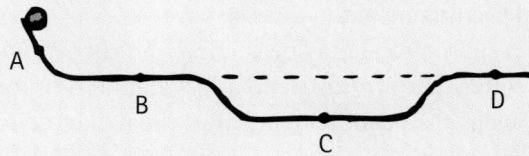

3. A block of iron is suspended by ropes in the positions shown below. Scales measure the tension (stretching force) in the ropes. Rank the scale readings from greatest to least.

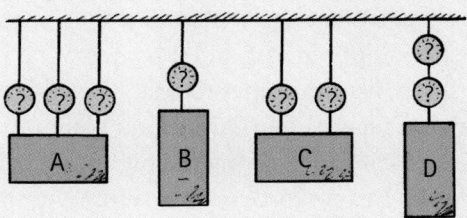

THINK AND EXPLAIN

1. Galileo found that a ball rolling down one incline will pick up enough speed to roll up another. How high will it roll compared to its initial height?

2. Correct your friend who says, "The race-car driver rounded the curve at a constant velocity of 100 km/h."

3. If the speedometer of a car reads a constant speed of 50 km/h, can you say that the car has a constant velocity? Why or why not?

4. A hungry mosquito watches you resting in a hammock in a 3-m/s breeze. With what velocity relative to the air should the mosquito hover above you for lunch?

5. If a huge bear were chasing you, its enormous mass would be very threatening. But if you ran in a zigzag pattern, the bear's mass would be to your advantage. Why?

6. Joshua gives his skateboard a push and it rolls across the classroom floor. Sophia says that after it leaves Joshua's hand, his *force* remains with it, keeping it going. Lillian disagrees and says that Joshua's push gives the skateboard *speed*, not force, and that when his hand no longer makes contact the force is no more. Which one do you agree with? Discuss this with your classmates.

7. A space probe may be carried by a rocket into outer space. Your friend asks what kind of force keeps the probe moving after the rocket no longer pushes it. What is your answer?

8. Consider a ball at rest in the middle of a toy wagon. When the wagon is pulled forward, the ball rolls to the back of the wagon. Interpret this observation in terms of Newton's first law.

9. Why do you lurch forward in a bus that suddenly slows? Why do you lurch backward when it picks up speed? What law applies here?

10. Push a shopping cart and it moves. When you stop pushing, it soon comes to rest. Does this violate Newton's law of inertia? Defend your answer.

11. When a car moves along the highway at constant velocity, the net force on it is zero. Why, then, do you continue running your engine? (*Hint:* How does friction enter into your answer?)

12. Consider a pair of parallel forces, one having a magnitude of 20 N, and the other 12 N. What maximum net force is possible for these two forces? What is the minimum net force possible?

13. The sketch shows a painting scaffold in mechanical equilibrium. The person in the middle weighs 250 N, and the tensions in each rope are 200 N. What is the weight of the scaffold?

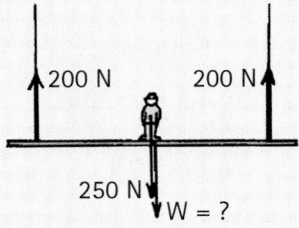

14. A different scaffold that weighs 300 N supports two painters, one of whom weighs 250 N and the other 300 N. The reading on the left scale is 400 N. What is the reading on the right-hand scale?

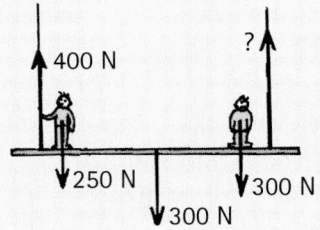

15. Nellie Newton hangs at rest from the ends of the rope, as shown. How does the reading on the scale compare to her weight?

16. Harry the painter swings year after year from his bosun's chair. His weight is 500 N and the rope, unknown to him, has a breaking point of 300 N. Why doesn't the rope break when he is supported as shown at the left below? One day Harry is painting near a flagpole, and, for a change, he ties the free end of the rope to the flagpole instead of to his chair, as shown at the right. Why did Harry end up taking his vacation early?

17. Latisha stands on a bathroom scale and reads her weight. Does the reading change if she raises one foot and leaves only one foot on the scale? How about if she stands with one foot on the scale and the other foot on the floor?

18. A child learns in school that Earth is traveling faster than 100,000 km/h around the Sun and in a frightened tone asks why we aren't swept off. What is your explanation?

19. If you toss a coin straight upward while riding in a train, where does the coin land when the motion of the train is uniform along a straight-line track? When the train slows while the coin is in the air? When the train is turning?

20. As Earth rotates about its axis, it takes 3 hours for the United States to pass beneath a point above Earth that is stationary relative to the Sun. What is wrong with the following scheme? To travel from Washington D.C. to San Francisco and use very little fuel, simply ascend in a helicopter high over Washington D.C. and wait 3 hours until San Francisco passes below.

THINK AND SOLVE

1. A tennis ball travels the full length of the court, 24 m, in 0.5 s. What is its average speed?

2. What is your average speed if you run 50 m in 10 s? Show that the distance you'd travel at this speed for 1 minute would be 300 m.

3. Find the net force produced by a 30-N and 20-N force in each of the following cases:
   (a) Both forces act in the same direction.
   (b) Both forces act in opposite directions.

4. A horizontal force of 100 N is required to push a crate of cell phones across a floor at a constant speed.
   (a) What is the net force acting on the crate?
   (b) What is the force of friction acting on the crate?

5. Phil Physicer weighs 600 N (132 lb) and stands on two bathroom scales. He stands so one scale reads twice as much as the other. What are the scale readings?

If you have a good handle on this chapter—if you really do—then you should be able to score 7 out of 10 on this RAT. Check your answers with your teacher. If you score less than 7, study further before moving on.

Choose the best answer to each of the following.

1. Aristotle distinguished between natural motion and unnatural motion. Galileo stated that all motion
   (a) is natural.
   (b) is unnatural.
   (c) follows the same laws.
   (d) is an illusion.

2. The difference between speed and velocity most involves
   (a) direction.
   (b) amount.
   (c) different units.
   (d) acceleration.

3. The first scientist to discover the concept of inertia was
   (a) Aristotle.   (b) Galileo.
   (c) Newton.   (d) Einstein.

4. According to Newton's first law of motion
   (a) objects at rest tend to remain at rest.
   (b) objects in motion tend to remain in motion.
   (c) Both of these.
   (d) None of these.

5. A ball rolling along a bowling alley moves at constant speed
   (a) without the need of a horizontal force.
   (b) due to a slight force in its direction of motion.
   (c) due to slight friction that slows it.
   (d) due to gravity.

6. If your textbook is pulled to the right with 8 N while being pulled to the left with 5 N, the net force on the book is
   (a) 3 N to the left.
   (b) 5 N to the right.
   (c) 13 N to the right.
   (d) None of these.

7. The equilibrium rule, $\Sigma F = 0$, applies to
   (a) objects or systems at rest.
   (b) objects or systems in uniform motion in a straight line.
   (c) Both of these.
   (d) None of these.

8. When you stand on two bathroom scales, one foot on each scale with weight evenly distributed, each scale will read
   (a) your weight.
   (b) half your weight.
   (c) zero.
   (d) actually more than your weight.

9. If you push your desk along the floor with a force of 80 N and it slides at constant speed, then the force of friction acting on the desk is
   (a) also 80 N in the direction of your push.
   (b) 80 N in the direction opposite to your push.
   (c) zero.
   (d) None of these.

10. Suppose you stand on the floor of a train moving at constant speed in a straight line. If you jump straight upward, you will land
    (a) at your original location.
    (b) slightly in front of your original location.
    (c) slightly in back of your original location.
    (d) in back of your original location a distance that depends on how high you jump.

3
NEWTON'S SECOND LAW OF MOTION— FORCE AND ACCELERATION

THE MAIN IDEA

An object accelerates when a net force acts on it.

Kick a soccer ball and it moves. Does the ball's motion depend on its mass? Does its motion depend on how hard it's kicked? Or does how far the ball moves depend both on its mass and on how hard you kick it? What about air resistance? Does that also affect how far the soccer ball goes?

Consider some more examples. Suppose that you drop two soccer balls from a cliff, one ball with air inside and the other filled with heavy sand. Which hits the ground first? Does air resistance make a difference? Suppose you and a much heavier friend skydive from a high-flying plane. You both open your same-size parachutes at the same time. Who reaches the ground first? Does air resistance make a difference in this case? What rules guide the answers to these questions?

Explore!

What Effect Does Air Resistance Have on Falling Objects?

1. Hold a sheet of paper and a book, side by side, over the top of a table. Then drop them at the same time.
2. Place the paper underneath the book and repeat.
3. Place the paper inside the book and repeat.

Analyze and Conclude

1. **Observing** Not surprisingly, when dropped side by side, the book fell to the tabletop first. When the paper was placed below and in the book, they fell at the same speed. Would you have expected otherwise?

2. **Predicting** If the sheet of paper is placed on *top* of the book, not sticking over its edge, how will the accelerations of each compare when dropped? (Prepare to be surprised!)
3. **Making Generalizations** Air resistance slows a falling object, providing the air is in the path of the object.

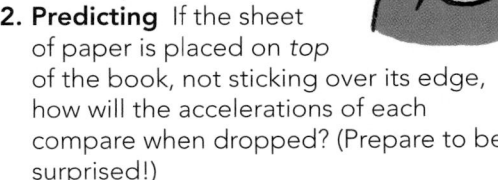

3.1 Galileo Developed the Concept of Acceleration

In addition to speed and velocity (previous chapter), Galileo experimented with inclined planes to developed the concept of *acceleration*. He found that balls rolling down inclines rolled faster and faster. Their velocity changed as they rolled. Further, the balls gained the same amount of velocity in equal time intervals. In other words, their velocity increased by a given amount each second.

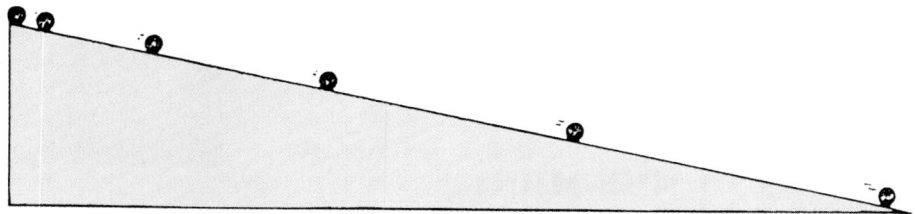

FIGURE 3.1 ▲
A ball gains the same amount of velocity each second as it rolls down an incline.

Galileo defined the rate of change of velocity as **acceleration:***

$$\textbf{Acceleration} = \frac{\textbf{change of velocity}}{\textbf{time interval}}.$$

You observe acceleration every time you ride in a bus. When the driver steps on the gas pedal the bus gains speed. We say it accelerates. We see why the gas pedal is called the "accelerator"!

When the brakes are applied, the car slows. This is also acceleration because the velocity of the car is changing. When something slows down, we often call this *deceleration*.

When a car makes a turn, even if its speed does not change, it is accelerating. Can you see why? Acceleration occurs because the car's direction is changing. Acceleration is a change in velocity. ✔ So acceleration is a change in speed, a change in direction, or a change in both speed *and* direction. Figure 3.2 illustrates this.

Can you see that a car has 3 controls that change velocity: the gas pedal (accelerator), brakes, and steering wheel?

FIGURE 3.2 ▶
The vehicle undergoes acceleration when there is a *change* in its state of motion.

---

* The Greek letter Δ (delta) is often used as a symbol for "change in" or "difference in." In "delta" notation, $a = \frac{\Delta v}{\Delta t}$, where Δv is the change of velocity, and Δt is the change in time (the time interval).

FIGURE 3.3 ▶
Rapid deceleration is sensed by the driver who lurches forward (inertia in action!).

Suppose we are driving. In 1 second, we steadily increase our velocity from 30 kilometers per hour to 35 kilometers per hour. In the next second, we go from 35 kilometers per hour to 40 kilometers per hour, and so on. We change our velocity by 5 kilometers per hour each second. We see that

$$\text{Acceleration} = \frac{\text{change of velocity}}{\text{time interval}} = \frac{5 \text{ km/h}}{1 \text{ s}} = 5 \text{ km/h·s.}$$

In this example, the acceleration is 5 kilometers per hour per second (abbreviated as 5 km/h·s). Note that a unit for time enters twice: once for the unit of velocity, and again for the interval of time in which the velocity is changing. Also note that acceleration is not just the change in velocity, it is the *change per second* of velocity. If either speed or direction (or both) changes, then the velocity changes.

Hold an apple above your head and drop it. It accelerates during its fall. When air resistance doesn't affect its motion, the apple is in **free fall.** Interestingly, the amount of acceleration is the same for all freely falling objects in the same vicinity. Experiment shows that a freely falling object gains speed at the rate of 10 m/s each second:

$$\text{Acceleration} = \frac{\text{change of speed}}{\text{time interval}} = \frac{10 \text{ m/s}}{1 \text{ s}} = 10 \text{ m/s}^2.$$

We read the acceleration of free fall as 10 meters per second squared. This is the same as saying that acceleration is 10 meters per second per second. Note again that the unit of time, the second, appears twice—once for the unit of velocity and again for the time during which velocity changes.

In Figure 3.4, we imagine a freely falling boulder with a speedometer attached. As the boulder falls, the speedometer shows that the boulder falls 10 m/s faster each second. This 10 m/s gain each second is the boulder's acceleration.

The acceleration of free fall is further developed in the Practice Book.

If you drop a ball, it gains speed because of the force of gravity. How about a ball thrown straight upward? Once it leaves your hand, the ball continues moving upward for a while and then comes back down. While going up, it moves against gravity and loses speed. Guess how much speed it loses each second while going upward? And guess how much speed it gains each second while coming down? That's right: The change in speed per second is 10 m/s—whether moving upward or downward! At the highest point, when the ball

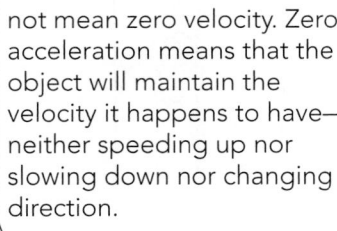
It's important to emphasize that zero acceleration does not mean zero velocity. Zero acceleration means that the object will maintain the velocity it happens to have—neither speeding up nor slowing down nor changing direction.

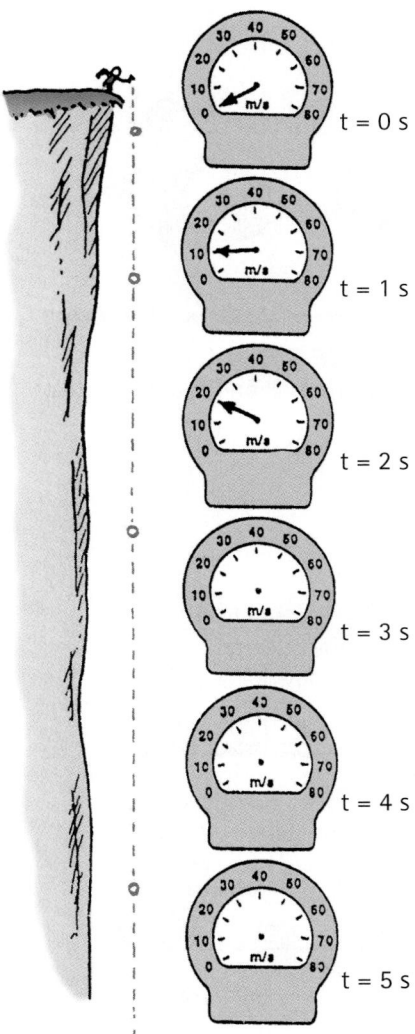

t = 0 s
t = 1 s
t = 2 s
t = 3 s
t = 4 s
t = 5 s

FIGURE 3.4 ▲
Imagine a falling boulder equipped with a speedometer. In each succeeding second of fall, you'd find the boulder's speed increasing by the same amount; 10 m/s. Sketch in the missing speedometer needle at t = 3 s, 4 s, and 5 s.

✓ **READING CHECK**

Name three kinds of changes in motion that occur with acceleration.

PhysicsPlace.com

Video
Force Causes Acceleration

FIGURE 3.5 ▲
Kick the ball and it accelerates.

Force of hand accelerates the brick

Twice as much force produces twice as much acceleration

Twice the force on twice the mass gives the same acceleration

FIGURE 3.6 ▲
Acceleration is directly proportional to net force.

changes direction from upward to downward, its instantaneous speed is zero. Then it starts downward *just as if it had been dropped from rest at that height.* It will return to its starting point with the same speed it had when thrown.

3.2 Force Causes Acceleration

Any object that accelerates is acted on by a push or a pull—a force of some kind. It may be a sudden push, like hitting a punching bag, or the steady pull of gravity when falling. Acceleration is caused by applying force.

Most often, more than one force acts on an object. Recall from the previous chapter that the combination of forces that act on an object is the *net force.* Acceleration depends on the *net force.* For example, if you push with 25 N on an object, and somebody else pushes in the opposite direction with 15 N, the net force applied to the object is 10 N. The object will accelerate as if a single 10-N force acts on it.

Suppose you pull a toy wagon with a net force of 20 N. The wagon accelerates. Now double the force and pull with 40 N. How much more will the wagon accelerate now? There is a general rule here. If the net force is doubled, the acceleration also doubles. Three times the net force produces three times the acceleration. We say that the acceleration is directly proportional to the net force. We write

Acceleration ~ net force.

The symbol ~ stands for "is directly proportional to." That means any change in one quantity matches the same amount of change in the other quantity.

✓ The direction of acceleration is always in the direction of the net force. When a force is applied in the direction of the object's motion, the speed increases. When a force is applied in the opposite direction, the speed decreases. When a force acts at right angles, it will change the direction of the object.

Force *changes* motion; it doesn't *cause* motion.

When one thing is directly proportional to another, then as one gets bigger the other gets bigger too.

CONCEPT CHECK

1. If you push on a shopping cart, it will accelerate. If you apply four times the net force, how much greater will the acceleration be?
2. If the net force acting on a sports car is reduced to half, how will the acceleration change?

Check Your Answers
1. Acceleration will be four times as much.
2. Acceleration will be half as much.

So we see that net force produces acceleration. How much acceleration, however, also depends on something else. It depends on the mass of the object being pushed or pulled.

3.3 Mass Is a Measure of Inertia

What happens when you kick a tin can? It accelerates—changes its state of motion. Now kick the same can filled with rocks. What happens? It doesn't accelerate as much as when it was empty. If the can is full of something really heavy like lead, it will hardly move. Ouch!

The more massive full can has more **inertia** than the empty can. In other words, it is more resistant to a change in motion. This suggests that the **mass** of an object relates to its inertia. The greater an object's mass, the greater its inertia. This is why powerful engines are required in tractor trailers that pull massive loads—and why they have powerful brakes for stopping. Heavy loads have lots of inertia.

Mass is more than an indication of an object's inertia. Mass is also a measure of how much material an object contains. Mass depends on the number and kinds of atoms making up the object. A dense material such as lead is made up of many tightly packed atomic particles. So it has a lot of mass for a given volume.

Mass Is Not Volume

Do not confuse mass and volume. **Volume** is a measure of space. It is measured in units such as cubic centimeters, cubic meters, or liters. Mass is measured in **kilograms.** If an object has a large mass, it may or may not have a large volume. For example, equal-size bags of popcorn and jellybeans may have equal volumes but very unequal masses. How many kilograms of matter an object contains and how much space the object occupies are two different things. Mass is different from volume.

Mass Is Not Weight

Do not confuse mass and weight. They are different from each other. As already mentioned, mass is a measure of the amount of matter in an object and depends on its number and kinds of atoms. **Weight,** however, depends on gravity. You would weigh less on the Moon, for example, than you do on Earth. Why? The Moon's gravity is weaker than Earth's so you'd be pulled to the Moon's surface with less force than on Earth. On the other hand, the numbers and kinds of atoms in your body are the same on the Moon as on Earth. There is just as much material in your body no matter where you are. So, unlike weight, your mass doesn't change if gravity varies.

You can sense how much mass is in an object by feeling its inertia. When you shake an object back and forth, you can sense its inertia. If it has a lot of mass, it's hard to change the object's direction. If it has a

✔ **READING CHECK**

How does the direction of acceleration compare with the direction of the net force that produces it?

Video
Definition of a Newton
Interactive Tutorial
Parachutes and Newton's Second Law

FIGURE 3.7 ▲
The greater the mass, the greater the force needed for a given acceleration.

FIGURE 3.8 ▲
An anvil in outer space, between the Earth and Moon for example, may be weightless, but it is not massless.

FIGURE 3.9 ▲
The astronaut in space finds it is just as difficult to shake the "weightless" anvil as if it were on Earth. If the anvil is more massive than the astronaut, which shakes more—the anvil or the astronaut?

small mass, shaking the object is easier. To-and-fro shaking requires the same force even in regions where gravity is different—on the Moon, for example. The rock's inertia, or mass, is a property of the object itself and not its location.

We define mass and weight as follows:

Mass is the amount of matter in an object; also, it is a measure of the inertia, or "laziness," that an object shows when you try to change its state of motion.

Weight is the force due to gravity on an object.

✓ Although mass and weight are different from each other, they are directly *proportional* to each other. Objects with a great mass have a great weight; objects with little mass have little weight. In the same location, twice the mass weighs twice as much. That's what we mean when we say that mass and weight are proportional to each other. Remember that mass has to do with the amount of matter in the object and its inertia, while weight has to do with how strongly that matter is attracted by gravity.

A pillow is bigger than an auto battery, but which has more matter—more *inertia*—more *mass*?

CONCEPT CHECK

1. Does a 2-kilogram iron block have twice as much *inertia* as a 1-kilogram iron block? Twice as much *mass*? Twice as much *volume*? Twice as much *weight* when weighed in the same location?
2. Does a 2-kilogram iron block have twice as much *inertia* as a 1-kilogram bunch of bananas? Twice as much *mass*? Twice as much *volume*? Twice as much *weight* when weighed in the same location?
3. How does the mass of a bar of gold vary with location?

Check Your Answers

1. The answer is yes to all questions. A 2-kilogram block of iron has twice as many iron atoms and therefore twice the amount of matter, mass, and weight. The blocks are made of the same material, so the 2-kilogram block also has twice the volume.
2. Two kilograms of *anything* have twice the inertia and twice the mass of 1 kilogram of anything else. Since mass and weight are proportional in the same location, 2 kilograms of anything will weigh twice as much as 1 kilogram of anything. Except for volume, the answer to all the questions is yes. Volume and mass are proportional only when the materials are the same—when they have the same *density*. Iron is much more dense than bananas, so 2 kilograms of iron must occupy less volume than 1 kilogram of bananas.
3. Not at all! It is made of the same number of atoms no matter what the location. Although its weight may vary with location, it has the same mass everywhere. This is why mass is preferred to weight in scientific studies.

1 Kilogram Weighs 10 Newtons

The standard unit of mass is the kilogram, abbreviated kg. The standard unit of force is the newton, as discussed in Chapter 2. The standard symbol for the newton is N. The abbreviation is written with a capital letter because the unit is named after a person. A 1-kg bag of any material has a weight of 10 N in standard units. Away from the Earth's surface where the force of gravity is less, the bag would weigh less.

So 1 kilogram of something weighs about 10 newtons. If you know the mass in kilograms and want weight in newtons, multiply the number of kilograms by 10. Or, if you know the weight in newtons, divide by 10 and you'll have the mass in kilograms. Weight and mass are proportional to each other. In cases where precision is needed—for example in some lab experiments—1 kilogram of matter actually weighs 9.8 N. For convenience, we round 9.8 N off to 10 N.

The relationship between kilograms and pounds is that 1 kg weighs 2.2 lb at Earth's surface. (That means 1 lb is the same as 4.45 N.)

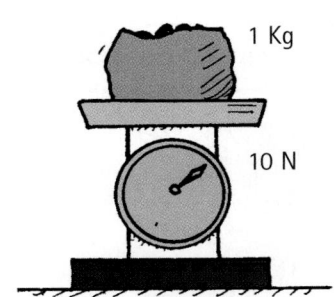

FIGURE 3.10 ▲
1 kilogram of nails weighs 10 newtons, which is equal to 2.2 pounds.

CONCEPT CHECK

Why is it okay to say a 1-kg bag of sand weighs 10 N, but a 1-kg bag of gold weighs 9.8 N? Don't they weigh the same?

Check Your Answers

Both 1-kg bags have the same weight, and both weigh 10 N. However, since gold is more valuable, saying precisely 9.8 N rather than 10 N is usually a good idea. Except for in the lab, we won't make a big deal in this book about whether or not rounding off is okay. It's more important to learn the concepts.

✔ **READING CHECK**

In what way are mass and weight related?

3.4 Mass Resists Acceleration

More massive objects are more difficult to accelerate. Experiments show that for the same force, twice as much mass results in half as much acceleration; three times the mass results in one-third the acceleration, and so forth. In other words,  the acceleration produced by a given force is *inversely* proportional to the mass. We write

$$\text{Acceleration} \sim \frac{1}{\text{mass}}.$$

PhysicsPlace.com

Tutorial
Parachutes and Newton's Second Law"

Force of hand accelerates the brick

The same force accelerates 2 bricks 1/2 as much

3 bricks, 1/3 as much acceleration

✔ **READING CHECK**

For a given force, how does the acceleration of an object relate to its mass?

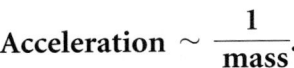
FIGURE 3.11 ▲
Acceleration is inversely proportional to mass.

Here's directly proportional.

Here's inversely proportional.

By **inversely** we mean that the two values change in opposite ways. When one value gets larger, the other gets smaller. Or when one gets smaller, the other gets larger. More mass means less acceleration, because more mass means more resistance to changes in motion. (Mathematically, we see that as the denominator increases, the whole quantity decreases. For example, the quantity 1/100 is less than the quantity 1/10.)

CONCEPT CHECK

1. Suppose you're offered either $\frac{1}{4}$ of an apple pie or $\frac{1}{8}$ of the pie. Which piece is larger?
2. Suppose you apply the same amount of force to two carts, one cart with a mass of 4 kg and the other with a mass of 8 kg.
   a. Which cart will accelerate more?
   b. How much greater will the acceleration be?

Check Your Answers

1. The larger piece is $\frac{1}{4}$, one-quarter of the pie. If you choose the $\frac{1}{8}$ piece, you'll have half as much pie as $\frac{1}{4}$.
2. a. The 4-kg cart will have more acceleration.
   b. The 4-kg cart will have *twice* the acceleration because it has half as much mass—which means half as much resistance to changes in motion!

3.5 Newton's Second Law Links Force, Acceleration, and Mass

PhysicsPlace.com

Video
Newton's Second Law
Interactive Tutorial
Parachutes and Newton's
Second Law

Isaac Newton was the first to realize the connection between force and mass in producing acceleration. He discovered one of the most important rules of nature ever proposed—his *second law of motion.* **Newton's second law** states:

> **The acceleration produced by a net force on an object is directly proportional to the net force, is in the same direction as the net force, and is inversely proportional to the mass of the object.**

Or in shorter notation,

$$\text{Acceleration} \sim \frac{\text{net force}}{\text{mass}}.$$

By using consistent units such as newtons (N) for force, kilograms (kg) for mass, and meters per second squared (m/s^2) for acceleration, we get the exact equation.

$$\text{Acceleration} = \frac{\text{net force}}{\text{mass}}$$

In briefest form, where a is acceleration, F is net force, and m is mass:

$$a = \frac{F}{m}.$$

FIGURE 3.12 ▲
Acceleration depends on the mass being pushed.

CALCULATION CORNER

Consider a 1000-kg car pulled by a cable with 2000 N of force. What will be the acceleration of the car? Using Newton's second law, we find

$$a = \frac{F}{m} = \frac{2000\ N}{1000\ kg} = \frac{2000\ kg \cdot m/s^2}{1000\ kg} = 2\ m/s^2.$$

Here we see that the units of force, the N, are the same as the units kg · m/s². From now on, we'll take a short-cut and simply say that the ratio N/kg equals m/s².

Suppose that the force were 4000 N. What would be the acceleration of the car?

$$a = \frac{F}{m} = \frac{4000\ N}{1000\ kg} = 4\ m/s^2$$

Doubling the force on the same mass simply doubles the acceleration.

✓ The acceleration of an object equals the net force divided by the mass. If the net force acting on an object is doubled, the object's acceleration will be doubled. Suppose instead that the mass is doubled. Then the acceleration will be halved. If both the net force and the mass are doubled, then the acceleration will be unchanged.

> When one thing is **inversely proportional** to another, then, as one gets bigger, the other gets smaller.

CONCEPT CHECK

If you push on a shopping cart, it will accelerate.
1. If you push the same, but the cart is loaded with groceries so it has five times as much mass, what happens to the acceleration?
2. If you push five times harder on the loaded cart, what happens to the acceleration?

Check Your Answers
1. Acceleration will be less—only one-fifth as much.
2. It will have the same acceleration as it had without the load.

✓ **READING CHECK**

How does the acceleration of an object relate to the net force on it and its mass?

3.6 Friction Is a Force That Affects Motion

✓ **Friction** occurs when one object tends to rub or does rub against something else. Friction occurs for solids, liquids, and gases. An important rule of friction is that it acts in a direction to oppose motion. If you drag a bag of cement along a floor to the left, the force of friction on the bag will be to the right. A boat propelled to the east by its motor experiences water friction to the west. When an object falls downward through the air, the friction due to air resistance (**air drag**) acts upward. Friction acts in a direction to oppose motion.

Video
Friction
Tutorial
Parachutes and Newton's Second Law

1. A crate filled with delicious peaches rests on a horizontal floor. The only forces acting on the crate are gravity and the support force of the floor. We show these forces with vectors. $\mathbf{F_w}$ is the weight vector, and $\mathbf{F_N}$ is the support force. We use the subscript $_N$ to indicate the force is **normal** to the floor. *Normal* means at a right angle with respect to the floor. Support forces, interestingly, are always normal to the surfaces of support.

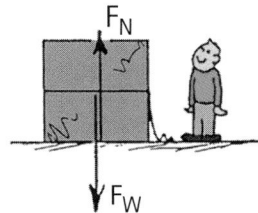

Choose the correct words.
   a. The net force on the crate is (zero) (greater than zero).
   b. Evidence for this is that there is no (velocity) (acceleration).

2. A small pull $\mathbf{P}$ is exerted on the crate, but not enough to move it.

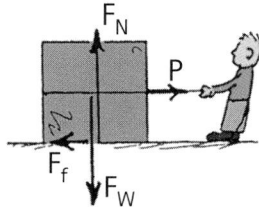

   a. A force of friction acts, which is (less than $\mathbf{P}$) (equal to $\mathbf{P}$) (greater than $\mathbf{P}$).
   b. Net force on the crate is (zero) (greater than zero).

3. Pull $\mathbf{P}$ is increased until the crate begins to slide. It is pulled so that it moves at constant velocity across the floor.

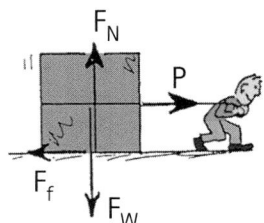

   a. Friction force $\mathbf{F_f}$ is (less than $\mathbf{P}$) (equal to $\mathbf{P}$) (greater than $\mathbf{P}$).
   b. Constant velocity means acceleration is (zero) (greater than zero).
   c. Net force on the crate is (less than zero) (zero) (greater than zero).

4. Pull $\mathbf{P}$ is further increased and is now greater than friction $\mathbf{F_f}$.

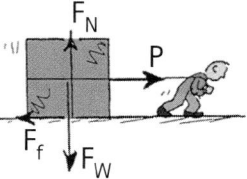

   a. Net force on the crate is (zero) (greater than zero).
   b. The net force acts to the right, so acceleration acts toward the (left) (right).
   We see here that the friction force varies with different pulling strengths. When the pull is small enough not enough to move the crate, friction equals the pull and net force is zero. As the pull is increased, friction increases the same and net force remains zero. When the pulling force is greater than the friction, the net force is greater than zero and the crate accelerates. Once the crate is moving, the pull can be reduced until it matches the friction and the crate moves steadily. Then acceleration is zero and the net force is zero.
   Discuss this with your classmates until you understand it. When you do, you're understanding good physics!

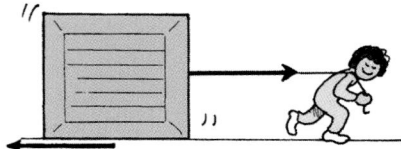

FIGURE 3.13 ▲
Applied force just overcomes friction so the crate slides at constant velocity.

The amount of friction between two surfaces depends on the kinds of material and how much they are pressed together. Friction between a crate and a rough wooden floor is greater than between the same crate and a polished linoleum floor. And if the surface is inclined, friction is less because the crate doesn't press as much on the inclined surface.

When you push horizontally on your textbook and it slides across the table, both your force and the opposite force of friction affect the motion. When you push hard enough to match friction, the net force on

the book is zero and it slides at constant velocity. Notice that we are talking about what we learned in the previous chapter—no change in motion occurs when $\Sigma F = 0$.

Learn your concepts now! If problem solving comes later, it will be much more meaningful!

CONCEPT CHECK

1. Two forces act on a bowl resting on a table: the bowl's weight and the support force from the table. Does a force of friction also act on the bowl?
2. Suppose a high-flying jumbo jet flies at constant velocity when the thrust of its engines is a constant 80,000 N. What is the *acceleration* of the jet? What is the force of air drag acting on the jet?

Check Your Answers

1. No, not unless the bowl tends to slide or does slide across the table. For example, if you push it toward the left, then friction between the bowl and table will act toward the right. Friction forces occur only when an object tends to slide or is sliding.
2. The acceleration is zero because the velocity is constant (not changing). Since the acceleration is zero, it follows from $a = \frac{F}{m}$ that the net force is zero. This says that the force of air drag must be equal to the thrusting force of 80,000 N and act in the opposite direction. So the air drag is 80,000 N.

 **READING CHECK**

When does friction occur?

3.7 Objects in Free Fall Have Equal Acceleration

When Galileo developed the concept of acceleration on inclined planes, he was interested in falling objects. Because there were no suitable timing devices at that time, he used inclined planes to effectively slow down acceleration. He found that as the planes were tipped at greater angles, the acceleration of the balls was greater. When tipped all the way vertical, acceleration was that of free fall. ✓ We define free fall as falling only under the influence of gravity, where other forces such as air drag can be neglected.

Galileo further discovered that the acceleration didn't depend on mass. On any incline all balls have the same acceleration. Likewise, the acceleration of free fall doesn't depend on mass.

PhysicsPlace.com
Interactive Tutorial
Parachutes and Newton's Second Law

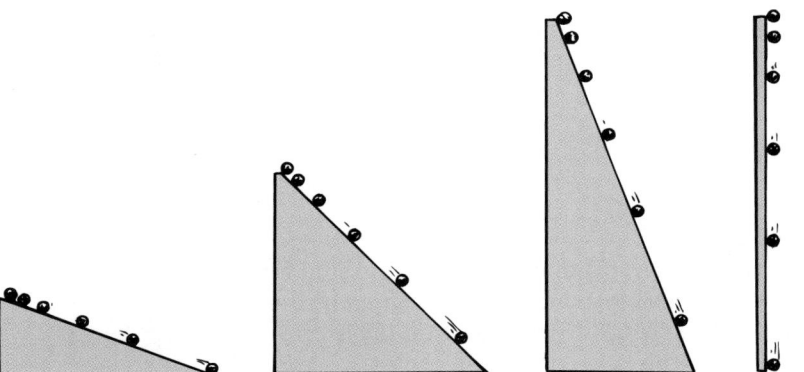

◀ **FIGURE 3.14**
The greater the slope of the incline, the greater the acceleration of the ball. What is the acceleration when the incline is vertical?

✔ **READING CHECK**

How is "free fall" defined?

PhysicsPlace.com

Video
Free-Fall Acceleration Explained
Interactive Tutorial
Parachutes and Newton's
Second Law

fyi

- We see in free fall that *weight/mass = g*. So we can say that *weight = mg!*

A 10-kilogram boulder and a 1-kilogram stone dropped from an elevated position at the same time will fall together and strike the ground at practically the same time. This experiment, said to be done by Galileo from the Leaning Tower of Pisa, destroyed the Aristotelian idea that heavy objects always fall faster than lighter ones in the same time. Galileo's experiment and many others demonstrated the same result. But Galileo couldn't say *why* the accelerations were equal. The explanation comes from Newton's second law.

3.8 Newton's Second Law Explains Why Objects in Free Fall Have Equal Acceleration

A falling 10-kg boulder "feels" ten times the force of gravity (weight) as a 1-kg stone. Followers of Aristotle believed the boulder should therefore fall ten times as fast as the stone—because they considered only the greater weight. But Newton's second law tells us to also consider the mass. Can you see that ten times as much force acting on ten times as much mass produces the same acceleration as the smaller force acting on the smaller mass? In symbolic notation,

$$\frac{F}{m} = \frac{F}{m},$$

where F stands for the force (weight) acting on the boulder and m stands for its correspondingly large mass. The small F and m stand for the smaller weight and mass of the stone. We see that the *ratio* of weight to mass is the same for these or any objects. All freely falling objects have the same force/mass ratio and undergo the same acceleration at the same location. The acceleration due to gravity is represented by the symbol g.

We can show the same result with numerical values. The weight of a 1-kg stone (or 1 kg of *anything*) is 10 N at the Earth's surface. The weight of 10 kg of matter, such as the boulder, is 100 N. The force acting on a falling object is its weight. The acceleration of the stone is

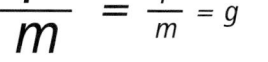

◀ **FIGURE 3.15**
The ratio of weight (F) to mass (m) is the same for the 10-kg cannonball and the 1-kg stone.

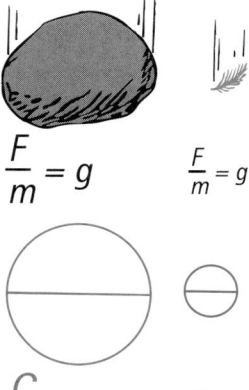

FIGURE 3.16 ▶
The ratio of weight (F) to mass (m) is the same for the large rock and the small feather; similarly, the ratio of circumference (C) to diameter (D) is the same for both the large and small circles.

CALCULATION CORNER

Fill in the blanks.

1. A 5-kg bag of sand has a weight of 50 N. When dropped, its acceleration is

$$a = \frac{50 \text{ N}}{5 \text{ kg}} = \underline{\hspace{1cm}} \text{ m/s}^2.$$

2. A 10-kg bag of sand has a weight of 100 N. When dropped its acceleration is

$$a = \frac{}{10 \text{ kg}} = \underline{\hspace{1cm}} \text{ m/s}^2.$$

3. Calculate the free-fall acceleration of a 20-kg bag of sand.

$$a = \underline{\hspace{1cm}} = \underline{\hspace{1cm}} \text{ m/s}^2.$$

$$a = \frac{F}{m} = \frac{weight}{m} = \frac{10 \text{ N}}{1 \text{ kg}} = 10 \text{ m/s}^2 = g,$$

and for the boulder,

$$a = \frac{F}{m} = \frac{weight}{m} = \frac{100 \text{ N}}{10 \text{ kg}} = 10 \text{ m/s}^2 = g.$$

We all know that a feather drops more slowly than a coin when both are dropped in air. Air drag more greatly affects the feather. But in a vacuum where air drag isn't present, a feather and coin dropped together will fall side by side (Figure 3.17). A vacuum pump removes air from the glass tube. We can see why acceleration is the same for both. With no air drag both have the same force/mass ratio.

It's of historical interest to note that although Galileo spoke about force and was the first to propose the concepts of acceleration and inertia, he did not make the connection between these three concepts. ✔ It took the genius of Isaac Newton to show the connection—namely, $a = \frac{F}{m}$. This rule of mechanics is one of the most profound in physics. With it, scientists and engineers have been able to put people on the Moon.

◀ **FIGURE 3.17**
A feather and a coin fall at equal accelerations in a vacuum.

✔ **READING CHECK**

What connection among force, mass, and acceleration did Galileo fail to realize, but Newton did?

3.9 Acceleration of Fall Is Less When Air Drag Acts

Most often, air drag is not negligible for falling objects. When this is the case, acceleration of fall is less than *g*. ✔ Air drag depends on two things: speed and surface area. When a sky diver steps from a high-flying plane, air drag builds up as speed increases. The result is reduced

For an object starting from rest, the free-fall distance *d* is given by $d = 1/2gt^2$.

In free fall, only a single force acts—the force of gravity. Whenever the force of air drag also occurs, the falling object is not in free fall.

◀ **FIGURE 3.18**
The flying squirrel increases its frontal area by spreading out. This increases air drag and decreases the speed of its fall.

PhysicsPlace.com

Videos
Falling and Air Resistance; Free Fall: How Fast?; $v = gt$; Air Resistance and Falling Objects; Free Fall: How Far?

Interactive Tutorial
Parachutes and Newton's Second Law

acceleration. More reduction can occur by increasing surface area. A diver does this by orienting the body so more air is encountered—by spreading out like a flying squirrel. So air drag depends on speed and the surface area encountered by the air (called the *frontal area*).

For free fall, the downward net force is weight. Only weight! But when air is present, the downward net force = weight − air drag. Can you see that the presence of air drag reduces net force? And that less net force means less acceleration? So as a diver falls faster and faster, the acceleration of fall gets less and less. What happens to the net force if air drag builds up to equal weight? The answer is, net force becomes zero. Here we see $\Sigma F = 0$ again. Then acceleration becomes zero. Does this mean the diver comes to a stop? No! What it means is the diver no longer picks up speed. Acceleration terminates—it no longer occurs. We say the diver has reached **terminal speed.** If we are concerned with direction—down for falling objects—we say the diver has reached **terminal velocity.**

Terminal speed for a human skydiver varies from about 150 to 200 km/h, depending on weight and orientation of the body. A heavier person falls faster for air drag to balance his or her weight. The greater weight is more effective in "plowing through" air. This means more terminal speed for a heavier person. Increasing frontal area reduces terminal speed. That's where a parachute comes in. A parachute greatly increases air drag, and terminal speed can be reduced to a safe 15 to 25 km/h.

Air drag

Air drag

Weight

Weight

FIGURE 3.19 ▲
The heavier parachutist must fall faster than the lighter parachutist for air drag to cancel her greater weight.

FIGURE 3.20 ▶
A stroboscopic study of a golf ball (left) and a Styrofoam ball (right) falling in air. The air drag is negligible for the heavier golf ball, and its acceleration is nearly equal to *g*. Air drag is not negligible for the lighter Styrofoam ball, which reaches its terminal velocity sooner.

When we previously discussed the interesting demonstration of the falling coin and feather in the glass tube, we found that the feather falls more slowly because of air drag. The feather's weight is very small so it reaches terminal speed very quickly. The feather doesn't have to fall very far or fast before air drag builds up to equal its small weight. The coin, on the other hand, doesn't have a chance to fall fast enough for air drag to build up to equal its weight.

What two things does air drag depend upon?

CONCEPT CHECK

Consider two parachutists, a heavy person and a light person, who jump from the same altitude with parachutes of the same size.
1. Which person reaches terminal speed first?
2. Which person has the greatest terminal speed?
3. Which person gets to the ground first?
4. If there were no air drag, as on the Moon, how would your answers to these questions differ?

Check Your Answers
To answer these questions, think of a coin and feather falling in air.
1. Just as a feather reaches terminal speed very quickly, the lighter person reaches terminal speed first.
2. Just as a coin falls faster than a feather through air, the heavy person falls faster and reaches a higher terminal speed.
3. Just like the race between a falling coin and feather in air, the heavier person falls faster and will reach the ground first.
4. If there were no air drag, there would be no terminal speed at all. Both would be in free fall and hit the ground at the same time.

fyi

- Skydivers and flying squirrels are not alone in increasing their surface areas when falling. When the Paradise Tree Snake (*Chrysopelea paradisi*) jumps from a tree branch, it doubles its width by flattening itself. It acquires a slightly concave shape and maneuvers itself by undulating in a graceful S-shape, traveling more than 20 meters in a single leap.

When Galileo reportedly dropped objects of different weights from the Leaning Tower of Pisa, they didn't actually hit at the same time. They almost did, but because of air drag, the heavier one hit a split second before the other. But this contradicted the much longer time difference expected by the followers of Aristotle. The behavior of falling objects was never really understood until Newton announced his second law of motion. Isaac Newton truly changed our way of seeing the world.

When Galileo tried to explain why all objects fall with equal accelerations, wouldn't he have loved to have known the rule $a = F/m$?

3 CHAPTER REVIEW

SUMMARY OF TERMS

Acceleration The rate at which velocity changes with time; the change may be in magnitude or direction or both.

Air drag Frictional resistance due to motion through air.

Free fall Motion under the influence of gravitational pull only.

Friction The resistive force that opposes the motion or attempted motion of an object past another with which it is in contact, or through a fluid.

Inertia The property of things to resist changes in motion.

Inversely When two values change in opposite directions, so that if one increases and the other decreases by the same amount, they are said to be inversely proportional to each other.

Kilogram The fundamental SI unit of mass.

Newton's second law The acceleration produced by a net force on an object is directly proportional to the net force, is in the same direction as the net force, and is inversely proportional to the mass of the object.

Mass The quantity of matter in an object. More specifically, it is the measure of the inertia or sluggishness that an object exhibits in response to any effort made to start it, stop it, deflect it, or change in any way its state of motion.

Terminal speed The speed at which the acceleration of a falling object terminates because air resistance balances its weight.

Terminal velocity Terminal speed with direction of motion (down for falling objects).

Volume The quantity of space an object occupies.

Weight The force due to gravity on an object.

REVIEW QUESTIONS

Galileo Developed the Concept of Acceleration

1. Distinguish between velocity and acceleration.
2. When are you most aware of motion in a moving vehicle—when it is moving steadily in a straight line or when it is accelerating?
3. What is the acceleration of free fall?

Force Causes Acceleration

4. Is acceleration proportional to net force or does acceleration equal net force?

Mass Is a Measure of Inertia

5. What relationship does mass have to inertia?
6. What relationship does mass have to weight?
7. Fill in the blanks: Shake something to and fro and you're measuring its_____. Lift it against gravity and you're measuring its_____.
8. What is the weight of a 1-kilogram brick?

Mass Resists Acceleration

9. Is acceleration *directly* proportional to mass, or is it *inversely* proportional to mass? Give an example.

Newton's Second Law Links Force, Acceleration, and Mass

10. If the net force acting on a sliding block is somehow tripled, by how much does the acceleration increase?
11. If the mass of a sliding block is somehow tripled at the same time the net force on it is tripled, how does the resulting acceleration compare to the original acceleration?

Friction Is a Force That Affects Motion

12. Suppose you exert a horizontal push on a crate that rests on a level floor and it doesn't move. How much friction acts compared with your push?
13. As you increase your push, will friction on the crate increase also?
14. Once the crate is sliding, how hard do you push to keep it moving at constant velocity?

Objects in Free Fall Have Equal Acceleration

15. What is meant by *free fall*?

Newton's Second Law Explains Why Objects in Free Fall Have Equal Acceleration

16. Why doesn't a heavy object accelerate more than a light object when both are freely falling?

17. The ratio of circumference/diameter for all circles is π. What is the ratio of force/mass for freely falling objects?

Acceleration of Fall Is Less When Air Drag Acts

18. What two principal factors affect the force of air drag on a falling object?

19. What is the acceleration of a falling object when it reaches its terminal velocity?

20. If two objects of the same size fall through air at different speeds, which encounters the greater air drag?

THINK AND DO

1. Drop a sheet of paper and a coin at the same time. Which reaches the ground first? Why? Now crumple the paper into a small, tight wad and again drop it with the coin. Explain the difference observed. Will they fall together if dropped from a second-, third-, or fourth-story window? Try it and explain your observations.

2. Drop two balls of different weights from the same height, and at small speeds they practically fall together. Will they roll together down the same inclined plane? If each is suspended from an equal length of string, making a pair of pendulums, and displaced through the same angle, will they swing back and forth in unison? Try it and see; then explain using Newton's laws.

3. The net force acting on an object and the resulting acceleration are always in the same direction. You can demonstrate this with a spool. If the spool is pulled horizontally to the right, in which direction will it roll?

4. Write a letter to grandpa, similar to the one to grandma in the previous chapter. Tell him that Galileo introduced the concepts of acceleration and inertia and was familiar with forces, but didn't see the connection among these three concepts. Tell him how Isaac Newton did and how the connection explains why heavy and light objects in free fall gain the same speed in the same time. In this letter, it's okay to use an equation or two, as long as you make it clear to grandpa that an equation is a shorthand notation of ideas you've explained.

PLUG AND CHUG

These are simply to familiarize you with the equations of a chapter—in this case, the definition of acceleration and its cause as stated by Newton's second law:

$$\text{Acceleration} = \frac{\text{change in velocity}}{\text{time interval}} = \frac{\Delta v}{\Delta t}.$$

1. Calculate the acceleration of a ball rolling down a hill that gains a speed of 4 m/s in 2 s.

2. Calculate the acceleration of a bike that goes from 8 m/s to 20 m/s in 4 s.

$$\text{Acceleration} = \frac{\text{net force}}{\text{mass}}; a = \frac{F}{m}$$

3. Calculate the acceleration of a 40-kg crate of softball gear when pulled sideways with a net force of 200 N.

4. Calculate the acceleration of a 2000-kg, single-engine airplane just before takeoff when the thrust of its engine is 500 N.

THINK AND COMPARE

1. Compare from greatest to least, the accelerations of these objects:
   (a) Net force 20 N acts on a mass of 10 kg.
   (b) Net force 30 N acts on a mass of 18 kg.
   (c) Net force 40 N acts on a mass of 30 kg.
   (d) Net force 50 N acts on a mass of 32 kg.
   Greatest _____ _____ _____ _____ Least

2. Boxes of chocolates of various masses are on a friction-free level table.

A   5 N [5 kg] 10 N   B 10 N [10 kg] 15 N

C 10 N [5 kg] 15 N   D 5 N [20 kg] 15 N

   Compare from greatest to least:
   (a) The net forces on the boxes.
   Greatest _____ _____ _____ _____ Least
   (b) The acceleration of the boxes.
   Greatest _____ _____ _____ _____ _____ Least

3. Compare from greatest to least the accelerations of these skydivers.
   (a) 1000-N man with 800 N of air drag.
   (b) 800-N woman with 700 N of air drag.
   (c) 700-N woman with 600 N of air drag.
   (d) 100-N dog with 90 N of air drag.
   Greatest _____ _____ _____ _____ Least

THINK AND EXPLAIN

1. What is the net force on a bright red Mercedes convertible traveling along a straight road at a steady speed of 100 km/h?

2. On a long alley a bowling ball slows down as it rolls. Is any horizontal force acting on the ball? How do you know?

3. In the orbiting space shuttle you are handed two identical boxes, one filled with sand and the other filled with feathers. How can you tell which is which without opening the boxes?

4. Your empty hand is not hurt when it bangs lightly against a wall. Why is it hurt if it does so while carrying a heavy load? Which of Newton's laws most applies here?

5. What happens to your weight when your mass increases?

6. When a junked car is crushed into a compact cube, does its mass change? Its weight? Its volume? Explain.

7. What is the net force on a 1-N apple when you hold it at rest above your head? What is the net force on the apple after you release it?

8. On which of these hills does the ball roll down with increasing speed and decreasing acceleration along the path? (Use this example if you wish to explain to someone the difference between speed and acceleration.)

9. If it takes 1 N to push horizontally on your book to make it slide at constant velocity, how much force of friction acts on the book?

10. A bear that weighs 4000 N grasps a vertical tree and slides down at constant velocity. What is the friction force that acts on the bear?

11. A crate remains at rest on a factory floor while you push on it with a horizontal force F. How big is the friction force exerted on the crate by the floor? Explain.

12. What is the acceleration of a ball at the top of its trajectory when thrown straight upward? Explain whether or not your answer is zero by using the equation $a = F/m$ as a guide to your thinking. (Interesting, almost everyone gets the wrong answer unless they choose F to be mg and let Newton's second law guide their thinking! Use this question if you're helping someone make the distinction between speed and acceleration.)

13. Aristotle claimed the speed of a falling object depends on its weight. We now know that objects in free fall, whatever their weights, undergo the same gain in speed. Why does weight not affect acceleration?

14. Two pumpkins are dropped from a high building through the air. One pumpkin is hollow and the other filled with rocks. Which accelerates more? Defend your answer.

15. In a vacuum, a coin and a feather fall equally, side by side. Would it be correct to say that in a vacuum equal forces of gravity act on both the coin and the feather? Defend your answer.

16. How does the force of gravity on a raindrop compare with the air drag it encounters when it falls at constant velocity?

17. Upon which will air drag be greater; a sheet of falling paper or the same paper wadded into a ball that falls at a faster terminal speed? (Careful!)

18. Why is it that a cat that accidentally falls from the top of a 50-story building hits the safety net below no faster than if it falls from the 20th story?

19. How does the terminal speed of a parachutist before opening a parachute compare with terminal speed after? Why is there a difference?

20. How does the gravitational force on a falling body compare with the air drag it encounters before it reaches terminal velocity? After?

THINK AND SOLVE

1. One pound is the same as 4.45 newtons. What is the weight in pounds of 1 newton?

2. What is your own mass in kilograms? Your weight in newtons?

3. Consider a 40-kg block of cement when pulled sideways with a net force of 200 N. Show that its acceleration is 5 m/s^2.

4. Consider a mass of 1 kg accelerated 1 m/s^2 by a force of 1 N. Show that the acceleration would be the same for 2 kg acted on by a force of 2 N.

5. Consider a jumbo jet of mass 30,000 kg in takeoff when the thrust from each of four engines is 30,000 N. Show that its acceleration is 4 m/s^2.

6. Leroy, who has a mass of 100 kg, is skateboarding at 9.0 m/s when he smacks into a brick wall and comes to a dead stop in 0.2 seconds.
   (a) Show that his deceleration is 45 m/s^2.
   (b) Show that the force of impact is 4500 N. (ouch!)

READINESS ASSURANCE TEST (RAT)

If you have a good handle on this chapter—if you really do—then you should be able to score 7 out of 10 on this RAT. Check your answers with your teacher. If you score less than 7, study further before moving on.

Choose the best *answer to each of the following.*

1. Consider a ball that picks up a speed of 2 m/s each second when it rolls from rest down an inclined plane. If the ball takes 5 seconds to reach the bottom, its speed at the bottom will be
   (a) 5 m/s.
   (b) 7.5 m/s.
   (c) 10 m/s.
   (d) None of the above.

2. When the ball rolls down the inclined plane in the previous question, as its speed increases its acceleration
   (a) increases.
   (b) decreases.
   (c) remains unchanged.

3. An object in free fall undergoes an increase in
   (a) speed.
   (b) acceleration.
   (c) both speed and acceleration.

4. An object with a huge mass also must have a huge
   (a) weight.
   (b) volume.
   (c) size.
   (d) surface area.

5. If gravity between the Sun and Earth suddenly vanished, Earth would continue moving in
   (a) a curved path.
   (b) a straight-line path.
   (c) an outward spiral path.
   (d) an inward spiral path.

6. A heavy rock and a light rock in free fall have the same acceleration. The *reason* the heavy rock does not have more acceleration is
   (a) the force of gravity on each is the same.
   (b) there is no air resistance.
   (c) the inertia of both rocks is the same.
   (d) All of these.
   (e) None of these.

7. When a 15-N falling object encounters 10 N of air resistance, its acceleration is
   (a) less than *g*.
   (b) *g*.
   (c) more than *g*.
   (d) Not enough information.

8. The amount of air resistance on a 0.8-N flying squirrel for terminal speed is
   (a) less than 0.8 N.
   (b) 0.8 N.
   (c) more than 0.8 N.
   (d) dependent on the orientation of its body.

9. If you drop at the same time a regular soccer ball and a soccer ball filled with sand from the top of a tall building, the ball to land first will be the
   (a) regular soccer ball.
   (b) ball filled with sand.
   (c) They will hit at the same time.

10. The scientist who found the connection between force, mass, and acceleration was
    (a) Galileo.
    (b) Newton.
    (c) Both.
    (d) Neither.

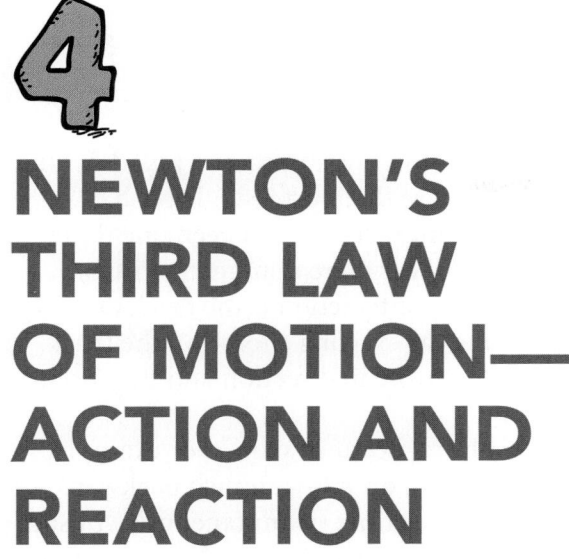

4
NEWTON'S THIRD LAW OF MOTION— ACTION AND REACTION

THE MAIN IDEA

For every force, there is an equal force in the opposite direction.

Here we see author Paul and wife Lillian touching. Who is touching whom? Can Paul touch Lillian without Lillian at the same time touching Paul? Newton's third law says no—that you can't touch without being touched! Consider a heavy truck hitting a small car in a head-on collision. Can the heavy truck hit the car without the car also hitting the truck? And does the truck hit the car with more force, or the same amount of force with which the car hits back? Can the heavyweight champion of the world punch a piece of paper in midair any harder than the paper hits back? In this chapter, we'll see how Newton's third law of motion guides our answers to these and other intriguing questions.

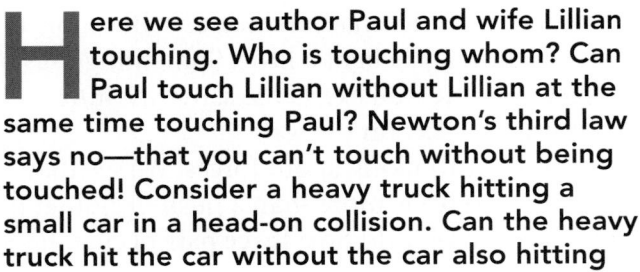

Explore!

Can There Be Only One Force in an Interaction?

1. With a classmate, hold two bathroom scales back to back. Then each of you push on the scales at the same time.
2. Record the readings on the scale, when the pushes are steady.
3. Repeat with a big person on one side and a smaller person on the other side.

Analyze and Conclude

1. **Observing** How did the readings on each scale compare?
2. **Predicting** Is there some way in which one person exerts more force than the other?
3. **Making Generalizations** Why do we say that forces only occur in pairs?

FIGURE 4.1 ▲
In the interaction between the car and the truck, is the force of impact the same on each? Is the damage the same?

4.1 A Force Is Part of an Interaction

In the previous chapters we've looked at force as a push or a pull. Looking closer, Newton realized that a force is more than just a single push or pull. A force is part of a mutual action—an **interaction**—between one thing and another. When a truck crashes into a car there is an interaction between the truck and the car. Part of the interaction is the truck exerting a force on the car. The other part is the car exerting a force on the truck. The forces are equal in strength and opposite in direction and they occur at exactly the same time.

✔ In every interaction, forces always occur in pairs. For example, you interact with the floor when you walk on it—you push backward against the floor, and the floor at the same time (simultaneously) pushes forward on you. Likewise, the tires of a car interact with the road—the tires push against the road, and the road pushes back on the tires. In swimming you interact with the water—you push the water backward, and the water pushes you forward. There is a pair of forces acting in each interaction. The interactions in these examples depend on friction. But a person or car on slippery ice, by contrast, may not be able to exert a force against the ice to produce the needed opposite force. Then the ice cannot push back and no change in motion occurs.

Can things like floors, car tires, and water exert forces? Your friends (not taking this course) may think that only living things like people and animals can exert forces. For example, when you push on a wall, how can the wall push back? It's not alive. It doesn't have muscles. But look at your fingers as you push on a wall. They're bent a little. Something must have pushed on them. Interestingly, the wall exerts the same amount of force on your fingers as your fingers exert on the wall.

So in this chapter we expand our thinking about forces. We see that nonliving things can exert them. A force is more than a push or pull. It is part of a mutual interaction between objects. So a **force pair** occurs in all interactions that involve forces.

Does a speeding baseball have force? The answer is no. Force is not something an object possesses, like mass. A speeding baseball *exerts*

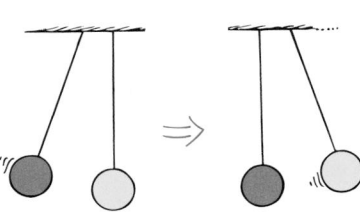

FIGURE 4.2 ▲
The impact forces between the blue and yellow balls move the yellow ball and stop the blue ball.

FIGURE 4.3 ▲
In the interaction between the hammer and the stake, each exerts the same amount of force on the other.

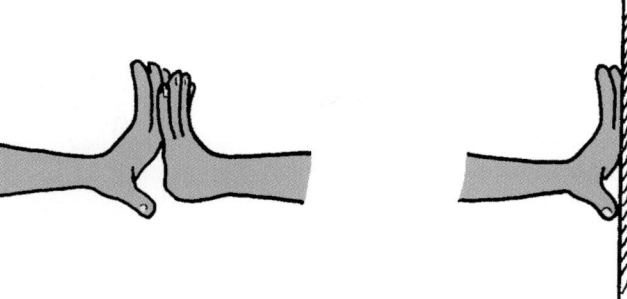

◀ **FIGURE 4.4**
FIGURE 4.4
You can feel your fingers being pushed by your friend's fingers. You also feel the same amount of force when you push on a wall and it pushes back on you. As a point of fact, you can't push on the wall *unless* it pushes back on you!

✔ **READING CHECK**

In what specific way do forces occur in every interaction?

a force when it hits something. How much force it exerts depends on how quickly the ball decelerates. Objects don't possess force as a thing in itself. As we will see in the following chapters, a speeding object possesses *momentum* and *kinetic energy*—but not force. A force is an interaction between one object and another.

CONCEPT CHECK
A car accelerates along a horizontal road. Strictly speaking, exactly what is it that pushes the car?

Check Your Answer
It is the road that pushes the car along. Really! Except for some road friction and air drag, only the road provides a horizontal force on the car. How? The rotating tires push back on the road (action). The road simultaneously pushes forward on the tires (reaction). The next time you see a car moving along a road, tell your friends that the road pushes the car along. If at first they don't believe you, convince them that there is more to the physical world than meets the eye of the casual observer. Turn them on to some physical science.

PhysicsPlace.com
Tutorial
Newton's Third Law

When pushing my fingers together I see the same discoloration on each of them. Aha—evidence that each experiences the same amount of force!

4.2 Newton's Third Law— Action and Reaction

In his investigation of many interactions, Newton discovered an underlying principle, called **Newton's third law:**

> **Whenever one object exerts a force on a second object, the second object exerts an equal and opposite force on the first.**

We can call one force the *action force,* and the other the *reaction force.* Then we can express Newton's third law in the form

> **To every action there is always an opposed equal reaction.**

It doesn't matter which force we call *action* and which we call *reaction.* The important thing is that they are co-parts of a single interaction and that neither force exists without the other. ✔ Action and reaction forces are equal in strength and opposite in direction.

FIGURE 4.5 ▶
When you lean against a wall, you exert a force on the wall. At the same time, the wall exerts an equal and opposite force on you. That's why you don't topple over.

HANDS-ON EXPLORATION

Playing with magnets is fun. Applying Newton's third law to magnets is also fun. Hold a toy magnet near another magnet. Notice that when one magnet moves another, it is also moved by the other. For equal-mass magnets, the effect is most noticeable. That's because the changes in motion (acceleration) are the same for each. For different-size magnets, the smaller magnets move more. Can you see how this ties into Newton's second law? (Newton's second law tells us that the acceleration of the magnet depends not only on force but on mass.)

> You can't pull on something unless that something pulls on you at the same time. That's the law!

✔ **READING CHECK**

Compare the relative strengths and directions of action and reaction forces.

CONCEPT CHECK

1. Which exerts more force, the Earth pulling on the Moon, or the Moon pulling on the Earth?
2. When a heavy football player and a light one run into each other, does the light player *really* exert as much force on the heavy player as the heavy player exerts on the light one? Is the damage to the heavy player the same as the damage to the light one?

Check Your Answers

1. This is like asking which is greater, the distance between New York and San Francisco, or the distance between San Francisco and New York. Both distances are the same, but in opposite directions; likewise for the pulls between the Earth and the Moon.
2. Yes. In the interaction between the two players, the forces each exert on the other have equal strengths. Although the forces are the same on each, the *effects* of these equal forces are quite unequal! The low-mass player may be completely unharmed. There is a difference between the *force* and the *effect* of the force.

PhysicsPlace.com

Tutorial
Newton's Third Law

4.3 A Simple Rule Helps Identify Action and Reaction

Here's a simple rule for identifying action and reaction forces. First, identify the interaction: One thing—say, object A—interacts with another—say, object B. Then action and reaction forces can be stated in the form

Action: Object A exerts a force on object B.

Reaction: Object B exerts a force on object A.

This is easy to remember. ✔ If the action is A on B, the reaction is B on A. We see that A and B are simply switched around. Consider the

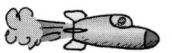

CALCULATION CORNER

Below we see two vectors on the sketch of the hand pushing the wall. The wall also pushes back on the hand. Note the others show only the action force.

Draw appropriate vectors showing the reaction forces. Can you specify the action-reaction pairs in each case?

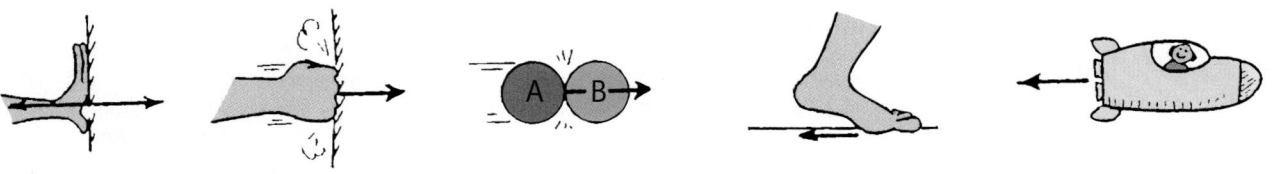

case of your hand pushing on the wall. The interaction is between your hand and the wall. We'll say the action is your hand (object A) exerting a force on the wall (object B). Then the reaction is the wall exerting a force on your hand.

Action: tire pushes on road   Reaction: road pushes on tire

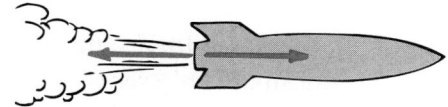

Action: rocket pushes on gas   Reaction: gas pushes on rocket

Action: man pulls on spring   Reaction: spring pulls on man

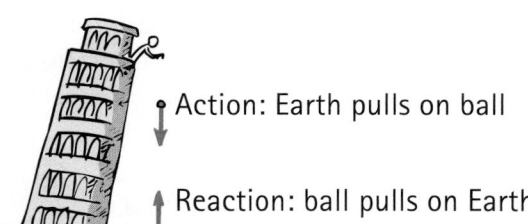

Action: Earth pulls on ball

Reaction: ball pulls on Earth

FIGURE 4.6 ▲
Action and reaction forces. Note that when the action is "A exerts force on B," the reaction is simply "B exerts force on A."

✓ **READING CHECK**

If the action is A acting on B, what is the reaction?

CONCEPT CHECK

Can you identify the action and reaction forces in the case of an object that is falling in a vacuum? (A vacuum is a region of space that is completely empty—no air.)

Check Your Answer

To identify a pair of action-reaction forces in any situation, first identify the pair of interacting objects. In this case the Earth is interacting with the falling object through the force of gravity. So the Earth pulls the falling object downward (call it *action*). Then *reaction* is the falling object pulling the Earth upward. (Hmm . . . you say, the falling object pulls the Earth upward? Yes, this may be hard to imagine at first but it is true. You don't notice the Earth being pulled upward by a falling object because of the Earth's large mass. More on this in Section 4.4.)

PhysicsPlace.com
Video
Action and Reaction on Different Masses
Tutorial
Newton's Third Law

FIGURE 4.7 ▲
The dog wags the tail and the tail wags the dog.

Careful: A pair of equal and opposite forces acting on the same object do not make up an action-reaction pair. Action and reaction act on different objects.

4.4 Action and Reaction on Objects of Different Masses

When a cannon is fired, there is an interaction between the cannon and the cannonball. The sudden force that the cannon exerts on the cannonball is exactly equal and opposite to the force the cannonball exerts on the cannon. This is why the cannon recoils (kicks). But the effects of these equal forces are very different. This is because the forces act on different masses. Recall Newton's second law:

$$a = \frac{F}{m}$$

Let F represent both the action and reaction forces, m the mass of the cannon, and m the mass of the cannonball. Different-sized symbols are used to indicate the differences in relative masses and resulting accelerations. Then the acceleration of the cannonball and cannon are

$$\textbf{Cannonball: } \frac{F}{m} = \boldsymbol{a}$$

$$\textbf{Cannon: } \frac{F}{m} = a$$

Do you see why the change in velocity of the cannonball is so large compared to the change in velocity of the cannon? ✓ A given force exerted on a small mass produces a large acceleration, while the same force exerted on a large mass produces a small acceleration.

✓ **READING CHECK**

What is the relationship between force and mass when acceleration is produced?

◀ **FIGURE 4.8**
The force exerted against the recoiling cannon is just as great as the force that drives the cannonball along the barrel. Why, then, does the cannonball undergo more acceleration than the cannon?

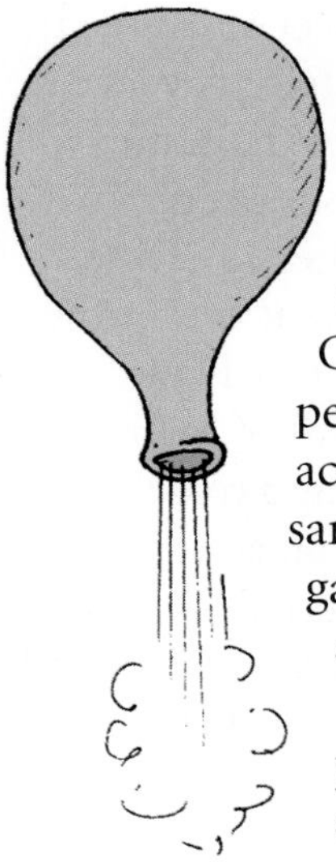

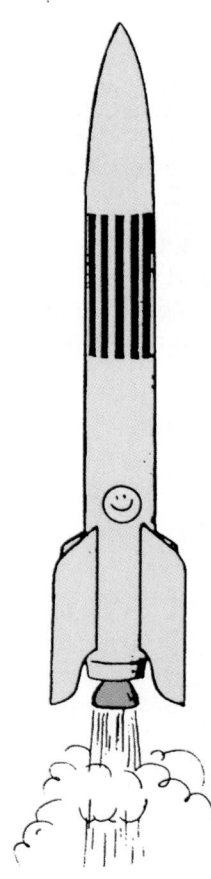

◀ **FIGURE 4.9**
The balloon recoils from the escaping air and climbs upward.

We can extend the idea of a cannon recoiling from the ball it fires to understanding rocket propulsion. Consider an inflated balloon recoiling when air is expelled. If the air is expelled downward, the balloon accelerates upward (Figure 4.9). A rocket accelerates the same way. It continually "recoils" from the ejected exhaust gas. Each molecule of exhaust gas is like a tiny cannonball shot from the rocket (Figure 4.10).

A common misconception is that a rocket is propelled by the impact of exhaust gases against the atmosphere. In fact, before the advent of rockets, it was commonly thought that sending a rocket to the Moon was impossible. Why? Because there is no air above the Earth's atmosphere for the rocket to push against. But this is like saying a cannon wouldn't recoil unless the cannonball had air to push against. Not true! Both the rocket and recoiling cannon accelerate because of the reaction forces of the material they fire—not because of any pushes on the air. In fact, a rocket works better above the atmosphere where there is no air drag.

FIGURE 4.10 ▲
The rocket recoils and rises from the "molecular cannonballs" it fires.

CONCEPT CHECK

A high-speed bus and an innocent bug have a head-on collision. The force of the bus on the bug splatters the poor bug all over the windshield. Is the corresponding force of the bug on the bus greater, less, or the same? Is the resulting deceleration of the bus greater than, less than, or the same as that of the bug?

Check Your Answers
The magnitudes of both forces are the same, for they make up an action-reaction force pair in the bus-bug interaction. The accelerations, however, are very different because of the different masses! The bug undergoes an enormous and lethal deceleration, while the bus undergoes a very tiny deceleration—so tiny that the very slight slowing of the bus is unnoticed by its passengers. But if the bug were more massive—as massive as another bus, for example—the slowing down would be quite evident.

PhysicsPlace.com

Video
Action and Reaction on Rifle and Bullet
Tutorial
Newton's Third Law

FIGURE 4.11 ▲
A acts on B, and B accelerates.

4.5 Action and Reaction Forces Act on Different Objects

Since action and reaction forces are equal and opposite, why don't they cancel to zero? They don't cancel out because they act on different objects. Consider kicking a football (Figure 4.11). Call the force exerted by your foot action. That's the only horizontal force on the football, so the football accelerates. Reaction is the football exerting a force on your foot, which tends to slow your foot down a bit. You can't cancel the force

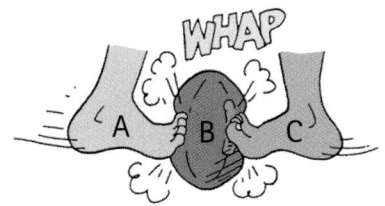

FIGURE 4.12 ▲
Both A and C act on B. They can cancel each other, so B does not accelerate.

on the football with a force on your foot. ✔ Forces cancel only when they act on the *same* object. Now, what would happen if two players kicked the same football with opposite and equal forces at the same time, as shown in Figure 4.12? In this case, two interactions occur. Two different forces act on the football and these forces together cancel to zero. Is there a reaction force to each of these two forces that act on the football? The answer is yes, in accordance with Newton's third law.

Interestingly, inside the football are trillions and trillions of interatomic forces that hold the ball together. But they don't play a role in accelerating the ball. Although every one of these forces is an action-reaction pair within the ball, they combine to zero. An external force, such as a kick, is needed to accelerate the ball. If this is confusing, it may be well to point out that Newton himself had difficulties with the third law.

CONCEPT CHECK

Why does a football or baseball never accelerate "spontaneously" in response to the trillions of interatomic forces acting within it?

Check Your Answer

Every one of these interatomic forces is part of an action-reaction pair within the ball. These forces add up to zero, no matter how many of them there are. A ball has zero acceleration unless an *external* force acts on it.

✔ READING CHECK

When can forces cancel each other?

PhysicsPlace.com

Tutorial
Newton's Third Law

4.6 The Classic Horse-Cart Problem— A Mind Stumper

A situation similar to the kicked football is shown in the comic strip "Horse Sense." Here we pretend that the horse believes its pull on the cart will be canceled by the opposite and equal pull by the cart on the horse, making acceleration impossible. This is the classic horse-cart problem that is a stumper for many students at the university level. By thinking carefully, you can understand it here.

The horse-cart problem can be looked at from different points of view. One is the farmer's point of view, where his only concern is getting his cart (the cart system) to market. Then, there is the point of view of the horse (the horse system). Finally, there is the point of view of the horse and cart together (the horse-cart system).

First look at the farmer's point of view—the cart system. The net force on the cart, divided by the mass of the cart, will produce an acceleration. The farmer doesn't care about the reaction on the horse.

Now look at the horse's point of view—the horse system. It's true that the opposite reaction force by the cart on the horse restrains the horse. This force tends to hold the horse back. Without this force the horse could freely gallop to the market. So how does the horse move forward? By interacting with the ground. At the same time the horse pushes backward against the ground, the ground pushes forward on the horse. If the horse pushes the ground with a greater force than its pull on the

HORSE SENSE

GIDDIUP! PULL THE CART SO WE CAN GET GOING.

FOR ME TO PULL THE CART WOULD BE A FUTILE EFFORT.

YOU SEE, IF I PULL ON THE CART, THE CART WILL PULL BACK ON ME. BY NEWTON'S 3rd LAW, THE FORCES ARE EQUAL AND OPPOSITE-SO THEY'LL CANCEL OUT. A ZERO NET FORCE WON'T GET US MOVING.

PHYSICS PHYSICS PHYSICS

$$a_{CART} = \frac{F_{CART}}{m_{CART}}$$

PHYSICS PHYSICS

I DON'T CARE ABOUT THE FORCE EXERTED ON YOU. I'M INTERESTED IN THE FORCE YOU EXERT ON THE CART! YOU PULL THE CART AND I GUARANTEE IT WILL MOVE!

BUT HOW CAN I MOVE FORWARD WHEN THE CART PULLS BACKWARD ON ME?

PHYSICS PHYSICS PHYSICS

JUST PUSH BACKWARD ON THE GROUND. BY NEWTON'S 3rd LAW, THE GROUND WILL PUSH FORWARD EQUALLY ON YOU --- THEN I'LL SIMPLY FOLLOW ALONG!

THAT GROUND IS DOING A VERY GOOD JOB!

PHYSICS GRUMBLE PHYSICS

cart, then there will be a net force on the horse. Acceleration occurs. When the cart is up to speed, the horse needs only to push against the ground with enough force to offset the friction between the cart's wheels and the ground.

Finally, look at the horse-cart system as a whole. From this viewpoint, the pull of the horse on the cart and the reaction of the cart on the horse are internal forces—forces that act and react within the system. They contribute nothing to the acceleration of the horse-cart system. The forces cancel and can be neglected. To move across the ground, there must be an interaction between the horse-cart system and

◀ **FIGURE 4.13**
All the pairs of forces that act on the horse and cart are shown: (1) the pull P of the horse and the cart on each other; (2) the push F of the horse and the ground on each other; and (3) the friction f between the cart wheels and the ground. Can you see that the acceleration is due to the net force $F - f$ on the horse-cart system (while opposite $F - f$ acts on the ground)?

the ground. This is similar to moving a car while you're sitting in it. You can push and push against the dashboard, but no acceleration occurs. To get a car moving, you must get outside and make the ground push you and the car. The horse-cart system is similar. ✓ It is the outside reaction force by the ground that pushes the system.

CONCEPT CHECK

1. What is the net force that acts on the cart in Figure 4.13? On the horse? On the horse-cart system?
2. Once the horse gets the cart moving at the desired speed, must the horse continue to exert a force on the cart?

Check Your Answers

1. The net force on the cart is $P - f$; on the horse, $F - P$; on the horse-cart system, $F - f$.
2. Yes, but only enough to counteract wheel friction and air resistance. Interestingly, air resistance would be absent if there were a wind blowing in the same direction and just as fast as the horse and cart. If the wind blows fast enough to provide a force to counteract friction, the horse could wear rollerblades and simply coast with the cart all the way to the market.

✓ **READING CHECK**

Specifically, what force pushes a system across the ground?

PhysicsPlace.com
Tutorial
Newton's Third Law

4.7 Action Equals Reaction

When you tie a rope to a wall and pull on it, you produce a tension in the rope. Your pull on the rope and the pull by the supporting wall are equal and opposite. Otherwise, there would be a net force on the rope and it would accelerate. The same is true if a friend holds one end of the rope and you have a tug-of-war. Rope tension when pulled at opposite ends is the same as the force provided by each end. Both pulls are the same in magnitude. This leads to a fascinating discovery for people who play tug-of-war. ✓ The team to win is not the team to exert the greatest force on the rope, but the greatest force *against the ground!* In this way a greater net force acts on the winning team.

FIGURE 4.14 ▶
Arnold and Suzie pull on opposite ends of the rope. Can Arnold pull any harder on the rope than Suzie pulls on it? If Suzie lets go, could Arnold provide tension in the rope?

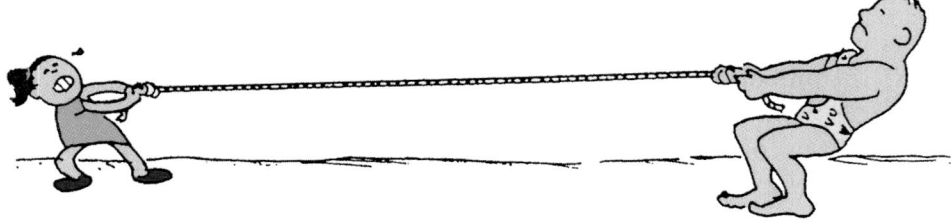

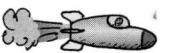

HANDS-AND-FEET-ON EXPLORATION

Perform a tug-of-war between boys and girls. Do it on a polished floor (that's somewhat slippery). Have the boys wear socks and the girls wear rubber-soled shoes. Who will surely win, and why?

CONCEPT CHECK

1. We said earlier that a car accelerates along a road because the road pushes it. Can we say that a team wins in a tug-of-war when the ground pushes harder on them than on the other team?
2. Does the scale read 100 N, 200 N, or zero?

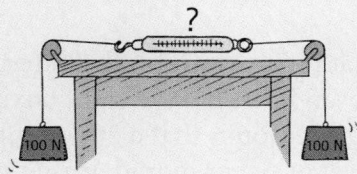

Check Your Answers

1. Yes!
2. Although the net force on the system is zero (as evidenced by no acceleration), the scale reading is 100 N, the tension in the string. Note that the string tension is 100 N in all the positions shown.

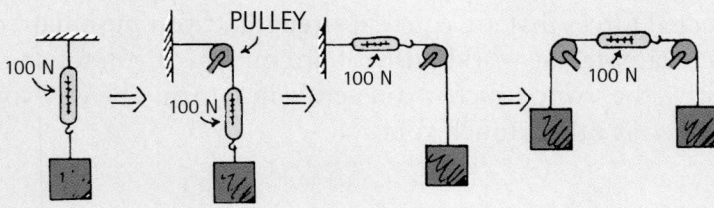

Newton's third law tells us how a helicopter gets its lifting force. The whirling blades are shaped to force air particles down (action), and the air forces the blades up (reaction). This upward reaction force is called *lift*. When lift equals the weight of the craft, the helicopter hovers in midair. When lift is greater, the helicopter rises.

The same is true for airplanes and birds. Birds fly by pushing air downward. The air simultaneously pushes the bird upward. Interestingly, when some birds deflect the air downward, the moving air meets air below and is swirled upward, mostly near the edges of the wings. Off to the side another bird can position itself to get added lift from this updraft. This bird, in turn, creates an updraft for a following bird and so on. This is the physics of why geese and ducks fly in a V formation!

Have you ever heard the expression that someone "can't fight their way out of a paper bag?" There's some interesting physics within this statement. According to Newton, you can't hit a piece of paper any harder than the paper can hit you back. Hold a sheet of paper in midair

FIGURE 4.15 ▲
The pair of vectors represents the force each wing exerts on the air. What forces act on the bird?

FIGURE 4.16 ▶
When a duck pushes air downward with its wings, air at the tips of its wings swirls upward, creating an updraft that is strongest off to its side. A trailing bird gets added lift by positioning itself in this updraft. This bird pushes air down and creates an updraft for the next bird, and so on. The result is a flock flying in a V formation.

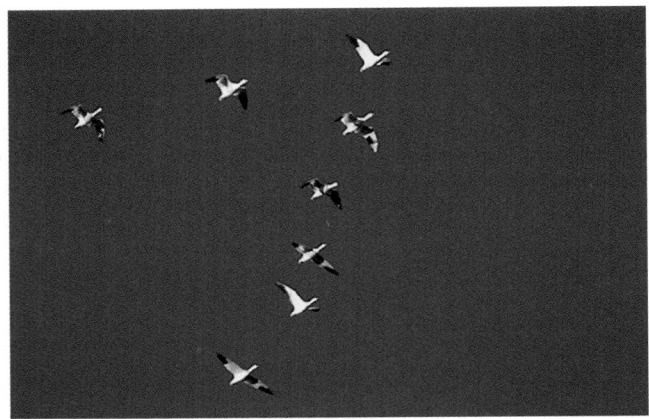

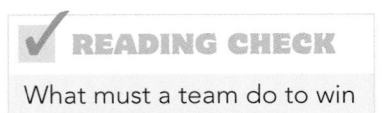

✔ **READING CHECK**

What must a team do to win in a tug-of-war?

and tell your friends that nobody can hit the paper with a force of 200 N (45 lb). You're correct even if the heavyweight boxing champion of the world hits the paper. The reason is that a 200-N interaction between the champ's fist and the sheet of paper in midair isn't possible—the paper is not capable of exerting a reaction force of 200 N. You cannot have an action force without its reaction force. Now, if you hold the paper against the wall, that is a different story. The wall will easily assist the paper in providing 200 N of reaction force, and more if needed!

For every interaction between things, there is always a pair of oppositely directed forces that are equal in strength. If you push hard on the world, for example, the world pushes hard on you. If you touch the world gently, the world touches you gently in return. The way you touch others is the way others touch you.

FIGURE 4.17 ▲
He can hit the massive bag with considerable force. But with the same punch he can exert only a tiny force on the tissue paper in midair.

◀ **FIGURE 4.18**
You cannot touch without being touched—Newton's third law.

4.8 Summary of Newton's Three Laws

Newton's first law, the law of inertia: An object at rest tends to remain at rest; an object in motion tends to remain in motion at constant speed along a straight-line path. This property of objects to resist change in motion is called *inertia*. Mass is a measure of inertia. Objects will undergo changes in motion only when acted on by a net force.

Newton's second law, the law of acceleration: When a net force acts on an object, the object will accelerate. The acceleration is directly proportional to the net force and inversely proportional to the mass. Symbolically, $a \sim F/m$. Acceleration is always in the direction of the net force. When objects fall in a vacuum, the net force is simply the weight, and the acceleration is g (the symbol g denotes that acceleration is due to gravity alone.) When objects fall in air, the net force is equal to the weight minus the force of air drag, and the acceleration is less than g. If and when the force of air drag equals the weight of a falling object, acceleration terminates, and the object falls at constant speed (called *terminal speed*).

Newton's third law, the law of action-reaction: Whenever one object exerts a force on a second object, the second object exerts an equal and opposite force on the first. Forces come in pairs, one action and the other reaction, both of which occur simultaneously and comprise the interaction between one object and the other. Action and reaction always act on different objects. Neither force exists without the other.

There has been a lot of new and exciting physics since the time of Isaac Newton. ✔ Nevertheless, and quite interestingly, it was primarily Newton's laws that got us to the Moon.

PhysicsPlace.com

Tutorial
Newton's Third Law

What laws of physics got humans to the Moon?

4 CHAPTER REVIEW

SUMMARY OF TERMS

Force pair The action and reaction pair of forces that occur in an interaction.

Interaction Mutual action between objects where each object exerts an equal and opposite force on the other.

Newton's third law Whenever one object exerts a force on a second object, the second object exerts an equal and opposite force on the first. Or put another way, "To every action there is always an opposed equal reaction."

REVIEW QUESTIONS

A Force Is Part of an Interaction

1. In the simplest sense, a force is a push or a pull. In a deeper sense, what is a force?
2. How many forces are required for an interaction?
3. When you push against a wall with your fingers, they bend because they experience a force. Identify this force.
4. Why do we say a speeding object doesn't *possess* force?

Newton's Third Law—Action and Reaction

5. State Newton's third law of motion.
6. Consider hitting a baseball with a bat. If we call the force on the bat against the ball the *action* force, identify the *reaction* force.
7. If a bat hits a ball with 1000 N of force, with how much force does the ball hit back on the bat?

A Simple Rule Helps Identify Action and Reaction

8. If the world pulls you downward, what is the reaction force?

Action and Reaction on Objects of Different Masses

9. If the forces that act on a cannonball and the recoiling cannon from which it is fired are equal in magnitude, why do the cannonball and cannon have very different accelerations?
10. Identify the force that propels a rocket.

Action and Reaction Forces Act on Different Objects

11. When can the net force on a ball be zero when you kick it?
12. Why do the interatomic forces inside a ball not accelerate the ball?

The Classic Horse-Cart Problem— A Mind Stumper

13. Referring to Figure 4.13, how many forces are exerted on the cart? What is the horizontal net force on the cart?
14. How many forces are exerted on the horse? What is the net force on the horse?
15. How many forces are exerted on the horse-cart system? What is the net force on the horse-cart system?

Action Equals Reaction

16. Which is most important in winning a tug-of-war—pulling harder on the rope, or pushing harder on the floor?
17. How does a helicopter get its lifting force?
18. A boxer can hit a heavy bag with great force. Why can't he hit a sheet of newspaper in midair with the same amount of force?
19. Can you physically touch another person without that person touching you with the same amount of force?

Summary of Newton's Three Laws

20. Fill in the blanks: Newton's first law is often called the law of _____; Newton's second law highlights the concept of _____; and Newton's third law is the law of _____ and _____.

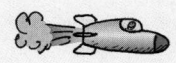

THINK AND DO

Hold your hand like a flat wing outside the window of a moving automobile. Then slightly tilt the front edge upward and notice the lifting effect. Can you see Newton's third law at work here?

THINK AND EXPLAIN

1. The photo shows Steve Hewitt and daughter Gretchen. Is Gretchen touching Dad, or is Dad touching her? Explain.

2. For each of the following interactions, identify action and reaction forces. (a) A hammer hits a nail. (b) Earth gravity pulls down on you. (c) A helicopter blade pushes air downward.
3. You hold an apple over your head. (a) Identify all the forces acting on the apple and their reaction forces. (b) When you drop the apple, identify all the forces acting on it as it falls and the corresponding reaction forces. Neglect air drag.
4. Identify the action-reaction pairs of forces for the following situations: (a) You step off a curb. (b) You pat your tutor on the back. (c) A wave hits a rocky shore.
5. Consider a tennis player hitting a ball. (a) Identify the action-reaction pairs when the ball is being hit and (b) while the ball is in flight.

6. A small car bumps into a van at rest in a parking lot. Upon which vehicle is the force of contact greater? Which vehicle undergoes the most acceleration? Defend your answer.

7. Your teacher challenges you and your best friend to pull on a pair of scales attached to the ends of a horizontal rope, in tug-of-war fashion. Can you each pull in such a way that the readings on the scales will differ? Defend your answer.

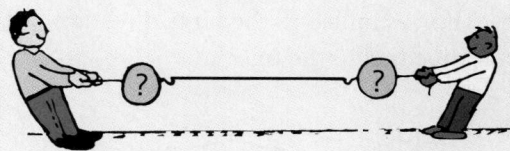

8. You push a heavy car by hand. The car, in turn, pushes back with an opposite but equal force on you. Doesn't this mean the forces cancel one another, making acceleration impossible? Why or why not?
9. A farmer urges his horse to pull a wagon. The horse refuses, saying to try would be futile for it would contradict Newton's third law. The horse concludes that she can't exert a greater force on the wagon than the wagon exerts on her, and therefore won't be able to accelerate the wagon. What is your explanation to convince the horse to pull?
10. Suppose two carts, one twice as massive as the other, fly apart when the compressed spring that joins them is released. How fast does the heavier cart roll compared to the lighter cart?

11. If you exert a horizontal force of 200 N to slide a desk across a classroom floor at constant velocity, how much friction does the floor exert on the desk? Is the force of friction equal and oppositely directed to your 200-N push? If the force of friction isn't the reaction force to your push, what is?

12. If a massive truck and small sports car have a head-on collision, upon which vehicle is the collision force greater? Which vehicle experiences the greater acceleration? Explain your answers.

13. Ken and Joanne are astronauts floating some distance apart in space. They are joined by a safety cord, the ends of which are tied around their waists. If Ken starts pulling on the cord, will he pull Joanne toward him, or will he pull himself toward Joanne, or will both astronauts move? Explain.

14. The strong man can withstand the tension force exerted by the two horses pulling in opposite directions. How would the tension compare if only one horse pulled and the left rope were tied to a tree? How would the tension compare if the two horses pulled in the same direction, with the left rope tied to the tree?

15. In a tug-of-war between two physics types, each pulls on the rope with a force of 250 N. What is the tension in the rope? If both remain motionless, what horizontal force does each exert against the ground?

16. A stone is shown at rest on the ground. (a) The vector shows the weight of the stone. Complete the vector diagram showing another vector that results in zero net force on the stone. (b) What is the conventional name of the vector you have drawn?

17. Here a stone is suspended at rest by a string. (a) Draw force vectors for all the forces that act on the stone. (b) Should your vectors have a zero resultant? (c) Why, or why not?

18. Here the same stone is being accelerated vertically upward. (a) Draw force vectors to some suitable scale showing relative forces acting on the stone. (b) Which is the longer vector, and why?

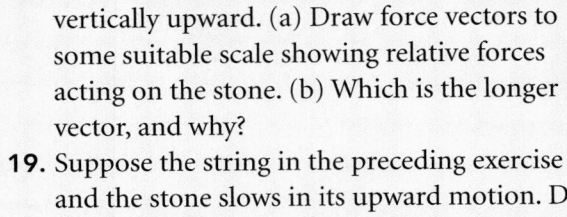

19. Suppose the string in the preceding exercise breaks and the stone slows in its upward motion. Draw a force vector diagram of the stone when it reaches the top of its path.

20. What is the acceleration of the stone in Think and Explain 19 at the top of its path?

THINK AND SOLVE

1. If you apply a net force of 5 N on a cart with a mass 5 kg, show that the acceleration is 1 m/s^2.

2. If you increase the speed of a 2.0-kg air puck by 3.0 m/s in 4.0 s, show that the force you exert on it is 1.5 N.

3. A boxer punches a sheet of paper in midair and brings it from rest up to a speed of 25 m/s in 0.05 s. If the mass of the paper is 0.003 kg, show that force the boxer exerts on it is only 1.5 N (about one-third of 1 lb).

4. Joshua has a mass of 60 kg and stands next to a wall on a frictionless skateboard. If he pushes the wall with a force of 30 N, how hard does the wall push on him? Show that Joshua accelerates at 0.5 m/s^2 away from the wall.

5. A 7.00-kg bowling ball moving at 8.0 m/s strikes a 1/0-kg bowling pin and slows to 7.0 m/s in 0.040 s.
   (a) Show that the force on the bowling ball is 175 N.
   (b) How much force acts on the bowling pin?

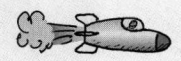

READINESS ASSURANCE TEST (RAT)

If you have a good handle on this chapter—if you really do—then you should be able to score 7 out of 10 on this RAT. Check your answers with your teacher. If you score less than 7, study further before moving on.

Choose the best answer to each of the following.

1. When walking, your foot pushes backward on the floor and
   (a) also pushes the floor forward.
   (b) the floor also pushes backward on your foot.
   (c) the floor pushes forward on your foot.
   (d) None of these.

2. A batted baseball soaring through the air with negligible air resistance has
   (a) speed.
   (b) force.
   (c) Both of these.
   (d) None of these.

3. A karate chop delivers a force of 3000 N to a board that breaks. The force that acts on the hand during this event is
   (a) less than 3000 N.
   (b) 3000 N.
   (c) more than 3000 N.
   (d) Not enough information to say.

4. A soccer ball is kicked to a 30-m/s speed. While being kicked, the amount of force that the player's foot exerts on the ball is
   (a) less than the amount of force on the foot.
   (b) the same as the amount of force on the foot.
   (c) more than the amount of force on the foot.
   (d) None of these.

5. When a cannonball is fired from a cannon, both the cannonball and the cannon experience
   (a) equal amounts of force.
   (b) equal accelerations.
   (c) Both of these.
   (d) None of these.

6. The force that propels a rocket is provided by
   (a) gravity.
   (b) Newton's laws of motion.
   (c) its exhaust gases.
   (d) the atmosphere against which the rocket pushes.

7. The team that wins in a tug-of-war is the team that
   (a) produces more tension in the rope than the opponent.
   (b) pushes hardest on the floor.
   (c) Both of these.
   (d) None of these.

8. The net force on a kicked football can be zero when
   (a) action and reaction both act on the ball.
   (b) it is kicked by two feet with equal and opposite amounts of force.
   (c) it is kicked in the same direction by two feet.
   (d) None of these.

9. The force that moves the horse and cart in the comic strip "Horse Sense" is
   (a) friction by the ground on the horse's feet.
   (b) the reaction to the horse's pull on the wagon.
   (c) the reaction to the cart's pull on the horse.
   (d) All of these.

10. The heavyweight boxing champion of the world can't hit a piece of tissue paper in midair with much force because
    (a) an interaction isn't possible in this case.
    (b) the paper can't produce much opposite force.
    (c) the paper is too light in weight.
    (d) Newton's third law can't occur.

5

MOMENTUM

THE MAIN
IDEA

Momentum is inertia in motion and is conserved for all interactions where external forces don't interfere.

How does the speed of the athlete affect the clearing of hurdles in his path? How can a karate expert break a stack of cement bricks with the blow of her bare hand? And why is her blow stronger if her hand bounces off the bricks? But doesn't bouncing reduce the impact on a trapeze artist when he falls into a circus net? And why is an extended swing "follow-through" important in golf, tennis, and boxing? A related question involves the game of pool. In playing pool, why does a cue ball stop short when it hits another one at rest head-on? And why does that struck ball continue with the speed of the first? The answers to these questions are related. They involve more than the concepts of inertia and force covered in previous chapters. Now we concern ourselves with a new concept—*momentum.*

Explore!

How Does a Collision Affect the Motion of Marbles?

1. Place four or five marbles, all identical in size and shape, in the center groove of a ruler. Launch another identical marble toward the stationary ones. Note the changes in the motion of the marbles.
2. Repeat, but with two marbles launched at about the same speed. Note any changes in the motion of the marbles.
3. Remove all but one marble. Launch an identical marble toward it. Note how the first marble stops and the one at rest moves.

Analyze and Conclude

1. **Observing** How did the approximate speeds of the marbles before and after collision compare?
2. **Drawing Conclusions** Did the number of moving marbles remain the same before and after collision?
3. **Predicting** What would occur if three marbles rolling to the right hit two marbles rolling to the left? Is a general rule at play here?

Check Your Answers

1. The force will be three times less than if he didn't pull back.
2. The force will be two times greater than if he held his head still. Forces of this kind account for many knockouts.
3. There is no contradiction because the best results for each are quite different. The best result for the boxer is reduced force, accomplished by maximizing time. The best result for the karate expert is increased force delivered in minimum time.

✔ **READING CHECK**

What is the impulse–momentum relationship?

5.3 Momentum Change Is Greater When Bouncing Occurs

You know that if a flowerpot falls from a shelf onto your head, you may be in trouble. If it bounces from your head, you may be in even more trouble. Why? Because impulses are greater when an object bounces. ✔ The impulse required to bring an object to a stop and then to "throw it back again" is greater than the impulse required merely to bring it to a stop. Suppose, for example, that you catch the falling pot with your hands. You provide an impulse to reduce its momentum to zero. If you throw the pot upward again, you have to provide additional impulse. It takes more impulse to catch it and throw it back up than merely to catch it. This increased amount of impulse is supplied by your head if the pot bounces from it.

An interesting application of the greater impulse that occurs with bouncing was employed with much success in California during the gold rush days. The water wheels used in gold-mining operations were not very effective. A man named Lester A. Pelton saw that the problem had to do with their flat paddles. He designed curved-shape paddles that would cause the incident water to make a U-turn—to "bounce." In this way the impulse exerted on the water wheels was increased. Pelton patented his idea and made more money from his invention, the Pelton wheel, than most gold miners made from gold.

FIGURE 5.9 ▲
Teacher Howie Brand shows that the block topples when the swinging dart bounces from it. When he changes the head of the dart so it doesn't bounce when it hits the block, no tipping occurs.

A flowerpot dropped onto your head bounces quickly. Ouch! If bouncing took a longer time, as with a safety net, then the force of the bounce would be much smaller.

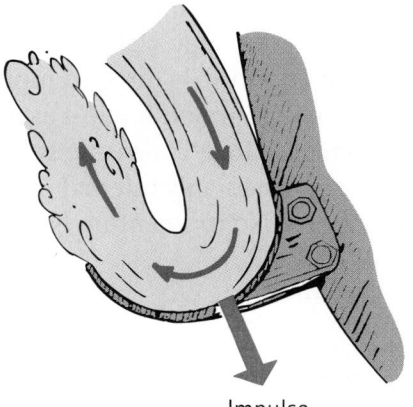

Impulse

◄ **FIGURE 5.10**
The Pelton wheel. The curved blades cause water to bounce and make a U-turn, which produces a greater impulse to turn the wheel.

Compare the impulses in catching a dropped ball to catching it and tossing it upward.

CONCEPT CHECK

1. Refer to Figure 5.8. How does the force that Cassy exerts on the bricks compare with the force exerted on her hand?
2. How will the impulse differ if her hand bounces back when striking the bricks?

Check Your Answers

1. In accord with Newton's third law, the forces are equal. Only the resilience of the human hand and the training she has undergone to toughen her hand allow her to do this without breaking bones.
2. The impulse will be greater if her hand bounces from the bricks. If the time of contact is not increased, a greater force is then exerted on the bricks (and her hand!).

5.4 When No External Force Acts, Momentum Doesn't Change—It Is Conserved

Newton's second law tells us that to accelerate an object, you apply a net force. We say much the same thing, but in different language. To change the momentum of an object, exert a net impulse on it.

In either case, the force or impulse must be exerted on the object by something outside the object. Internal forces won't work. For example, the molecular forces within a baseball have no effect upon the momentum of the baseball, just as your push against the dashboard of a car you're sitting in does not change the momentum of the car. Molecular forces within the baseball and a push on the dashboard are internal forces. They come in balanced pairs that cancel within the object. To change the momentum of the baseball or car, an outside force is required. Without an outside force, no change in momentum is possible.

When a cannon fires a cannonball, the explosive forces are internal forces. That means the total momentum of the cannon–cannonball system doesn't change (Figure 5.11). Can you see that the impulses must be the same, only in opposite directions? Think of Newton's third law of action and reaction. Then we see that the force on the cannonball is equal and opposite to the force on the cannon. Since these forces act for the same time, equal but oppositely directed impulses are produced. That means equal and oppositely directed momenta (the plural form of momentum). The recoiling cannon has just as much momentum as the speeding cannonball.* Together as one system, there is no net momentum change. In summary, when only internal forces act on a system, no change in momentum occurs. No momentum is gained and no momentum is lost.

Most of the cannonball's momentum is in speed; most of the recoiling cannon's momentum is in mass. So $mV = Mv$.

---

* Here we are neglecting the momentum of the ejected gases from the exploding gunpowder. Firing a gun with blanks at close range is a definite no-no because of the momentum of ejected gases. People have been killed by the firing of close-range blanks. Although no slug emerges from the gun, exhaust gases do—enough to be lethal.

FIGURE 5.11 ▲
The momentum before firing is zero. After firing, the net momentum is still zero, because the momentum of the cannon is equal and opposite to the momentum of the cannonball.

Two important ideas are to be learned from the cannon-and-cannonball example. The first is that momentum, like velocity, is a vector quantity that is described by both magnitude and direction; we measure both "how much" and "which direction." Therefore, like velocity, when momenta act in the same direction, they are simply added. When they act in opposite directions, they are subtracted.

The second important idea is *conservation.* For the cannon-cannonball system, no momentum was gained; none was lost. ✓ When a physical quantity remains unchanged during a process, we say that quantity is *conserved.* We say momentum is *conserved.* The concept that momentum is conserved when no external force acts is so important it is considered a law of mechanics. It is called the **law of conservation of momentum:**

In the absence of an external force, the momentum of a system remains unchanged.

When a system undergoes changes in which all forces are internal, the net momentum remains unchanged. This happens, for example, when cannons fire cannonballs, cars collide, or stars explode. In such cases, the net momentum of the system before and after the event remains the same.

Can you see how Newton's laws relate to momentum conservation?

CONCEPT CHECK
A high-speed bus and an innocent bug have a head-on collision. The sudden change of momentum for the bug spatters it all over the windshield. Is the change in momentum of the bus greater, less, or the same as the change in momentum of the unfortunate bug?

Check Your Answer
The momentum of both bug and bus change by the same amount because both the amount of force and the time, and therefore the amount of impulse, is the same on each. Momentum is conserved. Speed is another story. Because of the huge mass of the bus, its reduction of speed is very tiny—too small for the passengers to notice.

✓ READING CHECK

What does it mean to say that a physical quantity is conserved?

Stand at rest on a skateboard and throw a massive object to the front or the rear. Note that you recoil in the opposite direction. This is understandable if you understand momentum. The net momentum before the throw was zero. The net momentum just after is also zero—because your recoil momentum is equal and opposite to the momentum of the tossed object. You'll see that momentum is conserved. Now repeat, but don't let go when you "throw" the object. Do you still recoil when you go through the motions of throwing but don't really release the object? Defend your answer.

5.5 Momentum Is Conserved in Collisions

Momentum is conserved in collisions because the forces that act are internal forces—acting and reacting within the system itself. There is only a redistribution or sharing of whatever momentum exists before the collision.

In any collision, we can say

$$\textbf{Net momentum}_{\textbf{before collision}} = \textbf{net momentum}_{\textbf{after collision}}$$

Elastic Collisions

When a moving billiard ball hits another ball at rest head-on, the first ball comes to rest and the second ball moves away with a velocity equal to the initial velocity of the first ball. We see that momentum is transferred from the first ball to the second ball. When objects collide without being permanently deformed and without generating heat, we say the collision is an **elastic collision.** Colliding objects bounce perfectly in perfectly elastic collisions (Figure 5.12).

Inelastic Collisions

Momentum is conserved even when the colliding objects become distorted and generate heat during the collision. Whenever colliding objects become tangled or coupled together, we have an **inelastic collision.** In a perfectly inelastic collision, both objects stick together. Consider, for example, the case of a freight car moving along a track and colliding with another freight car at rest (Figure 5.13). If the freight cars are of equal mass and are coupled by the collision, can we predict the velocity of the coupled cars after impact?

Suppose the single car is moving at 10 meters per second, and we consider the mass of each car to be *m*. Then, from the conservation of momentum,

$$\textbf{Net } mv_{\textbf{before}} = \textbf{net } mv_{\textbf{after}}$$

$$(m \times 10 \text{ m/s})_{\textbf{before}} = (2m \times v)_{\textbf{after}}$$

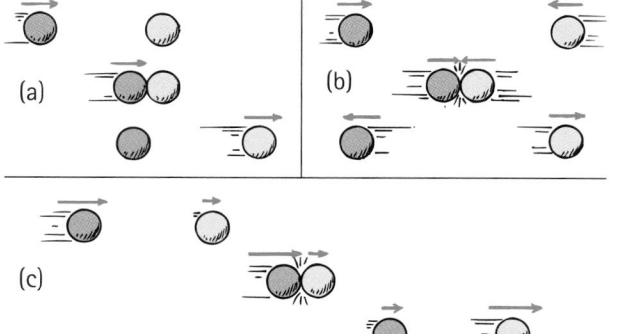

FIGURE 5.12 ▲
Elastic collisions of equally massive balls. (a) A green ball strikes a yellow ball at rest. (b) A head-on collision. (c) A collision of balls moving in the same direction. In each case, momentum is transferred from one ball to the other.

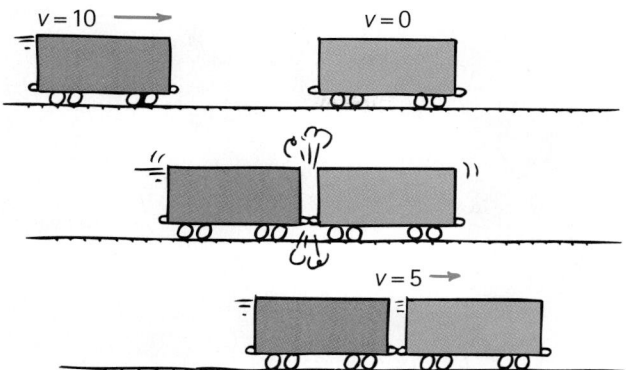

◀ **FIGURE 5.13**
Inelastic collision. The momentum of the freight car on the left is shared with the freight car on the right after collision.

By simple algebra, $v = 5$ m/s. This makes sense, for twice as much mass moves after the collision, with half as much as the velocity as before the collision. Then both sides of the equation are equal.

Note the inelastic collisions in Figure 5.14. Can you see the net momentum after collisions is the same? ✔ We see that when external forces don't play a role, momentum is conserved for both elastic and inelastic collisions.

If your teacher has an air track similar to that shown in Figure 5.15, you may be treated to fascinating demonstrations of momentum conservation. The air that spurts from the tiny holes in the track lets you see an almost friction-free performance. In the everyday world, friction usually shows itself. Ideally, the net momentum of a couple of freight cars that collide is the same before and just after collision. But as the combined cars move along the track, friction provides an impulse to decrease momentum.

Another thing: Perfectly elastic collisions are not common in the everyday world. Usually some heat is generated in collisions. Drop a ball, and after it bounces from the floor, both the ball and the floor are a bit warmer. Even a dropped "superball" will not bounce to its initial height. At the microscopic level, however, perfectly elastic collisions are commonplace. For example, gas molecules bounce off one another without generating heat; they don't even touch in the classic sense of the word. (As later chapters show, the notion of touching at the atomic level is different from touching at the everyday level.)

FIGURE 5.14 ▲
Inelastic collisions. The net momentum of the trucks before and after collision is the same.

FIGURE 5.15 ▶
An air track. Blasts of air from tiny holes provide a friction-free surface for the carts to glide upon.

Conservation of momentum and, as the next chapter will discuss, conservation of energy are the two most powerful tools of mechanics. These laws let us understand the details of interactions among subatomic particles to those of entire galaxies.

Galileo worked hard to produce smooth surfaces to minimize friction. How he would have loved to experiment with today's air tracks!

CONCEPT CHECK

Refer to the gliders on the air track in Figure 5.15 to answer these questions.

1. Suppose both gliders have the same mass. They move toward each other at the same speed and experience an elastic collision. Describe their motion after the collision.
2. Suppose both gliders have the same mass and move toward each other at equal speed. This time they stick together when they collide. Describe their motion after the collision.
3. Suppose one of the gliders is at rest and is loaded so that it has twice the mass of the moving glider. Again, the gliders stick together when they collide. Describe their motion after the collision.

Check Your Answers

1. Since the collision is elastic, the gliders reverse directions upon colliding and move away from each other at the same speed as before.
2. Before the collision, the gliders had equal and opposite momenta, since their equal masses were moving in opposite directions at the same speed. The net momentum of both was zero. Since momentum is conserved, their net momentum after they stick must also be zero. They slam to a dead halt.
3. Before the collision, the net momentum equals the momentum of the unloaded, moving glider. After the collision, the net momentum is the same as before, but now the gliders are stuck together and moving as a single unit. The mass of the stuck-together gliders is three times that of the unloaded glider. Thus, the velocity is one-third of the unloaded glider's velocity before collision and in the same direction as before, since the direction as well as the amount of momentum is conserved.

✔ **READING CHECK**

Is momentum conserved in elastic collisions, inelastic collisions, or both?

CALCULATION CORNER

For a numerical example of momentum conservation, consider a pair of carts, A and B, on the air track shown in Figure 5.15. Suppose that cart B is at rest and that A slides against it and they stick together. Let cart A have a mass of 5 kg and move 1 m/s toward cart B, which has a mass of 1 kg. What will be the velocity of both carts when they link together?

$$\text{Net } mv_{\text{before}} = \text{net } mv_{\text{after}}$$

$$[m_A v_A + m_B v_B]_{\text{before}} = [(m_A + m_B)v]_{\text{after}}$$

$$(5 \text{ kg})(1 \text{ m/s}) + (1 \text{ kg})(0 \text{ m/s}) = (5 \text{ kg} + 1 \text{ kg}) v$$

$$5 \text{ kg m/s} = (6 \text{ kg}) v$$

$$v = 5/6 \text{ m/s}$$

Here we see that the small cart has no momentum before collision because its velocity is zero. After collision, the combined mass of both carts moves at velocity v, which by simple algebra is seen to be 5/6 m/s. This velocity is in the same direction as that of cart A.

Now suppose cart B is not at rest, but moves toward the left at a velocity of 4 m/s, as shown in the sketch. The carts are moving toward each other ready for a head-on collision. Let the direction of cart A be +, and cart B − (because it moves in a direction negative relative to cart A). Then we see that

$$\text{Net } mv_{\text{before}} = \text{net } mv_{\text{after}}$$

$$[m_A v_A + m_B v_B]_{\text{before}} = [(m_A + m_B)v]_{\text{after}}$$

$$(5 \text{ kg})(1 \text{ m/s}) + (1 \text{ kg})(-4 \text{ m/s}) = (5 \text{ kg} + 1 \text{ kg}) v$$

$$(5 \text{ kg m/s}) + (-4 \text{ kg m/s}) = (6 \text{ kg}) v$$

$$1 \text{ kg m/s} = 6 \text{ kg } v$$

$$v = 1/6 \text{ m/s}$$

Note that the negative momentum of cart B before collision has more effect in slowing cart A after collision. If cart B were moving twice as fast, then

$$\text{Net } mv_{\text{before}} = \text{net } mv_{\text{after}}$$

$$[m_A v_A + m_B v_B]_{\text{before}} = [(m_A + m_B)v]_{\text{after}}$$

$$(5 \text{ kg})(1 \text{ m/s}) + (1 \text{ kg})(-8 \text{ m/s}) = (5 \text{ kg} + 1 \text{ kg}) v$$

$$(5 \text{ kg m/s}) + (-8 \text{ kg m/s}) = (6 \text{ kg}) v$$

$$-3 \text{ kg m/s} = 6 \text{ kg } v$$

$$v = -1/2 \text{ m/s}$$

Here we see the final velocity is −1/2 m/s. What is the significance of the minus sign? It means that the final velocity of the two-cart system is *opposite* to the initial velocity of cart A. After collision, the two-cart system moves toward the left.

Here we have discussed carts on an air track. Our example could well have been football players or even swimming fish. We leave as a chapter-end problem finding the initial velocity of a small fish halting the motion of a larger fish. Try the same type of calculations!

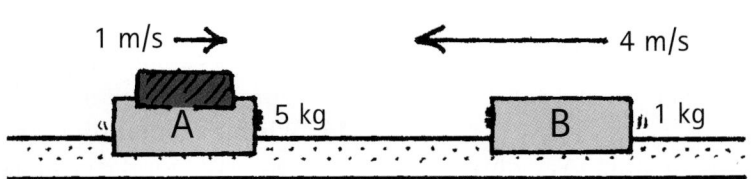

5 CHAPTER REVIEW

KEY TERMS

Elastic collision A collision in which colliding objects rebound without lasting deformation or the generation of heat.

Impulse The product of the force acting on an object and the time during which it acts. In an interaction, impulses are equal and opposite.

Impulse–momentum relationship Impulse is equal to the change in the momentum of the object that the impulse acts on. In symbol notation, $Ft = \Delta mv$.

Inelastic collision A collision in which the colliding objects become distorted, generate heat, and possibly join together.

Law of conservation of momentum When no external net force acts on an object or a system of objects, no change of momentum takes place. Hence, the momentum before an event involving only internal forces is equal to the momentum after the event: $mv_{\text{(before event)}} = mv_{\text{(after event)}}$.

Momentum The product of the mass of an object and its velocity.

REVIEW QUESTIONS

Momentum Is Inertia in Motion

1. Which has a greater momentum, a heavy truck at rest or a moving automobile?
2. How can a supertanker have a huge momentum when it moves relatively slowly?

Impulse Changes Momentum

3. Why is it incorrect to say that impulse equals momentum?
4. To impart the greatest momentum to an object, what should you do in addition to exerting the largest force possible?

5. For the same force, which cannon imparts the greater speed to a cannonball—a long cannon or a short one? Explain.
6. If you're in a car with faulty brakes and you have to hit something to stop, the momentum will change to zero whether you hit a brick wall or a haystack. So why is hitting a haystack a safer bet?
7. Why is it less damaging if you fall on a mat than if you fall on a solid floor?
8. Why is it a good idea to extend your hand forward when catching a fast-moving baseball with your bare hand?
9. In boxing, why is it advantageous to roll with the punch?
10. In karate, why is a short time for the applied force advantageous?

Momentum Change Is Greater When Bouncing Occurs

11. Which is the greater change in momentum, stopping something dead in its tracks or stopping it and then reversing its direction?
12. Which requires the greater impulse, stopping something dead in its tracks or stopping it and then reversing its direction?

When No External Force Acts, Momentum Doesn't Change—It Is Conserved

13. When can the momentum of two moving objects be canceled?
14. What does it mean to say that momentum (or any quantity) is *conserved*?
15. When a cannonball is fired, its momentum does change. Is momentum conserved for the cannonball?
16. When a cannonball is fired, the cannon recoils. Is momentum conserved for the cannon?
17. When a cannonball is fired, is momentum conserved for the cannon–cannonball system as a whole? (Why is your answer different than in the previous two questions?)

Momentum Is Conserved in Collisions

18. Distinguish between an *elastic* collision and an *inelastic* collision. For which type of collision is momentum conserved?

19. Railroad car A rolls at a certain speed and makes a perfectly elastic collision with car B of the same mass. After the collision, car A is observed to be at rest. How does the speed of car B compare with the initial speed of car A?

20. If the equally massive cars of the previous question stick together after colliding inelastically, how does their speed after the collision compare with the initial speed of car A?

THINK AND DO

1. Write a letter to a favorite relative reporting how nice it is to learn that nature follows rules, which means that predictions can be made. Give a couple of examples of how the conservation of momentum makes the behavior of colliding things more sensible.

2. If your teacher will let you, play around with an air track or air table. Predict what will happen before you initiate collisions of carts or air pucks.

THINK AND COMPARE

1. The balls have different masses and speeds.

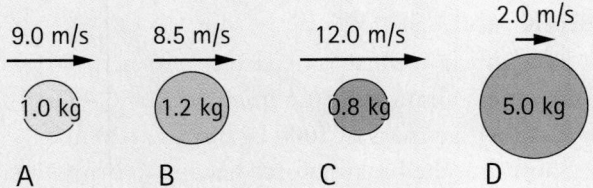

(a) Rank them in terms of momentum, from greatest to least.

(b) Rank them in terms of the impulse needed to stop them, from greatest to least.

2. Evan pushes crates starting at rest across a floor for 4 seconds with a net force as shown (including friction). Rank the three situations from biggest to smallest.

(a) Change in momentum
(b) Final speed
(c) Momentum in 4 seconds

PLUG AND CHUG

Become familiar with the equations for momentum and impulse.

Momentum = *mv*

1. Calculate the momentum of a 10-kg bowling ball rolling at 3 m/s.

2. Calculate the momentum of a 50-kg carton that slides at 3 m/s across an icy surface.

Impulse = *Ft*

3. Calculate the impulse that occurs when an average force of 10 N is exerted on a cart for 5 s.

4. Calculate the impulse that occurs when the same force of 10 N acts on the cart for twice the time.

THINK AND EXPLAIN

For answers below and solutions to Think and Solve, you may express momentum with the symbol *p*. Then $p = mv$.

1. When rollerblading, why is a fall less harmful on a wooden floor than on a concrete floor that has less give? Explain in terms of impulse and momentum.

2. In terms of impulse and momentum, why do air bags in cars reduce the chances of injury in accidents?

3. Many years ago, automobiles were manufactured to be as rigid as possible. Today's autos are designed to crumple upon impact. Why?

4. In terms of impulse and momentum, why are nylon ropes (which stretch considerably under tension) favored by mountain climbers?

5. If you throw an egg against a wall, the egg will break. But if you throw it at the same speed into a sagging sheet, it may not break. Why?

6. A lunar vehicle is tested on Earth at a speed of 10 km/h. When it travels as fast on the Moon, is its momentum more, less, or the same?

7. Which has the greater momentum when they move at the same speed—an automobile or a skateboard? Which requires the greatest stopping force?

8. In answering the preceding exercise, perhaps you stated that the automobile requires more stopping force. Make an argument that the skateboard could require more stopping force, depending on how quickly you want to stop it.

9. Jacob tries to jump from his canoe to the dock. He lands in the water, delighting his companions. What's your explanation for his mishap?

10. Why do 6-ounce boxing gloves hit harder than 16-ounce gloves?

11. Which undergoes the greatest change in momentum: (1) a baseball that is caught, (2) a baseball that is thrown, or (3) a baseball that is caught and then thrown back, if the baseball has the same speed just before being caught and just after being thrown?

12. In the preceding question, in which case is the greatest impulse required?

13. If a fully loaded shopping cart and an empty one traveling at the same speed have a head-on collision, which cart will experience the greater force of impact? The greater impulse? The greater change in momentum? The greater acceleration?

14. A fully dressed person is at rest in the middle of a pond on perfectly frictionless ice and must get to shore. How can this be accomplished?

15. Michael throws a ball horizontally while standing on roller skates. He rolls backward with a momentum that matches that of the ball. Will he end up rolling backward if he goes through the motions of throwing the ball, but instead holds onto it? Explain.

16. Two football players have a head-on collision and both stop short in their paths. If one player is twice as heavy as the other, how does his speed compare to the smaller player?

17. In the previous chapter, rocket propulsion was explained in terms of Newton's third law. That is, the force that propels a rocket is from the exhaust gases pushing against the rocket—the reaction to the force the rocket exerts on the exhaust gases. Explain rocket propulsion in terms of momentum conservation.

18. When you are traveling in a car at highway speed, the momentum of a grasshopper is suddenly changed as it splatters onto your windshield. Compared to the change in momentum of the grasshopper, by how much does the momentum of the car change?

19. If an 18-wheeler tractor-trailer and a sports car have a head-on collision, which vehicle will experience the greater force of impact? The greater impulse? The greater change in momentum? The greater acceleration?

20. Your friend says that the law of momentum conservation is violated when a ball rolls down a hill and gains momentum. What is your response?

THINK AND SOLVE

1. How much impulse is needed to stop a 10-kg bowling ball moving at 6 m/s?

2. A car with a mass of 1000 kg moves at 20 m/s. Show that the braking force needed to bring the car to a halt in 10 s is 2000 N.

3. A car carrying a 75-kg test dummy crashes into a wall at 25 m/s and is brought to rest in 0.1 s. Show that the average force exerted by the seat belt on the dummy is 18,750 N.

4. Jane (mass 40.0 kg), standing on slippery ice, catches her leaping dog (mass 15 kg) moving horizontally at 3.0 m/s. Show that the speed of Jane and her dog after the catch is 0.8 m/s.

5. A 2-kg ball of putty moving to the right has a head-on inelastic collision with a 1-kg putty ball moving to the left. If the combined blob doesn't move just after the collision, what can you conclude about the relative speeds of the balls before they collided?

6. A railroad diesel engine weighs four times as much as a freight car. If the diesel engine coasts at 5 km/h into a freight car that is initially at rest, show that the speed of the coupled cars is 4 km/h.

7. A 5-kg fish swimming 1 m/s swallows an absentminded 1-kg fish swimming toward it at a speed that brings both fish to a halt immediately after lunch. Show that the speed of the approaching smaller fish before lunch must have been 5 m/s.

8. A 1-kg ostrich egg is thrown at 2 m/s at a bedsheet and is brought to rest in 0.2 s. Show that the average amount of force on the egg is 10 N.

9. Can you run fast enough to have the same momentum as an automobile rolling at 1 mi/h? Make up reasonable figures to justify your answer.

READINESS ASSURANCE TEST (RAT)

If you have a good handle on this chapter—if you really do—then you should be able to score 7 out of 10 on this RAT. Check your answers with your teacher. If you score less than 7, study further before moving on.

Choose the best answer to each of the following.

1. If Fast Freddy doubles his running speed, what else doubles?
   (a) His momentum.
   (b) His inertia.
   (c) His time of running.
   (d) His acceleration.

2. A 1-kg ball has the same speed as a 10-kg ball. Compared with the 1-kg ball, the 10-kg ball has
   (a) one-tenth the momentum.
   (b) the same momentum.
   (c) 10 times as much momentum.
   (d) 100 times as much momentum.

3. Two iron balls, one twice the mass of the other, are dropped from rest from the top of a one-story building. Compared with the lighter ball, the twice-as-heavy ball when hitting the ground below has
   (a) half the momentum.
   (b) the same momentum.
   (c) twice the momentum.
   (d) four times the momentum.

4. If the mass of a cart full of groceries decreases to half and its speed doubles, the momentum of the cart
   (a) remains unchanged.
   (b) is doubled.
   (c) is quadrupled.
   (d) decreases.

5. Your friend says the impulse equals momentum. You disagree and say the friend is missing an important word, which is
   (a) inertia.
   (b) acceleration.
   (c) time.
   (d) change.

6. The impulse-momentum relationship is a direct result of Newton's
   (a) first law.
   (b) second law.
   (c) third law.
   (d) law of gravity.

7. Which of the following equations best illustrates the usefulness of automobile air bags?
   (a) $a = F/m$
   (b) $Ft = \Delta mv$
   (c) $d = 1/2\, gt^2$
   (d) $d = vt$

8. When you jump from an elevated position to the ground below, the force you experience when landing depends on
   (a) the jumping height.
   (b) the softness or hardness of the ground.
   (c) how much your knees bend.
   (d) All of the above.

9. Standing on a skateboard, you toss a ball horizontally away from you. The mass of the ball is one-tenth your mass. Compared with the speed you give to the ball, your ideal recoil speed will be
   (a) one-tenth as much.
   (b) the same.
   (c) ten times as much.
   (d) one hundred times as much.

10. A big fish swims upon and swallows a small fish at rest. After lunch, the momentum of the big fish is
    (a) the same as before.
    (b) less than before.
    (c) more than before.
    (d) zero.

ENERGY

THE MAIN IDEA

Energy can change from one form to another without a net loss or gain.

At an amusement park, why is the first summit on a roller coaster always the highest one? And why are following summits lower in height? In ancient times, people had great difficulty lifting huge loads, yet today a child can lift much heavier loads with a simple pulley system. How does a pulley so greatly increase output force? The answers to these questions involve the concept of *energy*—what this chapter is about.

Energy is the most central concept in physical science. Energy comes from the Sun in the form of sunlight. Plants capture it and make hydrocarbons like sugar. Solar energy is being tapped more and more for our energy needs. It produces hot water and electricity for our homes. We find energy in the foods we eat, which sustains life. Our study of energy begins with a related concept: *work.*

Explore!

What Happens When You Do Work on Sand?

1. Pour a handful of dry sand into a can.
2. Measure the temperature of the sand with a thermometer.
3. Remove the thermometer from the sand, then cover the can.
4. Shake the can vigorously for a minute or so. Now remove the cover and measure the temperature of the sand again.

Analyze and Conclude

1. **Observing** What happened to the temperature of the sand after you shook it?
2. **Predicting** Do you think temperature would increase for longer periods of shaking?
3. **Conclusion** How can you explain the change in temperature of the sand in terms of work and energy?

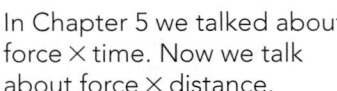

6.1 Work—Force × Distance

It takes energy to push something and make it move. How much energy depends on the force exerted and how long it is exerted. If "how long" means time, then we're talking about impulse, as in the previous chapter. Recall that impulse equals a change in *momentum*. In this chapter, by "how long," we mean distance. The quantity "force × distance" is equal to the change in *energy*. We call this quantity **work.**

In Chapter 5 we talked about force × time. Now we talk about force × distance.

Work = force × distance

◀ **FIGURE 6.1**
Compared with the work done in lifting a load one story high, twice as much works is done in lifting the same load two stories high. Twice the work is done because the *distance* is twice as much.

FIGURE 6.2 ▶
When twice the load is lifted to the same height, twice as much work is done because the *force* needed to lift it is twice as much.

We do work when we lift a load against Earth's gravity. The heavier the load or the higher we lift it, the more work we do. ✔ The amount of work done on an object depends on (1) how much force is applied and (2) how far the force causes the object to move. *

When a weight lifter raises a heavy barbell, he does work on the barbell. He gives energy to the barbell. Interestingly, when a weight lifter simply holds a barbell overhead, he does no work on it. He may get tired holding the barbell still, but if the barbell is not moved by the force he exerts, he does no work *on the barbell.* Work may be done on his muscles as they stretch and contract, which is force × distance on a biological scale. But this work is not done *on the barbell. Lifting* the barbell is different from *holding* the barbell.

The unit of work combines the unit of force (N) with the unit of distance (m), the newton-meter (N • m). We call a newton-meter the *joule* (J) (rhymes with *cool*). One joule of work is done when a force of 1 newton is exerted over a distance of 1 meter, as in lifting an apple over your head. For larger values, we speak of kilojoules (kJ)—thousands of joules—or megajoules (MJ)—millions of joules. The weight lifter in Figure 6.3 does work in kilojoules. The work done to vertically raise a heavily loaded truck can be in megajoules.

FIGURE 6.3 ▲
Work is done in lifting the barbell. Lifting it twice as high requires twice as much work.

✔ READING CHECK

What two things does the amount of work done depend upon?

---

* Force and distance must be in the same direction. When force is not along the direction of motion, then work equals the *component* of force in the direction of motion × distance moved.

FIGURE 6.4 ▲
He may expend energy when he pushes on the wall, but if it doesn't move, no work is done on the wall.

CONCEPT CHECK

1. How much work is needed to lift an object that weighs 500 N to a height of 4 m?
2. How much work is needed to lift it twice as high?
3. How much work is needed to lift a 1000-N object to a height of 8 m?

Check Your Answers

1. $W = F \times d = 500$ N $\times 4$ m $= 2000$ J.
2. Twice the height requires twice the work. That is, $W = F \times d = 500$ N $\times 8$ m $= 4000$ J.
3. Lifting twice the load twice as high requires four times the work. That is, $F \times d = 1000$ N $\times 8$ m $= 8000$ J.

6.2 Power—How Quickly Work Gets Done

Lifting a load quickly is more difficult than lifting the same load slowly. If equal loads are lifted to the same height, one quickly and the other slowly, the *work* done is the same. What's different is the *power*. ✔ **Power** is equal to the amount of work done per the time it takes to do it:

$$\text{Power} = \frac{\text{work done}}{\text{time interval}}$$

A high-power auto engine does work rapidly. An engine that delivers twice the power of another, however, does not necessarily go twice as fast. Twice the power means the engine can do twice the work in the same amount of time—or it can do the same amount of work in half the time. A more powerful engine can speed up a car more quickly.

FIGURE 6.5 ▶
The space shuttle can develop 33,000 MW of power when fuel is burned at the enormous rate of 3400 kg/s. This is like emptying an average-size swimming pool in 20 s!

The unit of power is the joule per second, called the **watt,** in honor of James Watt, the eighteenth-century developer of the steam engine. One watt (W) of power is used when 1 joule of work is done in 1 second. One kilowatt (kW) is 1000 watts. One megawatt (MW) is 1 million watts.

> ✔ **READING CHECK**
>
> What does work and time have to do with power?

CONCEPT CHECK

1. You do work when you do push-ups. If you do the same number of push-ups in half the time, how does your power output compare?

2. How many watts of power are needed when a force of 1 N moves a book 2 m in a time of 1 s?

Check Your Answers

1. Your power output is twice as much.
2. The power expended is 2 watts:

$$P = \frac{W}{t} = \frac{F \times d}{t} = \frac{1\,N \times 2\,m}{1\,s} = \textbf{2 W.}$$

6.3 Mechanical Energy

Work is done in lifting the heavy ram of a pile driver (Figure 6.6). When raised, the ram then has the ability to do work on a piling beneath it when it falls. When work is done by an archer in drawing a bow, the bent bow has the ability to do work on the arrow. When work is done to wind a spring mechanism, the spring then has the ability to do work on various gears to run a clock, ring a bell, or sound an alarm.

In each case, the ability to do work has been acquired by an object. This ability to do work is **energy.** Like work, energy is measured in joules.

Energy appears in many forms, such as heat, light, sound, electricity, and radioactivity. It even takes the form of mass, as celebrated in Einstein's famous $E = mc^2$ equation. In this chapter, we focus on potential energy and kinetic energy. **Potential energy** is energy that arises because of an object's position. **Kinetic energy** is energy possessed by moving objects. Potential and kinetic energy are both considered to be kinds of mechanical energy. ✔ So mechanical energy may be in the form of either potential energy, kinetic energy, or both.

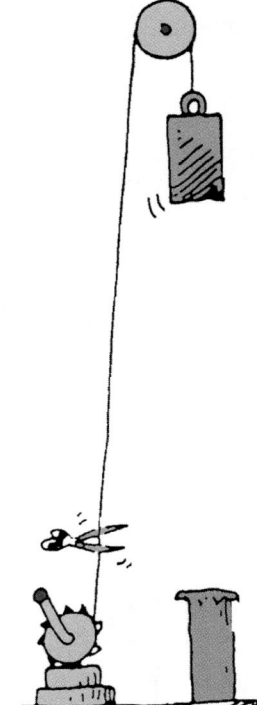

FIGURE 6.6 ▲
Work is required to lift the massive ram of the pile driver.

> What tells you whether or not work is done on something is a change in its energy. No change in energy means that no net work was done on it.

CONCEPT CHECK

1. She pushes the block five times farther up the incline than the man lifts it to the same height. How much more force does the man exert when he lifts the ice?

2. Who does more work on the ice?

3. If both jobs take the same time, who expends more power?

Check Your Answers

1. The man exerts five times as much force as the woman exerts.
2. Although he exerts more *force*, both do the same amount of work on the ice.
3. They both do the same amount of work in the same time, so both expend the same power.

READING CHECK

What are the two major forms of mechanical energy?

PE is relative to some reference level. If you're in a third-story classroom and a ball rests on the floor, you can say the ball is at height 0. Lift it and it has positive PE relative to the floor. Toss it out the window and it has negative PE relative to the floor.

6.4 Potential Energy Is Stored Energy

An object can store energy because of its position. In the stored state, energy has the potential to do work. This energy is stored and held in readiness. Therefore, it is called *potential energy* (PE). For example, when an archer draws an arrow with a bow, energy is stored in the bow. When released, energy is transferred to the arrow.

There are various kinds of potential energy. The potential energy that is easiest to visualize is in objects elevated against Earth's gravity. The potential energy due to an elevated position is called *gravitational potential energy.* The elevated ram of a pile driver and of water in an elevated reservoir both have gravitational potential energy.

The amount of gravitational potential energy that an elevated object has is equal to the work done to lift it—the force required to move it upward multiplied by the vertical distance moved ($W = F \times d$). Once upward motion begins, the upward force to keep an object moving at constant speed equals its weight. So the work done in lifting an object is its weight $\times$ height. We say:

Gravitational potential energy = weight $\times$ height

An object's weight is its mass m multiplied by the acceleration of gravity g. We write weight as mg. So the work done in lifting mg through a height h is equal to its gain in gravitational potential energy (PE):

$$PE = mgh$$

Note that the height h is the distance above some base level, such as the ground or the floor of a building. The potential energy is relative to that level and depends only on weight and height h. You can see in Figure 6.8 that the potential energy of the ball at the top of the structure depends on height only, and not on the path taken to get it there.

◄ **FIGURE 6.7**

The potential energy of Tenny's drawn bow equals the work (average force $\times$ distance) she did in drawing the arrow into position.

FIGURE 6.8 ▲

The PE of the 10-N ball is the same (30 J) in all three cases. That's because the work done in elevating it 3 m is the same whether it is (a) lifted with 10 N of force, (b) pushed with 6 N of force up the 5-m incline, or (c) lifted with 10 N up each 1-m step. No work is done in moving it horizontally (neglecting friction).

> ✓ **READING CHECK**
>
> How does the gravitational potential energy of an object relate to the work required to elevate it?

CONCEPT CHECK

1. How much work is done in lifting the 200-N block of ice shown in Figure 6.9 a vertical distance of 2.5 m?
2. How much work is done in pushing the same block of ice up the 5-m-long ramp? The force needed is only 100 N (which is why inclines are used).
3. What is the increase in the block's potential energy in each case?

Check Your Answers

1. 500 J (we get this either by *Fd* or *mgh*).
2. 500 J (she pushes with half the force over twice the distance).
3. Either way increases the block's potential energy by 500 J. The ramp simply makes this work easier to perform.

FIGURE 6.9 ▲

The man raises a block of ice by lifting it vertically. The girl pushes an identical block of ice up the ramp. When both blocks are raised to the same height, both have the same potential energy.

6.5 Kinetic Energy Is Energy of Motion

When you push on an object, you can make it move. Then a moving object is capable of doing work. It has energy of motion, or kinetic energy (KE). The kinetic energy of an object depends on its mass and speed. ✓ Specifically, kinetic energy is equal to the mass of an object multiplied by the square of its speed, multiplied by the constant 1/2:

$$\text{Kinetic energy} = 1/2 \text{ mass} \times \text{speed}^2$$

$$\text{KE} = 1/2\ mv^2$$

Since kinetic energy depends on mass, heavy objects have more kinetic energy than light ones moving at the same speed. For example, a car moving along the road has a certain amount of kinetic energy. A twice-as-massive car moving at the same speed has twice as much kinetic energy.

> Understanding the distinction between momentum and kinetic energy is high-level physics.

◀ **FIGURE 6.10**
The downhill "fall" of the roller coaster results in its roaring speed in the dip, and this kinetic energy sends it up the steep track to the next summit.

Kinetic energy also depends on speed. In fact, kinetic energy depends on speed more than it depends on mass. Why? Look at the equation. Kinetic energy depends on speed multiplied by speed, or speed squared. So if a car moving along the road has a certain amount of kinetic energy, a twice-as-fast car (with the same mass) has 2^2 or *four* times as much kinetic energy! The same car moving with three times the speed has 3^2 or nine times as much kinetic energy. So we see that small changes in speed produce large changes in kinetic energy.

READING CHECK

How does the kinetic energy of an object relate to its mass and speed?

CONCEPT CHECK

1. A car travels at 30 km/h and has kinetic energy of 1 MJ. If it travels twice as fast, 60 km/h, how much kinetic energy will it have?
2. If it travels three times as fast, at 90 km/h, what will be its kinetic energy?
3. If it travels four times as fast, at 120 km/h, what will be its kinetic energy?

Check Your Answers
1. Twice as fast means (2^2) four times the kinetic energy, or 4 MJ.
2. Three times as fast means (3^2) nine times the kinetic energy, or 9 MJ.
3. Four times as fast means (4^2) sixteen times the kinetic energy, or 16 MJ.

6.6 Work–Energy Theorem

To increase the kinetic energy of an object, work must be done on it. Or, if an object is moving, work is required to bring it to rest. In either case, the change in kinetic energy is equal to the work done. ✔ This important relationship between work and changes in energy is called the **work–energy theorem.** We abbreviate "change in" with the delta symbol, Δ, and say:

$$\textbf{Work} = \Delta \textbf{E}$$

When the energy that is changed is kinetic energy, we say:

$$\textbf{Work} = \Delta \textbf{KE}$$

Work equals change in kinetic energy. The work in this equation is the *net* work—that is, the work based on the net force.

The work–energy theorem emphasizes the role of *change.* If there is no change in an object's energy, then we know no work was done on it. This theorem applies to changes in potential energy also. Recall our previous example of the weight lifter raising the barbell. When work was

Which of these does a speeding ball not possess: force, momentum, energy? (*Hint:* The correct answer begins with an *F.*)

FIGURE 6.18 ▲
Work done on one end equals the work done on a load at the other end.

FIGURE 6.19 ▲
Force is multiplied. Note that a small input force × large distance = large output force × small distance.

CONCEPT CHECK

If a lever is arranged so that input distance is twice output distance, can we predict that energy output will be doubled?

Check Your Answer

No, no, a thousand times no! We can predict output *force* will be doubled, but never *energy*. Work and energy stay the same, which means force × distance stays the same. Shorter distance means greater force, and vice versa. Be careful to distinguish between the concepts of *force* and *energy!*

Another simple machine is a pulley. Can you see that it is a lever "in disguise"? When used as in Figure 6.20, it changes only the direction of the force. But when used as in Figure 6.21, the output force is doubled. Force is increased and distance moved is decreased.

Forces can be nicely multiplied with a system of pulleys. Such pulley arrangements are common wherever heavy loads are lifted, as in automobile service centers or machine shops. An ideal pulley system is shown in Figure 6.22. The man pulls 7 meters of rope with a force of 50 newtons and lifts 500 newtons through a vertical distance of 0.7 meter. The work the man does when pulling the rope is numerically equal to the increased potential energy of the 500-N bag.

Any machine that multiplies force does so at the expense of distance. Likewise, any machine that multiplies distance does so at the expense of force. No machine or device can put out more energy than is put into it. ✓ No machine can create energy; it can only transfer it or transform it from one form to another.

Energy is nature's way of keeping score.

✓ **READING CHECK**

A machine can transfer or transform energy, but can a machine create energy?

HANDS-ON EXPLORATION

Rub your hands briskly together. The friction between them multiplied by the distance of rubbing produces work that becomes heat. Note how quickly your palms are warmed.

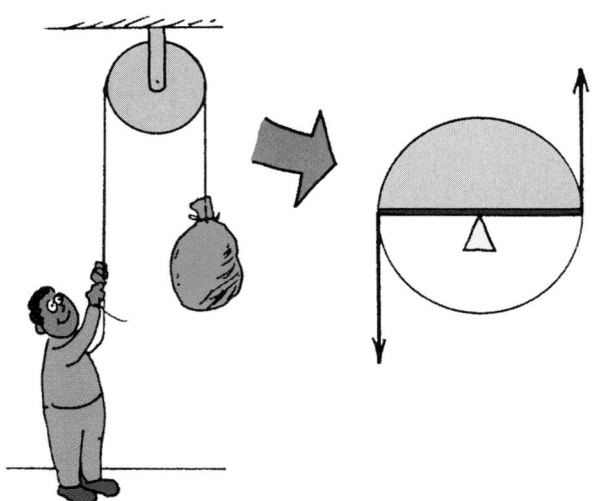

FIGURE 6.20 ▲
This pulley acts like a lever. It changes only the direction of the input force.

FIGURE 6.21 ▲
In this arrangement, a load can be lifted with half the input force.

$$F\mathsf{d} = F\mathsf{d}$$

◄ **FIGURE 6.22**
Input force × input distance = output force × output distance. Note the load is supported by 7 strands of rope. Each strand supports 1/7 the load. The tension in the rope pulled by the man is likewise 1/7 the load.

6.9 Efficiency—A Measure of Work Done for Energy Spent

The previous examples of machines were considered to be *ideal*. All the work input was transferred to work output. An ideal machine would have 100% efficiency. No real machine can be 100% efficient. ✓ In any machine, some energy is transformed into atomic or molecular kinetic energy—making the machine warmer. We say this wasted energy is dissipated as thermal energy, or roughly speaking, as heat. **Efficiency** can be expressed by the ratio

$$\text{Efficiency} = \frac{\textbf{work done}}{\textbf{energy used}}$$

Even a lever converts a small fraction of input energy into heat when it rotates about its fulcrum. We may do 100 joules of work but output 98 joules. The lever is then 98% efficient, and we waste 2 joules of work input on heat. In a pulley system, a larger fraction of input energy goes into heat. If we do 100 joules of work, the forces of friction acting through the distances through which the pulleys turn and rub about their axles may dissipate 60 joules of energy as heat. So the work output is only 40 joules, and the pulley system has an efficiency of 40%. The lower the efficiency of a machine, the greater the amount of energy wasted as heat.

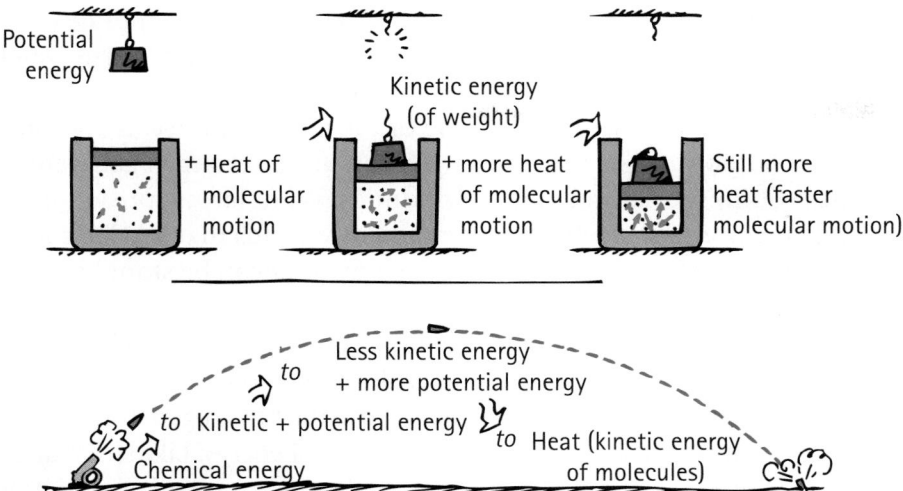

▲ **FIGURE 6.23**
Energy transitions. The graveyard of kinetic energy is thermal energy.

CONCEPT CHECK

Consider an imaginary miracle car that has a 100% efficient engine and burns fuel that has an energy content of 40 megajoules per liter. If the air drag plus frictional forces on the car traveling at highway speed is 500 N, what is the maximum distance the car can go on 1 liter of fuel?

Check Your Answer

From the definition work = force × distance, simple rearrangement gives distance = work/force. If all 40 million J of energy in 1 liter is used to do the work of overcoming the air drag and frictional forces, the distance covered is

$$\text{Distance} = \frac{\text{work}}{\text{force}} = \frac{40,000,000 \text{ J}}{500 \text{ N}} = 80,000 \text{ m} = 80 \text{ km.}$$

The important point here is that even with a perfect engine, there is an upper limit of fuel economy dictated by the conservation of energy.

An automobile engine is a machine that transforms chemical energy stored in gasoline into mechanical energy. But only a fraction of the energy in the gas is used by the car to move forward. Some of the fuel energy in the gas goes out in the hot exhaust gases and is wasted. Also, nearly half of the energy stored in the gas is wasted in the friction of the moving engine parts. In addition to these inefficiencies, some of the gas doesn't even burn completely. So the energy in the unburned gasoline also goes unused.

✔ **READING CHECK**

Why can't a machine be 100% efficient?

6.10 Sources of Energy

Except for nuclear power, the source of practically all our energy is the Sun. ✔ Even the energy we obtain from petroleum, coal, natural gas, and wood comes from the Sun. That's because these fuels are created by photosynthesis, the process by which plants trap solar energy and store it as plant tissue.

Sunlight is also directly transformed into electricity by photovoltaic cells, like those found in solar-powered calculators, or more recently, in the flexible solar shingles on the roofs of buildings. We use the energy in

FIGURE 6.24 ▲
Except for nuclear power, all the Earth's energy comes from the Sun.

Just as electricity requires an energy source, acquiring hydrogen requires an energy source. It's important to emphasize that hydrogen is *not* an energy source. It can carry and store energy produced elsewhere.

Watch for the growth of fuel-cell technology. The major hurdle for this technology is not the device itself, but with acquiring hydrogen fuel economically. One way is via solar cells.

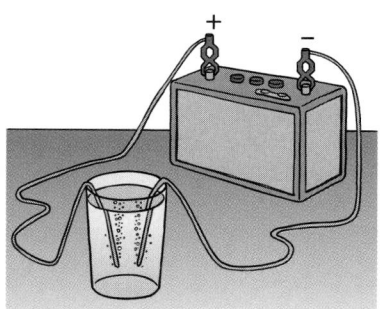

sunlight to generate electricity indirectly as well. Sunlight evaporates water, which later falls as rain; rainwater flows into rivers and turns water wheels, or it flows into modern generator turbines as it returns to the sea.

Wind, caused by unequal warming of the Earth's surface, is another form of solar power. The energy of wind can be used to turn generator turbines within specially equipped windmills. Wind farms spread out in windy regions feed electrical energy to power grids that service consumers. Harnessing the wind produces energy that can be stored in the form of free hydrogen.

Hydrogen is the least polluting of all fuels. Although hydrogen is an abundant element, it is locked up in molecules such as those in water and carbon compounds. Because it takes energy to extract hydrogen from water and carbon compounds, hydrogen is not a source of energy. Like steam and electricity, hydrogen can carry or store energy. A simple method of extracting hydrogen from water is shown in Figure 6.25. Place two platinum wires that are connected to the terminals of an ordinary battery into a glass of water (with an electrolyte dissolved in the water for conductivity). Be sure the wires don't touch each other. Bubbles of hydrogen form on one wire, and bubbles of oxygen form on the other. Electricity splits water into its constituent parts.

If you make the electrolysis process run backward, you have a fuel cell. In a fuel cell, hydrogen and oxygen gas are compressed at electrodes to produce water and electric current. The space shuttle uses fuel cells to meet its electrical needs while producing drinking water for the astronauts. Here on Earth, fuel-cell researchers are developing fuel cells for buses, automobiles, and trains.

The most concentrated form of usable energy is stored in uranium and plutonium, which are nuclear fuels. Interestingly, Earth's interior is kept hot because of a form of nuclear power—radioactivity—which has been with us since Earth formed. A by-product of nuclear power in Earth's interior is geothermal energy. Geothermal energy is held in underground reservoirs of hot water. Geothermal energy is predominantly limited to areas of volcanic activity, such as Iceland, New Zealand, Japan, and Hawaii. In these places, heated water near the Earth's surface is tapped to provide steam for running turbo-generators.

In locations where heat from volcanic activity is near the ground surface and groundwater is absent, another method holds promise for producing electricity: dry-rock geothermal power (Figure 6.26). With this method, water is put into the cavities in deep, dry, hot rock. When the water turns to steam, it is piped to a turbine at the surface. After turning the turbine, it is returned to the cavity for reuse. In this way, electricity is produced cheaply and cleanly.

◀ **FIGURE 6.25**
A simple electrical arrangement for separating hydrogen from water molecules. Hydrogen gas collects at one extended battery terminal and oxygen at the other. A fuel cell does the reverse and produces electricity from hydrogen and oxygen gases.

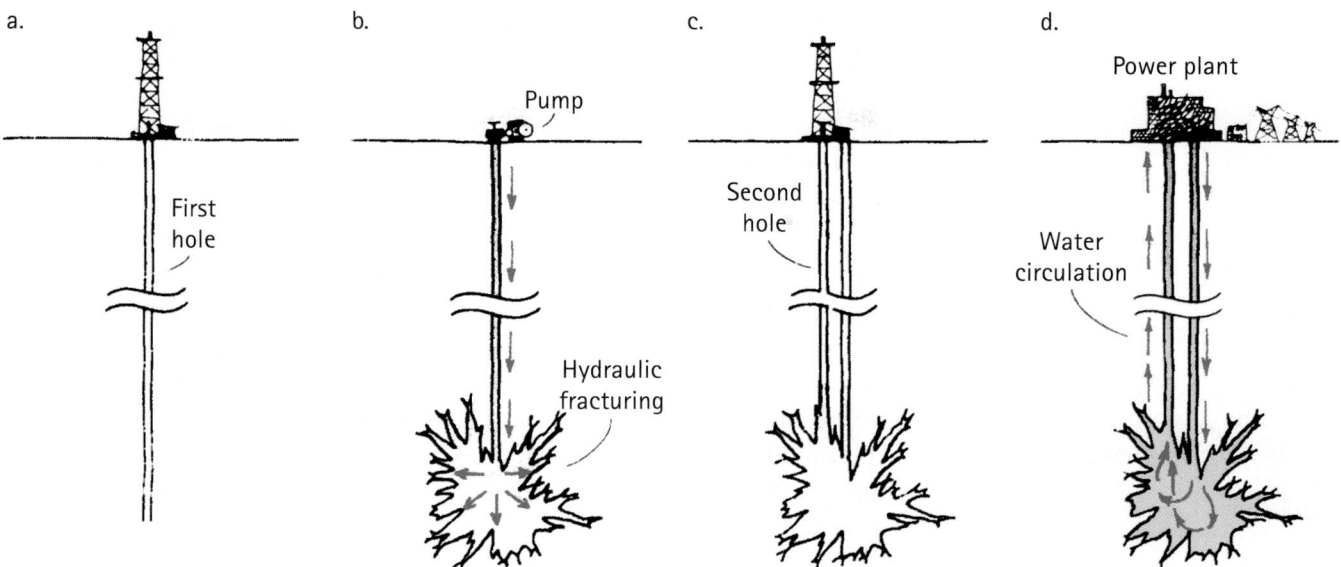

FIGURE 6.26 ▲

Dry-rock geothermal power. (a) A hole is sunk several kilometers into dry granite. (b) Water is pumped into the hole at high pressure and fractures surrounding rock to form a cavity with increased surface area. (c) A second hole is sunk to penetrate the cavity. (d) Water is circulated down one hole and through the cavity, where it is superheated before rising through the second hole. After driving a turbine, it is recirculated into the hot cavity again, making a closed cycle.

As the world population increases, so does our need for energy. With the rules of physics to guide them, technologists are presently researching newer and cleaner ways to develop energy sources. But they race to keep up with world population and greater demand in the developing world. Common sense dictates that in the meantime we should continue to optimize present sources and use what we consume efficiently and wisely.

READING CHECK

From where did fossil fuels originally acquire their energy?

6.11 Energy for Life

Your body is a machine—a wonderful machine. It is made up of smaller machines, the living cells. Like any machine, a living cell needs a source of energy. Most living organisms on this planet feed on molecules called hydrocarbons. Hydrocarbon compounds release energy when they react with oxygen. Like gasoline burned in an auto engine, ✓ there is more energy stored in the hydrocarbons than there is in the molecules that result after the hydrocarbons are "burned" for food. The energy difference is what sustains life.

Combustion, or burning, takes place when certain molecules combine with oxygen in air and release carbon dioxide molecules plus large amounts of energy. Combustion occurs when food is processed, or metabolized, in your body's digestive system. The combustion in your body is similar to the combustion that occurs in an engine when it burns gasoline. The main difference is the rate of combustion. During metabolism, the reaction rate is much slower and energy is released as needed by the body. Once the reaction starts, it is self-sustaining, similar to burning fossil fuels.

Each of us has a fundamental responsibility to treat our planet with respect and a sense of stewardship.

The reverse process is more difficult. Only green plants and certain one-celled organisms can make carbon dioxide combine with water to produce hydrocarbons like sugar. This process is *photosynthesis*. It requires an energy input, which normally comes from sunlight. Sugar is the simplest food. All other foods, such as carbohydrates, proteins, and fats, are made up of combinations of carbon, hydrogen, oxygen, and other elements. So green plants use the energy of sunlight to make food that gives us and all other organisms energy.

When you look at a tall tree, you might wonder where the atoms came from that compose it. Interestingly, the atoms that make up a tree come mostly from the air. Carbon dioxide molecules in the air make their way into tiny pores in the leaves. By photosynthesis, the carbon is separated from the oxygen and incorporated into the tree. What happens to the oxygen? It's expelled by the tree. And that's nice for us, for we need oxygen to live!

✔ **READING CHECK**

Energy is stored in hydro-carbons. After being burned for food, how does the energy in the resulting molecules compare?

JUNK SCIENCE

Scientists have to be open to new ideas. That's how science grows. But there is a body of established knowledge that can't be easily overthrown. That includes energy conservation, which is woven into every branch of science and supported by countless experiments from the atomic to the cosmic scale. Yet no concept has inspired more "junk science" than energy. Wouldn't it be wonderful if we could get energy for nothing, to possess a machine that gives out more energy than is put into it? That's what many practitioners of junk science offer. Gullible investors put their money into some of these schemes. But none of them passes the test of being real science. Perhaps some day a flaw in the law of energy conservation will be discovered. If it ever is, scientists will rejoice at the breakthrough. But so far, energy conservation is as solid as any knowledge we have. Don't bet against it.

6 CHAPTER REVIEW

KEY TERMS

Conservation of energy Energy cannot be created or destroyed; it may be transformed from one form into another, but the total amount of energy never changes. In an ideal machine, where no energy is transformed into heat,

$$\text{work}_{\text{input}} = \text{work}_{\text{output}} \text{ and } (Fd)_{\text{input}} = (Fd)_{\text{output}}.$$

Efficiency The percent of the work put into a machine that is converted into useful work output.

Energy The property of a system that enables it to do work.

Kinetic energy Energy of motion, described by the relationship kinetic energy $= 1/2\ m\text{v}^2$.

Machine A device such as a lever or pulley that increases (or decreases) a force or simply changes the direction of a force.

Potential energy The stored energy that a body possesses because of its position.

Power The time rate of doing work: Power = work/time.

Watt The unit of power, the joule per second.

Work The product of the force and the distance through which the force moves: $W = Fd$.

Work–energy theorem The work done on an object is equal to the energy gained by the object. Work $= \Delta E$. The energy change can be PE or KE.

REVIEW QUESTIONS

Work—Force × Distance

1. A force sets an object in motion. When the force is multiplied by the time of its application, we call the quantity *impulse,* which changes the *momentum* of that object. What do we call the quantity *force × distance*?

2. Cite an example where a force is exerted on an object without doing work on the object.

3. Which requires more work—lifting a 50-kg sack a vertical distance of 2 m or lifting a 25-kg sack a vertical distance of 4 m?

Power—How Quickly Work Gets Done

4. If both sacks in the preceding question are lifted their respective distances in the same time, how does the power required for each compare? How about for the case where the lighter sack is lifted its distance in half the time?

Mechanical Energy

5. What are the two main forms of mechanical energy?

6. Exactly what is it that a body having energy is capable of doing?

Potential Energy Is Stored Energy

7. A car is lifted a certain distance in a service station and therefore has potential energy relative to the floor. If it were lifted twice as high, how much potential energy would it have?

8. Two cars are lifted to the same elevation in a service station. If one car is twice as massive as the other, how do their potential energies compare?

9. How many joules of potential energy does a 1-N book gain when it is elevated 4 m? When it is elevated 8 m?

Kinetic Energy Is Energy of Motion

10. A moving car has kinetic energy. If it speeds up until it is going four times as fast, how much kinetic energy does it have in comparison?

Work–Energy Theorem

11. Compared to some original speed, how much work must the brakes of a car supply to stop a car moving four times as fast? How will the stopping distance compare?

Conservation of Energy

12. What will be the kinetic energy of a pile driver ram when it undergoes a 10-kJ decrease in potential energy? (Assume no energy goes to heat.)

Machines—Devices to Multiply Forces

13. Can a machine multiply input force? Input distance? Input energy? (If your three answers are the same, seek help, for the last question is especially important.)
14. If a machine multiplies force by a factor of 4, what other quantity is diminished, and by how much?
15. If the man in Figure 6.22 pulls 1 m of rope downward with a force of 100 N, and the load rises 1/7 as high, what is the maximum load that can be lifted?

Efficiency—A Measure of Work Done for Energy Spent

16. What is the efficiency of a machine that miraculously converts all the input energy to useful output energy?
17. Is a machine that has an efficiency greater than 100% physically possible? Discuss.

Sources of Energy

18. What is the ultimate source of the energy of fossil fuels, dams, and windmills?
19. What is the source of geothermal energy?

Energy for Life

20. The energy we require for existence comes from the chemically stored potential energy in food, which is transformed into other forms when it is metabolized. What happens to a person whose work output is less than the energy he or she consumes? Whose work output is greater than the energy he or she consumes? Can an undernourished person perform extra work without extra food? Briefly discuss.

THINK AND DO

1. Fill two mixing bowls with water from the cold tap and take their temperatures. Then run an electric or hand beater in the first bowl for a few minutes. Compare the temperatures of the water in the two bowls.

2. Pour some dry sand into a tin can with a cover. Compare the temperature of the sand before and after vigorously shaking the can for a couple of minutes.

PLUG AND CHUG

These one-step calculations are to familiarize you with the equations of the chapter.

Work $W = Fd$ (where F and d are in same direction)

1. Calculate the work done when a force of 2 N moves a book 3 m.
2. Calculate the work done when a 15-N force pushes a cart 3 m.

Power = work/time: $P = W/t$

3. Calculate the watts of power expended when a force of 1 N moves a book 2 m in a time interval of 1 s.
4. Calculate the power expended when a 20-N force pushes a cart 3.5 m in a time of 0.5 s.

Gravitational potential energy = weight × height: $PE = mgh$

5. How many joules of potential energy does a 1.5-kg book gain when it is elevated 2 m? When it is elevated 4 m? (Let $g = 10$ N/kg.)
6. Calculate the increase in potential energy when a 20-kg block of ice is lifted a vertical distance of 3 m.

Kinetic energy: $KE = 1/2\, mv^2$

7. Calculate the number of joules of kinetic energy a 1-kg parrot has when it flies at 6 m/s.
8. Calculate the kinetic energy of a 3-kg dog that runs at a speed of 4 m/s.

Work–energy theorem: Work = ΔKE; $Fd = \Delta 1/2\, mv^2$

9. How much work is required to increase the kinetic energy of a motor scooter by 4000 J?
10. What change in kinetic energy does a model airplane experience on takeoff if it is moved a distance of 5 m by a sustained net force of 10 N?

THINK AND COMPARE

Compare amounts for each of the following situations. If some are equal, put them in parentheses. For example, if A and B are tied, say (A-B tied).

1. The mass and speed of three vehicles is shown below.

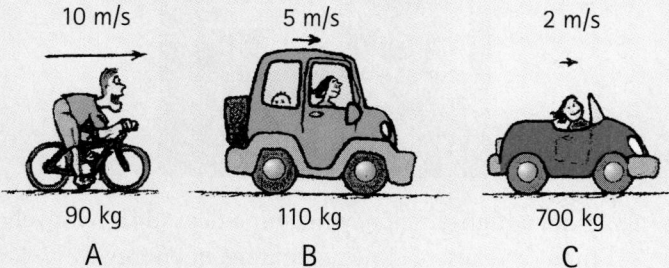

10 m/s   5 m/s   2 m/s

90 kg   110 kg   700 kg
A   B   C

   (a) Compare the momentum, from most to least.
   (b) Compare the kinetic energy, from most to least.

2. A ball is released at the left end of the metal track. Assume it has only enough friction to roll, but not to lessen its speed.

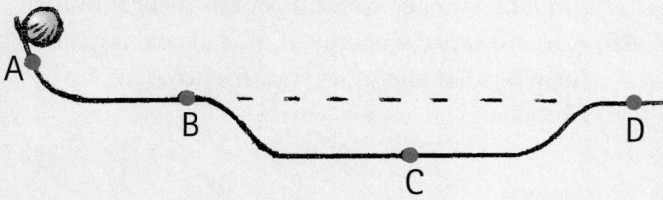

   (a) Compare the KE, from most to least.
   (b) Compare the PE.

3. The roller coaster starts from rest at point A, then proceeds down the incline. Compare the following for points A–E from most to least:

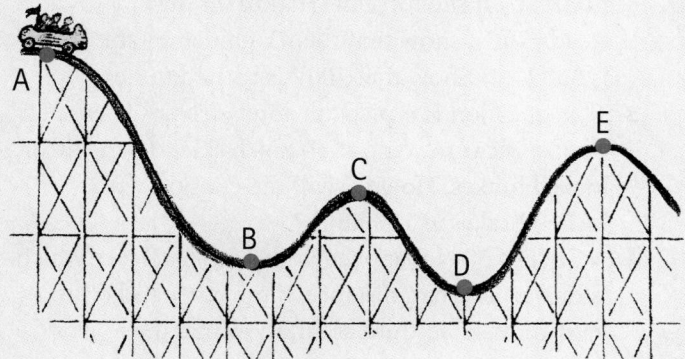

   (a) Speed.
   (b) KE.
   (c) PE.

4. Consider the efficiency of these four machines:
   (a) Energy in 100 J; energy out 60 J.
   (b) Energy in 100 J; energy out 50 J.
   (c) Energy in 200 J; energy out 80 J.
   (d) Energy in 200 J; energy out 120 J.
   Compare the efficiencies, from highest to lowest.

THINK AND EXPLAIN

1. When the mass of a moving object is doubled with no change in speed, by what factor is its momentum changed? Its kinetic energy?

2. When the velocity of an object is doubled, by what factor is its momentum changed? Its kinetic energy?

3. Consider a ball thrown straight up in the air. At what position is its kinetic energy a maximum? Where is its gravitational potential energy a maximum?

4. At what point in its motion is the KE of a pendulum bob a maximum? At what point is its PE a maximum? When its KE is half its maximum value, how much PE does it have?

5. A physical science teacher demonstrates energy conservation by releasing a heavy pendulum bob, as shown in the sketch, allowing it to swing to-and-fro. What would happen if in his exuberance he gave the bob a slight shove as it left his nose? Explain.

6. Discuss the design of the roller coaster shown in the sketch in terms of the conservation of energy.

7. Suppose that you and two classmates are discussing the design of a roller coaster. One classmate says that each summit must be lower than the previous one. Your other classmate says this is nonsense, for as long as the first one is the highest, it doesn't matter what height the others are. What do you say?

8. Consider a mouse and a dog running along a road have the same kinetic energy. Which has the greater average *speed*? (Use the equation for KE to guide your thinking.)

9. Consider molecules of hydrogen (tiny ones) and oxygen (bigger ones) in a gas mixture. If they have the same average kinetic energy (they will at the same temperature), which molecules have the greatest average speed?

10. According to the work-energy theorem, in the absence of friction, if you do 100 J of work on a cart while pushing it across a horizontal

surface, how much will you increase its kinetic energy?

11. When a driver applies brakes to keep a car going downhill at constant speed and constant kinetic energy, the potential energy of the car decreases. Where does this energy go? Where does most of it appear in a hybrid vehicle?

12. On a slide a child has potential energy that decreases by 1000 J while her kinetic energy increases by 900 J. What other form of energy is involved, and how much?

13. Consider the identical balls released from rest on tracks A and B as shown. When they reach the right ends of the tracks, which will have the greater speed? (*Hint:* Will their KEs be the same at the end?) Which will get to the end in the shortest time? (*Hint:* Considering the extra speed in the lower part of track B, which ball has the greatest average speed on the ramps?)

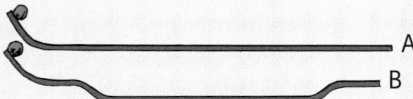

14. You tell your friend that no machine can possibly put out more energy than is put into it, and your friend states that a nuclear reactor puts out more energy than is put into it. What do you say?

15. Two lumps of clay with equal and opposite momenta have a head-on collision and come to rest. Is momentum conserved? Is kinetic energy conserved? Why are your answers different?

16. An automobile engine runs on compressed air. This means no pollution in the vicinity of the car. Does it also mean no energy is required? (*Hint:* Is energy required to compress air?)

17. Consider the swinging-balls apparatus. If two balls are lifted and released, momentum is conserved as two balls pop out the other side with the same speed as the released balls at impact. But momentum would also be conserved if one ball popped out at twice the speed. Can you explain why this never happens? (*Hint:* If the collision is perfectly elastic, what besides momentum would have to be conserved? Can you see

why this exercise is here rather than in the previous chapter on momentum?)

18. Does a high-efficiency machine degrade a relatively high or relatively low percentage of energy to thermal energy?

19. If an automobile had a 100% efficient engine, transferring all of the fuel's energy to work, would the engine be warm to your touch? Would its exhaust heat the surrounding air? Would it make any noise? Would it vibrate? Would any of its fuel go unused?

20. A friend says the energy of oil and coal is actually a form of solar energy. Is your friend correct, or mistaken?

THINK AND SOLVE

1. How many joules of work are done when a force of 1 N moves a book 2 m?

2. (a) How much work is done when you push a crate horizontally with 100 N across a 10-m factory floor? (b) If the force of friction on the crate is a steady 70 N, show that the KE gained by the crate is 300 J. (c) Show that 700 J is turned to heat.

3. This question is typical on some driver's license exams: A car moving at 50 km/h skids 15 m with locked brakes. How far will the car skid with locked brakes at 150 km/h?

4. A force of 50 N is applied to the end of a lever, which is moved a certain distance. If the other end of the lever moves one-third as far, show that the force it exerts is 150 N.

5. Consider an ideal pulley system. If you pull one end of the rope 1 m downward with a 50-N force, show that you can lift a 200-N load one-quarter of a meter high.

6. How many watts of power are expended when a force of 1 N moves a book 2 m in a time interval of 1 s?

READINESS ASSURANCE TEST (RAT)

If you have a good handle on this chapter—if you really do—then you should be able to score 7 out of 10 on this RAT. Check your answers with your teacher. If you score less than 7, study further before moving on.

Choose the best answer to the following.

1. How much work is done on a 100-kg crate that is hoisted 2 m in a time of 4 s?
   (a) 200 J
   (b) 500 J
   (c) 800 J
   (d) 2000 J

2. How much power is required to raise a 100-kg crate a vertical distance of 2 m in a time of 4 s?
   (a) 200 W
   (b) 500 W
   (c) 800 W
   (d) 2000 W

3. Raising an auto in a service station increases its potential energy. Raising it twice as high increases its potential energy by
   (a) half.
   (b) the same amount.
   (c) twice.
   (d) four times.

4. A model airplane moves three times as fast as another identical model airplane. Compared to the kinetic energy of the slower airplane, the kinetic energy of the faster airplane is
   (a) the same for level flight.
   (b) twice as much.
   (c) four times as much.
   (d) more than four times as much.

5. Which of the following equations is most useful for solving a problem that asks for the distance a fast-moving box slides across a post office floor and comes to a stop?
   (a) $F = ma$
   (b) $Ft = \Delta mv$
   (c) $KE = 1/2\ mv^2$
   (d) $Fd = \Delta 1/2\ mv^2$

6. A shiny sports car at the top of a vertical cliff has a potential energy of 100 MJ relative to the ground below. Unfortunately, a mishap occurs and it falls over the edge. When it is halfway to the ground, its kinetic energy is
   (a) the same as its potential energy at that point.
   (b) negligible.
   (c) about 60 MJ.
   (d) more than 60 MJ.

7. When a hybrid car brakes to a stop, much of its kinetic energy is transformed to
   (a) heat.
   (b) work.
   (c) electric potential energy.
   (d) gravitational potential energy.

8. In an ideal pulley system, a woman lifts a 100-N crate by pulling a rope downward with a force of 25 N. For every 1-meter length of rope she pulls downward, the crate rises
   (a) 50 centimeters.
   (b) 45 centimeters.
   (c) 25 centimeters.
   (d) None of these.

9. When 100 J are put into a device that puts out 40 J, the efficiency of the device is
   (a) 40%.
   (b) 50%.
   (c) 60%.
   (d) 140%.

10. A machine cannot multiply
    (a) forces.
    (b) distances.
    (c) energy.
    (d) All of these.

7

GRAVITY, PROJECTILES, AND SATELLITE MOTION

THE MAIN IDEA

Everything pulls on everything else; satellites are freely falling projectiles in the grip of gravity.

Are the space shuttle and the astronaut taking a "space walk" in the grips of Earth gravity, or beyond it? Was gravity discovered by Isaac Newton? Or was gravity discovered by earlier people who fell from trees or from their caves? If Newton didn't discover gravity, what did he discover about gravity? Does gravity reach to the Moon? Does it reach to the planets? To the stars? How far does gravity reach? How does gravity affect the motion of projectiles? Of satellites? Is it true that satellites, including planets, are freely falling projectiles? We'll learn the answers to these questions in this chapter.

Explore!

How Much Does the Size of Something Shrink with Increased Distance?

1. Hold your hands outstretched with one hand twice as far from your eyes as the other.
2. Make a judgment about which hand looks bigger.
3. Try it again, only this time overlap your hands slightly and view them with one eye closed.

Analyze and Conclude

1. **Observing** Many people see their hands in this position as the same size, while many see the nearer hand as slightly bigger. Very few people see the nearer hand as four times as big. But as

you will soon see by the inverse-square law, the nearer hand should appear twice as tall and twice as wide.

2. **Predicting** Twice times twice means four times as big! That's four times as much of your visual field as the farther hand.
3. **Making Generalizations** A person's belief that their hands are the same size is so strong that they likely overrule what is seen. That's why overlapping the hands better shows that the nearer hand is bigger. What other illusions do you have that are not so easily checked?

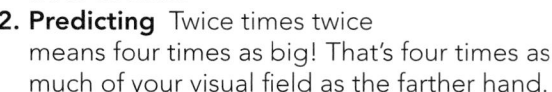

7.1 The Legend of the Falling Apple

Legend tells us that when Newton was a young man sitting under an apple tree, he made a connection that changed the way we see the world. He saw an apple fall. Perhaps he looked up through the tree branches toward the origin of the falling apple and noticed the Moon. In any event, Newton had the insight to realize that the force pulling on a falling apple is the same force that pulls on the Moon. Newton realized that Earth's gravity reaches to the Moon.

A force by Earth on the Moon would explain why the Moon does not follow a straight-line path, but instead circles about Earth. Circular motion is accelerated motion, which requires a force. ✓ Newton reasoned that the Moon is falling for the same reason the apple falls— they are both pulled by Earth's gravity.

FIGURE 7.1 ▲
Newton realizes that Earth's gravity affects both the apple *and* the Moon.

7.2 The Fact of the Falling Moon

Newton developed this idea further. How can the Moon fall and not hit Earth? He realized that the Moon must be falling *around* Earth—falling in the sense that it *falls beneath the straight line it would follow if no force acted on it*. He hypothesized that the Moon was simply a freely falling object circling Earth under the attraction of gravity.

As the Moon traces out its orbit around the Earth, it maintains a **tangential velocity**—a velocity parallel to the Earth's surface (Figure 7.2). ✓ Newton realized that the Moon's tangential velocity keeps it continually falling *around* the Earth instead of directly into it. Newton further realized that the Moon's path around the Earth is similar to the paths of the planets around the Sun.

This concept is illustrated in an original drawing by Newton, shown in Figure 7.3. Newton compared motion of the Moon with a cannonball fired from the top of a high mountain. He imagined that the mountaintop was above Earth's atmosphere, so air resistance would not slow the motion of the cannonball. If the cannonball were fired with a small horizontal speed, it would follow a curved path and soon hit the ground below. If it were fired faster, its path would be less curved and it would hit the ground farther away. If the cannonball were fired fast enough, Newton reasoned, its path would become a circle and the cannonball would circle indefinitely. It would be in orbit.

These ideas have very much changed the way people think.

✓ **READING CHECK**

What similarity did Newton see between a falling apple and the Moon?

FIGURE 7.2 ▲
The tangential velocity of the Moon allows it to fall around the Earth rather than directly into it.

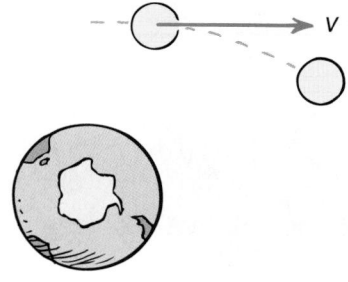

◀ **FIGURE 7.3**
This original drawing by Isaac Newton shows how an object projected fast enough would fall around Earth and become an Earth satellite.

☑ **READING CHECK**

Since the Moon falls toward Earth, why doesn't it crash into Earth's surface?

CONCEPT CHECK

1. In Figure 7.2, we see that the Moon falls around the Earth rather than straight into it. If the tangential velocity was zero, how would the Moon move?
2. If a rocket were fired from the surface of the Moon at a velocity equal to and opposite in direction to Moon's tangential velocity, what would be the fate of the rocket?

Check Your Answers

1. If the Moon's tangential velocity was zero, it would fall straight down and crash into the Earth!
2. The rocket would then have zero tangential velocity and would fall straight down and crash into the Earth. Ouch!

7.3 Newton's Grandest Discovery— The Law of Universal Gravitation

Newton further realized that everything pulls on everything else. He discovered that a force of gravity acts between all things in a beautifully simple way—involving only mass and distance. ☑ According to Newton, every mass attracts every other mass with a force that is directly proportional to the product of the two interacting masses. This statement is known as the **law of universal gravitation.** The force is inversely proportional to the square of the distance separating them:

$$\text{Force} \sim \frac{\text{mass}_1 \times \text{mass}_2}{\text{distance}^2}$$

Expressed in symbolic shorthand,

$$F \sim \frac{m_1 m_2}{d^2}$$

where m_1 and m_2 are the masses and d is the distance between their centers. Thus, the greater the masses m_1 and m_2, the greater the force of attraction between them. The greater the distance of separation d, the weaker is the force of attraction—weaker as the inverse square of the distance between their centers.

Just as sheet music guides a musician playing music, equations guide a physical science student to see how concepts are connected.

CONCEPT CHECK

1. According to the equation for gravity, what happens to the force between two bodies if the mass of one body is doubled?
2. What happens if instead the mass of the other body is doubled?
3. What happens if the masses of both bodies are doubled?
4. What happens if the mass of one body is doubled and the other tripled?

Check Your Answers

1. When one mass is doubled, the force between them doubles.
2. The force is still doubled, for it doesn't make any difference *which* mass doubles. ($2 \times 1 = 1 \times 2$; same product either way!)
3. The force is four times as much.
4. Double $\times$ triple = six. So the force is six times as much. (If you don't see why, discuss this with a friend before going further.)

 **READING CHECK**

State Newton's law of gravity in words.

7.4 Gravity and Distance— The Inverse-Square Law

Gravity gets weaker with distance the same way a light gets dimmer as you move farther away from it. Consider the candle flame in Figure 7.4. Light from the flame travels in all directions in straight lines. A patch is shown 1 meter from the flame. Notice that at a distance of 2 meters away, the light rays that fall on the patch spread to fill a patch twice as tall and twice as wide. The same light falls on a patch with four times the area. The same light 3 meters away spreads to fill a patch three times as tall and three times as wide. The light would fill a patch with nine times the area.

PhysicsPlace.com

Video
Inverse-Square Law

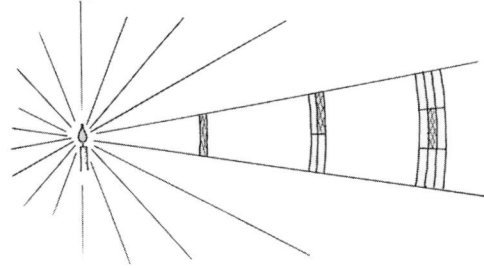

◀ **FIGURE 7.4**
Light from the flame spreads in all directions. At twice the distance, the same light is spread over four times the area; at three times the distance, it is spread over nine times the area.

As the light spreads out, its brightness decreases. Can you see that when you're twice as far away, it appears $\frac{1}{4}$ as bright? And can you see that when you're three times as far away, it appears $\frac{1}{9}$ as bright? There is a rule here: The intensity of the light gets less as the inverse square of the distance. This is the **inverse-square law.**

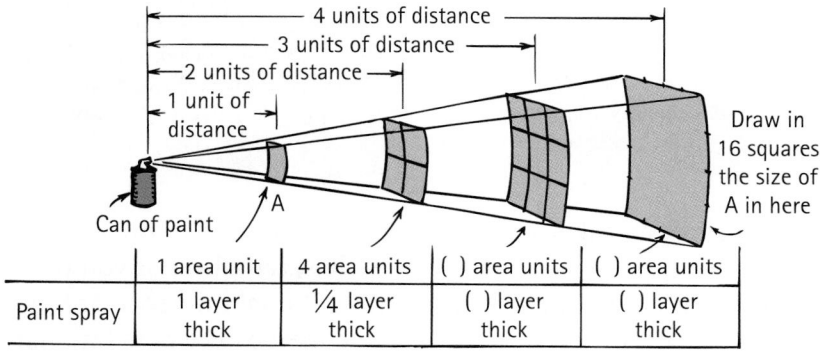

FIGURE 7.5 ▲
The inverse-square law. Paint spray travels in straight lines away from the nozzle of the can. Like gravity, the "strength" of the spray obeys the inverse-square law. Fill in the blanks.

Saying that if *F* is inversely proportional to the **square** of *d* means, for example, that if *d* gets bigger by 5, *F* gets *smaller* by 25. Get it?

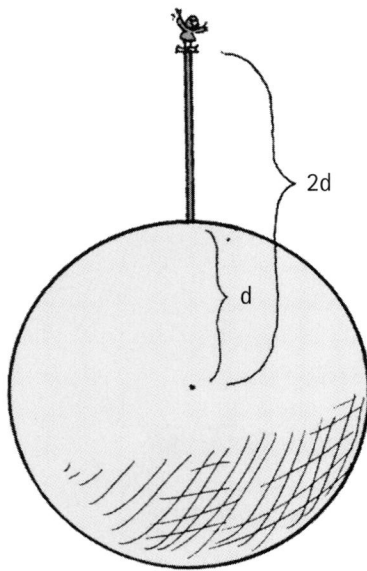

FIGURE 7.6 ▲
At the top of the ladder, the girl is twice as far from the Earth's center and weighs only $\frac{1}{4}$ as much as at the bottom of the ladder.

The inverse-square law also applies to a paint sprayer. Pretend you hold a paint gun at the center of a sphere with a radius of 1 meter (Figure 7.5). Suppose that a burst of paint produces a square patch of paint 1 millimeter thick. How thick would the patch be if the experiment were done in a sphere with twice the radius—that is, with the spray gun twice as far away? The answer is not half as thick, because the paint would spread to a patch twice as tall *and* twice as wide. It would spread over an area *four times* as big, and its thickness would be only $\frac{1}{4}$ millimeter. Can you see that for a sphere of radius 3 meters the thickness of the paint patch would be only $\frac{1}{9}$ millimeter? Do you see that the thickness of paint decreases as the *square* of the distance? The inverse-square law holds for light, for paint spray, and for gravity. It holds for all phenomena where something from a localized source spreads uniformly throughout the surrounding space. We'll see this to be true of the electric field about an electron, light from a match, radiation from a piece of uranium, and sound from a cricket.

✓ So we see that gravity decreases according to the inverse-square law. The force of gravity weakens as the square of the distance. In using Newton's equation for gravity, the distance term d is the distance between the *centers* of the masses of objects attracted to each other. Note in Figure 7.6 that the girl at the top of the ladder weighs only $\frac{1}{4}$ as much as she weighs at the Earth's surface. That's because she is twice the distance from the Earth's *center*.

CONCEPT CHECK

1. How much does the force of gravity change between the Earth and a receding rocket when the distance between them is doubled? Tripled? Ten times as much?

2. Consider an apple at the top of a tree. The apple is pulled by Earth's gravity with a force of 1 N. If the tree were twice as tall, would the force of gravity be only $\frac{1}{4}$ as strong? Defend your answer.

Check Your Answers

1. When the distance is doubled, the force is $\frac{1}{4}$ as much. When tripled, $\frac{1}{9}$ as much. When ten times, $\frac{1}{100}$ as much.

2. No, because the twice-as-tall apple tree is not twice as far from the Earth's center. The taller tree would have to be 6,370 km tall (the Earth's radius) for the apple's weight to reduce to $\frac{1}{4}$ N. For a decrease in weight by 1%, an object must be raised 32 km—nearly four times the height of Mt. Everest. So as a practical matter, we disregard the effects of everyday changes in elevation for gravity. The apple has practically the same weight at the top of the tree as at the bottom.

FIGURE 7.7 ▲
As the rocket gets farther from the Earth, gravitation between the rocket and the Earth gets less.

Earth's gravity weakens with distance, but no matter how far away, Earth's gravity approaches, but never reaches, zero. Even if you traveled to the far reaches of the universe, the gravitational influence of home would still be with you. It may be overwhelmed by the gravitational influences of nearer and/or more massive bodies, but it is there. The gravitational influence of every material object, however small or far, is present through all of space.

CONCEPT CHECK

1. Light from the Sun, like gravity, obeys the inverse-square law. If you were on a planet twice as far from the Sun, how bright would the Sun look?
2. How bright would the Sun look if you were on a planet twice as close to the Sun?

Check Your Answers

1. One-quarter as bright.
2. Four times brighter.

READING CHECK

How does the force of gravity change with distance?

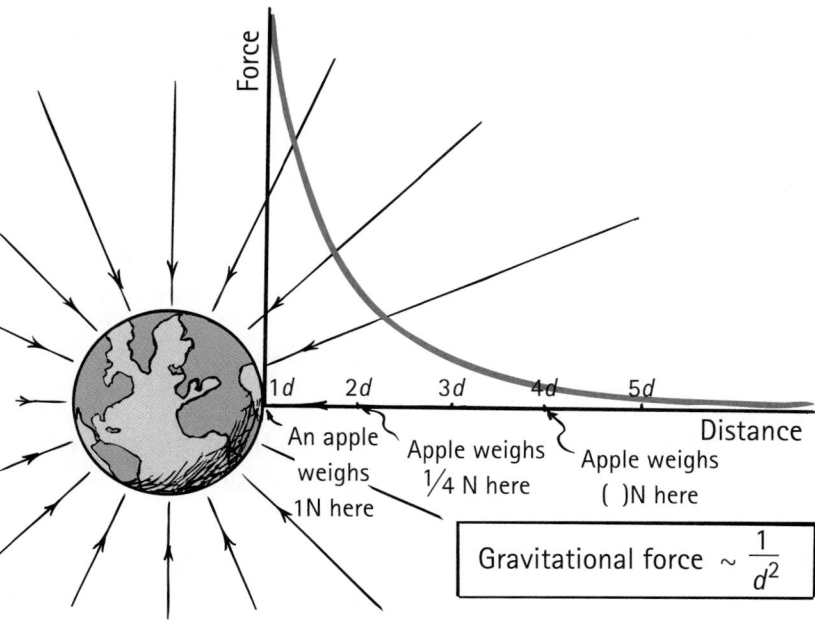

$$\text{Gravitational force} \sim \frac{1}{d^2}$$

◀ **FIGURE 7.8**

If an apple weighs 1 N at the Earth's surface, it weighs only $\frac{1}{4}$ N twice as far from the Earth's center. At three times the distance, it weighs only $\frac{1}{9}$ N. What would it weigh at four times the distance? Five times?

7.5 The Universal Gravitational Constant, *G*

The universal law of gravitation can be written as an exact equation when the **universal gravitational constant, *G*,** is used. Then we have

$$F = G\frac{m_1\,m_2}{d^2}$$

The units of *G* make the force come out in newtons. The magnitude of *G* is the same as the gravitational force between two 1-kilogram masses that are 1 meter apart: 0.0000000000667 newton.

$$G = 6.67 \times 10^{-11}\ \text{N} \cdot \text{m}^2/\text{kg}^2$$

This is an extremely small number. It shows that gravity is a very weak force compared with electrical forces. The large net gravitational force

Just as π relates circumference and diameter for circles, *G* relates gravitational force with mass and distance.

$$\frac{C}{D} = \pi$$

we feel as weight is because of the enormity of atoms in planet Earth that are pulling on us.

To better understand the constant *G*, consider the analogous case of the geometry constant, π, in the equation for the circumference of a circle:

$$C = \pi D$$

The equation tells you that the circumference of a circle, *C*, is equal to π multiplied by the diameter *D*. If we didn't know π, we could say

$$C \sim D$$

This expression tells us only that the circumference of a circle is *proportional* to its diameter. This means a small-circumference circle will have a small diameter and a large-circumference circle will have a large diameter. How much smaller or larger requires that we know π. We find this by dividing *C* by *D*. That is,

$$\frac{C}{D} = 3.14\ldots = \pi$$

Similarly with the proportion form of Newton's gravitational law on the previous page: The constant of proportionality *G* is found by

$$G = \frac{F}{m_1 m_2/d^2} = 6.67 \times 10^{-11}\,\text{N} \cdot \text{m}^2/\text{kg}^2$$

Regardless of the masses and distance between them, the gravitational force will have a value that results in the same value for *G*.

✔ **READING CHECK**

How do the units for *G* affect the unit of force?

7.6 The Mass of the Earth Is Measured

The value of *G* wasn't measured until a century after the publication of Newton's theory of universal gravitation. One method of measuring it, though not the first, is shown in Figure 7.9. Once we know the value of *G*, we have enough information to calculate the mass of the Earth! Here's how: The force that the Earth exerts on a 1-kilogram mass at its surface is 10 newtons (more precisely, 9.8 newtons). The distance between the 1-kilogram mass and the center of the Earth is the Earth's radius, 6.4×10^6 meters. Using

$$F = G\frac{m_1 m_2}{d^2}$$

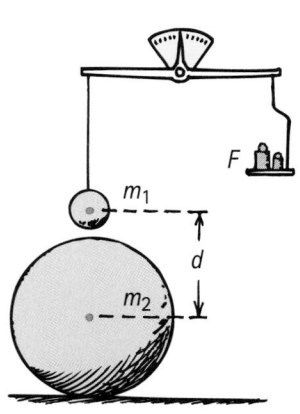

FIGURE 7.9 ▲
Philipp von Jolly's method of measuring *G*. A small ball of known mass is attracted to a 6-ton sphere rolled beneath it. The force of attraction can be measured by the weights needed to restore balance.

and using *F* as 9.8 N, m_1 is the mass of the 1-kilogram mass and m_2 is the mass of the Earth:

$$9.8\,\text{N} = 6.67 \times 10^{-11}\,\frac{\text{N} \cdot \text{m}^2}{\text{kg}^2} \times \frac{1\,\text{kg} \times m_2}{(6.4 \times 10^6\text{m})^2}$$

The only unknown quantity is m_2, the mass of the Earth. Solving, we find $m_2 = 6 \times 10^{24}$ kilograms.

✓ When *G* was first measured in the 18th century, people all over the world were excited about it. That's because newspapers everywhere announced the discovery as one that measured the mass of planet Earth. How exciting that Newton's formula gives the mass of the entire world, with all its oceans, mountains, and inner parts yet to be discovered! *G* and the mass of the Earth were measured when much of the Earth's surface was still undiscovered.

Myth: There is no gravity in space.
Fact: Gravity is everywhere!

CONCEPT CHECK

1. What value will result if you let your mass be m_1, the mass of the Earth m_2, and *d* the Earth's radius in the equation for gravity?
2. If your mass increases, does your weight increase also?

Check Your Answers

1. Your weight.
2. Yes, in direct proportion. That is, if you double your mass, your weight also doubles.

✓ **READING CHECK**

Why were people all over the world excited about the measurement of *G*?

7.7 Projectile Motion

A tossed stone, a cannonball, or any object projected by any means that continues in motion is called a **projectile.** A very simple projectile is a falling stone, Figure 7.10. This is a version of Figure 3.4, which we studied in Chapter 3. The stone gains speed as it falls straight downward, as indicated by a speedometer. Remember that a freely falling object gains 10 meters/second during each second of fall. This is the acceleration due to gravity, 10 m/s². If it begins its fall from rest, 0 m/s, then at the end of the first second of fall its speed is 10 m/s. At the end of 2 seconds, its speed is 20 m/s, and at the end of 3 seconds, it is 30 m/s—and so on. It keeps gaining 10 m/s each second it falls.

Although the change in speed is the same each second, the *distance* of fall keeps increasing. That's because the average speed of fall increases each second. Let's apply this to a new situation—throwing the stone horizontally from a high cliff.

First, imagine that gravity doesn't act on the stone (of course it *does* act, but we're just pretending for now). In Figure 7.11 we see the positions that a horizontally thrown stone would have with *no gravity*. Note that the positions each second are the same distance apart. That's because there is no force acting on it. The motion of the stone is like the motion of a bowling ball rolling along a bowling lane. Horizontal motion is constant because no horizontal force acts. Both the stone and the ball move without accelerating. They move at constant velocities, covering equal distances in equal times.

In the real world there is gravity, and the thrown stone falls beneath the straight line it would follow with no gravity (Figure 7.13). The stone curves as it falls. Interestingly, this familiar curve is the result of *two* kinds

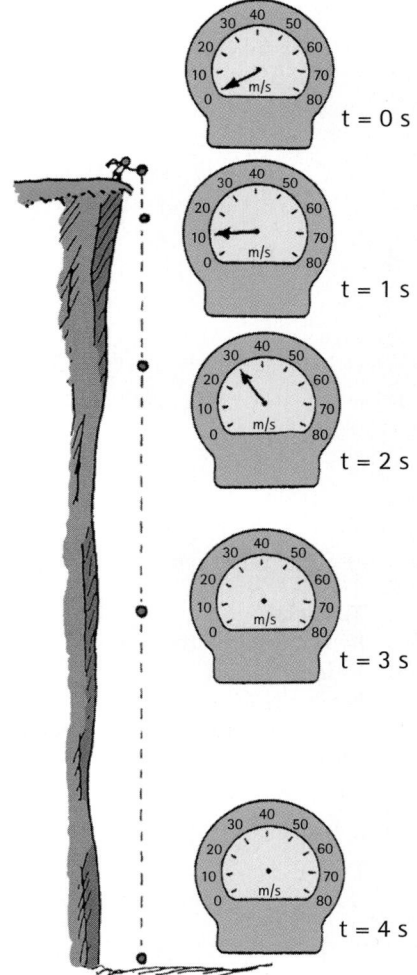

FIGURE 7.10 ▲
The falling stone gains a speed of 10 m/s each second. Fill in the speedometer readings for the times 3 and 4 seconds.

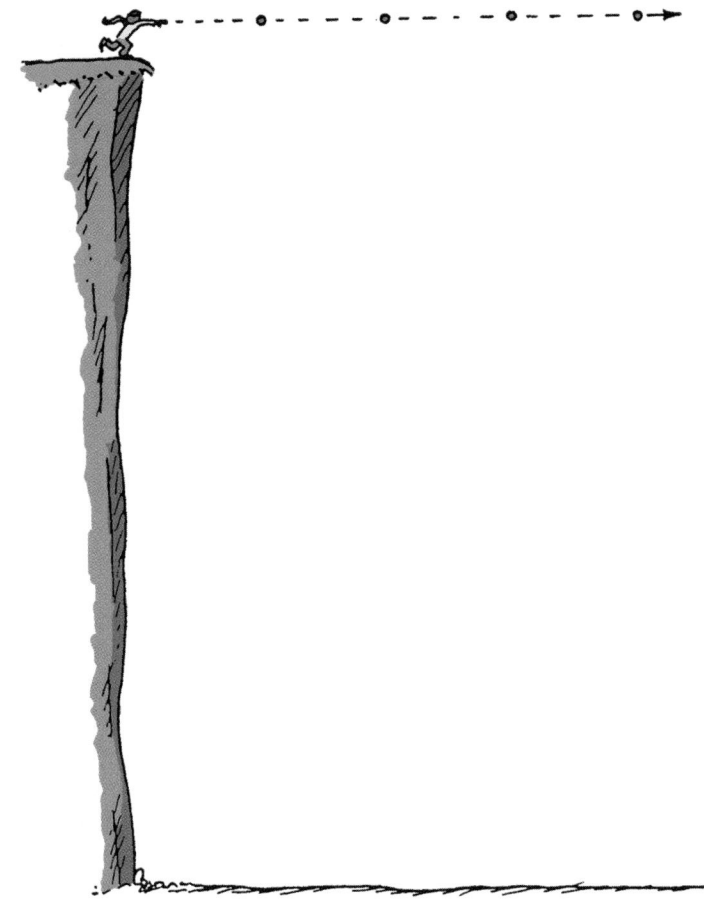

FIGURE 7.11 ▲
If there were no gravity, a stone thrown horizontally would move in a straight-line path and cover equal distances in equal time intervals.

of motion occurring at the same time. One kind is the straight-down vertical motion shown in Figure 7.10. The other is the horizontal motion of constant velocity, as imagined in Figure 7.11. Both occur simultaneously. As the stone moves horizontally, it also falls straight downward—beneath the place it would be if there were no gravity. This is indicated in Figure 7.13.

The curved path of a projectile is the result of constant motion horizontally and accelerated motion vertically under the influence of gravity. This curve is a **parabola.**

FIGURE 7.12 ▶
A bowling ball rolling along a lane similarly covers equal distances in equal times. It rolls at constant velocity.

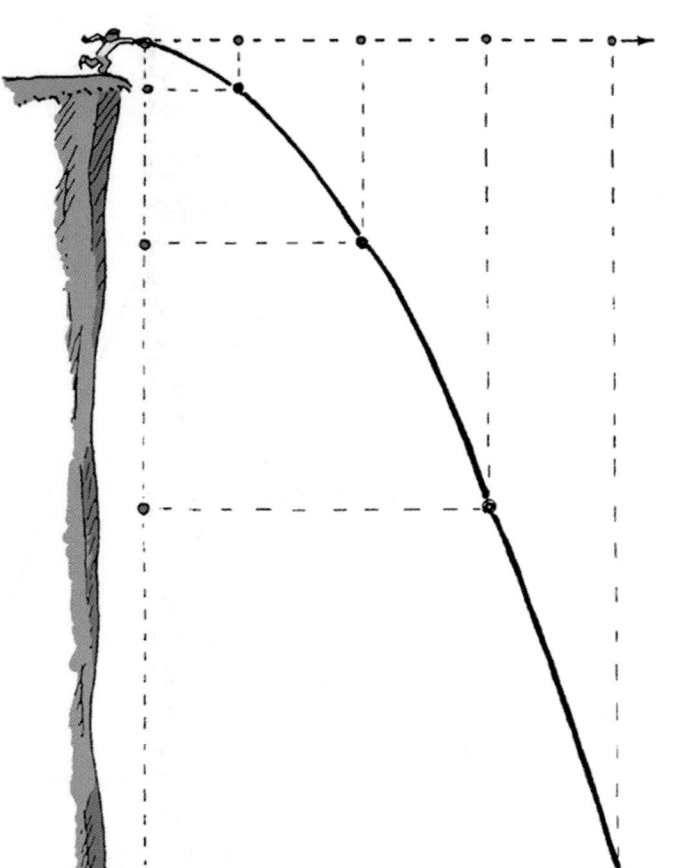

◀ **FIGURE 7.13**
The vertical path (dashed line) is for a stone dropped at rest. The horizontal path (dashed line) would occur with no gravity. The solid line shows the path that results from both the vertical and horizontal motions.

A projectile's path is called its *trajectory*.

CONCEPT CHECK

At the instant a horizontal cannon fires a cannonball from atop a high cliff, another cannonball is simply dropped from the same height. Which hits the ground below first, the one fired downrange, or the one that drops straight down?

Check Your Answer

Both cannonballs hit the ground at the same time, for both fall *the same vertical distance*. Can you see that the physics is the same as the physics of the figure above? We can reason this another way by asking which one would hit the ground first if the cannon were pointed at an *upward* angle. Then the dropped cannonball would hit first, while the fired ball is still in the air. Now consider the cannon pointing *downward*. In this case the fired ball hits first. So projected upward, the dropped one hits first; downward, the fired one hits first. Is there some angle at which there is a dead heat—where both hit at the same time? Can you see that this occurs when the cannon is horizontal?

In Figure 7.14, we consider a stone thrown upward at an angle. If there were no gravity, the path of the stone would be along the dashed line with the arrow. Positions of the stone at 1-second intervals along the line are shown by red dots. Because of gravity, the actual positions

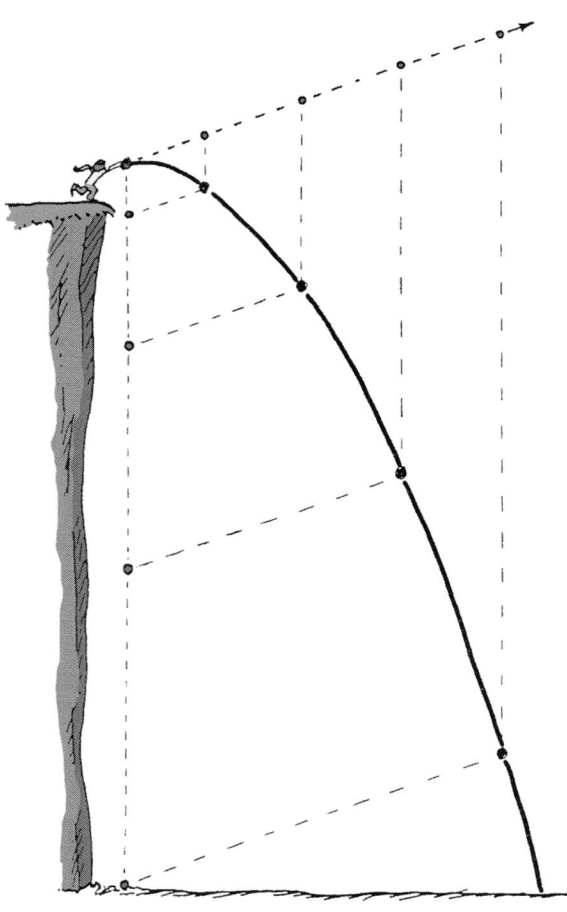

FIGURE 7.14 ▲
A stone thrown at an upward angle would follow the dashed line in the absence of gravity. Because of gravity, it falls beneath this line and describes the parabola shown by the solid curve.

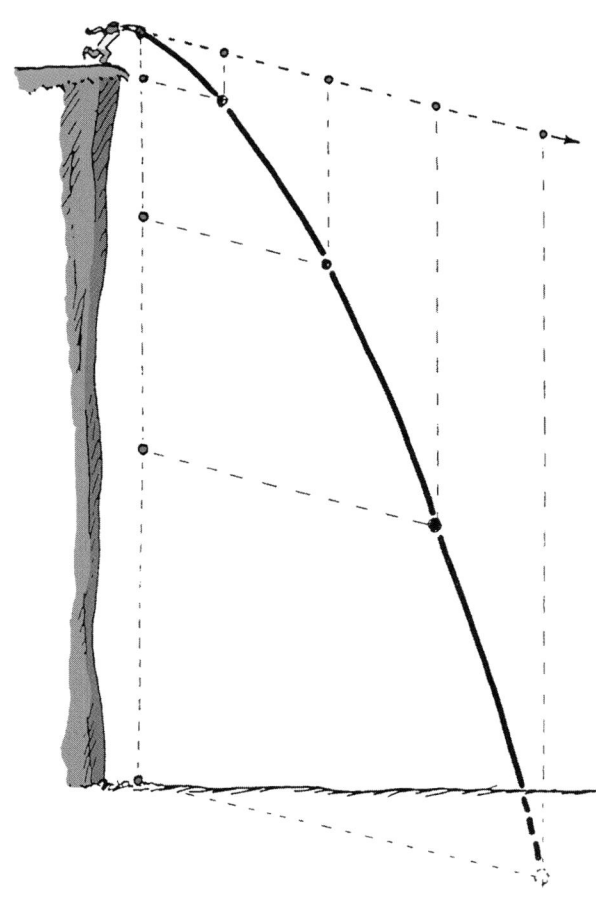

FIGURE 7.15 ▲
A stone thrown at a downward angle follows a somewhat different parabola.

(black dots) are below these points. How far below? The answer is, the same distance an object would fall if dropped from the red-dot positions. When we connect the black dots to plot the path, we get a different parabola. So we see that ✓ a projectile falls below an imaginary straight-line path a distance equal to the distance it would fall from rest at each point along the line.

In Figure 7.15, we consider a ball thrown at a downward angle. The physics is the same. If there were no gravity, it would follow the dashed line with the arrow. Because of gravity, it falls beneath this line, just as in the previous cases. The path is a somewhat different parabola.

The curved path of a projectile is a combination of horizontal and vertical motions. Consider the girl throwing the ball in Figure 7.16.

◀ **FIGURE 7.16**
The velocity of the ball (bold blue vector) has vertical and horizontal components. The vertical component relates to how high the ball will go. The horizontal component relates to the horizontal range of the ball.

HANDS-ON DANGLING BEADS EXPLORATION

Make your own model of projectile paths. On a ruler or a stick, at position 1, hang a bead from a string 1-cm long as shown. At position 2, hang a bead from a string 4-cm long. At position 3, do the same with a 9-cm length of string. At position 4, use 16 cm of string, and for position 5, 25 cm of string. Hold the stick horizontally and you have a version of Figure 7.13. Hold it at a slight upward angle to show a version of Figure 7.14. Held at an angle downward, you have Figure 7.15.

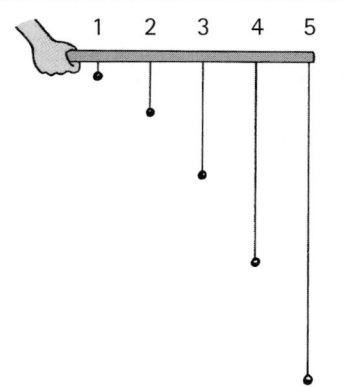

The velocity she gives the ball is shown by the bold blue vector. Notice that this vector has horizontal and vertical *components*. These components, interestingly, are completely independent of each other. The horizontal component is completely independent of the vertical component; they act as if the other doesn't exist. Their combined effects produce the curved paths of projectiles.

A typical projectile path in Figure 7.17 shows velocity vectors and their components. Notice that the horizontal component remains the same at all points. That's because no horizontal force exists to change this component of velocity (assuming negligible air drag). The vertical component, however, changes because of the vertical influence of gravity.

If you toss a ball vertically it will lose speed going up and gain the same speed in coming down, if air drag is negligible. Without air drag, time going up is the same as time coming down. For fast-moving balls, air drag is usually not negligible.

Air drag greatly affects the range of balls batted and thrown in baseball games. Without air drag, a ball normally batted to the middle of center field would be a home run. If baseball were played on the Moon (not scheduled in the near future!), the range of balls would be considerably farther—about six times the ideal range on Earth. This is because there is no atmosphere on the Moon, so air drag is completely absent. Second, gravity is one-sixth as strong on the Moon, which allows higher and longer paths.

The speed of a vertically thrown ball at the top of its path is zero. Is the acceleration there zero also? (Answer begins with an *N*.)

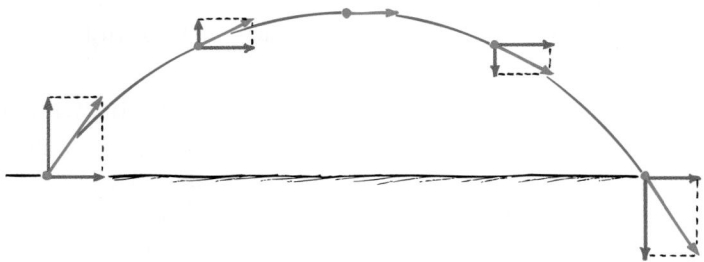

◀ **FIGURE 7.17**
The velocity of a projectile at various points. Note that the vertical component changes whereas the horizontal component is the same everywhere.

FIGURE 7.18 ▶
In the presence of air resistance, a high-speed projectile falls short of a parabolic path. The dashed line shows an ideal path with no air resistance. The solid line indicates an actual path.

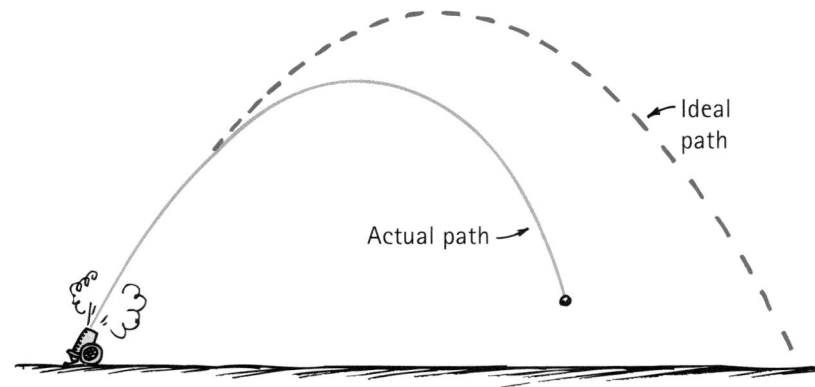

FIGURE 7.18 ▶
In the presence of air resistance, a high-speed projectile falls short of a parabolic path. The dashed line shows an ideal path with no air resistance. The solid line indicates an actual path.

Back here on Earth, baseball games normally take place on level ground. Baseballs curve over a flat playing field. The speeds of baseballs are not great enough for the Earth's curvature to affect the ball's path. For very long range projectiles, however, the curvature of the Earth's surface must be taken into account. As we will now see, when an object is projected fast enough, it can fall all the way around the Earth and become a **satellite.**

CONCEPT CHECK

1. At what part of its trajectory does a projectile have minimum speed?
2. (Challenge Question) A tossed ball changes speed along its parabolic path. When the Sun is directly overhead, does the shadow of the ball across the field also change speed?

Check Your Answers

1. The speed of a projectile is at minimum at the top of its path. If it is launched vertically, its speed at the top is zero. If it is projected at an angle, the vertical component of speed is zero at the top, leaving only the horizontal component. So the speed at the top is equal to the horizontal component of the projectile's velocity at any point.
2. No, for the shadow moves at constant velocity across the field, showing exactly the motion due to the horizontal component of the ball's velocity.

✓ **READING CHECK**

How far below an imaginary straight-line path does a projectile fall?

7.8 Fast-Moving Projectiles—Satellites

Suppose a cannon fires a cannonball so fast that its curved path matches the curvature of the Earth. Then without air drag, it would be an Earth satellite! The same would be true if you could throw a stone fast enough. ✓ Any **satellite** is simply a projectile moving fast enough to fall continually around the Earth.

In Figure 7.19, we see the curved paths of a stone thrown horizontally at different speeds. Whatever the pitching speed, in each case the stone drops the same vertical distance in the same time. For a 1-second drop, that distance is 5 meters (perhaps by now you have made use of this fact in lab). So if you simply drop a stone from rest, it will fall

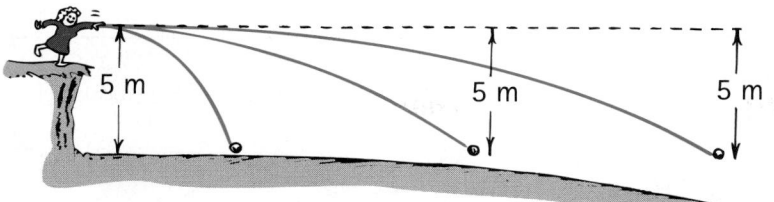

◄ **FIGURE 7.19**
Throw a stone at any speed
and 1 second later it falls
5 meters below where it
would have been if there were
no Earth gravity.

5 meters in 1 second of fall. Toss the stone sideways, and in 1 second it will be 5 meters below where it would have been without gravity. To be an Earth satellite, the stone's horizontal velocity must be great enough for its falling distance to match the Earth's curvature.

7.9 Earth Satellites in Circular Orbits

It is a geometrical fact that the surface of the Earth drops a vertical distance of 5 meters for every 8000 meters tangent to the surface. A tangent to a circle or to the Earth's surface is a straight line that touches the circle or surface at only one place. (So the tangent is parallel to the circle or sphere at the point of contact). We live on a round Earth.

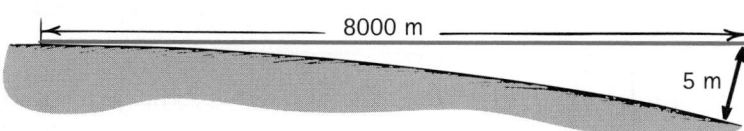

FIGURE 7.20 ▲
The Earth's curvature drops a vertical distance of 5 m for each 8000-m tangent (not to scale).

✓ A projectile that moves fast enough to travel a horizontal distance of 8 kilometers during 1 second is an Earth satellite. Neglecting air drag, it would follow the curvature of the Earth. A little thought tells you that this speed is 8 kilometers per second. If this doesn't seem fast, convert it to kilometers per hour and you get an impressive 29,000 kilometers per hour (18,000 mi/h). Fast, indeed! At this speed, atmospheric friction would incinerate the projectile. This happens to grains of sand and other meteorites that graze the Earth's atmosphere, burn up, and appear as "falling stars." That is why satellites like the space shuttles are launched to altitudes higher than 150 kilometers—to be above the atmosphere.

It is a common misconception that satellites orbiting at high altitudes are free from gravity. Nothing could be farther from the truth. The force of gravity on a satellite 150 kilometers above the Earth's surface is nearly as great as at the surface. If there were no gravity, motion would be along a straight-line path instead of curving around the Earth. High altitude puts the satellite beyond the Earth's *atmosphere,* but not beyond Earth's *gravity.* As mentioned earlier, Earth gravity goes on forever, getting weaker with distance, but never reaching zero.

✓ **READING CHECK**

Exactly what is a satellite?

PhysicsPlace.com
Video
Circular Orbits

FIGURE 7.21 ▲
If the stone is thrown fast enough so that its curve matches the Earth's curvature, it will be a satellite.

Newton calculated the speed that a cannonball fired high above the atmosphere would have to have for circular orbit about the Earth. No cannons could fire cannonballs at 8 km/s so he did not foresee humans launching satellites. Quite likely, he didn't foresee multistage rockets. Both the cannonball and Moon have a tangential ("sideways") velocity, parallel to the Earth's surface. This velocity is enough to ensure motion *around* the Earth rather than *into* it. Without air drag to reduce speed, the Moon or any Earth satellite "falls" around and around the Earth indefinitely. Similarly with the planets that continually fall around the Sun in closed paths.

Why don't the planets crash into the Sun? They don't because of their tangential velocities. What would happen if their tangential velocities were reduced to zero? The answer is simple enough: Their motion would be straight toward the Sun and they would indeed crash into it. Any objects in the solar system without sufficient tangential velocities have long ago crashed into the Sun. What remains is the harmony we observe.

Astronauts inside orbiting space facilities lack a support force. That's why they have no weight. Interestingly, the force of gravity between them and planet Earth is only slightly reduced from what it is at ground level.

CONCEPT CHECK

Can we also say a satellite stays in orbit because it is above the Earth's main pull of gravity?

Check Your Answer

No, no, no! No satellite is completely "above" the Earth's gravity. If the satellite were not in the grip of Earth's gravity, it would not orbit and would follow instead a straight-line path.

Think of the International Space Station that orbits Earth as Earth's lifeboat.

Satellites are payloads carried above the atmosphere by rockets. Putting a payload into orbit requires control over the speed and direction of the rocket. A rocket initially fired vertically is intentionally tipped from the vertical course as it rises. Then, once above the drag of the atmosphere, it is aimed horizontally, whereupon the payload is given a final thrust to orbital speed.

For a satellite close to the Earth, the period (the time for a complete orbit about the Earth) is about 90 minutes. For higher altitudes, gravitation is less and the orbital speed is less—the period is longer. For example,

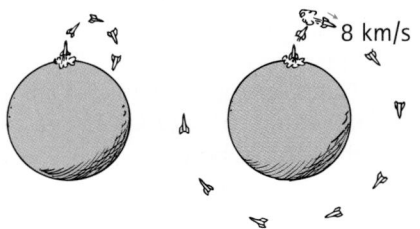

8 km/s

FIGURE 7.22 ▲
The initial thrust of the rocket pushes it above the atmosphere. Another thrust to a tangential speed of at least 8 km/s is needed if it is to fall around rather than into the Earth.

PRETEND CORNER: A CANNON SHOOT

Imagine you could be safely fired from a circus cannon horizontally at 8 kilometers per second. Pretend there is no air drag and the cannon is several meters above the ground. In the first second you'd travel 8 kilometers, fall vertically 5 meters, and still be the same distance above the ground. Your curved path would match the Earth's curvature. After another second you'd be another 8 kilometers down range, but still the same distance above the ground. If you didn't run into any obstacles you'd fall continually while remaining at a constant altitude—you'd be in low-Earth orbit! This is shown in Figure 7.24.

HANDS-ON EXPLORATION: THE WATER BUCKET SWING

Swing a bucket of water in a vertical circle, as shown in Figure 7.23. If you swing it sufficiently fast, the water won't spill. The explanation is similar to why satellites don't "fall" to Earth. Actually, both water in the bucket and satellites *are* falling. The water doesn't spill at the top of the swing because the bucket swings downward at least as fast as the water falls. Similarly, a satellite doesn't get closer to Earth because it falls a distance that matches the Earth's curvature. Analogies are the way to understand concepts!

FIGURE 7.23 ▲
Teacher Marshall Ellenstein whirls a bucket of water in a vertical circle and asks his class why water doesn't spill at the top of the swing. How does this relate to satellite motion?

Grains of sand and other meteorites that graze Earth's atmosphere burn up and appear as "falling stars." That's why space vehicles are launched to altitudes above the atmosphere.

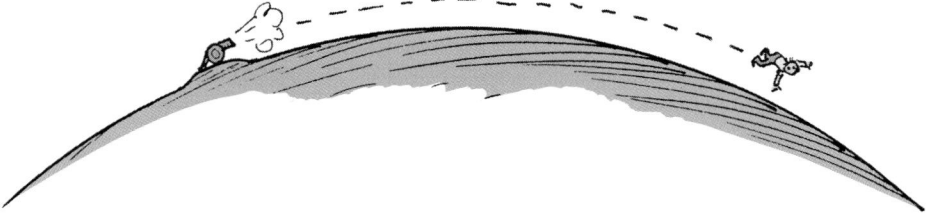

FIGURE 7.24 ▲
If you were shot from a make-believe circus cannon at 8 km/s, with no air drag, you'd be an Earth satellite!

communication satellites located at an altitude of 5.5 Earth radii have a period of 24 hours. This period matches the period of daily Earth rotation. For an orbit around the equator, these satellites stay above the same point on the ground. The Moon is even farther away and has a period of 27.3 days. The higher the orbit of a satellite, the less its speed and the longer its period.

✔ READING CHECK

What name is given to a projectile that travels a horizontal distance of 8 km in 1 s?

CONCEPT CHECK

What would be the fate of a rocket launched vertically that remains vertical as it rises?

Check Your Answer

After the rocket reaches its highest point it would fall back to its launching site—not a good idea!

7.10 Satellites in Elliptical Orbits

If a payload above the drag of the atmosphere is given a horizontal speed somewhat greater than 8 kilometers per second, it will overshoot a circular path and trace an elliptical path.

An **ellipse** is a specific curve. It is an oval-like path along which the sum of the distances of a point from two fixed points (called *foci*) is a constant. For a satellite orbiting Earth, one focus is at the Earth's center; the other focus could be inside or outside the Earth.

An ellipse can be easily constructed by using a pair of tacks, one at each focus, a loop of string, and a pencil. As the foci get farther apart, the path gets more elongated, or "eccentric." ✓ The closer the foci are to each other, the closer the ellipse is to a circle. When both foci are together, the ellipse is a circle. So we see that a circle is a special case of an ellipse.

Unlike the constant speed of a satellite in a circular orbit, speed varies in an elliptical orbit. In a circular orbit, the satellite path is always parallel to the Earth's surface. As if it were a bowling ball traveling on a lane parallel to the Earth's surface, the satellite's speed doesn't change. Both satellite and bowling ball go neither with nor against gravity.

A satellite following an elliptical orbit is different. Half the time the satellite moves away from the Earth, and half the time it moves toward the Earth. When it moves away, against the force of gravity, it loses speed. Like a stone thrown into the air, it slows to a point where it no longer recedes and then begins to fall back toward the Earth. Then it gains speed. The speed lost in receding is regained as it falls back. Then the satellite rejoins its original path with the same speed it had initially. The procedure repeats over and over, and an ellipse is traced in each cycle.

Satellite motion is nicely seen from a conservation of energy point of view. When the satellite is farthest from Earth its potential energy is maximum. As it falls back toward Earth its potential energy decreases and its kinetic energy increases. Kinetic energy is maximum when it passes closest to Earth. Total energy, potential plus kinetic, remains the same.

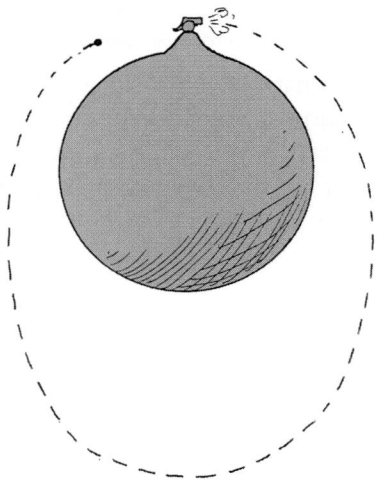

FIGURE 7.25 ▲
A cannonball fired at 9 km/s will overshoot a circular path and follow an ellipse.

✓ READING CHECK

What shape occurs when the foci of an ellipse are moved closer and closer together?

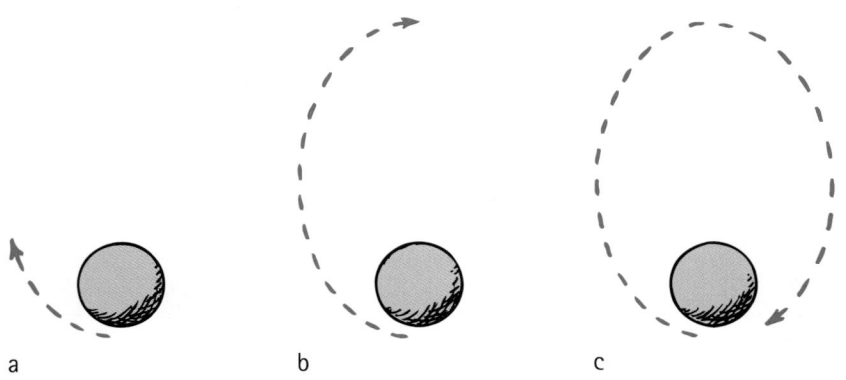

a   b   c

◀ FIGURE 7.26
The satellite loses speed in receding from the Earth, and regains it when falling back toward the Earth. The cycle is repeated and the satellite remains in an elliptical orbit.

Construct your own ellipses with string and tacks as shown. Note that the closer your tacks are, the more circular your ellipse. When the tacks are farther apart, the ellipse is more *eccentric*.

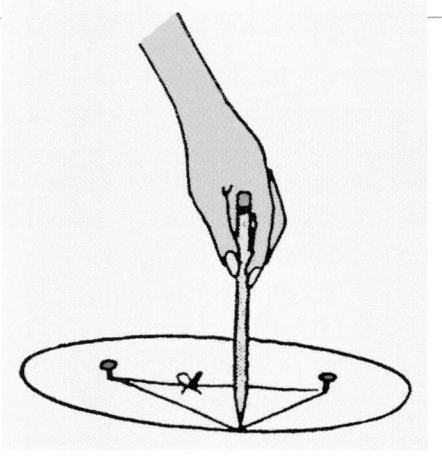

7.11 Escape Speed—Getting "Out There"

Theory in science differs from that in common usage. The theory of gravity, for example, is universally accepted by scientists, based on enormous evidence and the success of the model. The term *theory* does not imply doubts about a phenomenon's fundamental existence.

We know that a cannonball fired horizontally at 8 kilometers per second from Newton's mountain would be in orbit. But what would happen if the cannonball were instead fired at the same speed *vertically*? It would rise to some maximum height, reverse direction, and then fall back to Earth. Then the old saying "What goes up must come down" would hold true, just as surely as a stone tossed skyward will be returned by gravity (unless, as we shall see, its speed is too great).

Today it is more accurate to say, "What goes up *may* come down," because there is a critical speed at which a projectile can outrun gravity and escape the Earth. This critical speed is called **escape speed** or, if direction is involved, *escape velocity*. ✔ From the surface of the Earth, escape speed is 11.2 kilometers per second. A projectile launched at any greater speed will leave the Earth, traveling slower and slower due to Earth's gravity, but never stopping. The projectile will outrun the Earth's influence. Although it never escapes the tug of Earth's gravitation, it escapes the Earth itself.

FIGURE 7.27 ▶
Project anything away from the Earth at 11.2 km/s and its trip is a one-way affair. It won't return.

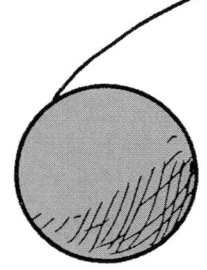

MECHANICS REVIEW

Several positions of a satellite in an elliptical orbit are shown. At which position does the satellite have the greatest

1. speed?
2. velocity?
3. mass?
4. gravitational attraction to Earth?
5. kinetic energy?
6. potential energy?
7. total energy?
8. acceleration? (Let the equation $a = F/m$ guide you.)

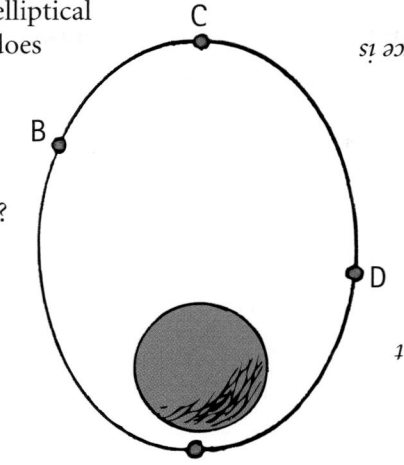

Answers

1. *Greatest speed at A.*
2. *Greatest velocity where greatest speed occurs, A.*
3. *Same mass everywhere. Mass does not depend on location.*
4. *Greatest at A, where it is closest to the Earth.*
5. *Greatest at A, where speed is greatest.*
6. *Greatest at C, where distance is greatest and speed least.*
7. *Same everywhere, in accord with energy conservation.*
8. *Greatest at A, where gravitational force is greatest.*

CONCEPT CHECK

If a flight mechanic drops a wrench from a high-flying airplane, it crashes to Earth. But if an astronaut outside the orbiting space shuttle drops a wrench, it doesn't crash to Earth. Explain.

Check Your Answer

If a wrench or anything else is "dropped" from an orbiting space vehicle, it has much the same tangential speed as the vehicle and remains in orbit. If a wrench is dropped from a high-flying airplane, it too has the tangential speed of the airplane. But this speed is too small for the wrench to orbit the Earth. Instead, it soon crashes to the Earth's surface.

 READING CHECK

What is escape speed from planet Earth?

PhysicsPlace.com

Tutorial
Detecting Dark Matter in a Spiral Galaxy
Video
Discovery of Neptune

7.12 Gravitation Is Universal

Why are Earth and other planets round? Why is the Sun round? They are round because everything attracts everything else, and all parts of these bodies have attracted themselves together as much as they can! Any "corners" of them have been pulled in, making them spheres. Rotational effects make some of them bulge slightly at their equators.

The shapes of distant galaxies show that the law of gravitation applies to large distances. Gravity underlies the fate of the entire universe.  Current scientific speculation is that the universe originated in the explosion of a primordial fireball some 13.7 billion years ago. This is the **Big Bang** theory of the origin of the universe. The explosion was space itself, with all the matter of the universe hurled outward. Space is still stretching out, carrying the galaxies with it. Evidence for this expansion includes precise measurement of the earlier remnants of the Big Bang: its cosmic microwave background.

The mind that encompasses the universe is as marvelous as the universe that encompasses the mind.

 READING CHECK

How old is the universe according to present evidence?

Your authors wonder about readers of this book who will continue in their study of science and help to decipher the nature of dark matter, dark energy, and other wonders of the universe yet to be discovered.

More recent evidence suggests the universe is not only expanding but *accelerating* outward. It is pushed by an antigravity *dark energy* that makes up an estimated 73% of the universe. Yet-to-be discovered particles of exotic *dark matter* compose 23% of the universe. Ordinary matter—the stuff of stars, cabbages, and kings—makes up only 4%. The concepts of dark matter and dark energy will continue to inspire exciting research throughout this century. They may hold clues to how the cosmos began and where it is headed and may be the key to understanding the fate of the universe. Our present view of the universe has progressed appreciably beyond the universe as Newton perceived it.

Our universe is still young. But humankind is younger by far.

7 CHAPTER REVIEW

KEY TERMS

Big Bang The primordial explosion that is thought to have resulted in the expanding universe.

Ellipse The oval path followed by a satellite. The sum of the distances from any point on the path to two points called foci is a constant. When the foci are together at one point, the ellipse is a circle. As the foci get farther apart, the path gets more "eccentric."

Escape speed The speed that a projectile, space probe, or similar object must reach to escape the gravitational influence of the Earth or celestial body to which it is attracted.

Inverse-square law A law relating the intensity of an effect to the inverse square of the distance from the cause:

$$\text{Intensity} \sim \frac{1}{\text{distance}^2}$$

Law of universal gravitation Every body in the universe attracts every other body with a mutually attracting force. For two bodies, this force is directly proportional to the product of their masses and inversely proportional to the square of the distance separating them:

$$F = G\frac{m_1 m_2}{d^2}$$

Parabola The curved path followed by a projectile near the Earth under the influence of gravity only.

Projectile Any object that moves through the air or through space under the influence of gravity.

Satellite A projectile or small body that orbits a larger body.

Tangential velocity Velocity that is parallel (tangent) to a curved path.

Universal gravitational constant, G The proportionality constant in Newton's law of universal gravitation.

REVIEW QUESTIONS

The Legend of the Falling Apple

1. What connection did Newton make between a falling apple and the Moon?

The Fact of the Falling Moon

2. What does it mean to say something moving in a curve has a tangential velocity?
3. In what sense does the Moon "fall?"

Newton's Grandest Discovery—The Law of Universal Gravitation

4. State Newton's law of universal gravitation in words. Then state it again in one equation.

Gravity and Distance—The Inverse-Square Law

5. How does the force of gravity between two bodies change when the distance between them is doubled?
6. How does the thickness of paint sprayed on a surface change when the sprayer is held twice as far away?
7. How does the brightness of light change when a point source of light is moved twice as far away?

The Universal Gravitational Constant, G

8. What is the magnitude of gravitational force between two 1-kilogram bodies that are 1 meter apart?
9. What is the magnitude of the gravitational force between the Earth and a 1-kilogram body?
10. What do we call the gravitational force between the Earth and your body?

The Mass of the Earth Is Measured

11. When G was first measured, the experiment was called the "weighing the Earth experiment." Why?

Projectile Motion

12. With no gravity, a horizontally moving projectile follows a straight-line path. With gravity, how far below the straight-line path does it fall compared with the distance of free fall?

13. A ball is batted upward at an angle. What happens to the vertical component of its velocity as it rises? As it falls?

14. With no air drag, what happens to the horizontal component of velocity for the batted baseball?

Fast-Moving Projectiles—Satellites

15. How can a projectile "fall around the Earth?"

Earth Satellites in Circular Orbits

16. Are the planets of the solar system simply projectiles falling around and around the Sun?

Satellites in Elliptical Orbits

17. Why does the speed of a satellite undergo change in an elliptical orbit?

18. At what part of an elliptical orbit does a satellite have the greatest speed? The least speed?

Escape Speed—Getting "Out There"

19. What is the minimum speed for orbiting the Earth in close orbit? The maximum speed? What happens above this speed?

Gravitation Is Universal

20. What makes the Earth round?

THINK AND DO

Do a version of the Explore! activity at the beginning of this chapter, only this time use two dollar bills—one regular, and the other folded in half lengthwise and again widthwise, so it has 1/4 the area. Now hold the two in front of your eyes. Where do you hold the folded bill so that it looks the same size as the unfolded one? Share this with your friends!

THINK AND COMPARE

1. The planet and its much less massive moon gravitationally attract each other. Compare the gravitational attractions between them, from greatest to least.

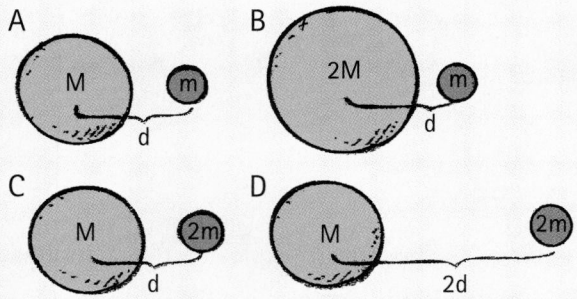

2. Three boys on three towers of the same height toss a ball at the velocities shown. Compare the horizontal distances from the tower that the balls travel when hitting the ground, from greatest range to least.

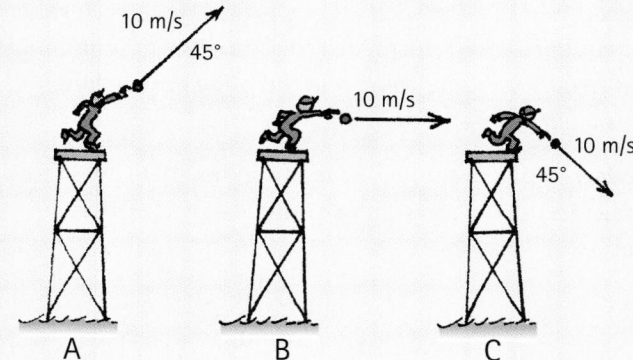

THINK AND EXPLAIN

1. Comment on whether or not this label on a consumer product should be cause for concern. *CAUTION: The mass of this product pulls on every other mass in the universe, with an attracting force that is proportional to the product of the masses and inversely proportional to the square of the distance between them.*

2. Gravitational force acts on all bodies in proportion to their masses. Why, then, doesn't a heavy body fall faster than a light body?

3. What would be the path of the Moon if somehow all gravitational forces on it vanished to zero?

4. Is the force of gravity stronger on a piece of iron than on a piece of wood if both have the same mass? Defend your answer.

5. Is the force of gravity stronger on a piece of paper when it is crumpled? Defend your answer.

6. What is the magnitude and direction of the gravitational force that acts on a teacher who weighs 1000 N at the surface of the Earth?

7. The Earth and the Moon are attracted to each other by gravitational force. Does the more massive Earth attract the less massive Moon with a force that is greater, smaller, or the same as the force with which the Moon attracts the Earth?

8. What do you say to a friend who says that if gravity follows the inverse-square law, when you are on the 20th floor of a building gravity on you should be one-fourth as much as when you're on the 10th floor?

9. A friend says that astronauts in orbit are weightless because they are beyond the pull of Earth's gravity. Correct your friend's ignorance.

10. Is the horizontal component of velocity for a projectile affected by the vertical component? Defend your answer.

11. A heavy crate accidentally falls from a high-flying airplane just as it flies directly above a shiny red sports car parked in a car lot. Relative to the car, where will the crate crash?

12. At what point in its trajectory does a batted baseball have its minimum speed? If air drag can be neglected, how does this compare with the horizontal component of its velocity at other points?

13. A park ranger shoots a monkey hanging from a branch of a tree with a tranquilizing dart. The ranger aims directly at the monkey, not realizing that the dart will follow a parabolic path and thus fall below the monkey. The monkey, however, sees the dart leave the gun and lets go of the branch to avoid being hit. Will the monkey be hit anyway? Defend your answer.

14. Since Earth is gravitationally attracted to the Sun, why don't we simply crash into the Sun?

15. Consider a high-orbiting spaceship that travels at 7 km/s with respect to the Earth. Suppose it projects a capsule rearward at 7 km/s with respect to the ship. Describe the path of the capsule with respect to the Earth.

16. The orbital velocity of the Earth about the Sun is 30 km/s. If the Earth were suddenly stopped in its tracks, it would simply fall directly into the Sun. Devise a plan whereby a rocket loaded with radioactive wastes could be fired into the Sun for permanent disposal. How fast and in what direction with respect to the Earth's orbit should the rocket be fired?

17. If a cannonball is fired from a tall mountain, gravity changes its speed all along its trajectory. But if it is fired fast enough to go into circular orbit, gravity does not change its speed at all. Why?

18. When an Earth satellite is placed into a higher orbit, what happens to its time for making a complete orbit? In other words, what happens to its period?

19. If a space vehicle circled Earth at a distance equal to the Earth–Moon distance, how long would it take to make a complete orbit?

20. Some people dismiss the validity of scientific theories by saying they are "only" theories. The law of universal gravitation is a theory. Does this mean that scientists still doubt its validity? Explain.

THINK AND SOLVE

1. If you stood atop a ladder that was so tall you were three times as far from the Earth's center as you were standing on the Earth's surface, how would your weight compare with its present value?

2. Find the change in the force of gravity between two planets when the masses of both planets are doubled but the distance between them stays the same.

3. Find the change in the force of gravity between two planets when their masses remain the same but the distance between them is increased by ten times.

4. Find the change in the force of gravity between two planets when the distance between them is *decreased* by ten times.

5. Find the change in the force of gravity between two planets when the masses of the planets don't change but the distance between them is decreased by five times.

6. Find the change in the force of gravity between two objects when both masses are doubled and the distance between them is also doubled.

7. Consider a bright point of light located 1 m from a square opening of area 1 square meter. Light passing through the opening illuminates an area of 4 m² on a wall 2 m from the opening. (a) Find the area illuminated if the wall is moved to a distance of 3 m, 5 m, or 10 m. (b) How can the same amount of light illuminate more area as the wall is moved farther away?

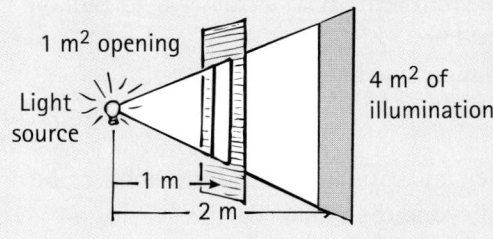

1 m² opening

Light source

4 m² of illumination

1 m

2 m

8. Calculate the force of gravity between the Earth (6×10^{24} kg) and the Sun (2×10^{30} kg). The average distance between the two is 1.5×10^{11} m.

9. A 3-kg newborn baby at the Earth's surface is gravitationally attracted to Earth with a force of about 30 N. (a) Calculate the force of gravity with which the baby on Earth is attracted to the planet Mars when Mars is closest to Earth. (The mass of Mars is 6.4×10^{23} kg and its closest distance is 5.6×10^{10} m.) (b) Calculate the force of gravity between the baby and the physician who delivers it. Assume the physician has a mass of 100 kg and is 0.5 m from the baby. (c) How do the forces compare?

10. Students in a lab measure the speed of a steel ball launched horizontally from a tabletop to be 4.0 m/s. If it takes 0.5 s for a ball to fall from the table to the floor below, where should they place a small piece of paper so that the ball will hit it when it lands?

READINESS ASSURANCE TEST (RAT)

If you have a good handle on this chapter—if you really do—then you should be able to score at least 7 out of 10 on this RAT. Check your answers with your teacher. If you score less than 7, consider studying further before moving on.

Choose the best answer to each of the following.

1. The force of gravity between two planets depends on their
   (a) masses and distance apart.
   (b) planetary atmospheres.
   (c) rotational motions.   (d) All of these.

2. When the distance between two stars is reduced by $\frac{1}{2}$, the force between them
   (a) decreases by $\frac{1}{2}$.   (b) decreases by $\frac{1}{4}$.
   (c) increases twice times as much.
   (d) increases four times as much.

3. If the Sun were twice as massive, its pull on Mars would be
   (a) unchanged.   (b) twice.
   (c) half.   (d) four times as much.

4. When an astronaut in orbit is weightless, he or she is
   (a) beyond the pull of Earth's gravity.
   (b) still in the grip of Earth's gravity.

   (c) in the grip of interstellar gravity.
   (d) None of these.

5. When no air resistance acts on a fast-moving baseball, its acceleration is
   (a) downward, *g*.
   (b) due to a combination of constant horizontal motion and accelerated downward motion.
   (c) opposite to the force of gravity.
   (d) zero.

6. When you toss a projectile sideways, it curves as it falls. It will be an Earth satellite if the curve it makes
   (a) matches the curve of Earth's surface.
   (b) results in a straight line.
   (c) spirals out indefinitely.
   (d) None of these.

7. In circular orbit, the gravitational force on a satellite is
   (a) constant in magnitude.
   (b) at right angles to satellite motion.
   (c) Both of these.
   (d) Neither of these.

8. The speed of a satellite in an elliptical orbit
   (a) remains constant.
   (b) acts at right angles to its motion.
   (c) varies.   (d) All of these.

9. A satellite in elliptical orbit about Earth travels fastest when it moves
   (a) close to Earth.   (b) far from Earth.
   (c) in either direction—the same everywhere.
   (d) between the near and far points from Earth.

10. A satellite in Earth orbit is above Earth's
    (a) atmosphere.   (b) gravitational field.
    (c) Both of these.   (d) Neither of these.

8
FLUID MECHANICS

THE MAIN
IDEA

Laws of mechanics applied to liquids and gases explain flotation of boats that are denser than water and the flight of birds and aircraft that are denser than air.

Liquids and gases have the ability to flow; hence, they are called *fluids.* Because they are both fluids, we find that they obey similar mechanical laws. Why is it easier to float in saltwater than in freshwater? How is it that iron boats don't sink in water or that helium balloons don't sink from the sky? Why is gas compressible while liquids are not? Why do your ears pop when riding an elevator?

Why does the sheet of paper rise when Maitreya blows across its top surface? Is the physics of the lift the same that accounts for the lift of hydrofoils, birds, and airplanes? Flotation, gas compressibility, and lift are properties of fluids. To answer these questions about fluids, it is important to introduce two concepts—*density* and *pressure.*

Explore!

When Will an Egg Float in Water?

1. Place an egg in a small container of water. Note that it rests on the bottom.
2. Gradually pour salt in the water. Let it dissolve.
3. Observe what happens for the egg.

Analyze and Conclude

1. **Observing** Note that the egg floats in the saltwater.

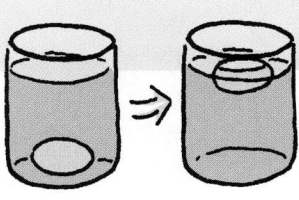

2. **Predicting** Are there other objects that will float in saltwater and sink in freshwater?
3. **Making Generalizations** How does the density of an egg compare to that of tap water? To saltwater? If other objects behave in the same way, how do their densities compare with the fluids in which they're immersed?

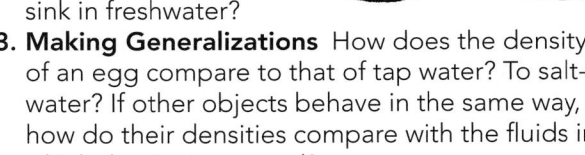

FIGURE 8.1 ▲
When the volume of the bread is reduced, its density increases.

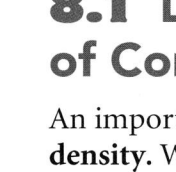

Our bodies are composed of 60% water.

8.1 Density—A Measure of Compactness

An important property of materials is the measure of compactness: **density.** We think of density as the "lightness" or "heaviness" of materials of the same size. ✔ Density is a measure of how much mass occupies a given space, whether solid, liquid, or gas. Density is the amount of matter per unit volume:

$$\text{Density} = \frac{\text{mass}}{\text{volume}}$$

The densities of a few materials are listed in Table 8.1. Mass is measured in grams or kilograms, and volume in cubic centimeters (cm^3) or cubic meters (m^3).* A gram of any material has the same mass as 1 cubic centimeter of water at a temperature of 4°C. So water has a density of 1 gram per cubic centimeter. Mercury's density is 13.6 grams per cubic centimeter, which means that it has 13.6 times as much mass as an equal volume of water. Iridium, a hard, brittle, silvery-white metal in the platinum family, is the densest substance on Earth.

A quantity known as weight density, commonly used when discussing liquid pressure, is expressed by the amount of weight per unit volume:†

$$\text{Weight density} = \frac{\text{weight}}{\text{volume}}$$

CONCEPT CHECK

1. Which has the greater density—1 kg of water or 10 kg of water?
2. Which has the greater density—5 kg of lead or 10 kg of aluminum?
3. Which has the greater density—an entire candy bar or half a candy bar?

Check Your Answers

1. The density of any amount of water is the same: 1 g/cm^3 or, equivalently, 1000 kg/m^3, which means that 1 gram of water would exactly fill a thimble of volume 1 cubic centimeter (or equivalently, 1000 kg would fill a 1-cubic-meter tank). One kg of water would fill a tank only a thousandth as large, 1 liter. This means that 10 kg would fill a 10-liter tank. Nevertheless, the important concept is that the ratio of mass/volume is the same for *any* amount of water.
2. Density is a *ratio* of weight or mass per volume, and this ratio is greater for any amount of lead than for any amount of aluminum—see Table 8.1.
3. Both the entire candy bar and half of it have the same density.

How do mass and volume relate to the density of solids, liquids, and gases?

---

* A cubic meter is a sizable volume and contains a million cubic centimeters, so there are a million grams of water in a cubic meter (or, equivalently, a thousand kilograms of water in a cubic meter). Hence, 1 g/cm^3 = 1000 kg/m^3.

† Weight density is common to the United States Customary System (USCS) units in which 1 cubic foot of freshwater (nearly 7.5 gallons) weighs 62.4 pounds. So freshwater has a weight density of 62.4 lb/ft^3. Saltwater is slightly denser, 64 lb/ft^3.

TABLE 8.1 Densities of Some Materials

Material	Grams per Cubic Centimeter (g/cm³)	Kilograms per Cubic Meter (kg/m³)
Liquids		
Mercury	13.6	13,600
Glycerin	1.26	1260
Seawater	1.03	1025
Water at 4°C	1.00	1000
Benzene	0.90	899
Ethyl alcohol	0.81	806
Solids		
Iridium	22.6	22,650
Osmium	22.6	22,610
Platinum	21.1	21,090
Gold	19.3	19,300
Uranium	19.0	19,050
Lead	11.3	11,340
Silver	10.5	10,490
Copper	8.9	8920
Brass	8.6	8600
Iron	7.8	7874
Tin	7.3	7310
Aluminum	2.7	2700
Ice	0.92	919
Gases (atmospheric pressure at sea level)		
Dry air		
0°C	1.29	
10°C	1.25	
20°C	1.21	
30°C	1.16	
Helium	0.178	
Hydrogen	0.090	
Oxygen	1.43	

fyi

- The metals lithium, sodium, and potassium (not in Table 8.1) are all less dense than water and will float in water.

FIGURE 8.2 ▲
Although the weight of both books is the same, the upright book exerts greater pressure against the table.

PhysicsPlace.com
Video
Dam Keeps Water in Place

FIGURE 8.3 ▲
This water tower does more than store water. The depth of water above ground level insures reliable water pressure to the many homes it serves.

8.2 Pressure—Force per Area

Place a book on a bathroom scale and whether you place it on its back, on its side, or balanced on a corner, it still exerts the same force. The weight reading is the same. Now balance the book on the palm of your hand and you sense a difference—the *pressure* of the book depends on the area of contact (Figure 8.2). There is a difference between force and pressure. **Pressure** is defined as the force exerted over a unit of area, such as a square meter or square foot:*

$$\textbf{Pressure} = \frac{\textbf{force}}{\textbf{area}}$$

CONCEPT CHECK

Does a bathroom scale measure weight, pressure, or both?

Check Your Answer

A bathroom scale measures weight, the force that compresses an internal spring or equivalent. The weight reading is the same whether you stand on one or both feet (although the pressure on the scale is twice as much when standing on one foot).

Pressure in a Liquid

When you swim under water, you can feel the water pressure acting against your eardrums. The deeper you swim, the greater the pressure. What causes this pressure? ✔ Pressure in the water is due to the weight of the fluid directly above you—both the weight of water plus the weight of the atmosphere. More water is above you when you swim deeper. That means more pressure. If you swim twice as deep, twice the weight of water pushes on you. So the water's contribution to the pressure you feel is doubled. Added to the water pressure is the pressure of the atmosphere, which is equivalent to an extra 10.3-meter depth of water. Because atmospheric pressure at Earth's surface is nearly constant, pressure *differences* you feel under water depend only on changes in the depth of the water.

The pressure due to a liquid is precisely equal to the product of weight density and depth:†

Liquid pressure = weight density × depth

---

* Pressure may be measured in any unit of force divided by any unit of area. The standard international (SI) unit of pressure, the newton per square meter, is called the *pascal* (Pa), after the 17th-century theologian and scientist Blaise Pascal. A pressure of 1 Pa is very small and approximately equals the pressure exerted by a dollar bill resting flat on a table. Science types prefer kilopascals (1 kPa = 1000 Pa).

† This is derived from the definitions of pressure and density. Consider an area at the bottom of a vessel that contains liquid. The weight of the column of liquid directly above this area produces pressure. From the definition *weight density = weight/volume*, we can express the amount of liquid weight as *weight = weight density × volume*. The volume of the column is simply the area multiplied by the depth. Then we get

$$\text{Pressure} = \frac{\text{force}}{\text{area}} = \frac{\text{weight}}{\text{area}} = \frac{\text{weight density} \times \text{volume}}{\text{area}} = \frac{\text{weight density} \times (\text{area} \times \text{depth})}{\text{area}}$$

$$= \text{weight density} \times \text{depth}$$

For the total pressure, we should add to this equation the pressure due to the atmosphere on the surface of the liquid.

It is important to note that pressure does not depend on the *volume* of liquid. You feel the same pressure a meter deep in a small pool as you do a meter deep in the middle of the ocean. This is illustrated by the connecting vases shown in Figure 8.4. Pressure depends on depth, not volume.

Water seeks its own level. This can be demonstrated by filling a garden hose with water and holding the two ends upright. The water levels will be equal whether the ends are held close together or far apart. Pressure depends on depth, not volume. So we see there is an explanation for why water seeks its own level.

FIGURE 8.4 ▲
Liquid pressure is the same for any given depth below the surface, regardless of the shape of the containing vessel.

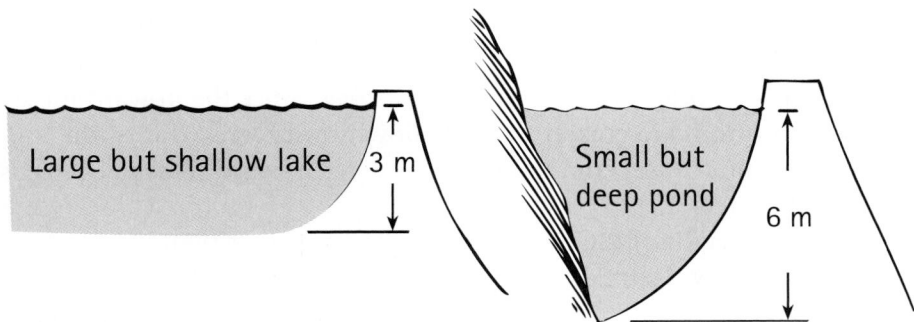

FIGURE 8.5 ▲
The average water pressure acting against the dam depends on the average depth of the water—*not* on the volume of water held back. The large shallow lake exerts only one-half the average pressure that the small deep pond exerts.

In addition to pressure depending on depth, liquid pressure is exerted equally in all directions. For example, if you are submerged in water, it makes no difference which way you tilt your head—your ears feel the same amount of water pressure. Because a liquid can flow, the pressure isn't only downward. We know pressure acts upward when we try to push a beach ball beneath the water's surface. The bottom of a boat is certainly pushed upward by water pressure. And we know water pressure acts sideways when we see water spurting sideways from a leak in an upright can. Pressure in a liquid at any point is exerted in equal amounts in all directions.

When liquid presses against a surface, there is a net force directed perpendicular to the surface (Figure 8.6). If there is a hole in the surface, the liquid spurts at right angles to the surface before curving downward due to gravity (Figure 8.7). At greater depths the pressure is greater and the speed of the exiting liquid is greater.*

Blood pressure is measured in your upper arm—level with your heart.

FIGURE 8.6 ▲
The forces due to liquid pressure against a surface combine to produce a net force that is perpendicular to the surface.

FIGURE 8.7 ▶
The force vectors act in a direction perpendicular to the inner container surface and increase with increasing depth.

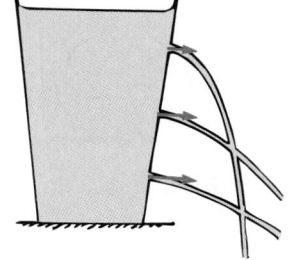

✔ **READING CHECK**

What causes the pressure you feel when you are beneath the surface of water?

---

* The speed of liquid exiting the hole is $\sqrt{2gh}$, where h is the depth below the free surface. Interestingly, this is the same speed that water or anything else would have if freely falling the same distance h.

8.3 Buoyancy in a Liquid

Lifting a large boulder off the bottom of a riverbed is easy if the boulder is submerged. When it is lifted above the surface, however, greater lifting force is needed. This is because, when the boulder is submerged, the water exerts an upward force on it—opposite in direction to gravity. This upward force is called the **buoyant force.** Figure 8.8 shows why the buoyant force acts upward. Pressure is exerted everywhere on the submerged object. At each point it produces a force that is perpendicular to its surface. The vector arrows show the magnitude and direction of forces at different places. Forces against the sides cancel one another at equal depths. Pressure is greatest against the bottom of the boulder simply because the bottom of the boulder is deeper. Since the upward forces against the bottom are greater than the downward forces against the top, the forces do not cancel, and there is a net force upward. This net force is the buoyant force.

✓ If the weight of the submerged object is greater than the buoyant force, the object will sink. If its weight is equal to the buoyant force, the object will remain at any level, like a fish. If the buoyant force is greater than the weight of the completely submerged object, it will rise to the surface and float.

To have a good understanding of buoyancy, you need to understand the meaning of the expression "volume of water displaced." If a stone is placed in a container that is already filled to its brim with water, some water will overflow (Figure 8.9). Water is *displaced* by the stone. A little thought will tell us that the *volume of the stone*—that is, the amount of space it occupies or its number of cubic centimeters—is equal to the *volume of water displaced.* Place any object in a container partially filled with water, and the level of the surface rises (Figure 8.10). How high? Water will rise to exactly the level that would be reached by pouring in a volume of water equal to the volume of the submerged object. This is a good method for determining the volume of irregularly shaped objects: *A completely submerged object always displaces a volume of liquid equal to its own volume.*

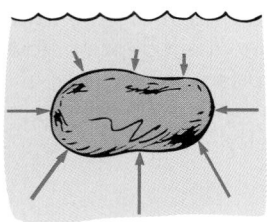

FIGURE 8.8 ▲
The greater pressure against the bottom of a submerged object produces an upward buoyant force.

When you use a measurement cup in the kitchen to find something's volume, you're using the displacement method.

FIGURE 8.9 ▲
When a stone is submerged, it displaces a volume of water equal to the volume of the stone.

Water displaced

FIGURE 8.10 ▲
The raised level due to placing a stone in the container is the same as if a volume of water equal to the volume of the stone were poured in.

8.4 Archimedes' Principle— Sink or Swim

The relationship between buoyancy and displaced liquid was first discovered in the 3rd century BC by the Greek scientist Archimedes. This relationship is called **Archimedes' principle.** It applies to liquids and gases, which are both fluids. It is stated as follows:

> **An immersed body is buoyed up by a force equal to the weight of the fluid it displaces.**

Immersed means "either *completely* or *partially submerged.*" If an immersed body displaces 10 N of fluid, the buoyant force on it will be 10 N. If we immerse a sealed 1-L container halfway into the water, it will displace 1 half-liter of water and be buoyed up by the weight of 1 half-liter of water, 5 N. If we immerse it completely (submerge it), it will be buoyed up by the weight of a full liter (or 10 N) of water. Unless the completely submerged container is compressed, the buoyant force will equal 10 N at *any* depth. This makes sense because the submerged object can't displace more water than its own volume. And the weight of this volume of water (not the weight of the submerged object!) is equal to the buoyant force.

The apparent weight of an object decreases when immersed in a fluid. Notice in Figure 8.12 that the 3-N block has an apparent weight of 1 N when submerged. The apparent weight of a submerged object is its weight out of water minus the buoyant force.

FIGURE 8.11 ▲
A liter of water occupies a volume of 1000 cm³, has a mass of 1 kg, and weighs 10 N (more precisely, 9.8 N). Its density may therefore be expressed as 1 kg/L and its weight density as 10 N/L. (Seawater is slightly denser than 10 N/L).

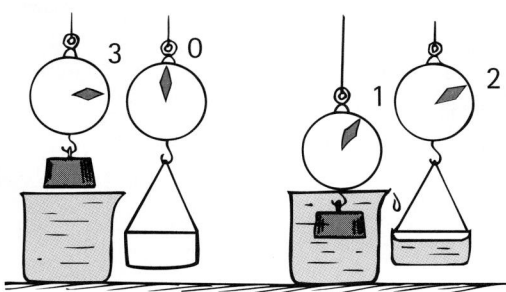

FIGURE 8.12 ▲
A 3-N block weighs more in air than it does in water. When submerged in water, the block's loss in weight is the buoyant force, which equals the weight of water displaced.

Stick your foot in a swimming pool and your foot is immersed. Jump in and sink and immersion is total—you're submerged.

Section 8.4 may need multiple and careful readings. It is a stumbling block for many students. If you understand it, help classmates who don't—yet.

CONCEPT CHECK

1. Does Archimedes' principle tell us that if an immersed block displaces 10 N of fluid, the buoyant force on the block is 10 N?
2. A 1-L container completely filled with lead has a mass of 110 N and is submerged in water. What is the buoyant force acting on it?
3. A boulder is thrown into a deep lake. As it sinks deeper and deeper into the water, does the buoyant force on it increase? Decrease?

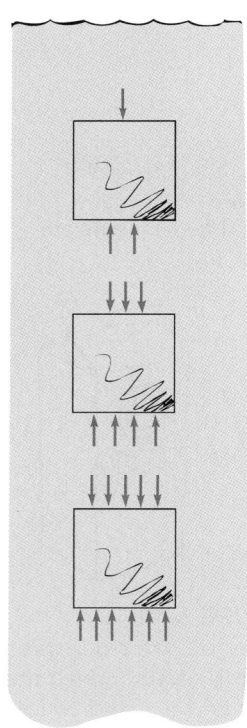

FIGURE 8.13 ▲
The difference in the upward and downward forces acting on the submerged block is the same at any depth.

Check Your Answers

1. Yes. Looking at it in a Newton's-third-law way, when the immersed block pushes 10 N of fluid aside, the fluid reacts by pushing back on the block with 10 N.
2. The buoyant force is 10 N because 10 N of water is displaced (the weight of 1 L of water). The 110-N weight of the lead is irrelevant; 1 L of anything submerged in water will displace 1 L and be buoyed upward with a force of 10 N. (Get this straight before going further!)
3. Buoyant force remains the same. It doesn't change as the boulder sinks because the boulder displaces the same volume of water at any depth. So the buoyant force remains the same at all depths.

Flotation

Iron is much denser than water and therefore sinks. But an iron ship floats. Why is this so? Consider a solid 1-ton block of iron. Iron is nearly eight times as dense as water, so when the block is submerged it will displace only $\frac{1}{8}$ ton of water—not enough to prevent it from sinking. Suppose we reshape the same iron block into a bowl, as shown in Figure 8.14. It still weighs 1 ton. When placed in water, the iron displaces a greater volume of water than before. The deeper it is immersed, the more water it displaces and the greater the buoyant force acting on it. When the buoyant force equals 1 ton, it will sink no further. ✓ Only in the special case of floating does the buoyant force acting on an object equal the object's weight.

Video
Flotation

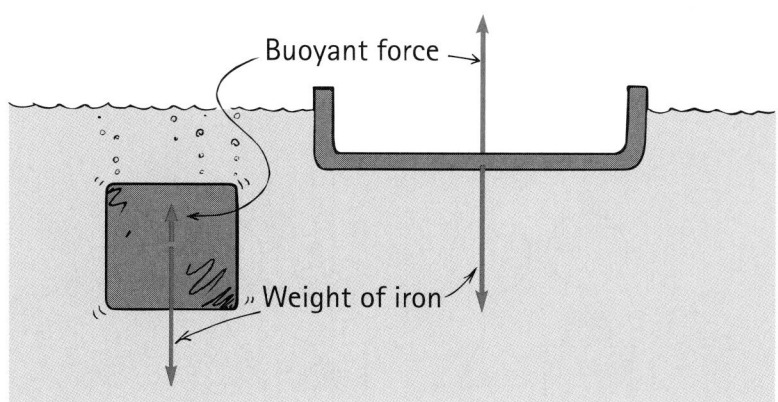

FIGURE 8.14 ▲
An iron block sinks, but the same quantity of iron shaped like a bowl floats.

PEOPLE IN HISTORY: ARCHIMEDES AND THE GOLD CROWN

According to legend, Archimedes (287–212 BC) was given the task of determining whether a crown made for King Hiero II of Syracuse was of pure gold or whether it contained some less expensive metal such as silver. Archimedes' problem was to determine the density of the crown without destroying it. He could weigh the gold, but determining its volume was a problem. The story tells us that Archimedes came to the solution when he noted the rise in water level while immersing his body in the public baths of Syracuse. Legend reports that he excitedly rushed naked through the streets shouting "Eureka! Eureka!" ("I have found it! I have found it!").

What Archimedes discovered was a simple and accurate way of finding the volume of an irregular object—the displacement method we use today with measuring cups. Once he knew both the weight and volume, he could calculate the density, which could be compared with the density of gold. Archimedes found that the crown indeed was not pure gold.

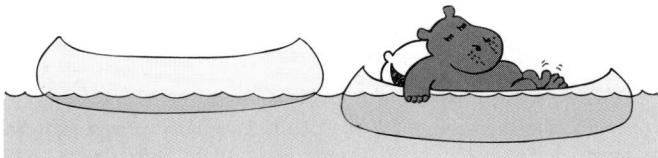

FIGURE 8.15 ▲

The weight of a floating object equals the weight of the water displaced by the submerged part.

FIGURE 8.16 ▲

A floating object displaces a weight of fluid equal to its own weight.

When the iron boat displaces a weight of water equal to its own weight, it floats. This is called the **principle of flotation,** which states:

> **A floating object displaces a weight of fluid equal to its own weight.**

Every ship, submarine, or dirigible airship must be designed to displace a weight of fluid equal to its own weight. Thus, a 10,000-ton ship must be built wide enough to displace 10,000 tons of water before it immerses too deep in the water. The same applies to vessels in air. A dirigible or huge balloon that weighs 100 tons displaces at least 100 tons of air. If it displaces more, it rises; if it displaces less, it descends. If it displaces exactly its weight, it hovers at constant altitude.

Since the buoyant force upon a body equals the weight of the fluid it displaces, denser fluids will exert more buoyant force upon a body than less-dense fluids of the same volume. A ship therefore floats higher in saltwater than in freshwater because saltwater is slightly denser than freshwater. In the same way, a solid chunk of iron will float in mercury even though it sinks in water.

> People who can't float are, nine times out of ten, males. Most males are more muscular and slightly denser than females. Also, cans of diet soda float whereas cans of regular soda sink in water. What does this tell you about their relative densities?

FIGURE 8.17 ▲

The same ship empty and loaded. How does the weight of its load compare to the weight of additional water displaced?

> Again for emphasis: Only in the special case of floating does the buoyant force acting on an object equal the object's weight.

CONCEPT CHECK

Fill in the blanks for these statements:
1. The volume of a submerged body is equal to the _____ of the fluid displaced.
2. The weight of a floating body is equal to the _____ of the fluid displaced.
3. Why is it easier to float in saltwater than in freshwater?

✔ READING CHECK

According to Archimedes' principle, the buoyant force on an object equals the weight of water displaced, and not necessarily the weight of the object. What is the special case where buoyant force equals both the weight of water displaced and the weight of the object?

Check Your Answers

1. volume
2. weight
3. When you're floating, the weight of water you displace equals your weight. Saltwater is denser, so you don't "sink" as far to displace your weight. You'd float even higher in mercury (density 13.6 g/cm^3), and you'd sink completely in alcohol (density 0.8 g/cm^3).

Notice in our discussion of liquids that Archimedes' principle and the law of flotation were stated in terms of *fluids*, not liquids. That's because although liquids and gases are different phases of matter, they are both fluids, with much the same mechanical principles. Let's turn our attention to the mechanics of gases in particular.

8.5 Pressure in a Gas

The main difference between a gas and a liquid is the distance between molecules. In a gas, the molecules are far apart and free from intermolecular forces. Molecular motions in a gas are less restricted. A gas expands, fills all space available to it, and exerts a pressure against its container. Only when the quantity of gas is very large, such as Earth's atmosphere or a star, do the gravitational forces limit the size or determine the shape of the mass of gas.

Liquids and gases are both fluids. A gas takes the shape of its container. A liquid does so only below its surface.

Boyle's Law

The air pressure inside the inflated tires of an automobile is much greater than the atmospheric pressure outside. The density of air inside is also greater than the density of the air outside. To understand the relation between pressure and density, think of the molecules of air (primarily nitrogen and oxygen) inside the tire. The air molecules behave like tiny billiard balls, randomly moving and banging against the inner walls, which produces a jittery force that appears to our coarse senses as a steady push. This pushing force, averaged over the wall area, provides the pressure of the enclosed air.

LINK TO EARTH SCIENCE: FLOATING MOUNTAINS

Mountains float on Earth's semiliquid mantle just as icebergs float in water. Both the mountains and icebergs are less dense than the material they float upon. Just as most of an iceberg is below the water surface (90%), most of a mountain (about 85%) extends into the dense, semiliquid mantle. If you could shave off the top of an iceberg, the iceberg would be lighter and be buoyed up to nearly its original height before its top was shaved. Similarly, when mountains erode they are lighter and are pushed up from below to float to nearly their original heights. So when a kilometer of mountain erodes away, some 85% of a kilometer of mountain returns. That's why it takes so long for mountains to weather away. Mountains, like icebergs, are bigger than they appear to be. The concept of floating mountains is *isostacy*—Archimedes' principle for rocks.

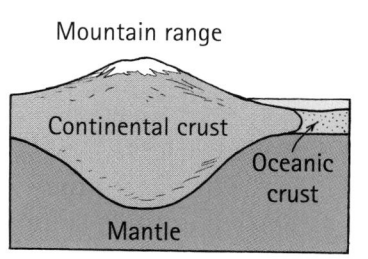

Suppose there are twice as many molecules in the same volume (Figure 8.18). Then the air density is doubled. If the molecules move at the same average speed—or, equivalently, if they have the same temperature—then the number of collisions will be doubled. This means that the pressure is doubled. So pressure is proportional to density.

We double the density of air in the tire by doubling the amount of air. We can also double the density of a *fixed* amount of air by compressing it to half its volume. Consider the cylinder with the movable piston in Figure 8.19. If the piston is pushed downward so that the volume is half the original volume, the density of molecules is doubled, and the pressure is correspondingly doubled. Decrease the volume to a third of its original value, and the pressure is increased by three, and so forth (provided the temperature remains the same).

Notice in these examples with the piston that the product of pressure and volume remains the same. For example, a doubled pressure multiplied by a halved volume gives the same value as a tripled pressure multiplied by a one-third volume. In general, we can state that the product of pressure and volume for a given mass of gas is a constant *as long as the temperature does not change.* ✓ "Pressure × volume" for a sample of gas at some initial time is equal to any "different pressure × different volume" of the same sample of gas at some later time. In shorthand notation,

$$P_1V_1 = P_2V_2.$$

where P_1 and V_1 represent the original pressure and volume, respectively, and P_2 and V_2 the second pressure and volume. This relationship is called **Boyle's law,** after Robert Boyle, the 17th-century physicist who is credited with its discovery.*

Boyle's law applies to ideal gases. An ideal gas is one in which the forces between molecules and the size of the individual molecules can be neglected. Air and other gases under normal pressures and temperatures approach ideal gas conditions.

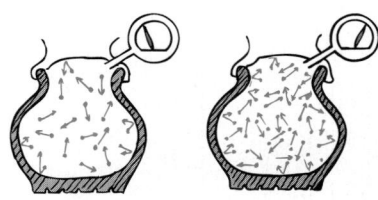

FIGURE 8.18 ▲
When the density of gas in the tire is increased, pressure is increased.

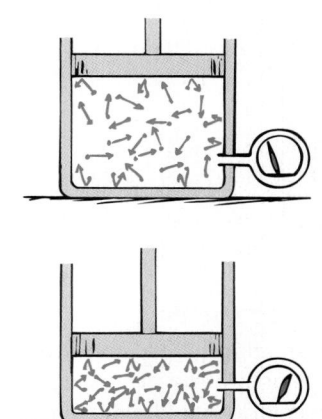

FIGURE 8.19 ▲
When the volume of gas is decreased, density and therefore pressure are increased.

CONCEPT CHECK

1. A piston in an airtight pump is withdrawn so that the volume of the air chamber is increased three times. What is the change in pressure?
2. A scuba diver breathes compressed air beneath the surface of water. If she holds her breath while returning to the surface, what happens to the volume of her lungs?

Check Your Answers

1. The pressure in the piston chamber is reduced to one-third. This is the principle that underlies a mechanical vacuum pump.
2. When she rises toward the surface, the surrounding water pressure on her body decreases, allowing the volume of air in her lungs to increase—ouch! A first lesson in scuba diving is to not hold your breath when ascending. To do so can be fatal.

✓ **READING CHECK**

State Boyle's law in words.

So $PV = PV$.

---

* A general law that takes temperature changes into account is $P_1V_1/T_1 = P_2V_2/T_2$, where T_1 and T_2 represent the initial and final *absolute* temperatures, measured in SI units called kelvins (Chapter 9).

8.6 Atmospheric Pressure Is Due to the Weight of the Atmosphere

We live at the bottom of an ocean of air. The atmosphere, much like the water in a lake, exerts a pressure. One of the most celebrated experiments demonstrating the pressure of the atmosphere was conducted in 1654 by Otto von Guericke, burgermeister of Magdeburg and inventor of the vacuum pump. Von Guericke placed together two copper hemispheres about $\frac{1}{2}$ meter in diameter to form a sphere, as shown in Figure 8.20. He inserted a gasket made of a ring of leather soaked in oil and wax between them to make an airtight joint. When he evacuated the sphere with his vacuum pump, two teams of eight horses each were unable to pull the hemispheres apart.

Interestingly, von Guericke's demonstration preceded knowledge of Newton's third law. The forces on the hemispheres would have been the same if he used only one team of horses and tied the other end of the rope to a tree!

FIGURE 8.20 ▲
The famous "Magdeburg hemispheres" experiment of 1654, demonstrating atmosphere pressure. Two teams of horses couldn't pull the evacuated hemispheres apart. Were the hemispheres sucked together or pushed together? By what?

When the air pressure inside a cylinder like that shown in Figure 8.21 is reduced, there is an upward force on the piston. This force is large enough to lift a heavy weight. If the inside diameter of the cylinder is 12 centimeters or greater, a person can be lifted by this force.

What do the experiments of Figures 8.20 and 8.21 demonstrate? Do they show that air exerts pressure or that there is a "force of suction"? If we say there is a force of suction, then we assume that a vacuum can exert a force. But what is a vacuum? It is an absence of matter; it is a condition of nothingness. How can nothing exert a force? The hemispheres are not sucked together, nor is the piston holding the weight

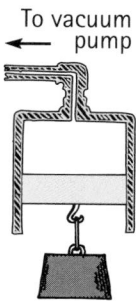

To vacuum
◄— pump

FIGURE 8.21 ▲
Is the piston pulled up or pushed up?

sucked upward. The hemispheres and the piston are being pushed against by the pressure of the atmosphere.

✓ Just as water pressure is caused by the weight of water, **atmospheric pressure** is caused by the weight of air. We have adapted so completely to the invisible air that we sometimes forget it has weight. Perhaps a fish "forgets" about the weight of water in the same way. The reason we don't feel this weight crushing against our bodies is that the pressure inside our bodies equals that of the surrounding air. There is no net force for us to sense.

At sea level, 1 cubic meter of air at 20°C has a mass of about 1.2 kilograms. To estimate the mass of air in your room, estimate the number of cubic meters there, multiply by 1.2 kg/m^3, and you'll have the mass. Don't be surprised if it's heavier than your kid sister. If your kid sister doesn't believe air has weight, maybe it's because she's always surrounded by air. Hand her a plastic bag of water and she'll tell you it has weight. But hand her the same bag of water while she's submerged in a swimming pool, and she won't feel the weight. We don't notice that air has weight because we're submerged in air.

The density of water is the same at any level (assuming constant temperature). Not so for air. The density of air in the atmosphere becomes less with altitude. Although 1 cubic meter of air at sea level has a mass of about 1.2 kg, at 10 kilometers 1 cubic meter of air has a mass of about 0.4 kilograms. Air is thinner at higher altitudes. To compensate for this, airplanes are pressurized; the additional air needed to fully pressurize a 777 jumbo jet, for example, is more than 1000 kilograms. Air is heavy, if you have enough of it.

Consider the mass of air in an upright 30-kilometer-tall hollow bamboo pole that has an inside cross-sectional area of 1 square centimeter. If the density of air inside the pole matches the density of air outside, the enclosed mass of air would be about 1 kilogram. The weight of this much air is about 10 newtons. So air pressure at the bottom of the bamboo pole would be about 10 newtons per square centimeter (10 N/cm^2). Of course, the same is true without the bamboo pole. There are 10,000 square centimeters in 1 square meter, so a column of air 1 square meter in cross section that extends up through the atmosphere has a mass of about 10,000 kilograms. The weight of this air is about 100,000 newtons (10^5 N). This weight produces a pressure of 100,000 newtons per square meter.

Recall that 1 N/m^2 is called 1 pascal. So atmospheric pressure at Earth's surface is 100,000 pascals, or 100 kilopascals. To be more precise, the average atmospheric pressure at sea level is 101.3 kilopascals (101.3 kPa).*

The pressure of the atmosphere is not uniform. Besides altitude variations, there are variations in atmospheric pressure at any one locality due to moving fronts and storms. Measurement of changing air pressure is important to meteorologists in predicting weather.

---

* The pascal is the SI unit of measurement. The average pressure at sea level (101.3 kPa) is often called 1 atmosphere. In British units, the average atmospheric pressure at sea level is 14.7 lb/in^2 (psi).

FIGURE 8.22 ▲
You don't notice the weight of a bag of water while you're submerged in water. Similarly, you don't notice that the air around you has weight.

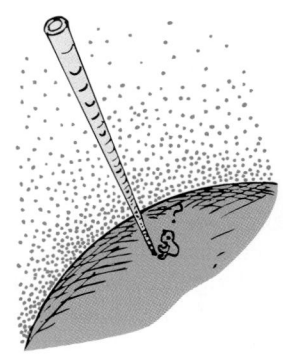

FIGURE 8.23 ▲
The mass of air that would occupy a bamboo pole that extends to the "top" of the atmosphere is about 1 kg. This air has a weight of about 10 N.

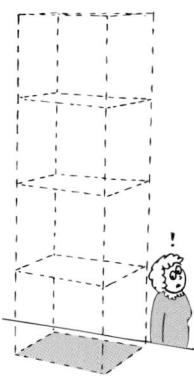

FIGURE 8.24 ▲
The weight of air that presses down on a 1-square-meter surface at sea level is about 100,000 newtons. So atmospheric pressure is about 10^5 N/m^2, or about 100 kPa.

Workers in underwater construction work in an environment of compressed air. The air pressure in their underwater chambers is at least as much as the combined pressure of water and atmosphere outside.

PhysicsPlace.com

Video
Air is Matter: Pouring Air from One Glass to Another

fyi

- Could water be used to make a barometer? Yes, but the glass tube would have to be much longer— 13.6 times as long, to be exact. The density of mercury is 13.6 the density of water. That's why a tube of water 13.6 times longer than one of mercury (of the same cross section) is needed to provide the same weight as mercury in the tube. A water barometer would have to be 13.6 × 0.76 meter, or 10.3 meters high—too tall to be practical.

CONCEPT CHECK

1. Estimate the mass of air in kilograms in a classroom that has a 200-m² floor area and a 4-m-high ceiling. (Assume a chilly 10°C temperature).
2. Why doesn't the pressure of the atmosphere break windows?

Check Your Answers

1. The mass of air is 1000 kg. The volume of air is 200 m² × 4 m = 800 m³; each cubic meter of air has a mass of about 1.25 kg, so 800 m³ × 1.25 kg/ m³ = 1000 kg (about a ton).
2. Atmospheric pressure is exerted on *both* sides of a window, so no net force is exerted on the window. If for some reason the pressure is reduced or increased on one side only, as in a strong wind, then watch out!

Barometers

An instrument used for measuring atmospheric pressure is called a **barometer.** A simple mercury barometer is illustrated in Figure 8.25. A glass tube, longer than 76 centimeters and closed at one end, is filled with mercury and tipped upside down in a dish of mercury. The mercury in the tube flows out of the submerged open bottom until the difference in the mercury levels in the tube and the dish is 76 centimeters. The empty space trapped above, except for some mercury vapor, is a pure vacuum.

In a way, a barometer acts like children balancing on a seesaw. The barometer "balances" when the weight of liquid in the tube exerts the same pressure as the atmosphere outside. Whatever the width of the tube, a 76-centimeter column of mercury weighs the same as the air that would fill a vertical 30-kilometer tube of the same width. If the atmospheric pressure increases, then the atmosphere pushes down harder on the mercury in the dish. This in turn pushes mercury higher in the tube. A balancing equilibrium is reached when the pressure of the mercury column equals atmospheric pressure.

What happens in a barometer is similar to what happens when you drink through a straw. By sucking, you reduce the air pressure in the straw when it is placed in a drink. Atmospheric pressure on the drink then pushes the liquid up into the reduced-pressure region. Strictly speaking, the liquid is not sucked up; it is pushed up the straw by the

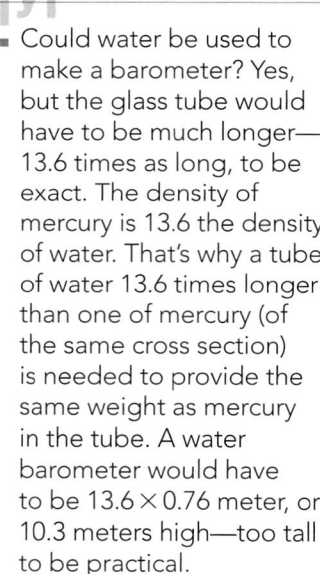

760 mm

◀ **FIGURE 8.25**
A simple mercury barometer. Mercury is pushed up into the tube by atmospheric pressure.

FIGURE 8.26 ▶
Strictly speaking, they do not suck the soda up the straws. They instead reduce pressure in the straws, which allows the weight of the atmosphere to press the liquid up into the straws. Could they drink a soda this way on the Moon?

When the pump handle is raised, air in the pipe is "thinned" as it expands to fill a larger volume. Atmospheric pressure on the well surface pushes water up into the pipe, causing water to overflow at the spout.

FIGURE 8.27 ▲
The atmosphere pushes water from below up into a pipe that is evacuated of air by the pumping action.

pressure of the atmosphere. If the atmosphere is prevented from pushing on the surface of the drink, as in the party-trick bottle with the straw through an air-tight cork stopper, one can suck and suck and get no drink.

If you understand these ideas, you can understand why there is a 10.3-meter limit on the height water can be lifted with vacuum pumps. The old fashioned farm-type pump, shown in Figure 8.27, operates by producing a partial vacuum in a pipe that extends down into the water below. Atmospheric pressure on the surface of the water simply pushes the water up into the region of reduced pressure inside the pipe. Can you see that, even with a perfect vacuum, the maximum height to which water can be lifted in this way is 10.3 meters?

A small portable instrument that measures atmospheric pressure is the *aneroid barometer* (Figure 8.28). A metal box has a partial vacuum inside, and a slightly flexible lid bends in or out with changes in atmospheric pressure. A spring-and-lever system activates a scale and indicates atmospheric pressure. At higher altitudes, atmospheric pressure is less, so a barometer can be used to determine elevation. An aneroid barometer calibrated for altitude is called an *altimeter* (altitude meter). Some of these instruments are sensitive enough to indicate a change in elevation as you walk up a flight of stairs.*

Reduced air pressures are produced by pumps, which operate on the principle of a gas tending to fill its container. If a space with less pressure is provided, gas will flow from the region of higher pressure to the one of lower pressure. A vacuum pump simply provides a region of lower pressure

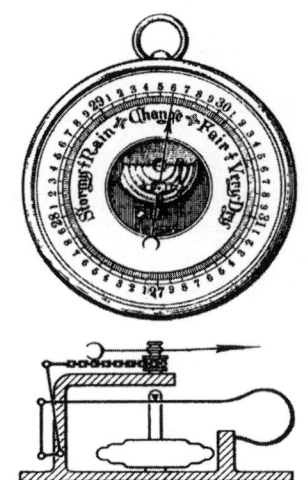

FIGURE 8.28 ▲
The aneroid barometer.

* Evidence of slightly greater atmospheric pressure over a few centimeters in elevation is any small helium-filled balloon that rises in air. The atmosphere really does push with more force against the lower bottom than against the higher top!

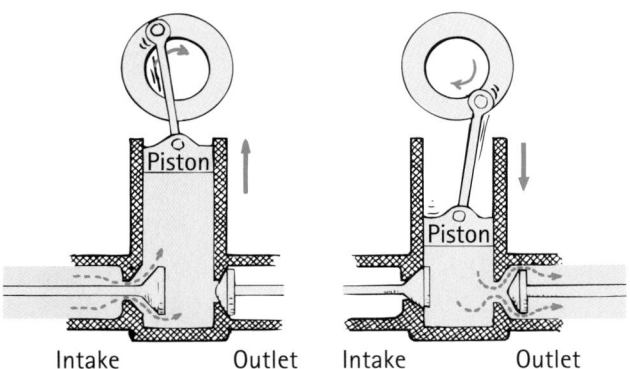

FIGURE 8.29 ▲

A mechanical vacuum pump. When the piston is lifted, the intake valve opens and air moves in to fill the empty space. When the piston is moved downward, the outlet valve opens and the air is pushed out. What changes would you make to convert this pump into an air compressor?

Intake   Outlet   Intake   Outlet

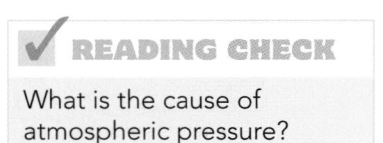

READING CHECK

What is the cause of atmospheric pressure?

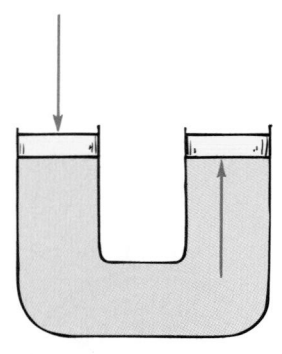

FIGURE 8.30 ▲

The force exerted on the left piston increases the pressure in the liquid and is transmitted to the right piston.

into which normally fast-moving gas molecules randomly move. The air pressure is repeatedly lowered by piston and valve action (Figure 8.29).

8.7 Pascal's Principle—The Transmission of Pressure in a Fluid

An important fact about fluid pressure is that a change in pressure at one part of the fluid will be transmitted undiminished to other parts. For example, if the pressure of city water is increased at the pumping station by 10 units of pressure, the pressure everywhere in the pipes of the connected system will be increased by 10 units of pressure (providing the water is at rest). This rule is called **Pascal's principle:**

> **A change in pressure at any point in an enclosed fluid at rest is transmitted undiminished to all points in the fluid.**

Pascal's principle was discovered in the 17th century by theologian and scientist Blaise Pascal, for whom the SI unit of pressure, the pascal ($1 \text{ Pa} = 1 \text{ N/m}^2$), is named.

Fill a U-tube with water and place pistons at each end, as shown in Figure 8.30. Pressure exerted against the left piston is transmitted throughout the liquid and against the bottom of the right piston. (The pistons are simply "plugs" that can slide freely but snugly inside the tube.) The pressure that the left piston exerts against the water will be exactly equal to the pressure the water exerts against the right piston. This is nothing to write to grandma about. But suppose you make the tube on the right side wider and use a piston of larger area; then the result is impressive. In Figure 8.31, the piston on the right has fifty times the area of the piston on the left (say the left has 100 square centimeters and the right 5000 square centimeters). Suppose a 10-kg load is placed on the left piston. Additional pressure due to the weight of the load is transmitted throughout the liquid and up against the larger piston. Here

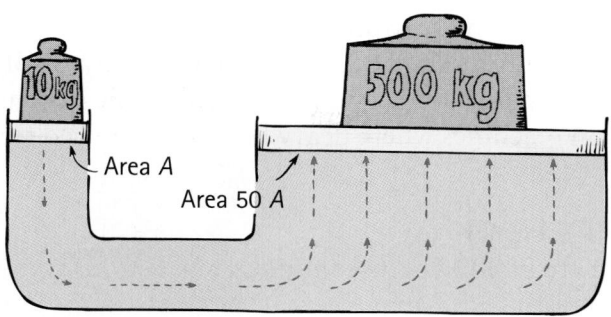

◀ **FIGURE 8.31**
A 10-kg load on the left piston
will support 500 kg on the
right piston.

is where the difference between force and pressure comes in. The additional pressure is exerted against every square centimeter of the larger piston. Since there is fifty times as much area, fifty times as much force is exerted on the larger piston. Thus, the larger piston will support a 500-kg load—fifty times the load on the smaller piston!

This *is* something to write to grandma about, for we can use such a device to multiply forces. One newton input produces 50 newtons output. By further increasing the area of the larger piston (or reducing the area of the smaller piston), we can multiply force, in principle, by any amount. Pascal's principle underlies the operation of the hydraulic press.

✓ The hydraulic press does not violate energy conservation, because a decrease in the distance moved compensates for the increase in force. Force is increased only at the expense of distance. When the small piston in Figure 8.31 is moved downward 10 centimeters, the large piston will be raised only one-fiftieth of this, or 0.2 centimeter. The input force multiplied by the distance moved by the smaller piston is equal to the output force multiplied by the distance moved by the larger piston (recall the same physics with mechanical levers back in Section 6.8).

Pascal's principle applies to all fluids, whether gases or liquids. A typical application of Pascal's principle for gases and liquids is the automobile lift seen in many service stations (Figure 8.32). Increased air pressure produced by an air compressor is transmitted through the air to the surface of oil in an underground reservoir. The oil in turn transmits the pressure to a piston, which lifts the automobile. The relatively low pressure that exerts the lifting force against the piston is about the same as the air pressure in automobile tires.

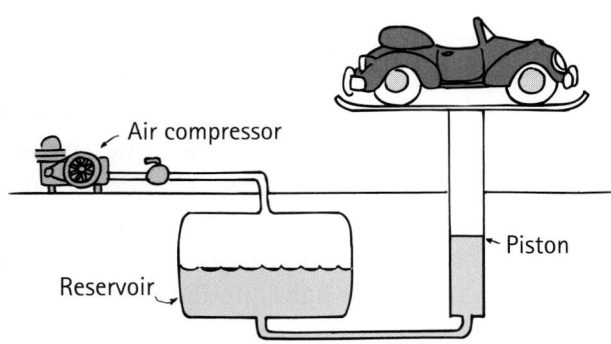

◀ **FIGURE 8.32**
Pascal's principle in a service station.

✔ **READING CHECK**

Forces can be greatly multiplied with a hydraulic press. Why does this not violate the law of conservation of energy?

More common than the hydraulics in automobile service stations are the hydraulics in industry and construction. Relatively small pressures can produce enormous forces. Look for the hydraulic pistons in almost all construction machines where heavy loads are involved (Figure 8.33).

CONCEPT CHECK

1. If a pressure of 30 N/cm^2 (common in auto tires) is exerted against a piston with an area of 200 cm^2 (about the area of the palm of your hand), what force can be produced?
2. If a friend commented that a hydraulic device is a common way of multiplying energy, what would you say?

Check Your Answers

1. $P = F \times A = 30$ N/cm^2 $\times$ 200 cm^2 = 6000 N (more than 1300 lb). No wonder hydraulic devices are common in industry!
2. No, no, no! Although a hydraulic device, like a mechanical lever, can multiply *force*, it always does so at the expense of distance. Energy is the product of force and distance. Increase one, decrease the other. *No device has ever been found that can multiply energy!*

FIGURE 8.33 ▲
Pascal's principle at work in the hydraulic devices on this common but incredible machine. We can only wonder whether Pascal envisioned the extent to which his principle would allow huge loads to be so easily lifted.

8.8 Buoyancy in a Gas— More Archimedes' Principle

A crab lives at the bottom of its ocean floor and looks upward at jellyfish and other lighter-than-water marine life drifting above it. Similarly, we live at the bottom of our ocean of air and look upward at balloons and other lighter-than-air objects drifting above us. Balloons and jellyfish are suspended in fluids because they are buoyed upward by displaced weights of fluid equal to their weights. Objects in air are buoyed upward just as are objects in water. Archimedes' principle applies to air just as it does for water:

An object surrounded by air is buoyed up by a force equal to the weight of the air displaced.

Table 8.1 tells us that the mass of 1 cubic meter of air at ordinary atmospheric pressure and room temperature is about 1.2 kilograms—a weight of about 12 newtons. ✔ Therefore, any 1-cubic-meter object in air is buoyed up with a force of 12 newtons. If the weight of the 1-cubic-meter object is greater than 12 newtons, it falls to the ground when released. If an object of this size weighs less than 12 newtons, buoyant force is greater than weight, and it rises in the air. Any object that weighs less than an equal volume of air rises in the air. Stated another way, any object less dense than air will rise in air. Gas-filled balloons that rise in air are less dense than air.

Low-density gas is used in a balloon for the same reason that plastic foam is used in life preservers. The plastic foam possesses no strange tendency to be drawn toward the water's surface, and the gas possesses no strange tendency to rise. Plastic foams and gases are buoyed upward like anything else. They are simply light enough for the buoyancy to be significant.

FIGURE 8.34 ▲
All bodies are buoyed up by a force equal to the weight of air they displace. Why, then, don't all objects float like this balloon?

Unlike water, there is no sharp surface at the "top" of the atmosphere. Furthermore, unlike water, the atmosphere becomes less dense with altitude. Whereas a piece of wood will float to the surface of water, balloons do not rise to any atmospheric surface. How high will a balloon rise? A gas-filled balloon will rise until the weight of displaced air equals its weight. Then the sum of the forces on the balloon is zero. Rising rubber balloons expand when rising. They usually reach a height where expansion stretches the rubber until it ruptures.

CONCEPT CHECK

Is there a buoyant force acting on you? If there is, why are you not buoyed up by this force?

Check Your Answer

There *is* a buoyant force acting on you, and you *are* buoyed upward by it. You aren't aware of it only because your weight is so much greater.

Large helium-filled dirigible airships are designed to rise slowly in air; that is, their total weight is a little less than the weight of air displaced. When in motion, the ship may be raised or lowered by means of horizontal "elevators."

Thus far we have treated pressure only as it applies to stationary fluids. Motion produces an additional influence.

8.9 Bernoulli's Principle— Flying with Physics

When a fluid continuously flows through a pipe, the volume that flows past any cross section of the pipe in a given time is the same that flows past other sections of the pipe—even if the pipe widens or narrows. For continuous flow, a fluid speeds up when it goes from a wide to a narrow part of the pipe. This is evident in a wide, slow-moving river that flows more swiftly as it enters a narrow gorge. It is also evident when water flowing from a garden hose speeds up when you squeeze the end of the hose to make the stream narrower.

The motion of a fluid in steady flow follows imaginary *streamlines*, represented by thin lines in Figure 8.36 and in other figures that follow. Streamlines are the smooth paths of bits of fluid. The lines are closer together in narrower regions, where the flow speed is greater. (Streamlines are visible when smoke or other visible fluids are passed through evenly spaced openings, as in a wind tunnel.)

fyi

- If a balloon is free to expand when rising, it gets larger. If a balloon is not free to expand, buoyancy decreases as the balloon rises because of the less dense displaced air. It then settles where buoyancy matches weight.

PhysicsPlace.com

Video
Buoyancy of Air

✓ **READING CHECK**

How much buoyant force acts on a 1-cubic-meter refrigerator in your school cafeteria?

FIGURE 8.35 ▲
Because the flow is continuous, water speeds up when it flows through the narrow and/or shallow part of the brook.

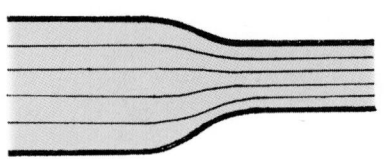

FIGURE 8.36 ▶
Water speeds up when it flows into the narrower pipe. The close-together streamlines indicate increased speed and decreased internal pressure.

- Because the volume of water flowing through a pipe of different cross-sectional areas *A* remains constant, speed of flow *v* is high where the area is small, and the speed is low where the area is large. This is stated in the equation of continuity:

$$A_1 v_1 = A_2 v_2.$$

The product $A_1 v_1$ at point 1 equals the product $A_2 v_2$ at point 2.

The friction of both liquids and gases sliding over one another is called *viscosity* and is a property of all fluids.

Recall from Chapter 5 that a large change in momentum is associated with a large impulse. So when water from a firefighter's hose hits you, the impulse can knock you off your feet. Interestingly, the pressure *within* that water is relatively small!

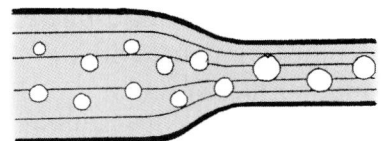

◀ **FIGURE 8.37**
Internal pressure is greater in slower-moving water in the wide part of the pipe, as evidenced by the squeezed-down air bubbles. The bubbles are bigger in the narrow part because internal pressure there is less.

Daniel Bernoulli, an 18th-century Swiss scientist, studied fluid flow in pipes. His discovery, now called **Bernoulli's principle**, can be stated as follows:

Where the speed of a fluid increases, internal pressure in the fluid decreases.

Where streamlines of a fluid are closer together, flow speed is greater and pressure within the fluid is less. Changes in internal pressure are evident in water containing air bubbles. The volume of an air bubble depends on the surrounding water pressure. Where water gains speed, pressure is lowered and bubbles become bigger. In water that slows, pressure is greater and bubbles are squeezed to a smaller size.

Bernoulli's principle is a consequence of the conservation of energy, although, surprisingly, he developed it long before the concept of energy was formalized. The full energy picture for a fluid in motion is quite complicated. ✓ Simply stated, more speed and kinetic energy mean less internal pressure, and more internal pressure means less speed and kinetic energy. Bernoulli's principle applies to a smooth, steady flow (called *laminar* flow) of constant-density fluid. At speeds above some critical point, however, the flow may become chaotic (called *turbulent* flow) and follow changing, curling paths called *eddies*. This exerts friction on the fluid and dissipates some of its energy. Then Bernoulli's equation doesn't apply well.

The decrease of fluid pressure with increasing speed may at first seem surprising, particularly if you fail to distinguish between the pressure *within* the fluid, internal pressure, and the pressure *by* the fluid on something that interferes with its flow. Internal pressure within flowing water is not the same as the external pressure it can exert on whatever it encounters. Internal pressure and external pressure are two different pressures. When the momentum of moving water or anything else is suddenly reduced, the impulse it exerts is relatively huge. A dramatic example is the use of high-speed jets of water to cut steel in modern machine shops. The water has very little internal pressure, but the pressure the stream exerts on the steel interrupting its flow is enormous.

Applications of Bernoulli's Principle

Hold a sheet of paper in front of your mouth, as shown in the chapter opener photo and also in Figure 8.38. When you blow across the top surface, the paper rises. That's because the internal pressure of moving air against the top of the paper is less than the atmospheric pressure beneath it.

Anyone who has ridden in a convertible car with the canvas top up has noticed that the roof puffs upward as the car moves. This is

Bernoulli's principle again. Pressure outside is less on top of the fabric where air is moving than static atmospheric pressure is on the inside.

Consider wind blowing across a peaked roof. The wind gains speed as it flows over the roof, as the crowding of streamlines in the sketch indicates. Pressure along the streamlines is reduced where they are closer together. The greater pressure inside the roof can lift it off the house. During a severe storm, the difference in outside and inside pressure doesn't need to be very much. A small pressure difference over a large area can produce a huge force.

If we think of the blown-off roof as an airplane wing, we can better understand the lifting force that supports a heavy aircraft. In both cases, a greater pressure below pushes the roof or the wing into a region of lesser pressure above. Wings come in a variety of designs. What they all have in common is that air is made to flow faster over the wing's top surface than under its lower surface. This is mainly accomplished by a tilt in the wing, called its *angle of attack*. Then air flows faster over the top surface for much the same reason that air flows faster in a narrowed pipe or in any other constricted region. Most often, but not always, different speeds of airflow over and beneath a wing are enhanced by a difference in the curvature (*camber*) of the upper and lower surfaces of the wing. The result is more crowded streamlines along the top wing surface than along the bottom.

When the average pressure difference over the wing is multiplied by the surface area of the wing, we have a net upward force—lift. Lift is greater when there is a large wing area and when the plane is traveling fast. A glider has a very large wing area relative to its weight, so it does not have to be going very fast for sufficient lift. At the other extreme, a fighter plane designed for high-speed flight has a small wing area relative to its weight. Consequently, it must take off and land at high speeds.

We all know that a baseball pitcher can throw a ball in such a way that it will curve off to one side as it approaches home plate. The pitcher does this by imparting a large spin to the ball. Similarly, a tennis player can hit a ball so it curves. A thin layer of air is dragged around the spinning ball by friction, which is enhanced by the baseball's threads or the tennis ball's fuzz. The moving layer of air produces a crowding of streamlines on one side. Note, in Figure 8.41b, that the streamlines are more crowded at B than at A for the direction of spin shown. Air pressure is greater at A, and the ball curves as shown.

FIGURE 8.38 ▲
The paper rises when Tim blows air across its top surface.

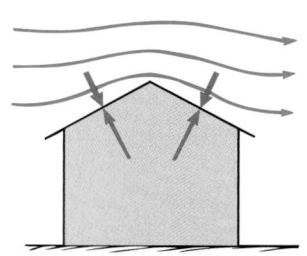

FIGURE 8.39 ▲
Air pressure above the roof is less than air pressure beneath the roof.

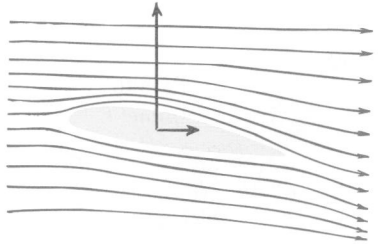

FIGURE 8.40 ▲
The vertical vector represents the net upward force (lift) that results from more air pressure below the wing than above the wing. The horizontal vector represents air drag.

◀ **FIGURE 8.41**
(a) The streamlines are the same on either side of a non-spinning baseball. (b) A spinning ball produces a crowding of streamlines. The resulting "lift" (red arrow) causes the ball to curve as shown by the blue arrow.

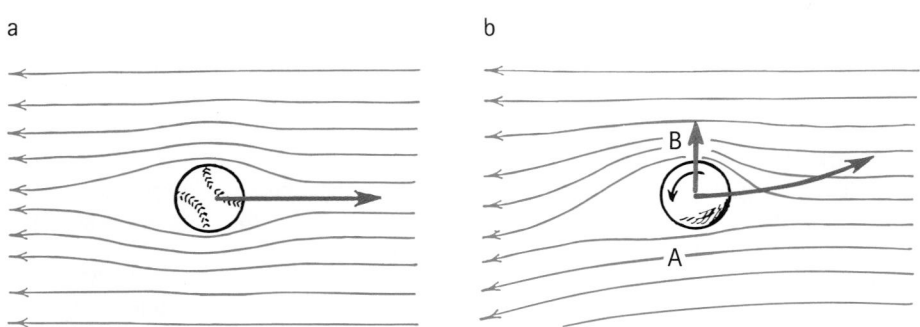

a   b

Motion of air relative to ball

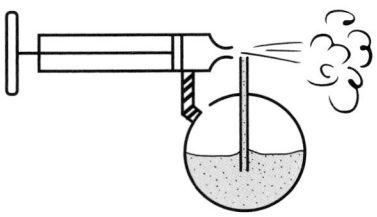

FIGURE 8.42 ▲
Why does the liquid in the reservoir go up the tube?

The familiar sprayer shown in Figure 8.42 uses Bernoulli's principle. When you force air across the open end of a tube inserted into the liquid, pressure is reduced. Atmospheric pressure on the liquid below pushes it up into the tube, where it is carried away by the stream of air.

Bernoulli's principle explains why trucks passing closely on the highway are drawn to each other, and why passing ships run the risk of a sideways collision. Water flowing between the ships travels faster than water flowing past the outer sides. Water pressure acting against the hulls is reduced between the ships. Unless the ships are steered to compensate for this, the greater pressure against the outer sides of the ships forces them together. Figure 8.43 shows how to demonstrate this in your kitchen sink or bathtub.

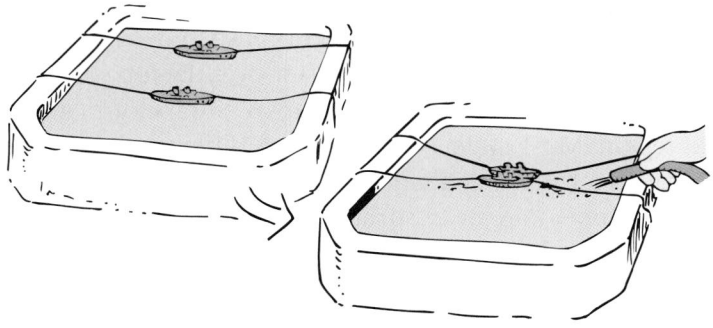

FIGURE 8.43 ▲
Try this in your sink. Loosely moor a pair of toy boats side by side. Then direct a stream of water between them. The boats will draw together and collide. Why?

✔ READING CHECK

What is the relationship between the speed of a fluid and its internal pressure? Or equivalently, between its kinetic energy and its internal pressure?

Rats to you too, Daniel Bernoulli!

FIGURE 8.44 ▲
The curved shape of an umbrella can be disadvantageous on a windy day.

CONCEPT CHECK

1. On a windy day, waves in a lake or the ocean are higher than their average height. How does Bernoulli's principle contribute to the increased height?

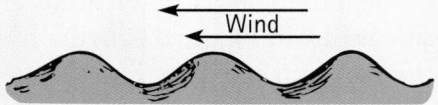

2. Blimps, airplanes, and rockets operate under three very different principles. Which operates by way of buoyancy? Bernoulli's principle? Newton's third law?

Check Your Answers

1. The troughs of the waves are partially shielded from the wind, so air travels faster over the crests. Pressure there is more reduced than down below in the troughs. The greater pressure in the troughs pushes water into the even higher crests.

2. Blimps operate by way of buoyancy, airplanes by the Bernoulli effect, and rockets by way of Newton's third law. Interesting, Newton's third law also plays a significant role in airplane flight—wing pushes air downward; air pushes wing upward.

8 CHAPTER REVIEW

KEY TERMS

Archimedes' principle An immersed body is buoyed up by a force equal to the weight of the fluid it displaces (applies to both liquids and gases).

Atmospheric pressure The pressure exerted against bodies immersed in the atmosphere resulting from the weight of air pressing down from above. At sea level, atmospheric pressure is about 101 kPa.

Barometer Any device that measures atmospheric pressure.

Bernoulli's principle The pressure in a fluid moving steadily without friction or outside energy input decreases when the fluid velocity increases.

Boyle's law The product of pressure and volume is a constant for a given mass of confined gas regardless of changes in either pressure or volume individually, so long as temperature remains unchanged:

$$P_1 V_1 = P_2 V_2$$

Buoyant force The net upward force that a fluid exerts on an immersed object.

Density The amount of matter per unit volume:

$$\text{Density} = \frac{\text{mass}}{\text{volume}}$$

Weight density is expressed as weight per unit volume.

Pascal's principle A change in pressure at any point in an enclosed fluid at rest is transmitted undiminished to all points in the fluid.

Pressure The ratio of force to the area over which that force is distributed:

$$\text{Pressure} = \frac{\text{force}}{\text{area}}$$

$$\text{Liquid pressure} = \text{weight density} \times \text{depth}$$

Principle of flotation A floating object displaces a weight of fluid equal to its own weight.

REVIEW QUESTIONS

Density—A Measure of Compactness

1. Distinguish between mass density and weight density. What are the mass density and the weight density of water?

Pressure—Force per Area

2. How does the pressure exerted by a liquid change with depth in the liquid? How does the pressure exerted by a liquid change as the density of the liquid changes?

3. How does water pressure 1 meter below the surface of a small pond compare to water pressure 1 meter below the surface of a huge lake?

4. If you punch a hole in the side of a container filled with water, in what direction does the water initially flow outward from the container?

Buoyancy in a Liquid

5. Why does buoyant force act upward on an object submerged in water?

6. How does the volume of a fully submerged object compare with the volume of water displaced?

Archimedes' Principle—Sink or Swim

7. How does the buoyant force on a fully submerged object compare with the weight of water displaced?

8. What is the mass in kilograms of 1 L of water? What is its weight in newtons?

9. If a 1-L container is immersed halfway in water, what is the volume of water displaced? What is the buoyant force on the container?

10. Does the buoyant force on a floating object depend on the weight of the object or on the weight of the fluid displaced by the object? Or are these two weights the same for the special case of floating? Defend your answer.

11. What weight of water is displaced by a 100-ton floating ship? What is the buoyant force that acts on this ship?

Pressure in a Gas

12. By how much does the density of air increase when it is compressed to half its volume? Its pressure?

Atmospheric Pressure Is Due to the Weight of the Atmosphere

13. How does the downward pressure of the 76-cm column of mercury in a barometer compare with the air pressure at the surface of the dish of mercury that it stands in?

14. How does the weight of mercury in a barometer tube compare with the weight of an equal cross section of air from sea level to the top of the atmosphere?

15. Consider two tubes of equal cross-sectional area. One is a 76-cm-tall tube of mercury and the other a 10.3-m-tall tube of water. Which weighs more? Which exerts more pressure at the bottom of the tube?

Pascal's Principle—The Transmission of Pressure in a Fluid

16. What happens to the pressure in all parts of a confined fluid when the pressure in one part is increased?

17. Does Pascal's principle provide a way to get more energy from a machine than is put into it? Defend your answer.

Buoyancy in a Gas—More Archimedes' Principle

18. A balloon that weighs 1 N is suspended in air, drifting neither up nor down. How does the buoyant force acting on it compare with its weight? What happens if the buoyant force decreases? Increases?

Bernoulli's Principle—Flying with Physics

19. What are streamlines? Is pressure greater or less in regions of crowded streamlines?

20. What do peaked roofs, convertible tops, and airplane wings have in common when air moves faster across their top surfaces?

THINK AND DO

1. You ordinarily pour water from a full glass into an empty glass simply by placing the full glass above the empty glass and tipping. Have you ever poured air from one glass to another? The procedure is similar. Lower two glasses in water, mouths downward. Let one fill with water by tilting its mouth upward. Then hold the water-filled glass mouth downward above the air-filled glass. Slowly tilt the lower glass and let the air escape, filling the upper glass. You will be pouring air from one glass into another!

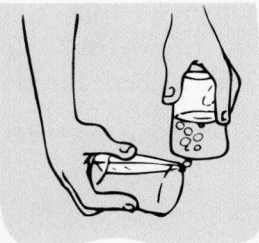

2. Raise a filled glass of water above the waterline in a body of water, but with its mouth beneath the surface. Why does the water not run out? How tall would a glass have to be before water began to run out? (You won't be able to do this indoors unless you have a ceiling that is at least 10.3-m higher than the waterline.)

3. Place a card over the open top of a glass filled to the brim with water, and then invert it. Why does the card stay in place? Try it sideways.

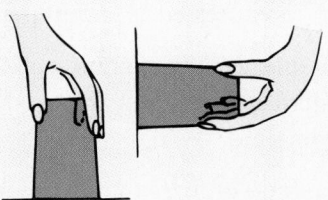

4. Lower a narrow glass tube or drinking straw into water and place your finger over the top of the tube. Lift the tube from the water and then lift your finger from the top of the tube. What happens? (You'll do this often in chemistry experiments.)

5. Blow across the top of a sheet of paper as Tim does in Figure 8.35. Try this with those of your friends who are not taking a physical science course. Then explain it to them!

6. Push a pin through a small card and place it over the hole of a thread spool. Try to blow the card from the spool by blowing through the hole. Try it in all directions.

PLUG AND CHUG

Density = mass/volume

1. Calculate the density of a block of mass 22 kg that has a volume of 2.6 cm^3.
2. Calculate the density of 3 L of a liquid that has a mass of 2.6 kg.
3. Calculate the density of a gas with a mass of 4.3 kg and a volume of 3.0 m^3.

Water Pressure = weight density × depth (where weight density = density × 10 N/kg)

4. Calculate the pressure in N/m^2 at the bottom of the Hoover dam. The depth of the water behind the dam is 220 m. (Neglect the pressure due to the atmosphere.)
5. Calculate the water pressure at the base of a 30-m-tall water tower.

THINK AND COMPARE

1. The three metal balls are held suspended and submerged in water. Rank the buoyant force on each from greatest to least.

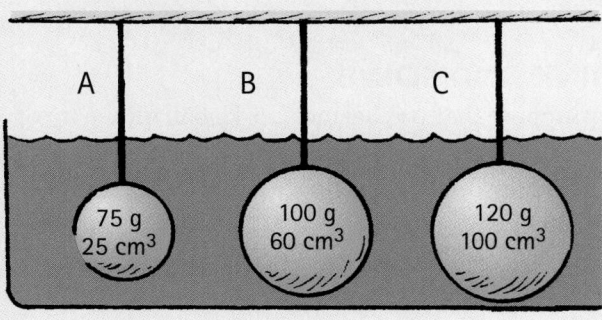

2. A buoyant force acts on each of the three gas-filled balloons. Rank the buoyant force from greatest to least.

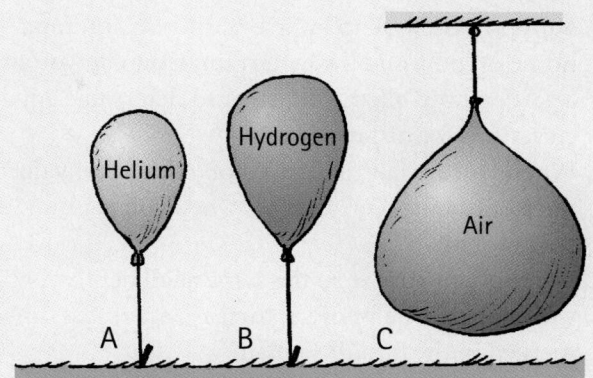

3. Rank the amount of force the right-hand piston can lift, from greatest to least.

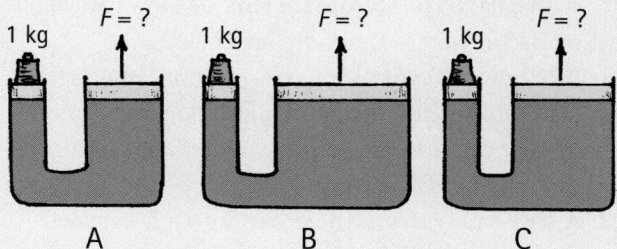

THINK AND EXPLAIN

1. Stand on a bathroom scale and read your weight. When you lift one foot up so that you're standing on one foot, does the reading change? Does a scale read force or pressure?
2. In a deep dive, a whale is appreciably compressed by the pressure of the surrounding water. What happens to the whale's density?
3. A balloon is weighted so that it is barely able to float in water. If it is pushed beneath the surface, will it rise back to the surface, stay at the depth to which it is pushed, or sink? Explain. (*Hint:* Does the balloon's density change?)

4. The density of a rock doesn't change when it is submerged in water. Does your density change when you are submerged deep in water? Defend your answer.

5. If water faucets upstairs and downstairs are turned fully on, will more water per second flow out the downstairs faucet? Or will the water flowing from the faucets be the same?

6. Suppose you wish to lay a level foundation for a home on hilly and bushy terrain. How can you use a garden hose filled with water to determine equal elevations for distant points?

7. When you are bathing on a stony beach, why do the stones hurt your feet less when you get in deep water?

8. If liquid pressure were the same at all depths, would there be a buoyant force on an object submerged in the liquid? Explain.

9. Why is it inaccurate to say that heavy objects sink and that light objects float? Give exaggerated examples to support your answer.

10. Compared to an empty ship, would a ship loaded with a cargo of Styrofoam sink deeper into water or rise in water? Defend your answer.

11. A barge filled with scrap iron is in a canal lock. If the iron is thrown overboard, does the water level at the side of the lock rise, fall, or remain unchanged? Explain.

12. A ship sailing from the ocean into a freshwater harbor sinks slightly deeper into the water. Does the buoyant force on it change? If so, does it increase or decrease?

13. Suppose you are given the choice between two life preservers that are identical in size, the first a light one filled with Styrofoam and the second a very heavy one filled with lead pellets. If you submerge these life preservers in the water, upon which will the buoyant force be greater? Upon which will the buoyant force be ineffective? Why are your answers different?

14. A half-filled bucket of water is on a spring scale. Will the reading of the scale increase or remain the same if a fish is placed in the bucket? Will your answer be different if the bucket is initially filled to the brim?

15. There is much concern about rising sea levels due to melting of ice in the polar regions. What happens to sea levels when enormous blocks of ice presently on land slip into the ocean?

16. Your friend says that the buoyant force of the atmosphere on an elephant is significantly greater than the buoyant force of the atmosphere on a small helium-filled balloon. What do you say?

17. When you replace helium in a balloon with hydrogen, which is less dense, does the buoyant force on the balloon change if the balloon remains the same size? Explain.

18. A steel tank filled with helium gas doesn't rise in air, but a balloon containing the same helium easily does? Why?

19. When a steadily flowing gas moves from a larger-diameter pipe to a smaller-diameter pipe, what happens to (a) its speed, (b) its pressure, and (c) the spacing between its streamlines?

20. What physics principle underlies the following three observations? When passing an oncoming truck on the highway, your car tends to sway toward the truck. The canvas roof of a convertible automobile bulges upward when the car is traveling at high speeds. The windows of older passenger trains sometimes break when a high-speed train passes by on the next track.

PHYSICS
YUM

THINK AND SOLVE

1. Suppose that you balance a 5-kg ball on the tip of your finger, which has an area of 1 cm². Show that the pressure on your finger is 50 N/cm², which is 500 kPa.

2. A 6-kg piece of metal displaces 1 L of water when submerged. Show that its density is 6000 kg/m³. How does this compare with the density of water?

3. A 1-kg rock suspended above water weighs 10 N. When the rock is suspended beneath the surface of the water, the scale reads 8 N.

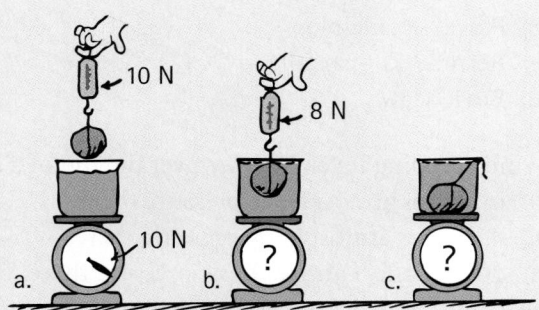

(a) What is the buoyant force on the rock?

(b) If the container of water weighs 10 N on the weighing scale, what is the scale reading when the rock is suspended beneath the surface of the water?

(c) What is the scale reading when the rock is released and rests at the bottom of the container?

4. A merchant in Katmandu sells you a solid-gold 1-kg statue for a very reasonable price. When you get home, you wonder whether or not you got a bargain, so you lower the statue into a container of water and measure the volume of displaced water. Show that, for pure gold, the volume of water displaced will be 51.8 cm³.

5. In the hydraulic pistons shown in the sketch, the small piston has a diameter of 2 cm; the large piston has a diameter of 6 cm. How much more force can the larger piston exert compared with the force applied to the smaller piston?

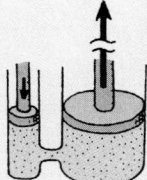

6. The wings of a certain airplane have a total surface area of 100 m². At a particular speed, the difference in air pressure below and above the wings is 4% of atmospheric pressure. Show that the lift on the airplane is 4 × 10⁵ N.

READINESS ASSURANCE TEST (RAT)

If you have a good handle on this chapter—if you really do—then you should be able to score 7 out of 10 on this RAT. Check your answers with your teacher. If you score less than 7, study further before moving on.

Choose the best answer to each of the following.

1. Pumice is a volcanic rock that floats in water. The density of pumice is
(a) less than the density of water.
(b) equal to the density of water.
(c) more than the density of water.
(d) But being a rock, it sinks!

2. The pressure at the bottom of a pond does NOT depend on
(a) the acceleration due to gravity.
(b) water density.
(c) the depth of the pond.
(d) the surface area of the pond.

3. A completely submerged object always displaces its own
(a) weight of fluid.
(b) volume of fluid.
(c) density of fluid.
(d) All of these.

4. A rock suspended by a weighing scale weighs 5 N out of water and 3 N when submerged in water. What is the buoyant force on the rock?
(a) 3 N
(b) 5 N
(c) 8 N
(d) None of these.

5. A block of wood with a flat rock tied to its top floats in a bucket of water. If the wood and rock are turned over so the rock is submerged beneath the wood, the water level at the side of the bucket
   (a) rises.
   (b) falls.
   (c) remains the same.
   (d) But with the rock beneath the wood, both rock and wood sink.

6. In a vacuum, an object has no
   (a) buoyant force.
   (b) mass.
   (c) weight.
   (d) All of these.

7. Consider two mercury barometers, one having a cross-sectional area of 1 cm² and the other 2 cm². Mercury in the smaller tube will rise
   (a) to the same height as the other.
   (b) twice as high as the other.
   (c) four times as high as the other.
   (d) None of these.

8. In a hydraulic press operation, it is impossible for the
   (a) output piston to move farther than the input piston.
   (b) force output to exceed the force input.
   (c) output piston's speed to exceed the input piston's speed.
   (d) energy output to exceed energy input.

9. The flight of a blimp best illustrates
   (a) Archimedes' principle.
   (b) Pascal's principle.
   (c) Bernoulli's principle.
   (d) Boyle's law.

10. Wind speeding up as it blows over the top of a hill
    (a) increases atmospheric pressure there.
    (b) decreases atmospheric pressure there.
    (c) doesn't affect atmospheric pressure there.
    (d) brings atmospheric pressure to zero.

Learning colors your life.

PART TWO
FORMS OF ENERGY

Although the temperature of these sparks exceeds 2000°C, the heat they impart when striking my skin is very small—which illustrates that **temperature** and **heat** are different concepts. Learning to distinguish between closely related concepts is the challenge and essence of *Conceptual Physical Science—Explorations*.

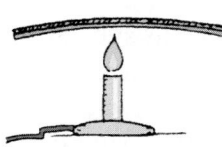

9

HEAT

THE MAIN
IDEA

Heat is the movement of thermal energy from high to low temperature. It transfers from one place to another by the processes of conduction, convection, and radiation.

Why can co-author John Suchocki walk barefoot without harm on red-hot coals? Holding one of these coals in your hands would burn you, but when white-hot sparks from a holiday sparkler hit your skin, they don't burn you—why is this so? Why does a bite of the filling of a hot apple pie burn your tongue, while the crust does not, especially since both parts come from the same oven with the same temperature?

Why does air in a balloon, the concrete of a sidewalk, and almost everything else expand as it heats up? Why does ice water do the opposite—contract instead of expand as its temperature rises? And why does a tile floor feel cooler than a carpeted floor when both have the same temperature? Is it true that everything continually emits radiation? If so, then why doesn't everything get colder with time? Let's explore!

Explore!

Can You Trust Your Senses?

1. Put some hot water, some warm water, and some cold water in three open containers.
2. Place a finger in the hot water and a finger of the other hand in the cold water. How do they feel?
3. After a few seconds, place both fingers in the warm water. How do they feel now?

Analyze and Conclude

1. **Observing** The differences in temperature experienced by your two immersed fingers is not surprising—until they are immersed in the warm water.

2. **Predicting** Can you trust your sense of hot and cold? Will both fingers feel the same temperature when they are put in the warm water?
3. **Making Generalizations** Our senses are a good guide at times—but not at all times. To measure the hotness or coldness of things, use a thermometer.

FIGURE 9.1 ▶
Count Rumford (1753–1814).
Besides studying projectiles,
setting up an early system of
welfare, breeding horses, and
acting as a spy, Count Rumford
showed that heat is related to
the motion of matter.

FIGURE 9.2 ▲
Push down on the piston and
you do work on the air inside.
Air molecules gain kinetic
energy and move faster, and
the air is warmed.

✔ **READING CHECK**

How does thermal energy
relate to the kinetic and
potential energies of its
particles?

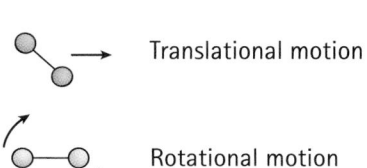
→ Translational motion

Rotational motion

↔ Vibrational motion

9.1 Thermal Energy— The Total Energy in a Substance

Until a couple hundred years ago, people thought heat was a mysterious substance called *caloric* that passed from warmer objects to cooler ones. In 1798, scientist Count Rumford observed the drilling of a metal cylinder to make a cannon. He noticed that the metal got very hot as the drill turned. Rumford realized that the drill's motion must be producing the heat. He realized that motion produces heat and that heat and the energy of motion must be connected.

To understand how motion, heat, and energy are related, you need to know a bit about the particles of matter. Matter is made of tiny particles called *atoms.* Atoms join together to make clusters called *molecules.* Atoms and molecules constantly wiggle and jiggle. When jiggling slowly, they form solids. When they jiggle faster so they slide over one another, we have a liquid. When atoms and molecules move so fast that they disconnect and fly loose, we have a gas. Heated further and electrons are shaken off atoms and we have a plasma. So whether a substance is a solid, liquid, gas, or plasma depends on the motion of its particles. All this is included in the *kinetic theory of matter,* which states:

Matter is composed of tiny particles called *atoms* and *molecules.* These particles are always moving. This random motion of atoms and molecules is present in different degrees in solids, liquids, gases, and plasmas.

The warmer an object is, the faster its particles move. That means they have more kinetic energy. For example, when you strike a penny with a hammer, the penny becomes warm. Why? Because the hammer's blow causes atoms in the coin to jiggle faster. When you put a flame to a liquid, the liquid becomes warmer as its particles move faster. When you rapidly compress air in a tire pump, the air gets warmer. In these cases, the molecules are made to move faster. They gain kinetic energy and become warmer.

The warmer an object is, the more *thermal energy* it contains. The **thermal energy** in a substance is the total energy of all of its atoms and molecules. ✔ Thermal energy consists of all the potential *and* kinetic energy of the particles in a substance as they travel about, twist and turn, and vibrate back and forth (Figure 9.3).

◀ **FIGURE 9.3**
Types of motion in matter. Increases in translational motion increase kinetic energy and therefore temperature. Increases in rotational and vibrational motion increase potential energy.

9.2 Temperature—Average Kinetic Energy per Molecule in a Substance

To tell how warm or cold an object is we measure its **temperature.** A common thermometer measures temperature by expansion or contraction of a liquid, usually colored alcohol.

The most common thermometer in the world is the *Celsius thermometer,* named after the Swedish astronomer Anders Celsius (1701–1744). Celsius was the first person to suggest the scale of 100 degrees between the freezing point and boiling point of water. The number 0 represents the temperature at which water freezes, and the number 100 represents the temperature at which water boils (at standard atmospheric pressure). In between are 100 equal parts called *degrees.*

In the United States, the number 32 represents the temperature of freezing water, and the number 212 represents the temperature at which water boils. This temperature scale makes up a Fahrenheit thermometer, named after its originator, the German physicist Gabriel D. Fahrenheit (1686–1736). The Fahrenheit scale is still popular in the United States.

Arithmetic formulas are used for converting from one temperature scale to the other and are common in classroom exams. Because such arithmetic exercises are not really physical science, we won't be concerned with these conversions. (This may be important in a math class, but not here.) Besides, the conversion between Celsius and Fahrenheit temperatures is closely approximated in the side-by-side scales of Figure 9.5.*

✓ As Figure 9.3 indicates, temperature is proportional to the average translational kinetic energy per particle that makes up a substance. By "translational" we mean to-and-fro linear motion. A substance with a high temperature has atoms or molecules with high average translational kinetic energies. (To be brief, from now on in this chapter, we'll simply say "molecules" to mean "atoms and molecules.")

Interestingly, a thermometer actually registers its own temperature. When a thermometer is in contact with something whose temperature we wish to know, thermal energy flows between the two until their temperatures are equal. At this point, thermal equilibrium is established. So when we look at the temperature of the thermometer, we learn about the temperature of the substance with which it reaches thermal equilibrium.

---

* Okay, if you really want to know, the formulas for temperature conversion are: $C = \frac{5}{9}(F - 32)$; $F = \frac{9}{5}C + 32$, where C is the Celsius temperature and F is the corresponding Fahrenheit temperature.

FIGURE 9.4 ▲
A testament to Fahrenheit outside his home (now in Gdansk, Poland).

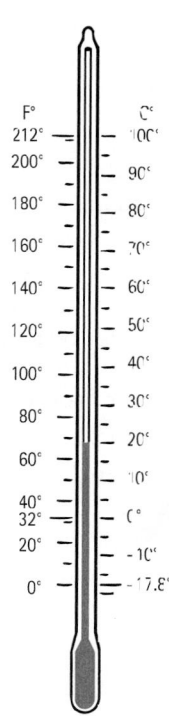

FIGURE 9.5 ▲
Fahrenheit and Celsius scales on a thermometer.

fyi

- Thermal contact is not required with infrared thermometers that show digital temperature readings by measuring the infrared radiation emitted by all bodies.

READING CHECK

How does temperature relate to the average kinetic energy of molecular motion?

CONCEPT CHECK

True or false: Temperature is a measure of the total kinetic energy in a substance.

Check Your Answer

False. Temperature is a measure of the *average* (not the *total!*) kinetic energy of the molecules in a substance. For example, there is twice as much total molecular kinetic energy in 2 liters of boiling water as in 1 liter—but the temperatures of the two volumes of water are the same because the *average* kinetic energy per molecule in each is the same.

PhysicsPlace.com

Video
Low Temperatures with
Liquid Nitrogen

Plasmas are found in stars, where the temperature is many millions of degrees Celsius. Cooler plasmas are in fluorescent lamps and in the pixels of some TV screens.

9.3 Absolute Zero—Nature's Lowest Possible Temperature

In principle, there is no upper limit of temperature. As thermal motion keeps increasing, a solid object melts to a liquid and then evaporates to a gas. As mentioned earlier, further heating of a gas breaks molecules up into atoms that lose some or all of their electrons—a cloud of electrically charged particles called a plasma. Temperature has no upper limit.

In contrast, there is a definite lower limit to temperature. Here's how experimenters in the 19th century found that limit. Gases expand when heated and contract when cooled. All gases were found to shrink by $\frac{1}{273}$ of their volume at 0°C for each Celsius degree lowering in temperature. This occurs when the gas pressure is held constant. So if a gas at 0°C were cooled down by 273°C, it would contract $\frac{273}{273}$ volumes and be reduced to zero volume.

The same occurs with pressure. The pressure of a gas of fixed volume decreases by $\frac{1}{273}$ for each Celsius degree lowering of temperature. If it is cooled 273°C below zero, it would have no pressure at all. In practice, every gas turns to a liquid before it gets this cold. Nevertheless,

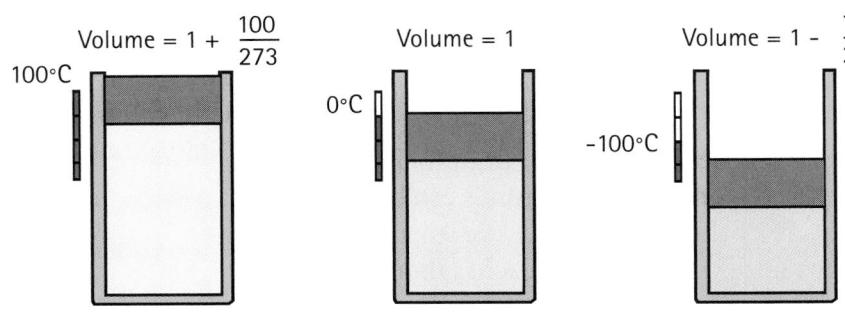

FIGURE 9.6 ▲

When pressure is held constant, the volume of a gas at 0°C changes by $\frac{1}{273}$ of its volume with each 1°C change in temperature. At 100°C, the volume is $\frac{100}{273}$ greater than it is at 0°C. When the temperature is reduced to −100°C, the volume is reduced by $\frac{100}{273}$. At −273°C, the volume of the gas would be reduced by $\frac{273}{273}$ and so would be zero.

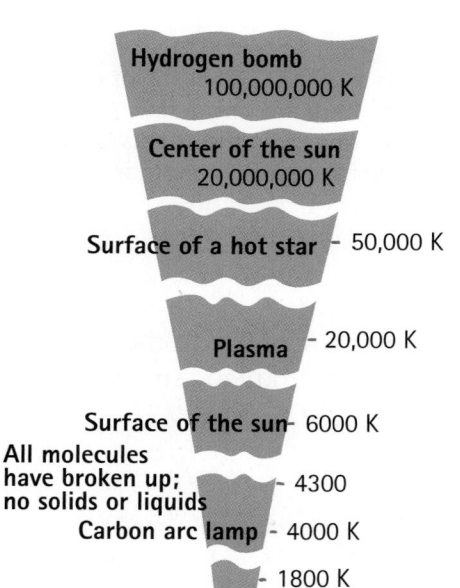

- Hydrogen bomb 100,000,000 K
- Center of the sun 20,000,000 K
- Surface of a hot star – 50,000 K
- Plasma – 20,000 K
- Surface of the sun – 6000 K
- All molecules have broken up; no solids or liquids – 4300
- Carbon arc lamp – 4000 K
- 1800 K Iron melts
- +200°C – 500 K Tin melts
- 400 K
- +100°C – Water boils
- 300 K
- 0°C – 273 K – Ice melts Ammonia boils
- 200 K
- –100°C – Dry ice vaporize
- 100 K
- –200°C – Oxygen boils Helium boils
- –273°C – 0 K

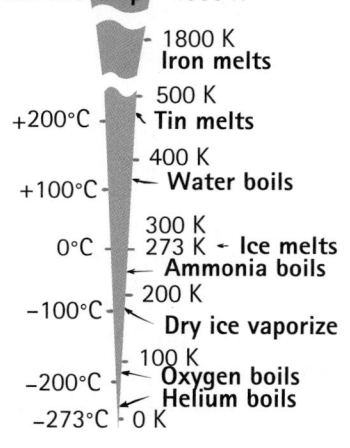

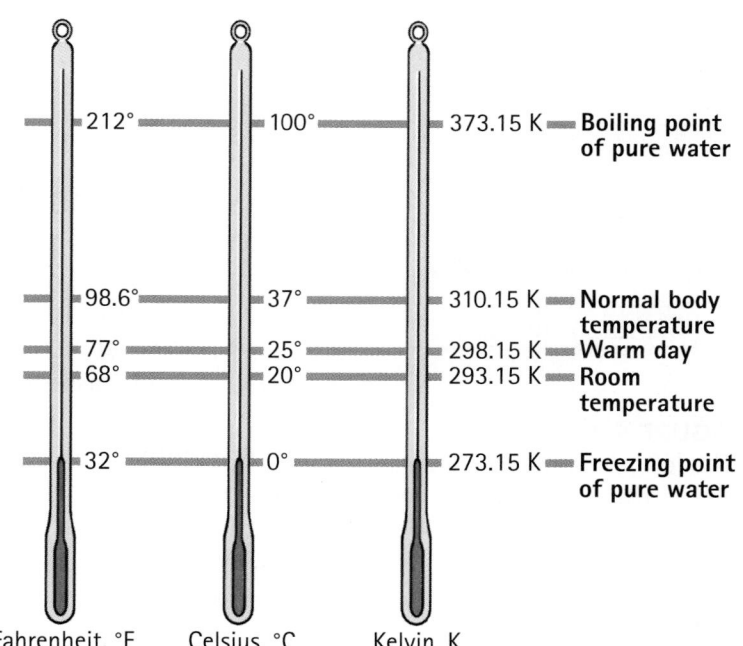

FIGURE 9.8 ▶
Some familiar temperatures measured on the Fahrenheit, Celsius, and Kelvin scales.

Fahrenheit, °F	Celsius, °C	Kelvin, K	
212°	100°	373.15 K	Boiling point of pure water
98.6°	37°	310.15 K	Normal body temperature
77°	25°	298.15 K	Warm day
68°	20°	293.15 K	Room temperature
32°	0°	273.15 K	Freezing point of pure water

these decreases by $\frac{1}{273}$ increments suggested the idea of a lowest temperature: −273°C. We call this lower limit of temperature **absolute zero.** At this temperature molecules have lost all available kinetic energy. No more energy can be taken from a substance at absolute zero. It can't get any colder.

The absolute temperature scale is called the Kelvin scale, named after the famous British physicist Lord Kelvin. Absolute zero is 0 K (short for "0 kelvin"; note that the word *degrees* is not used with Kelvin temperatures). There are no negative numbers on the Kelvin scale. ✓ Degrees on the Kelvin scale are the same size as divisions on the Celsius scale. Thus the melting point of ice is 273 K, and the boiling point of water is 373 K.

Absolute zero isn't the coldest you can get. It's the coldest you can hope to approach.

◀ **FIGURE 9.9**
The temperature of the sparks is very high, about 2000°C. That's a lot of thermal energy per molecule of spark. Because there are only a few molecules per spark, however, the total amount of thermal energy in the sparks is safely small. Temperature is one thing; transfer of thermal energy is another.

CONCEPT CHECK

1. Which is larger, a Celsius degree or a kelvin?
2. A sample of hydrogen gas has a temperature of 0°C. If the gas is heated until its molecules have doubled their kinetic energy, what is its temperature?

Check Your Answers

1. Neither. They are equal.
2. The 0°C gas has an absolute temperature of 273 K. Twice as much kinetic energy means that it has twice the absolute temperature, or two times 273 K. This would be 546 K, or 273°C.

✔ **READING CHECK**

How do the size of divisions compare on the Kelvin and Celsius scales?

Hot stove

FIGURE 9.10 ▲
The left pot contains 1 liter of water. The right one contains 3 liters. Although both pots absorb the same quantity of heat, the temperature increases three times as much in the pot with the smaller amount of water.

9.4 Heat Is the Movement of Thermal Energy

If you touch a hot stove, thermal energy enters your hand because the stove is warmer than your hand. When you touch a piece of ice, however, thermal energy passes out of your hand and into the ice. The direction of thermal energy flow is normally from a warmer substance to a cooler one. A scientist defines **heat** as the thermal energy transferred from one substance to another due to a temperature difference between the two substances.

According to this definition, matter does not *contain* heat. Matter contains *thermal energy*. Heat is *thermal energy in transit.* ✔ After heat has been transferred to an object or substance, it ceases to be heat—and becomes thermal energy.

For substances in thermal contact, thermal energy flows from the higher-temperature substance into the lower-temperature one until thermal equilibrium is reached. Energy flow depends on temperature differences, not thermal energy differences. For example, there is more thermal energy in a bowl of warm water than there is in a red-hot thumbtack. If the tack is put into the water, thermal energy doesn't flow from the warm water to the tack. Instead, it flows from the hot tack to the cooler water. Thermal energy never flows by itself from a low-temperature substance into a higher-temperature one.

We Know What Heat Is—What Is Cold?

Heat actually exists. Heat is thermal energy that transfers in a direction from hot to cold. But what is cold? Does a cold substance contain something opposite to thermal energy? The answer is no.

Just as dark is the absence of light, cold is the absence of thermal energy.

An object is cold not because it contains something, but because it *lacks* something. Something that is cold lacks thermal energy. On a near-zero winter day when you're waiting at the bus stop, you feel cold not because something called cold gets to you. You feel cold because you lose heat. Cold is not a thing in itself, but the result of lowered thermal energy.

CONCEPT CHECK
1. Suppose you apply a flame to 1 L of water and its temperature rises by 3°C. If you apply the same flame for the same length of time to 3 L of water, by how much does its temperature rise?
2. When you touch a cold surface, does cold travel from the surface to your hand or does energy travel from your hand to the cold surface?

Check Your Answers
1. Its temperature rises by only 1°C. This is because there are three times as many molecules in 3 L of water and each molecule receives only one-third as much energy on the average. So the average kinetic energy, and thus the temperature, increases by one-third as much. See Figure 9.10.
2. The direction of energy travel is from hot to cold—from your hand to the cold surface. There is no "cold" that travels in the other direction.

Heat Units Are Energy Units

Heat is a form of energy and is measured in joules. It takes 4.18 joules of heat to change 1 gram of water by 1 Celsius degree. A unit of heat still common in the United States is the *calorie*. A calorie is defined as the amount of heat needed to change the temperature of 1 gram of water by 1 Celsius degree. (The relationship between calories and joules is that 1 calorie = 4.18 joules.)

The energy ratings of foods and fuels are measured by the energy released when they are burned. (Metabolism is really "burning" at a slow rate.) The heat unit for labeling foods is the kilocalorie, which is 1000 calories (the heat needed to change the temperature of 1 kilogram of water by 1°C). To tell the difference between this unit and the smaller calorie, the food unit is usually called a *Calorie* with a capital C. So 1 C = 1000 calories.

Temperature is measured in degrees. Heat is measured in joules (or calories).

FIGURE 9.11 ▲
To the weight watcher, the peanut contains 10 Calories; to the physicist, it releases 10,000 calories (41,800 joules) of energy when burned or digested.

In the United States, we speak of low-calorie foods and drinks. Most of the world speaks of low-joule foods and drinks.

CONCEPT CHECK
Which will raise the temperature of water more, adding 4.18 joules or 1 calorie?

Check Your Answer
Both the same. This is like asking which is longer, a 1-mile-long track or a 1.6-kilometer-long track. They're the same in different units.

FIGURE 9.12 ▲
The filling of hot apple pie may be too hot to eat, even though the crust is not.

9.5 Specific Heat Capacity— A Measure of Thermal Inertia

When you're eating, have you noticed that some foods remain hotter much longer than others? For example, just after an apple pie has been taken out of an oven, the filling burns your tongue while the crust doesn't. Or you can take a bite of a piece of hot toast a few seconds after it is out of the hot toaster, but you have to wait several minutes before eating soup from a stove as hot as the toaster.

Different substances have different capacities for storing thermal energy. When you heat a pot of water on a stove, you find that it takes about 15 minutes to bring it to a boil. If you put an equal mass of iron on the same stove, you'd find it rising through the same temperature range in only about 2 minutes. For silver, the time would be less than a minute. Different materials require different amounts of thermal energy to raise temperature. This is because different materials absorb energy in different ways. The added energy may increase the translational motion of molecules, which raises the temperature. Or added energy may increase the amount of internal vibration or rotation within the molecules and therefore become potential energy. ✓ It's mainly the translation motion of its atoms and molecules that raises the temperature of a substance. Each substance has its own characteristic **specific heat capacity.***

> **The specific heat capacity of any substance is defined as the quantity of heat required to change the temperature of a unit mass of the substance by 1 degree.**

Specific heat capacity is a measure of thermal inertia. Recall in our study of Newton's laws that inertia is the property of matter to resist changes in motion. Specific heat capacity is a similar property of matter to resist a change in temperature.

Water has a much higher capacity for storing energy than most all other substances. A lot of heat energy is needed to change the temperature of water. This explains why water is very useful in the cooling system of automobiles and other engines. It absorbs a great quantity of heat for small rises in temperature. Water also takes longer to cool.

CONCEPT CHECK
Which has a higher specific heat capacity, water or sand? In other words, which takes longer to warm in sunlight (or longer to cool at night)?

---

* If we know the specific heat capacity c of a substance, the formula for the quantity of heat Q involved when a mass m of the substance undergoes a change in temperature ΔT is $Q = cm\Delta T$. In words, heat transferred = specific heat capacity × mass × temperature change.

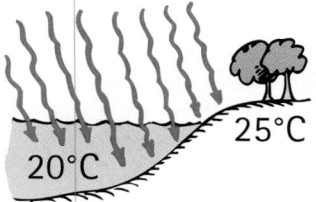

Check Your Answer
Water has the higher specific heat capacity. In the same sunlight, the temperature of water increases more slowly than the temperature of sand. And water will cool more slowly at night. Sand and soil's low specific heat capacity, as evidenced by how quickly it warms in the morning sun and how quickly it cools at night, affects local climates.

Water's high specific heat capacity affects the world's climate. Look at a world globe and notice the high latitude of Europe. Water's high specific heat keeps climate there milder than regions of the same latitude in northeastern Canada. Both Europe and Canada receive about the same amount of sunlight per square kilometer. What happens is that the Atlantic Ocean current known as the Gulf Stream carries warm water northeast from the Caribbean. It holds much of its thermal energy long enough to reach the North Atlantic Ocean off the coast of Europe. Then it cools, releasing 4.18 joules of energy for each gram of water that cools 1°C. The released energy is carried by westerly winds over the European continent.

A similar effect occurs in the United States. The winds in North America are mostly westerly. On the West Coast, air moves from the Pacific Ocean to the land. In winter months, the ocean water is warmer than the air. Air blows over the warm water and then moves over the coastal regions. This warms the climate. In summer, the opposite occurs. The water cools the air and the coastal regions are cooled. Temperature changes are moderate rather than extreme. The East Coast does not benefit from the moderating effects of water because the direction of air is from the land to the Atlantic Ocean. Land, with a lower specific heat capacity, gets hot in the summer but cools rapidly in the winter.

Islands and peninsulas do not have the extremes of temperatures that are common in interior regions of a continent. The high summer

FIGURE 9.13 ▲
Because water has a high specific heat capacity and is transparent, it takes more energy to warm the water than to warm the land. Solar energy striking the land is concentrated at the surface, but solar energy striking the water extends beneath the surface and so is "diluted."

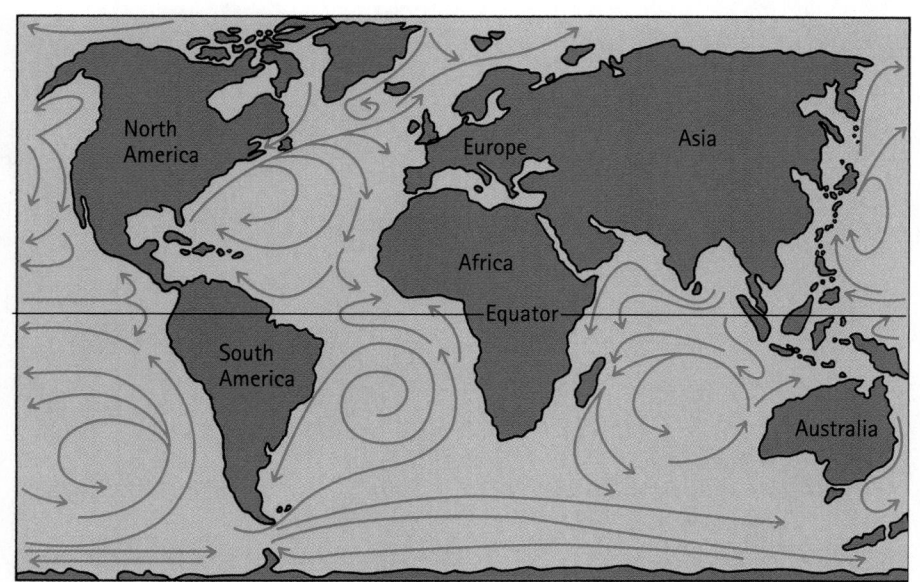

◄ **FIGURE 9.14**
Many ocean currents, shown in blue, distribute heat from the warmer equatorial regions to the colder polar regions.

READING CHECK

What kind of molecular motion affects the temperature of substances?

and low winter temperatures common in Manitoba and the Dakotas, for example, are largely due to the absence of large bodies of water. Europeans, islanders, and people living near ocean air currents should be glad that water has such a high specific heat capacity. San Franciscans are!

CONCEPT CHECK

Bermuda is close to North Carolina, but unlike North Carolina, it has a tropical climate year-round. Why?

Check Your Answer

Bermuda is an island. The surrounding water warms it when it might be too cold, and cools it when it might be too warm.

PhysicsPlace.com

Video
How a Thermostat Works

9.6 Thermal Expansion

Molecules in a hot substance jiggle faster and move farther apart. The result is **thermal expansion.** Most substances expand when heated and contract when cooled. Sometimes the changes are too small to be noticed, and sometimes not. Telephone wires are longer and sag more on a hot summer day than in winter. Railroad tracks that are laid on cold winter days expand and buckle in the summer (Figure 9.15).

Thermal expansion must be taken into account in structures and devices of all kinds. A dentist uses filling material with the same rate of expansion as teeth. A civil engineer uses reinforcing steel with the same expansion rate as concrete. A long steel bridge usually has one end fixed while the other rests on rockers (Figure 9.16).

FIGURE 9.15 ▶
Thermal expansion. Extreme heat on a July day caused the buckling of these railroad tracks.

◀ **FIGURE 9.16**
One end of the bridge is fixed, but the end shown rides on rockers to allow for thermal expansion.

FIGURE 9.17 ▶
This gap in the roadway of a bridge is called an expansion joint; it allows the bridge to expand and contract. (Was this picture taken on a warm or a cold day?)

Notice also tongue-and-groove gaps called *expansion joints* on bridges (Figure 9.17). We see the effects of expansion all around us.

We can see that different substances expand at different rates with a bimetallic strip (Figure 9.18). This device is made of two strips of different metals welded together, one of brass and the other of iron. When heated, the greater lengthening of the brass bends the strip. This bending may be used to turn a pointer, regulate a valve, or close a switch.

 Brass
Iron

Room temperature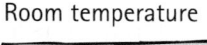

Brass  Iron

FIGURE 9.18 ▲
A bimetallic strip. Brass expands more when heated than iron does and contracts more when cooled. Because of this behavior, the strip bends as shown.

A practical application of a bimetallic strip wrapped into a coil is the thermostat (Figure 9.19). When a room becomes too cold, the coil bends toward the brass side and activates an electrical switch that turns on the heater. When the room gets too warm, the coil bends toward the iron side, which breaks the electrical circuit and turns off the heater. Bimetallic strips are used in oven thermometers, refrigerators, electric toasters, and various other devices.

☑ Liquids expand more than solids with increases in temperature. We notice this when gasoline overflows from a car's tank on a hot day. If the tank and contents expanded at the same rate, no overflow would occur. This is why a gas tank being filled shouldn't be "topped off" on a hot day.

Thermal expansion accounts for the creaky noises often heard in the attics of old houses on cold nights.

To furnace →

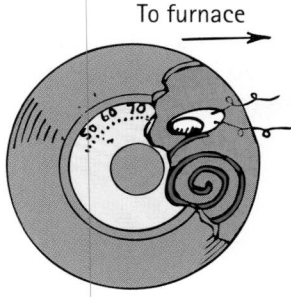

FIGURE 9.19 ▲
A thermostat. When the bimetallic coil expands, the drop of liquid mercury rolls away from the electrical contacts and breaks the electrical circuit. When the coil contracts, a small blob of mercury rolls against the contacts and completes the circuit.

CONCEPT CHECK

1. When you can't loosen a metal lid on a glass jar, how can you use the concept of thermal expansion to rescue the situation?
2. Supersonic aircraft can be 20 cm longer when in flight than when parked on the ground. Offer an explanation.

Check Your Answers

1. Hold the lid of the jar under hot water for a few seconds. The metal should expand more than the glass, making it easier to loosen.
2. At cruising speed (faster than the speed of sound), air friction against the aircraft raises its temperature dramatically, resulting in this significant thermal expansion.

☑ **READING CHECK**

Which expands more with increases in temperature, liquids or solids?

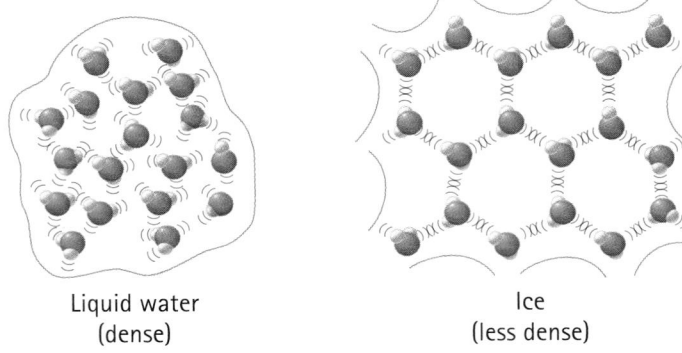

FIGURE 9.20 ▶
Water molecules in a liquid are denser than water molecules frozen in ice, where they have an open crystalline structure.

Liquid water
(dense)

Ice
(less dense)

FIGURE 9.21 ▲
The six-sided structure of snow crystals is a result of the six-sided ice crystals that make them up.

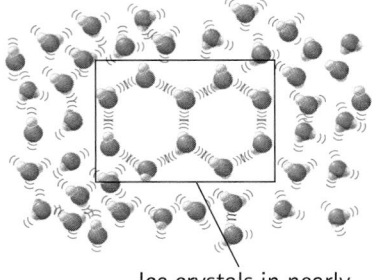

Ice crystals in nearly frozen liquid water

FIGURE 9.22 ▲
Close to 0°C, liquid water contains crystals of ice. The open structure of these crystals increases the volume of water slightly.

Expansion of Water

Water, like most other substances, expands when heated. But interestingly, it doesn't expand in the temperature range between 0°C and 4°C. Something quite fascinating happens in this range. Ice has a crystalline structure, with six-sided open-structured crystals. Water molecules in this open structure occupy a greater volume than they do in the liquid phase (Figure 9.20). This means that ice is less compact (less dense) than water.

When ice melts, not all the six-sided crystals collapse. Some remain in the ice-water mixture, making up a microscopic slush that slightly "bloats" the water—increases its volume slightly. This results in ice water being less dense than slightly warmer water. As the temperature of water near 0°C is increased, more of the remaining ice crystals collapse. This further decreases the volume of the water. This contraction continues only up to 4°C. ✓ That's because two things happen at the same time; volume decreases due to ice crystal collapse, and volume increases due to greater molecular motion. The collapsing effect dominates until the temperature reaches 4°C. After that, expansion overrides contraction because most of the ice crystals have melted. (Figure 9.23).

When ice water freezes to become solid ice, its volume increases tremendously—and its density is much lower. That's why ice floats on water. Like most other substances, solid ice contracts with further cooling.

CONCEPT CHECK

What's inside the open spaces of the water crystals shown in Figures 9.20 and 9.22 and the cartoon to the right? Is it air, water vapor, or nothing?

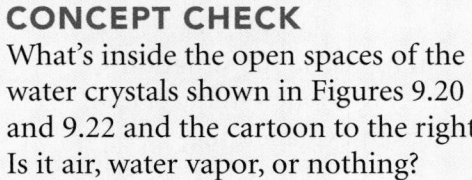

Check Your Answer

There's nothing at all in the open spaces. It's empty space—a void.
If there were air in the spaces, the illustration would have to show air molecules. If there were water vapor in the spaces, H$_2$O molecules would have to be shown.

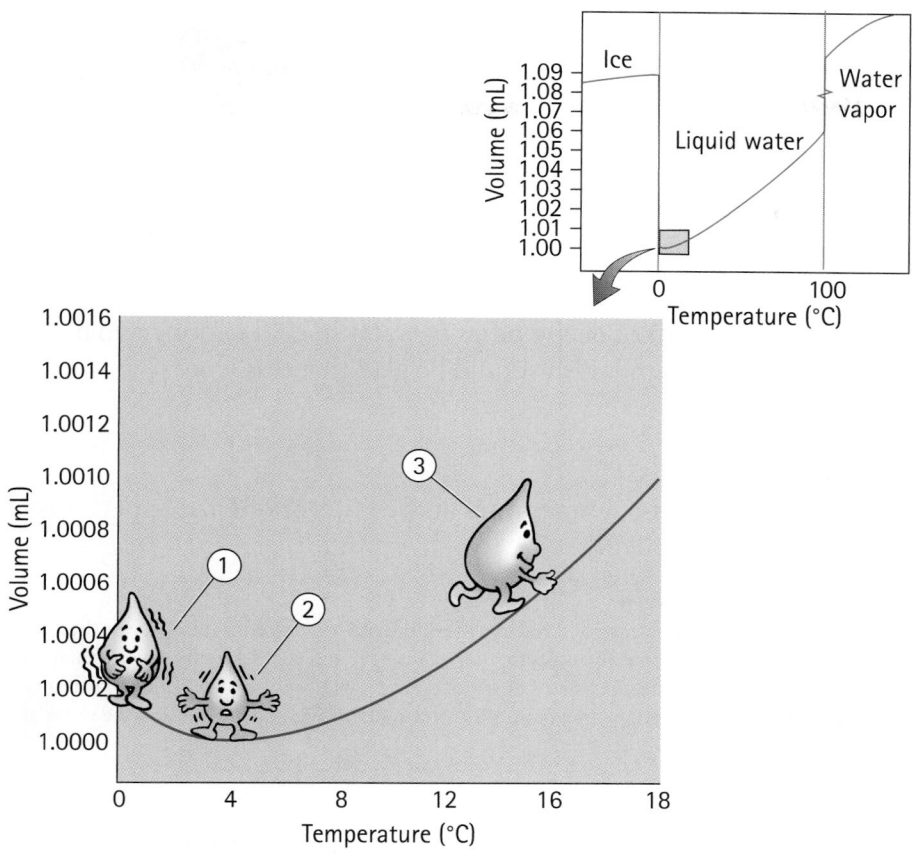

FIGURE 9.23 ▲

Between 0°C and 4°C, the volume of liquid water decreases as temperature increases. Above 4°C, water behaves the way other substances do. Its volume increases as its temperature increases. The volumes shown here are for a 1-gram sample.

This behavior of water is very important in nature. If water were most dense at 0°C, it would settle to the bottom of a pond or lake. Water at 0°C is less dense and "floats" at the surface. That's why ice forms at the surface. So a pond freezes from the surface downward. In a cold winter, the ice will be thicker than in a milder winter. Water at the bottom of an ice-covered pond remains at 4°C, relatively warm for organisms that live there.

Interestingly, very deep bodies of water are not ice-covered even in the coldest of winters. This is because all the water must be cooled to 4°C before lower temperatures can be reached. For deep water, the winter is not long enough to reduce an entire pond to 4°C. Any 4°C water lies at

FIGURE 9.24 ▶
As water cools, it sinks until the entire pond is at 4°C. Then, as water at the surface is cooled further, it floats on top and can freeze. Once ice is formed, temperatures lower than 4°C can extend down into the pond.

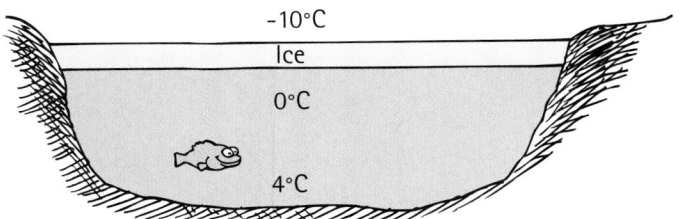

Because water is most dense at 4°C, colder water rises and freezes on the surface. This means that fish remain in relative warmth!

✔ **READING CHECK**
What two processes occur for water as its temperature increases from 0°C?

PhysicsPlace.com
Videos
The Secret to Walking on Hot Coals; Air is a Poor Conductor

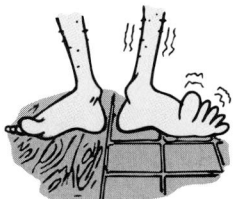

the bottom. Because of water's high specific heat and poor ability to conduct heat, the bottom of deep bodies of water in cold regions remains at a constant 4°C year round. Fish should be glad that this is so.

CONCEPT CHECK
What was the precise temperature at the bottom of Lake Superior on New Year's Eve in 2000?

Check Your Answer
The temperature at the bottom of any body of water with 4°C water in it is 4°C, for the same reason that rocks are at the bottom. Both 4°C water and rocks are more dense than water at any other temperature. Water does not conduct heat well, so if the body of water is deep and in a region of long winters and short summers, as with Lake Superior, the water at the bottom is 4°C year round.

There are three main ways that heat is conducted from one substance to another or from one place to another. These are *conduction, convection,* and *radiation.* We investigate each in turn.

9.7 Conduction—Heat Transfer via Particle Collision

If you hold one end of an iron nail in a flame, the nail quickly becomes too hot to hold. If you hold one end of a short glass rod in a flame, the rod takes much longer before it becomes too hot to hold. In both cases, heat at the hot end travels along the entire length. This method of heat transfer is called **conduction.** Thermal conduction occurs by collisions between particles and their immediate neighbors. Because the heat travels quickly through the nail, we say that it is a good *conductor* of heat. Materials that are poor conductors are called *insulators.*

Solids (such as metals) whose atoms or molecules have loosely held electrons are good conductors of heat. These mobile electrons move quickly and transfer energy to other electrons, which migrate quickly throughout the solid. Poor conductors (such as glass, wool, wood, paper, cork, and plastic foam) are made up of molecules that hold tightly to

◀ **FIGURE 9.25**
The tile floor feels colder than the wooden floor, even though both are at the same temperature. Tile is a better heat conductor than wood, and it more quickly conducts internal energy from your feet.

their electrons. In these materials—insulators—molecules vibrate in place and don't transfer energy well. Since the electrons in insulators are not mobile, energy is transferred much more slowly.

Wood is a good insulator and is used for cookware handles. Even when a pot is hot, you can quickly grasp the wooden handle with your bare hand without harm. An iron handle of the same temperature would surely burn your hand. Wood is a good insulator even when it's red hot. This explains how firewalking co-author John Suchocki can walk barefoot on red-hot wooden coals without burning his feet (see the photo at the beginning of this chapter). (CAUTION: Don't try this on your own; even experienced firewalkers sometimes receive bad burns when conditions aren't just right.) The main factor here is the poor conductivity of wood—even red-hot wood. Although its temperature is high, very little heat is conducted to the feet. A firewalker must be careful that no iron nails or other good conductors are among the hot coals. Ouch!

Air is a very poor conductor. That's why you can briefly put your hand in a hot pizza oven without harm. The hot air doesn't conduct heat well. But don't touch the metal in the hot oven. Ouch again! ✓ The good insulating properties of such things as wool, fur, and feathers are largely due to the air spaces they contain. Porous substances are also good insulators because of their many small air spaces. Be glad that air is a poor conductor; if it weren't, you'd feel quite chilly on a 20°C (68°F) day!

Snow is a poor conductor of heat. Snowflakes are formed of crystals that trap air and provide insulation. That's why a blanket of snow keeps the ground warm in winter. Animals in the forest find shelter from the cold in snow banks and in holes in the snow. The snow doesn't provide them with thermal energy—it simply slows down the loss of body heat generated by the animals. Then there are the Arctic dwellings, igloos, that are shielded from the cold by their snow covering.

Homes are insulated with rock wool or fiberglass. Interestingly, insulation doesn't prevent the flow of heat. Insulation simply slows down the rate at which heat flows. Even a well-insulated warm home gradually cools. Insulation merely delays the rate at which heat conducts from a warmer region to a cooler one. In winter, we wish to slow conduction from inside to outside. But on hot summer days we wish to slow down conduction in the other direction, from outside to inside. Insulation slows conduction in either direction.

What can be both good and poor at the same time? Answer: Any good insulator is a poor conductor. Or any good conductor is a poor insulator.

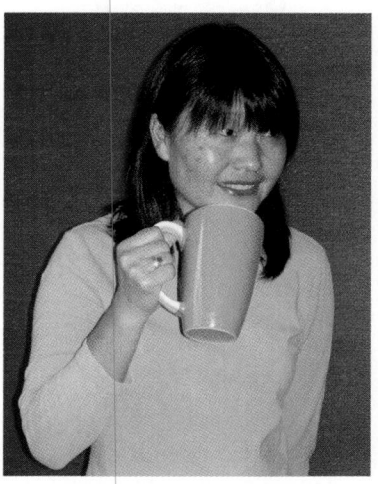
FIGURE 9.26 ▲
Lil can comfortably hold the handle of a cup of hot chocolate because the cup is a good insulator.

Insulators can only slow the flow of heat. They cannot prevent or stop heat flow.

◄ **FIGURE 9.27**
Snow patterns on the roof of a house show areas of conduction and insulation. Bare parts show where heat from inside has conducted through the roof and melted the snow.

What accounts for the good insulating properties of wool, fur, feathers, and snow?

CONCEPT CHECK

1. In desert regions that are hot in the day and cold at night, the walls of houses are often made of mud. Why is it important that the mud walls be thick?
2. Wood is a better insulator than glass. Yet fiberglass is commonly used to insulate buildings. Why?

Check Your Answers

1. A wall of correct thickness keeps the house warm at night by slowing heat conduction from warmer inside to cooler outside. In the daytime, conduction is slowed from warmer outside to cooler inside. Such a wall has "thermal inertia." Thick walls hold up the roof better, too!
2. Fiberglass is a good insulator, many times better than glass, because of the air that is trapped among its fibers.

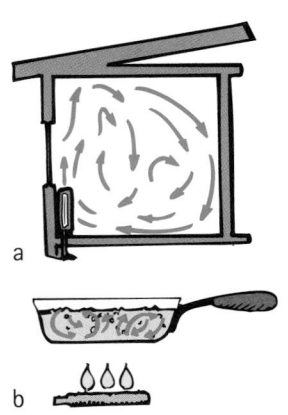

FIGURE 9.28 ▲
Convection currents in a gas (air) and a liquid.

fyi

- Convection ovens are simply ovens with a fan inside. Cooking is speeded up by the circulation of heated air.

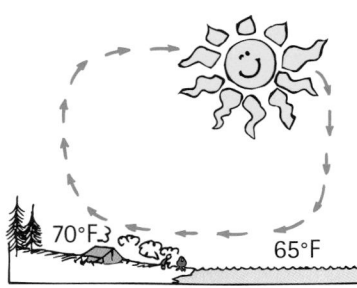

9.8 Convection—Heat Transfer via Movements of Fluid

Liquids and gases transfer heat mainly by **convection**, which is transfer by motion of a fluid—by currents. Convection occurs in all fluids. Whether we heat water in a pot or warm air in a room, the process is the same (Figure 9.28). ✔ As the fluid is heated from below, the molecules at the bottom begin moving faster. They spread apart and become less dense. Then they are buoyed upward. Denser, cooler fluid migrates to the bottom. In this way, convection currents keep the fluid stirred up. Warmer fluid moves away from the heat source and cooler fluid moves toward the heat source and is warmed.

Warm air expands, becomes less dense, and rises in the cooler surrounding air—like a balloon buoyed upward. When the rising air reaches an altitude where air density is the same, it no longer rises. We see this occurring when smoke from a fire rises and then settles off as it cools and its density matches that of the surrounding air. As air expands, it cools. Cooling by expansion is the opposite of what occurs when air is compressed. If you've ever compressed air with a tire pump, you probably noticed that both air and pump became quite hot.

Convection currents stir the atmosphere and produce winds. Some parts of the Earth's surface absorb heat from the Sun more readily than others. This results in uneven heating of the air near the ground. We see this at the seashore, as Figure 9.29 shows. In daytime, the ground warms up more than the water, and air above the ground that is warmed then rises. It is replaced by cooler air that moves in from above the water. The

◄ **FIGURE 9.29**
Convection currents produced by unequal heating of land and water. During the day, warm air above the land rises, and cooler air over the water moves in to replace it. At night, the direction of air flow is reversed because now the water is warmer than the land.

BREATH-ON EXPLORATION

Do the following experiment right now. With your mouth open, blow on your hand. Your breath is warm. Now repeat, but this time pucker your lips to make a small hole so your breath expands as it leaves your mouth. Note that your breath is appreciably cooler! Expanding air cools.

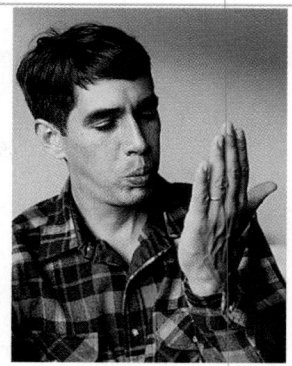

result is a sea breeze. At night, the process reverses because the shore cools off more quickly than the water, and then the warmer air is over the sea. Build a fire on the beach and you'll notice that the smoke sweeps inward during the day and seaward at night.

9.9 Radiation—Heat Transfer via Radiant Energy

Thermal energy from the Sun passes through space and then through the atmosphere before it warms the Earth's surface. This heat transfer is not by conduction and convection, for there's no material between the Sun and Earth. Heat transfer must be by some other way—by **radiation.*** The energy transferred this way is called *radiant energy.*

✓ Radiant energy is in the form of *electromagnetic waves.* It includes a wide span of waves that begin with radio waves and infrared waves, continue through to visible-light waves, and end with gamma rays. The lengths of waves differ. The waves of infrared (below-the-red) waves, for instance, are longer than those of visible-light waves. The longest visible wavelengths are for red light, and the shortest are for violet light. Shorter

> ✔ **READING CHECK**
>
> Why does a warm fluid, such as heated air, rise?

fyi

- Opening a refrigerator door lets warm air in, which then takes energy to cool. The more empty your frig, the more cold air is swapped with warm air. So keep your frig full for lower operating costs—especially if you're an excessive open-and-close-the-door type.

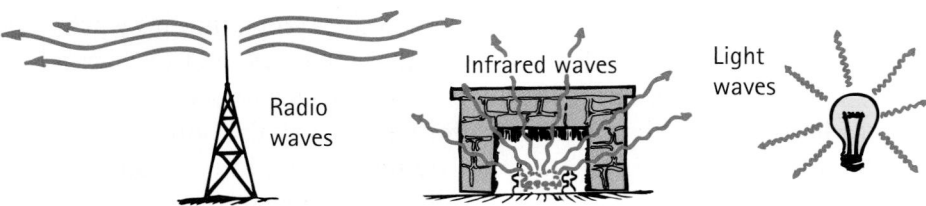

FIGURE 9.30 ▲
Types of radiant energy (electromagnetic waves).

> Radiation by Earth is *terrestrial radiation.* Radiation by the Sun is *solar radiation.* Both are regions in the electromagnetic spectrum. (What do you call radiation from that special someone?)

---

* Do not confuse radiation with radioactivity—reactions that involve the atomic nucleus and are characteristic of nuclear power plants and the like. Radiation here is electromagnetic radiation—"heat" waves of low-frequency light—which we will study in detail in Parts 3 and 4.

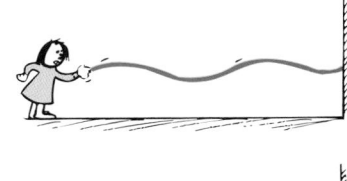

◀ **FIGURE 9.31**
A wave of long wavelength is produced when the rope is shaken gently (at a low frequency). When shaken more vigorously (high frequency), a wave of shorter wavelength is produced.

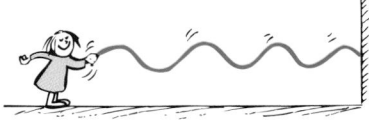

waves can't be seen by the eye. We'll treat waves further in Chapters 12 and 13, and electromagnetic waves in Chapters 11 and 13.

The wavelength of radiation is related to the frequency of radiation. Frequency is the rate of vibration of a wave source. The girl in Figure 9.31 shakes a rope at a low frequency (top) and a higher frequency (bottom). Note that shaking at a low frequency produces a long, lazy wave, and the higher-frequency shake produces shorter waves. Likewise with electromagnetic waves. We will see in Chapter 13 that vibrating electrons emit electromagnetic waves. High-frequency vibrations produce short waves and low-frequency vibrations produce longer waves.

✔ **READING CHECK**

In what form does radiant energy travel?

(a) Cool

(b) Medium

(c) Hot

FIGURE 9.32 ▲
(a) A low-temperature (cool) source emits primarily low-frequency, long-wavelength waves. (b) A medium-temperature source emits primarily medium-frequency, long-wavelength waves. (c) A high-temperature source (hot) emits primarily high-frequency, short-wavelength waves.

Emission of Radiant Energy

All substances at any temperature above absolute zero emit radiant energy. The average frequency f of the radiant energy is directly proportional to the absolute temperature T of the emitter:

$$f \sim T$$

The Sun has a very high temperature and therefore emits radiant energy at a high frequency—high enough to stimulate our sense of sight. The Earth, in comparison, is relatively cool. So the radiant energy it emits has a frequency lower than that of visible light. The radiation emitted by the Earth is in the form of infrared waves. ✔ Radiant energy emitted by the Earth is called **terrestrial radiation.**

Most people know that the Sun glows and emits radiant energy. And many educated people know that the source of the Sun's radiant energy involves nuclear reactions in its deep interior. However, relatively few people know that the Earth also "glows" and emits radiant energy of the same nature. If you visit the depths of any mine, you'll find it's warm down there—year-round. Radioactivity in the Earth's interior warms the Earth. Much of this heat conducts to the surface to become terrestrial radiation. So radiant energy is emitted by both the Sun and the Earth and differs only in the range of frequencies and the amount. In Chapter 31 we'll learn how the atmosphere is transparent to the

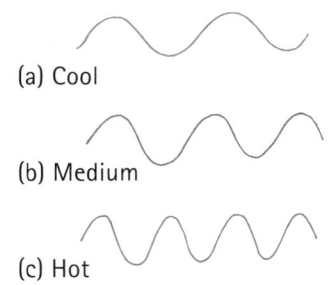

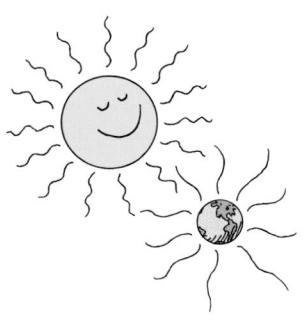

◀ **FIGURE 9.33**
Both the Sun and the Earth emit the same kind of radiant energy. The Sun's glow is visible to the eye; the Earth's glow consists of longer waves and so is not visible to the eye.

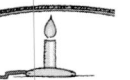

high-frequency solar radiation but opaque to much of the lower-frequency terrestrial radiation. This produces a "greenhouse effect" and plays a role in global warming.

All objects—you, your teacher, and everything in your surroundings—continually emit radiant energy in a mixture of frequencies (because temperature corresponds to a mixture of molecular kinetic energies). Objects of everyday temperatures mostly emit low-frequency infrared waves. When the higher-frequency infrared waves are absorbed by your skin, you feel the sensation of heat. So it is common to refer to infrared radiation as *heat radiation.*

Heat radiation underlies infrared thermometers. You simply point the thermometer at something whose temperature you want, press a button, and a digital temperature reading appears. The radiation emitted by the object in question provides the reading. Typical classroom infrared thermometers operate in the range of about −30°C to 200°C.

Common hotter sources that give the sensation of heat are the burning embers in a fireplace, a lamp filament, and the Sun. All of these emit both infrared radiation and visible light. When this radiant energy falls on other objects, it is partly reflected and partly absorbed. The part that is absorbed increases the thermal energy of the objects.

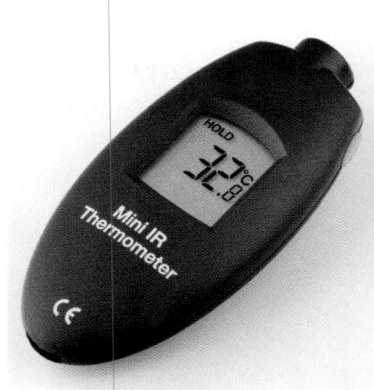

FIGURE 9.34 ▲
An infrared thermometer measures the infrared radiant energy emitted by a body and converts it to temperature.

CONCEPT CHECK

Which of the following do *not* give off radiant energy? (a) The Sun; (b) Lava from a volcano; (c) Red-hot coals; (d) This book that you're reading.

Check Your Answer

Did you answer (d), the book? Sorry, wrong answer. None of the choices is correct. The book, like the other things listed, has temperature—though not as high as the others listed. By the rule $f \sim T$, it therefore emits radiation. Since its temperature is low, the frequency of radiation is also low. Everything with any temperature above absolute zero emits electromagnetic radiation. That's right—*everything!*

Everything around you both radiates and absorbs energy continuously!

✔ **READING CHECK**

What is terrestrial radiation?

Absorption of Radiant Energy

If everything is emitting energy, why doesn't everything finally run out of it? The answer is, all things also *absorb* energy. Good emitters of radiant energy are also good absorbers; poor emitters are poor absorbers. For example, a radio antenna constructed to be a strong emitter of radio waves is also, by its very design, a strong receiver (absorber) of them. A poorly designed transmitting antenna is also a weak receiver. An object that absorbs radiant energy looks dark. If it absorbs all the radiant energy on it, it looks perfectly black.

✔ Every surface, hot or cold, both absorbs and emits radiant energy. If the surface is hotter than its surroundings, the surface will be a net emitter and will cool. If it's colder than its surroundings, it will be a net absorber and will become warmer.

A hot pizza put outside on a winter day is a net emitter. The same pizza placed in a hotter oven is a net absorber.

READING CHECK

Does the surface of a substance absorb, or does it emit, radiant energy?

CONCEPT CHECK
If a good absorber of radiant energy were a poor emitter (instead of a good emitter), how would its temperature compare with the surroundings?

Check Your Answer
There would be a net absorption of radiant energy and the temperature would be above that of the surroundings. Things around us approach a common temperature because good absorbers are, by their very nature, also good emitters.

fyi

- Emission and absorption in the visible part of the spectrum are affected by color. But not so in the infrared part of the spectrum, where surface texture has more effect. In the infrared, a dull finish emits/absorbs better than a polished one, whatever the color.

Reflection of Radiant Energy

Radiant energy can be reflected. There are no perfect reflectors, so some absorption always occurs. When you look at the open ends of pipes in a stack, the holes appear black. Look at open doorways or windows of distant houses in the daytime, and they, too, look black. ✔ Openings appear black because the light that enters them is reflected back and forth on the inside walls many times and is partly absorbed at each reflection. As a result, very little or none of the light returns back out the opening (Figure 9.35).

The pupil of your eye is another example. The pupil is a hole in your iris that allows light to enter with no reflection. That's why it appears black. (An exception occurs in flash photography when pupils appear red or pink, which occurs when very bright light is reflected off the eye's inner pink surface and back through the pupil.)

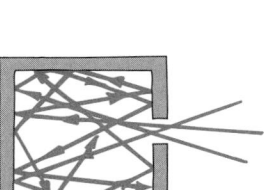

FIGURE 9.35 ▲
Radiation that enters the opening has little chance of leaving because most of it is absorbed. For this reason, the opening to any cavity looks black to us.

FIGURE 9.36 ▶
The hole looks perfectly black and indicates a black interior, when in fact the interior has been painted a bright white.

READING CHECK

Why do window openings in the daytime look black?

CONCEPT CHECK
Which would be more effective in heating the air in a room, a heating radiator painted black or one painted silver?

9.10 Energy Changes with Changes of Phase

As mentioned earlier, matter exists in solid, liquid, gas, and plasma phases of matter. When the phase of a substance changes to another phase (ice melting to water, for example), a transfer of thermal energy occurs.

Thermal energy must be added to a solid to transform it to a liquid. For example, energy is needed to melt ice to form water. Energy is needed to change a liquid to a gas; boiling water to become steam, for example. Likewise, energy is needed to transform a gas into plasma. The direction of energy changes is shown in Figure 9.37.

Energy is taken away from a substance when the phase change is in the opposite direction. When steam turns into water, thermal energy is taken from the steam. That's why steam burns are more damaging than burns from water at the same temperature. The energy that was in the steam is now on your skin. Ouch! And to change liquid water to ice, energy has to be taken from the water.

✓ Whenever a substance changes phase, energy is either given to the substance or taken from it.

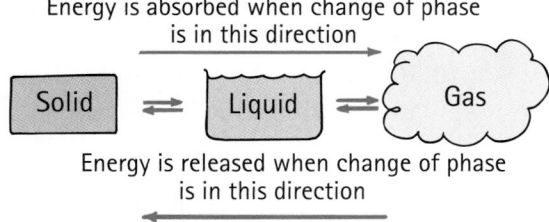

◀ **FIGURE 9.37**
Energy changes with change of phase.

CONCEPT CHECK

1. Is energy taken away or given to ice when part of it melts?
2. Is energy taken away or given to water when some of it evaporates?
3. Is energy taken away or given to steam when part of it turns to water?

Check Your Answers
1. The phase change is from solid to liquid, so energy is given to the ice.
2. The phase change is from liquid to gas, so energy is given to the water.
3. The phase change is from gas to liquid, so energy is taken from the steam.

✓ **READING CHECK**

What role does energy play when a substance changes from one phase to another?

9 CHAPTER REVIEW

KEY TERMS

Absolute zero The theoretical temperature at which a substance possesses no kinetic energy.

Conduction The transfer of internal energy by molecular and electronic collisions within a substance (especially a solid).

Convection The transfer of internal energy in a gas or liquid by means of currents in the heated fluid. The fluid flows, carrying energy with it.

Heat The thermal energy that flows from a substance of higher temperature to a substance of lower temperature, commonly measured in calories or joules.

Radiation The transfer of energy by means of electromagnetic waves.

Specific heat capacity The quantity of heat required to raise the temperature per unit mass of a substance by 1 degree Celsius.

Temperature A measure of the hotness or coldness of substances, related to the average kinetic energy per molecule in a substance, measured in degrees Celsius, or in degrees Fahrenheit, or in kelvins.

Terrestrial radiation The radiant energy emitted by Earth.

Thermal energy The total energy (kinetic plus potential) of the submicroscopic particles that make up a substance.

Thermal expansion The expansion of a substance due to increased molecular motion in that substance.

REVIEW QUESTIONS

Thermal Energy—The Total Energy in a Substance

1. Why does a penny become warmer when struck by a hammer?

Temperature—Average Kinetic Energy per Molecule in a Substance

2. What are the temperatures for freezing water on the Celsius and Fahrenheit scales? For boiling water at sea level?
3. What is meant by the statement "a thermometer measures its own temperature"?

Absolute Zero—Nature's Lowest Possible Temperature

4. What is the temperature of melting ice on the Kelvin scale? Of boiling water at atmospheric pressure?

Heat Is the Movement of Thermal Energy

5. When you touch a cold surface, does "coldness" travel from the surface to your hand or does thermal energy travel from your hand to the cold surface? Explain.
6. Distinguish between temperature and heat.
7. Is cold the opposite of thermal energy or the lack of it?
8. Distinguish between a joule and a calorie.

Specific Heat Capacity—A Measure of Thermal Inertia

9. Does a substance that heats up quickly have a high or a low specific heat capacity?
10. Why is the West Coast of the United States warmer in winter than the East Coast?

Thermal Expansion

11. Why does a bimetallic strip bend with changes in temperature?
12. When the temperature of ice-cold water is increased slightly, does it undergo a net expansion or net contraction?
13. At what temperature do the combined effects of contraction and expansion produce the smallest volume for water?

Conduction—Heat Transfer via Particle Collision

14. What is the explanation for a barefoot firewalker being able to walk safely on red-hot wooden coals?

15. Does a good insulator prevent heat from getting through it, or does it simply slow its passage?

Convection—Heat Transfer via Movements of Fluid

16. What happens to the temperature of air when it expands?

17. Why does the direction of coastal winds change from day to night?

Radiation—Heat Transfer via Radiant Energy

18. How does the frequency of radiant energy relate to the absolute temperature of the radiating source?

19. What is terrestrial radiation? How does it differ from solar radiation?

Energy Changes with Changes of Phase

20. Is energy added, or is it released, when boiling water is changed to steam?

THINK AND DO

1. Write a letter to your grandparents and tell them how you're learning to see the connections in nature. Also tell them how you're learning to distinguish between closely related ideas. Use temperature and heat as an example.

2. Hold the bottom end of a test tube full of cold water in your hand. Heat the top part in a flame until the water boils. The fact that you can still hold the bottom shows that water is a poor conductor of heat. This is even more dramatic when you wedge chunks of ice at the bottom; then the water above can be brought to a boil without melting the ice. Try it and see.

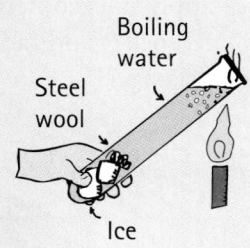
Boiling water
Steel wool
Ice

3. Wrap a piece of paper around a thick metal bar and place it in a flame. Note that the paper will not catch fire. Can you figure out why? (Paper generally will not ignite until its temperature reaches 233°C.)

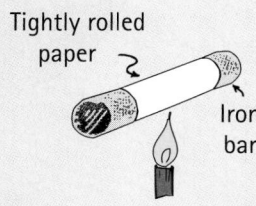
Tightly rolled paper
Iron bar

THINK AND EXPLAIN

1. In your room there are things such as tables, chairs, other people, and so forth. Which of these things has a temperature (1) lower than, (2) greater than, and (3) equal to the temperature of the air?

2. Which is greater, an increase in temperature of 1 Celsius degree or an increase of 1 Fahrenheit degree? 1 Celsius degree or an increase of 1 kelvin?

3. Which has the greater amount of thermal energy, an iceberg or a cup of hot tea? Explain.

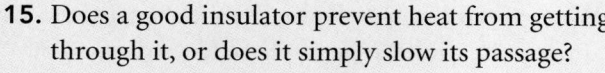

4. Adding the same amount of heat to two different objects does not necessarily produce the same increase in temperature. Why not?

5. Why will a watermelon stay cool for a longer time than sandwiches when both are removed from a cooler on a hot day?

6. Iceland, so named to discourage conquest by expanding empires, is not at all ice-covered like Greenland and parts of Siberia, even though it is nearly on the Arctic Circle. The average winter temperature of Iceland is considerably higher than regions at the same latitude in eastern Greenland and central Siberia. Why is this so?

7. If the winds at the latitude of San Francisco and Washington, D.C., were from the east rather than from the west, why might San Francisco be able to grow only cherry trees and Washington, D.C., only palm trees?

8. An old remedy for a pair of nested drinking glasses that stick together is to run water at different temperatures into the inner glass and over the surface of the outer glass. Which water should be hot, and which cold?

9. A metal ball is just able to pass through a metal ring. When the ball is heated, however, it will not pass through the ring. What would happen if the

ring, rather than the ball, were heated? Does the size of the hole increase, stay the same, or decrease?

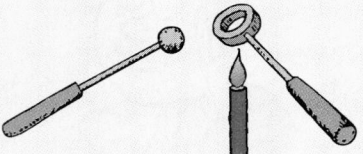

10. Suppose you cut a small gap in a metal ring. If you heat the ring, will the gap become wider or narrower?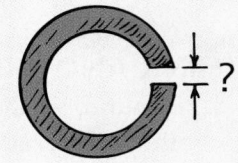
11. State whether water at the following temperatures will expand or contract when warmed a little: 0°C; 4°C; 6°C.
12. Why is it important to protect water pipes so they don't freeze?
13. If you hold one end of a metal nail against a piece of ice, the end in your hand soon becomes cold. Does cold flow from the ice to your hand? Explain.
14. Many tongues have been injured by licking a piece of metal on a very cold day. Why would no harm result if a piece of wood were licked on the same day even when it has the same temperature?
15. All objects continuously emit radiant energy. Why, then, doesn't the temperature of all objects continuously decrease?
16. All objects continuously absorb energy from their surroundings. Why, then, doesn't the temperature of all objects continuously increase?
17. What determines whether an object is a net emitter or a net absorber of radiant energy?
18. You can comfortably hold your fingers close beside a candle flame, but not very close above the flame. Why?
19. In a mixture of hydrogen and oxygen gases at the same temperature, which molecules move faster? Defend your answer, and compare it to the fact that if a mouse and a cat run with the same kinetic energy, the mouse runs faster.

20. Water vapor changes phase when it changes into snow. Does this change of phase tend to warm or to cool the surrounding air?

THINK AND SOLVE

1. What would be the final temperature of a mixture of 50 gram of 20°C water and 50 g of 40°C water?
2. (Challenge!) Consider a 40,000-km steel pipe that forms a ring to fit snugly all around the circumference of the world. Suppose people along its length breathe on it so as to raise its temperature 1°C. The pipe gets longer. It also is no longer snug. How high does it stand above ground level? (To simplify, consider only the expansion of its radial distance from the center of the Earth, and apply the geometry formula that relates circumference C and radius r: $C = 2\pi r$. The result is surprising!)

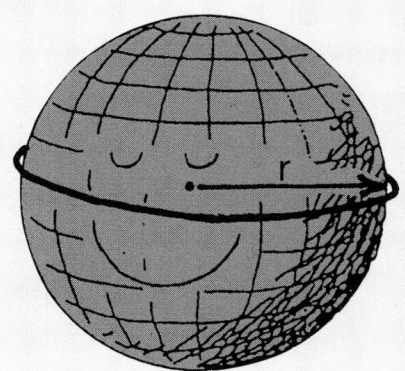

READINESS ASSURANCE TEST (RAT)

If you have a good handle on this chapter—if you really do—then you should be able to score 7 out of 10 on this RAT. Check your answers with your teacher. If you score less than 7, study further before moving on.

Choose the best *answer to each of the following.*

1. Temperature is generally proportional to a substance's
   (a) thermal energy.
   (b) vibrational kinetic energy.
   (c) average translational kinetic energy.
   (d) rotational kinetic energy.

2. When three-quarters of a container of hot water is poured into a second empty container, the second container then has
   (a) 3/4 the thermal energy.
   (b) 3/4 the original volume of water.
   (c) the same temperature.
   (d) All of these.

3. Water freezes at a temperature of
   (a) 0°C.
   (b) 273 K.
   (c) Both of these.
   (d) None of these.

4. *Heat* is simply another word for
   (a) temperature.
   (b) thermal energy.
   (c) thermal energy that flows from hot to cold.
   (d) All of these.

5. To say that water has a high specific heat capacity is to say that water
   (a) requires a lot of energy in order to increase in temperature.
   (b) gives off a lot of energy in cooling.
   (c) absorbs a lot of energy for a small increase in temperature.
   (d) All of these.

6. A bimetallic strip used in thermostats relies on the fact that different metals have different
   (a) specific heat capacities.
   (b) thermal energies at different temperatures.
   (c) rates of thermal expansion.
   (d) All of these.

7. The density of water at 4°C will slightly increase when it is
   (a) cooled.
   (b) warmed.
   (c) Both.
   (d) Neither.

8. A firewalker walking barefoot across hot wooden coals depends on wood's
   (a) good conduction.
   (b) poor conduction.
   (c) low specific heat capacity.
   (d) low radiation.

9. Thermal convection is linked mostly to
   (a) radiant energy.
   (b) fluids.
   (c) insulators.
   (d) All of these.

10. A high-temperature source radiates relatively
   (a) short wavelengths.
   (b) long wavelengths.
   (c) low frequencies of radiation.
   (d) None of these.

10
ELECTRICITY

THE MAIN
IDEA

The amount of electric current in an electrical device depends on the voltage that "pushes it" and the resistance that "resists it."

Why do we sometimes get a shock when we scuff our shoes on the rug? Why does a balloon rubbed on our hair stick to a wall? What causes shock—electric current or electric voltage? And what's the difference between current and voltage? What's the difference between direct current and alternating current? We'll discuss the answers to these questions and more in this chapter.

Electricity is a part of just about everything around us. It's in the lightning from the sky and in what holds atoms together. Harnessing electricity has enormously changed the world. It's worth serious study. Let's begin with the concept of electric charge.

Explore!

What Does It Take to Light a Lightbulb?

1. Try to light a lightbulb with just a battery and two pieces of wire.
2. Now try to get the same result with different arrangements of the wires, battery, and lightbulb.

Analyze and Conclude

1. **Observing** Describe both successful and unsuccessful attempts to light the lightbulb.
2. **Predicting** How many possible arrangements of the wires, battery, and lightbulb will produce a lit bulb?
3. **Making Generalizations** What conditions are necessary in order for the bulb to light?

10.1 Electric Charge Is a Basic Characteristic of Matter

When you rub an inflated rubber balloon on your hair, you charge the balloon. In charging it, you scrape electrons off the atoms of your hair onto the balloon. The rubber apparently has more "grab" on electrons than your hair does. If you do the same with two balloons, you'll find that when you bring them close, they repel.

The first rule of electricity is: Like charges repel one another. For historical reasons we say that the charge on an electron is negative. Because there are extra electrons on the balloons rubbed with hair, the balloons are negatively charged. A pair of negatively charged balloons, or anything else negatively charged, repels. Again, for emphasis:

Rule 1: Like charges repel one another.

If you instead charge one balloon by rubbing it with a piece of kitchen plastic wrap, the plastic grabs electrons from the balloon. The balloon then has an opposite charge to the one rubbed on your hair. When you bring the two oppositely charged balloons together, they attract. Here's the second rule of electricity:

Rule 2: Unlike charges attract one another.

In Figure 10.1, we see a pair of like charges attracting and a pair of unlike charges repelling. In Figure 10.2, we see a simple model of an atom. Particles called *protons* in the nucleus carry the positive charge. The protons attract the whirling electrons and hold them in orbit.

To understand electric charge, we look at a preview of atomic parts—what we'll learn more about in Part 5. Here are some basic facts about atoms:

1. Every atom has a positively charged *nucleus* surrounded by negatively charged electrons.
2. The electrons of all atoms are identical. Each has the same quantity of negative charge and the same mass.
3. Protons and neutrons make up the nucleus. (The common form of hydrogen has no neutron and is the only exception.) Protons are about 1800 times more massive than electrons but carry an amount of positive charge equal to the negative charge of electrons. Neutrons have slightly more mass than protons and have no charge.

PhysicsPlace.com

Tutorial
Electrostatics

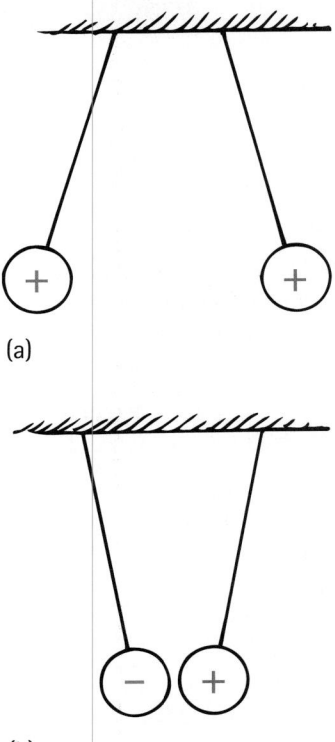

(a)

(b)

FIGURE 10.1 ▲
(a) Like charges repel. (b) Unlike charges attract.

Negative and *positive* are just the **names** given to opposite charges. The names chosen could just as well have been "east and west" or "up and down" or "Mary and Larry."

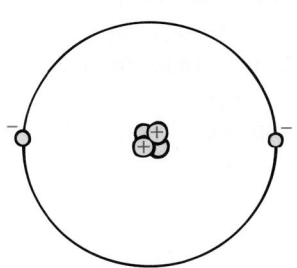

◀ **FIGURE 10.2**
Model of a helium atom. The atomic nucleus is made up of two protons and two neutrons. The positively charged protons attract two negative electrons. What is the net charge of this atom?

◀ **FIGURE 10.3**
Electrons are transferred from the fur to the rod. The rod is then negatively charged. Is the fur charged? How much compared to the rod? Positively or negatively?

An atom in its normal state has the same number of electrons as protons. When an atom loses one or more electrons, it has a positive net charge. When it gains one or more electrons, it has a negative net charge. A charged atom is called an *ion*. A *positive ion* has a net positive charge. A *negative ion,* with one or more extra electrons, has a net negative charge.

It is important to note that ✓ when we charge something, no electrons are created or destroyed. They are simply transferred from one material to another. Charge is *conserved.* In every event, whether on a large scale or small scale, the principle of *conservation of charge* has always proved true. No case of the creation or destruction of net electric charge has ever been found. Conservation of charge ranks with conservation of energy and momentum as a significant fundamental principle in physics.

When you charge the balloon negatively, you also charge your hair positively.

A flow of electric charge is *current*. Electric charge at rest is *static charge.*

CONCEPT CHECK
If you walk across a rug and scuff electrons from your feet, are you negatively or positively charged?

Check Your Answer
You have fewer electrons after you scuff your feet, and so you are positively charged (and the rug is negatively charged).

✓ **READING CHECK**
What does it mean to say that charge is conserved?

LINK TO ELECTRONICS TECHNOLOGY: ELECTROSTATIC CHARGE

Electric charge can be dangerous. Two hundred years ago, young boys called *powder monkeys* ran below the decks of warships to bring sacks of black gunpowder to the cannons above. It was ship law that this task be done barefoot. Why? Because it was important that no static charge build up on the powder on their bodies as they ran to and fro. Bare feet scuffed the decks much less than shoes and assured no charge buildup that might produce an igniting spark and an explosion.

Static charge is a danger in many industries today.

Not because of explosions, but because delicate electronic circuits may be destroyed by static charge. Some sensitive circuit components can be "fried" by static electric sparks. Electronics technicians often wear clothing of special fabrics with ground wires between their sleeves and their socks. Some wear special wrist bands that are connected to a grounded surface to prevent static charge buildup—when moving a chair, for example. The smaller the electronic circuit, the more hazardous are sparks that may short-circuit their elements.

10.2 Coulomb's Law—The Force Between Charged Particles

The electrical force has a pattern much like gravitational force. It depends on the quantity of charge and is inversely proportional to the square of the distance between charged particles. This relationship was discovered by Charles Coulomb in the 18th century and so is called **Coulomb's law.** ✔ It states that the force between two charged particles varies directly as the product of their charges and inversely as the square of the separation distance. The force acts along a straight line between the particles. Coulomb's law can be expressed as

$$F = k\frac{q_1 q_2}{d^2}$$

where k is the proportionality constant, q_1 represents the quantity of charge of one particle, q_2 represents the quantity of charge of the other particle, and d is the distance between the charged particles.

The unit of charge is the **coulomb,** abbreviated C. It turns out that a charge of 1 C is the charge on 6.25 billion billion electrons. This might seem like a great number of electrons, but it represents only the amount of charge that passes through a common 100-watt lightbulb in little more than a second.

The proportionality constant k in Coulomb's law is similar to G in Newton's law of gravity. Instead of being a very small number like G, k is a very large number, approximately

$$k = 9{,}000{,}000{,}000 \; \text{N·m}^2/\text{C}^2$$

In scientific notation, $k = 9 \times 10^9 \; \text{N·m}^2/\text{C}^2$. The unit $\text{N·m}^2/\text{C}^2$ is not important to learn here. It simply converts the right-hand side of the equation to the unit of force, the newton (N). What *is* important is the large magnitude of k. If, for example, a pair of like-charged particles each carrying a charge of 1 coulomb were 1 meter apart, the force of repulsion between them would be 9 billion newtons. That would be about 10 times the weight of a battleship! Obviously, such amounts of net charge do not usually exist in our everyday environment. If such charges were common, we'd see things attracting and repelling quite often. Like charges repel and separate from one another before much charge can build up. The electrical force between even slightly charged objects greatly overwhelms the gravitational force between them.

So Newton's law of gravitation for masses is similar to Coulomb's law for electrically charged objects. The most important difference between gravitational and electrical forces is that electrical forces may be either attractive or repulsive, whereas gravitational forces are only attractive. Coulomb's law underlies the bonding forces between molecules that will be covered in chemistry (Part 3).

PhysicsPlace.com

Tutorial
Electrostatics

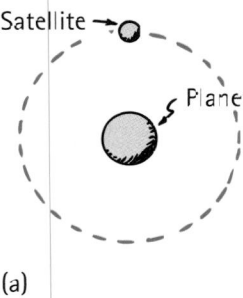

(a)

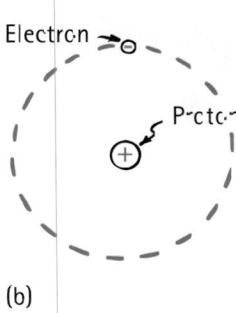

(b)

FIGURE 10.4 ▲
(a) A gravitational force holds the satellite in orbit about the planet and (b) an electrical force holds the electron in orbit about the proton. In both cases, the force follows the inverse-square law.

Coulomb's law is like Newton's law of gravity. But unlike gravity, forces can be attractive or repulsive.

✔ **READING CHECK**

State Coulomb's law in words.

PhysicsPlace.com

Tutorial
Electrostatics
Video
Van de Graaff Generator

10.3 Charge Polarization

We began this chapter by discussing an inflated balloon rubbed on your hair. If you place the balloon against a wall, you'll see that it sticks. This is because the negative charge on the balloon pulls the positive part of atoms in the wall closer to it. The balloon has the effect of "inducing" an opposite charge in the wall. Although the atoms in the wall don't move, their "centers of charge" are moved. The positive part of each atom is attracted toward the balloon and the negative part is repelled. We say the distorted atoms (Figure 10.6) are **electrically polarized.**

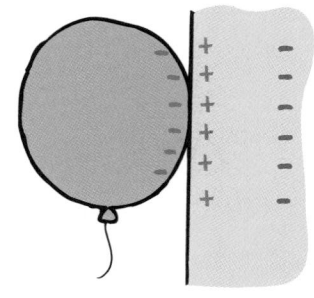

FIGURE 10.5 ▲
Because the negatively charged balloon polarizes atoms in the wall and creates a positively charged surface, the balloon sticks to the wall.

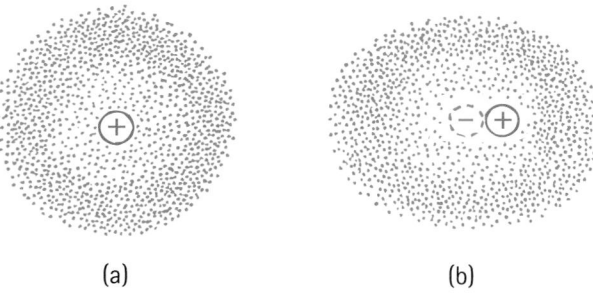

(a)   (b)

FIGURE 10.6 ▲
(a) The center of the negative "cloud" of electrons coincides with the center of the positive nucleus in an atom. (b) When an external negative charge is brought nearby to the right, as on a charged balloon, the electron cloud is distorted, so the centers of negative and positive charge no longer coincide. The atom is electrically polarized.

✔ A polarized object has no net charge. Only the distribution of charge in the material is altered. We will return to electrical polarization in Part 3 and see how it causes the stickiness between many kinds of molecules.

✔ **READING CHECK**

When an object is electrically polarized, does it have a net charge, or is its distribution of charge altered?

10.4 Electric Current—The Flow of Electric Charge

Recall from Chapter 9 that loose electrons in metals are responsible for the good heat conduction in metals. The same is true for electrical conduction. Loosely held outer electrons in the atoms of a metal are called *conduction electrons.* ✔ Protons don't move about in a metal because they are bound inside the atomic nucleus. Conduction electrons, however, can freely migrate through a metal. In Part 3, we'll learn that in addition to electron flow, both positive and negative ions can make up the flow of electric charge in fluids. In this chapter, we'll focus on currents made of flowing electrons.

Electrons flow in a way similar to water flow. Just as water current is the flow of H_2O molecules, **electric current** is the flow of electrons. But there are differences between water flow and electron flow. If you buy a water pipe at a hardware store, the clerk doesn't sell you the water to go with it. You provide that yourself. By contrast, when you buy "an electron pipe," an electric wire, you get the electrons too. Every bit of matter, wires included, contains enormous numbers of electrons that swarm about in random directions. When they are set in motion in one direction, a *net* direction, we have an electric current.

The *rate* of electrical flow is measured in *amperes* (abbreviation A). An **ampere** is the rate of flow of 1 coulomb of charge per second. (That's a flow of 6.25 billion billion electrons per second.) In a wire that carries 5 amperes, 5 coulombs of charge pass any cross section in the wire each second. In a wire that carries 10 amperes, twice as many coulombs pass any cross section each second.

✔ **READING CHECK**

Why do electrons in a wire, rather than protons, make up the flowing charge that gives us electric current?

10.5 An Electric Current Is Produced by Electrical Pressure—Voltage

When water flows in a pipe, there is more pressure at one end than the other. There must be a pressure difference to keep the water flowing. Also recall from our study of heat in Chapter 9 that heat flow depends on a temperature difference. Heat flows from zones of high temperature to those of low temperature. Similarly for electric current. ✔ Electrons flow in a wire only when a difference in "electrical pressure" exists. The name for this pressure is *voltage*.

So what, more specifically, is voltage? Voltage is directly proportional to electric potential energy. Recall the concept of potential energy discussed in

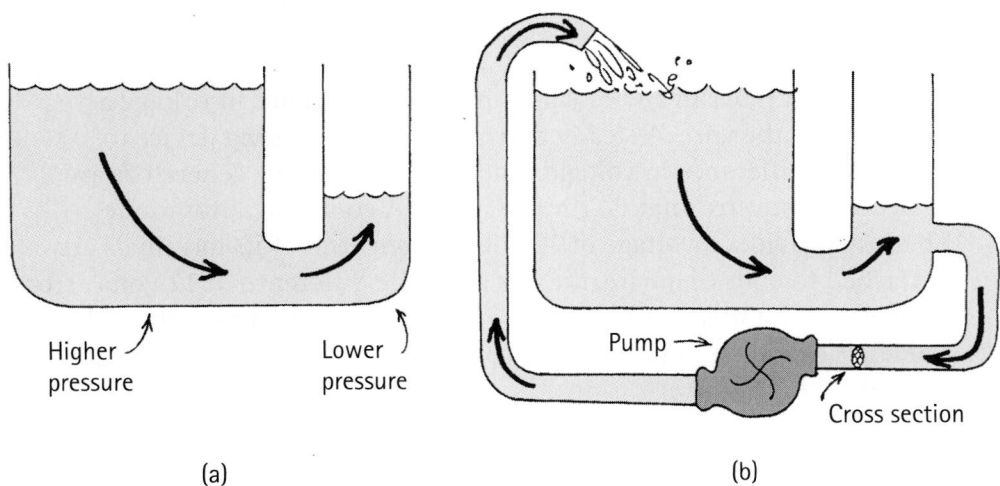

Higher pressure   Lower pressure   Pump →   Cross section

(a)   (b)

FIGURE 10.7 ▲
(a) Water flows from the reservoir of higher pressure to the reservoir of lower pressure. The flow ceases when the difference in pressure ceases.
(b) Water continues to flow because a difference in pressure is maintained with the pump.

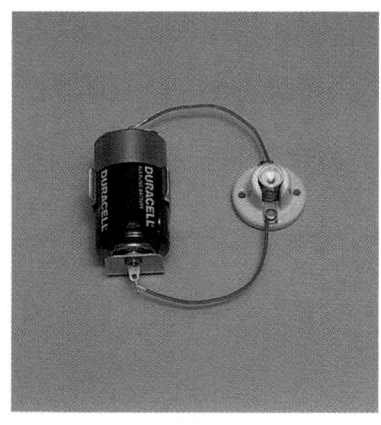

◀ **FIGURE 10.8**
Each coulomb of charge that is made to flow in the circuit that connects the ends of this 1.5-V flashlight cell is energized with 1.5 J.

Chapter 6. Things have energy due to their positions (Figure 10.9). Likewise with electrons and other charged particles. In Figure 10.10, we see that pushing a spring gives it potential energy, for the compressed spring can do work on something when released. We also see that pushing a charged particle toward another charge can increase its electrical potential energy. This potential energy compared with the quantity of charge is what we mean by **voltage.**

$$\text{Voltage} = \frac{\text{potential energy}}{\text{charge}}$$

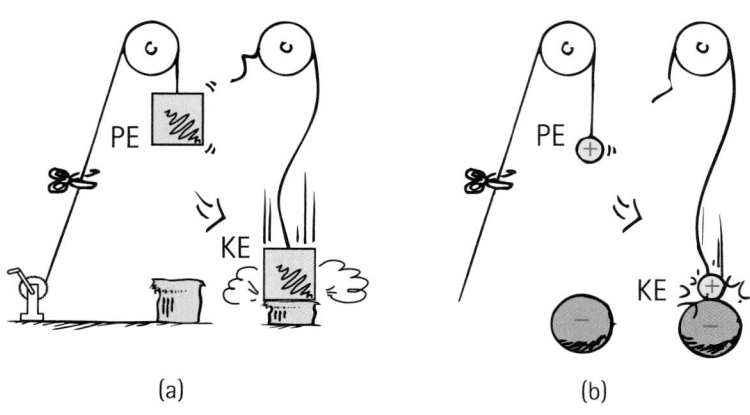

(a) (b)

FIGURE 10.9 ▲
(a) Gravitational potential energy converts to kinetic energy. (b) Similarly, electric potential energy converts to kinetic energy of electrical charge.

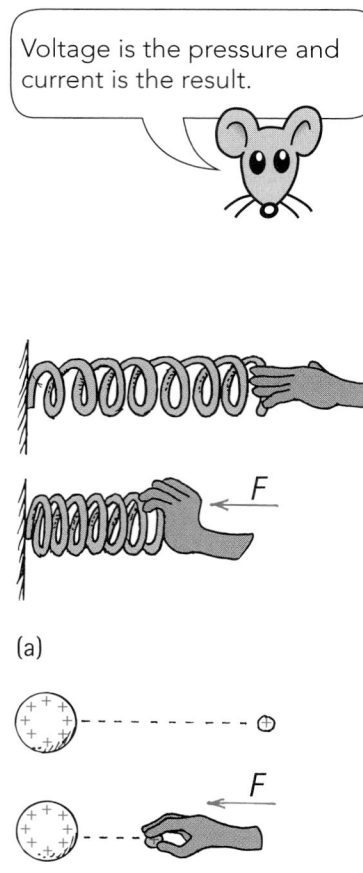

Voltage is the pressure and current is the result.

(a)

(b)

FIGURE 10.10 ▲
(a) The spring has more mechanical PE when compressed. (b) The charged particle similarly has more electrical PE when pushed closer to the charged sphere. In both cases, the increased PE is the result of work input.

Current flows in a wire when there is a difference in voltage across the ends of the wire. A steady current needs a pumping device to provide a difference in voltage. Chemical batteries or generators are "electrical pumps" that do the job nicely. A common automobile battery provides a voltage of 12 volts. When each of its terminals are attached to ends of a wire, there is a voltage difference of 12 volts across the wire. That means 12 joules of energy are supplied to each coulomb of charge flowing in the wire. The wire is usually a part of an electric circuit.

There is often some confusion about charge flowing *through* a circuit and voltage *across* a circuit. We can see the difference by thinking of a long pipe filled with water. Water flows *through* the pipe if there is a difference in pressure *across,* or between, its ends; it flows from the high-pressure to the low-pressure end. Only the water flows, not the pressure. Similarly, electrons flow because of a difference in electrical pressure (voltage difference). Electrons flow *through* a circuit because of an applied voltage *across* the circuit. Voltage doesn't flow through a circuit—it doesn't go anywhere, for it is the electrons that flow. Voltage produces current (if there is a complete circuit).

10.6 Electrical Resistance

A battery or generator of some kind moves electrons in a circuit. How much current there is depends on the voltage and also on the **electrical resistance** of the circuit. Just as narrow pipes resist water flow more than wide pipes, narrow wires resist electrical current more than wider wires. And length contributes to resistance also. Just as long pipes have more resistance than short ones, long wires offer more electrical resistance. And most important is the kind of material. Copper has a low electrical resistance, while a strip of rubber has an enormous resistance. ✔ So we see that the electrical resistance of a wire depends on the kind of material used in the wire and the thickness and length of the wire.

Temperature also affects electrical resistance. The greater the jostling of atoms within a conductor (in other words, the higher the temperature), the greater resistance a conductor has. The resistance of some materials reaches zero at very low temperatures. These are *superconductors* (see Link to Technology box on page 201).

(see Link to Technology box on page 201).

What condition is necessary for the flow of charge in a wire?

✔ **READING CHECK**

What condition is necessary for the flow of charge in a wire?

FIGURE 10.11 ▲
An unusual source of voltage. The electric potential between the head and tail of the electric eel (*Electrophorus electricus*) can be up to 600 V.

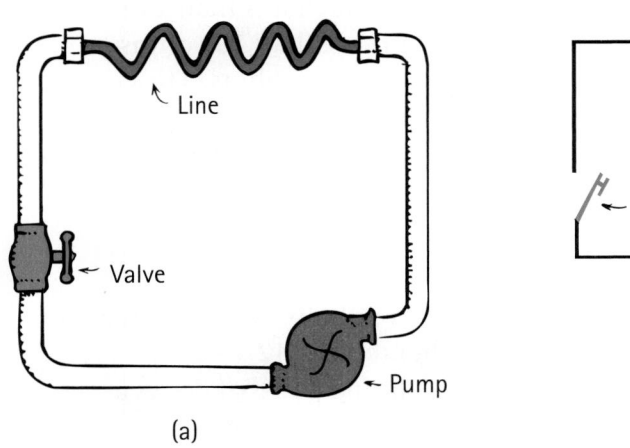

Line

Valve

Pump

(a)

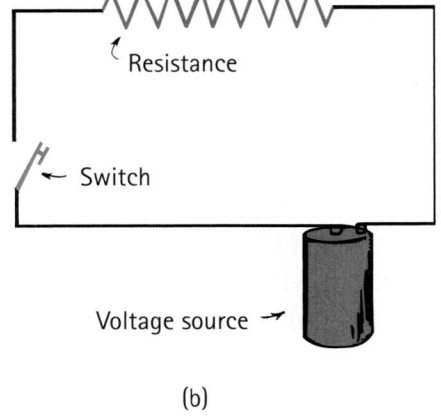

Resistance

Switch

Voltage source →

(b)

FIGURE 10.12 ▲
(a) In the hydraulic circuit, the narrow pipe (green) offers resistance to water flow. (b) In the electric circuit, a lightbulb or other device (shown by the zigzag symbol for resistance) offers resistance to electron flow.

Electrical resistance is measured in units called *ohms*. The Greek letter *omega*, Ω, is commonly used as the symbol for the ohm, which is named after Georg Simon Ohm, a German physicist who in 1826 discovered a simple and very important relationship among voltage, current, and resistance.

✔ **READING CHECK**

Cite three factors that determine the electrical resistance of a wire.

10.7 Ohm's Law—The Relationship Among Current, Voltage, and Resistance

PhysicsPlace.com

Videos
Caution on Handling Electrical Wires; Birds & High Voltage Wires; Ohm's Law

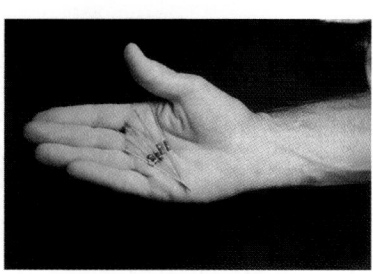

FIGURE 10.13 ▲
Resistors. The symbol of resistance in an electric circuit is ─\\/\\/\\─.

FIGURE 10.14 ▲
More water flows through a thick hose than through a thin one connected to a city's water system (same water pressure). Likewise for electric current in thick and thin wires connected across the same voltage.

✓ **READING CHECK**

What three concepts are tied together in Ohm's law?

The relationship among current, voltage, and resistance is **Ohm's law.** Ohm discovered that ✓ the amount of current in a circuit is directly proportional to the voltage across the circuit and inversely proportional to the resistance of the circuit:

$$\text{Current} = \frac{\text{voltage}}{\text{resistance}}$$

Or, in units form:

$$\text{Amperes} = \frac{\text{volts}}{\text{ohms}}$$

Ohm's law tells us that 1 volt across a circuit with a resistance of 1 ohm produces a current of 1 ampere. If there are 12 volts across the same circuit, the current is 12 amperes. So for a given circuit of constant resistance, current and voltage are proportional to each other.* This means we get twice the current for twice the voltage. The greater the voltage, the greater the current. But if the resistance is doubled for a circuit, the current is reduced to half. The higher the resistance, the lower the current. Ohm's law makes good sense.

The resistance of a typical lamp cord is much less than 1 ohm, and a typical lightbulb has a resistance of more than 100 ohms. An iron or electric toaster has a resistance of 15 to 20 ohms. The current inside these and all other electrical devices is regulated by circuit elements called *resistors* (Figure 10.13), whose resistance may be a few ohms or millions of ohms.

CONCEPT CHECK

1. How much current flows through a lightbulb with a resistance of 60 Ω when the voltage across the lightbulb is 12 V?
2. What is the resistance of an electric frying pan that draws a current of 12 A when connected to a 120-V circuit?

Check Your Answers

1. This is calculated from Ohm's law: Current $= \dfrac{\text{voltage}}{\text{resistance}}$, so $\dfrac{12 \text{ V}}{60 \text{ }\Omega} = 0.2$ A.

2. Rearrange Ohm's law to read

$$\text{Resistance} = \frac{\text{voltage}}{\text{current}} = \frac{120 \text{ V}}{12 \text{ A}} = 10 \text{ }\Omega$$

---

* If we use the symbol *V* for voltage, *I* for current, and *R* for resistance, Ohm's law is expressed $I = V/R$. It can also be written, $V = IR$, or $R = V/I$, so if any two variables are known, the third can be found. Units are abbreviated V for volts, A for amperes, and Ω for ohms.

In ordinary conductors, such as household wiring, the moving electrons that make up current often collide with atomic nuclei in the wire. The colliding electrons transfer their kinetic energy to the wire and the wire heats up. Energy is wasted. However, experiments show that certain metals lose all their electrical resistance when placed in a bath of liquid helium at a temperature of 4 K. All of it! The electrons in these extremely cold conductors travel pathways that avoid collisions. Hence the electrons can flow indefinitely. The materials that work this way are called *superconductors*. **Superconductors** have zero electrical resistance to the flow of charge. In superconductivity, no current is lost and no heat is generated.

Various ceramic oxides are superconducting at temperatures above 100 K. Steady currents have been observed to persist for years in some superconductors without any apparent loss in energy. There is presently enormous interest in the technology of superconductors. Imagine the energy-saving potential of superconducting devices! Explanations of superconductivity involve the wave nature of matter (quantum mechanics) and are being vigorously researched.

10.8 Electric Shock

Which causes electric shock in the human body—current or voltage? The damaging effects of shock result from current through the body. From Ohm's law, we see current in a body depends on the voltage applied and also on the body's electrical resistance. A person's resistance ranges from about 100 ohms if the body is soaked with saltwater to about 500,000 ohms if the skin is very dry. If you touch the two electrodes of a battery with dry fingers, you make up a circuit with a resistance of about 100,000 ohms. You usually cannot feel 12 volts. Even 24 volts just barely tingles. If your skin is moist, 24 volts can be quite uncomfortable. Table 10.1 describes the effects of different amounts of current on the human body.

TABLE 10.1 Effect of Electric Current on the Body

Current (A)	Effect
0.001	Can be felt
0.005	Is painful
0.010	Causes involuntary muscle contractions (spasms)
0.015	Causes loss of muscle control
0.070	Goes through the heart; serious damage, probably fatal for if current lasts for more than 1 s

CONCEPT CHECK

1. At 100,000 Ω, how much current will flow through your body if you touch the terminals of a 12-V battery?
2. If your skin is very moist, so that your resistance is only 1000 Ω, and you touch the terminals of a 12-V battery, how much current do you receive?

FIGURE 10.15 ▲
The bird can stand harmlessly on one wire of high voltage, but it had better not reach over and grab a neighboring wire! Why not?

FIGURE 10.16 ▲
The round prong connects the body of the appliance directly to ground (the Earth). Any charge that builds up on an appliance is therefore conducted to the ground—preventing accidental shock.

Check Your Answers

1. $\dfrac{12 \text{ V}}{100,000 \ \Omega} = 0.00012 \text{ A}.$

2. $\dfrac{12 \text{ V}}{1000 \ \Omega} = 0.012 \text{ A}.$ Ouch!

✓ To receive a shock, there must be a *difference* in voltage between one part of your body and another part. Electron flow will pass along the path of least electrical resistance connecting these two points. Suppose you fell from a bridge and grabbed onto a high-voltage power line, halting your fall. If you touch nothing else of different voltage, you receive no shock. Even if the wire is a few thousand volts and even if you hang by two hands, no significant electron flow will occur between your hands. This is because there is no voltage difference between your hands. If, however, you reach over with one hand and grab onto a wire of different voltage . . . zap! We have all seen birds perched on high-voltage wires. Every part of their bodies is at the same high voltage as the wire, and so they have no problem.

Most electric plugs and sockets are wired with three connections. The two flat prongs on a plug are for the current-carrying double wire inside the socket. One part is "live" (energized) and the other is neutral. The larger round prong connects to a wire in the electrical system that is grounded—connected directly to the ground (Figure 10.16). If the live wire of the plugged-in appliance accidentally comes in contact with the metal surface of the appliance, and you touch the appliance, you could receive a dangerous shock. This won't occur when the appliance casing is grounded via the ground wire. Then the voltage of the appliance is the same voltage as the ground—relatively speaking, zero. You can't get shocked unless there is a voltage difference.

CONCEPT CHECK

1. So which causes electric shock—current or voltage?
2. What is the source of electrons that produce a shock in your body?
3. What is the source of a simple battery-powered circuit?

Check Your Answers

1. Electric shock *occurs* when current is produced in the body, which is *caused* by an applied voltage. Voltage is the cause; current is the effect.
2. The source of electrons are those already in your body. Like any conductor, the electrons are already there. A voltage across your body will set them in motion.
3. The circuit elements themselves. Just as a water pump moves water in pipes, a battery supplies energy to move electrons already in a circuit.

✓ **READING CHECK**

What role does voltage play in getting an electric shock?

Many people are killed each year by current from common 120-V electric circuits. If you touch a faulty 120-V light fixture with your hand while your feet are on the ground, there may be a 120-V "electrical pressure" between your hand and the ground. Resistance to current is usually greatest between your feet and the ground, and so the current is usually not enough to do serious harm. But if your feet and the ground are wet, there is a low-resistance electrical path between you and the ground. The 120 V across this lowered resistance may produce a current greater than your body can stand.

Pure water is not a good conductor. But the ions normally found in water make it a fair conductor. More dissolved materials, especially small amounts of salt, lower the resistance even more. There is usually a layer of salt left from perspiration on your skin, which when wet lowers your skin

resistance to a few hundred ohms or less. Handling electrical devices while taking a bath is a definite no-no.

Injury by electric shock comes in three forms: (1) burning of tissues by heating, (2) muscle contraction, and (3) disruption of cardiac rhythm. These conditions are caused by too much electric power delivered in critical body regions for too long a time.

Electric shock can upset the nerve center that controls breathing. In rescuing shock victims, the first thing to do is clear them from the electric supply. Use a dry wooden stick or some other nonconductor so that you don't get electrocuted yourself. Then apply artificial respiration. It is important to continue artificial respiration. There have been cases of victims of lightning who did not breathe for several hours, but were eventually revived and completely regained good health.

10.9 Direct Current and Alternating Current

Electric current may be *dc* or *ac*. By *dc,* we mean **direct current.** Direct current is current made up of electrons that flow in *one direction.* A battery produces direct current in a circuit because the terminals of the battery always have the same opposite signs. Electrons move from the repelling negative terminal toward the attracting positive terminal, always moving through the circuit in the same direction.

It is interesting to note that the speed of electrons as they drift through a wire is surprisingly slow. This is because electrons continually bump into atoms in the wire. ✓ The *drift speed* of electrons in a typical circuit is much less than 1 centimeter per second. The electric signal, however, travels at nearly the speed of light. That's the speed at which the electric *field* in the wire is established. (An electric field exists around charged particles like a gravitational field exists around massive bodies—both fields travel from one point to another at the speed of light.)

Alternating current (ac) acts as the name implies. Electrons in the circuit flow initially in one direction and then in the opposite direction. This is done by switching the sign at the terminals of the power-station generator (next chapter). The net drift speed of electrons in an ac circuit is zero. They vibrate back and forth about

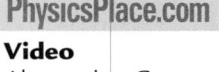

Video
Alternating Current

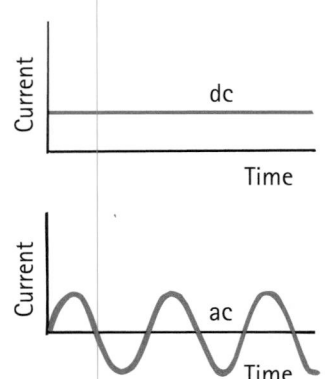

FIGURE 10.17 ▲
Time graphs of *dc* and *ac.*

✔ READING CHECK

How does electron drift speed differ from the speed of an electric signal?

relatively fixed positions and travel nowhere. Nearly all commercial ac circuits involve currents that alternate back and forth at a frequency of 60 cycles per second. This is 60-hertz (Hz) current (a cycle per second is called a *hertz*).

10.10 Electric Power—The Rate of Doing Work

FIGURE 10.18 ▲
The power and voltage on the lightbulb read "100 W 120 V." How many amperes will flow through the bulb?

Moving charges in an electric current can do work. They can heat a circuit or turn a motor. Recall from Chapter 6 how we defined power as the rate of using energy. Power is the energy transformed divided by the elapsed time. Electrical energy may be transformed to mechanical energy (as in a motor), to light (as in a lamp), to thermal energy (as in a heater), or to other forms. ✔ In electrical terms, power is equal to current multiplied by voltage.

$$\textbf{Power} = \textbf{current} \times \textbf{voltage}$$

When current is in amperes and voltage is in volts, then power is expressed in watts. So in units form,

$$\textbf{Watts} = \textbf{amperes} \times \textbf{volts}$$

If a lamp rated at 120 watts operates on a 120-volt line, it draws a current of 1 ampere (120 W = 1 A × 120 V). A 60-watt lightbulb draws $\frac{1}{2}$ ampere on a 120-volt line.

✔ READING CHECK

Mechanical power is the rate at which work is done, energy per charge. How is electrical power expressed?

CONCEPT CHECK
What power is needed to operate a clock if it draws a current of 0.05 amperes from your household circuit?

Check Your Answer
Power = current × voltage = 0.05 A × 120 V = 6 W.

LINK TO HISTORY OF TECHNOLOGY: 110 VOLTS ●

In the early days of electricity, high voltages burned out electric light filaments, and so low voltages were more practical. The hundreds of power plants built in the United States prior to 1900 adopted 110 volts (or 115 or 120 volts) as their standard. Tradition has it that 110 volts was agreed upon because it made bulbs of the day glow as brightly as a gas lamp. By the time electricity became popular in Europe, engineers had figured out how to make lightbulbs that would not burn out so fast at higher voltages. Because power transmission is more efficient at higher voltages, Europe adopted 220 volts as its standard. The United States remained with 110 volts (today, officially 120 volts) because of the installed base of 110-volt equipment.

10.11 Electric Circuits—Series and Parallel

Any path along which electrons can flow is a *circuit*. For a steady current, there must be a complete circuit with no gaps. A gap is usually provided by an electric switch that can be opened or closed. Then there is control to either stop or allow energy flow.

Most circuits contain more than one device that receives electric energy in the circuit. These may be several lamps, for example. These devices are commonly connected in one of two ways, *series* or *parallel*. When connected in series, the devices and the wires connecting them form a single pathway for electron flow between the terminals of the battery, generator, or wall socket. When connected in parallel, the devices and wires connecting them form branches, each providing separate paths for electron flow. Series and parallel connections each have their own distinctive characteristics.

Series Circuits

A simple **series circuit** is shown in Figure 10.20, where three lightbulbs are connected in series with a battery. When the switch is closed, the same current exists almost immediately in all three lightbulbs. The current does not "pile up" in any lightbulb but flows *through* each lightbulb. Remember that the electrons in a circuit are in the conductors themselves. They flow in a direction from the negative terminal of the battery, through each lightbulb, then toward the positive terminal of the battery. Inside the battery, the electrons move in a direction toward the negative terminal. This means that ✓ the amount of current passing through the battery is the same as the amount passing through the rest of the circuit. This is the only path of the electrons through the circuit. If a break occurs anywhere in the path, the flow of electrons stops. Burning out of one of the lightbulb filaments or simply opening the switch causes such a break. The circuit shown in Figure 10.20 illustrates the following important characteristics of series connections:

1. Electric current has but one pathway. This means that the current is the same in every part of the circuit.
2. This current is resisted by the resistance of the first device, the resistance of the second, and that of the third also. So the total resistance to current in the circuit is the sum of the individual resistances along the circuit path (assuming the resistance of the connecting wires is negligible).
3. The current in the circuit is numerically equal to the voltage supplied by the source divided by the total resistance of the circuit. This is in accord with Ohm's law.
4. The total voltage established across a series circuit divides among the electrical devices in the circuit so that the sum of the "voltage drops" across each device is equal to the total voltage supplied by the source.

PhysicsPlace.com
Tutorial
Circuits
Video
Electric Currents

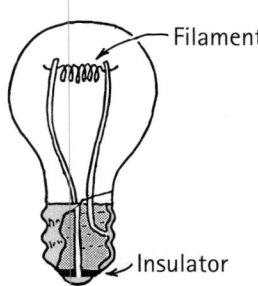

FIGURE 10.19 ▲
The conduction electrons that surge back and forth in the filament of a lightbulb do not come from the voltage source. They are in the filament to begin with. The voltage source simply provides them with surges of energy. Likewise for modern fluorescent bulbs.

An open circuit (open switch) is one without current. A closed circuit (closed switch) is one with current. This is opposite of a closed door or closed highway.

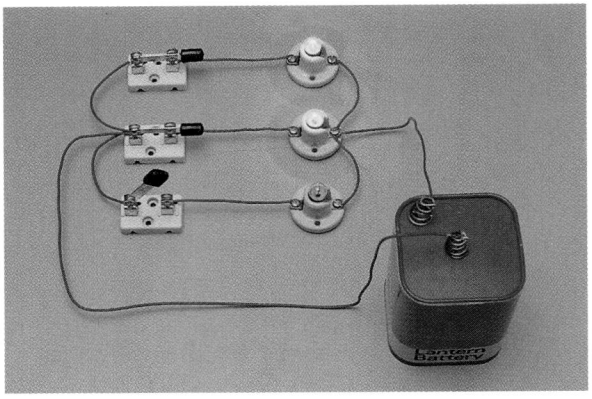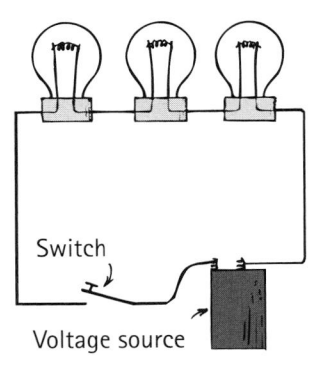

FIGURE 10.20 ▲
A simple series circuit. The 6-V battery provides 2 V across each lightbulb.

> A series circuit is like a single-lane road with no alternate path. If there is a roadblock or cave-in, traffic will stop.

(This follows from the fact that the amount of energy given to the total current is equal to the sum of energies given to each device.)

5. The voltage drop across each device is proportional to its resistance. (This follows from the fact that more energy is wasted as heat when a current passes through a high-resistance device than through a low-resistance device.)

CONCEPT CHECK

1. What happens to the current in other lightbulbs in a series circuit if one bulb burns out?
2. What happens to the light intensity of each lightbulb in a series circuit when more bulbs are added to the circuit?

Check Your Answers

1. If one bulb burns out, the circuit path is broken and current ceases. All bulbs go out.
2. The addition of more bulbs in a series circuit results in more circuit resistance. This lowers the current in the circuit and therefore in each bulb. That's why the bulbs dim. The same amount of energy is divided among more lightbulbs, which means less energy per bulb—less voltage drop across each bulb.

✓ **READING CHECK**

How does the amount of current in a battery compare with the current in the rest of the circuit?

It is easy to see the main disadvantage of a series circuit. If one device fails, current in the entire circuit ceases. Some inexpensive Christmas tree lights are connected in series. When one bulb burns out, it's fun and games (or frustration) trying to locate which bulb to replace.

Most circuits are wired so that it is possible to operate several electrical devices, each independently of the others. In your home, for example, a light switch can be turned on or off without affecting other electrical appliances on the same circuit. This is because these devices are connected not in series, but in parallel.

> A parallel circuit is like a road with alternate routes. If there is a roadblock or cave-in, the traffic along alternate routes keeps moving.

You electrify me!

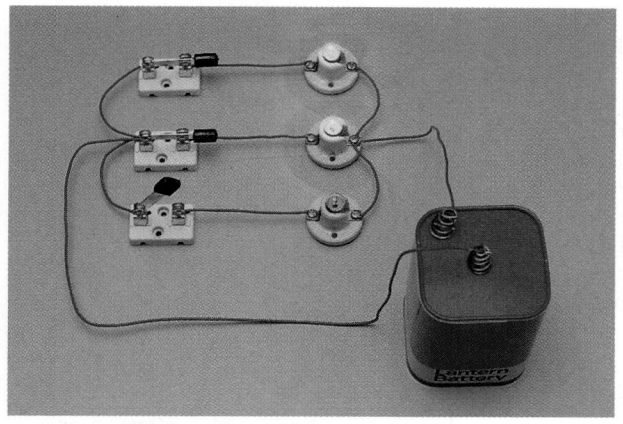

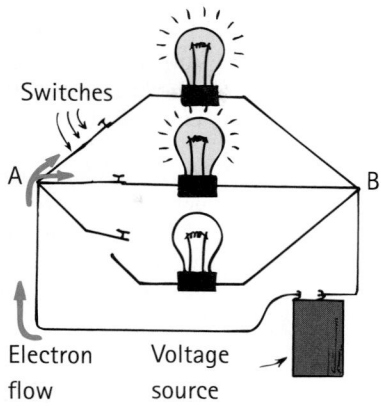

FIGURE 10.21 ▲
A simple parallel circuit. A 6-V battery provides 6 V across each lightbulb.

Parallel Circuits

A simple **parallel circuit** is shown in Figure 10.21. Three lightbulbs are connected to the same two points, A and B. Electrical devices connected to the same two points of an electrical circuit are *connected in parallel.* Electrons leaving the negative battery terminal need to travel through only *one* lightbulb filament before returning to the positive terminal of the battery. In this case, current branches into three separate pathways from A to B. ✔ How much current is in each branch depends on the resistance in that branch. A break in any one path does not interrupt the flow of charge in the other paths. The current in one branch isn't affected by the current in any other branch. Each device operates independently of the others.

The circuit shown in Figure 10.21 illustrates the following major characteristics of parallel connections:

1. Each device connects the same two points, A and B, of the circuit. The voltage is therefore the same across each device.
2. The total current in the circuit divides among the parallel branches. Since the voltage across each branch is the same, the amount of current in each branch is inversely proportional to the resistance of the branch.
3. The total current in the circuit equals the sum of the currents in its parallel branches.
4. As the number of parallel branches is increased, the overall resistance of the circuit is lowered (just as more checkout cashiers at a supermarket lowers people-flow resistance). With each added parallel path, the overall circuit resistance is lowered. This means the overall resistance of the circuit is less than the resistance of any one of the branches.

CONCEPT CHECK
1. What happens to the current in other lightbulbs in a parallel circuit when one bulb burns out?
2. What happens to the light intensity of each lightbulb in a parallel circuit when more bulbs are added?

After failing more than 6000 times before perfecting the first electric lightbulb, Thomas Edison stated that his trials were not failures, because he successfully discovered 6000 ways that don't work.

fyi

- Switching a lightbulb on and off shortens its life by generating mechanical stress in its filament. Filament resistance in a 120-V 60-W bulb increases about 15 times from room temperature to its nearly 3000 K operating temperature in a time of about 100 milliseconds. The initial 10-A current drawn quickly decreases to a steady 0.7 A.

✔ **READING CHECK**

In a three-bulb series circuit, current in one lightbulb is the same in the other lightbulbs. How does current differ in the branches of a three-bulb parallel circuit?

Check Your Answers

1. When one bulb burns out, the others are unaffected. The current in each branch, according to Ohm's law, is equal to voltage/resistance. Because neither voltage nor resistance is affected in the other branches, the current in those branches is unaffected. The total current in the overall circuit (the current through the battery), however, is lowered. It is reduced by an amount equal to the current drawn by the bulb in question before it burned out. But the current in any other single branch is unchanged.

2. The light intensity of each bulb is unchanged as other bulbs are introduced (or removed). Although changes of resistance and current occur for the circuit as a whole, no changes occur in any individual branch in the circuit.

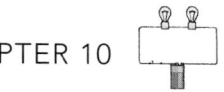

ELECTRICAL ENERGY AND TECHNOLOGY

Try to imagine life before electrical energy was something that humans could control. Imagine home life without electric lights, refrigerators, heating and cooling systems, the telephone, TV, and computers. We may romanticize a better life without these, but only if we overlook the hours of daily toil doing laundry, cooking, and heating homes. We'd also have to overlook how difficult it was getting a doctor in times of emergency before the advent of the telephone—when all the doctor had in his bag were laxatives, aspirins, and sugar pills—and when infant death rates were staggering.

We have become so accustomed to the benefits of technology that we are only faintly aware of

our dependency on dams, power plants, mass transportation, electrification, modern medicine, and modern agricultural science for our very existence. When we dig into a good meal, we give little thought to the technology that went into growing, harvesting, and delivering the food on our table. When we turn on a light, we give little thought to the power grid that links the widely separated power stations by long-distance transmission lines. These lines serve as the productive life force of industry, transportation, and the electrification of civilization. Anyone who thinks of science and technology as "inhuman" fails to grasp the ways in which they make our lives more human.

Parallel Circuits and Overloading

Electricity is usually fed into a home by way of two wires called *lines*. These lines are very low in resistance and are connected to wall outlets in each room—sometimes through two or more separate circuits. About 110–120 volts are applied across these lines by a transformer in the neighborhood. (A transformer is a device that steps down the higher voltage supplied by the power utility.) As more devices are connected to a circuit, more pathways for current result. ✓ More pathways lower the combined resistance of the parallel circuit. Therefore, more current exists in the circuit, which is sometimes a problem. Circuits that carry more than a safe amount of current are said to be *overloaded*.

We can see how overloading occurs in Figure 10.22. The supply line is connected to an

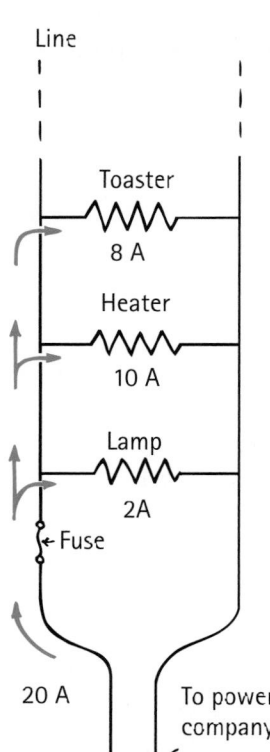

In practice, the lines in your home are not perfect conductors. For large currents used to operate a vacuum cleaner, the connecting wires do warm up (touch the connecting wire after the device has been running a while and you'll feel its warmth). But for most cases, the resistance of the lines can be neglected.

FIGURE 10.22 ▲
Circuit diagram for appliances connected to a household circuit.

✔ **READING CHECK**

What is the effect of more pathways on the amount of current in a household circuit?

FIGURE 10.23 ▶
A safety fuse.

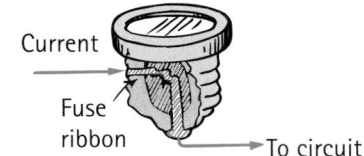

Current

Fuse ribbon

To circuit

electric toaster that draws 8 amperes, an electric heater that draws 10 amperes, and an electric lamp that draws 2 amperes. When only the toaster is operating and drawing 8 amperes, the total line current is 8 amperes. When the heater is also operating, the total line current increases to 18 amperes (8 amperes to the toaster plus 10 amperes to the heater). If you turn on the lamp, the line current increases to 20 amperes. Connecting any more devices increases the current still more. Connecting too many devices into the same circuit results in overheating the wires, which can cause a fire.

To prevent overloading in circuits, fuses are connected in series along the supply line. In this way the entire line current must pass through the fuse. The fuse shown in Figure 10.23 is constructed with a wire ribbon that heats up and melts at a given current. If the fuse is rated at 20 amperes, it will pass 20 amperes but no more. A current above 20 amperes melts the fuse ribbon, which "blows out" and breaks the circuit. Before a blown fuse is replaced, the cause of overloading should be determined and remedied. Often, insulation that separates the wires in a circuit wears away and allows the wires to touch each other. This is called a *short circuit*.

In modern buildings, fuses have been largely replaced by circuit breakers, which use magnets or bimetallic strips to open a switch when the current is too much. Utility companies use circuit breakers to protect their lines all the way back to the generators.

10 CHAPTER REVIEW

KEY TERMS

Alternating current (ac) Electrically charged particles that repeatedly reverse direction, vibrating about relatively fixed positions. In the United States, the vibrational rate is 60 Hz.

Ampere The unit of electric current, equivalent to 1 coulomb per second (the flow of 6.25×10^{18} electrons per second). In symbols, 1 A = 1 C/s.

Coulomb The unit of electrical charge. It is equal in magnitude to the total charge of 6.25×10^{18} electrons.

Coulomb's law The electrical force between two charged bodies is directly proportional to the product of the charges and inversely proportional to the square of the distance between them:

$$F = k\frac{q_1 q_2}{d^2}$$

Direct current (dc) Electrically charged particles flowing in one direction only.

Electric current The flow of electric charge that transports energy from one place to another. Measured in amperes.

Electric power The rate of energy transfer, or rate of doing work. Measured by

$$\text{Power} = \text{current} \times \text{voltage}$$

Measured in watts (or kilowatts), where 1 A × 1 V = 1 W.

Electrical resistance The property of a material that resists the flow of charged particles through it. Measured in ohms (Ω).

Electrically polarized Term applied to an atom or molecule in which the charges are aligned so that one side has a slight excess of positive charge and the other side a slight excess of negative charge.

Ohm's law The statement that the current in a circuit varies in direct proportion to the voltage across the circuit and inversely with the circuit's resistance:

$$\text{Current} = \frac{\text{voltage}}{\text{resistance}}$$

Parallel circuit An electric circuit in which electrical devices are connected so that the same voltage acts across each one and any single one completes the circuit independently of all the others.

Series circuit An electric circuit in which electrical devices are connected so that the same electric current exists in all of them.

Voltage A form of "electrical pressure":

$$\text{Voltage} = \frac{\text{potential energy}}{\text{charge}}$$

REVIEW QUESTIONS

Electric Charge Is a Basic Characteristic of Matter

1. Which part of an atom is *positively* charged and which part is *negatively* charged?
2. How does the charge of one electron compare with that of another electron?
3. How does the number of protons in the atomic nucleus normally compare with the number of electrons that orbit the nucleus?
4. What is meant by saying charge is *conserved*?

Coulomb's Law—The Force Between Charged Particles

5. How is Coulomb's law similar to Newton's law of gravitation? How is it different?

Charge Polarization

6. How does an electrically *polarized* object differ from an electrically *charged* object?

Electric Current—The Flow of Electric Charge

7. Why do electrons rather than protons make up the flow of charge in a metal wire?

An Electric Current Is Produced by Electrical Pressure—Voltage

8. How much energy is given to each coulomb of charge passing through a 6-V battery?
9. Does electric charge flow *across* a circuit or *through* a circuit? Does voltage *flow* across a circuit or is it *impressed* across a circuit? Explain.

Electrical Resistance

10. Which has the greater resistance, a thick wire or a thin wire of the same length?

Ohm's Law—The Relationship Among Current, Voltage, and Resistance

11. When the voltage across the ends of a piece of wire is doubled, what effect does this have on the current in the wire?
12. When the resistance of a circuit is doubled, and no other changes occur, what effect does this have on the current in the circuit?

Electric Shock

13. What is the function of the third prong on the plug of an electric appliance?

Direct Current and Alternating Current

14. Distinguish between *dc* and *ac*.
15. Does a battery produce *dc* or *ac*? Does the generator at a power station produce *dc* or *ac*?

Electric Power—The Rate of Doing Work

16. Which draws more current, a 40-W bulb or a 100-W bulb?

Electric Circuits—Series and Parallel

17. In a circuit consisting of two lightbulbs connected in series, if the current through one bulb is 1 A, what is the current through the other bulb?
18. In a circuit consisting of two lightbulbs connected in parallel, if there is 6 V across one bulb, what is the voltage across the other bulb?
19. How does the total current through the branches of a parallel circuit compare with the current through the voltage source?

20. Are household circuits normally wired in series or in parallel?

THINK AND DO

1. Demonstrate charging by friction and discharging from points with a friend who stands at the far end of a carpeted room. Scuff your shoes across the rug until your noses are close together. This can be a delightfully tingling experience, depending on how dry the air is (and how pointed your noses are).

2. Briskly rub a comb on your hair or a woolen garment and bring it near a small but smooth stream of running water. Is the stream of water charged? (Before you say yes, note the behavior of the stream when an opposite charge is brought nearby.)

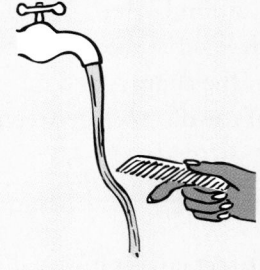

3. An electric cell is made by placing two plates made of different materials that have different affinities for electrons in a conducting solution. (A battery is actually a series of cells.) You can make a simple 1.5-V cell by placing a strip of copper and a strip of zinc in a tumbler of saltwater. The voltage of a cell depends on the materials used and the solution they are placed in, not on the size of the plates.

4. An easy cell to construct is the citrus cell. Stick a paper clip and a piece of copper wire into a lemon. Hold the ends of the wire close together, but not touching, and place the ends on your tongue. The slight tingle you feel and the metallic taste you experience result from a slight current of electricity pushed by the citrus cell through the wires when your moist tongue closes the circuit.

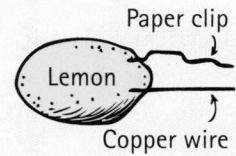

Paper clip

Lemon

Copper wire

PLUG AND CHUG

Ohm's Law: Current = voltage/resistance; $I = V/R$

1. Calculate the current in the 240-Ω filament of a lightbulb connected to a 120-V line.
2. Calculate the current in a toaster that has a heating element of 15 Ω when connected to a 120-V outlet.
3. Calculate how much current warms your feet from electric socks that have a 90-Ω heating element powered by a 9-V battery.
4. Calculate the current that moves through your fingers (resistance 1000 Ω) when you touch them to the terminals of a 6-V battery.

Power = current $\times$ voltage; $P = IV$

5. Calculate the power of a device that carries 0.5 A when impressed with 120 V.
6. Calculate the power of a hair dryer that operates on 120 V and draws a current of 10 A.

THINK AND COMPARE

1. Shown below are three separate pairs of point charges. Assume the pairs only interact with each other. Compare and rank the magnitudes of the force between each pair from highest to lowest.

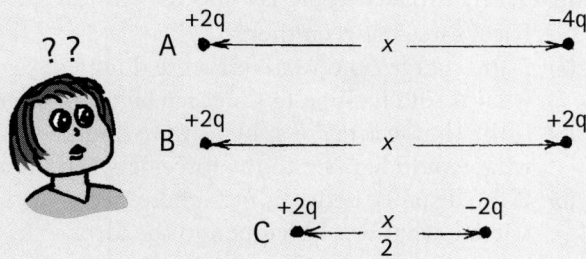

2. Compare the circuits below according to the brightness of the bulbs. Rank them from brightest to dimmest. (Or, do they have the same brightness?)

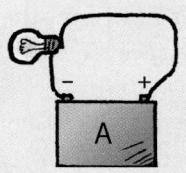

3. Rank the current in the three lightbulbs and the current in the battery, from greatest to least.

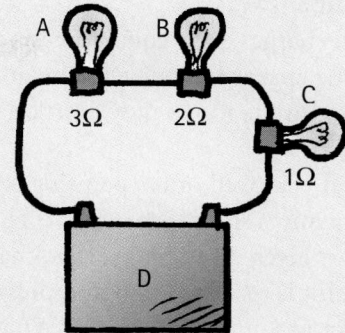

4. Rank the current in the three branches and the current in the battery, from greatest to least.

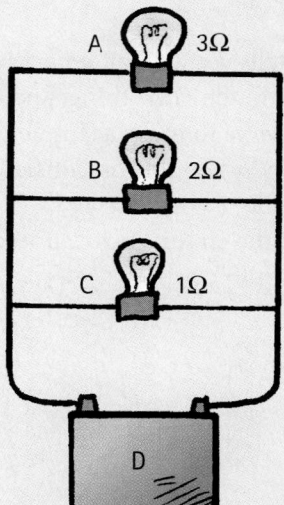

THINK AND EXPLAIN

1. When combing the fur of a cat, electrons transfer from its fur onto the comb. Is the cat's fur then positively or negatively charged? How about the comb?
2. If electrons were positive and protons were negative, would Coulomb's law be written the same or differently?
3. The 5000 billion billion freely moving electrons in a penny repel one another. Why don't they fly out of the penny?
4. How does the magnitude of electrical force between a pair of charged objects change when the objects are moved twice as far apart? Three times as far apart?
5. How does the magnitude of electric force compare between a pair of charged particles when they are brought to half their original distance of

separation? To one-quarter their original distance? To four times their original distance? (What law guides your answers?)

6. Two equal charges exert equal forces on each other. What if one charge has twice the magnitude of the other. How do the forces they exert on each other compare?

7. When a car is moved into a painting chamber, a mist of paint is sprayed around it. Then the body of the car is given a sudden electric charge and the mist of paint is attracted to it and presto—the car is quickly and uniformly painted. What does the concept of polarization have to do with this?

8. What happens to the brightness of light emitted by a lightbulb when the current that flows in it increases?

9. Your tutor tells you that an *ampere* and a *volt* really measure the same thing, and the different terms only serve to make a simple concept seem confusing. Why should you consider getting a different tutor?

10. In which of the circuits below does a current exist to light the bulb?

11. Does more current flow out of a battery than into it? Does more current flow into a lightbulb than out of it? Explain.

12. Only a small percentage of the electric energy going into a common lightbulb is transformed into light. What happens to the rest?

13. Why are thick wires rather than thin wires usually used to carry large currents?

14. What is the effect on current in a wire if both the voltage across it and its resistance are doubled? If both are halved?

15. Will the current in a lightbulb connected to a 220-V source be greater or less than when the same bulb is connected to a 110-V source?

16. Which will do less damage—plugging a 110-V appliance into a 220-V circuit or plugging a 220-V appliance into a 110-V circuit? Explain.

17. The damaging effects of electric shock result from the amount of current that flows in the body.

Why, then, do we see signs that read "Danger—High Voltage" rather than "Danger—High Current"?

18. If several bulbs are connected in series to a battery, they may feel warm to the touch but not visibly glow. What is your explanation?

19. In the circuit shown, how does the brightness of each identical lightbulb compare? Which bulb draws the most current? What will happen if bulb A is unscrewed? If C is unscrewed?

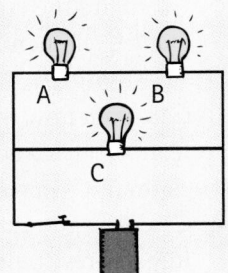

20. As more and more bulbs are connected in series to a flash-light battery, what happens to the brightness of each bulb? Assuming heating inside the battery is negligible, what happens to the brightness of each bulb when more and more bulbs are connected in parallel?

THINK AND SOLVE

1. Two point charges are separated by 6 cm. The attractive force between them is 20 N. Find the force between them when the distance between them is 12 cm. (Why can you solve this problem without knowing the magnitudes of the charges?)

2. Make use of Coulomb's law here. Suppose you have a pair of electrically charged metal spheres suspended a certain distance from each other by insulating threads. There is a specific amount of electrical force between them.
   (a) If the charge on one sphere were doubled, what would happen to the force between them?
   (b) If the charge on *both* spheres were doubled, what would happen to the force between them?
   (c) If the distance between the spheres were tripled, what would happen to the force between them?
   (d) If the distance between them were reduced to one-fourth the original distance, what would happen to the force between them?
   (e) If the charge on each sphere were doubled and the distance between them were doubled, what would happen to the force between them?

3. A radio speaker has a resistance of 8 Ω. Show that when 12 V is impressed across the speaker, the current is 1.5 A.

4. Rearrange the equation current = voltage/resistance to express *resistance* in terms of current and voltage.

Then consider the following: A certain device in a 120-V circuit has a current rating of 20 A. Show that the resistance of the device is 6 Ω.

5. What is the effect on current through a circuit of steady resistance when the voltage is doubled? What if both voltage and resistance are doubled?

6. Begin with the definition of power (current × voltage) and then show that the current in a 6-W clock radio operating in a common 120-V circuit is 0.05 A.

READINESS ASSURANCE TEST (RAT)

If you have a good handle on this chapter—if you really do—then you should be able to score 7 out of 10 on this RAT. Check your answers with your teacher. If you score less than 7, study further before moving on.

Choose the best answer to each of the following.

1. When you brush your hair and scrape electrons from your hair, the charge of your hair becomes
   (a) positive.
   (b) negative.
   (c) Actually both of these.
   (d) Actually neither of these.

2. According to Coulomb's law, a pair of particles that are placed twice as close to each other will experience forces
   (a) twice as strong.
   (b) four times as strong.
   (c) half as strong.
   (d) one-quarter as strong.

3. When you buy a water pipe in a hardware store, the water isn't included. When you buy copper wire, electrons
   (a) must be supplied by you, just as water must be supplied for a water pipe.
   (b) are already in the wire.
   (c) may fall out, which is why wires are insulated.
   (d) None of these.

4. To receive an electric shock, there must be
   (a) current in one direction.
   (b) moisture in an electrical device being used.
   (c) high voltage and low body resistance.
   (d) a difference in potential across part or all of your body.

5. A 10-Ω resistor carries 10 A. The voltage across the resistor is
   (a) zero.
   (b) more than zero but less than 10 V.
   (c) 10 V.
   (d) more than 10 V.

6. You can touch and discharge a 10,000-V Van de Graaff generator with little harm because, although the voltage is high, there is relatively little
   (a) resistance.   (b) energy.
   (c) grounding.   (d) All of these.
   (e) None of these.

7. Compared with the amount of current in the filament of a lamp, the amount of current in the connecting wire is
   (a) definitely less.   (b) often less.
   (c) actually more.   (d) the same.
   (e) Incredibly, all of these.

8. As more lamps are connected to a series circuit, the overall current in the power source
   (a) increases.   (b) decreases.
   (c) remains the same.

9. As more lamps are connected to a parallel circuit, the overall current in the power source
   (a) increases.   (b) decreases.
   (c) remains the same.

10. The difference between *dc* and *ac* in electrical circuits is that in *dc*
    (a) charges flow steadily in one direction only.
    (b) charges flow in one direction.
    (c) charges steadily flow to and fro
    (d) charges flow to and fro.

11

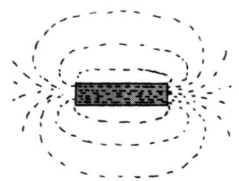

MAGNETISM

THE MAIN IDEA

A moving electric charge is surrounded by a magnetic field. Magnetism can produce electric current, which in turn can produce magnetism.

Megan not only shows how a magnet attracts nails but how nails stuck to the magnet attract nails below. In the previous chapter, we discussed a charged balloon that sticks to a wall. Is the physics similar for the sticking nails? And what causes magnetism—is it electrical? What are electro-magnets? Does the fact that a compass needle points north tell us that Earth is a giant magnet? Does magnetism about Earth shield us from harmful cosmic rays? Is Earth's magnetism responsible for the spectacular colors of the aurora borealis? These questions and more will be answered in this chapter.

Explore!

What Does a Magnetic Field Look Like?

1. Place a bar magnet on a level surface. Cover the magnet with a thin sheet of glass, clear plastic, or cardboard.
2. Sprinkle iron filings onto the sheet in the area above the magnet. Gently tap the sheet and observe the pattern formed by the filings.
3. Repeat Steps 1 and 2 using two bar magnets. Arrange the magnets in a straight line with opposite poles facing each other. Leave a gap of 4–6 cm between the poles.
4. Rotate one of the magnets 180 degrees and observe any change in the pattern formed by the filings.

Analyze and Conclude

1. **Observing** Make sketches of the patterns produced by the single magnet and by the pair of magnets in both orientations.
2. **Predicting** What pattern do you think would be formed if you used a horseshoe magnet in Step 1?
3. **Making Generalizations** In some places the patterns were of bunched-up filings and other places the filings were more spread out. What do you think this indicates about the magnetic field strengths?

The term *magnetism* comes from the region of Magnesia, a province of Greece. Certain magnetic stones were found by the Greeks who lived there more than 2000 years ago. These stones, called *lodestones,* could attract pieces of iron. When lodestones were rubbed on pieces of iron, magnets were made. Magnets were first made into compasses and used for navigation by the Chinese in the 12th century. Today, much of modern technology—from motors to computers—relies on magnetism.

11.1 Magnetic Poles—Attraction and Repulsion

If you've played around with magnets, you know they exert forces on one another. **Magnetic forces** are similar to electrical forces, for both kinds of forces cause objects to attract and repel. Also, both electrical and magnetic forces act between objects that are not touching. And similar to electrical force, the strength of a magnetic interaction depends on the distance between the two magnets. Whereas electric charges produce electrical forces, regions called *magnetic poles* produce magnetic forces.

Hang a bar magnet at its center by a piece of string and you've got a compass. One end, called the *north-seeking pole,* points northward. The opposite end, called the *south-seeking pole,* points southward. More simply, these are called the *north* and *south poles.* All magnets have both a north and a south pole (some have more than one of each). Refrigerator magnets have narrow strips of alternating north and south poles. These magnets are strong enough to hold sheets of paper against a refrigerator door. But they have a very short range because the north and south poles cancel a short distance from the magnet. In a simple bar magnet, the magnetic poles are located at the two ends. A common horseshoe magnet is a bar magnet bent into a U shape. Its poles are also at its two ends.

When the north pole of one magnet is brought near the north pole of another magnet, they repel each other.* The same is true of a south pole near a south pole. If opposite poles are brought together, however, attraction occurs. The rule is

Like poles repel each other; opposite poles attract.

This rule is similar to the rule for electric charges, where like charges repel one another and unlike charges attract. But there is a very

FIGURE 11.1 ▲
A horseshoe magnet.

You've got a magnetic personality.

---

* The force of interaction between magnetic poles is given by $F \sim p_1 p_2/d^2$, where p_1 and p_2 represent magnetic pole strengths and d represents the separation distance between the poles. Note the similarity of this relationship to Coulomb's law, and to Newton's law of gravity.

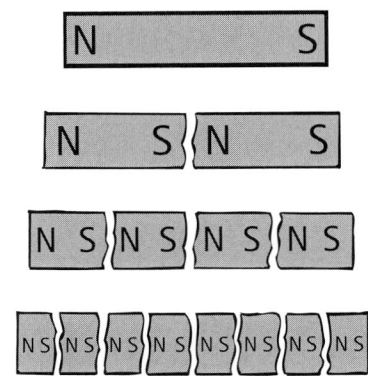

◀ **FIGURE 11.2**
Break a magnet in half, and you have two magnets. Break these in half, and you have four magnets, each with a north and south pole. Continue breaking the pieces further and further, and you find the same results. Magnetic poles always exist in pairs.

important difference between magnetic poles and electric charges. Electric charges can be isolated, but magnetic poles cannot. Electrons and protons are entities by themselves. A cluster of electrons doesn't need the company of a cluster of protons, and vice versa. ✓ But a north magnetic pole never exists without the company of a south pole, and vice versa. The north and south poles of a magnet are like the head and tail of the same coin.

If you break a bar magnet in half, each half still behaves as a complete magnet. Break the pieces in half again, and you have four complete magnets. You can continue breaking the pieces in half and never isolate a single pole. Even when your piece is one atom thick, there are two poles. This suggests that atoms themselves are magnets.

✓ READING CHECK

In electricity, a negative charge doesn't need a positive companion. In magnetism, can a north pole exist without a south pole?

CONCEPT CHECK
Must every magnet have a north and south pole?

Check Your Answer
Yes, just as every coin has two sides, a head and a tail. Some trick magnets may have more than one pair of poles, but nevertheless poles occur in pairs.

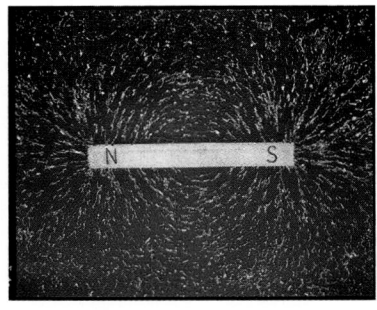

FIGURE 11.3 ▲
Top view of iron filings sprinkled around a magnet. The filings trace out a pattern of *magnetic field lines* in the space surrounding the magnet. Interestingly, the magnetic field lines continue inside the magnet (not revealed by the filings) and form closed loops.

11.2 Magnetic Fields—Regions of Magnetic Influence

When you sprinkle some iron filings on a sheet of paper placed on a magnet, you'll see that the filings trace out an orderly pattern of lines that surround the magnet. The space around the magnet contains a **magnetic field.** The shape of the field is shown by the filings. Note that the filings line up with the magnetic field lines that spread out from one pole and return to the other.

The direction of the field outside a magnet is from the north to the south pole. Where the lines are closer together, the field is stronger. The concentration of iron filings at the poles of the magnet in Figure 11.3 shows the magnetic field strength is greater there. If we place a small compass anywhere in the field, it will line up with the magnetic field.

✓ A magnetic field is produced by the motion of electric charge. Where, then, is this motion in a common bar magnet? The answer is, in

FIGURE 11.4 ▶
(Left) When the compass needle is not aligned with the magnetic field, the oppositely directed forces on the needle produce a pair of torques (twisting forces called a *couple*) that (right) twist the needle into alignment.

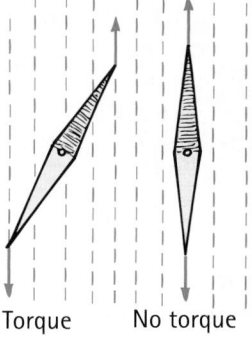

Torque   No torque

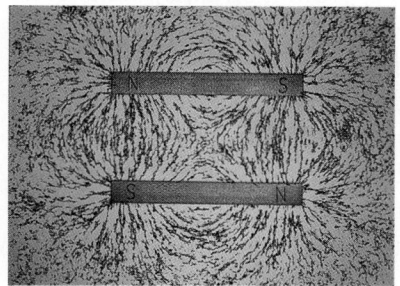 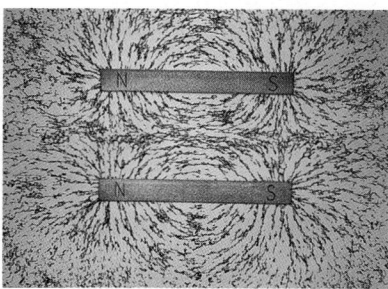

FIGURE 11.5 ▲
The magnetic field patterns for a pair of magnets. Opposite poles are nearest each other (left), and like poles are nearest each other (right).

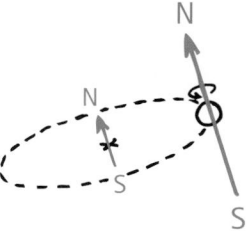

the electrons of the atoms that make up the magnet. These electrons are in constant motion. Two kinds of electron motion make magnetism: electron spin and electron revolution. Electrons spin about their own axes like tops, and they revolve about the atomic nucleus like planets revolving around the Sun. In most common magnets, electron spin is the main contributor to magnetism.

Every spinning electron is a tiny magnet. A pair of electrons spinning in the same direction makes a stronger magnet. A pair of electrons spinning in opposite directions, however, work against each other. The magnetic fields cancel. This is why most substances are not magnets. In most atoms, the various fields cancel one another because the electrons spin in opposite directions. In materials such as iron, nickel, and cobalt, however, the fields do not cancel each other entirely. Each iron atom has four electrons whose spin magnetism is uncanceled. Each iron atom, then, is a tiny magnet. The same is true to a smaller extent for the atoms of nickel and cobalt. Most common magnets are therefore made from alloys containing iron, nickel, and cobalt in various proportions.

FIGURE 11.6 ▲
Both the spinning motion and the revolving (orbital) motion of every electron in an atom produce magnetic fields. The field due to spin (large vector) combines with the field due to revolution (small vector) to produce the magnetic field of the atom. The resulting field is greater for iron atoms.

✔ **READING CHECK**

What, exactly, produces a magnetic field?

HANDS-ON EXPLORATION

Most iron objects around you are magnetized to some degree. A filing cabinet, a refrigerator, or even cans of food on your pantry shelf have north and south poles induced by the Earth's magnetic field.

Pass a compass from their bottoms to their tops and their poles are easily identified. Turn cans upside down and see how many days it takes for the poles to reverse themselves!

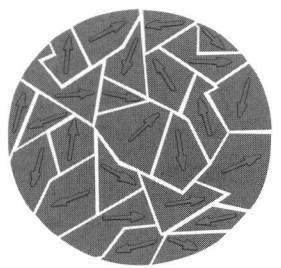

FIGURE 11.7 ▲
A microscopic view of magnetic domains in a crystal of iron. Each domain consists of billions of aligned iron atoms. In this view, the domains are not aligned.

Why are some pieces of iron magnetic, while most pieces of iron are not magnetic?

11.3 Magnetic Domains—Clusters of Aligned Atoms

Large clusters of iron atoms line up with one another. These clusters of aligned atoms are **magnetic domains.** Each domain is made up of billions of aligned atoms. The domains are microscopic (Figure 11.7), and there are many of them in a crystal of iron. Domains themselves tend to align with one another.

☑ In a strong magnet, most domains are aligned. The reason most pieces of iron aren't magnets is because the domains in ordinary iron are not lined up. In a common iron nail, for example, the domains are randomly oriented. But when you bring a magnet nearby, they can be induced into alignment. (It is interesting to listen with an amplified stethoscope to the clickety-clack of domains aligning in a piece of iron when a strong magnet approaches.) The domains align themselves much as electrical charges in a piece of paper align themselves (become polarized) in the presence of a charged rod. When you remove the nail from the magnet, ordinary thermal motion causes most or all of the domains in the nail to return to a random arrangement.

Permanent magnets can be made by placing pieces of iron or other magnetic materials in strong magnetic fields. Iron alloys differ in their ability to become magnetized; soft iron is easier to magnetize than steel. It helps to tap the material to nudge any stubborn domains into alignment. Another way is to stroke the material with a magnet. The stroking

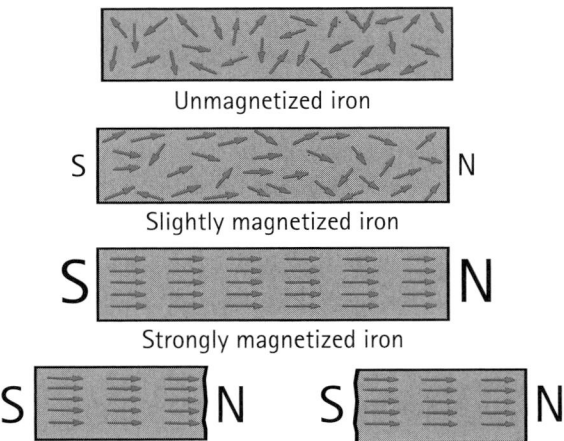

Unmagnetized iron

S ⟶ N
Slightly magnetized iron

S ⟶ N
Strongly magnetized iron

S ⟶ N   S ⟶ N
When a magnet is broken into two pieces, each piece is an equally strong magnet

FIGURE 11.8 ▲
Pieces of iron in successive stages of becoming magnetized. The arrows represent domains; each arrowhead is a north pole and each tail a south pole. Poles of neighboring domains neutralize each other's effects, except at the two ends of a piece of iron.

motion lines up the domains. If a permanent magnet is dropped or heated, some of the domains are jostled out of alignment and the magnet becomes weaker.

CONCEPT CHECK

1. How can a magnet attract a piece of iron that is not magnetized?
2. Why will a magnet not pick up a penny or a piece of wood?

Check Your Answers

1. Like the compass needle in Figure 11.4, domains in the unmagnetized piece of iron are induced into alignment by the magnetic field of the magnet. One domain pole is attracted to the magnet and the other domain pole is repelled. Does this mean that the net force is zero? No, because the force is slightly greater on the domain pole closest to the magnet than on the farther pole. That's why there is a net attraction. In this way a magnet attracts nonmagnetized pieces of iron (Figure 11.9).
2. A penny and a piece of wood have no magnetic domains that can be induced into alignment.

FIGURE 11.9 ▲
The iron nails become induced magnets.

11.4 The Interaction Between Electric Currents and Magnetic Fields

A single moving charge produces a magnetic field. A current of charges, then, also produces a magnetic field. The magnetic field that surrounds a current-carrying wire can be demonstrated by arranging an assortment of compasses around the wire (Figure 11.10). ✓ The magnetic field about the current-carrying wire makes up a pattern of concentric circles. When the current reverses direction, the compass needles turn around, showing that the direction of the magnetic field also changes.*

If the wire is bent into a loop, the magnetic field lines bunch together inside the loop (Figure 11.11). If the wire is bent into

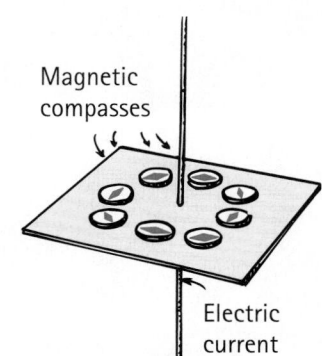

FIGURE 11.10 ▲
The compasses show the circular shape of the magnetic field surrounding the current-carrying wire.

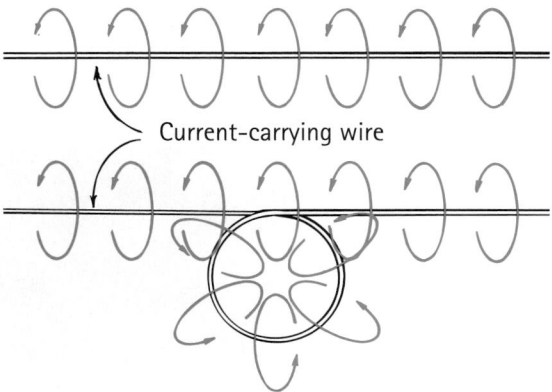

Current-carrying wire

◀ **FIGURE 11.11**
Magnetic field lines about a current-carrying wire crowd up when the wire is bent into a loop.

---

* Earth scientists think that the Earth's magnetism is the result of electric currents that accompany thermal convection in the molten parts of the Earth's interior. Evidence shows that the Earth's poles periodically reverse places—more than 20 reversals in the past 5 million years. This is perhaps the result of changes in the direction of electric currents within the Earth. More about this in Chapter 32.

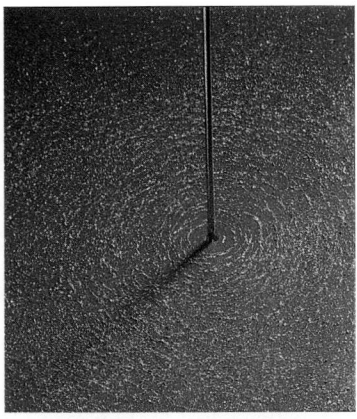

◀ **FIGURE 11.12**
Iron filings sprinkled on paper reveal the magnetic field configurations about (a) a current-carrying wire, (b) a current-carrying loop, and (c) a current-carrying coil of loops.

another loop, overlapping the first, the bunching of field lines in the loops is doubled. More loops mean more magnetic field intensity. The magnetic field intensity is strong for a current-carrying coil of many loops.

Electromagnets

If a piece of iron is placed in a current-carrying coil of wire, the alignment of magnetic domains in the iron produces a stronger magnet. Then we have an **electromagnet.** Its strength is increased by increasing the current through the coil. Strong electromagnets are used to control charged-particle beams in high-energy accelerators.

Electromagnets powerful enough to lift automobiles are a common sight in junkyards. The strength of these electromagnets is limited mainly by overheating of the current-carrying coils. The most powerful electromagnets omit the iron core and use superconducting coils through which large electrical currents easily flow.

Superconducting Electromagnets

Superconductors (previous chapter) have the interesting property of expelling magnetic fields. Because magnetic fields cannot penetrate the surface of a superconductor, magnets levitate above them. The reasons for this behavior are beyond the scope of this book and involve quantum mechanics. One of the hot applications of superconducting electromagnets is the levitation of high-speed trains for transportation. Prototype trains have already been demonstrated in the United States, Japan, and Germany. Watch for the growth of this relatively new technology.

 **READING CHECK**

What pattern occurs in the magnetic field about a current-carrying wire?

FIGURE 11.13 ▶
A permanent magnet levitates above a superconductor because the magnet's magnetic field cannot penetrate the superconducting material.

FIGURE 11.14 ▲

This passenger train is a magnetically levitated vehicle—a *magplane.*
Whereas conventional trains vibrate as they ride on rails at high speeds,
magplanes can travel vibration-free at high speeds because they make no
physical contact with the guideway they float above.

11.5 Magnetic Forces Are Exerted on Moving Charges

A charged particle has to be moving to interact with a magnetic field.
Charges at rest don't respond to magnets. But when moving, charged
particles experience a deflecting force.* ✔️ The force is greatest when
the particles move at right angles to the magnetic field lines. At other
angles, the force is less and becomes zero when the particles move parallel to the field lines. The force is always perpendicular to the magnetic
field lines and perpendicular to the velocity of the charged particle
(Figure 11.15). So a moving charge is deflected when it crosses through
a magnetic field. Since only direction changes, there is no change in
kinetic energy. If a particle travels parallel to the field, no deflection
occurs.

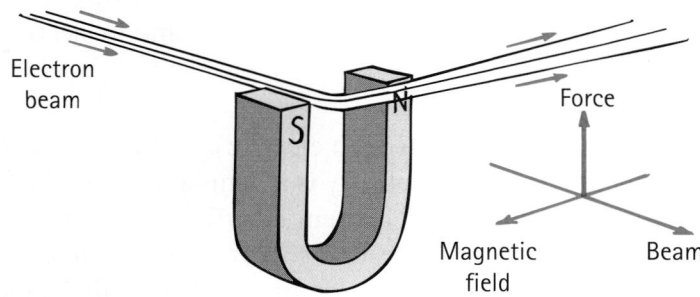

FIGURE 11.15 ▲

A beam of electrons is deflected by a magnetic field.

---

* When particles of electric charge q and velocity v move perpendicularly into a magnetic
  field of strength B, the force F on each particle is $F = qvB$. For nonperpendicular angles, v
  in this relationship must be the component of velocity perpendicular to B.

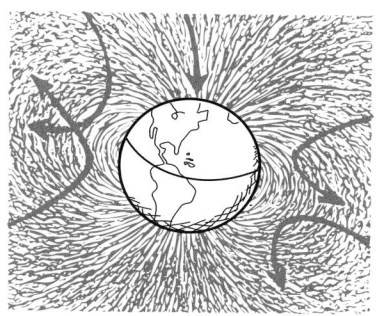

FIGURE 11.16 ▲
The magnetic field of the Earth deflects many charged particles that make up cosmic radiation.

This deflecting force is very different from the forces that occur in other interactions. Gravitation acts in a direction parallel to the line between masses, and electrical force acts in a parallel direction between charges. But magnetic force acts at right angles to the magnetic field and the velocity of the charged particle.

We are fortunate that charged particles are deflected by magnetic fields. This fact was useful in the first TV picture tubes where electrons were guided to provide a picture. More interesting, charged particles from outer space are deflected by the Earth's magnetic field. The intensity of harmful cosmic rays bombarding the Earth's surface would be stronger otherwise.

Magnetic Force on Current-Carrying Wires

Simple logic tells you that if a charged particle moving through a magnetic field experiences a deflecting force, then a current of charged particles moving through a magnetic field also experiences a deflecting force. If the particles are forced while moving inside a wire, the wire is also forced (Figure 11.17).

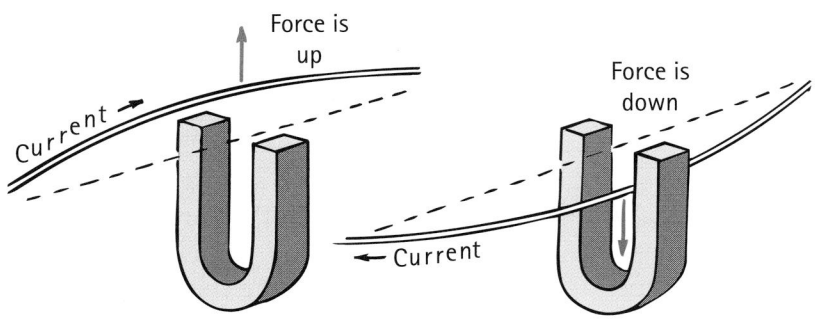

FIGURE 11.17 ▲
A current-carrying wire experiences a force in a magnetic field. (Can you see this is a follow-up of what happens in Figure 11.15?)

If we reverse the direction of current, the deflecting force acts in the opposite direction. This is useful in electric meters and in electric motors.

Electric Meters

The simplest meter that can detect an electric current is a magnetic compass. The next level of complexity is a compass in a coil of wires (Figure 11.18). When an electric current is in the coil, each loop produces its own effect on the needle, and so a very small current can be detected. A current-indicating instrument is called a *galvanometer*.

In an advanced course you'll learn the "simple" right-hand rule!

FIGURE 11.18 ▶
A very simple galvanometer.

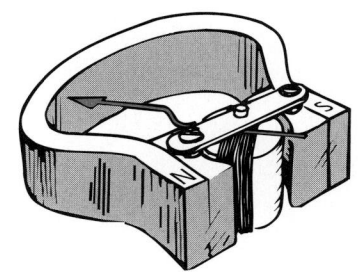

A more common galvanometer design is shown in Figure 11.19. It uses more loops of wire and is therefore more sensitive. The coil is free to move, and the magnet is held stationary. The greater the current in its windings, the greater its deflection. The coil turns against a spring, with an attached needle showing the amount of deflection. A galvanometer may be calibrated to measure current (amperes), in which case it is called an *ammeter*. Or it may be calibrated to measure voltage, in which case it is called a *voltmeter*.*

FIGURE 11.19 ▲
A common galvanometer design.

Electric Motors

If we change the design of the galvanometer slightly, so that deflection makes a complete turn rather than a partial rotation, we have an electric **motor**. The principal difference is that in a motor the current is made to change direction each time the coil makes a half rotation. This happens in cyclic fashion to produce continuous rotation, which has been used to run clocks, operate gadgets, and lift heavy loads.

◀ **FIGURE 11.20**
Both the ammeter and the voltmeter are basically galvanometers.

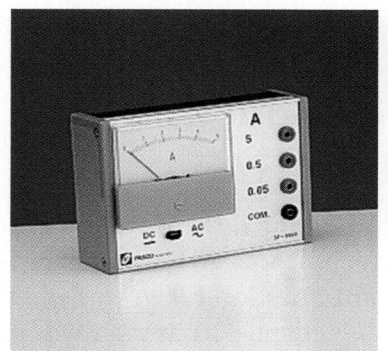 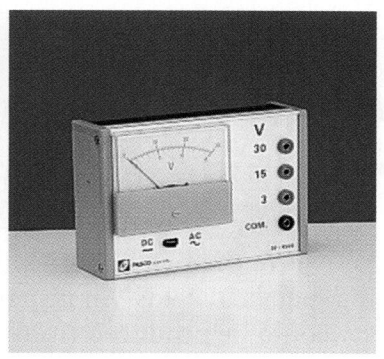

In Figure 11.21, we see the principle of the electric motor in bare outline. A permanent magnet produces a magnetic field in a region where a rectangular loop of wire is mounted to turn about the dashed axis shown. Can you see that any current in the loop has one direction in the upper side of the loop and the opposite direction in the lower side? If the upper side of the loop is forced to the left by the magnetic field, the lower side is forced to the right, as if it were a galvanometer. But unlike in a galvanometer, the current in a motor is reversed during each half revolution by stationary contacts on the shaft. The parts of the wire that rotate and brush against these contacts are called brushes. In this way, the current in the loop alternates so that the forces on the upper and lower regions do not change directions as the loop rotates. The rotation is continuous as long as current is supplied.

We have described here only a very simple dc motor. Larger motors, dc or ac, are usually made by replacing the permanent magnet by an electromagnet that is energized by some power source. Of course, more than

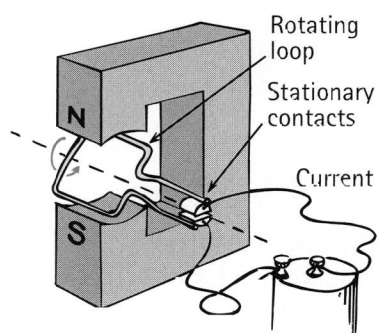

Rotating loop
Stationary contacts
Current

FIGURE 11.21 ▲
A simplified electric motor.

---

* To some degree, measuring instruments change what is being measured—ammeters and voltmeters included. Because an ammeter is connected in series with the circuit it measures, its resistance is made very low. That way it doesn't appreciably lower the current it measures. Because a voltmeter is connected in parallel, its resistance is made very high and it draws very little current for its operation.

a single loop is used. Many loops of wire are wound about an iron cylinder, called an *armature,* which then rotates when the wire loops carry current.

The advent of electric motors brought to an end much human and animal toil in many parts of the world. Electric motors have greatly changed the way people live.

CONCEPT CHECK
What is the major similarity between a galvanometer and a simple electric motor? What is the major difference?

Check Your Answers
A galvanometer and a motor are similar in that they both use coils positioned in a magnetic field. When a current passes through the coils, forces on the wires rotate the coils. The major difference is that the maximum coil rotation in a galvanometer is one half turn, whereas in a motor the coil (wrapped on an armature) rotates through many complete turns. This is accomplished by alternating the direction of the current with each half turn of the armature.

11.6 Electromagnetic Induction— How Voltage Is Created

In the early 1800s, the only current-producing devices were voltaic cells. These were the first batteries, which produced small currents by dissolving metals in acids. An important question asked was whether electricity could be produced from magnetism. The answer was provided in 1831 by two physicists, Michael Faraday in England and Joseph Henry in the United States—each working without knowledge of the other. Their discovery changed the world. The technology that followed made electricity commonplace, powering industries by day and lighting up cities at night.

Faraday and Henry both discovered that electric current can be produced in a wire simply by moving a magnet in or out of a coiled part of the wire (Figure 11.22). No battery or other voltage source is needed— only the motion of a magnet in a wire coil. ✔ They discovered that voltage is caused, or *induced,* by the relative motion between a wire and a magnetic field. Whether the magnetic field moves near a stationary conductor or the conductor moves in a stationary magnetic field— voltage is induced either way (Figure 11.23).

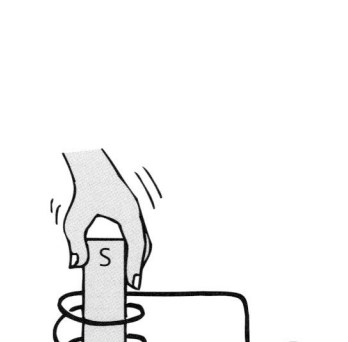

FIGURE 11.22 ▲
When the magnet is plunged into the coil, charges in the coil are set in motion; voltage is induced in the coil.

FIGURE 11.23 ▶
Voltage is induced in the wire loop either when the magnetic field moves past the wire or when the wire moves through the magnetic field.

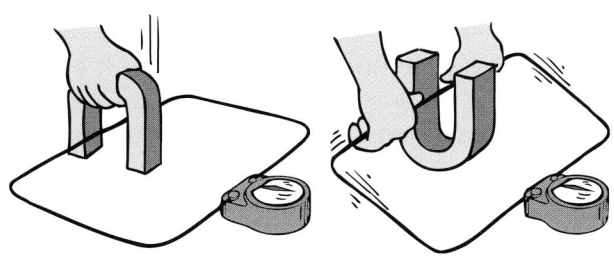

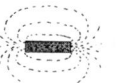

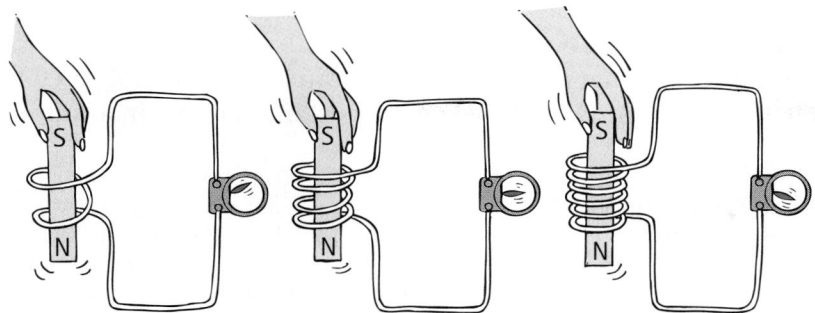

FIGURE 11.24 ▲
When a magnet is plunged into a coil of twice as many loops as another, twice as much voltage is induced. If the magnet is plunged into a coil with three times as many loops, three times as much voltage is induced.

FIGURE 11.25 ▲
It is more difficult to push the magnet into a coil made up of many loops because the magnetic field of each current loop resists the motion of the magnet.

The greater the number of loops of wire moving in a magnetic field, the greater the induced voltage (Figure 11.24). Pushing a magnet into a coil with twice as many loops induces twice as much voltage; pushing into a coil with ten times as many loops induces ten times as much voltage; and so on. It may seem that we get something (energy) for nothing by simply increasing the number of loops in a coil of wire. But we don't: We find it is more difficult to push the magnet into a coil made up of more loops. This is because the induced voltage produces a current, which makes an electromagnet, which repels the magnet in our hand. So we do more work against this "back force" to induce more voltage (Figure 11.25).

The amount of voltage induced depends on how fast the magnetic field lines are entering or leaving the coil. Very slow motion produces hardly any voltage at all. Quick motion induces a greater voltage. This phenomenon of inducing voltage by changing the magnetic field in a coil of wire is **electromagnetic induction.**

Faraday's Law

Electromagnetic induction is summarized by **Faraday's law:**

> **The induced voltage in a coil is proportional to the number of loops multiplied by the rate at which the magnetic field changes within those loops.**

The amount of *current* produced by electromagnetic induction depends on more than the induced voltage. It also depends on the resistance in both the coil and the circuit to which it's connected. For example, we can plunge a magnet in and out of a closed rubber loop and produce no current. Doing the same in a copper loop easily produces current. So the current in each is quite different. The electrons in the rubber sense the same voltage as those in the copper, but their bonding to the fixed atoms prevents the movement of electrons that move so freely in copper.

Shake flashlights need no batteries. Shake the flashlight for 30 seconds or so and generate up to 5 minutes of bright illumination. E & M induction in action! When brightness diminishes, shake again. You provide the energy to operate the device.

✓ READING CHECK

What was the discovery of Michael Faraday and Joseph Henry in 1831?

Changing a magnetic field in a closed loop induces voltage. If the loop is in an electrical conductor, then current is induced.

CONCEPT CHECK

If you push a magnet into a coil, as shown in Figure 11.25, you'll feel a resistance to your push. Why is this resistance greater in a coil with more loops?

Check Your Answer

Simply put, more work is required to provide more energy. You can also look at it this way: When you push a magnet into a coil, you cause the coil to become an electromagnet. The more loops on the coil, the stronger the electromagnet that you produce and the stronger it pushes back against you. (If the coil's electromagnet attracted your magnet instead of repelling it, energy would be created from nothing and the law of energy conservation would be violated. So the coil has to repel your magnet.)

We have mentioned two ways in which voltage can be induced in a loop of wire: by moving the loop near a magnet or by moving a magnet near the loop. There is a third way, by changing a current in a nearby loop. All three cases possess the same essential ingredient—a changing magnetic field in the loop.

We see electromagnetic induction all around us. On the road we see it operate when a car drives over buried coils of wire to activate a nearby traffic light. When iron parts of a car move over the buried coils, the Earth's magnetic field in the coils is changed, inducing a voltage to trigger the changing of the traffic lights. Similarly, when you walk through the upright coils in the security system at an airport, any metal you carry slightly alters the magnetic field in the coils. This change induces voltage and sounds an alarm. When the magnetic strip on the back of a credit card is scanned, induced voltage pulses identify the card. Similarly with the recording head of a tape recorder. Magnetic domains in the tape are sensed as the tape moves past a current-carrying coil. Electromagnetic induction is everywhere. As we shall see in Chapter 13, it underlies the electromagnetic waves we call light.

A magstripe on a credit card contains millions of tiny magnetic domains held together by a resin binder. Data are encoded in binary, with 0s and 1s distinguished by the frequency of domain reversals. It's quite amazing how quickly your name pops up when your card is swiped.

11.7 Generators and Alternating Current

A **generator** is a motor in reverse. The device is much the same, but with the roles of input and output reversed. ✔ In a motor, electrical energy is the input and mechanical energy the output. In a generator, mechanical energy is the input and electric energy is the output. Both devices simply transform energy from one form to another.

A striking example of a device functioning as both motor and generator is found in hybrid automobiles. When extra power for accelerating or hill climbing is needed, this device draws current from a battery and acts as a motor to assist the gasoline engine. When braking or rolling downhill causes the wheels to exert a torque on the device, it acts as a generator and recharges the battery. The electrical part of the hybrid engine is both a motor and a generator.

✔ READING CHECK

How do the inputs and outputs of a motor and a generator differ?

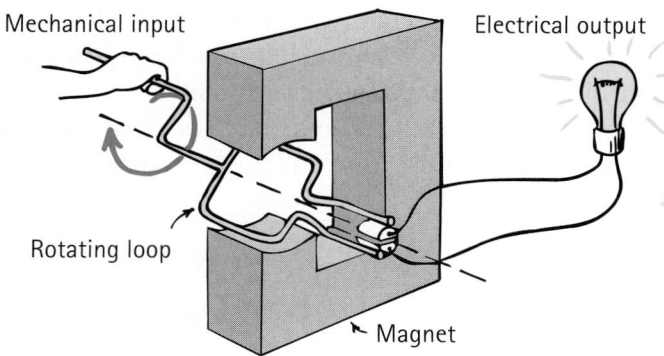

FIGURE 11.26 ▲
A simple generator. Voltage is induced in the loop when it is rotated in the magnetic field. (Compare this figure with Figure 11.21 on page 225.)

11.8 Power Production— A Technological Extension of Electromagnetic Induction

Fifty years after Faraday and Henry discovered electromagnetic induction, Nikola Tesla and George Westinghouse put those findings to practical use. They showed the world that electricity could be generated reliably and in sufficient quantities to light entire cities.

Tesla built generators much like those still in use but more complicated than the simple model we have discussed. His generators had armatures made up of bundles of copper wires. The armatures were forced to spin within strong magnetic fields by a turbine, which in turn was spun by the energy of either falling water or steam. The rotating loops of wire in the armature cut through the magnetic field of the surrounding electromagnets. In this way they induced alternating voltage and current.

✓ It's important to know that generators don't produce energy— they simply convert energy from some other form to electric energy.

FIGURE 11.27 ▲
Steam drives the turbine, which is connected to the armature of the generator.

Beware of junk scientists who sell magnets to cure physical ailments. Claims for cures are bogus. We each need a *knowledge filter* to tell the difference between what is true and what seems to be true. The best knowledge filter ever invented is science.

READING CHECK

Does a generator produce energy? Defend your answer.

As we discussed in Chapter 6, energy from a source, whether fossil or nuclear fuel or wind or water, is converted to mechanical energy to drive the turbine. The attached generator converts most of this mechanical energy to electrical energy. Some people think that electricity is a primary source of energy. It is not. It is a carrier of energy that must have a source.

11.9 The Induction of Fields— Both Electric and Magnetic

Electromagnetic induction explains the induction of voltages and currents. Actually, the more basic *fields* are at the root of both voltages and currents. The modern view of electromagnetic induction states that electric and magnetic fields are induced. These, in turn, produce the voltages we have considered. So induction takes place whether or not a conducting wire or any material medium is present. In this more general sense, Faraday's law states:

> **An electric field is induced in any region of space in which a magnetic field is changing with time.**

There is a second effect, an extension of Faraday's law. The effect is the same, except that the roles of electric and magnetic fields are interchanged. It is one of nature's many symmetries. This effect was advanced by the British physicist James Clerk Maxwell in about 1860 and is known as **Maxwell's counterpart to Faraday's law:**

> **A magnetic field is induced in any region of space in which an electric field is changing with time.**

At only one speed will the linkage between electric and magnetic fields be in perfect balance with no gain or loss of energy—exactly the speed of light!

In both cases, the strengths of the induced field are proportional to the rates of change of the inducing field. The induced electric and magnetic fields are at right angles to each other.

Maxwell saw the link between electromagnetic waves and light. If electric charges are set into vibration in the range of frequencies that match those of light, waves are produced that *are* light! ✓ Maxwell discovered that light is simply electromagnetic waves in the range of frequencies to which the eye is sensitive.

On the eve of his discovery, Maxwell had a date with a young woman he was later to marry. While walking in a garden, his date remarked about the beauty and wonder of the stars. Maxwell asked how she would feel to know that she was walking with the only person in the world who knew what the starlight really was. For it was true. At that time, James Clerk Maxwell was the only person in the world to know that light of any kind is energy carried in waves of electric and magnetic fields that continually regenerate each other.

READING CHECK

What relationship did Maxwell discover about electromagnetic waves?

11 CHAPTER REVIEW

KEY TERMS

Electromagnet A magnet whose field is produced by an electric current. Electromagnets are usually in the form of a wire coil with a piece of iron inside the coil.

Electromagnetic induction A magnetic field is induced in any region of space in which an electric field is changing with time. The magnitude of the induced magnetic field is proportional to the rate at which the electric field changes.

Faraday's law The induction of voltage when a magnetic field changes with time. If the magnetic field within a closed loop changes in any way, a voltage is induced in the loop:

$$\text{Voltage induced} \sim \text{number of loops} \times \frac{\text{magnetic field change}}{\text{time}}$$

This is a statement of Faraday's law. The induction of voltage is the result of a more fundamental phenomenon: the induction of an electric *field*.

Generator An electromagnetic induction device that produces electric current by rotating a coil within a stationary magnetic field. A generator converts mechanical energy to electrical energy.

Magnetic domains Clustered regions of aligned magnetic atoms. When these regions are aligned with one another, the substance containing them is a magnet.

Magnetic field The region of magnetic influence around either a magnetic pole or a moving charged particle.

Magnetic force (1) Between magnets, it is the attraction of unlike magnetic poles for each other and the repulsion between like magnetic poles. (2) Between a magnetic field and a moving charged particle, it is a deflecting force due to the motion of the particle. It is perpendicular to the velocity of the particle and perpendicular to the magnetic field lines. Also, it is greatest when the particle moves perpendicular to the field lines and zero when the particle moves parallel to the field lines.

Maxwell's counterpart to Faraday's law An electric field is induced in any region of space in which a magnetic field is changing with time. The magnitude of the induced electric field is proportional to the rate at which the magnetic field changes.

Motor A device employing a current-carrying coil that is forced to rotate in a magnetic field. A motor converts electrical energy to mechanical energy.

REVIEW QUESTIONS

Magnetic Poles—Attraction and Repulsion

1. In what way is the rule for the interaction between magnetic poles similar to the rule for the interaction between electric charges?
2. In what way are *magnetic poles* very different from *electric charges*?

Magnetic Fields—Regions of Magnetic Influence

3. An electric field surrounds an electric charge. What additional field surrounds a moving electric charge?
4. What two kinds of motion are exhibited by electrons in an atom?

Magnetic Domains—Clusters of Aligned Atoms

5. Most metals are not magnetic, but iron is. Why?
6. Why will dropping an iron magnet on a hard floor make it a weaker magnet?

The Interaction Between Electric Currents and Magnetic Fields

7. What is the shape of magnetic field lines about a current-carrying wire?

8. What happens to the direction of the magnetic field about an electric current when the direction of the current is reversed?

Magnetic Forces Are Exerted on Moving Charges

9. In what direction relative to a magnetic field does a charged particle move in order to experience maximum deflecting force? Minimum deflecting force?

10. Both gravitational and electrical forces act along the direction of the force fields. How is the direction of the magnetic force on a moving charge different?

11. Since a magnetic force acts on a moving charged particle, does it make sense that a magnetic force also acts on a current-carrying wire? Defend your answer.

12. What happens to the direction of the force on a wire when the current in it is reversed?

13. What is a galvanometer called when calibrated to read current? Voltage?

Electromagnetic Induction—How Voltage Is Created

14. What must change in order for electromagnetic induction to occur?

15. What are the three ways that voltage can be induced in a wire?

Generators and Alternating Current

16. What is the basic difference between a generator and an electric motor?

17. What is the basic similarity between a generator and an electric motor?

Power Production—A Technological Extension of Electromagnetic Induction

18. What commonly supplies the energy input to a turbine?

The Induction of Fields—Both Electric and Magnetic

19. What is induced by the rapid alternation of a magnetic field?

20. What is induced by the rapid alternation of an electric field?

THINK AND DO

1. Find the direction and dip (slant from the vertical) of the Earth's magnetic field lines in your locality. Magnetize a large steel needle or straight piece of steel wire by stroking it a couple of dozen times with a strong magnet. Run the needle or wire through a cork in such a way that when the cork floats, the magnet remains horizontal (parallel to the water surface). Float the cork in a plastic or wooden container of water. The needle points toward one of the Earth's magnetic poles. Then press unmagnetized common pins into the sides of the cork. Rest the pins on the rims of a pair of drinking glasses so that the needle or wire points toward the magnetic pole. It should dip in line with the Earth's magnetic field.

2. An iron bar can be easily magnetized by aligning it with the magnetic field lines of the Earth and striking it lightly a few times with a hammer. This works best if the bar is tilted down to match the dip of the Earth's field. The hammering jostles the domains so they can better fall into alignment with the Earth's field. The bar can be demagnetized by striking it when it is in an east-west direction.

THINK AND COMPARE

The magnets are moved into the wire coils in identical quick fashion. Voltage induced in each coil causes a current to flow, as indicated on the galvanometer. Neglect the electrical resistance of the loops in the coil.

1. Rank from most to least the reading on the galvanometer.

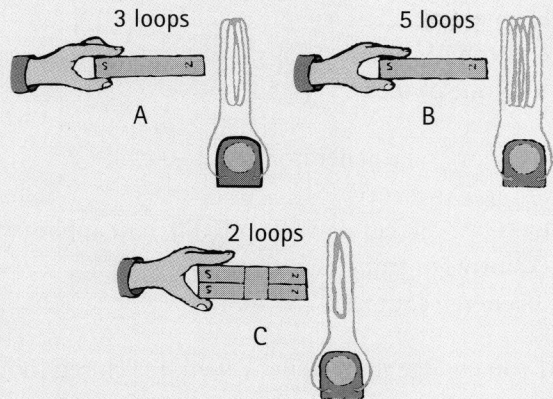

3 loops

5 loops

A

B

2 loops

C

Most current _____ _____ _____ Least current
Or, induced current the same for each. _____

2. Make the same ranking, only this time for each coil having twice as many loops as in Question 1.

THINK AND EXPLAIN

1. A paper clip near the north pole of a magnet is attracted to the magnet. Will the paper clip be attracted to the south pole of the magnet? Explain.

2. What is different about the magnetic poles of common refrigerator magnets compared with common bar magnets?

3. What surrounds a stationary electric charge? A moving electric charge?

4. "An electron always experiences a force in an electric field, but not always in a magnetic field." Defend this statement.

5. Cans of food in your kitchen pantry are likely magnetized. Why?

6. A friend tells you that a refrigerator door, beneath its layer of white painted plastic, is made of aluminum. How could you check to see if this is true (without any scraping)?

7. Magnet A has twice the magnetic field strength of magnet B (at equal distance) and at a certain distance pulls on magnet B with a force of 50 N. With how much force, then, does magnet B pull on magnet A?

8. A strong magnet attracts a paper clip to itself with a certain force. Does the paper clip exert a force on the strong magnet? If not, why not? If so, does it exert as much force on the magnet as the magnet exerts on it? Defend your answers.

9. A common pickup for an electric guitar consists of a coil of wire around a small permanent magnet. The magnetic field of the magnet magnetizes the nearby guitar string. When the string is plucked, the rhythmic oscillations of the string produce the same rhythmic changes in the magnetic field through the coil, which in turn induce the same rhythmic voltages in the coil, which when amplified and sent to a speaker produce music! Why will this type of pickup not work with nylon strings?

10. Why is a generator armature harder to rotate when it is connected to a circuit and supplying electric current?

11. If your metal car moves over a wide, closed loop of wire embedded in a road surface, will the magnetic field of the Earth within the loop be altered? Will this produce a current pulse? Can you think of a practical application for this at a traffic intersection?

12. At the security area of an airport, you walk through a weak ac magnetic field inside a coil of wire. What is the result of a small piece of metal on your person that slightly alters the magnetic field in the coil?

13. A piece of plastic tape coated with iron oxide is magnetized more in some parts than in others. When the tape is moved past a small coil of wire, what happens in the coil? What is a practical application of this?

14. Joseph Henry's wife tearfully sacrificed part of her wedding gown for silk to cover the wires of Joseph's electromagnets. What was the purpose of the silk covering?

15. If you place a metal ring in a region in which a magnetic field is rapidly alternating, the ring may become hot to your touch. Why?

16. If you force the armature of a motor to spin, will it generate an electric current? Why or why not?

17. Your friend says that if you crank the shaft of a dc motor manually, the motor becomes a dc generator. Do you agree or disagree?

18. A length of wire is bent into a closed loop and a magnet is plunged into it, inducing a voltage and, consequently, a current in the wire. A second length of wire, twice as long, is bent into two loops of wire and a magnet is similarly plunged into it. Twice the voltage is induced, but the current is the same as that produced in the single loop. Why?

19. Two separate but similar coils of wire are mounted close to each other, as shown below. The first coil is connected to a battery. The second coil is connected to a galvanometer. How does the galvanometer respond when the switch in the first circuit is closed? After being closed when the current is steady? When the switch is opened?

Primary   Secondary

20. A friend says that changing electric and magnetic fields generate one another, and this gives rise to visible light when the frequency of change matches the frequencies of light. Do you agree? Explain.

READINESS ASSURANCE TEST (RAT)

If you have a good handle on this chapter—if you really do—then you should be able to score 7 out of 10 on this RAT. Check your answers with your teacher. If you score less than 7, study further before moving on.

Choose the best *answer to each of the following.*

1. Moving electric charged particles can interact with a(n)
   (a) electric field.
   (b) magnetic field.
   (c) Both of these.
   (d) None of these.

2. The magnetic field lines about a current-carrying wire form
   (a) circles.
   (b) radial lines.
   (c) eddy currents.
   (d) energy loops.

3. A magnetic force can act on an electron even when it
   (a) is at rest.
   (b) moves parallel to magnetic field lines.
   (c) Both of these.
   (d) None of these.

4. A magnetic force acting on a beam of electrons can change its
   (a) direction.
   (b) energy.
   (c) Both of these.
   (d) None of these.

5. A motor and a generator are
   (a) similar devices.
   (b) very different devices with different applications.
   (c) forms of transformers.
   (d) energy sources.

6. If you change the magnetic field in a closed loop of wire, you induce in the loop a(n)
   (a) current.
   (b) voltage.
   (c) electric field.
   (d) All of these.

7. A voltage will be induced in a wire loop when the magnetic field within that loop
   (a) changes.
   (b) aligns with the electric field.
   (c) is at right angles to the electric field.
   (d) converts to magnetic energy.

8. A galvanometer can be calibrated to read electric
   (a) current.
   (b) voltage.
   (c) Either of these.
   (d) Neither of these.

9. An electric field is induced in any region of space in which
   (a) a magnetic field's orientation is at right angles to the electric field.
   (b) the accompanying electric field undergoes changes in time.
   (c) a magnetic field changes with time.
   (d) All of these.

10. Electricity and magnetism connect to form
   (a) mass.
   (b) energy.
   (c) ultra-high-frequency sound.
   (d) light.

12
WAVES AND SOUND

THE MAIN IDEA

Sound is a form of energy that travels in waves that spread out through space and time.

Michelle and Miriam produce vibrations in their violins when they draw their bows across the strings. These vibrations cause air inside the violins and the surrounding air to vibrate in the same way—hence the sound of music. Vibrations are wiggles in time; wiggles through space and time are waves. Sound is produced by vibrations, and likewise with light, which will be emphasized in the following chapter.

How do the natures of sound and light differ from each other? Why does sound travel so much slower than light? Is an echo simply sound that's reflected? Does the speed of sound differ in various materials? Can sound travel in a vacuum? Can one sound wave cancel another? And how does an airplane create a sonic boom? Let's begin our study of wave motion and find the answers to these questions.

Explore!

What Is Forced Vibration?

1. Strike a tuning fork with a rubber hammer or the heel of your shoe. (Do not strike the tuning fork on the edge of a table.)
2. Listen to the sound produced by the vibrating fork.
3. Hold the base of the fork against the top of a table or against other surfaces.

Analyze and Conclude

1. **Observing** How did the loudness of sound compare before and when the tuning fork was held against a surface?
2. **Predicting** How would the loudness compare if the tuning fork were held against the body of a guitar or other stringed instrument?
3. **Making Generalizations** How do forced vibrations affect the loudness of sound? How does the area of a "sounding board" affect loudness?

Be clear about the distinction between *frequency* and *speed*. How frequently a wave vibrates is altogether different from how fast it moves from one location to another.

12.1 Special Wiggles— Vibrations and Waves

When something moves periodically back and forth, side to side, or up and down, we say it vibrates. When a vibration is carried through space and time it is a **wave.** A wave is a pattern of matter or energy that extends from one place to another. Light and sound are both vibrations that move through space as waves. But they are two very different kinds of waves. Sound is the movement of vibrations of matter—through solids, liquids, or gases. If there is no matter to vibrate, then no sound is possible. Sound cannot travel in a vacuum. But light can, for light is a vibration of electric and magnetic fields—a vibration of pure energy. Light can pass through many materials, although it requires none to travel. Light travels quite nicely from the Sun and stars to the Earth, for example.

You can see the relationship between a vibration and a wave in Figure 12.1. A marking pen on a bob attached to a vertical spring vibrates up and down. Notice that it traces a waveform on a sheet of paper that moves horizontally at constant speed. The waveform is a *sine curve,* with a shape like rolling hills and valleys and a precise mathematical description. As in a water wave, the high points of a sine wave are called *crests* and the low points are *troughs.* The dashed green line shows the "home" position, or midpoint of the vibration.

✔ The term **amplitude** refers to the distance from the midpoint to the crest (or trough) of the wave. The amplitude equals the maximum displacement from the home position—from equilibrium.

✔ The **wavelength** of a wave is the distance from one crest to the next one. Equivalently, it is the distance between any two successive identical parts of the wave. The wavelengths of waves at the beach are measured in meters, the wavelengths of ripples in a pond in centimeters, and the wavelengths of light waves in billionths of a meter (nanometers).

✔ How frequently a vibration occurs is described by its **frequency.** The frequency of any vibrating object is the number of to-and-fro vibrations the object makes in a given time (usually 1 second). If a complete to-and-fro vibration (one cycle) occurs in 1 second, the frequency is one vibration per second. If two vibrations occur in 1 second, the frequency is two vibrations per second.

FIGURE 12.1 ▶
When the bob vibrates up and down, a marking pen traces out a sine curve on paper that moves horizontally at constant speed.

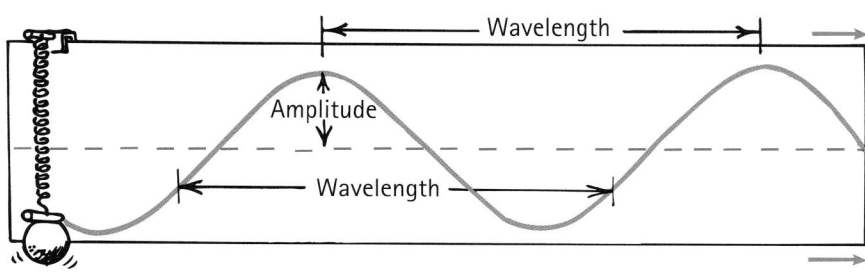

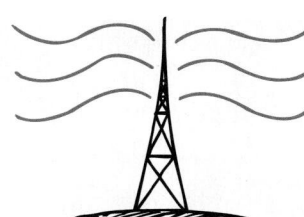

The unit of frequency is the **hertz** (Hz), as was mentioned briefly in the previous chapter. It is named after Heinrich Hertz, who demonstrated the existence of radio waves in 1886. We call one vibration per second 1 hertz; two vibrations per second is 2 hertz, and so on. Higher frequencies are measured in kilohertz (kHz), and still higher frequencies in megahertz (MHz). AM radio waves are usually measured in kilohertz, and FM radio waves are measured in megahertz. A station at 960 kHz on the AM radio dial, for example, broadcasts radio waves that have a frequency of 960,000 vibrations per second. A station at 101.7 MHz on the FM dial broadcasts radio waves that have a frequency of 101,700,000 hertz. These radio-wave frequencies are the frequencies at which electrons in the broadcasting antenna are forced to vibrate.

FIGURE 12.2 ▲
Electrons in the transmitting antenna vibrate 940,000 times each second and produce 940-kHz radio waves.

✔ The **period** of a wave or vibration is the time it takes for a complete vibration. Period can be calculated from frequency, and vice versa. Suppose, for example, that a pendulum makes two vibrations in 1 second. Its frequency is 2 Hz, and the time needed to complete one vibration—that is, the period—is $\frac{1}{2}$ second. If the vibration frequency is 3 Hz, then the period is $\frac{1}{3}$ second. Frequency and period are the inverse of each other:

$$\text{Frequency} = \frac{1}{\text{period}} \qquad \text{Period} = \frac{1}{\text{frequency}}$$

Cell phones operate in the GHz range—where electrons inside jiggle in unison billions of times per second!

CONCEPT CHECK

1. An electric toothbrush completes 90 cycles every second. What are (a) its frequency and (b) its period?
2. Gusts of wind cause the Sears Building in Chicago to sway back and forth, completing a cycle every 10 seconds. What are (a) its frequency and (b) its period?

Check Your Answers
  1. (a) 90 cycles per second is 90 vibrations per second, or 90 Hz; (b) $\frac{1}{90}$ s.
  2. (a) $\frac{1}{10}$ Hz; (b) 10 s.

✔ **READING CHECK**

Distinguish between the amplitude, wavelength, frequency, and period of a wave.

12.2 Wave Motion— Transporting Energy

When you drop a stone into a quiet pond, waves travel outward in expanding circles. Energy is carried by the wave, moving from place to place. The water itself goes nowhere. This can be seen by watching a leaf floating in the water. The leaf bobs up and down but doesn't travel with the waves. When you speak, the energy of your voice travels across the room at about 340 meters per second. ✔ Wave energy travels through the air. The air goes nowhere. The air itself doesn't travel at this speed. If it did, then speaking would be a windy experience.

Wave speed can be expressed as $v = f\lambda$, where f is the wave frequency and λ (the Greek letter lambda) is the wavelength of the wave.

FIGURE 12.3 ▲
Water waves.

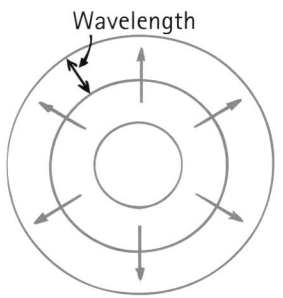

Wavelength

FIGURE 12.4 ▲
A top view of water waves.

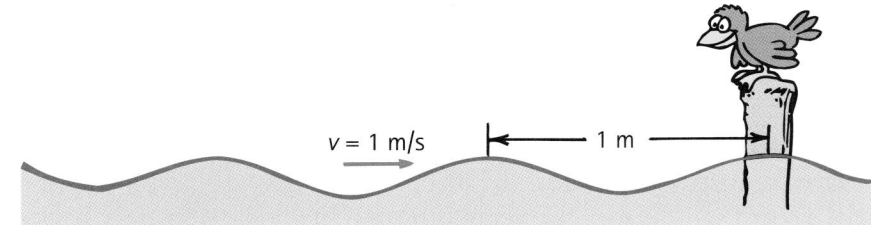

✔ **READING CHECK**

Exactly what is it that moves through the air in the case of sound? What doesn't travel in a sound wave?

Wave Speed

The speed of a wave is related to the frequency and wavelength of the waves. You can understand this by considering the simple case of water waves (Figures 12.3 and 12.4). Imagine that you fix your eyes at a stationary point on the surface of water and observe the waves passing by this point. You can measure the amount of time that passes between the arrival of one crest and the arrival of the next one (this time interval is the period). And you can also estimate the distance between crests (the wavelength). You know that speed is defined as distance divided by time. In this case, the distance is one wavelength and the time is one period. That means wave speed = wavelength/period.

For example, if the wavelength is 10 meters and the time between crests at a point on the surface is 0.5 second, the wave moves 10 meters in 0.5 seconds. So its speed is 10 meters divided by 0.5 seconds = 20 meters per second.

Since period is the inverse of frequency, the formula wave speed = wavelength/period can also be written

$$\textbf{Wave speed} = \textbf{frequency} \times \textbf{wavelength}$$

This relationship applies to all kinds of waves, whether they are water waves, sound waves, or light waves.

$v = 1$ m/s   1 m

FIGURE 12.5 ▲
If the wavelength is 1 m, and one wavelength per second passes the pole, then the speed of the wave is 1 m/s.

CONCEPT CHECK

1. If a train of freight cars, each 10 m long, rolls by you at the rate of three cars each second, what is the speed of the train?
2. If a water wave vibrates up and down three times each second and the distance between wave crests is 2 m, what are (a) the wave's frequency? (b) its wavelength? (c) its wave speed?

Check Your Answers

1. 30 m/s. We can see this in two ways. (1) According to the speed definition from Chapter 2, $v = \frac{d}{t} = (3 \times 10 \text{ m})/1 \text{ s} = 30$ m/s, because 30 m of train passes you in 1 s. (2) If we compare the train to wave motion, with wavelength corresponding to 10 m and frequency 3 Hz, then speed = frequency × wavelength = 3 Hz × 10 m = 30 m/s.
2. (a) 3 Hz; (b) 2 m; (c) wave speed = frequency × wavelength = 3/s × 2 m = 6 m/s. (Note that 3 Hz is 3 vibrations/s, and because "vibrations" has no unit, we write 3 Hz = 3/s.) It is customary to express the equation for wave speed as $v = f\lambda$, where v is wave speed, f is wave frequency, and λ (the Greek letter lambda) is wavelength.

12.3 Two Types of Waves— Transverse and Longitudinal

Fasten one end of a Slinky to a wall and hold the free end in your hand. Shake it up and down and you produce vibrations that are at right angles to the direction of wave travel. The right-angled, or sideways, motion is called *transverse motion.* ✓ This type of wave—in which the direction of wave travel is perpendicular to the direction of the vibrating source—is called a **transverse wave.** Waves in the stretched strings of musical instruments are transverse. Electromagnetic waves are also transverse.

✓ A **longitudinal wave** is one in which the direction of wave travel is *along* the direction in which the source vibrates. You produce a longitudinal wave with your Slinky when you shake it back and forth along the Slinky's axis (Figure 12.6a). The vibrations are then parallel to the direction of energy transfer. Part of the Slinky is compressed, and a wave of *compression* travels along it. In between successive compressions is a stretched region, called a *rarefaction.* Both compressions and rarefactions travel in the same direction along the Slinky. Together they make up the longitudinal wave.

READING CHECK

Distinguish between a transverse wave and a longitudinal wave.

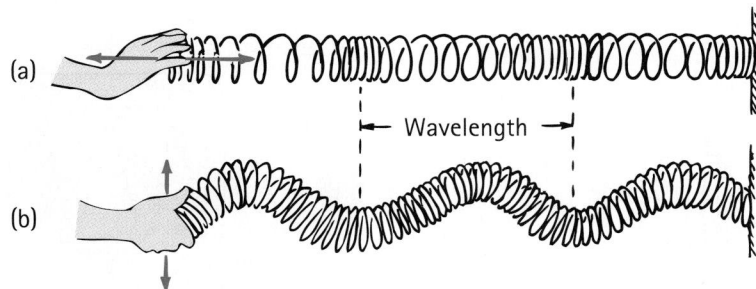

(a)

├── Wavelength ──┤

(b)

◀ **FIGURE 12.6**
Both waves transfer energy from left to right. (a) When the Slinky is pushed and pulled rapidly along its length, a longitudinal wave is produced. (b) When the end of the Slinky is shaken up and down, a transverse wave is produced.

12.4 Sound Travels in Longitudinal Waves

Think of the air molecules in a room as tiny, randomly moving Ping-Pong balls. If you vibrate a Ping-Pong paddle in the midst of the balls, you'll set them vibrating to and fro. In some regions the balls are momentarily bunched up (compressions) and in other regions in between they are momentarily spread out (rarefactions). The same is done to air molecules by the vibrating prongs of a tuning fork. Vibrations made up of compressions and rarefactions spread from the tuning fork throughout the air. A **sound wave** is produced.

The wavelength of a sound wave is the distance between successive compressions or, equivalently, the distance between successive rarefactions. Each molecule in the air vibrates to-and-fro about some equilibrium position as the waves move by.

FIGURE 12.7 ▲
Vibrate a Ping-Pong paddle in the midst of a lot of Ping-Pong balls, and you make the balls vibrate also.

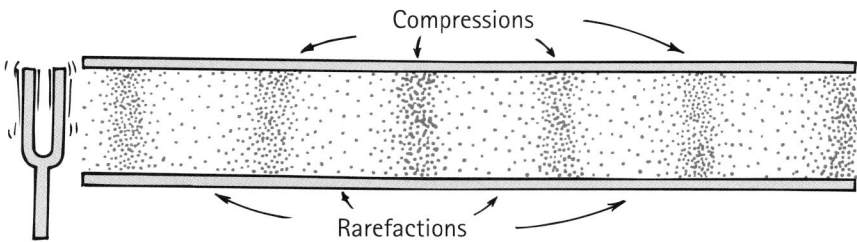

FIGURE 12.8 ▲
Compressions and rarefactions travel (both at the same speed in the same direction) from the tuning fork through the air in the tube.

Our subjective impression about the frequency of sound is described as **pitch.** ✓ A high-pitch sound like that from a tiny bell has a high vibration frequency. Sound from a large bell has a low pitch because its vibrations are of a low frequency. A young human ear can normally hear pitches from sound in a range from about 20 to 20,000 Hz. As we age, this range shrinks. So by the time you can afford to trade in your old sound system for an expensive hi-fi one, you may not be able to tell the difference. Sound waves of frequencies below 20 Hz are called *infrasonic waves,* and those of frequencies above 20,000 Hz are called *ultrasonic waves.* We cannot hear infrasonic or ultrasonic sound waves. But dogs and some other animals can.

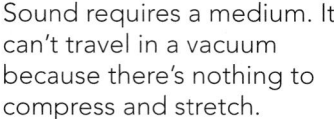

Sound requires a medium. It can't travel in a vacuum because there's nothing to compress and stretch.

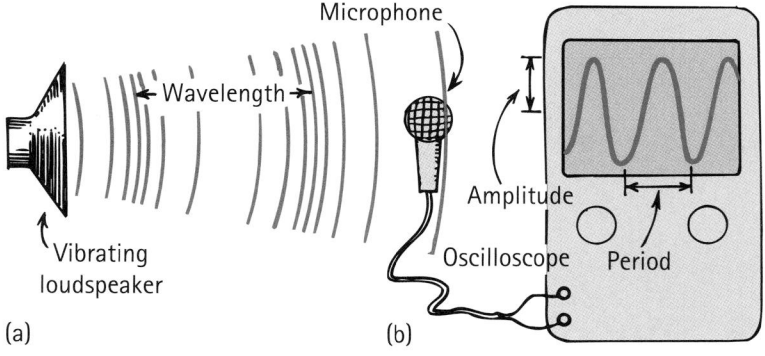
(a)   (b)

FIGURE 12.9 ▲
The radio loudspeaker is a paper cone that vibrates in rhythm with an electric signal. The sound produced sets up similar vibrations in the microphone, which are displayed on an oscilloscope. The shape of the waveform on the oscilloscope screen reveals information about the sound.

FIGURE 12.10 ▲
Waves of compressed and rarefied air, produced by the vibrating cone of the loudspeaker, make up the pleasing sound of music.

Most sound travels through air, but any elastic substance—solid, liquid, or gas—can transmit sound. An elastic substance is "springy," has resilience, and can transmit energy with little loss. Steel, for example, is elastic, whereas lead or putty is not. Many solids and liquids conduct sound better than air. You can hear the sound of a distant train clearly by placing your ear against the rail. When swimming, have a friend some

LINK TO TECHNOLOGY: LOUDSPEAKERS

The loudspeaker of your radio or other sound-producing systems changes electrical signals into sound waves. The electrical signals pass through a coil wound around the neck of a paper cone. This coil acts as an electromagnet, which is located near a permanent magnet. When the current direction is one way, magnetic force pushes the electromagnet toward the permanent magnet, pulling the cone inward. When the current direction reverses, the cone is pushed outward. Vibrations in the electric signal then cause the cone to vibrate. Vibrations of the cone produce sound waves in the air.

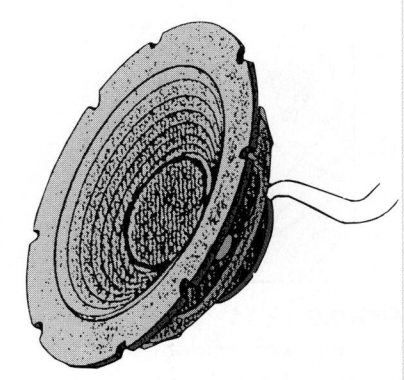

distance away click two rocks together beneath the water surface while you're submerged. Observe how well water conducts the sound.

Pause to reflect on the physics of sound while you are quietly listening to your radio sometime. The radio loudspeaker is a paper cone that vibrates in rhythm with an electrical signal. Air molecules next to the vibrating cone are set into vibration. These in turn vibrate against neighboring molecules, which in turn do the same, and so on. As a result, rhythmic patterns of compressed and rarefied air emanate from the loudspeaker and vibrate the air in the whole room. This sets your eardrum into vibration, which in turn sends cascades of rhythmic electrical impulses along the cochlea nerve canal and into your brain. And you listen to the sound of music.

Your two ears are so sensitive to the differences in sound reaching them that you can tell from what direction a sound is coming with almost pinpoint accuracy. With only one ear you would have no idea (and in an emergency might not know which way to jump).

Speed of Sound

If you watch a person at a distance chopping wood or hammering, you can easily see that the blow occurs before you hear it. Likewise, you see a flash of lightning before you hear thunder. Sound takes time to travel from one location to another. The speed of sound depends on wind conditions, temperature, and humidity. But it doesn't depend on the loudness or the frequency of sound. All sounds, loud or soft, high- or low-pitched, travel at the same speed in a given medium. The speed of sound in dry air at 0°C is about 330 meters per second (nearly 1200 km/h). It travels slightly faster if water vapor is in the air. Sound travels faster through warm air than cold air. This makes sense because faster-moving molecules in warm air bump into each other more often and transmit a pulse in less time. For each degree rise in temperature above 0°C, the speed of sound in air increases by 0.6 meter per second. So in air at a normal 20°C room temperature, sound travels at about 340 meters per second. In water, sound speed is about 4 times its speed in air; in steel it's about 15 times.

✔ **READING CHECK**

How do the pitches of a tiny bell and a large bell compare?

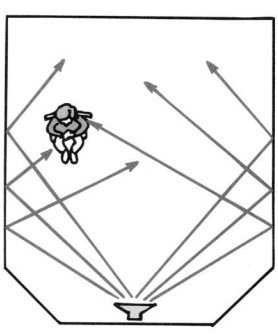

FIGURE 12.11 ▲
The angle of incident sound is equal to the angle of reflected sound.

FIGURE 12.12 ▲
Reflecting surfaces above the orchestra nicely reflect sound.

Compared with the way in which light reflects, how does sound reflect?

12.5 Sound Can Be Reflected

We call the reflection of sound an *echo*. A large fraction of sound energy is reflected from a surface that is rigid and smooth. Less sound is reflected if the surface is soft and irregular. Sound energy that is not reflected is either transmitted or absorbed.

✓ Sound reflects from a smooth surface the same way light does—the angle of incidence is equal to the angle of reflection (Figure 12.11). Reflected sound in a room makes it sound lively and full, as you have probably noticed while singing in the shower. Sometimes when the walls, ceiling, and floor of a room are too reflective, the sound becomes garbled. This is due to multiple reflections called *reverberations*. On the other hand, if the reflective surfaces are too absorbent, the sound level is low and the room may sound dull and lifeless. In the design of an auditorium or concert hall, a balance must be achieved between reverberation and absorption. The study of sound properties is called *acoustics*.

In concert halls, good acoustics require highly reflective surfaces behind the stage to direct sound out to an audience. Sometimes, reflecting surfaces are suspended above the stage. In the San Francisco opera hall, the reflective surfaces are large, shiny, plastic plates that also reflect light. A listener can look up at these reflectors and see the reflected images of the members of the orchestra (the plastic reflectors are somewhat curved, which increases the field of view). Both sound and light obey the same law of reflection, so if a reflector is oriented in such a way that you can see a particular musical instrument, you'll also hear it. Sound from the instrument follows the line of sight to the reflector and then to you.

12.6 Sound Can Be Refracted

✓ Sound waves bend when parts of the waves travel at different speeds. This occurs in uneven winds or when sound is traveling through air of varying temperatures. This bending of sound is called **refraction**. On a warm day, air near the ground may be warmer than air above, and so the speed of sound near the ground is greater. Sound waves therefore tend

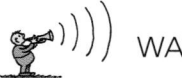 

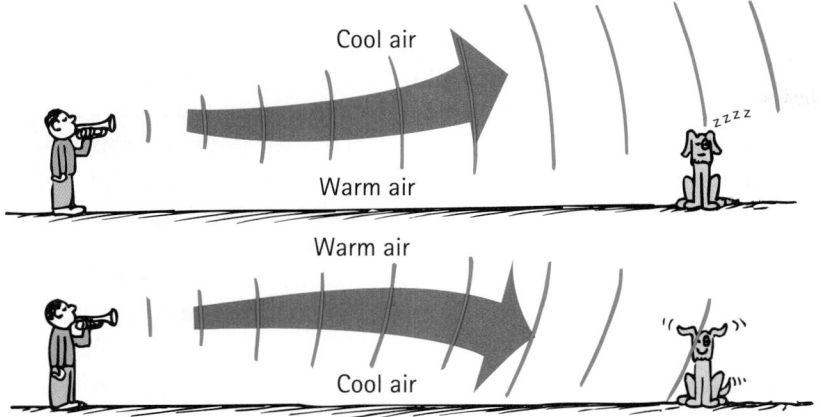

to bend away from the ground, resulting in sound that does not seem to carry well (Figure 12.13).

The refraction of sound occurs under water, too, where the speed of sound varies with temperature. This poses a problem for surface vessels that chart the bottom features of an ocean by bouncing ultrasonic waves off the bottom. This is a blessing for submarines that wish to escape detection. Layers of water at different temperatures (thermal gradients) result in refraction of sound that leaves gaps, or "blind spots," in the water. This is where submarines hide. If it weren't for refraction, submarines would be easier to detect.

The multiple reflections and refractions of ultrasonic waves are used by physicians in a technique for harmlessly "seeing" inside the body without the use of X-rays. When high-frequency sound (ultrasound) enters the body, it is reflected more strongly from the outside of organs than from their interior. A picture of the outline of the organs is obtained (Figure 12.14). This ultrasound echo technique has always been used by bats, which emit ultrasonic squeaks and locate objects by their echoes. Dolphins do this and much more.

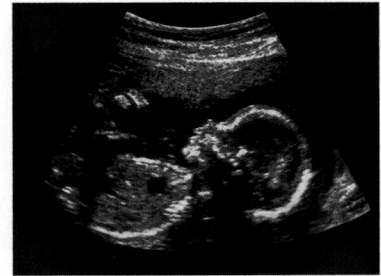

FIGURE 12.14 ▲
Megan Hewitt Abrams when she was a 14-week-old fetus.

◀ **FIGURE 12.15**
A dolphin emits ultra-high-frequency sound to locate and identify objects in its environment. Distance is sensed by the time delay between sending sound and receiving its echo, and direction is sensed by differences in time for the echo to reach its two ears. A dolphin's main diet is fish and since hearing in fish is limited to fairly low frequencies, they are not alerted to the fact they are being hunted.

LINK TO ZOOLOGY: DOLPHINS AND ACOUSTICAL IMAGING

The primary sense of the dolphin is acoustic, for sight is not a very useful sense in the often murky and dark depths of the ocean. Whereas sound is a passive sense for us, it is an active sense for dolphins when they send out sounds and then perceive their surroundings via echoes. The ultrasonic waves emitted by a dolphin enables it to "see" through the bodies of other animals and people. Because skin, muscle, and fat are almost transparent to dolphins, they "see" only a thin outline of the body, but the bones, teeth, and gas-filled cavities are clearly apparent. Physical evidence of cancers, tumors, and heart attacks can all be "seen" by dolphins—as humans have only recently been able to do with ultrasound.

What's more interesting, a dolphin can reproduce the sonic signals that paint the mental image of its surroundings. Thus the dolphin probably communicates its experience to other dolphins by communicating the full acoustic image of what is "seen," placing the image directly in the minds of other dolphins. It needs no word or symbol for "fish," for example, but communicates an image of the real thing. It is quite possible that dolphins highlight portions of the images they send by selective filtering, as we similarly do when communicating a musical concert to others via various means of sound reproduction. Small wonder that the language of the dolphin is very unlike our own!

READING CHECK

In terms of speed, what occurs when sound bends from a straight-line course?

CONCEPT CHECK
A depth-sounding vessel surveys the ocean bottom with ultrasonic sound that travels an average 1530 m/s in seawater. How deep is the water if the time delay of the echo from the ocean floor is 2 s?

Check Your Answer
The 2-s delay means it takes 1 s for the sound to reach the bottom (and another 1 s to return). Sound traveling at 1530 m/s for 1 s tells us the bottom is 1530 m deep.

12.7 Forced Vibrations and Natural Frequency

If you strike an unmounted tuning fork, its sound is rather faint. Repeat but hold the fork against a table after you strike it, and the sound is louder. This is because the table is forced to vibrate, and with its larger surface it sets more air in motion. The table can be made to vibrate by a fork of any frequency. This is a case of **forced vibration.** The vibration of a factory floor caused by the running of heavy machinery is another example of forced vibration. A more pleasing example is given by the sounding boards of stringed instruments.

Drop a wrench and a baseball bat on a concrete floor, and you can tell the difference between their sounds. This is because each vibrates differently when striking the floor. They are not forced to vibrate at a particular frequency, but instead each vibrates at its own special frequency. When disturbed, any object made of an elastic material vibrates at its own special set of frequencies, which together form its special sound. We speak of an object's **natural frequency,** which depends on factors such as the elasticity and shape of the object. Bells and tuning

forks, of course, vibrate at their own special frequencies. And interestingly, ✓ most other things, from planets to atoms and almost everything in between, have a springiness to them and vibrate at one or more natural frequencies.

✓ **READING CHECK**

What besides bells and tuning forks vibrate at their own natural frequencies?

12.8 Resonance and Sympathetic Vibrations

When the frequency of forced vibrations imposed on an object matches the object's natural frequency, a dramatic increase in amplitude occurs. This phenomenon is called **resonance** (which means "resounding" or "sounding again"). Putty doesn't resonate because it isn't elastic, and a dropped handkerchief is too limp. ✓ In order for something to resonate, it needs a force to pull it back to its starting position and enough energy to keep it vibrating.

A common experience illustrating resonance occurs on a swing. When pumping a swing, you pump in rhythm with the natural frequency of the swing. More important than the force with which you pump is the timing. Even small pumps or small pushes from someone else, if delivered in rhythm with the frequency of the swinging motion, produce large amplitudes.

A common classroom demonstration of resonance is illustrated with a pair of tuning forks adjusted to vibrate at the same frequency and spaced a meter or so apart. When one of the forks is struck, it sets the other fork into vibration. This is a small-scale version of pushing a friend on a swing—it's the timing that's important. When a series of sound waves impinge on the fork, each compression gives the prongs a tiny push. Because the frequency of these pushes matches the natural frequency of the fork, the succession of pushes increases the amplitude of vibration. This is because the pushes occur at the right time and in the same direction as the instantaneous motion of the fork. The motion of the second fork is called a *sympathetic vibration*.

Respect the knowledge that you are acquiring. It's the one thing that can never be taken away from you.

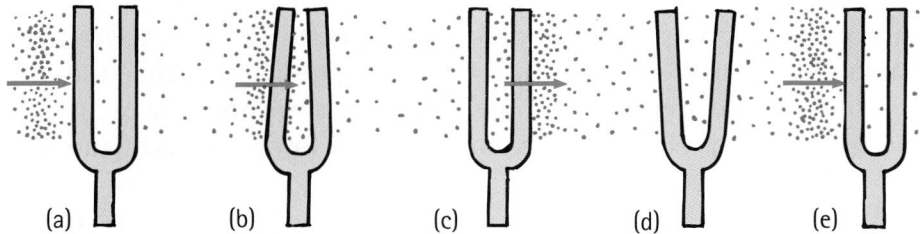

(a)   (b)   (c)   (d)   (e)

FIGURE 12.16 ▲

Stages of resonance. (a) The first compression meets the prong of the fork and gives it a tiny and momentary push; (b) the prong bends and then (c) returns to its initial position just at the time a rarefaction arrives. Still moving, the prong (d) overshoots in the opposite direction. Just when the prong restores to its initial position, (e) the next compression arrives to repeat the cycle. Now it bends farther because it is already moving.

FIGURE 12.17 ▲

In 1940, four months after being completed, the Tacoma Narrows Bridge in the state of Washington was destroyed by wind-generated resonance. A mild gale produced an irregular force in resonance with the natural frequency of the bridge, steadily increasing the amplitude of vibration until the bridge collapsed.

If the forks are not adjusted for matched frequencies, the timing of pushes is off and resonance does not occur. When you tune your radio, you adjust the natural frequency of the electronics in the radio to match one of the many surrounding signals in the air. The radio then resonates to one station at a time instead of playing all stations at once.

Resonance occurs whenever successive impulses are applied to a vibrating object in rhythm with its natural frequency. Cavalry troops marching across a footbridge near Manchester, England, in 1831 mistakenly caused the bridge to collapse when they marched in rhythm with the bridge's natural frequency. Since then, it is customary to order troops to "break step" when crossing bridges. A 20th-century bridge disaster was caused by wind-generated resonance (Figure 12.17).

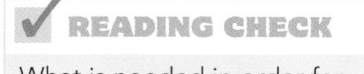
READING CHECK

What is needed in order for something to resonate?

12.9 Interference—The Addition and Subtraction of Waves

One of the most interesting properties of all waves is **interference.** Consider transverse waves. When the crest of one wave overlaps the crest of another, the crests add together. The result is a wave of increased amplitude. This is *constructive interference* (Figure 12.18). When the crest of one wave overlaps the trough of another, the opposite occurs. The high portions of one wave simply fill in the low portions of another. This is *destructive interference.*

Wave interference is easiest to see in water. Study Figure 12.19 and look at the interference pattern made when two vibrating objects touch

FIGURE 12.18 ▶

Constructive and destructive interference in a transverse wave.

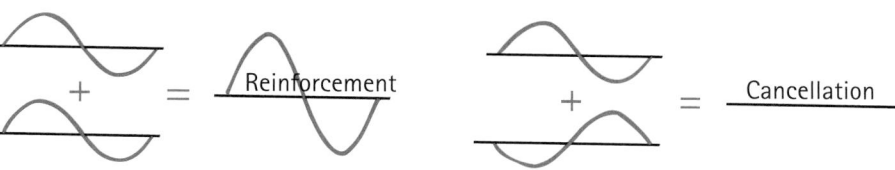

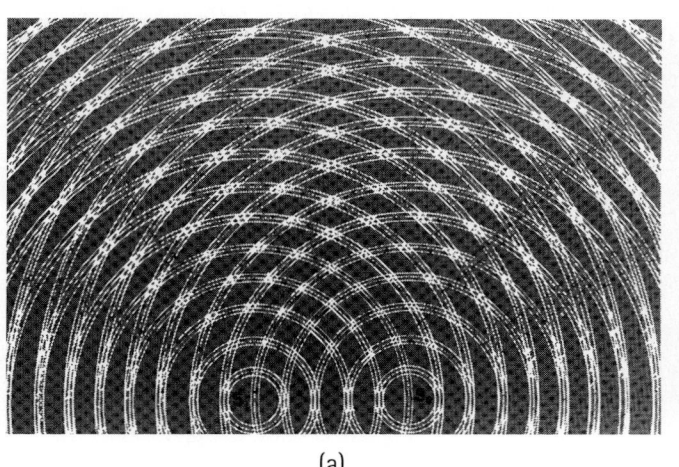
(a)

(b)

FIGURE 12.19 ▲
A drawing (left) and a photo (right) of two sets of overlapping water waves that produce interference patterns.

the surface of water. Notice the regions where a crest of one wave overlaps the trough of another. This results in regions of zero amplitude. At points along these regions, the waves arrive out of step. We say they are *out of phase* with one another.

☑ Interference is a property of all wave motion, whether the waves are water waves, sound waves, or light waves. We see a comparison of interference for both transverse and longitudinal waves in Figure 12.20. In the case of sound, the crest of a wave corresponds to a compression and the trough of a wave corresponds to a rarefaction.

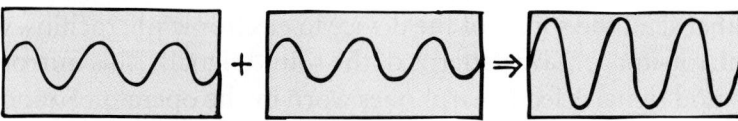

The superposition of two identical transverse waves in phase produces a wave of increased amplitude.

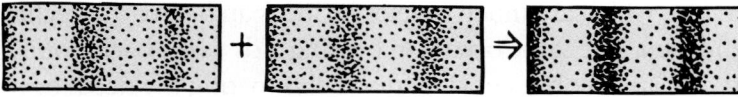

The superposition of two identical longitudinal waves in phase produces a wave of increased intensity.

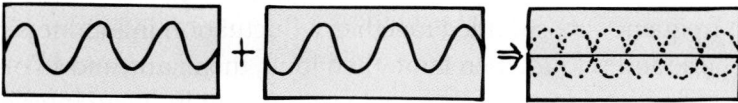

Two identical transverse waves that are out of phase destroy each other when they are superimposed.

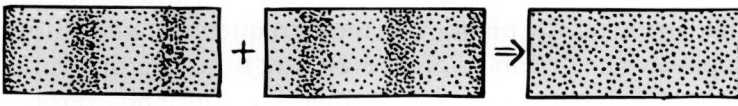

Two identical longitudinal waves that are out of phase destroy each other when they are superimposed.

◄ **FIGURE 12.20**
Constructive (top two panels) and destructive (bottom two panels) wave interference in transverse and longitudinal waves.

SOUND-OFF EXPLORATION

Get a sound system with a pair of detachable speakers and set it for monaural sound (not stereo). Reverse the wiring on one of the speakers by switching the positive and negative wire inputs. Your speakers are then out of phase. When both speakers emit the same signal, a compression emitted by one speaker occurs at the same time a rarefaction is being emitted by the other. The resulting sound is not as full and not as loud as from speakers properly connected in phase. That's because the waves, mainly the longer ones, are being canceled by interference.

Now here's the important step. Unhook the speakers so you can face them toward each other. First hold them apart, with music playing, then bring them closer together. You'll hear weaker sound because shorter waves are being canceled. The sound is tinny. When the pair of speakers are brought face to face and touching each other, you'll hear very little sound! You must try this to appreciate it.

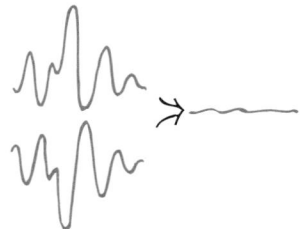

FIGURE 12.21 ▲
When a mirror image of a sound signal combines with original signal, the sound is canceled.

FIGURE 12.22 ▲
Ken Ford tows gliders in quiet comfort when he wears his noise-canceling earphones.

Destructive sound interference is a useful property in *antinoise technology*. Some noisy devices such as jackhammers are equipped with microphones that send the sound of the device to electronic microchips, which create mirror-image wave patterns of the sound signals. This mirror-image sound signal is fed to earphones worn by the operator. Sound compressions (or rarefactions) from the device are canceled by mirror image rarefactions (or compressions) in the earphones. The combination of signals cancels the jackhammer noise. Antinoise devices are also common in some aircraft, which are much quieter inside than before this technology was introduced. The same occurs in hearing aids when background noise is canceled. Are automobiles next, perhaps eliminating the need for mufflers?

Beats—An Effect of Sound Interference

A tone is a sound of distinct pitch and duration. When two tones of slightly different frequency are sounded together, a fluctuation in loudness may be heard. The sound is loud, then faint, then loud, then faint, and so on. This periodic variation in loudness is called *beats*. A **beat** is due to interference.

Beats can occur with any kind of wave and provide a practical way to compare frequencies. To tune a piano, for example, a piano tuner listens for beats produced between a standard tuning fork and the frequency of a particular string on the piano. When the frequencies are identical, the beats disappear. The members of an orchestra tune their instruments by listening for beats between their instruments and a standard tone produced by a piano or some other instrument.

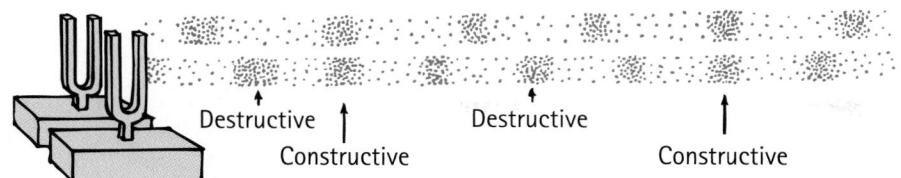

◄ **FIGURE 12.23**
The interference of two sound sources of slightly different frequencies produces beats.

Standing Waves—The Effect of Waves Passing Through Each Other

Another interesting effect of interference is *standing waves.* Tie a rope to a wall and shake the free end up and down. (A rubber tube works even better.) Waves that hit the wall are reflected back along the rope. By shaking the rope just right, you can cause the incident and reflected waves to interfere and form a **standing wave,** where parts of the rope, called the *nodes,* are stationary. You can hold your fingers on either side of the rope at a node, and the rope will not touch them. Thus nodes are positions of zero rope displacement. The distance between nodes is $\frac{1}{2}$ wavelength. Two loops make up a complete wave. The positions on a standing wave that have the largest displacements are known as *antinodes* and occur halfway between nodes.

Standing waves are produced when two sets of waves of equal amplitude and wavelength pass through each other in opposite directions. Then the waves are steadily in and out of phase with each other and produce stable regions of constructive and destructive interference (Figure 12.24).

Standing waves are set up in the strings of musical instruments when plucked, bowed, or struck. Standing waves can be set up in a tub of water or a bowl of soup by sloshing it back and forth with the right frequency. They can be produced with either transverse or longitudinal vibrations.

A life without music misses a richness of spirit. Likewise for a life without science.

✔ **READING CHECK**

What kind of waves exhibit the property of interference?

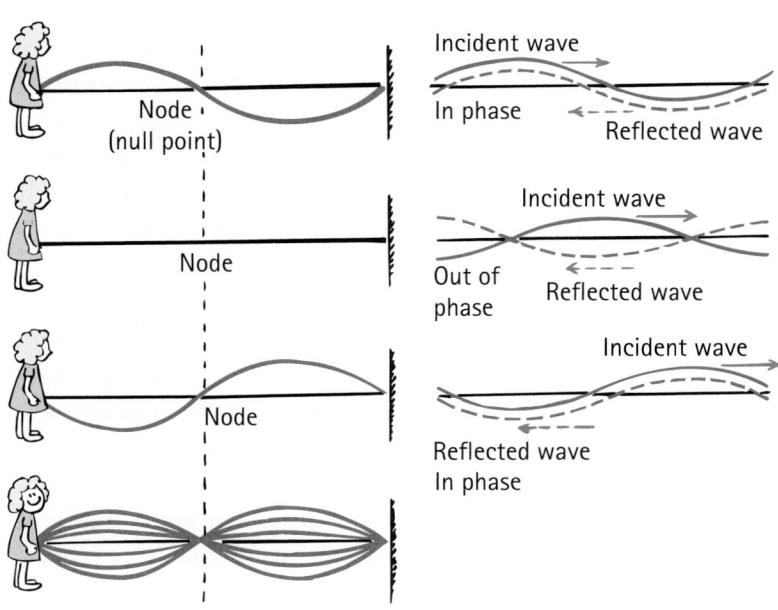

◄ **FIGURE 12.24**
The incident and reflected waves interfere to produce a standing wave.

FIGURE 12.25 ▶
(a) Shake the rope until you set up a standing wave of one loop ($\frac{1}{2}$ wavelength). (b) Shake with twice the frequency and produce a wave having two loops (1 wavelength). (c) Shake with three times the frequency and produce three loops ($\frac{3}{2}$ wavelengths).

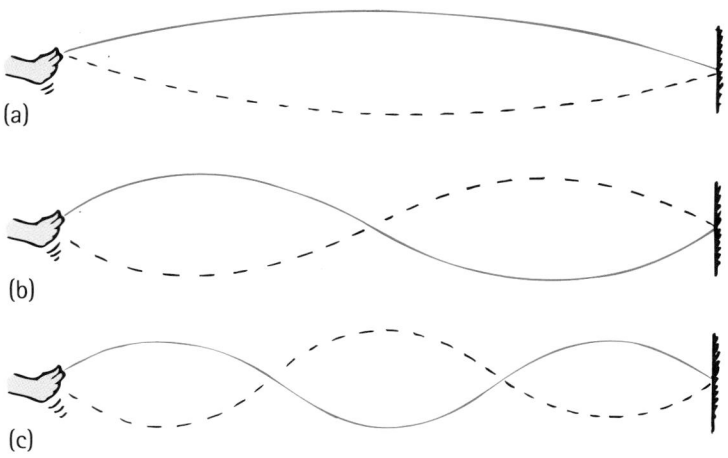

(a)

(b)

(c)

CONCEPT CHECK

Is it possible for one wave to cancel another so that no amplitude remains at some points?

Check Your Answer

Yes. This is destructive interference. In a standing wave in a rope, for example, parts of the rope—the nodes—have no amplitude because of destructive interference.

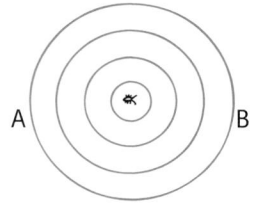

FIGURE 12.26 ▲
Top view of water waves made by a stationary bug jiggling in still water.

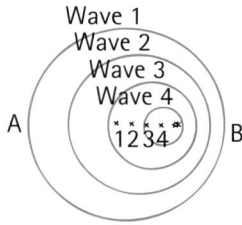

FIGURE 12.27 ▲
Water waves made by a bug swimming in still water toward point B.

12.10 The Doppler Effect—Changes in Frequency Due to Motion

A pattern of water waves produced by a bug jiggling its legs and bobbing up and down in the middle of a quiet puddle is shown in Figure 12.26. The bug is not going anywhere but is merely treading water in a fixed position. The waves it makes are concentric circles because wave speed is the same in all directions. If the bug bobs in the water at a constant frequency, the distance between wave crests (the wavelength) is the same in all directions. Waves encounter point A as frequently as they encounter point B. This means that the frequency of wave motion is the same at points A and B, or anywhere else near the bug. This wave frequency is the same as the bug's bobbing frequency.

Suppose now that the jiggling bug begins moving to the right at a speed less than the wave speed. In effect, the bug chases part of the waves it makes. The circular waves are no longer concentric (Figure 12.27). Each one has its center where the bug was previously. The outermost wave (wave 1 in Figure 12.27) was made when the bug was at the center of that circle. The center of wave 2 was made when the bug was at the center of that circle, and so on. The centers of the circular waves move in the direction of the swimming bug. Although the bug maintains the same bobbing frequency as before, an observer at B sees the waves coming more frequently. In other words, observer B measures a *higher* frequency. An observer at A, on the other hand, measures a *lower* frequency because of the longer time between wave-crest arrivals. ✓ This change in frequency due to the motion of the source (or receiver) is called the **Doppler effect** (after the Austrian scientist Christian Doppler).

Water waves spread over the two-dimensional surface of the water. Sound and light waves, on the other hand, travel in three-dimensional space in all directions like an expanding balloon. Just as circular waves are closer together in front of the swimming bug, spherical sound or light waves ahead of a moving source are closer together and reach a receiver more frequently. The Doppler effect applies to all types of waves.

The Doppler effect is evident when you hear the changing pitch of a siren as the vehicle drives by. When it approaches, the pitch is higher than normal. The wave crests reach your ear more frequently. When the vehicle passes and moves away, you hear a drop in pitch. Then the crests of the waves hit your ear less frequently.

FIGURE 12.28 ▲
The pitch (frequency) of sound increases when the source moves toward you and decreases when the source moves away.

The Doppler effect also occurs when the receiver moves. Move toward a stationary wave source and you encounter its waves more frequently. Move away, and you encounter waves less frequently. The Doppler effect results from relative motion between a wave source and a receiver.

The Doppler effect also occurs with light waves. When a source of light waves approaches, the increase in frequency is called a *blue shift*. That's because the increase is toward the high-frequency, blue end of the color spectrum. A decrease in frequency is called a *red shift*, referring to a shift toward the lower-frequency, red end of the color spectrum. The galaxies, for example, show a red shift in the light they emit. A measurement of this shift permits a calculation of the speeds at which they are receding from the Earth. A rapidly spinning star shows a red shift on the side turning away from us and a relative blue shift on the side turning toward us. This enables astronomers to calculate the star's spin rate.

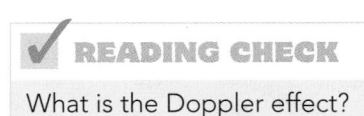

✔ READING CHECK

What is the Doppler effect?

12.11 Wave Barriers and Bow Waves

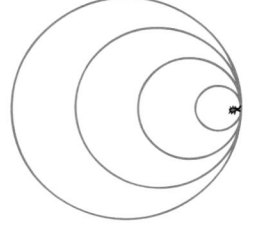

FIGURE 12.29 ▲
Wave pattern made by a bug swimming at wave speed.

✓ When a moving wave source travels as fast as the waves it produces, a *wave barrier* is produced. Consider the bug in our previous example. When it swims as fast as the wave speed, can you see that it keeps up with the waves it produces? Instead of the waves getting ahead of the bug, they pile up and overlap directly in front of it (Figure 12.29). The bug encounters a wave barrier. Much effort is required of the bug to swim over this barrier—before it can swim faster than wave speed.

The same thing happens when an aircraft travels at the speed of sound. The sound waves produced by the engines overlap to produce a barrier of compressed air on the leading edges of the wings and other parts of the craft. Considerable thrust is required for the aircraft to push through this barrier. Once through, the craft can fly faster than the speed of sound without similar opposition. It is now *supersonic.* It is like the bug, which once over its wave barrier finds the water ahead relatively smooth and undisturbed.

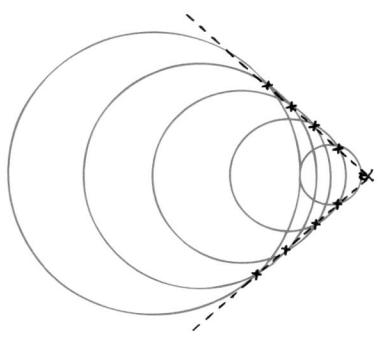

FIGURE 12.30 ▲
A bow wave, the pattern made by a bug swimming faster than wave speed. The points at which adjacent waves overlap (X) produce the V-shape.

✓ When the bug swims faster than wave speed, it produces a V-shaped wave pattern like the one shown in Figure 12.30. It outruns the waves it produces. The V-pattern caused by the overlapping waves is called a **bow wave,** which appears to be dragging behind the bug. The familiar bow (rhymes with cow) wave generated by a speedboat knifing through the water is produced by the overlapping of many periodic circular waves.

Some wave patterns made by sources moving at various speeds are shown in Figure 12.31. Note that after the speed of the source exceeds wave speed, increased speeds produce a narrower and narrower V-shape.*

---

* Bow waves generated by boats in water are more complex than indicated here. Our idealized treatment serves as an analogy for the production of the less complex shock waves in air.

v less than v_w v equals v_w v exceeds v_w v greatly exceeds v_w

FIGURE 12.31 ▲
Patterns made by a bug swimming at successively greater speeds. Overlapping at the edges occurs only when the bug swims faster than wave speed.

✓ **READING CHECK**

Compared to wave speed, how fast must an object move to produce a wave barrier? A V-shaped wave pattern?

12.12 Shock Waves and the Sonic Boom

A speedboat knifing through the water generates a two-dimensional bow wave. A supersonic aircraft similarly generates a three-dimensional **shock wave.** ✓ Just as a bow wave is produced by overlapping circles that form a V, a shock wave is produced by overlapping spheres that form a cone. And just as the bow wave of a speedboat spreads until it reaches the shore of a lake, the conical wake generated by a supersonic craft spreads until it reaches the ground. The bow wave of a speedboat that passes by can splash and douse you if you are at the water's edge. In a sense, you can say that you are hit by a "water boom." In the same way, when the conical shell of compressed air that sweeps behind a supersonic aircraft reaches listeners on the ground, the sharp crack they hear is described as a **sonic boom.**

We don't hear a sonic boom from slower-than-sound, or subsonic, aircraft because the sound waves reach our ears one at a time and make one continuous tone. Only when the craft moves faster than sound do the waves overlap to reach the listener in a single burst. The sudden increase in pressure is much the same in effect as the sudden expansion of air produced by an explosion. Both processes direct a burst of high-pressure air to the listener. The ear normally can't recognize a difference between an explosion and a sonic boom.

A water skier is familiar with the fact that next to the high hump of the bow wave is a V-shaped depression. The same is true of a shock wave. There are two cones: a high-pressure one generated at the bow of the supersonic aircraft and a low-pressure one that follows at the tail. You can see the edges of these cones in the photograph of the supersonic bullet in Figure 12.32. Between these two cones the air pressure rises sharply to above atmospheric pressure, then falls below atmospheric pressure before sharply returning to normal beyond the inner tail cone (Figure 12.33). This overpressure suddenly followed by underpressure intensifies the sonic boom.

Don't confuse *supersonic* with *ultrasonic*. Supersonic has to do with speed—faster than sound. Ultrasonic involves frequency—higher than we can hear.

FIGURE 12.32 ▲
Shock wave of a bullet piercing a sheet of Plexiglas. Light is deflected as it passes through the compressed air that makes up the shock wave, making it visible. Look carefully to see the second shock wave originating at the tail of the bullet.

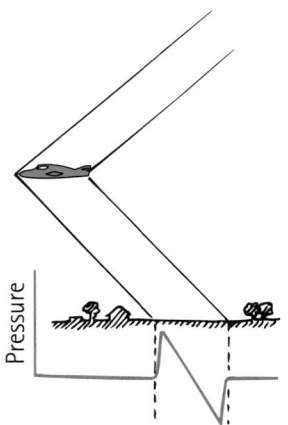

FIGURE 12.33 ▲
A shock wave and the air pressure differences it causes.

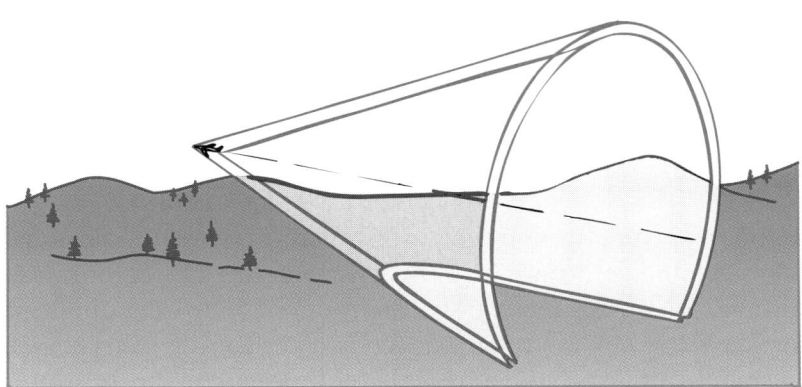

FIGURE 12.34 ▲
A shock wave is made up of two cones—a high-pressure cone with the apex at the bow of the aircraft and a low-pressure cone with the apex at the tail.

A   B   C

FIGURE 12.35 ▲
The shock wave has not yet reached listener A, but is now reaching listener B and has already reached listener C.

✔ **READING CHECK**

How are a bow wave and a shock wave similar?

In Figure 12.35, listener B is in the process of hearing a sonic boom. Listener C has already heard it, and listener A will hear it shortly. The aircraft that generated this shock wave may have broken through the sound barrier hours ago!

It is not necessary that the moving source be "noisy" to produce a shock wave. Once an object—even a silent one—is moving faster than the speed of sound, it *makes* sound. A supersonic bullet passing overhead produces a crack, which is a small sonic boom. If it were larger and disturbed more air in its path, the crack would be more boomlike. When a lion tamer cracks a whip, the cracking sound is a sonic boom produced by the tip when it travels faster than the speed of sound. Both the bullet and the whip are not in themselves sound sources. But when traveling at supersonic speeds, they produce their own sound as they generate shock waves.

12 CHAPTER REVIEW

KEY TERMS

Amplitude For a wave or vibration, the maximum displacement on either side of the equilibrium (midpoint) position.

Beats A series of alternate reinforcements and cancelations produced by the interference of two waves of slightly different frequency, heard as a throbbing effect in sound waves.

Bow wave The V-shaped wave produced by an object moving across a liquid surface at a speed greater than the wave speed.

Doppler effect The change in frequency of wave motion resulting from motion of the wave source or receiver.

Forced vibration The setting up of vibrations in an object by a vibrating force.

Frequency For a vibrating body or medium, the number of vibrations per unit time. For a wave, the number of crests that pass a particular point per unit time.

Hertz The SI unit of frequency. It equals one vibration per second.

Interference The pattern formed by superposition of different sets of waves that produces mutual reinforcement in some places and cancelation in others.

Longitudinal wave A wave in which the medium vibrates in a direction parallel (longitudinal) to the direction in which the wave travels. Sound is an example.

Natural frequency A frequency at which an elastic object naturally tends to vibrate, so that minimum energy is required to produce a forced vibration or to continue vibrating at that frequency.

Period The time required for a vibration or a wave to make a complete cycle; equal to 1/frequency.

Pitch The "highness" or "lowness" of a tone, as on a musical scale, which is mainly governed by frequency.

Refraction The bending of a wave through either a nonuniform medium or from one medium to another, caused by differences in wave speed.

Resonance The response of a body when a forcing frequency matches its natural frequency.

Shock wave The cone-shaped wave created by an object moving at supersonic speed through a fluid.

Sonic boom The loud sound resulting from the incidence of a shock wave.

Sound wave A longitudinal vibratory disturbance that travels in a medium, which a young ear can hear in the approximate frequency range 20–20,000 Hz.

Standing wave A stationary wave pattern formed in a medium when two sets of identical waves pass through the medium in opposite directions.

Transverse wave A wave in which the medium vibrates in a direction perpendicular (transverse) to the direction in which the wave travels. Light is an example.

Wave A disturbance or vibration propagated from point to point in a medium or in space.

Wave speed The speed with which waves pass a particular point:

$$\textbf{Wave speed} = \textbf{frequency} \times \textbf{wavelength.}$$

Wavelength The distance between successive crests, troughs, or identical parts of a wave.

REVIEW QUESTIONS

Special Wiggles—Vibrations and Waves

1. What is the source of all waves?
2. How do frequency and period relate to each other?

Wave Motion—Transporting Energy

3. What is it that moves from source to receiver in wave motion?
4. What is the relationship among frequency, wavelength, and wave speed?

Two Types of Waves—Transverse and Longitudinal

5. In a transverse wave, in what direction are the vibrations relative to the direction of wave travel?
6. In a longitudinal wave, in what direction are the vibrations relative to the direction of wave travel?

Sound Travels in Longitudinal Waves

7. Why will sound not travel in a vacuum?
8. How does the speed of sound in water compare with the speed of sound in air? How does the speed in steel compare with the speed in air?

Sound Can Be Reflected

9. How does the angle of incidence compare with the angle of reflection for sound?
10. What is a *reverberation*?

Sound Can Be Refracted

11. What causes refraction?
12. Does sound tend to bend upward or downward when its speed near the ground is greater than its speed at a higher level?

Forced Vibrations and Natural Frequency

13. Why does a struck tuning fork sound louder when it is held against a table?
14. Give three examples of forced vibration.

Resonance and Sympathetic Vibrations

15. Distinguish between *forced vibrations* and *resonance*.

16. When you listen to a radio, why are you able to hear only one station at a time rather than all stations at once?

Interference—The Addition and Subtraction of Waves

17. What kind of waves exhibit interference?
18. Distinguish between *constructive interference* and *destructive interference*.

The Doppler Effect—Changes in Frequency Due to Motion

19. In the Doppler effect, does frequency change? Does wavelength change? Does wave speed change?
20. Can the Doppler effect be observed with longitudinal waves, transverse waves, or both?

Wave Barriers and Bow Waves

21. How does the V-shape of a bow wave depend on the speed of the wave source?

Shock Waves and the Sonic Boom

22. How does the V-shape of a shock wave depend on the speed of the wave source?
23. True or false: A sonic boom occurs only when an aircraft is breaking through the sound barrier.
24. True or false: In order for an object to produce a sonic boom, it must be a sound source.

THINK AND DO

1. Tie a rubber tube, a spring, or a rope to a fixed support and produce standing waves. See how many nodes you can produce.
2. Test to see which ear has the better hearing by covering one ear and finding the distance away that your open ear can hear the ticking of a clock. Repeat for the other ear. Notice also how the sensitivity of your hearing improves when you cup your ears with your hands.
3. The Doppler shift is nicely heard with a buzzer of any kind that emits a steady tone. Put it in a plastic bag and swing it around your head in a circle. Your friends will hear the frequency shift as the buzzer alternately moves toward and away from them. (Why do *you* not hear a Doppler shift?)

PLUG AND CHUG

Frequency $f = \frac{1}{T}$; Period $T = \frac{1}{f}$

1. Find the frequency, in hertz, that corresponds to each of the following periods:
   (a) 0.10 s   (b) 5 s   (c) $\frac{1}{60}$ s
2. Find the period, in seconds, that corresponds to each of the following frequencies:
   (a) 10 Hz   (b) 0.2 Hz   (c) 60 Hz

Speed $v = f\lambda$

3. Find the speed of a wave with a frequency of 3 Hz and a wavelength of 2 m.
4. Find the speed of a wave with a frequency of 340 Hz and a wavelength of 1 m.

THINK AND COMPARE

1. The three waves below have the same frequency (and obviously, different wavelengths). Rank their speeds from greatest to least.

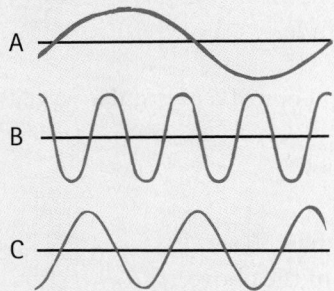

2. A fire engine sounds its siren as it approaches you. Rank the Doppler effect you hear from highest to lowest frequency.
   (a) The speed of the fire engine is 20 km/h and you are at rest.
   (b) The speed of the fire engine is 30 km/h and you are at rest.
   (c) The speed of the fire engine is 30 km/h while you drive away from it at 20 km/h.
   (d) The speed of the fire engine is 30 km/h while you drive toward it at 20 km/h.

THINK AND EXPLAIN

1. If we double the frequency of a vibrating object, what happens to its period?
2. If the frequency of a sound wave is doubled, what change occurs in its speed? In its wavelength?

3. Red light has a longer wavelength than blue light. Which has the greater frequency?
4. You dip your finger repeatedly into a puddle of water and make waves. What happens to the wavelength if you dip your finger more frequently?
5. Why will marchers at the end of a long parade following a band be out of step with marchers near the front?
6. What two physics mistakes occur in a science fiction movie that shows a distant explosion in outer space, where you see and hear the explosion at the same time?
7. A cat can hear sound frequencies up to 70,000 Hz. Bats send and receive ultra-high-frequency squeaks up to 120,000 Hz. Which hears shorter wavelengths, cats or bats?
8. You notice smoke from the starter's gun at a racetrack before you hear it fire. Explain.
9. In an Olympic competition, a microphone picks up the sound of the starter's gun and sends it electrically to speakers at every runner's starting block. Why?
10. Sound from Source A has twice the frequency of sound from Source B, both in air. Compare the speeds and wavelengths of sound from the two sources.
11. If the speed of sound depended on frequency, how would distant music sound?
12. Why is an echo weaker than the original sound?
13. What is the danger posed by people in the balcony of an auditorium stamping their feet in a steady rhythm?
14. What physical principle is used by Manuel when he pumps in rhythm with the natural frequency of the swing?

15. Would there be a Doppler effect if the source of sound were stationary and the listener in motion? Why or why not? In which direction should the listener move to hear a higher frequency? A lower frequency?

16. When you blow your horn while driving toward a stationary listener, the listener hears an increase in the horn frequency. Would the listener hear an increase in the horn frequency if he were in another car traveling at the same speed in the same direction as you? Explain.

17. Astronomers find that light coming from one edge of the Sun has a slightly higher frequency than light from the opposite edge. What do these measurements tell us about the Sun's motion?

18. Does the conical angle of a shock wave open wider, narrow down, or remain constant as a supersonic aircraft increases its speed?

19. If the sound of an airplane does not originate in the part of the sky where the plane is seen, does this imply that the airplane is traveling faster than the speed of sound? Explain.

20. Why is it that a subsonic aircraft, no matter how loud it may be, cannot produce a sonic boom?

THINK AND SOLVE

1. A weight suspended from a spring bobs up and down over a distance of 20 cm twice each second. What is its frequency? Its period? Its amplitude?

2. From far away you watch a woman driving nails into her front porch at a regular rate of one stroke per second. You hear the sound of the blows exactly synchronized with the blows you see. And then you hear one more blow after you see her cease hammering. How far away is she?

3. A skipper on a boat notices wave crests passing his anchor chain every 5 s. He estimates the distance between wave crests to be 15 m. He also correctly estimates the speed of the waves. Show that this speed is 3 m/s.

4. An oceanic depth-sounding vessel surveys the ocean bottom with ultrasonic sound that travels 1530 m/s in seawater. When the time delay of the echo from the ocean floor is 6 s, show that the depth is almost 4600 m.

5. A bat flying in a cave emits a sound and receives its echo 0.1 s later. Show that the cave wall is 21.5 m distant.

6. What frequency of sound produces a wavelength of 1 m in room-temperature air?

READINESS ASSURANCE TEST (RAT)

If you have a good handle on this chapter—if you really do—then you should be able to score 7 out of 10 on this RAT. Check your answers with your teacher. If you score less than 7, study further before moving on.

Choose the best answer to each of the following.

1. When we consider the distance a pendulum swings to and fro, we're talking about its
   (a) frequency.
   (b) period.
   (c) wavelength.
   (d) amplitude.

2. If the frequency of a particular wave is 30 Hz, its period is
   (a) $\frac{1}{30}$ s.
   (b) 30 s.
   (c) more than 30 s.
   (d) None of the above.

3. In Europe, alternating electric current vibrates to and fro 50 cycles in 1 s. The frequency of these vibrations is
   (a) 50 Hz with a period of $\frac{1}{50}$ s.
   (b) $\frac{1}{50}$ Hz with a period of 50 s.
   (c) 50 Hz with a period of 50 s.
   (d) $\frac{1}{50}$ Hz with a period of $\frac{1}{50}$ s.

4. If you dip your finger repeatedly onto the surface of still water, you produce waves. The more frequently you dip your finger, the
   (a) lower the wave frequency and the longer the wavelengths.
   (b) higher the wave frequency and the shorter the wavelengths.
   (c) Strangely, both of these.
   (d) Neither of these.

5. The vibrations along a longitudinal wave move in a direction
   (a) parallel to the wave direction.
   (b) perpendicular to the wave direction.
   (c) Both of these.
   (d) Neither of these.

6. A common example of a transverse wave is
   (a) sound.
   (b) light.
   (c) Both of these.
   (d) Neither of these.

7. When your radio set is tuned to an incoming radio signal, what is occurring?
   (a) Refraction
   (b) Forced vibration
   (c) Resonance
   (d) Diffraction

8. When sound or light undergoes interference, it can sometimes
   (a) build up to an amplitude greater than the sum of amplitudes.
   (b) cancel completely.
   (c) Both of these.
   (d) Neither of these.

9. What does NOT occur with the Doppler effect are changes in
   (a) frequency due to motion.
   (b) the speed of sound due to motion.
   (c) Both of these.
   (d) Neither of these.

10. A sonic boom is the result of wave
    (a) interference.
    (b) resonance.
    (c) superposition.
    (d) reflection and refraction.

13

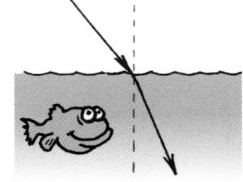

LIGHT, REFLECTION, AND COLOR

THE MAIN IDEA

Light interacting with matter can be reflected, transmitted, absorbed, or a combination of these. Transmitted light can be refracted.

Light is the only thing we see, but what exactly *is* light? How does it differ from sound? And why does light travel so fast? How does it get through glass? Does it lose speed when it passes through transparent materials like glass or when it reflects? How many mirrors are responsible for the multiple images above of physics teacher Fred Myers and his daughter McKenzie? How does the photo show you that light doesn't change frequency when it is reflected? And why do colored lights red, green, and blue mix to form white; but red, green, and blue paints mix to a muddy brown? Is there a physical science reason for why the sky is blue, sunsets are red, and clouds are white?

We begin our study of light by looking at its electromagnetic nature, how it interacts with materials, and how it appears so nicely as color.

Explore!

What Is the Color of a Candle's Reflection?

1. Hold the flame of a candle or match or any small source of white light between you and a piece of colored glass.
2. What is the color of the flame reflected from the front surface?
3. From the back surface?

Analyze and Conclude

1. **Observing** How many reflections do you see?
2. **Predicting** Why are the colors of the reflected light different?
3. **Making Generalizations** Would you expect the same differences for pieces of different-colored pieces of glass?

13.1 The Electromagnetic Spectrum— and the Tiny Bit That Is Light

If you shake the end of a stick back and forth in still water, you'll create waves on the water surface. If you similarly shake an electrically charged rod to-and-fro in empty space, you'll create electromagnetic waves in space. This is because the moving charge is in effect an electric current. Recall from Chapter 11 that a magnetic field surrounds an electric current and that the field changes as the current changes. Recall also from Chapter 11 that a changing magnetic field induces an electric field— electromagnetic induction. And what does the changing electric field do? It induces a changing magnetic field. The vibrating electric and magnetic fields regenerate each other to make up an **electromagnetic wave.** In a vacuum, all electromagnetic waves travel at the same speed.

FIGURE 13.1 ▲
Shake an electrically charged object to-and-fro, and you produce an electromagnetic wave.

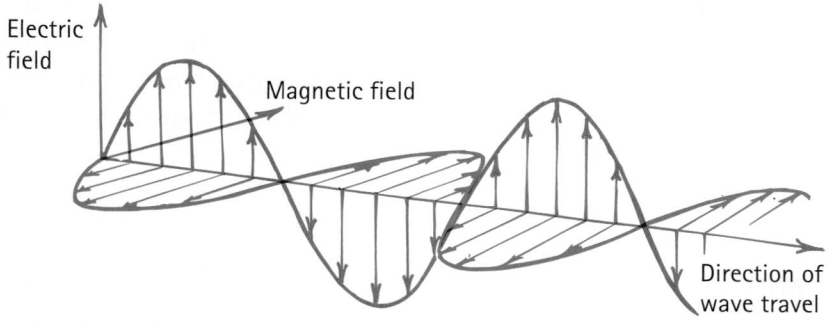

◀ **FIGURE 13.2**
The electric and magnetic fields of an electromagnetic wave are perpendicular to each other and to the direction of motion of the wave.

The classification of electromagnetic waves according to frequency is the **electromagnetic spectrum** (Figure 13.3). Electromagnetic waves of frequency as low as 0.01 hertz (Hz) have been detected. Others with frequencies of several thousand hertz (kHz) are classified as low-frequency

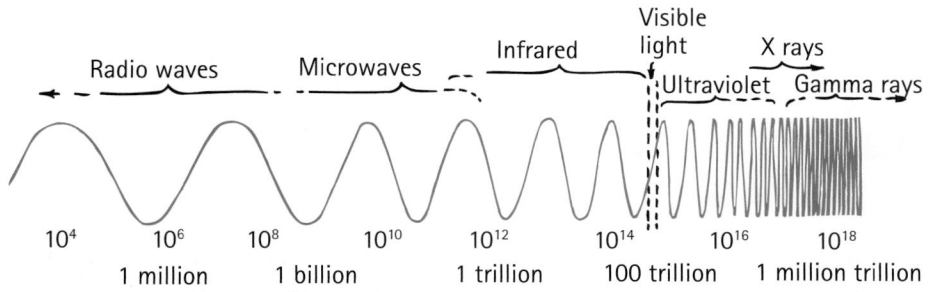

FIGURE 13.3 ▲
The electromagnetic spectrum is a continuous range of waves extending from radio waves to gamma rays. The descriptive names of the sections are merely a historical classification, for all waves are the same in nature, all traveling at the same speed. They differ mainly in frequency and wavelength.

Light is energy carried in an electromagnetic wave emitted by vibrating electrons in atoms.

radio waves. A frequency of 1 million hertz (1 MHz) lies in the middle of the AM radio band. The very-high-frequency (VHF) television band of waves starts at about 50 million Hz. FM radio ranges from 88 to 108 MHz. Then there are ultrahigh frequencies (UHF), followed by microwaves, beyond which are infrared waves, the "heat waves" we studied in Chapter 9. Farther to the right in Figure 13.3 is visible light. Surprisingly, visible light makes up less than a millionth of 1% of the electromagnetic spectrum.

The lowest frequency of light that our eyes can see appears red. The highest visible frequencies are nearly twice the frequency of red and appear violet. Still higher frequencies are ultraviolet, which are beyond our range of vision. These higher-frequency waves are more energetic and cause sunburns. Higher frequencies beyond ultraviolet extend into the X-ray and gamma-ray regions. There is no sharp boundary between these regions, which actually overlap each other. The spectrum is separated into regions merely for classification.

✓ The frequency of an electromagnetic wave as it vibrates through space is identical to the frequency of the vibrating electric charge that generates it. Different frequencies result in different wavelengths—low frequencies produce long wavelengths, and high frequencies produce short wavelengths. The higher the frequency of the vibrating charge, the shorter the wavelength of radiation.

CONCEPT CHECK

Is it correct to say that a radio wave is a low-frequency light wave? Is a radio wave also a sound wave?

Check Your Answers

Both a radio wave and a light wave are electromagnetic waves, and all electromagnetic waves originate in the vibrations of electrons. Because radio waves have lower frequencies than light waves, a radio wave may be considered a low-frequency light wave (and a light wave a high-frequency radio wave). A sound wave, however, is a *mechanical* vibration of matter and is *not* electromagnetic. A sound wave is fundamentally different from an electromagnetic wave. So a radio wave is definitely not a sound wave. (Don't confuse a radio wave with the sound that a loudspeaker emits.)

✓ **READING CHECK**

How does the frequency of emitted light compare with the frequency of the vibrating electron that generates it?

13.2 Why Materials Are Either Transparent or Opaque

When light passes through matter, some of the electrons in the matter are forced into vibration. In this way, vibrations in the emitter are transmitted to vibrations in the receiver. This is similar to the way sound is transmitted (Figure 13.4).

Materials such as glass and water allow light to pass through in straight lines. They are **transparent** to light. To understand how light penetrates through glass (or any transparent material), visualize the electrons in the atoms of glass as if they were connected to the nucleus

Just as a sound wave can force a sound receiver into vibration, a light wave can force electrons in materials into vibration.

by springs (Figure 13.5). When a light wave meets them, the electrons are set into vibration.

As we learned in the previous chapter, materials that are springy (elastic) respond more to vibrations at particular frequencies than to other frequencies. Bells ring at a particular frequency, tuning forks vibrate at a particular frequency, and so do the electrons of atoms and molecules. The natural vibration frequencies of an electron depend on how strongly it is attached to its atom or molecule. Different atoms and molecules have different "spring strengths." Electrons in atoms of glass vibrate in the ultraviolet range. So when ultraviolet waves shine on glass, resonance occurs. The vibration of electrons builds up to large amplitudes, just as pushing someone at the resonant frequency on a swing builds to a large amplitude. The energy received by a glass atom is either re-emitted or passed on to neighboring atoms by collisions. Resonating atoms in the glass can hold onto the energy of the ultraviolet light for quite a long time (about 100 millionths of a second). During this time, the atom makes about 1 million vibrations. This is time enough to collide with neighboring atoms and release its energy as heat (or more accurately, into thermal energy). Glass is therefore not transparent to ultraviolet. Ultraviolet is absorbed by glass.

At lower wave frequencies, such as those of visible light, electrons in the glass atoms are forced into vibration—but at less amplitude. The atoms retain the energy for a briefer time, with less chance of collisions with neighboring atoms. This means that less energy is transformed to heat. Instead, quite nicely, the energy of vibrating electrons is re-emitted as light. Glass is transparent to light of all the visible frequencies. The frequency of the re-emitted light that is passed from one atom to another atom is identical to the frequency of the light that initially produced the vibration. However, there is a slight time delay between absorption and re-emission.

✓ It is this time delay that results in a lower average speed of light through glass (Figure 13.6). Light travels at different average speeds through different materials. We say *average speeds* because the speed of light in a vacuum, whether in interstellar space or in the space between atoms in a piece of glass, is a constant 300,000 kilometers per second (km/s). [The presently accepted value is 299,792 km/s, rounded to 300,000 km/s (186,000 mi/s).] We call this speed of light c. The speed of

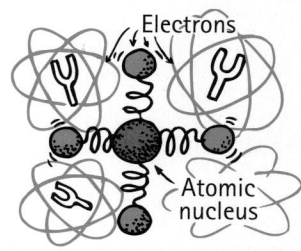

FIGURE 13.5 ▲
The electrons of atoms have certain natural frequencies of vibration and can be modeled as particles connected to the atomic nucleus by a spring. As a result, atoms and molecules behave somewhat like optical tuning forks.

In air, light travels a million times faster than sound.

Light slows when it enters glass?

FIGURE 13.6 ▲
A wave of visible light incident upon a pane of glass produces vibrations in the glass atoms, which cascade to a chain of absorptions and re-emissions that pass the light energy through the material and out the other side. Because of the time delay between absorptions and re-emissions, light travels through the glass more slowly than through empty space.

FIGURE 13.7 ▲
When the raised ball is released and hits the others, the ball that emerges from the opposite side is not the same ball that initiated the transfer of energy. Likewise, light that emerges from a pane of glass is not the same light that was incident on the glass. Both the emerging ball and emerging light are different from, though identical to, the incident ones.

light in the atmosphere is slightly less than in a vacuum but is usually rounded off as c. In water, light travels at 75% of its speed in a vacuum, or $0.75c$. In glass, light travels about $0.67c$, depending on the wavelength of light and the type of glass. In a diamond, light travels at less than half its speed in a vacuum, only $0.41c$. When light emerges from these materials into the air, it again travels at its original speed c.

Infrared waves, which have frequencies lower than those of visible light, vibrate entire molecules in the structure of glass and many other materials. This molecular vibration increases the thermal energy and temperature of the material, which is why infrared waves are often called *heat waves*. Thus glass is not transparent to infrared light.

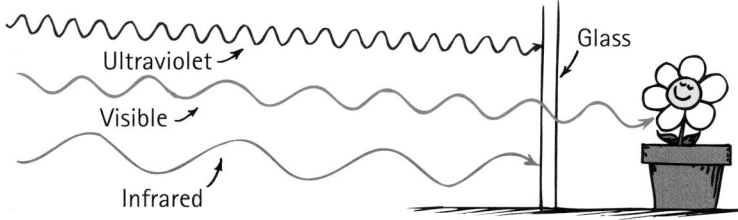

FIGURE 13.8 ▲
Glass blocks both infrared and ultraviolet waves but is transparent to all the frequencies of visible light.

CONCEPT CHECK
1. Why is glass transparent to visible light but not to ultraviolet and infrared?
2. Pretend that while you walk across a room you make several short stops along the way to greet people who are "on your wavelength." How is this analogous to visible light traveling through glass?
3. In what way is it not analogous?

 **READING CHECK**

What causes light to slow down when it passes through a transparent material?

Check Your Answers

1. The natural vibration frequency for electrons in glass matches the frequency of ultraviolet light, so resonance occurs when ultraviolet waves shine on glass. The absorbed energy is passed on to other atoms as heat, instead of being re-emitted as light. So glass is opaque at ultraviolet frequencies. But in the range of visible light, the forced vibrations of electrons in the glass are at smaller amplitudes. Vibrations are smaller and light is re-emitted instead of becoming thermal energy. So the glass is transparent. Lower-frequency infrared light causes entire molecules, rather than electrons, to resonate. So heat is generated and the glass is opaque.

2. Your average speed across the room is less than it would be in an empty room because of the time delays of your stops. In the same way, light takes a longer time to travel through glass. The result of interactions of light and atoms it encounters is lower average speed.

3. In walking across the room, it is you who begins and completes the walk. But it is different for light. The light that entered the glass is not the same light that emerges. The frequencies are the same, so like identical twins, they are indistinguishable.

Most things around us are **opaque**—they absorb light without re-emitting it. Books, desks, chairs, and people are opaque. Vibrations of their atoms and molecules produced by light are turned into random kinetic energy—into thermal energy. They become slightly warmer.

Metals are opaque. As we learned in Chapter 10, the outer electrons of atoms in metals are not bound to any particular atom. They are loose and free to wander throughout the material (which is why metal conducts electricity and heat so well). When light shines on metal and sets these free electrons into vibration, their energy does not "spring" from atom to atom in the material, but is instead reflected. That's why metals are shiny.

The Earth's atmosphere is transparent to some ultraviolet light, all visible light, and some infrared light. But the atmosphere is opaque to high-frequency ultraviolet light. If all ultraviolet light got through the atmosphere, we would be fried to a crisp. The small amount of ultraviolet that does get through causes sunburns. Clouds are semitransparent to ultraviolet light, which is why you can get a sunburn on a cloudy day. Ultraviolet light is not only harmful to your skin but also damaging to tar roofs. Now you know why tarred roofs are often covered with gravel.

Have you noticed that things look darker when they are wet than when they are dry? Light incident on a dry surface such as sand bounces

fyi

• Materials such as glass are transparent only for those creatures who see in the "visible" part of the spectrum. Other creatures who are tuned to different frequency ranges will see glass as opaque and other materials as transparent.

fyi

• Dark or black skin absorbs ultraviolet radiation before it can penetrate too far. In fair skin, it can travel deeper. Fair skin may develop a tan upon exposure to ultraviolet, which may afford some protection against further exposure. Ultraviolet radiation is also damaging to the eyes.

◀ **FIGURE 13.9**
Metals are shiny because light that shines on them forces free electrons into vibration, and these vibrating electrons then emit their "own" light waves as reflection.

 **READING CHECK**

Why do wet materials look darker than the same materials when dry?

directly to your eye. ✓ But light incident on a wet surface bounces around inside the transparent wet region before it reaches your eye. What happens with each bounce? Absorption! So sand and other things look darker when wet.

13.3 Reflection of Light

When sunlight or a lamp illuminates this page, electrons in the atoms of the paper and ink vibrate more energetically in response to the vibrating electric fields of the light. The energized electrons re-emit the light, which enables you to see the page. When illuminated by white light, the paper appears white, which tells you that the electrons re-emit all the visible frequencies. In short, light undergoes **reflection.** Very little absorption occurs. The ink is a different story. Except for a bit of reflection, it absorbs all the visible frequencies and therefore appears black.

Law of Reflection

Anyone who has played pool or billiards knows that when a ball bounces from a surface, the angle of rebound is equal to the angle of incidence. Likewise for light. This is the **law of reflection,** and it applies to all angles:

> **The angle of reflection equals the angle of incidence.**

The law of reflection is illustrated with arrows that represent light rays in Figure 13.10. Instead of measuring angles from the reflecting surface, it is customary to measure the angles of each ray from a line perpendicular to the reflecting surface. This imaginary line is called the *normal.* The incident ray, the normal, and the reflected ray all lie in the same plane.

FIGURE 13.10 ▶
The law of reflection.

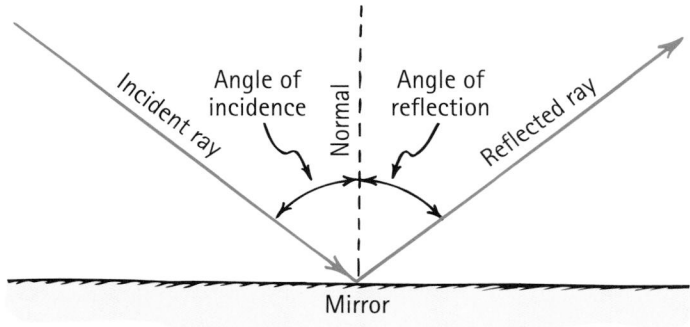

Look at a candle flame placed in front of a mirror. Rays of light radiate from the flame in all directions. Figure 13.11 shows only four of the infinite number of rays leaving one point on the flame. When these rays meet the mirror, they reflect at angles equal to their angles of incidence. The rays diverge (spread out) from the flame. Notice that they diverge also from the mirror. They appear to emanate from behind the mirror (dashed lines). You would see an image of the candle flame at this point. ✓ The image is as far behind the mirror as the flame is in front of the mirror. Both the flame and its image have the same size.

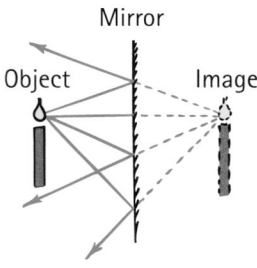

FIGURE 13.11 ▲
A virtual image is formed behind the mirror. It is located at the position where the extended reflected rays (dashed lines) converge.

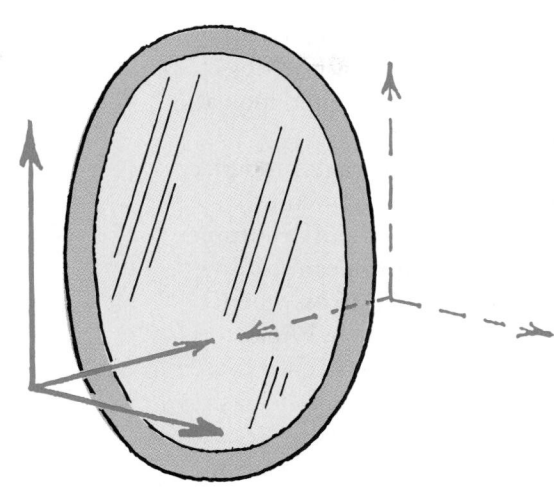

FIGURE 13.12 ▲

Marjorie's image is as far behind the mirror as she is in front. Note that she and her image have the same color of clothing—evidence that light doesn't change frequency upon reflection. Interestingly, her left-right axis is not reversed. Likewise for her up-down axis. The axis that *is* reversed, as shown to the right, is front-back. That's why it appears that her left hand faces the right hand of her image.

When you view yourself in a mirror, for example, the size of your image is the same as the size of your twin as if located as far behind the mirror as you are in front—as long as the mirror is flat. A flat mirror is called a *plane mirror.* When the mirror is curved, the sizes and distances of object and image are no longer equal. We shall not study curved mirrors in this book, except to say that the law of reflection still holds. A curved mirror behaves as a succession of flat mirrors, each tilted slightly different from the one next to it. At each point, the angle of incidence is equal to the angle of reflection (Figure 13.13). Note that in a curved mirror, the normals (dashed lines between the solid rays) at different points on the surface are not parallel to one another.

Whether a mirror is flat or curved, the human eye-brain system cannot ordinarily tell the difference between an object and its reflected

Your image is as far behind a mirror as you are in front of it— as if your twin stood behind it as far as you are in front.

HANDS-ON EXPLORATIONS: PLAYING WITH MIRRORS

1. Stand a pair of pocket-size mirrors on edge with the faces parallel to each other. Place an object such as a coin between the mirrors and look at the reflections in each mirror. Nice?
2. Now set up the mirrors at right angles and place a coin between them. You'll see four coins.
3. Change the angle of the mirrors and determine how many images of the coin you can see.
4. Look at yourself in the pair of mirrors when they're at right angles to each other. Wink. Notice that you see yourself as others see you.
5. Rotate the mirrors, still at right angles to each other. Does your image rotate also?
6. Now place the mirrors 60° apart so you again see your face. Again rotate the mirrors and see if your image rotates also. Amazing?

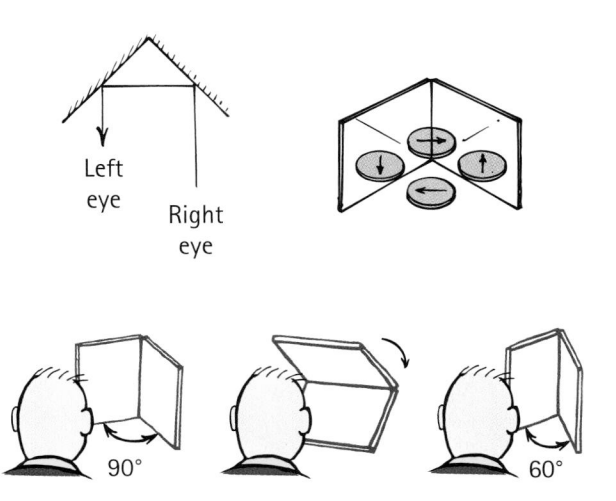

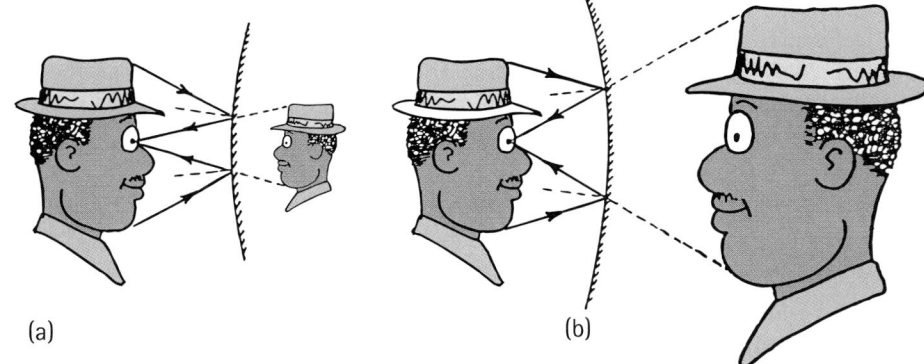

FIGURE 13.13 ▲
(a) The virtual image formed by a *convex* mirror (a mirror that curves outward) is smaller and closer to the mirror than the object. (b) When the object is near a *concave* mirror (a mirror that curves inward like a "cave"), the virtual image is larger and farther away than the object. In either case, the law of reflection holds for each ray.

image. So we see the illusion that an object exists behind a mirror (or, in some cases, in front of a concave mirror). Light reaches our eye in the same manner whether it comes from an object or a reflection of the object.

✔ READING CHECK

When you look at something in a plane mirror, how far behind the mirror is the image?

CONCEPT CHECK

1. What evidence can you cite to support the claim that the frequency of light does not change upon reflection?
2. If you wish to take a picture of your image while standing 2 m in front of a plane mirror, what distance should you set your camera to provide the sharpest focus?

Wanna get rich? Be the first to invent a surface that will reflect 100% of the light incident upon it.

Only part of the light that strikes a surface is reflected. On a pane of clear glass, for example, light perpendicular to the surface reflects only about 4% from each surface of the pane. On a polished aluminum or silver surface, however, about 90% of incident light is reflected.

Diffuse Reflection

We call reflection from a smooth or polished surface *specular* reflection. ✓ If a surface is so smooth that the distances between microscopic irregularities on the surface are less than about one-eighth the wavelength of the light, the surface is said to be *polished*. From rough surfaces, light reflects in many directions. This is not specular reflection, but **diffuse reflection** (Figure 13.14). There is very little diffuse reflection on a polished surface. Interestingly, an irregular surface for one range of wavelengths may be polished for other ranges. The wire-mesh "dish" shown in Figure 13.15 is very rough for light waves. We certainly don't see it as polished. But for long-wavelength radio waves, it is quite polished and is an excellent reflector.

Light reflecting from this page is diffuse. The page may be smooth to a radio wave, but to a light wave it is rough, as Figure 13.16 clearly indicates. Rays of light striking this page encounter millions of tiny surfaces facing in all directions. The incident light therefore reflects in all directions. This is good because it lets us see the page without glare. We don't

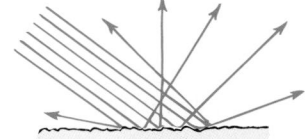

FIGURE 13.14 ▲
Diffuse reflection. Although each ray obeys the law of reflection, the many different surface angles that light rays encounter in striking a rough surface cause reflection in many directions.

◀ **FIGURE 13.15**
The open-mesh parabolic dish is a diffuse reflector for short-wavelength visible light waves but a polished reflector for long-wavelength radio waves.

FIGURE 13.16 ▶
A magnified view of the surface of ordinary paper.

have to hold our head in any special position to read the page. Similarly for other objects. In an automobile you can see the road ahead at night because of diffuse reflection by the road surface. When the road is wet, however, water provides a more mirrored surface and it is harder to see. Most of our environment is seen by diffuse reflection.

CONCEPT CHECK
Why is it more dangerous to drive a car on a rainy night?

Check Your Answer
In addition to the less diffuse road surface when wet, as described above, there's another reason. Headlights from oncoming cars reflect from the more mirrored surface full force into your eyes. Glare is more intense.

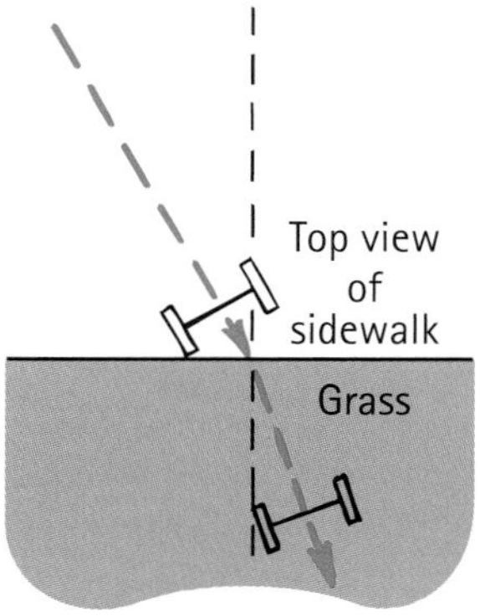

13.4 Refraction—The Bending of Light Due to Changing Speed

In Section 13.2, we learned that light slows down when it enters glass and other transparent materials. Light travels at different speeds in various materials. It travels at 300,000 km/s in a vacuum, at a slightly lower speed in air, and at about $\frac{3}{4}$ that speed in water. In a diamond, light travels at about 40% of its speed in a vacuum. When light bends as it passes from one medium to another, we call the process **refraction**. When light is not perpendicular to the surface of penetration, bending occurs.

To better understand the bending of light in refraction, look at the pair of toy cart wheels in Figure 13.17. The wheels roll from a smooth sidewalk onto a grass lawn. If the wheels meet the grass at an angle, as the figure

Top view of sidewalk

Grass

FIGURE 13.17 ▲
The direction of the rolling wheels changes when one wheel slows down before the other does.

FIGURE 13.18 ▶
When light is refracted, the direction of the waves changes when one part of the wave slows down before the other part.

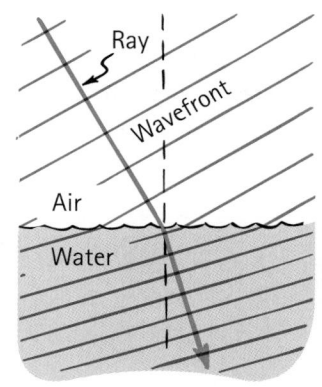

shows, they are deflected from their straight-line course. Note that the left wheel slows first when it interacts with the grass on the lawn. The right wheel maintains its higher speed while on the sidewalk. It pivots about the slower-moving left wheel because it travels farther in the same time. So the direction of the rolling wheels is bent toward the "normal," the black dashed line perpendicular to the grass-sidewalk border in Figure 13.17.

Figure 13.18 shows how a light wave bends in a similar way. Note the direction of light, indicated by the blue arrow (the light ray), and also note the *wavefronts* (red) drawn at right angles to the ray. In the

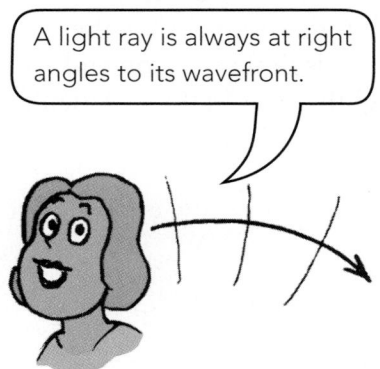

A light ray is always at right angles to its wavefront.

figure, the wave meets the water surface at an angle. This means that the left portion of the wave slows down in the water while the remainder in the air travels at speed *c*. The light ray remains perpendicular to the wavefront and therefore bends at the surface. It bends like the wheels bend when they roll from the sidewalk into the grass.

✔ In both cases, the bending is caused by a change in speed.

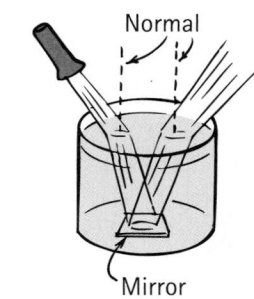

FIGURE 13.19 ▲
Refraction.

Figure 13.19 shows a beam of light entering water at the left and exiting at the right. The path would be the same if the light entered from the right and exited at the left. Light paths are reversible for both reflection and refraction. If you see someone's eyes by means of reflection or refraction, such as with a mirror or a prism, then that person can see you also.

Refraction causes many illusions. One of them is the apparent bending of a stick that is partially in water. The submerged part appears closer to the surface than it actually is. Likewise when you look at a fish in water. The fish appears nearer to the surface and closer than it really is (Figure 13.21). If we look straight down into water, an object submerged 4 m beneath the surface appears to be only 3 m deep. Because of refraction, submerged objects appear to be magnified.

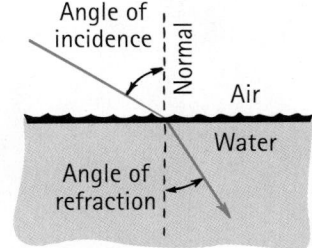

FIGURE 13.20 ▲
When light slows down in going from one medium to another, like going from air to water, it refracts toward the normal. When light speeds up in traveling from one medium to another, like going from water to air, it refracts away from the normal.

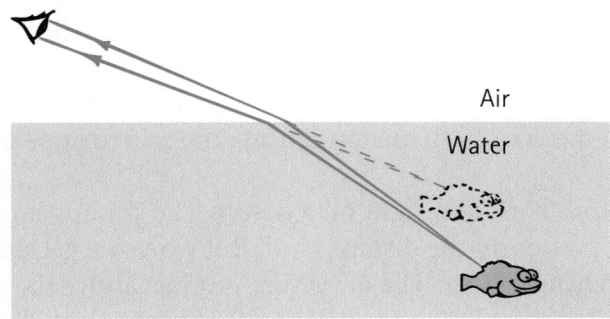

FIGURE 13.21 ▲
Because of refraction, a submerged object appears to be nearer to the surface than it actually is.

READING CHECK

What causes light to bend?

CONCEPT CHECK

If the speed of light were the same in all media, would refraction still occur when light passes from one medium to another?

Check Your Answer

No.

13.5 Illusions and Mirages Are Caused by Atmospheric Refraction

Refraction occurs in the Earth's atmosphere. Whenever we watch a sunset, we see the Sun for several minutes after it has sunk below the horizon (Figure 13.22). The Earth's atmosphere is thin at the top and dense at the bottom. Because light travels slightly faster in thin air than in

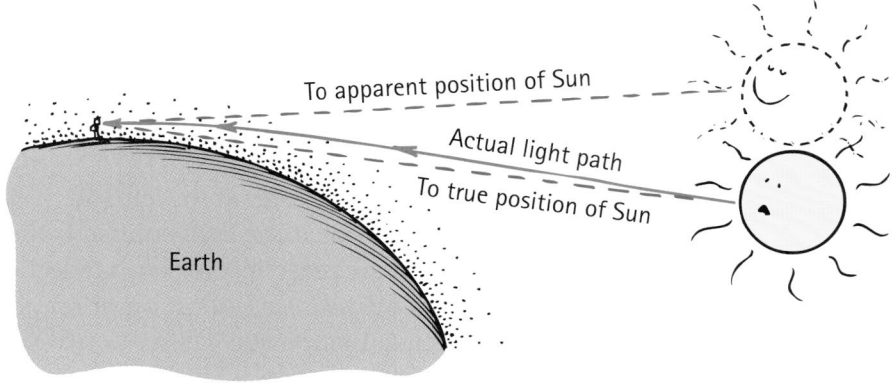

FIGURE 13.22 ▲
The Sun's shape is distorted by differential refraction.

FIGURE 13.23 ▲
Because of atmospheric refraction, when the Sun is near the horizon it appears elliptical.

dense air, parts of the wavefronts of sunlight higher up travel faster than parts of the wavefronts closer to the ground. Light rays bend. The density of the atmosphere changes gradually, so light rays bend gradually and follow a curved path. So we gain additional minutes of daylight each day. Furthermore, when the Sun (or Moon) is near the horizon, the rays from the lower edge are bent more than the rays from the upper edge. This shortens the vertical diameter, causing the Sun to appear elliptical (Figure 13.23).

Mirages are a common sight on a desert. The sky appears to be reflected from water on the distant sand. But when we get there, the sand is dry. Why is this so? The air is very hot just above the surface and cooler above. Light travels faster through the thinner hot air below than through the denser cool air above. So wavefronts near the ground travel faster than above. The result is upward bending (Figure 13.24). So we see

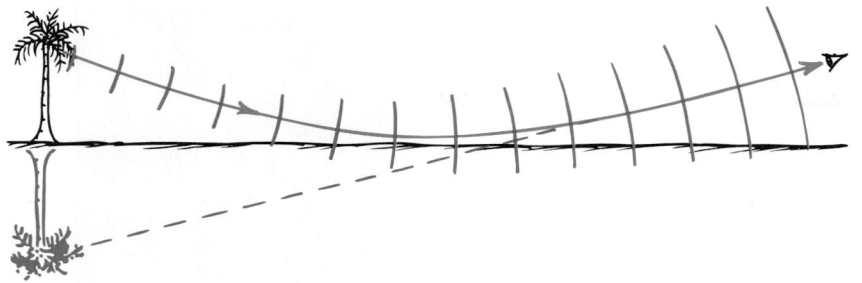

FIGURE 13.24 ▲
Light from the top of the tree gains speed in the warm and less-dense air near the ground. When the light grazes the surface and bends upward the observer sees a mirage.

FIGURE 13.25 ▲
A mirage. The apparent wetness of the road is not reflection by water, but rather refraction through the warmer and less-dense air near the road surface.

an upside-down view as if reflection were occurring from a water surface. We see a mirage, which is formed by real light and can be photographed (Figure 13.25). A mirage is not, as many people think, a trick of the mind.

When we look at an object over a hot stove or over a hot pavement, we see a wavy, shimmering effect. This is due to varying densities of air caused by changes in temperature. ✔ The twinkling of stars results from similar variations in the sky, where light passes through unstable layers in the atmosphere.

CONCEPT CHECK
If the speed of light were the same in air of various temperatures and densities, would there still be slightly longer daytimes, twinkling stars at night, mirages, and a slightly squashed Sun at sunset?

Check Your Answer
No.

READING CHECK
Why do stars twinkle?

13.6 Color Science

To the scientist, color is not in a material. Color is a physiological experience and is in the eye of the beholder. So when we say that light from a rose is red, in a strict sense we mean that it *appears* red. Many organisms, including people with defective color vision, do not see the rose as red at all.

The colors we see depend on the frequency of the light we see. Light of different frequencies is perceived as having different colors; the lowest frequency we detect appears to most people as the color red. The highest frequency appears as violet. In between is the infinite number of hues that make up the color spectrum of the rainbow. By convention, these hues are grouped into the seven colors: red, orange, yellow, green, blue, indigo, and violet. These colors all blended together appear white. The white light from the Sun is a blend of all the visible frequencies.

FIGURE 13.26 ▲
Sunlight passing through a prism separates into a color spectrum. The colors of things depend on the colors of the light that illuminates them.

FIGURE 13.27 ▶
The square on the left *reflects* all the colors illuminating it. In sunlight, it is white. When illuminated with blue light, it is blue. The square on the right *absorbs* all the colors illuminating it. In sunlight, it is warmer than the white square.

All the colors added together produce white. The absence of all color is black.

Selective Reflection

Except for sources of light such as lamps, most things around us reflect rather than emit light. They reflect only part of the light incident upon them. That's the part that gives them their color. A rose, for example, doesn't emit light; it reflects light. If we pass sunlight through a prism and then place a deep-red rose in various parts of the spectrum, the petals appear brown or black in all parts of the spectrum except in the red. In the red part of the spectrum, the petals appear red. But the green stem and leaves look black. This shows that the red petals have the ability to reflect red light but not other colors of light. Likewise, the green leaves have the ability to reflect green light but not other colors. When the rose is held in white light, the petals appear red and the leaves green because the petals reflect the red part of the white light and the leaves reflect the green part. To understand why objects reflect only particular colors of light, we turn again to the atom.

Light reflects from things similar to the way sound "reflects" from a tuning fork when another one sets it into vibration. One tuning fork can cause another to vibrate even when the frequencies are not matched, although at much less amplitude. The same is true of atoms and molecules. Electrons can be forced into vibration by the vibrating electric fields of electromagnetic waves. ✓ Once vibrating, these electrons emit their own electromagnetic waves just as vibrating tuning forks send out sound waves.

Usually a material absorbs light of some frequencies and reflects the rest. The color reflected is the color we see. An object that reflects light of all the visible frequencies—for example, the white part of this page—is the same color as the light that shines on it. If a material absorbs all the light that shines on it, it reflects none and is black.

Interestingly, the petals of most yellow flowers, like daffodils, reflect red and green as well as yellow. Yellow daffodils reflect a broad band of frequencies. The reflected colors of most objects are not pure single-frequency colors, but are a mixture of frequencies.

An object can reflect only those frequencies present in the illuminating light. The appearance of a colored object therefore depends on the kind of light that illuminates it. An incandescent lamp, for instance, emits light of lower average frequencies than sunlight, enhancing any reds viewed in this light. In a fabric having only a little bit of red in it, the red is more apparent under an incandescent lamp than under a fluorescent lamp. Fluorescent lamps are richer in the higher frequencies and

FIGURE 13.28 ▲
The rabbit's dark fur absorbs all the radiant energy in incident sunlight and therefore is black. Light fur on other parts of the body reflects light of all frequencies and therefore is white.

so blues are enhanced under them. For this reason, it is difficult to tell the true color of objects viewed in artificial light. What a color looks like depends on the light source (Figure 13.29).

Selective Transmission

The color of a transparent object depends on the color of the light it transmits. A red piece of glass appears red because it absorbs all the colors of white light, except red. So red is transmitted. Similarly, a blue piece of glass appears blue because it transmits primarily blue and absorbs the other colors that illuminate it. These pieces of glass contain dyes or *pigments*— fine particles that selectively absorb light of particular frequencies and selectively transmit others. From an atomic point of view, electrons in the pigment molecules are set into vibration by the illuminating light. Light of some of the frequencies is absorbed by the pigments. The rest is re-emitted from atom to atom in the glass. The energy of the absorbed light increases the kinetic energy of the atoms and the glass is warmed. Ordinary window glass doesn't have a color because it transmits light of all visible frequencies equally well.

FIGURE 13.29 ▲
Color depends on the light source.

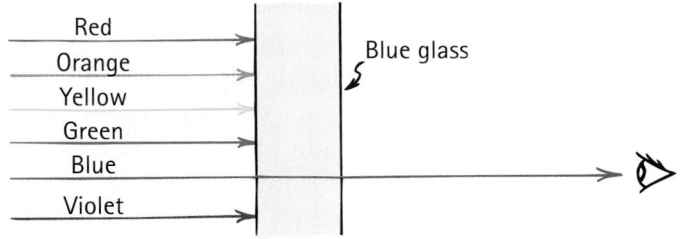

Red
Orange
Yellow
Green
Blue
Violet
Blue glass

◄ **FIGURE 13.30**
Only energy having the frequency of blue light is transmitted; energy of the other frequencies is absorbed and warms the glass.

CONCEPT CHECK

1. When red light shines on a red rose, why do the leaves become warmer than the petals?
2. When green light shines on a red rose, why do the petals look black?
3. If you hold any small source of white light between you and a piece of red glass, you'll see two reflections from the glass: one from the front surface and one from the back surface. What color is each reflection? (Did you try this Explore Activity at the beginning of this chapter?)

Check Your Answers

1. The leaves don't reflect red light, but absorb it. So the leaves become warmer.
2. The petals don't reflect the green light, but absorb it. Because green is the only color illuminating the rose and because green contains no red to be reflected, the rose reflects no color and appears black.
3. The reflection from the front surface is white because the light doesn't reach far enough into the colored glass to allow absorption of non-red light. Only red light reaches the back surface because the pigments in the glass absorb all the other colors, and so reflection from the back is red.

✔ **READING CHECK**

What two kinds of waves are emitted by vibrating electrons and vibrating tuning forks?

13.7 Mixing Colored Lights

You can see that white light from the Sun is composed of all the visible frequencies when you pass sunlight through a prism. The white light is dispersed into a rainbow-colored spectrum. The distribution of solar frequencies (Figure 13.31) is uneven and is most intense in the yellow-green part of the spectrum. How fascinating that our eyes have evolved to have maximum sensitivity in this range! That's why fire engines are painted yellow-green, particularly at airports, where visibility is vital. Our sensitivity to yellow-green light is also why we see better under the illumination of yellow sodium-vapor lamps at night than under incandescent lamps of the same brightness.

FIGURE 13.31 ▶
The radiation curve of sunlight is a graph of brightness versus frequency. Sunlight is brightest in the yellow-green region, in the middle of the visible range.

All the colors combined make white. Interestingly, we see white also from the combination of only red, green, and blue light. We can understand this by dividing the solar radiation curve into three regions, as in Figure 13.32. Three types of cone-shaped receptors in our eyes perceive color. Each is stimulated only by certain frequencies of light. ✔ Light of lower visible frequencies stimulates the cones sensitive to low frequencies and appears red. Light of middle frequencies stimulates the mid-frequency-sensitive cones and appears green. Light of higher frequencies stimulates the higherfrequency-sensitive cones and appears blue. When all three types of cones are stimulated equally, we see white.

Aha! Now I know why the color of tennis balls is yellow-green!

Additive Primary Colors—Red, Green, and Blue

When red, green, and blue lights are projected on a screen, they overlap to produce white. In the language of scientists, colored lights that overlap *add* to each other. So we say that red, green, and blue light *add to produce white light.* Any two of these colors of light add to produce another color (Figure 13.33). Various amounts of red, green, and blue produce any color in the spectrum. For this reason, red, green, and blue are called the **additive primary colors.** A close examination of the picture on a TV screen reveals that the picture is a mixture of tiny spots, each less than a millimeter across. When the screen is lit, some of the spots are red, some green, and some blue. The mixtures of these primary colors at a distance provide a complete range of colors, plus white.*

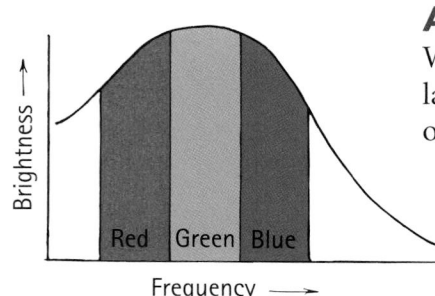

FIGURE 13.32 ▲
Radiation curve of sunlight divided into three regions—red, green, and blue. These are the additive primary colors.

---

* It's interesting to note that the "black" you see on the darkest scenes on a TV screen is simply the color of the face itself, which is more a light gray than black. Because our eyes are sensitive to the contrast with the illuminated parts of the screen, we see this gray as black.

13.8 Mixing Colored Pigments

Every artist knows that if you mix red, green, and blue paint, the result is not white but rather a muddy dark brown. Red and green paint certainly do not mix to form yellow, as is the rule for combining colored lights.

Mixing pigments in paints and dyes is entirely different from mixing lights. Pigments are tiny particles that absorb specific colors. For example, 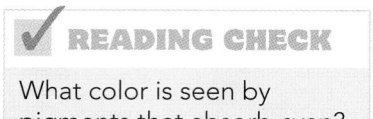 pigments that produce the color red absorb the complementary color cyan. So something painted red absorbs cyan, which is why it reflects red. In effect, cyan has been *subtracted* from white light. Something painted blue absorbs yellow, and so reflects all the colors except

✔ READING CHECK

What color is seen by pigments that absorb cyan?

(a) (b) (c) (d) (e) (f)

FIGURE 13.35 ▲
Only four colors of ink are used to print color illustrations and photographs—(a) magenta, (b) yellow, (c) cyan, and black. When magenta, yellow, and cyan are combined, they produce (d). Addition of black (e) produces the finished result (f).

yellow. Take yellow away from white and you've got blue. The colors magenta, cyan, and yellow are the **subtractive primary colors.**

The variety of colors you see in the colored photographs in this or any other book are the result of magenta, cyan, and yellow dots. Light illuminates the book, and light of some frequencies is subtracted from the reflected light. The rules of color subtraction differ from the rules of light addition. We'll not go deeper into this topic. Much information is available on the web.

FIGURE 13.36 ▲
Look through a magnifying glass and you will see that the color green on a printed page is composed of cyan and yellow dots.

◀ **FIGURE 13.37**
Dyes or pigments, as in the three transparencies shown, absorb (subtract) light of some frequencies and transmit only part of the spectrum. When white light passes through overlapping sheets of these colors, all light is blocked (subtracted). Then we have black. Where only yellow and cyan overlap, only green is not subtracted. Various proportions of yellow, cyan, and magenta dyes will produce nearly any color in the spectrum.

FIGURE 13.38 ▲
The rich colors of Sneezlee represent many frequencies of light. The photo, however, is a mixture of only yellow, magenta, cyan, and black.

13.9 Why the Sky Is Blue

Not all colors are the result of the addition or subtraction of light. Some colors, like the blue of the sky, are the result of selective scattering. Consider the analogous case of sound: If a beam of a particular frequency of sound is directed to a tuning fork of similar frequency, the tuning fork is set into vibration and redirects the beam in multiple directions. The tuning fork *scatters* the sound. A similar process occurs with the scattering of light from atoms and particles that are far apart from one another. This occurs in the atmosphere.

We know that atoms behave like tiny optical tuning forks and re-emit light waves that shine on them. Very tiny particles act the same.

There are no blue pigments in the feathers of a blue jay. They are blue due to scattering!

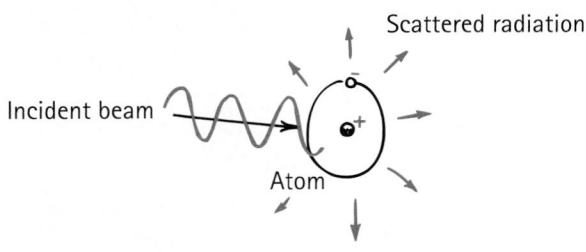

◀ **FIGURE 13.39**
A beam of light falls on an atom and increases the vibrational motion of electrons in the atom. The vibrating electrons re-emit the light in various directions. Light is scattered.

FIGURE 13.40 ▲

In clean air, the scattering of high-frequency light gives us a blue sky. When the air is full of particles larger than oxygen and nitrogen molecules, lower-frequency light is also scattered, which adds to give a whitish sky.

Atmospheric soot heats the Earth's atmosphere by absorbing light while cooling local regions by blocking sunlight from reaching the ground. Soot particles in the air can trigger severe rains in one region and droughts and dust storms in another.

READING CHECK

Why does the sky sometimes have a whitish appearance, rather than blue?

The tinier the particle, the higher the frequency of light it will re-emit. This is similar to the way small bells ring with higher notes than larger bells. The nitrogen and oxygen molecules that make up most of the atmosphere are like tiny bells that "ring" with high frequencies when energized by sunlight. Like sound from the bells, the re-emitted light is sent in all directions. When light is re-emitted in all directions, we say the light is *scattered*.

Of the visible frequencies of sunlight, violet is scattered the most by nitrogen and oxygen in the atmosphere. Then the other colors are scattered in order: blue, green, yellow, orange, and red. Red is scattered only a tenth as much as violet. Although violet light is scattered more than blue, our eyes are not very sensitive to violet light. Therefore, the blue scattered light is what predominates in our vision, so we see a blue sky!

The blueness of the sky varies in different locations under various conditions. A main factor is the content of water vapor in the atmosphere. On clear dry days, the sky is a much deeper blue than on clear humid days. Places where the upper air is exceptionally dry, such as Italy and Greece, have beautiful blue skies that have inspired painters for centuries. ✓ Where the atmosphere contains a lot of dust particles and other particles larger than oxygen and nitrogen molecules, light of the lower frequencies is also strongly scattered. This causes the sky to appear less blue, with a whitish appearance. After a heavy rainstorm when the airborne particles have been washed away, the sky becomes a deeper blue.

The grayish haze in the skies over large cities is the result of particles emitted by car and truck engines and by industry. Even when idling, a typical automobile engine emits more than 100 billion particles per second. Most are invisible but act as tiny centers to which other particles adhere. These are the primary scatterers of lower-frequency light. With the largest of these particles, absorption rather than scattering takes place, and a brownish haze is produced. Yuck!

Scattering is not limited to particles in the air. The intriguing vivid blue of lakes in the Canadian Rocky Mountains is due to scattering. The lakes are fed by runoff from melting glaciers that contain fine particles of silt, called rock flour, which remain suspended in the water. Light scatters from these tiny particles and gives the water its eerily vivid color (Figure 13.41). (Tourists who photograph these lakes are advised to inform their photo processors *not* to adjust the lake color to a "real" blue.)

FIGURE 13.41 ▶

The extraordinary blue of Canadian Rocky Mountain lakes is produced by scattering from extremely fine particles of glacial silt suspended in the water.

13.10 Why Sunsets Are Red

Light that isn't scattered is light that is transmitted. Because red, orange, and yellow light are the least scattered by the atmosphere, light of these low frequencies is better transmitted through the air. Red is scattered the least and passes through more atmosphere than any other color. So the thicker the atmosphere through which a beam of sunlight travels, the more time there is to scatter all the higher-frequency parts of the light. This means red light travels through it best. As Figure 13.42 shows, sunlight travels through more atmosphere at sunset, which is why sunsets are red.

One of the many beauties of physics is the redness of a fully eclipsed Moon—resulting from the refraction of sunsets and sunrises that completely circle the world. This refracted light shines on an otherwise dark Moon.

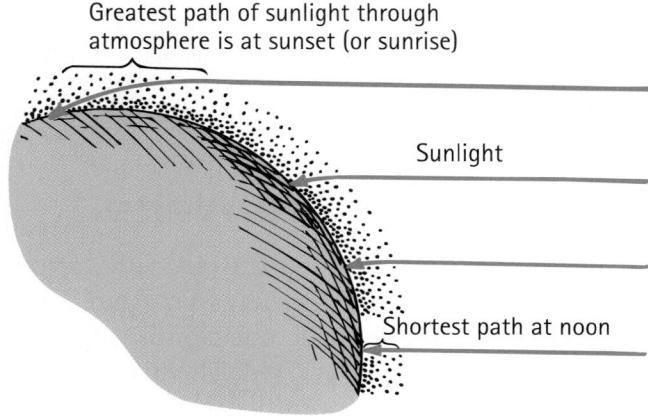

Greatest path of sunlight through atmosphere is at sunset (or sunrise)

Sunlight

Shortest path at noon

FIGURE 13.42 ▲
A sunbeam must travel through more atmosphere at sunset than at noon. As a result, more blue is scattered from the beam at sunset than at noon. By the time a beam of initially white light reaches the ground, only light of the lower frequencies survives to produce a red sunset.

At noon, sunlight travels through the least amount of atmosphere to reach the Earth's surface. Only a small amount of high-frequency light is scattered from the sunlight, enough to make the Sun look yellowish. As the day progresses and the Sun descends lower in the sky (Figure 13.42), the path through the atmosphere is longer, and more violet and blue are scattered from the sunlight. ✓ The removal of violet and blue leaves the transmitted light redder. The Sun becomes progressively redder, going from yellow to orange and finally to a red-orange at sunset. Sunsets and sunrises are unusually colorful following volcanic eruptions, because particles larger than atmospheric molecules are then more abundant in the air.

While watching a sunset, does an astronomer sense the Sun moving downward, or Earth turning away from a stationary Sun?

SHINE-ON EXPLORATION

You can simulate a sunset with a fish tank full of water by adding a few drops of milk. A few drops will do. Then shine a flashlight beam through the water and you'll notice that it looks bluish from the side.

Milk particles are scattering the higher frequencies of light in the beam. Light emerging from the far end of the tank will have a reddish tinge. That's the light that wasn't scattered.

For me, knowing why the sky is blue and why sunsets are red adds to their beauty—knowledge doesn't subtract.

✔ **READING CHECK**

What color results when violet and blue are removed from sunlight?

The colors of the sunset are consistent with our rules for color mixing. When blue is subtracted from white light, the complementary color that remains is yellow. When higher-frequency violet is subtracted, the resulting complementary color is orange. When medium-frequency green is subtracted, magenta is left. The combinations of resulting colors vary with atmospheric conditions, which change daily, giving us a variety of sunsets.

CONCEPT CHECK

If molecules in the sky scattered low-frequency light more than high-frequency light, what color would the sky be? What color would sunsets be?

Check Your Answers

If low-frequency light were scattered, the noontime sky would appear reddish-orange. At sunset, more reds would be scattered by the longer path of the sunlight, and the sunlight would be predominantly blue and violet. So sunsets would appear blue!

13.11 Why Clouds Are White

Clouds are made up of clusters of water droplets in a variety of sizes. The different-size clusters result in a variety of scattered colors. In effect, the tiniest clusters tend to make blue clouds; slightly larger clusters, green; and still larger clusters, red. The overall result is a white cloud. Electrons close to one another in a cluster vibrate in phase. This results in a greater intensity of scattered light than there would be from the same number of electrons vibrating separately. Hence, clouds are bright!

✔ Larger clusters of droplets absorb much of the light incident upon them, and so the scattered intensity is less. Therefore, clouds composed of larger clusters are darker. Further increase in the size of the clusters causes them to fall as raindrops, and we have rain.

The next time you find yourself admiring a crisp blue sky or delighting in the shapes of bright clouds or watching a beautiful sunset, think about all those ultratiny optical tuning forks vibrating away. You'll appreciate these everyday wonders of nature even more!

FIGURE 13.43 ▲
A cloud is composed of various sizes of water-droplet clusters. The tiniest clusters scatter blue light, slightly larger ones scatter green light, and still larger ones scatter red light. The result is a white cloud.

✔ **READING CHECK**

Why are clouds composed of larger clusters of water drops darker than bright white clouds?

FIGURE 13.44 ▲
Water is cyan because it absorbs red light. The froth in the waves is white because, like clouds, it is composed of a variety of tiny clusters of water droplets that scatter all the visible frequencies.

13 CHAPTER REVIEW

KEY TERMS

Additive primary colors The three colors—red, blue, and green—that when added in certain proportions produce any other color in the visible-light part of the electromagnetic spectrum.

Complementary colors Any two colors that when added produce white light.

Diffuse reflection Reflection in irregular directions from an irregular surface.

Electromagnetic wave A wave emitted by vibrating electrical charges (often electrons) and composed of vibrating electric and magnetic fields that regenerate one another.

Electromagnetic spectrum The range of electromagnetic waves extending in frequency from radio waves to gamma rays.

Law of reflection The angle of a reflection equals the angle of incidence. The reflected and incident rays lie in a plane that is normal to the reflecting surface.

Opaque The term applied to materials through which light cannot pass.

Reflection The return of light into the medium from which it came.

Refraction The bending of an oblique ray of light when it passes from one transparent medium to another.

Subtractive primary colors The three colors of absorbing pigments—magenta, yellow, and cyan—that when mixed in certain proportions reflect any other color in the visible-light part of the electromagnetic spectrum.

Transparent The term applied to materials through which light can pass in straight lines.

REVIEW QUESTIONS

The Electromagnetic Spectrum—and the Tiny Bit That Is Light

1. How do the speeds of various electromagnetic waves compare?
2. How does the frequency of a radio wave compare with the frequency of the vibrating electrons that produce it?

Why Materials Are Either Transparent or Opaque

3. In what region of the electromagnetic spectrum is the resonant frequency of electrons in glass?
4. What is the fate of the energy in ultraviolet light incident on glass?
5. How does the average speed of light in glass compare with its speed in a vacuum?

Reflection of Light

6. Relative to the distance of an object in front of a plane mirror, how far behind the mirror is the image?

Refraction—The Bending of Light Due to Changing Speed

7. When a wheel rolls from a smooth sidewalk onto grass, the interaction of the wheel with the blades of grass slows the wheel. What slows light when it passes from air into glass or water?
8. What is the angle between a ray of light and its wavefront?

Illusions and Mirages Are Caused by Atmospheric Refraction

9. Why does a setting Sun often appear elliptical instead of round?
10. What produces the wavy effect in air over a hot stove?

Color Science

11. Which has the higher frequency, red light or blue light?
12. What is the appearance of a material that absorbs all the light that shines on it and reflects none?

Mixing Colored Lights

13. What is the color of the peak frequency of solar radiation?
14. Why are red, green, and blue called the *additive primary colors*?

Mixing Colored Pigments

15. What are the subtractive primary colors? Why are they so called?
16. Why are red and cyan called *complementary colors*?

Why the Sky Is Blue

17. Is the blueness of the sky due to reflection, refraction, or scattering?
18. Why does the sky sometimes appear whitish?

Why Sunsets Are Red

19. Why does the Sun look reddish at sunrise and sunset but not at noon?

Why Clouds Are White

20. What is the evidence for a variety of particle sizes in a cloud?

THINK AND DO

1. Stare at a piece of colored paper for 45 seconds or so. Then look at a white surface. Because the cones in your retina receptive to the color of the paper have become fatigued, you see an afterimage of the complementary color when you look at the white area. This is because the fatigued cones send a weaker signal to the brain. All the colors produce white, but all the colors minus one produce the color complementary to the missing one.
2. Stare intently for a minute or so at the flag below. Then turn your view to a white area, such as a wall.

What colors do you see in the image of the flag that appears on the wall?

THINK AND COMPARE

1. A top view of wheels from a toy cart rolled from a concrete sidewalk onto the following surfaces. Surface A is slightly-rougher concrete, B is coarse grass, and C is cropped grass on a golf course. When they roll on the surfaces, they bend from their initial directions. Rank the bending from greatest to least on the surfaces.

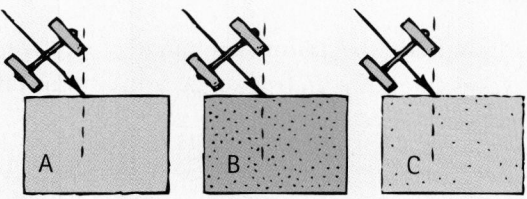

2. Identical rays of light enter three transparent blocks composed of different materials. Light slows when it enters them. Rank the blocks according to the speed of light in them, from greatest to least.

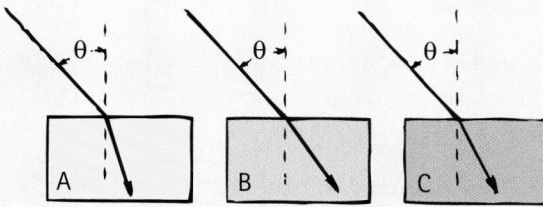

THINK AND EXPLAIN

1. Which waves have the longest wavelengths: light waves, X-rays, or radio waves? Which have the highest frequencies?
2. Is glass transparent or opaque to frequencies of light that match its own natural frequencies? Explain.
3. You can get a sunburn on a cloudy day, but you can't get a sunburn even on a sunny day if you are behind glass. Explain.

4. An eye at point P looks into the mirror. Which of the numbered cards is seen reflected in the mirror?

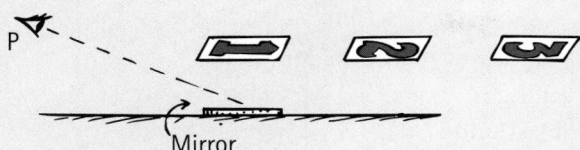

5. What must be the minimum length of a plane mirror in order for you to see a full view of yourself? What effect does your distance from the plane mirror have in the answer? (Try it and see!)

6. On a steamy mirror, wipe away just enough to see your full face. How tall will the wiped area be compared with the vertical dimension of your face?

7. Hold a pocket mirror at almost arm's length from your face and note the amount of your face you can see. To see more of your face, should you hold the mirror closer or farther, or would you need a larger mirror? (Try it and see!)

8. Why is the lettering on the front of some vehicles "backwards"?

AMBULANCE

9. Which kind of road surface is easier to see when driving at night, a pebbled uneven surface or a mirror-smooth surface? Explain.

10. A person in a dark room looking through a window can clearly see a person outside in the daylight, whereas the person outside cannot see the person inside. Explain.

11. A pair of toy cart wheels are rolled obliquely from a smooth surface onto two plots of grass, a rectangular plot and a triangular plot, as shown below. The ground is on a slight incline so that after slowing down in the grass, the wheels speed up again when they emerge on the smooth surface. Finish each sketch by showing some positions of the wheels inside the plots and on the other sides, thereby indicating the direction of travel.

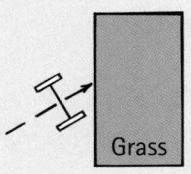
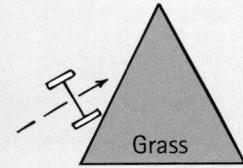

12. Peter Hopkinson stands astride a large mirror and boosts class interest with this zany demonstration. How does he accomplish his apparent levitation in midair? (Hint: Consider his left leg.)

13. What color besides white is common on today's table-tennis balls, and why?

14. A spotlight that has a white-hot filament is coated so that it won't transmit yellow light. What color is the emerging beam of light?

15. Below is a photo of science author Suzanne with son Tristan wearing red and daughter Simone wearing green. Below that is the negative photo, which shows these colors differently. What is your explanation?

16. Complete the following equations:

Yellow light + blue light = _____ light.
Green light + _____ light = white light.
Magenta light + yellow light + cyan light = _____ light.

17. If the sky were composed of atoms that predominantly scattered orange light rather than blue, what color would sunsets be?

18. Comment on the statement, "Oh, that beautiful red sunset is just the leftover colors that weren't scattered on their way through the atmosphere."

19. Volcanic emissions spew fine ashes in the air that scatter red light. What color does a full Moon appear through these ashes?

20. Why do white clouds in midday often turn dark?

THINK AND SOLVE

1. If you take a photo of your image in a plane mirror, how many meters away should you set your focus if you are 3 m in front of the mirror?

2. If you walk toward a mirror at 2 m/s, how fast do you and your image approach each other?

3. No glass is perfectly transparent. Consider a pane of glass that transmits 92% of the light incident upon it. How much light is transmitted through two of these sheets together?

READINESS ASSURANCE TEST (RAT)

If you have a good handle of this chapter—if you really do—then you should be able to score 7 out of 10 on this RAT. Check your answers with your teacher. If you score less than 7, study further before moving on.

Choose the best answer to each of the following.

1. The electromagnetic spectrum spans waves ranging from lowest to highest frequencies. The smallest portion of the electromagnetic spectrum is that of
   (a) radio waves.   (b) microwaves.
   (c) visible light.   (d) gamma rays.

2. Strictly speaking, the photons of light that shine on glass are
   (a) also the ones that travel through and exit the other side.
   (b) not the ones that travel through and exit the other side.
   (c) absorbed and transformed to thermal energy.
   (d) diffracted.

3. Window glass is normally transparent to
   (a) infrared.   (b) visible light.
   (c) ultraviolet.   (d) All of these.

4. The law of reflection applies to
   (a) light.
   (b) sound.
   (c) both light and sound.
   (d) neither light nor sound.

5. To see your full image in a mirror on the wall, the minimum height of the mirror is
   (a) at least one-quarter your height.
   (b) half your height.
   (c) the same as your height.
   (d) dependent on your distance from the mirror.

6. When light refracts, there is a change in
   (a) speed.
   (b) direction.
   (c) both speed and direction.
   (d) frequency.

7. When a light ray passes at an angle from air to water, the ray inside the water bends
   (a) toward the normal.
   (b) away from the normal.
   (c) either away or toward the normal.
   (d) parallel to the normal.

8. A red rose will not appear red when illuminated only with
   (a) red light.   (b) orange light.
   (c) white light.   (d) cyan light.

9. Red, green, and blue light overlap to form
   (a) red light.   (b) green light.
   (c) blue light.   (d) white light.

10. The redness of a sunset is due to light that
    (a) is scattered in the sky.
    (b) survives scattering in the sky.
    (c) is dispersed.
    (d) refracts in the air.

14

PROPERTIES OF LIGHT

THE MAIN IDEA

Light has both wave and particle properties: It travels as a wave and hits like a particle.

What causes the colors of rainbows and why are they bow-shaped? What causes the colors in soap bubbles—especially big ones, like the one shown here? How do lenses redirect light to form images? Does light bend around corners like sound can? Can light waves cancel other light waves? What causes the vivid colors of gasoline spilled on a wet street? Why does the color of some butterfly wings depend on your angle of sight? What's the difference between regular sunglasses and Polaroid sunglasses? Is light a wave or is light a stream of particles? Can it be both at the same time? We'll answer these questions and more in this chapter. Onward!

Explore!

Seeing the Image of the Sun Cast by a Pinhole

1. Poke a small hole with the tip of a pencil or ballpoint pen in the middle of a small piece of cardboard or index card.
2. Hold the card in bright sunlight about a meter or so above the floor.
3. The card casts a shadow on the floor. In the middle of the card's shadow is a circular spot of light—an image of the Sun. If the Sun is low in the sky, the circular image is lengthened to an ellipse.

Analyze and Conclude

1. **Observing** Note that the size of the Sun's image depends on the distance between the card and the floor. Hold the card higher and the image is bigger.

2. **Predicting** Repeat the procedure, but instead of a round hole poked in the card, carefully cut a diamond shape. Will the image be diamond-shaped?

3. **Making Generalizations** The fact that only a circular image occurs shows that the hole casts an image of the Sun—and not that of the opening in the card. Look at spots of light beneath sunlit trees. When the openings between leaves above are small compared with the distance to the ground below, images of the Sun are cast. What will be the shape of these images at the time of a partial solar eclipse?

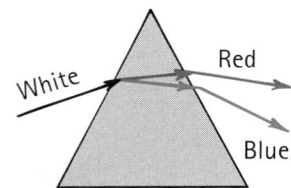

FIGURE 14.1 ▲
Dispersion by a prism makes the components of white light visible. (Only red and blue are indicated here.)

14.1 Light Dispersion and Rainbows

Recall from the previous chapter that light slows when it goes from air to glass or to any transparent material. Recall also that high-frequency light in a transparent medium travels slower than low-frequency light. Violet light travels about 1% slower in ordinary glass than red light. Light of colors between red and violet travel at their own respective speeds in glass.

✓ Because light of various frequencies travel at different speeds in transparent materials, it refracts by different amounts. When white light is refracted twice, as in a prism, the separation of colors of light is quite noticeable. This separation of light into colors arranged by frequency is called *dispersion* (Figure 14.1). Because of dispersion, there are rainbows!

To see a rainbow, the Sun must shine on water drops in a cloud or in falling rain. The drops act as prisms that disperse light. When you face a rainbow, the Sun is behind you, in the opposite part of the sky. Seen from an airplane near midday, the bow forms a complete circle.

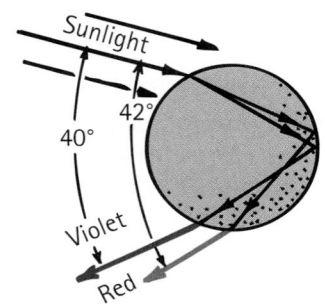

FIGURE 14.2 ▲
Dispersion of sunlight by a single raindrop.

You can see how a raindrop disperses light in Figure 14.2. Follow the ray of sunlight as it enters the drop near its top surface. Some of the light here is reflected (not shown), and the remainder is refracted into the water. At this first refraction, the light is dispersed into its spectrum colors, red being deviated the least and violet the most. When the light reaches the opposite side of the drop, each color is partly refracted out into the air (not shown) and partly reflected back into the water. Arriving at the lower surface of the drop, each color is again partly reflected (not shown) and partly refracted into the air. After two refractions and a reflection, light can leave the droplet at any angle up to some maximum angle, which is about 40° for violet light and about 42° for red light. Bunching up of light intensity at those maximum angles produces a rainbow.

Although each drop disperses a full spectrum of colors, an observer is in a position to see only a single color from any one drop (Figure 14.3). If violet light from a single drop reaches an observer's eye, red light from

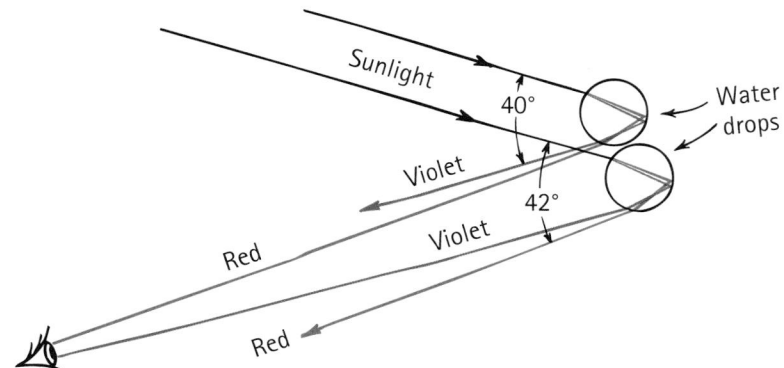

FIGURE 14.3 ▲
Sunlight incident on two raindrops, as shown, emerges as dispersed light. The observer sees the red light from the upper drop and the violet light from the lower drop. Millions of drops produce the whole spectrum of visible light.

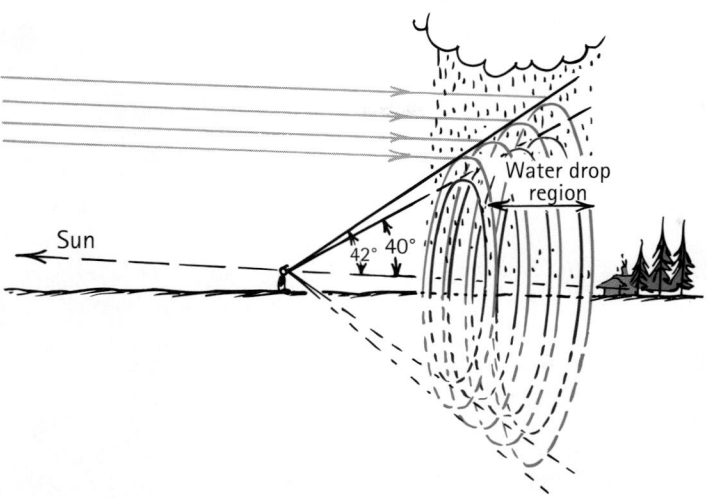

FIGURE 14.4 ▲
When your eye is located between the Sun (not shown off to the left) and a water-drop region, the rainbow you see is the edge of a three-dimensional cone that extends through the water-drop region. (Innumerable layers of drops form innumerable two-dimensional arcs like the four suggested here.)

the same drop travels a lower path (toward the observer's feet). To see red light, you should look to a higher drop in the sky. The color red is seen when the angle between a beam of sunlight and the dispersed light is 42°. The color violet is seen when the angle between the sunbeams and dispersed light is 40°.

A rainbow is not the flat, two-dimensional arc it appears to be. It appears flat for the same reason a spherical burst of fireworks high in the sky looks flat—because it's far away. A rainbow is actually part of a three-dimensional cone. The tip of the cone is at your eye. Think of a glass cone. If you held the pointed tip very close to your eye, what would you see? You'd see the glass as a circle; likewise with a rainbow. All the drops that disperse the rainbow's light toward *you* lie in the shape of a cone—a cone of different layers. Four of the innumerable layers are shown in Figure 14.4. Drops that deflect red to your eye are on the outside, orange is beneath the red, yellow is beneath the orange, and so on all the way to violet on the inner conical surface. The thicker the region containing water drops, the thicker the conical edge you see through, and the more vivid the rainbow.

Your cone of vision intersects the cloud of drops and creates your rainbow. It is ever so slightly different from the rainbow seen by a person nearby. So when a friend says, "Look at the pretty rainbow," you can reply, "Okay, move aside so I can see it, too." Everybody sees his or her own personal rainbow.

Another fact about rainbows: A rainbow always faces you squarely. When you move, your rainbow appears to move with you. So you can never approach the side of a rainbow or see it end-on as in the exaggerated view of Figure 14.4. You *can't* reach its end. Hence the saying

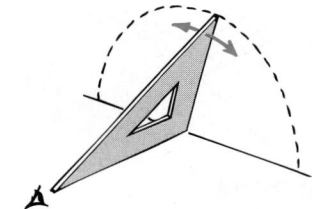

FIGURE 14.5 ▲
Only raindrops along the dashed line disperse red light to the observer at a 42° angle; hence, the light forms a bow.

The primary rainbow needs the Sun to be less than 42° above the horizon. A higher Sun won't do because any rainbow formed would be below the horizon.

FIGURE 14.6 ▲
Two refractions and a reflection in water droplets produce light at all angles up to about 42°, with the intensity concentrated where we see the rainbow at 40° to 42°. No light leaves the water droplet at angles greater than 42° unless it undergoes two or more reflections inside the drop. So the sky is brighter inside the rainbow than outside it. Notice the weak secondary rainbow to the right of the primary.

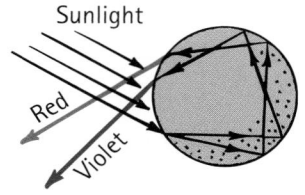

FIGURE 14.7 ▲
Double reflection in a drop produces a secondary bow.

"looking for the pot of gold at the end of the rainbow" means pursuing something you can never reach.

Often a larger, secondary rainbow can be seen arcing at a greater angle around the primary bow. We won't treat this secondary rainbow except to say that it is formed by similar circumstances and is a result of double reflection within the raindrops (Figure 14.7). Because of this extra reflection (and extra refraction loss), the secondary rainbow is much dimmer and its colors are reversed.

READING CHECK

Why do different colors of light bend differently in a transparent material?

CONCEPT CHECK

1. Suppose you point to a wall with your arm extended. Then you sweep your arm around, making an angle of about 42° to the wall. If you rotate your arm in a full circle while keeping the same angle, what shape does your arm describe? What shape on the wall does your finger sweep out?
2. If light traveled at the same speed in raindrops as it does in air, would we have rainbows?

14.2 Lenses

When you think of a lens, think of a set of glass prisms arranged as shown in Figure 14.8. They refract incoming parallel light rays so that they converge to (or diverge from) a point. The arrangement shown in Figure 14.8a converges the light, and we have a **converging lens.** Notice that it is thicker in the middle. In Figure 14.8b, the middle is thinner than the edges and makes a lens that diverges the light—we have a **diverging lens.** Note that the prisms in Figure 14.8b diverge incident rays in a way that makes them appear to originate from a single point in front of the lens.

(a)　　　　　(b)

FIGURE 14.8 ▲
Imagine a lens as a set of blocks and prisms.

HANDS-ON EXPLORATIONS: PLAYING AROUND WITH LENSES

Get your hands on some lenses. Not doing so is like taking swimming lessons away from water. Hold converging lenses in parallel light and note the point where the light converges. Note that different-shaped lenses have different focal lengths.

Determine the magnifying power of a lens by focusing on the lines of a ruled piece of paper. Count the spaces between the lines that fit into one magnified space, and you have the magnifying power of the lens. You can do the same with binoculars and a distant brick wall. Hold the binoculars so that only one eye looks at the bricks through the eyepiece while the other eye looks directly at the bricks. The number of bricks seen with the unaided eye that will fit into one magnified brick gives the magnification of the instrument.

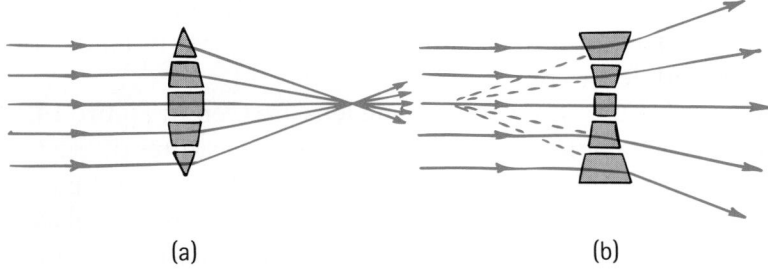

Magnified space

3 spaces fit into one magnified space

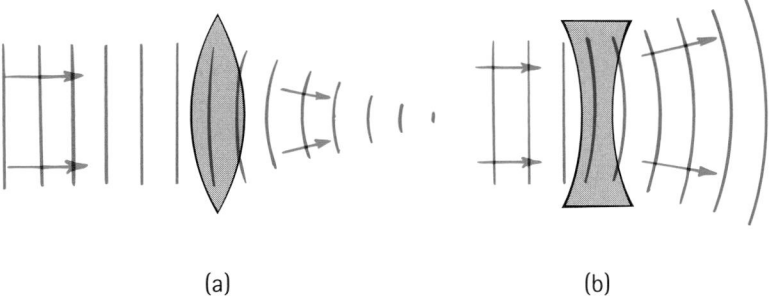

(a) (b)

FIGURE 14.9 ▲
Wavefronts travel more slowly in glass than in air. In (a), the waves are retarded more through the center of the lens, and convergence results. In (b), the waves are retarded more at the edges, and divergence results.

In both lenses, the greatest deviation of rays occurs at the outermost prisms because they have the greatest angle between the two refracting surfaces. ✔ No deviation occurs exactly in the middle, for in that region the two surfaces of the glass are parallel to each other. Real lenses are not made of prisms, of course. They are made of a solid piece of glass with surfaces ground usually to a circular curve. In Figure 14.9, we see how smooth lenses refract waves.

Some key features for a converging lens are seen in Figure 14.10. The *principal axis* is the line joining the centers of curvatures of the two lens surfaces. The *focal point* is the point of convergence for light parallel to the principal axis. Incident beams not parallel to the principal axis focus at points above or below the focal point. All such possible points make up a *focal plane* (not shown). Because a lens has two surfaces, it has two focal points and two focal planes. The *focal length* of the lens is the distance between the center of the lens and either focal point.

In a diverging lens, an incident beam of light parallel to the principal axis is not converged to a point, but is diverged—so the light appears to emerge from a point in front of the lens.

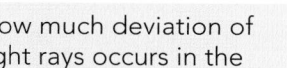

READING CHECK

How much deviation of light rays occurs in the exact middle of a lens?

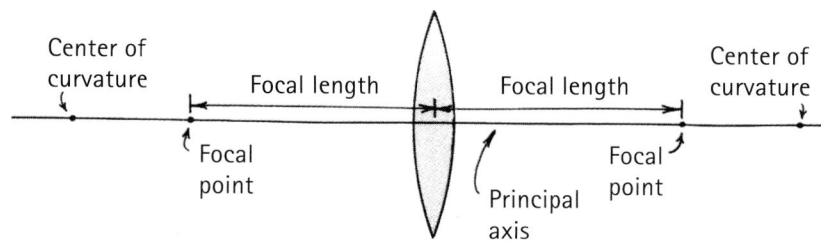

FIGURE 14.10 ▲
Key features of a converging lens.

LINK TO PHYSIOLOGY: YOUR EYE

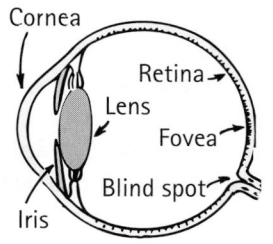

With all of today's technology, the most remarkable optical instrument known is your eye. Light enters through your *cornea*, which does about 70% of the necessary bending of the light before it passes through your *pupil* (the aperture—opening—in the iris). Light then passes through your *lens*, which provides the extra bending power needed to focus images of nearby objects on your extremely sensitive *retina*. (Only recently have artificial detectors been made with greater sensitivity to light than the human eye.) An image of the visual field outside your eye is spread over the retina. The retina is not uniform. There is a spot in the center of your field of view called the *fovea*—region of most distinct vision. You see greater detail here than at the side parts of your eye. There is also a spot in your retina where the nerves carrying all the information exit; this is your *blind spot*.

You can demonstrate that you have a blind spot in each eye. Simply hold this book at arm's length, close your left eye, and look at the round dot and X below with your right eye only. You can see both the dot and the X at this distance. Now move the book slowly toward your face, with your right eye fixed upon the dot, and you'll reach a position about 20–25 centimeters from your eye where the X disappears. When both eyes are open, one eye "fills in" the part to which your other eye is blind. Now repeat with only the left eye open, looking this time at the X, and the dot will disappear. But note that your brain fills in the two intersecting lines. Amazingly, your brain fills in the "expected" view even with one eye closed. Instead of seeing nothing, your brain graciously fills in the appropriate background. Repeat this for small objects on various backgrounds. You not only see what's there—you see what's not there!

The light receptors in your retina do not connect directly to your optic nerve but are instead interconnected to many other cells. Through these interconnections a certain amount of information is combined and "digested" in your retina. In this way, the light signal is "thought about" before it goes to the optic nerve and then to the main body of your brain. So some brain functioning occurs in your eye. Amazingly, your eye does some of your "thinking."

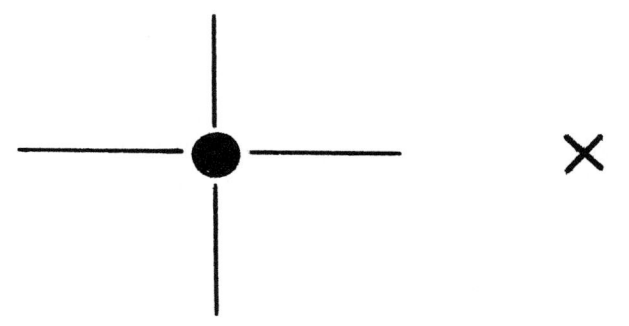

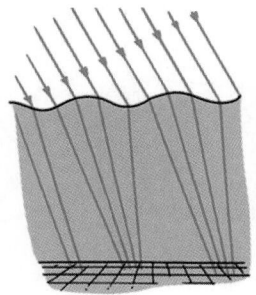

FIGURE 14.11 ▲
The moving patterns of bright and dark areas at the bottom of the pool result from the uneven surface of the water, which behaves like a blanket of undulating lenses. Just as we see the pool bottom shimmering, a fish looking upward at the Sun would see it shimmering, too. Because of similar irregularities in the atmosphere, we see the stars twinkle.

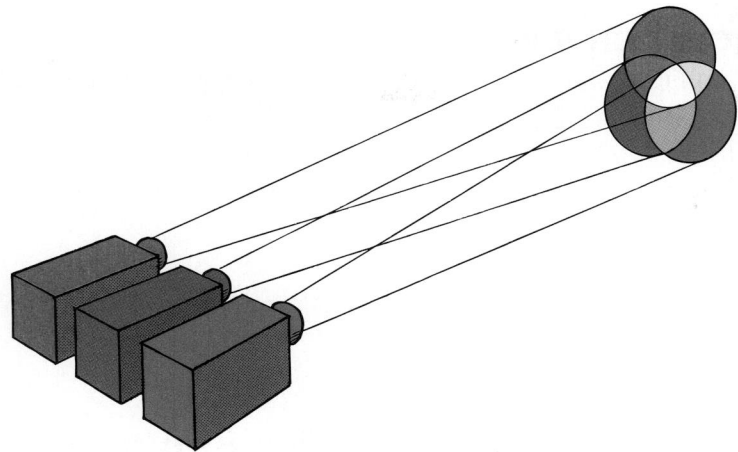

◀ **FIGURE 13.33**
Color addition by the mixing of colored lights. When three projectors shine red, green, and blue light on a white screen, the overlapping parts produce different colors. White is produced where all three overlap.

Complementary Colors

Here's what happens when two of the three additive primary colors are combined:

Red + Blue = Magenta

Red + Green = Yellow

Blue + Green = Cyan

We say that magenta is the opposite of green; cyan is the opposite of red; and yellow is the opposite of blue. When you add any color and its opposite color, you get white.

Magenta + Green = White (= Red + Blue + Green)

Yellow + Blue = White (= Red + Green + Blue)

Cyan + Red = White (= Blue + Green + Red)

When two colors are added together to produce white, they are called **complementary colors.** Every hue has some complementary color that when added to it makes white.

The fact that a color and its complement combine to produce white light is nicely used in lighting stage performances. Blue and yellow lights shining on performers, for example, produce the effect of white light—except where one of the two colors is absent, as in the shadows. The shadow of one lamp, say the blue, is illuminated by the yellow lamp and appears yellow. Similarly, the shadow cast by the yellow lamp appears blue. This is a most interesting effect. We see this effect in Figure 13.34, where red, green, and blue light shine on the golf ball. Note the shadows cast by the ball. The middle shadow is cast by the green spotlight and is not dark because it is illuminated by the red and blue lights, which make magenta. The shadow cast by the blue light appears yellow because it is illuminated by red and green light. Can you see why the shadow cast by the red light appears cyan?

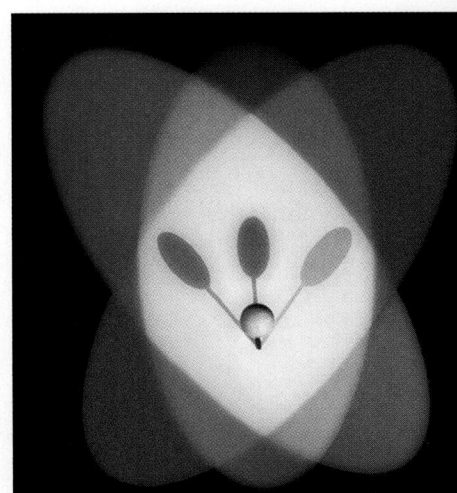

FIGURE 13.34 ▲
The white golf ball appears white when illuminated with red, green, and blue lights of equal intensities. Why are the shadows of the ball cyan, magenta, and yellow?

 **READING CHECK**

What three colors are the cones of our eyes sensitive to?

Make your own historic first camera. Cut out one end of a small cardboard box and cover the end with semi-transparent tracing or tissue paper. Make a clean-cut pinhole at the other end. (If the cardboard is thick, you can make the pinhole through a piece of tinfoil placed over a larger opening in the cardboard.) Aim the camera at a bright object in a darkened room, and you will see an upside-down image on the tracing paper. The tinier the pinhole, the dimmer and sharper the image. If, in a dark room, you replace the tracing paper with unexposed photographic film, cover the back so that it is light-tight, and cover the pinhole with a removable flap, you have a camera. You're ready to take a picture. Exposure times differ depending mostly on the type of film and amount of light. Try different exposure times, starting with about 3 seconds. Also, try boxes of various lengths. The lens on a commercial camera is much bigger than the pinhole and therefore admits more light in less time—hence the name *snapshots*.

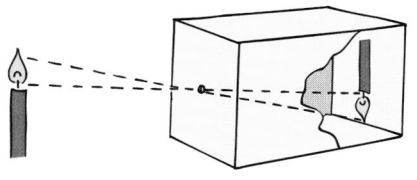

14.3 Image Formation by a Lens

At this moment, light is reflecting from your face onto this page. Light that reflects from your forehead, for example, strikes every part of the page. Likewise for the light that reflects from your chin. Every part of the page is illuminated with reflected light from your forehead, your nose, your chin, and every other part of your face. You don't see an image of your face on the page because there is too much overlapping of light. But place a barrier with a pinhole in it between your face and the page, and the light that reaches the page from your forehead does not overlap the light from your chin. Likewise for the rest of your face. Without this overlapping, an image of your face is formed on the page. It will be very dim, for very little light reflected from your face passes through the pinhole. To see the image, you'd have to shield the page from other light sources. The same is true of the vase and flowers in Figure 14.12b.

The first cameras had no lenses and admitted light through a small pinhole. Long exposure times were required because of the small

Can you see why the image of Figure 14.12 is upside down?

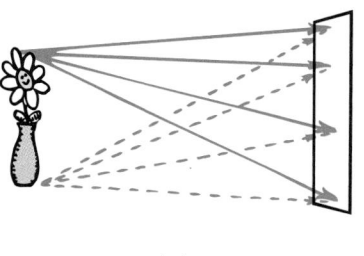

(a)

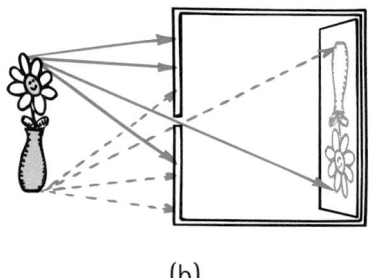

(b)

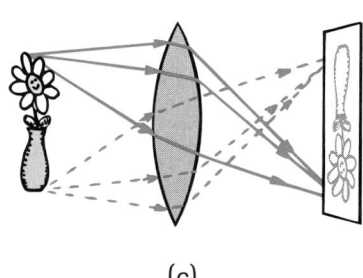

(c)

FIGURE 14.12 ▲

Image formation. (a) No image appears on the wall because rays from all parts of the object overlap all portions of the wall. (b) A single small opening in a barrier prevents overlapping rays from reaching the wall; a dim upside-down image is formed. (c) A lens converges the rays upon the wall without overlapping; more light makes a brighter image.

amount of light admitted by the pinhole. That meant subjects being photographed had to remain still. Motion would produce a blur. If the hole were a bit larger, exposure time would be shorter, but overlapping rays would produce a blurry image. Too large a hole would allow too much overlapping, resulting in no image. That's where a converging lens comes in (Figure 14.12c). The lens converges light onto the film without any overlapping of rays. Moving objects can be taken with the lens camera because of the short exposure time. As mentioned earlier, that's why early photographs taken with lens cameras were called snapshots.

The simplest use of a converging lens is a magnifying glass. To understand how it works, think about how you examine objects near and far. With unaided vision, you see a distant object through a relatively narrow angle of view and a close object through a wider angle of view (Figure 14.13). To see the details of a small object, you want to get as close to it as possible for the widest-angle view. But your eye can't focus when it's too close to the object. That's where the magnifying glass plays a role. When close to the object, the magnifying glass gives you a clear image that would be blurry otherwise.

◀ **FIGURE 14.13**

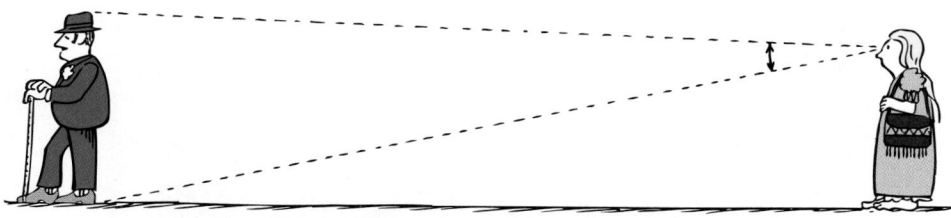
Object is viewed through a narrow angle

Object is viewed through a wide angle

When you use a magnifying glass, you hold it close to the object you wish to examine. This is because a converging lens provides an enlarged, right-side-up image only when the object is inside the focal point. If a screen is placed at the image distance, no image appears on it because no light is directed to the image position. The rays that reach your eye, however, behave virtually *as if* they originated at the image position. This is called a **virtual image.**

When the object is distant enough to be outside the focal point of a converging lens, a **real image** is formed instead of a virtual image, Figure 14.15 shows this case. A real image is upside-down. Likewise for projecting slides and motion pictures on a screen. To see a right-side image, the slide or film is upside-down. The same is true for the image

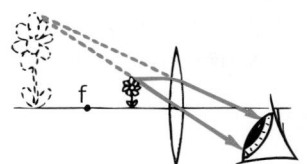

FIGURE 14.14 ▲
When an object is near a converging lens (inside its focal point *f*), the lens acts as a magnifying glass to produce a virtual image. The image appears larger and farther from the lens than the object.

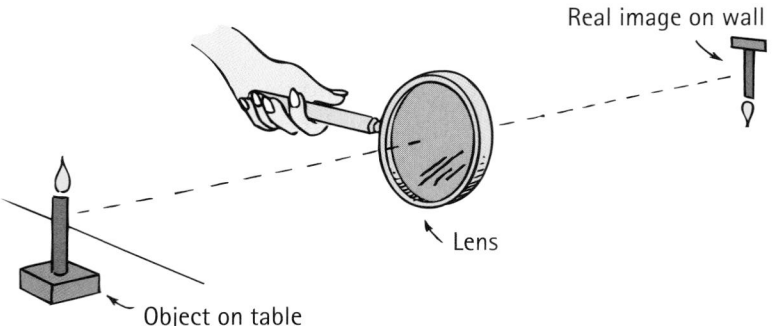

Real image on wall

Lens

Object on table

FIGURE 14.15 ▲
When an object is far from a converging lens (beyond its focal point), a real upside-down image is formed.

Oh my goodness, all my photographs are upside-down!

in a camera. ✔ Real images formed with a single lens are always upside-down.

A diverging lens used alone produces a reduced virtual image. It makes no difference how far or how near the object is. The image is always virtual, right-side-up, and smaller than the object. That's why a diverging lens is often used as a "finder" on a camera. When you look at the object to be photographed through such a lens, you see a virtual image that approximates the same proportions as the photograph.

✔ READING CHECK

Are real images formed with a single lens always, or sometimes, upside-down?

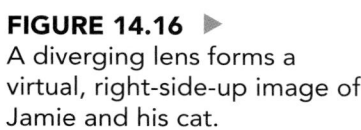
FIGURE 14.16 ▶
A diverging lens forms a virtual, right-side-up image of Jamie and his cat.

CONCEPT CHECK

Why is the greater part of the photograph in Figure 14.16 out of focus?

Check Your Answer

Both Jamie and his cat and the virtual image of Jamie and his cat are "objects" for the lens of the camera that took this photograph. Since the objects are at different distances from the lens, images are at different distances relative to the film in the camera. So only one object can be brought into focus. The same is true of your eyes. You cannot focus on near and far objects at the same time.

Lens Defects

No lens forms a perfect image. A distortion in an image is called an *aberration*. Aberrations can be minimized by combining lenses in particular ways. For this reason, most optical instruments use compound lenses instead of using single lenses. Each compound lens consists of several simple lenses.

✓ *Spherical aberration* results when light passing through the edges of a lens focuses at a slightly different place from where light passing through the center of the lens focuses (Figure 14.17). As a result, the image you see is blurred. This can be remedied by covering the edges of the lens. A camera does this with the use of a diaphragm. In good optical instruments, spherical aberration is corrected by a combination of lenses.

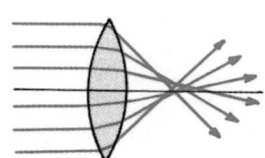

◀ **FIGURE 14.17**
Spherical aberration.

FIGURE 14.18 ▶
Chromatic aberration.

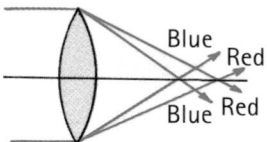

✓ *Chromatic aberration* is the result of different colors having different speeds. This means different refractions in the lens. Different colors of light focus in different places (Figure 14.18). As a result, some colors in the image you see may be in focus while other colors are out of focus. This is corrected with *achromatic lenses,* which combine simple lenses of assorted glass.

The pupil of your eye changes in size to regulate the amount of light that enters. Vision is sharpest when your pupil is smallest because light passes through only the center of your eye's lens. Through the center, spherical and chromatic aberrations are minimal. Also, the eye then functions like a pinhole camera, needing minimum focusing for a sharp image. Your vision is better in bright light because your pupils are smaller.

Astigmatism of the eye is a defect caused by an irregularity in the curvature of the cornea. It curves more in one direction than another—somewhat like the side of a barrel. Due to this defect, the eye doesn't form sharp images. The remedy is eyeglasses with cylindrical lenses that have more curvature in one direction than in another.

To read tiny print clearly, look at it through a pinhole in a piece of paper, close to the print. It works!

✓ **READING CHECK**

Distinguish between spherical and chromatic aberration.

One option for those with poor sight today is wearing eyeglasses. The advent of eyeglasses probably occurred in Italy in the late 1200s. (Curiously, the telescope wasn't invented until some 300 years later. If, in the meantime, anybody viewed objects through a pair of lenses separated along their axes, such as fixed at the ends of a tube, there is no record of it.) Another present-day option to wearing eyeglasses is contact lenses. And yet another is cornea surgery, where the cornea is shaved to a proper shape for normal vision. Soon the wearing of eyeglasses and contact lenses may be a thing of the past. We really do live in a rapidly changing world.

CONCEPT CHECK

1. If light traveled at the same speed in glass and in air, would glass lenses change the direction of light rays?
2. Why is chromatic aberration associated with a lens but not with a mirror?

Check Your Answers

1. No.
2. Light of different frequencies travels at different speeds in a transparent medium and therefore refracts at different angles. This is the cause of chromatic aberration. The angles of reflected light, however, have no relation to frequency. One color reflects the same as any other color. In telescopes, therefore, mirrors are preferred over lenses because of the absence of chromatic aberration for reflection.

14.4 Diffraction—The Spreading of Light

When you touch your finger to the surface of still water, circular ripples are produced. When you touch the surface with a straight edge, such as a horizontally held meterstick, you produce a plane wave. You can produce a series of plane waves by successively dipping a meterstick into the surface (Figure 14.19).

The photographs in Figure 14.20 are top views of water ripples in a shallow glass tank (called a ripple tank). A barrier with an adjustable opening is in the tank. When plane waves meet the barrier, they continue

FIGURE 14.19 ▲
The vibrating meterstick makes plane waves in the tank of water. These waves diffract through the opening.

FIGURE 14.20 ▶
Plane waves passing through openings of various sizes. The smaller the opening, the greater the bending of the waves at the edges—in other words, the greater the diffraction.

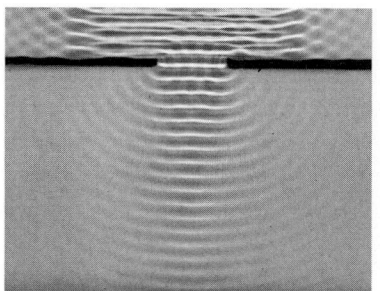
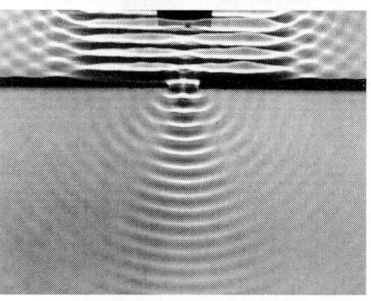
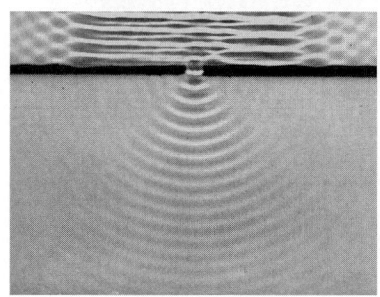

through with some distortion. In the top image, where the opening is wide, the waves continue through the opening almost without change. At the two ends of the opening, however, the waves bend. This bending is called **diffraction.** Any bending of light by means other than reflection and refraction is diffraction. As the width of the opening is narrowed, as in the center image in Figure 14.20, the waves spread more. When the opening is small relative to the wavelength of the incident wave, they spread even more. We see that smaller openings produce more diffraction. ✔ Diffraction is a property of all kinds of waves, including sound and light waves.

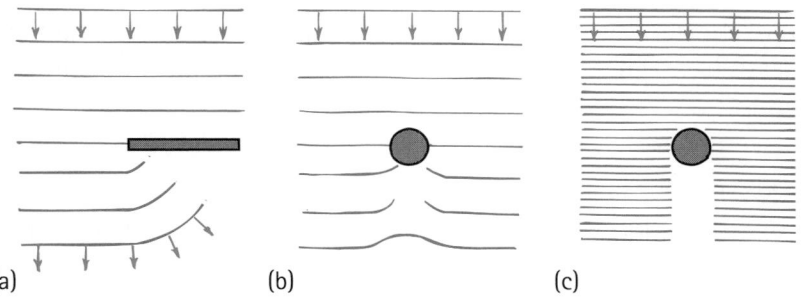
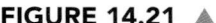

(a)   (b)   (c)

FIGURE 14.21 ▲
(a) Waves tend to spread into the shadow region. (b) When the wavelength is about the size of the object creating the barrier, the shadow is soon filled in. (c) When the wavelength is short relative to the object's size, a sharp shadow is cast.

Diffraction is not confined to narrow slits or to openings. Diffraction occurs around the edges of surfaces and it can be seen with all shadows (Figure 14.22). On close examination, even the sharpest shadow is blurred slightly at the edge .

The amount of diffraction depends on the wavelength of the wave compared with the size of the obstruction that casts the shadow. Long waves are better at filling in shadows, which is why foghorns emit low-frequency sound waves—to fill in any "blind spots." Likewise for radio waves. The wavelength of AM radio waves ranges from 180 to 550 meters. These waves are longer than most objects in their path. They readily bend around buildings and other objects. A long-wavelength radio wave doesn't "see" a relatively small building in its path—but a short-wavelength radio wave does. The radio waves of the FM band range from 2.8 to 3.4 meters and don't bend very well around buildings. This is one of the reasons FM reception is often poor in localities where AM comes in loud and clear. In the case of radio reception, we don't wish to "see" objects in the path of radio waves, and so diffraction is nice.

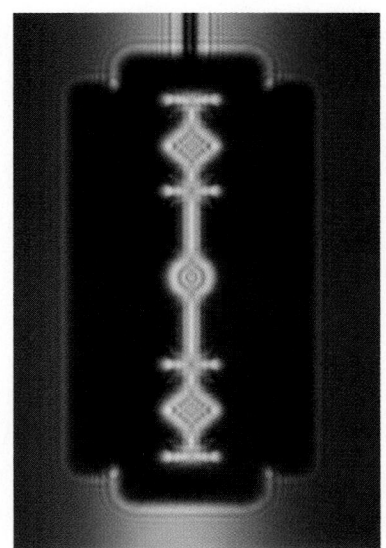

FIGURE 14.22 ▲
Diffraction fringes are evident in the shadows of monochromatic (single-frequency) laser light.

Bats hunt moths in darkness by echo location. Some moths are protected by a thick covering of fuzzy scales that deaden the echos.

Diffraction is not so nice for viewing very small objects with a microscope. If the size of an object is about the same as the wavelength of light, diffraction blurs the image. If the object is smaller than the wavelength of light, no structure can be seen. The entire image is lost due to diffraction. No amount of magnification or perfection of microscope design can defeat this fundamental diffraction limit.

To minimize this problem, microscopists illuminate tiny objects with electron beams rather than light. Compared with light waves, electron beams have extremely short wavelengths. *Electron microscopes* take advantage of the fact that all matter has wave properties. A beam of electrons has a wavelength smaller than those of visible light. In an electron microscope, electric and magnetic fields, rather than optical lenses, are used to focus and magnify images.

The fact that smaller details can be better seen with smaller wavelengths is useful to dolphins, who scan their environment with ultrasound. The echoes of long-wavelength sound give the dolphin an overall image of objects in its surroundings. To examine more detail, the dolphin emits sound of shorter wavelengths. The dolphin has always done naturally what physicians have only recently been able to do with an ultrasonic imaging devices.

READING CHECK

What kinds of waves diffract?

CONCEPT CHECK

Why does a microscopist use blue light rather than white light to illuminate objects being viewed?

Check Your Answer

There is less diffraction with blue light. This allows the microscopist to see more detail (just as a dolphin beautifully investigates fine detail in its environment by the echoes of ultra-short wavelengths of sound).

14.5 Interference—Constructive and Destructive

The fringes of brightness and darkness produced when light is diffracted are due to **interference.** Recall our study of interference in Chapter 12, where we learned waves can combine constructively or destructively. ✓ Interference is a property of all waves, including light. Constructive and destructive interference are reviewed in Figure 14.23.

FIGURE 14.23 ▶
Wave interference.

+ = Reinforcement   + = Cancellation   + = Partial cancellation

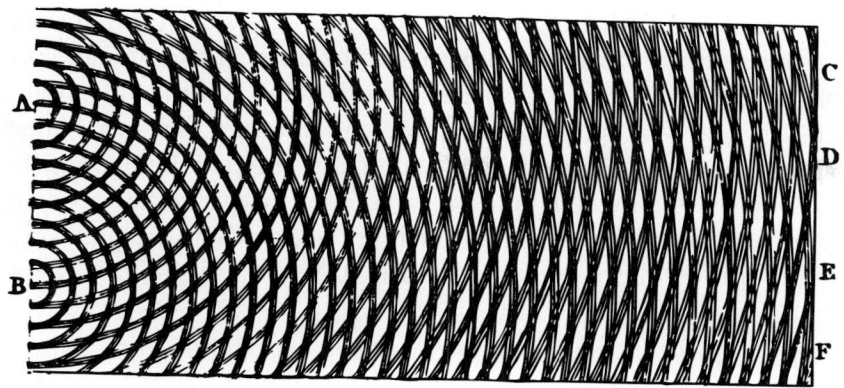

FIGURE 14.24 ▲
Thomas Young's original drawing of a two-source interference pattern.

We see that the adding, or *superposition*, of a pair of identical waves in phase produces a wave of the same frequency but twice the amplitude. If the waves are exactly one-half wavelength out of phase, their superposition results in complete cancellation. If they are out of phase by other amounts, partial cancellation occurs.

Wave interference was convincingly demonstrated by Thomas Young in 1801.* Young shone light through two closely spaced side-by-side pinholes. He found that the light recombines to produce fringes of brightness and darkness on a screen behind. The bright fringes form when a crest from the light wave through one hole meets a crest from the light wave through the other hole. Both waves arrive in phase at the screen. The dark fringes form when a crest from one wave meets a trough from the other. Figure 14.24 shows Young's drawing of the pattern of superimposed waves from the two sources. When his experiment is done with two closely spaced slits instead of pinholes, the fringe patterns are straight lines (Figure 14.25).

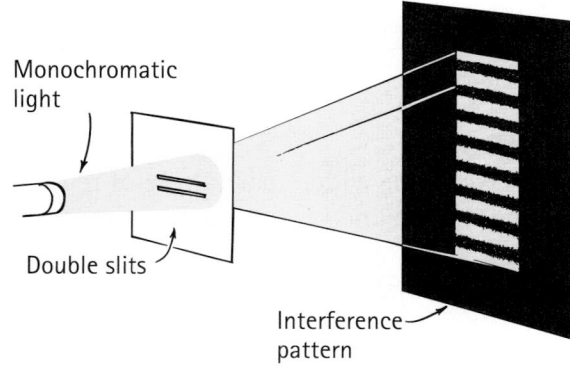

Monochromatic
light

Double slits

Interference
pattern

◀ **FIGURE 14.25**
When monochromatic light passes through two closely spaced slits, an interference pattern of fringes is produced.

---

*Thomas Young read fluently at the age of 2. By the age of 4, he had read the Bible twice. When he was 14, he knew eight languages. In his adult life he was a physician and physicist, contributing to an understanding of fluids, work and energy, and the elastic properties of materials. He was the first person to make progress in deciphering Egyptian hieroglyphics. No doubt about it—Thomas Young was a bright guy!

We see in Figures 14.26 and 14.27 how the series of bright and dark
fringes result from the different path lengths from slits to screen. For the
central bright fringe, the paths from the two slits are the same length and
so the waves arrive in phase and reinforce each other. The dark fringes
on either side of the central fringe result from one path being longer (or
shorter) by one-half wavelength, so waves there arrive half a wavelength
out of phase. The other sets of dark fringes occur where the paths differ
by odd multiples of one-half wavelength: 3/2, 5/2, and so on.

FIGURE 14.26 ▲
A bright area occurs when waves from both slits arrive in phase; a dark area
results from the overlapping of waves that are out of phase.

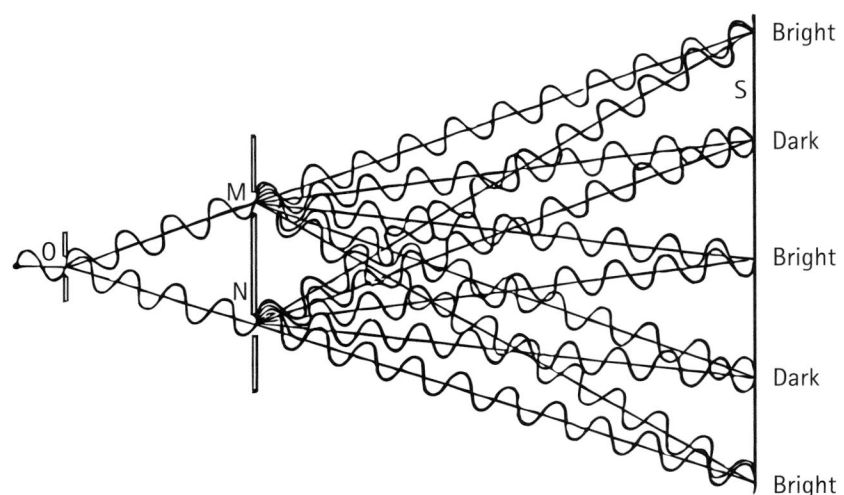

FIGURE 14.27 ▶
Light from O passes through
slits M and N and produces an
interference pattern on the
screen S.

CONCEPT CHECK

1. Which fringes would be wider apart, those of red light or blue
   light, assuming the red and blue are monochromatic (single
   frequency)?
2. Why is it important that monochromatic light be used?

Check Your Answers

1. Red fringes would be more widely spaced. Can you see that longer waves in
   Figure 14.27 would produce wider-apart fringes?
2. Light of various wavelengths would result in dark fringes for one wavelength filling in
   bright fringes for another. The result would be no distinct fringe pattern. If you haven't
   seen this, perhaps your teacher will demonstrate it.

LINK TO OPTOMETRY: SEEING STAR-SHAPED STARS

Have you ever wondered why stars are represented with spikes? The stars on the American flag have five spikes, and the Jewish Star of David has six spikes. All through the ages, stars have been drawn with spikes. Stars do not have spikes, for they are point sources of light in the night sky. The reason for the spikes is poor eyesight.

The surface of your eye, the cornea, becomes scratched by a variety of causes. These scratches make up a sort of diffraction grating. A scratched cornea is not a very good diffraction grating, but its effects are evident. Look at a bright point source against a dark background—like a star in the night sky. Instead of seeing a point of light, you may see a spiky shape. The spikes even shimmer and twinkle if there are some temperature differences in the atmosphere to produce some refraction. And if you live in a windy desert region where sandstorms are frequent, your cornea will be even more scratched and you'll see more vivid star spikes.

Interference patterns are not limited to one or two slits. A multitude of closely spaced slits make up a *diffraction grating*. These devices, like prisms, disperse white light into colors. These are used in devices called *spectrometers*, which we shall discuss in the next chapter. The feathers of some birds act as diffraction gratings and disperse colors. Likewise for the microscopic pits on the reflective surface of compact discs.

FIGURE 14.28 ▲
Because of the interference it causes, a diffraction grating disperses light into colors. It may be used in place of a prism in a spectrometer.

14.6 Interference Colors by Reflection from Thin Films

We have all noticed the beautiful spectrum of colors reflected from a soap bubble or from gasoline on a wet pavement. These colors are produced by the interference of light waves. This phenomenon is often called *iridescence* and is observed in thin, transparent films.

A soap bubble appears iridescent in white light when the thickness of the soap film is about the same as the wavelength of light. Light waves reflected from the outer and inner surfaces of the film to your eye travel different distances. When illuminated by white light, the film may be just the right thickness at one place to cause the destructive interference of, say, red light. When red light is subtracted from white light, the mixture left appears as the complementary color—cyan. At another place, where the film is thinner, perhaps blue is canceled. Then the light seen is the complement of blue—yellow. ✓ Whatever color is canceled by interference, the light seen is its complementary color.

This occurs for gasoline on a wet street (Figure 14.29). Light reflects from two surfaces. One is the upper, air-gasoline surface. The other is

Soap-bubble colors result from the interference of reflected light from the inside and outside surfaces of the soap film. When a color is canceled, what you see is its complementary color.

FIGURE 14.29 ▶
The thin film of gasoline is just the right thickness to cancel the reflections of blue light from the top and bottom surfaces. If the film were thinner, perhaps shorter-wavelength violet would be canceled. (One wave is drawn black to show how it is out of phase with the blue wave upon reflection.)

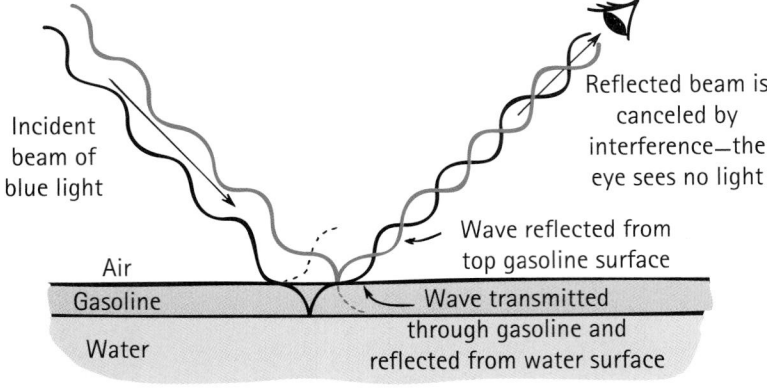

FIGURE 14.30 ▲
Magenta is seen in Emily's soap bubbles, which is due to the cancellation of green light. What primary color is canceled to produce cyan?

the lower, gasoline-water surface. If the thickness of the gasoline is such to cancel blue, as the figure suggests, then the gasoline surface appears yellow to the eye. As mentioned earlier, blue subtracted from white leaves yellow. Why are a variety of colors seen in the thin film of gasoline? The answer is that the film thickness is not uniform. Different film thicknesses show a "contour map" of microscopic differences in surface "elevations."

If you view the thin film of gasoline at a lower angle, you'll see different colors. That's because light through the film travels a longer path. A longer wave is canceled and a different color is seen. Different wavelengths of light are canceled for different angles.

Dishes washed in soapy water and poorly rinsed have a thin film of soap on them. Hold such a dish up to a light source so that *interference colors* can be seen. Then turn the dish to a new position, keeping your eye on the same part of the dish. Do you notice a change in color? Light reflecting from the bottom surface of the transparent soap film cancels light reflecting from the top surface.

Interference techniques can be used to measure the wavelengths of light and other regions of the electromagnetic spectrum. Interference provides a means of measuring extremely small distances with great accuracy. Instruments called *interferometers* use the principle of interference and are the most accurate instruments known for measuring small distances.

HANDS-ON EXPLORATION: SWIRLING COLORS

Dip a dark-colored cup (dark colors make the best background for viewing interference colors) in dishwashing detergent. Then hold the cup sideways and look at the reflected light from the soap film that covers its mouth. Swirling colors appear as the soap runs down to form a wedge that grows thicker at the bottom. The top becomes thinner—so thin that it appears black. This occurs when the film is thinner than 1/4 the wavelength of the shortest waves of visible light. The film soon becomes so thin it pops.

CONCEPT CHECK

1. What color appears to be reflected from a soap bubble in sunlight when its thickness is such that green light is canceled?

2. In the left column are the colors of certain objects. In the right column are various ways in which colors are produced. Match the right column to the left.

   (a) yellow daffodil   (1) interference

   (b) blue sky   (2) diffraction

   (c) rainbow   (3) selective reflection

   (d) peacock feathers   (4) refraction

   (e) soap bubble   (5) scattering

Check Your Answers

1. The composite of all the visible wavelengths except green is the complementary color, magenta. (Go back and see Figures 13.35 and 13.36.)
2. a-3; b-5; c-4; d-2; e-1.

> ✔ **READING CHECK**
>
> When white light illuminates a thin film and a particular color is canceled by interference, what color remains?

14.7 Polarization—Evidence for the Transverse Wave Nature of Light

Interference and diffraction provide the best evidence that light is wavelike. As we learned in Chapter 12, waves can be either longitudinal or transverse. Sound waves are longitudinal, which means the vibratory motion of the medium is *along* the direction of wave travel. ✔ The fact that light waves exhibit **polarization** demonstrates that they are transverse.

If you shake a rope either up and down or from side to side as shown in Figure 14.31, you'll produce a transverse wave along the rope. The plane of vibration is the same as the plane of the wave. If we shake it up and down, the wave vibrates in a vertical plane. If we shake it back and forth, the wave vibrates in a horizontal plane. We say that such a wave is *plane-polarized*—that the waves traveling along the rope are confined to a single plane. Polarization is a property of transverse waves. (Polarization does not occur among longitudinal waves—there is no such thing as polarized sound.)

A single vibrating electron can emit an electromagnetic wave that is plane-polarized. The plane of polarization matches the vibrational direction of the electron. That means a vertically accelerating electron emits light that is vertically polarized. A horizontally accelerating electron emits light that is horizontally polarized (Figure 14.32).

A common light source—such as an incandescent lamp, a fluorescent lamp, or a candle flame—emits light that is unpolarized. This is because electrons that emit the light are vibrating in many random directions. There are as many planes of vibration as the vibrating electrons producing them. A few planes are represented in Figure 14.33a. We can represent all these planes by radial lines, shown in Figure 14.33b. (Or, more simply, the planes can be represented by vectors in two mutually perpendicular directions, Figure 14.33c.) A vertical vector represents all the components of vibration in the vertical direction. The horizontal vector represents all

FIGURE 14.31 ▲
A vertically plane-polarized plane wave and a horizontally plane-polarized plane wave.

(a)

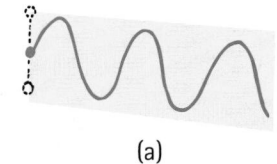

(b)

FIGURE 14.32 ▲
(a) A vertically plane-polarized wave from a charge vibrating vertically. (b) A horizontally plane-polarized wave from a charge vibrating horizontally.

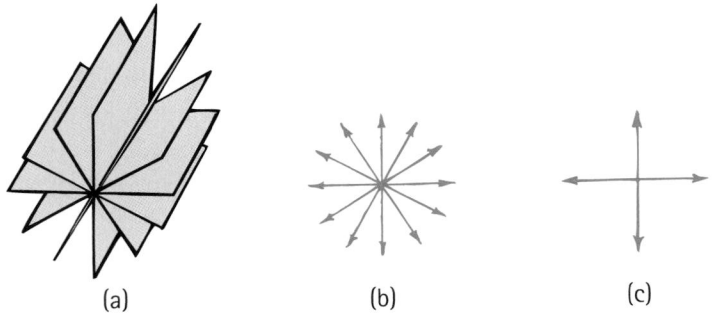

FIGURE 14.33 ▲

Representations of planes of waves. The center and right configurations show electric vectors that make up the electric part of electromagnetic waves.

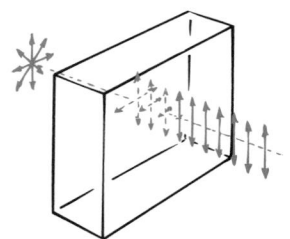

FIGURE 14.34 ▲

One component of the incident unpolarized light is absorbed, resulting in emerging polarized light.

the components of vibration horizontally. The simple model of Figure 14.33c represents unpolarized light. Polarized light would be represented by a single vector.

All transparent crystals having a noncubic natural shape have the property of polarizing light. These crystals divide unpolarized light into two internal beams polarized at right angles to each other. Some crystals strongly absorb one beam while transmitting the other (Figure 14.34). This makes them excellent polarizers. Herapathite is such a crystal. Microscopic herapathite crystals are aligned and embedded between cellulose sheets. They make up Polaroid filters, popular in sunglasses. Other Polaroid sheets consist of certain aligned molecules rather than tiny crystals.

If you look at unpolarized light through a Polaroid filter, you can rotate the filter in any direction and the light appears unchanged. But if the light

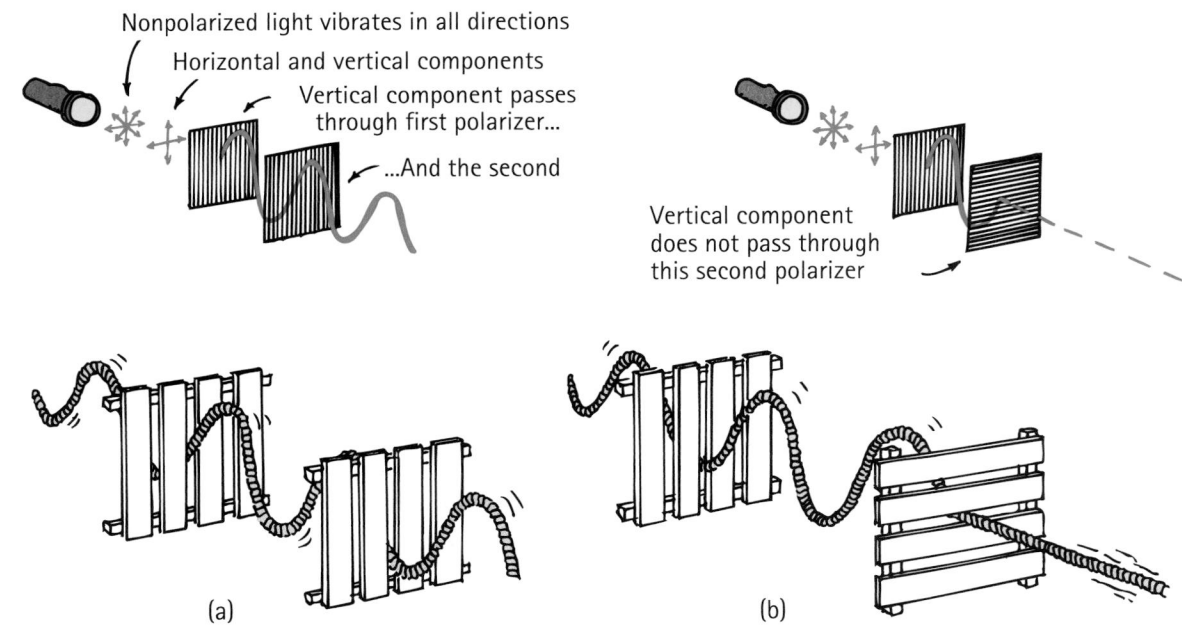

FIGURE 14.35 ▲

(a) Both light and the vibrations of the rope pass through aligned filters.
(b) Neither light nor vibrations of the rope pass through crossed filters. The blue dashed line shows the direction light would travel if the filters weren't crossed.

HANDS-ON EXPLORATION: INTERFERENCE COLORS WITH POLAROIDS

Place a source of white light on a table in front of you. Then place a sheet of Polaroid in front of the source, a bottle of corn syrup in front of the sheet, and a second sheet of Polaroid in front of the bottle. Look through the Polaroid sheets that sandwich the syrup and view spectacular colors as you rotate one of the sheets. (If you don't have sheets, use lenses from Polaroid sunglasses.)

is polarized, then as you rotate the filter, you progressively reduce more and more of the light until it is blocked out. An ideal Polaroid filter transmits 50% of incident unpolarized light. That 50% is polarized. When two Polaroid filters are arranged so that their polarization axes are aligned, light can pass through both (Figure 14.35a). If their axes are at right angles to each other (in this case we say the filters are *crossed*), almost no light penetrates the pair (Figure 14.35b). (A small amount of shorter wavelengths do get through.) When Polaroid filters are used in pairs like this, the first one is called the *polarizer* and the second one the *analyzer*.

Much of the light reflected from nonmetallic surfaces is polarized. The glare from glass or water is a good example. Except for light that hits vertically, the reflected ray has more vibrations parallel to the reflecting surface. The part of the ray that penetrates the surface has more vibrations at right angles to the surface (Figure 14.37). Skipping flat rocks off the surface of a pond is analogous. When the rocks hit parallel to the surface, they easily reflect. But when they hit with their faces at right angles to the surface, they "refract" into the water. The glare from reflecting surfaces can be dimmed a lot with the use of Polaroid sunglasses. The polarization axes of the lenses are vertical because most glare reflects from horizontal surfaces.

FIGURE 14.36 ▲
Polaroid sunglasses block out horizontally vibrating light. When the lenses overlap at right angles, no light gets through.

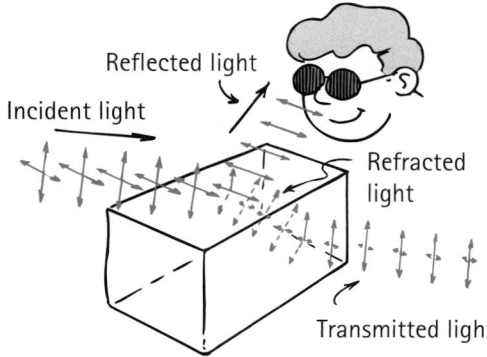

FIGURE 14.37 ▲
Most glare from nonmetallic surfaces is polarized. Notice that the components of incident light parallel to the surface are reflected. Also notice that the components perpendicular to the surface pass through the surface into the medium. Because most of the glare we encounter is from horizontal surfaces, the polarization axes of Polaroid sunglasses are vertical.

FIGURE 14.38 ▲

Light (a) passes through when the axes of the Polaroid filters are aligned but (b) absorbed when Ludmila rotates one filter so that the axes are at right angles to each other. (c) When she inserts a third Polaroid filter at an angle between the crossed ones, light again passes through. Why? (The answer is in Appendix C.)

CONCEPT CHECK

Which pair of glasses is best suited for automobile drivers? (The polarization axes are shown by straight lines.)

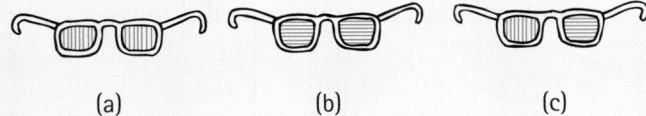

(a) (b) (c)

Check Your Answer

Pair A is best suited because the vertical axis blocks horizontally polarized light, which makes up much of the glare from horizontal surfaces. Pair C is suited for viewing 3-D movies.

What kind of waves can exhibit polarization?

Beautiful colors similar to interference colors can be seen when certain materials are placed between crossed Polaroid filters. Cellophane and transparent tape are wonderful. The explanation for these colors is another story—advanced study.

14.8 Wave-Particle Duality— Two Sides of the Same Coin

In ancient times, Plato and other Greek philosophers hypothesized that light was made up of tiny particles. And in the early 1700s, so did Isaac Newton, who first became famous for his experiments with light. Then a hundred years later, the wave nature of light was demonstrated by Thomas Young in his interference experiments. This wave view was validated in 1862 by James C. Maxwell's finding that light is energy carried in the vibrating electric and magnetic fields of electromagnetic waves. The wave view of light was confirmed experimentally by Heinrich Hertz 25 years later. The wave nature of light seemed to be established.

Then in 1905, Albert Einstein published a Nobel Prize–winning paper that challenged the wave theory of light. Einstein stated that, in its interactions with matter, light strikes not as a wave, but as a tiny particle of energy called a *photon.*

So science had come full circle in its view of light—particle to wave and back to particle. As we shall see, both views are correct! First let's look at Einstein's particle model of light, which explained a mystery to scientists in the early 1900s—the *photoelectric effect.*

When light shines on certain metal surfaces, electrons are ejected from the surfaces. This is the **photoelectric effect,** used in electric eyes, light meters, and some predigital motion-picture sound tracks. What puzzled investigators in the early 1900s was that only ultraviolet and violet light knock electrons from these surfaces. Lower-frequency light, like red, won't do it, even if the red light is very bright with much energy. They discovered that electrons are ejected by high-frequency light, not by bright light. Very dim violet light ejects electrons, but a bright red or green light doesn't. If the brightness of violet light is increased, more electrons are ejected—but not at greater energies.

Einstein's explanation was that the electrons in the metal are bombarded by "particles of light"—now called **photons.** Einstein stated that the energy of each photon is proportional to its frequency:

$$E \sim f$$

So Einstein viewed a beam of light as a hail of photons, each carrying energy proportional to its frequency. One photon is completely absorbed by each electron ejected from the metal.

All attempts by investigators at the time to explain the photoelectric effect by waves failed. A light wave is broad and its energy is spread out. For a light wave to eject a single electron from a metal surface, all the light's energy would have to be concentrated on that one electron. This is as unlikely as an ocean wave hitting a beach and knocking only one single seashell far inland with an energy equal to that of the whole wave. The photoelectric effect only makes sense by thinking of light as a succession of particle-like photons.

Einstein's idea was experimentally verified 11 years after he announced it. Interestingly, it was verified by an experimenter, the American physicist Robert Millikan, who thought he would discredit it (a not-too-uncommon feature of science). Every aspect of Einstein's concept was confirmed. The photoelectric effect proves convincingly that light has particle properties.

Recall Thomas Young's double-slit interference experiment, which we earlier discussed in terms of waves. When monochromatic light passes through a pair of closely spaced thin slits, an interference pattern is produced on photographic film (Figure 14.40). Now let's consider the experiment in terms of photons. Suppose we dim our light source so that in effect only one photon at a time reaches the thin slits. If the film behind the slits is exposed to the light for a very short time, the film becomes exposed as simulated in Figure 14.41a. Each spot shows where the film

Einstein won the Nobel Prize for his explanation of the photoelectric effect.

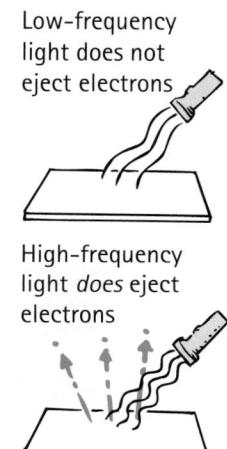
Low-frequency light does not eject electrons

High-frequency light *does* eject electrons

FIGURE 14.39 ▲
The photoelectric effect depends on frequency.

A photon is a lump of light.

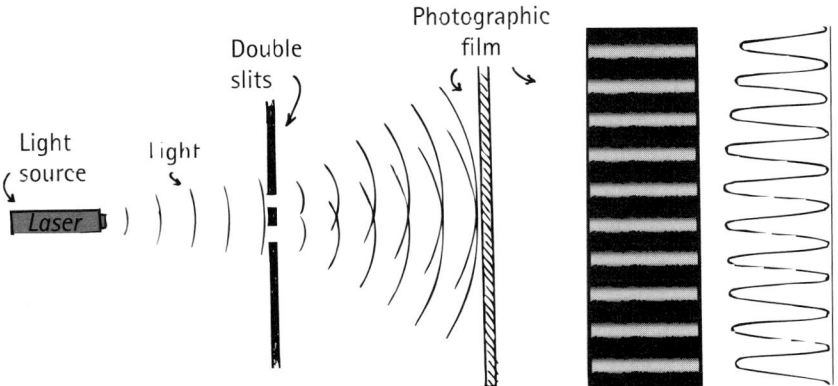

FIGURE 14.40 ▲
Arrangement for double-slit experiment. The black-and-white-striped tall rectangle is a photograph of the interference pattern. To the far right is a graphic of the pattern.

has been exposed to a photon. If the light is allowed to expose the film for a longer time, a pattern of fringes begins to emerge, as in Figure 14.41b and c. This is quite amazing. We see spots on the film progressing photon by photon to form the same interference pattern characterized by waves!

(a)   (b)   (c)

FIGURE 14.41 ▲
Stages of a two-slit interference pattern. The pattern of individual exposed grains progresses from (a) 28 photons to (b) 1000 photons to (c) 10,000 photons. As more photons hit the screen, a pattern of interference fringes appears.

Evidently, light has both a wave nature and a particle nature—a wave-particle duality. This duality is evident in the formation of optical images. Let's look carefully at the way a photographic image is formed, say on film, photon by photon (Figure 14.42). The photographic film consists of an emulsion that contains grains of silver halide crystal, each grain containing about 10^{10} silver atoms. Each photon that hits the film gives up its energy to a single grain in the emulsion. This energy activates surrounding crystals in the grain and develops the film. Many photons activating many grains produce the usual photographic exposure. When a photograph is taken with very feeble light, the image is built up by individual photons that arrive independently and at random. Figure 14.42 beautifully shows how an exposure progresses photon by photon.

What all this means is that light has both wave and particle properties. Simply put: ✓   *Light behaves as a stream of photons when it interacts with the photographic film or other detectors. But light behaves as a wave in traveling from a source to the place where it is detected.* In

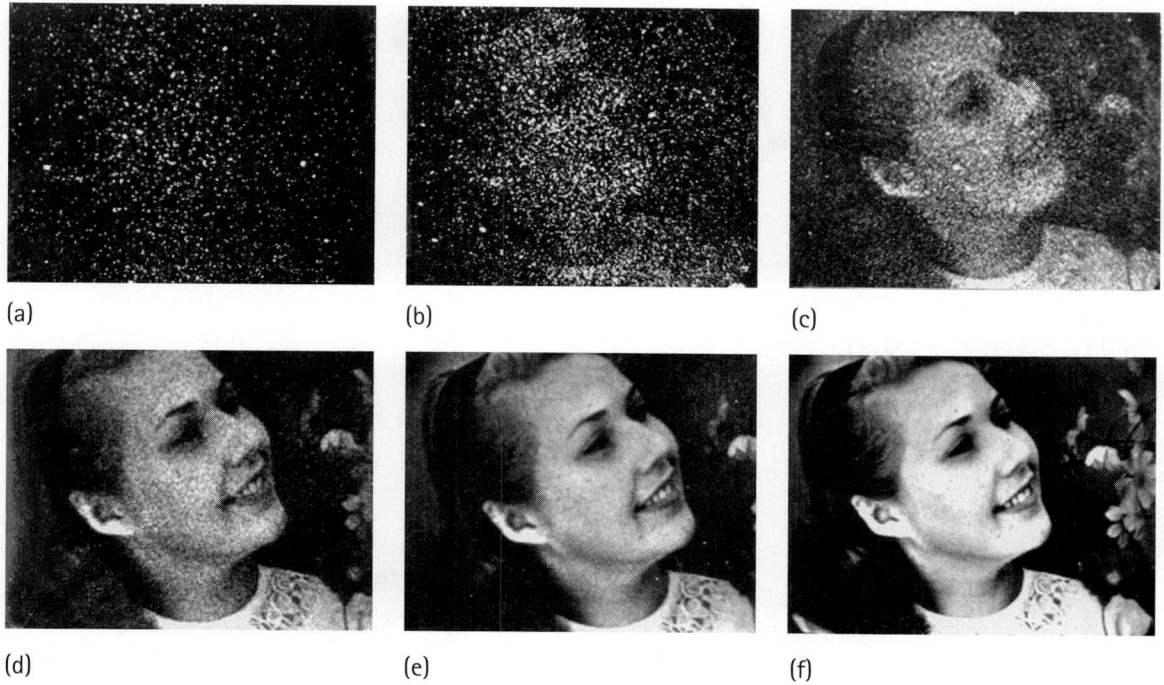

(a)   (b)   (c)

(d)   (e)   (f)

FIGURE 14.42 ▲
Stages of film exposure reveal the photon-by-photon production of a
photograph. The approximate numbers of photons at each stage are
(a) 3×10^3; (b) 1.2×10^4; (c) 9.3×10^4; (d) 7.6×10^5; (e) 3.6×10^6; and
(f) 2.8×10^7.

interference experiments, photons strike the film at places where we
would expect to see constructive interference of waves.

The fact that light shows both wave and particle behavior is one
of the most interesting surprises that scientists discovered in the previous
century. The finding that light comes in bundles, *photons*—sometimes
called *quanta*—led to a whole new way of looking at nature: wave
mechanics, or *quantum mechanics*. An outcome of this new mechanics is
that just as light has particle properties, particles have wave properties.
First, electrons were found to have wave properties; a beam of electrons
passing through slits exhibits the same type of diffraction pattern as light.
Then other particles—even baseballs and orbiting planets—could be
described by the new mechanics of waves. Quantum mechanics and
Newtonian physics overlap in the macroworld, and both are seen as
"correct." But only quantum mechanics, with its emphasis on waves, is
wholly accurate in the microworld of the atom. Let's explore the atom!

Light travels as a wave and
hits like a particle.

✔ **READING CHECK**

When does light behave as
a wave, and when does it
behave as a stream of
particles?

14 REVIEW

KEY TERMS

Converging lens A lens that is thicker in the middle than at the edges and refracts parallel rays passing through it to a focus.

Diffraction The bending of light as it passes around an obstacle or through a narrow slit, causing the light to spread and to produce light and dark fringes.

Diverging lens A lens that is thinner in the middle than at the edges, causing parallel rays passing through it to diverge as if from a point.

Interference The result of superposing two or more waves of the same wavelength.

Photoelectric effect The emission of electrons from a metal surface when light shines on it.

Photon A particle of light, or the basic packet of electromagnetic radiation.

Polarization The alignment of the transverse electric vectors that make up electromagnetic radiation.

Real image An image formed by light rays that converge at the location of the image—which can be displayed on a screen.

Virtual image An image formed by light rays that do not converge at the location on the image. Cannot be displayed on a screen.

REVIEW QUESTIONS

Light Dispersion and Rainbows

1. Which travels more slowly in glass, red light or violet light?
2. What color light bends most in a prism?
3. Where must the Sun be to view a rainbow?

Lenses

4. Distinguish between a *converging lens* and a *diverging lens*.
5. What is the *focal length* of a lens?

Image Formation by a Lens

6. Distinguish between a *virtual image* and a *real image*.
7. What kind of lens can be used to produce a real image? A virtual image?

Diffraction—The Spreading of Light

8. Is diffraction more pronounced through a small opening or through a large opening?
9. For an opening of a given size, is diffraction more pronounced for a longer wavelength or a shorter wavelength?

Interference—Constructive and Destructive

10. Is interference restricted to only some types of waves or does it occur for all types of waves?
11. What is monochromatic light?

Interference Colors by Reflection from Thin Films

12. What causes the variety of colors seen in gasoline splotches on a wet street?
13. What accounts for the variety of colors in a soap bubble?

Polarization—Evidence for the Transverse Wave Nature of Light

14. What phenomenon distinguishes between longitudinal and transverse waves?
15. How does the direction of polarization of light compare with the direction of vibration of the electrons that produced it?
16. How much unpolarized light does an ideal Polaroid filter transmit?

Wave-Particle Duality—Two Sides of the Same Coin

17. What evidence can you cite for the wave nature of light? For the particle nature of light?
18. Which are more successful in dislodging electrons from a metal surface, photons of violet light or photons of red light? Why?
19. Why won't a very bright beam of red light impart more energy to an electron than a feeble beam of violet light?
20. When does light behave as a wave? When does it behave as a particle?

THINK AND DO

With a razor blade, cut a slit in a card and look at a light source through it. You can vary the size of the opening by bending the card slightly. See the interference fringes? Try it with two closely spaced slits.

THINK AND EXPLAIN

1. A rainbow viewed from an airplane may form a complete circle. Where will the shadow of the airplane appear? Explain.
2. When you stand with your back to the Sun, you see a rainbow as a circular arc. Could you move off to one side and see it as a segment of an ellipse, as Figure 14.4 suggests? Defend your answer.
3. Why is a secondary rainbow dimmer than a primary bow?
4. In taking a photograph, what would happen to the image if you cover up the bottom half of the lens?
5. Why do radio waves diffract around buildings but light waves do not?
6. A pattern of fringes is produced when mono- chromatic light passes through a pair of thin slits.

Is such a pattern produced by three parallel thin slits? By thousands of such slits? Give an example to support your answer.

7. The colors of peacocks and hummingbirds are the result not of pigments but of ridges in the surface layers of their feathers. By what physical principle do these ridges produce colors?
8. The colored wings of many butterflies are due to pigmentation, but in some species, such as the Morpho butterfly, the colors do not result from any pigmentation. When the wing is viewed from different angles, the colors change. How are these colors produced?
9. Why do the iridescent colors seen in some seashells, especially abalone shells, change as the shells are viewed from different positions?
10. When dishes are not properly rinsed after washing, different colors are reflected from their surfaces. Explain.
11. If you notice the interference patterns of a thin film of oil or gasoline on water, you'll note that the colors form complete rings. How are these rings similar to the lines of equal elevation on a contour map?
12. Why aren't interference colors seen on films of gasoline on a dry street?
13. Because of wave interference, a film of oil on water is seen to be yellow to observers directly above in an airplane. What color does it appear to a scuba diver directly below?
14. Polarized light is a part of nature, but polarized sound is not. Why?
15. Why do Polaroid sunglasses reduce glare, whereas unpolarized sunglasses simply reduce the total amount of light reaching your eyes?
16. The digital displays of watches and other devices are normally polarized. What problem occurs when you look at them while wearing polarized sunglasses?
17. What percentage of light is transmitted by two ideal Polaroid filters atop each other with their polarization axes aligned? With their axes at right angles to each other?
18. A beam of red light and a beam of blue light have exactly the same energy. Which beam contains the greater number of photons?

19. Suntanning produces cell damage in the skin. Why is ultraviolet radiation capable of producing this damage, whereas visible radiation is not?

20. Does the photoelectric effect *prove* that light is made of particles? Do interference experiments *prove* that light is composed of waves? (Is there a distinction between what something *is* and how it *behaves*?)

READINESS ASSURANCE TEST (RAT)

If you have a good handle on this chapter—if you really do—then you should be able to score 7 out of 10 on this RAT. Check your answers with your teacher. If you score less than 7, study further before moving on.

Choose the best answer to each of the following.

1. When light incident on a prism separates into a spectrum, we call the process
   (a) polarization.
   (b) interference.
   (c) dispersion.
   (d) None of these.

2. A rainbow illustrates the properties of
   (a) reflection.
   (b) refraction.
   (c) dispersion.
   (d) All of these.

3. The properties of lenses depends mainly on
   (a) reflection.
   (b) refraction.
   (c) dispersion.
   (d) All of these.

4. A real image can be cast on a screen by a
   (a) converging lens.
   (b) diverging lens.
   (c) Either of these.
   (d) Neither of these.

5. The glasses of a nearsighted person are usually thicker at the
   (a) middle.
   (b) edges.
   (c) Either of these.
   (d) Neither of these.

6. When light undergoes interference, it can sometimes
   (a) build up to more than the sum of amplitudes.
   (b) cancel completely.
   (c) Both of these.
   (d) Neither of these.

7. When a beam of light reflects from a pair of closely spaced surfaces, color is produced because some of the reflected light is
   (a) converted to a different frequency.
   (b) deflected.
   (c) subtracted from the beam.
   (d) amplified.

8. Polarization occurs for waves that are
   (a) transverse.
   (b) longitudinal.
   (c) Both of these.
   (d) Neither of these.

9. Which of these photons is more likely to initiate the photoelectric effect?
   (a) Red
   (b) Green
   (c) Blue
   (d) Violet

10. In the equation $E = hf$, f stands for the
    (a) frequency of a photon with energy E.
    (b) wavelength of a photon with energy E.
    (c) Both of these.
    (d) Neither of these.

15

THE ATOM

THE MAIN IDEA

Atoms are the building blocks of most matter.

Imagine you are falling off your chair in slow motion. While falling, imagine you are also shrinking in size. By the time you reach the floor, you have shrunk to just one-billionth of a meter—you are small enough to have fallen into a single atom. What do you see from inside this atom? Occasionally, you may encounter tiny bits of matter, such as fast-flying electrons or the central atomic nucleus. But these particles are very small compared to the entire volume of the atom, which you come to realize is mostly empty space! So if the floor we stand on is made of atoms and atoms are mostly empty space, how can we say the floor is solid? How do we know atoms exist? What evidence do we have for saying that atoms are the fundamental building blocks of everything around us, and how many different kinds of atoms are there? In this chapter, we explore the bizarre world of atoms and the amazing chart that tells their story—the periodic table.

Explore!

How Fast Do Atoms Migrate?

1. Put a few drops of food coloring into a small container of cool or room temperature water.
2. Note the approximate time for the colored drops to spread to all parts of the water.
3. Repeat using warmer water.

Analyze and Conclude

1. **Observing** What difference in the spreading in cool and warm water did you see?

2. **Predicting** How would the time for spreading compare if you used hot water instead of warm water?

3. **Making Generalizations** What can you say about the motion of molecules that make up the food coloring?

15.1 Discovering the Invisible Atom

How small are atoms? Atoms are so small that the number of them in a baseball is roughly equal to the number of Ping-Pong balls that could fit inside a hollow sphere as big as the Earth. This is illustrated in Figure 15.1.

Atoms in a baseball

Ping-Pong balls in the Earth

FIGURE 15.1 ▲
If the Earth were filled with nothing but Ping-Pong balls, the number of balls would be huge and roughly equal to the number of atoms in a baseball. Said differently, if a baseball were the size of the Earth, one of its atoms would be the size of a Ping-Pong ball.

Atoms are so small, we could stack microscope on top of microscope and never "see" an atom. This is because atoms are smaller than the wavelengths of visible light. As illustrated in Figure 15.2, the diameter of an object visible under the highest magnification must be larger than the wavelengths of visible light.

Where do atoms come from? The origin of most atoms goes back to the birth of the universe. Hydrogen, H, the lightest atom, was probably

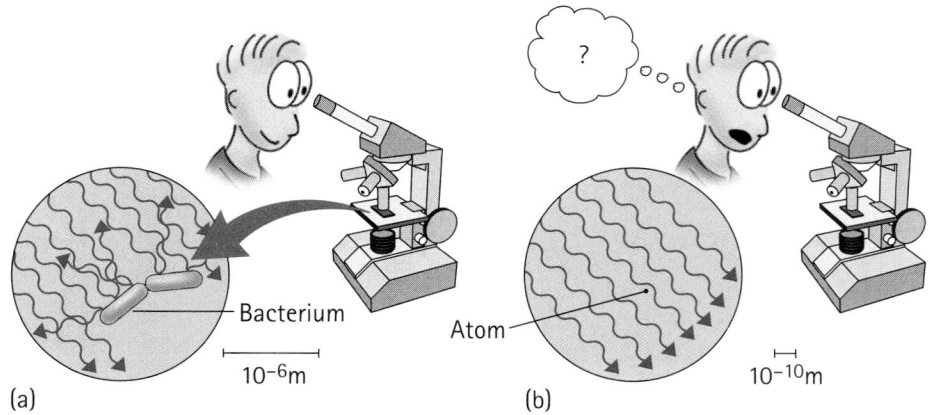

Bacterium

10^{-6}m

(a)

Atom

10^{-10}m

(b)

FIGURE 15.2 ▲
(a) A bacterium is visible because it is much larger than the wavelengths of visible light. We can see the bacterium through the microscope because the bacterium reflects visible light back towards the eye. (b) An atom is invisible because it is smaller than the wavelengths of visible light, which passes by the atom with no reflection.

the original atom, and hydrogen atoms make up more than 90% of the atoms in the known universe. Heavier atoms are produced in stars. There, enormous temperatures and pressures force light atoms to join, which makes heavier atoms. Anything you see on Earth today is made of atoms manufactured in stars that exploded billions of years before our solar system came into being. You are made of stardust, as is everything that surrounds you. We'll be exploring this idea further in Chapter 34.

So, atoms in your body have literally existed long before Earth came to be. They have been recycling throughout the universe for eons. In this way, you don't "own" the atoms that make up your body—you're simply the present caretaker. There will be many caretakers to follow.

There are about as many atoms of air in your lungs at any moment as there are breaths of air in the Earth's atmosphere. That's because about 10^{22} atoms make up a liter of air and there are about 10^{22} liters of air in the atmosphere. Here's what that means. Exhale a deep breath; the number of atoms exhaled approximately equals the number of breathfuls of air in the Earth's atmosphere. It will take several years for your breath to become uniformly mixed in the atmosphere. Then anyone, anywhere on Earth, who inhales a breath of air takes in, on average, one of the atoms you exhaled! Furthermore, you exhale many, many breaths, and so other people breathe in many, many atoms that were once in your lungs—that were once a part of you. And of course, vice versa: With each breath you take in, you recycle atoms that were once a part of everyone who ever lived. Some of the atoms breathed became part of our bodies, so it can be truly said that we are literally breathing one another. Science tells us, we're all one family.

Life is not measured by the number of breaths we take, but by the moments that take our breath away.— *George Carlin*

Evidence for Atoms

Because atoms are so tiny, scientists didn't hypothesize their existence until the early 1800s. They realized that if atoms are real, then it would make sense that some materials transform into other materials. Iron, for example, transforms into rust when iron *atoms* combine with oxygen *atoms*. Thinking in terms of atoms, scientists were soon able to create new materials, such as dyes and medicines, not found in nature. These successes led early scientists to accept atoms as real even though these tiny bits of matter could not be directly seen. As we explore in later chapters, modern chemistry is founded upon the idea that atoms exist.

☑ The first direct evidence for atoms was discovered in 1827 by a Scottish botanist, Robert Brown. He was studying grains of pollen in a drop of water under a microscope. He noticed that the grains were continually moving about. At first he thought the grains were moving life forms. But later he found that dust particles and grains of soot moved the same way. This perpetual jiggling of particles is called *Brownian motion* and it wasn't until the early 1900s that Einstein was able to provide the explanation. Einstein correctly noted that Brownian motion results from collisions between invisible atoms and visible particles. Brown's pollen grains were moving because they were constantly being jostled by groups of atoms that make up the water.

Are atoms big or small? The answer begins with an *s*.

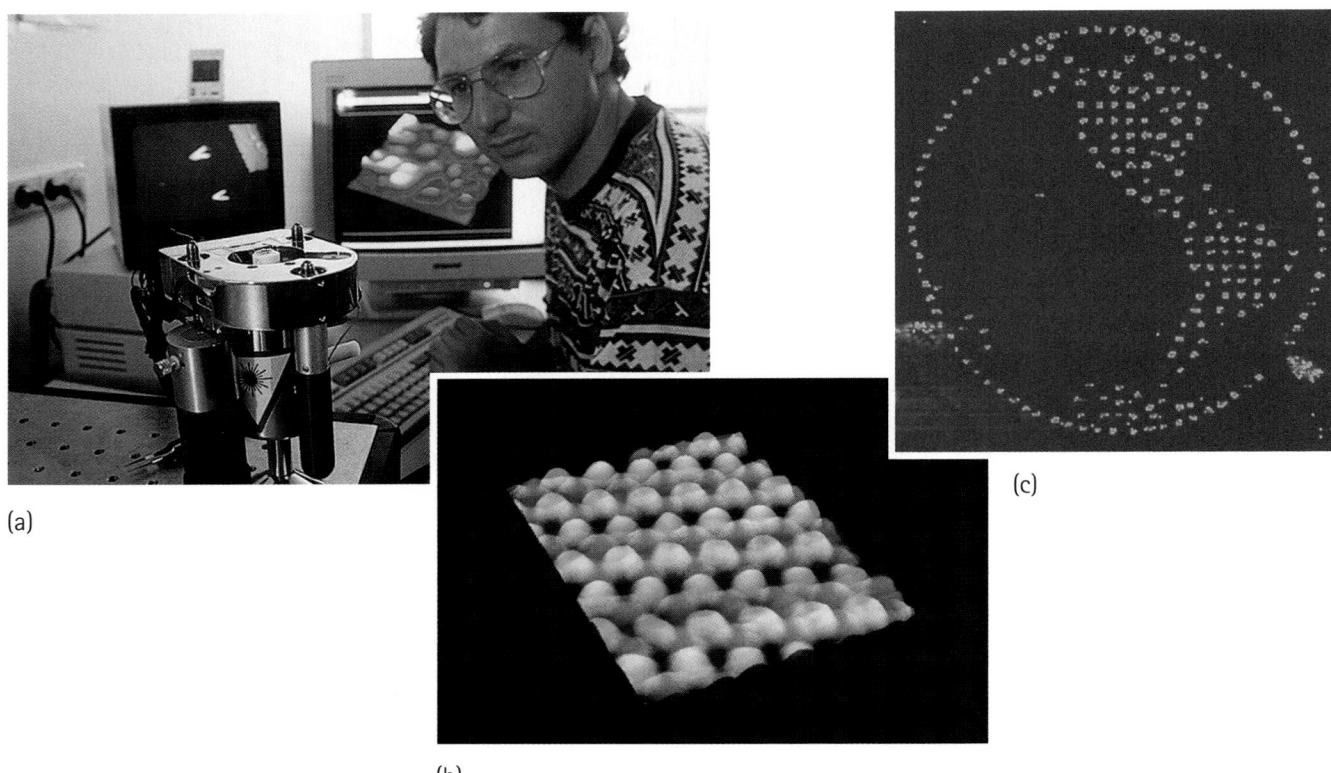

(a)

(b)

(c)

FIGURE 15.3 ▲
(a) Scanning probe microscopes are devices used to create submicroscopic imagery. (b) An image of gallium and arsenic atoms. (c) Each dot in the world's tiniest map consists of a few thousand gold atoms. Each dot was moved into its proper place by a scanning probe microscope.

Most of an atom is nothing at all.

Today, we still cannot see atoms directly. However, we can now generate images of them *indirectly*. In the mid-1980s, researchers developed the *scanning probe microscope*, which produces images by dragging an ultrathin needle back and forth over the surface of a sample, as illustrated in this chapter's opening photograph. Interactions with the surface atoms are detected and translated by a computer into a three-dimensional image that corresponds to the positions of atoms on the surface, as shown in Figure 15.3. A scanning probe microscope can also move individual atoms into desired positions. This ability ushered in the field of *nanotechnology,* which we discuss on page 324.

CONCEPT CHECK
Why are atoms invisible?

Check Your Answer
Atoms are invisible because visible light passes right by them. An individual atom is smaller than the wavelengths of visible light and so is unable to reflect that light. The atomic images generated by scanning probe microscopes are not photographs taken by a camera. Rather, they are computer-generated images produced by the movements of an ultrathin needle.

FIGURE 15.4 ▶

Electrons whiz around the atomic nucleus, forming what can be best described as a cloud. The cloud is most dense where the electrons tend to spend most of their time. Electrons, however, are invisible to us. Hence, such a cloud can only be imagined. Furthermore, if this illustration were drawn to scale, the atomic nucleus would be too small to be seen. Realistically, atoms are not well suited to graphical illustrations.

Electron cloud

Nucleus

During the early 1900s, scientists came to realize through their experiments that atoms are not the smallest particles of matter. Instead, atoms themselves are made of even smaller particles, which we call *subatomic particles.* Three examples include *protons, neutrons,* and *electrons.* Protons and neutrons are bound together at the atom's center to form the **atomic nucleus.** The nucleus is a relatively heavy core that makes up most of an atom's mass. Surrounding the nucleus are the tiny **electrons,** as indicated in Figure 15.4.

All atoms and all things made of atoms, including ourselves, are mostly empty space. How can this be? The answer has to do with the fact that the bulk of an atom's mass is concentrated within its nucleus. Surrounding that nucleus is empty space through which electrons are buzzing around. The size of an atom, therefore, is determined by how far away electrons move around the nucleus. Because electrons are even smaller than the nucleus, and because they are widely separated from each other (as well as from the nucleus), atoms are indeed mostly empty space—just as our solar system is mostly empty space.

So if atoms are mostly empty space, why don't they simply pass through one another? They don't because electrons repel the electrons of neighboring atoms. Therefore, two atoms can get only so close to each other before they start repelling. This explains why gravity doesn't pull you through the floor as you stand. While the force of gravity pulls you down, the electric force of repulsion between the atoms of the floor and your feet pushes you up. As you stand on the floor, these two forces are balanced and you find yourself neither falling nor rising—there's more to standing than most people realize!

When the atoms of your hand push against the atoms of a wall, repulsions between electrons in your hand and electrons in the wall prevent your hand from passing through the wall. You sense this repulsion as a pressure that pushes back. Also, our sense of touch comes from these electrical repulsions. Interestingly, when you touch someone, your atoms and those of the other person do not meet. Instead, atoms from the two of you get close enough so that you sense an electrical repulsion. There is still a tiny, though imperceptible, gap between the two of you (Figure 15.5).

If a typical atom were expanded to a diameter of 3 km, about as big as a medium-sized airport, the nucleus would be about the size of a basketball. Atoms are mostly empty space.

FIGURE 15.5 ▲

As close as Tracy and Ian are in this photograph, none of their atoms meet. The closeness between us is in our hearts.

✔ **READING CHECK**

What scientist discovered the first direct evidence for the existence of atoms?

15.2 Elements and the Periodic Table

You know that atoms make up the matter around you, from stars to steel to chocolate ice cream. Given all these different types of material, you might think that there must be many different kinds of atoms. Not true. The number of different kinds of atoms is surprisingly small. The great variety of substances results from the many ways a few kinds of atoms can be combined. Just as the three colors red, green, and blue can be combined to form any color on a television screen, or the 26 letters of the alphabet make up all the words in a dictionary, only a few kinds of atoms combine in different ways to produce all substances. To date, we know of slightly more than 100 distinct atoms. Of these, about 90 are found in nature. The remaining atoms have been created in the laboratory.

✓ Any material made of only one type of atom is classified as an **element.** A few examples are shown in Figure 15.6. Pure gold, for example, is an element because it contains only gold atoms. Nitrogen gas is also an element because it contains only nitrogen atoms. Likewise, the graphite in your pencil is an element—solely made up of carbon atoms. All of the elements are listed in a chart called the **periodic table,** shown in Figure 15.7.

As you can see from the periodic table, each element is designated by its **atomic symbol,** which comes from the letters of the element's name. For example, the atomic symbol for carbon is C, and that for chlorine is Cl. In many cases, the atomic symbol is derived from the element's Latin name. Gold has the atomic symbol Au after its Latin name, *aurum.* Lead has the atomic symbol Pb after its Latin name, *plumbum* (Figure 15.8). Elements having symbols derived from Latin names are usually those discovered earliest.

Note that only the first letter of an atomic symbol is capitalized. The symbol for the element cobalt, for instance, is Co, while CO is a combination of two elements: carbon, C, and oxygen, O.

The elements are organized within the periodic table in a remarkable fashion. We will be discussing this organization in much more detail in Chapter 16, which is the first chemistry chapter. For now, you should know that each vertical column of the periodic table is called a **group** of elements (or sometimes a *family* of elements), as shown in Figure 15.9. Elements within the same group (vertical column) have similar

Most materials are made from more than one kind of atom. Water, H_2O, for example, is made from the combination of hydrogen and oxygen atoms. These materials are called *compounds,* which we discuss further in Chapter 16.

FIGURE 15.6 ▶
Any element consists of only one kind of atom. Gold consists of only gold atoms, a flask of gaseous nitrogen consists of only nitrogen atoms, and the carbon of a graphite pencil consists of only carbon atoms.

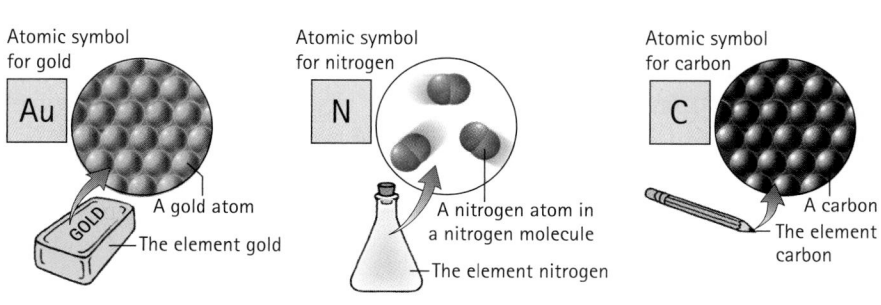

✔ The proton is heavy—nearly 2000 times as massive as the electron. The proton and electron have the same quantity of charge, but the charges are opposite. The number of protons in the nucleus of any atom is equal to the number of electrons whirling about the nucleus. So the proton's positive charge and the electron's negative charge cancel each other. Can you see that this guarantees that the atom has an overall net electric charge of zero? It is electrically neutral. For example, an electrically balanced oxygen atom has eight electrons and eight protons.

How do atoms of one element differ from another element? Each element is identified by its **atomic number.** This is the number of protons contained in each atomic nucleus. Hydrogen, with one proton per atom, has atomic number 1. Helium, with two protons in its nucleus, has atomic number 2; and so on. Look at the periodic table in Figure 15.7 on page 321. You will see that it lists the elements in order of increasing atomic number.

CONCEPT CHECK
How many protons are there in an iron atom? (Iron has the chemical symbol Fe and the atomic number 26.)

Check Your Answer
The atomic number of an atom and its number of protons are the same. Thus, there are 26 protons in an iron atom. Another way to say this is that all atoms that contain 26 protons are, by definition, iron atoms.

A neutron goes into a restaurant and asks the waiter, "How much for a drink?" The waiter replies, "For you, no charge."

In addition to the proton, the atomic nucleus also includes another particle—the *neutron*. Helium, for example, has twice the electric charge of hydrogen but four times the mass. The added mass is due to the neutrons in helium's nucleus. ✔ The **neutron** is a nuclear particle with about the same mass as the proton, but with no electric charge. Any object with no net electric charge is said to be electrically neutral (which is why the neutron got its name). We'll have much more to say about the neutron in the following chapter. Table 15.1 summarizes the basic facts about electrons, protons, and neutrons. Note from this

TABLE 15.1 Subatomic Particles

	Particle	Charge	Mass Compared to Electron	Actual Mass* (kg)
	Electron	−1	1	$9.11 \times 10^{-31\dagger}$
Nucleons	Proton	+1	1836	1.673×10^{-27}
	Neutron	0	1841	1.675×10^{-27}

* Not measured directly but calculated from experimental data.
† 9.11×10^{-31} kg = 0.00000000000000000000000000000911 kg.

Nanotechnology is the technology of tiny things. It began with microtechnology, which was ushered in some 60 years ago. Engineers learned to build electronic circuits within the size of a micron (10^{-6} meters), thus the term *micro*technology. Such circuits have had a major impact on society—from personal computers, to cell phones, to the Internet.

Today we are past the realm of microns to the realm of the nanometer (10^{-9} meters), which is the realm of individual atoms. Technology at this scale is called *nano*technology. Nanotechnology deals with incredibly tiny objects from 1 to 100 nanometers (nm). For perspective, a water molecule is less than 1 nm across. Nanotechnology is interdisciplinary, which means it requires the cooperative efforts of chemists, engineers, physicists, molecular biologists and many others. Interestingly, there are already many products on the market that contain components developed through nanotechnology. These include sunscreens, dental bonding agents, stain-free clothing, the heads to computer hard drives, and much more. Nanotechnology is still young and it may be decades before its potential is fully realized. Consider, for example, that personal computers didn't become widely available until the 1990s, some 40 years after the start of microtechnology.

Most experts agree, however, that the first big benefits will arise in computer science and medicine.

The ultimate expert on nanotechnology is nature. Living organisms, for example, are complex systems of interacting biomolecules all functioning within the realm of nanometers. In this sense, the living organism is nature's nanomachine. We need look no further than our own bodies to find evidence of the power of nanotechnology. With nature as our teacher, we have much to learn. In medicine, for example, nanotechnology can help us discover the exact causes of nearly any disease or disorder (aging included) and their cures.

What are the limits of nanotechnology? As a society, how will we deal with the impending changes nanotechnology may bring? Consider the possibilities. Wall paint that can change color or be used to display video. Solar cells that capture sunlight so efficiently that they render fossil fuels obsolete. Robots with so much processing power that we begin to wonder whether they experience consciousness. Nanomachines that can "photocopy" three-dimensional objects, including living organisms. Medicines that more than double the average human life span. Stay tuned for an exciting new revolution in human capabilities.

 **READING CHECK**

How do protons and neutrons compare in terms of mass and electric charge?

table that protons and neutrons are both examples of **nucleons,** which is the general term given to a subatomic particle found within the nucleus.

15.4 Isotopes and Atomic Mass

An element has a definite number of protons, but its number of neutrons may vary. For example, most hydrogen atoms (atomic number 1) have no neutrons. A small percentage of hydrogen atoms, however, have one neutron, and a smaller percentage have two neutrons. Similarly, most iron atoms (atomic number 26) have 30 neutrons, but a small percentage have 29 neutrons. Atoms of the same element that contain different numbers of neutrons are **isotopes** of one another.

How can we tell isotopes apart? ✓ We identify isotopes by their **mass number,** which is the total number of protons and neutrons they contain. In other words, mass number is the number of nucleons. As Figure 15.11 shows, a hydrogen isotope with only one proton is called

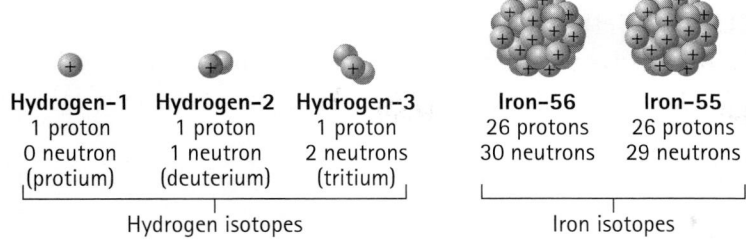

Isotopes of an element have the same number of protons but different numbers of neutrons. That means they have different mass numbers. The three hydrogen isotopes have special names: *protium* for hydrogen-1, *deuterium* for hydrogen-2, and *tritium* for hydrogen-3. Hydrogen-1 is most common. For most elements, such as iron, the isotopes have no special names and are indicated merely by mass number.

hydrogen-1, where 1 is the *mass number*. A hydrogen isotope with one proton and one neutron is therefore hydrogen-2, and a hydrogen isotope with one proton and two neutrons is hydrogen-3. Similarly, an iron isotope with 26 protons and 30 neutrons is called iron-56, and one with only 29 neutrons is iron-55.

CONCEPT CHECK
What is the mass number of iron-56? How about uranium-238?

Check Your Answers
The mass number of any element is the total number of protons and neutrons, which is shown as a whole number after the name of the element. For iron-56, the mass number is 56. For uranium-238, the mass number is 238.

There is another way to express isotopes. Write the mass number as a superscript and the atomic number as a subscript to the left of the atomic symbol. For example, an iron isotope with a mass number of 56 and atomic number of 26 is written

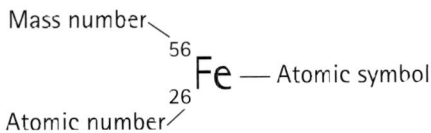

Mass number $\nearrow$
$^{56}_{26}\text{Fe}$ — Atomic symbol
Atomic number $\nearrow$

The total number of neutrons in an isotope can be calculated by subtracting its atomic number from its mass number:

> **mass number**
> **− atomic number**
> **number of neutrons**

For example, uranium-238 has 238 nucleons. The atomic number of uranium is 92, which tells us that 92 of these 238 nucleons are protons. The remaining 146 nucleons must be neutrons:

> **238 protons and neutrons**
> **− 92 protons**
> **146 neutrons**

We can't "see" an atom because they're too small. We can't see the farthest star either. There's much that we can't see. But that doesn't prevent us from thinking about such things or even collecting indirect evidence.

fyi

- Don't confuse atomic mass with mass number. *Mass number* is an integer that specifies an isotope and has no units—it's simply equal to the number of nucleons in a nucleus. The element with mass number 12, for example, is carbon. *Atomic mass* is an average of the isotopes of a given element, with units called *amu's* (atomic mass units). The average atomic mass of carbon is 12.011 amu.

CONCEPT CHECK

How many neutrons are in the lithium-7 isotope (Li, atomic number 3)?

Check Your Answer

Subtract the total number of protons from the total number of nucleons to obtain the total number of neutrons. For lithium-7, there are $7 - 3 = 4$ neutrons.

Isotopes of an element differ only by mass, not by electric charge. That means they cannot easily be distinguished from one another. For example, about 1% of the carbon we eat is the carbon-13 isotope with seven neutrons per nucleus. The remaining 99% of the carbon in our diet is the common carbon-12 isotope, with six neutrons per nucleus. Our bodies can't tell the difference.

The total mass of an atom is called its **atomic mass.** This is the sum of the masses of all the atom's electrons, protons, and neutrons. Because the mass of electrons is so little compared with the mass of protons and neutrons, electrons contribute practically nothing to atomic mass. A small portion of the periodic table in Figure 15.12 indicates some atomic masses of elements. The larger periodic table on the inside back cover of this textbook shows a full list of atomic masses.

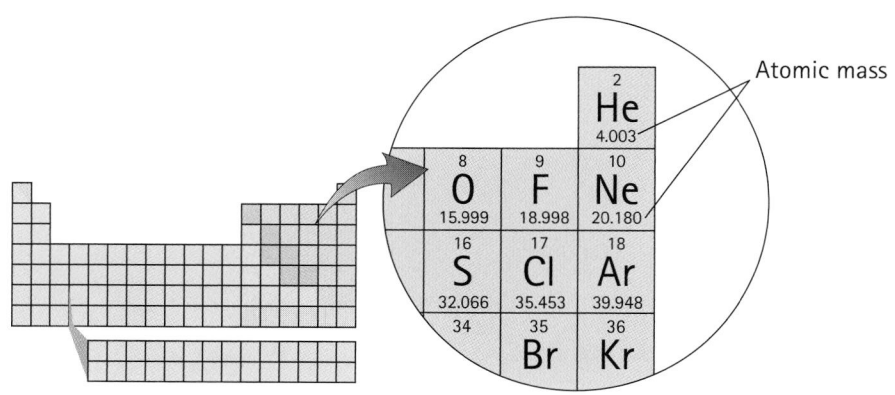

FIGURE 15.12 ▶
Helium, He, has an atomic mass of 4.003; and neon, Ne, has an atomic mass of 20.180.

The atomic mass of an element is defined as the *average* atomic mass of its various isotopes. This is why the atomic masses shown in the periodic table are not whole numbers. To understand why, consider the following analogy. Imagine 100 students take a test worth 100 points. If all students scored 80 points, then the average score would be 80%. Right? If instead one of the students scored a perfect 100 points, what would happen to the class average? The answer is that the class average would rise a little bit because of that one higher-scoring student. Similarly, we find that most all carbon atoms are the carbon-12 isotope. One out of every 100 carbon atoms, however, is the heavier carbon-13 isotope. This small amount of carbon-13 raises the *average* mass of carbon from exactly 12.0000 to the slightly higher value of 12.011, which is what you see listed in the periodic table.

Most water molecules, H_2O, consist of hydrogen atoms with no neutrons. The few that do, however, are heavier and because of this difference they can be isolated. Such water is appropriately called "heavy water."

CONCEPT CHECK
Distinguish between mass number and atomic mass.

Check Your Answer
Both terms include the word *mass* and so are easily confused. Focus your attention on the second word of each term, however, and you'll get it right every time. Mass number is a count of the *number* of nucleons in an isotope. An atom's mass number requires no units because it is simply a count. Atomic mass is a measure of the total *mass* of an atom, which is given in atomic mass units.

How does the mass number of an atom compare with the number of nucleons it contains?

Nucleon, neutron, atomic mass, mass number, atomic number, and more! Whew! Why all these new terms? Remember, we're describing an invisible world of atoms. Atoms are not part of our common language. To describe this tiny and very real world of the atom means that we expand our vocabulary. The same thing is true when you're introduced to the world of a new Internet game—you need new words to describe the new objects and the new experiences. The more you learn, the more words you need to describe what you learn!

Sometimes atomic mass is expressed as a tiny fraction of a kilogram.

15.5 Electron Shells—Regions About the Nucleus Where Electrons Are Located

The identity of a song is determined by the arrangement of its musical notes. In a similar way, the identity of an atom is determined by how its electrons are arranged. Within the atom, electrons behave as though they are arranged in **shells.** There are at least seven shells, and each shell can hold only a limited number of electrons. In Figure 15.13, the innermost shell can hold 2 electrons, the second and third shells, 8 electrons each; the fourth and fifth shells, 18 electrons each; and the sixth and seventh shells, 32 electrons each.

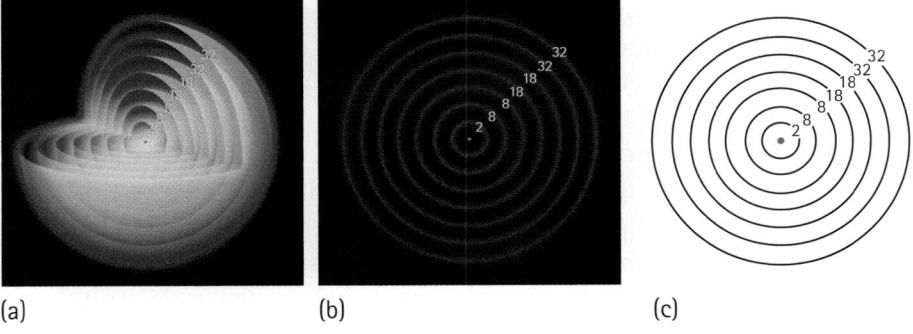

(a)   (b)   (c)

Sometimes a model is useful even when it's incorrect. Scotsman James Watt constructed a workable steam engine in the 18th century based on a model of heat that turned out to be quite incorrect.

FIGURE 15.13 ▲
(a) A cutaway view of the seven shells, indicating the number of electrons each shell can hold. (b) A two-dimensional, cross-sectional view of the shells. (c) A simple cross-sectional view resembling Bohr's planetary model of the atom.

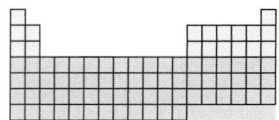

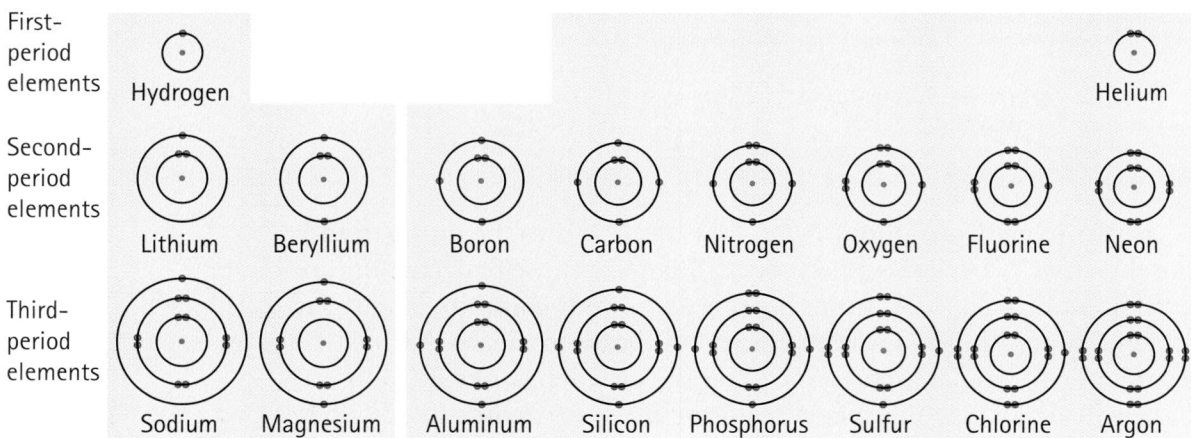

First-period elements — Hydrogen, Helium

Second-period elements — Lithium, Beryllium, Boron, Carbon, Nitrogen, Oxygen, Fluorine, Neon

Third-period elements — Sodium, Magnesium, Aluminum, Silicon, Phosphorus, Sulfur, Chlorine, Argon

FIGURE 15.14 ▲

The first three periods of the periodic table. Elements in the same period have electrons in the same shells. Elements in the same period differ from one another by the number of electrons in the outermost shell.

▲ Two-time Nobel laureate Linus Pauling (1901–1994) was an early proponent of teaching beginning chemistry students a shell model of the atom from which the organization of the periodic table could be described.

These seven shells tell us why there are seven periods in the periodic table. Also, the number of elements in each period is equal to the shell's capacity for electrons. The first shell, for example, can hold only two electrons. That's why we find only two elements, hydrogen and helium, in the first period (Figure 15.14). The second and third shells each can hold eight electrons, and so eight elements are found in both the second and third periods.

The electrons of the outermost shell in any atom are the ones directly exposed to their external environment. ✓ These important outermost electrons are called **valence electrons.** They are the first to interact with other atoms, and the ones that participate in chemical bonding. (We'll return to chemical bonding in Chapter 17.) The term *valence* comes from the Latin *valentia,* "strength," and it refers to the "combining power" of an atom. Atoms combine to form *molecules,* which are tightly held groups of atoms, by way of valence electrons.

Look carefully at Figure 15.14. Can you see that the valence electrons of atoms above and below one another (within the same group) are similarly organized? For example, atoms of the first group, which include hydrogen, lithium, and sodium, each have a single valence electron. The atoms of the second group, including beryllium and magnesium, each have two valence electrons. Similarly, atoms of the last group, including helium, neon, and argon, each have their outermost shells filled to capacity with valence electrons—two for helium, and eight for both neon and

argon. So we see that the valence electrons of atoms in the same group of the periodic table are similarly organized. This explains why elements directly above or below one another in the periodic table have similar properties, as was first stated on page 322.

as was first stated on page 322.

✓ READING CHECK

What are valence electrons and why are they important?

CONCEPT CHECK
Do atoms really look like the shells of Figure 15.13?

Check Your Answer
No! Remember, atoms are smaller than the wavelengths of visible light, and so they are not well suited to graphical illustrations. The shells of Figure 15.13 are not actual images of atoms. Instead, these are just pictures that help us to understand the behavior of atoms.

Electron shells are not what atoms look like, but they help us understand and predict the properties and behaviors of atoms. In Chapter 17, we will be using a simplified version of these shells, known as *electron-dot structures,* to show how atoms join together to form molecules. For the next chapter, however, we will be exploring in greater detail the nature of the atomic nucleus, which is a potential source of enormous amounts of energy.

15 CHAPTER REVIEW

KEY TERMS

Atomic mass The mass of an element's atoms listed in the periodic table as an average value based on the relative abundance of the element's isotopes.

Atomic nucleus The dense, positively charged center of every atom.

Atomic number A count of the number of protons in the atomic nucleus.

Atomic symbol An abbreviation for an element or atom.

Electron An extremely small, negatively charged subatomic particle found outside the atomic nucleus.

Element Any material that is made up of only one type of atom.

Group A vertical column in the periodic table, also known as a family of elements.

Isotopes Any member of a set of atoms of the same element whose nuclei contain the same number of protons but different numbers of neutrons.

Mass number The total number of protons and neutrons within an isotope.

Nucleon Any subatomic particle found in the atomic nucleus; another name for either proton or neutron.

Neutron An electrically neutral subatomic particle of the atomic nucleus.

Period A horizontal row in the periodic table.

Periodic table A chart in which all known elements are listed in order of atomic number.

Proton A positively charged subatomic particle of the atomic nucleus.

Shell A region of space around the atomic nucleus in which electrons may reside.

Valence electron An electron located in the outermost occupied shell in an atom that can participate in chemical bonding.

REVIEW QUESTIONS

Discovering the Invisible Atom

1. Which atom is most plentiful in the universe?
2. What was the source of motion in the particles that Brown investigated?
3. Is it possible to see an atom using visible light? Why or why not?
4. What type of microscope allows us to visualize the positions of individual atoms?
5. What is at the center of every atom?

Elements and the Periodic Table

6. Why is pure gold classified as an element?
7. What is the atomic symbol for the element cobalt?
8. How many periods are in the periodic table?
9. How many groups are in the periodic table?
10. What happens to the properties of elements across any period of the periodic table?

The Atomic Nucleus Consists of Protons and Neutrons

11. What role does atomic number play in the periodic table?
12. What is a nucleon?
13. Distinguish between a proton and a neutron in terms of electric charge.

Isotopes and Atomic Mass

14. What occurs for two atoms to be isotopes of each other?

15. Distinguish between atomic number and mass number.

16. Distinguish between mass number and atomic mass.

17. How many nucleons are in hydrogen-3?

Electron Shells—Regions About the Nucleus Where Electrons Are Located

18. How many shells are needed to explain the periodic table?

19. Which electrons are most responsible for the properties of an atom?

20. What is the relationship between the maximum number of electrons each shell can hold and the number of elements in each period of the periodic table?

THINK AND DO

About how many Ping-Pong balls can fit within an average-size room? Assume that a cubic meter can hold about 20,000 Ping-Pong balls. Use this estimate to figure the number of Ping-Pong balls that could fit within an average-size house containing 10 average-size rooms. How does this number compare to the number of atoms in a baseball? Consider this next time you're outside with a clear view of the Earth around you.

THINK AND COMPARE

1. Consider three 1-gram samples of matter: A, carbon-12; B, carbon-13; C, uranium-238. Rank them in terms of having the greatest number of atoms, from most to least.

2. Consider these atoms: potassium K, sodium Na, and lithium Li. Rank them in order of atomic size from smallest to largest.

3. Consider these atoms: hydrogen H, aluminum Al, argon, Ar. Rank them, from smallest to largest, in order of (a) size; (b) number of protons in the nucleus; (c) number of electrons.

4. Consider these atoms: helium He, chlorine Cl, and argon Ar. Rank them in terms of their atomic number, from smallest to largest.

THINK AND EXPLAIN

1. A cat strolls across your backyard. An hour later, a dog with its nose to the ground follows the trail of the cat. Explain what is occurring from a molecular point of view.

2. Which atoms are older, those that make up the body of an old woman, or those that make up the body of a young girl?

3. Where did the atoms that make up a newborn baby originate?

4. Where did the carbon atoms in Leslie's hair originate? (This is co-author Leslie at age 16.)

5. Does it make sense to say that this textbook is more than 99.9% empty space? Defend your answer.

6. The atoms that compose your body are mostly empty space, and structures such as the chair you're sitting on are composed of atoms that are also mostly empty space. So what prevents you from falling through the chair?

7. If two protons and two neutrons are removed from the nucleus of an oxygen-16 atom, a nucleus of which element remains?

8. If an atom has 43 electrons, 56 neutrons, and 43 protons, what is its approximate atomic mass? What is the name of this element?

9. The nucleus of an electrically neutral gold atom contains 79 protons. How many electrons does this atom normally have?

10. Evidence for the existence of neutrons was not discovered until many years after the discoveries of the electron and the proton. Give a possible explanation.

11. Why are the atomic masses listed in the periodic table not whole numbers?

12. Which contributes more to an atom's mass: electrons or protons? Which contributes more to an atom's size?

13. Which of the following diagrams best represents the size of the atomic nucleus relative to the size of the atom:

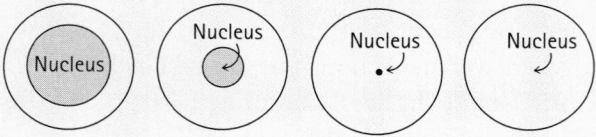

14. Might it be possible to see the individual shells of an atom using the scanning probe microscope? Defend your answer.

15. Must a shell contain electrons in order to exist? Defend your answer.

16. Using the idea of shells, explain why a potassium atom, K, is larger than a sodium atom, Na.

17. Using the idea of shells, explain why a lithium atom, Li, is larger than a beryllium atom, Be.

18. Why don't equal masses of golf balls and Ping-Pong balls contain the same number of balls?

19. Why don't equal masses of carbon atoms and oxygen atoms contain the same number of particles?

20. Which contains more atoms: 1 kg of lead, or 1 kg of aluminum?

READINESS ASSURANCE TEST (RAT)

If you have a good handle on this chapter—if you really do—then you should be able to score 7 out of 10 on this RAT. Check your answers with your teacher. If you score less than 7, study further before moving on.

Choose the best answer to each of the following.

1. Which are older, the atoms in the body of an elderly person or those in the body of a baby?
   (a) A baby because this is surely a trick question.
   (b) An elderly person because they have been on Earth for a longer time.
   (c) They are of the same age, which is appreciably older than the solar system.
   (d) It depends upon their diet.

2. If an atom were the size of a baseball, its nucleus would be about the size of
   (a) a walnut.   (b) a raisin.
   (c) a flea.   (d) an atom.

3. Someone argues that he or she doesn't drink tap water because it contains thousands of impure atoms in each glass. How would you respond in defense of the water's purity, if it indeed does contain thousands of atoms of some impurity per glass?
   (a) Impurities aren't necessarily bad; in fact, they may be good for you.
   (b) The water contains water molecules, each of which is pure.
   (c) There's no defense. If the water contains impurities, it should not be drunk.
   (d) Compared to the billions and billions of water molecules, a thousand atoms of something else is practically nothing.

4. You could swallow a capsule of germanium, Ge (atomic number 32), without significant ill effects. If a proton were added to each germanium nucleus, however, you would not want to swallow the capsule because the germanium would
   (a) become arsenic.
   (b) become radioactive.
   (c) expand and likely lodge in your throat.
   (d) have a change of flavor.

5. If two protons and two neutrons are removed from the nucleus of neon-20, a nucleus of which element remains?
   (a) Magnesium-22
   (b) Magnesium-20
   (c) Oxygen-18
   (d) Oxygen-16

6. The nucleus of an electrically neutral iron atom contains 26 protons. How many electrons does this iron atom have?
   (a) 52
   (b) 26
   (c) 24
   (d) None

7. Why are the atomic masses listed in the periodic table not whole numbers?
   (a) Scientists have yet to make the precise measurements.
   (b) That would be too much of a coincidence.
   (c) The atomic masses are average atomic masses.
   (d) Today's instruments are able to measure the atomic masses to many decimal places.

8. If, hypothetically, 80.0% of the atoms of an element had an atomic mass of 80.00, and the other 20.0% with an atomic mass of 82.00, what is the approximate atomic mass of the element?
   (a) 80.4 amu
   (b) 81.0 amu
   (c) 81.6 amu
   (d) 64.0 amu

9. How many electrons are there in the third shell of sodium, Na (atomic number 11)?
   (a) None
   (b) One
   (c) Two
   (d) Three

10. Strontium, Sr (number 38), is especially dangerous to humans because it tends to accumulate in calcium-rich bone marrow tissues (calcium, Ca, number 20). This fact relates to the organization of the periodic table in that strontium and calcium are both
    (a) metals.
    (b) in the same vertical column.
    (c) made of relatively large atoms.
    (d) isotopes of the same atom.

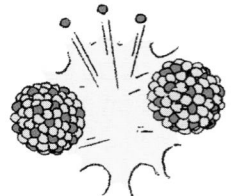

16
NUCLEAR ENERGY

THE MAIN IDEA

The atomic nucleus is the source of a tremendous amount of energy, which poses both risks and benefits.

Uranium-235 nucleus

Neutron

Nuclear energy arises from the atomic nucleus. It comes from a number of different sources. Nuclear energy is in the radioactivity that occurs naturally in the air we breathe, in the rocks around us, and even in the food we eat. Some nuclear energy is human-controlled, an important part of medical diagnostics and radiation therapy. Another source of nuclear energy is the nuclear power plant, which transforms nuclear energy into electricity. Of course, a nuclear bomb is another source of nuclear energy. The most significant source of nuclear energy, however, is the Sun. We'll see in this chapter how this energy comes from thermonuclear fusion, and how it relates to Einstein's famous equation $E = mc^2$. In this nuclear age, it makes good sense to have a basic understanding of nuclear energy.

Explore!

How Can You Model Exponential Growth and Decay?

1. Place a single paperclip on your desktop.
2. To the right of the clip, place two paperclips in a vertical column.
3. Repeat the process of arranging paperclips in columns, doubling the number each time. Continue the process until you have five columns.

Analyze and Conclude

1. **Observing** Look for a mathematical pattern in your columns. For example, how does the number of paperclips in a given column compare to the sum of paperclips in all the preceding columns?

2. **Predicting** Can you predict the total number of columns you could create with 127 paperclips?
3. **Making Generalizations** Starting with your greatest column, remove half until you have one paperclip left. As you read the chapter, relate this to radioactive half-life.

16.1 Radioactivity—The Disintegration of the Atomic Nucleus

As we learned in Chapter 15, atoms are made up of electrons, neutrons, and protons. The neutrons and protons are in the core of the atom—the nucleus. Most atoms in us and in the world around us have stable nuclei. This stability is due to a correct balance of protons and neutrons. Other atoms, however, have nuclei that are unstable because they contain an "off-balance" number of protons and neutrons. For example, they may contain too many neutrons and not enough protons. Atoms with unstable nuclei are said to be *radioactive.* Sooner or later, they break down and eject energetic particles. This process is **radioactivity.** It is often called *radioactive decay.*

A common misconception is that radioactivity is something new in the environment, something caused by humans. Actually, it has been around since the dawn of time and accounts for geothermal energy. ✔ Radioactive decay in Earth's interior heats the water that spurts from a geyser or wells up from a natural hot spring.

As Figure 16.1 shows, most of the radiation we encounter is natural background radiation that originates in Earth and in space. Even the cleanest air we breathe is radioactive as a result of bombardment by cosmic rays. These rays originate in the Sun and other stars. At sea level, the protective blanket of the atmosphere reduces this background radiation. But radiation is more intense at higher altitudes where the air is thinner. In Denver, the "mile-high city," a person receives more than twice as much cosmic radiation as someone at sea level. A couple of round-trip airplane flights between New York and San Francisco exposes us to as much radiation as we receive in a chest X-ray at the doctor's office. Because extended exposure to this level of radiation is dangerous, the flight time of pilots and flight crew is limited.

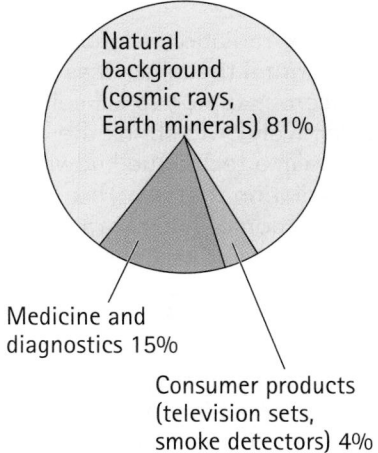

FIGURE 16.1 ▲
Origins of radiation exposure for an average individual in the United States.

✔ **READING CHECK**

What is the source of heat in geysers and hot springs?

SCIENCE AND SOCIETY: NUCLEAR TECHNOLOGY

When household electricity made its way across the country more than a century ago, it represented a new technology with great potential. But it was not without its hazards. While electric grids could provide cities with a quantity and quality of energy previously unheard of, they could also kill people in ways previously unheard of. Many people opposed the adoption of household electricity because of the inherent dangers. But safeguards were engineered, and society determined that the benefits of electricity outweighed the risks. Similar debates continue today; not over electricity, but over nuclear energy and radioactivity.

Without nuclear technology, we would not have medical X-rays, radiation treatments for fighting cancer, smoke detectors, nuclear power as a source of electricity, nor hundreds of other useful applications. Then again, we would not have nuclear bombs. Along with any technology comes responsibility. Our responsibility is to safeguard nuclear material and dispose of it in such a way as to reduce danger to future generations. As a member of society, you will have to make important decisions in these matters. You should do so with an adequate understanding of both the benefits and risks of nuclear technologies.

FIGURE 16.2 ▶
Nuclear radiation is focused on harmful tissue, such as a cancerous tumor, to kill cells selectively or to shrink the tissue in a technique known as radiation therapy. This application of nuclear radiation has saved millions of lives—a clear-cut example of the benefits of nuclear technology. The inset shows the international symbol indicating an area where radioactive material is being handled or produced.

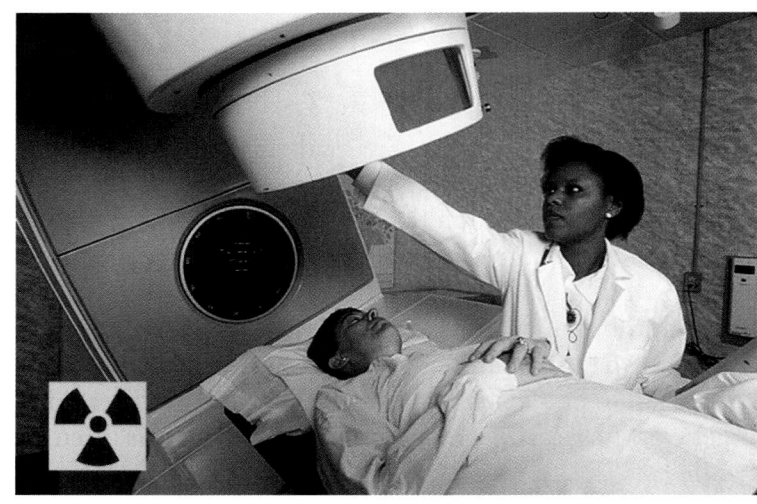

Radioactivity has been around since Earth's beginnings.

16.2 Alpha, Beta, and Gamma Rays

☑ The atoms of radioactive elements emit three distinct types of radiation, named by the first three letters of the Greek alphabet, α, β, γ—alpha, beta, and gamma. Alpha rays carry a positive electrical charge, beta rays carry a negative charge, and gamma rays carry no charge. The three rays can be separated by placing a magnetic field across their paths (Figure 16.3).

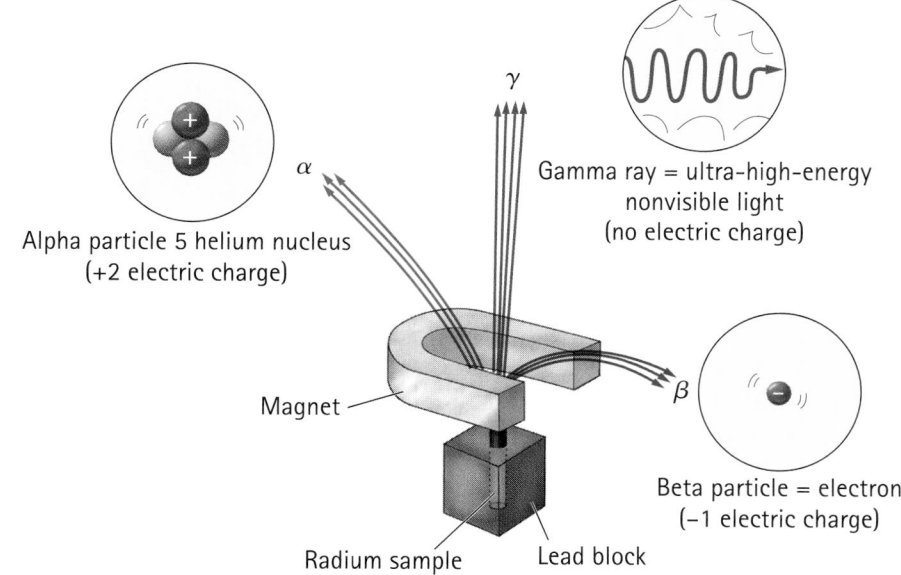

Alpha particle 5 helium nucleus (+2 electric charge)

γ

Gamma ray = ultra-high-energy nonvisible light (no electric charge)

Magnet

β

Radium sample   Lead block

Beta particle = electron (−1 electric charge)

FIGURE 16.3 ▲
In a magnetic field, alpha rays bend one way, beta rays bend the other way, and gamma rays don't bend at all. Note that the alpha rays bend less than beta rays. This occurs because alpha particles have more inertia (mass) than beta particles. The combined beam comes from a radioactive material placed at the bottom of a hole drilled in a block of lead.

Alpha radiation is a stream of alpha particles. An **alpha particle** is the combination of two protons and two neutrons (in other words, it is the nucleus of the helium atom, atomic number 2). Alpha particles are relatively easy to shield against because of their relatively large size and their double positive charge (+2). For example, they do not normally penetrate such lightweight materials as paper or clothing. Because of their great kinetic energies, however, alpha particles can cause significant damage to the surface of a material, especially living tissue. When traveling through only a few centimeters of air, alpha particles pick up electrons and become nothing more than harmless helium. As a matter of fact, that's where the helium in a child's balloon comes from—practically all Earth's helium atoms were at one time energetic alpha particles.

Beta radiation is a stream of beta particles. A **beta particle** is merely an electron ejected from a nucleus. In other words, a beta particle is a fast-flying electron. Interestingly, the neutron becomes a proton once it loses the beta particle, as shown in Figure 16.4.

A beta particle is normally faster than an alpha particle, and it carries only a single negative charge (−1). Beta particles are not as easy to stop as alpha particles are, and they are able to penetrate light materials such as paper or clothing. They can penetrate fairly deeply into skin, where they have the potential for harming or killing living cells. But they are unable to penetrate even thin sheets of denser materials, such as aluminum. Beta particles, once stopped, simply become a part of the material they are in, like any other electron.

Gamma rays are the high-frequency electromagnetic radiation emitted by radioactive nuclei. Like photons of visible light, a gamma ray photon is pure energy. The amount of energy in a gamma ray, however, is much greater than in visible light, ultraviolet light, or even X-rays. Because they have no mass or electric charge, and because of their high energies, gamma rays are able to penetrate through most materials. However, they cannot easily penetrate very dense materials, such as lead. Lead is commonly used as a shielding material in laboratories or hospitals where there can be much gamma radiation. Delicate molecules inside cells throughout our bodies that are zapped by gamma rays suffer structural damage. Hence, gamma rays generally cause more damage to our cells than do alpha or beta rays.

Figure 16.6 shows the relative penetrating power of the three types of radiation. Figure 16.7 shows an interesting practical use for gamma radiation.

FIGURE 16.4 ▲
A neutron has zero net charge. As the neutron releases a beta particle (an electron) it transforms into a proton, so the net charge remains zero.

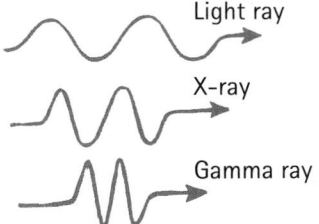
Once a beta particle slows by collisions, it becomes just another electron in the environment.

FIGURE 16.5 ▲
A gamma ray is simply electromagnetic radiation, much higher in frequency than light and X-rays.

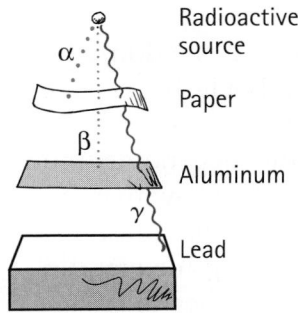

◄ **FIGURE 16.6**
Alpha particles are the least penetrating and can be stopped by a few sheets of paper. Beta particles will readily pass through paper, but not through a sheet of aluminum. Gamma rays penetrate several centimeters into solid lead.

✔ **READING CHECK**

What types of radiation are emitted by the nuclei of radioactive elements?

FIGURE 16.7 ▶
The shelf life of fresh strawberries and other perishables is markedly increased when the food is subjected to gamma rays from a radioactive source. The strawberries on the right were treated with gamma radiation, which kills the microorganisms that normally lead to spoilage. The food is only a receiver of radiation and is not transformed into an emitter of radiation, as can be confirmed with a radiation detector.

CONCEPT CHECK

Pretend that you are given three radioactive rocks. One is an alpha emitter, one is a beta emitter, and one is a gamma emitter, and you know which is which. You can throw away one, but, of the remaining two, you must hold one in your hand and place one in your shirt pocket. What can you do to minimize your exposure to radiation?

Check Your Answer
Hold the alpha emitter in your hand because the skin on your hand will shield you. Put the beta emitter in your pocket because beta particles will likely be stopped by the combined thickness of your clothing and skin. Throw away the gamma emitter because it would penetrate your body from any of these locations. (Ideally, of course, you should distance yourself as much as possible from all of the rocks.)

FIGURE 16.8 ▲
A commercially available radon test kit.

16.3 Environmental Radiation

✓ Most radioactivity we encounter originates in nature. Common rocks and minerals in our environment, for example, contain significant quantities of radioactive isotopes, because most of them contain trace amounts of uranium. As a matter of fact, people who live in brick, concrete, or stone buildings are exposed to greater amounts of radiation than people who live in wooden buildings.

A common source of naturally occurring radiation is radon-222, an inert gas arising from uranium deposits. Radon is a heavy gas that tends to accumulate in basements after it seeps up through cracks in the floor. Levels of radon vary from region to region, depending upon local geology. You can check the radon level in your home with a radon detector kit (Figure 16.8). If levels are high, corrective measures, such as sealing the basement floor and walls and maintaining adequate ventilation, should be taken. Radon gas poses a serious health risk.

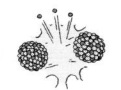

About 20% of our annual exposure to radiation comes from sources outside of nature, primarily medical procedures. Various medical devices, fallout from nuclear testing, and the coal and nuclear power industries are also contributors. Interestingly, the coal industry outranks the nuclear power industry as a source of radiation. The global combustion of coal annually releases about 13,000 tons of radioactive thorium and uranium into the atmosphere (in addition to other environmentally damaging molecules, including greenhouse gases). Both of these elements are found naturally in coal deposits, so their release is a natural consequence of burning coal. Nuclear power plants also produce radioactive by-products. Worldwide, the nuclear power industries generate about 10,000 tons of radioactive waste each year. Almost all of this waste, however, is contained and *not released into the environment.*

Most of the radiation we receive is from natural sources and from medical procedures. The human body, however, is a significant source of natural radiation, primarily from the potassium we ingest. Our bodies contain about 200 grams of potassium. Of this quantity, about 20 milligrams is the radioactive isotope potassium-40, which is a beta-ray emitter. Between every heartbeat, about 5000 potassium-40 isotopes in the average human body undergo spontaneous radioactive decay. Radiation is indeed everywhere.

When radiation encounters our intricately structured cells, it can create chaos. Cells are able to repair most kinds of damage caused by radiation, if the radiation is not too severe. A cell can survive an otherwise lethal dose of radiation if the dose is spread over a long period of time to allow intervals for healing. When radiation is sufficient to kill cells, the dead cells can be replaced by new ones. Sometimes a radiated cell will survive with damaged DNA. This can alter the genetic information contained in a cell, producing one or more mutations (Chapter 23). Although the effects of many mutations are insignificant, others affect the functioning of cells. For example, certain genetic mutations lead to cancer. In addition, if mutations occur in an individual's reproductive cells, they can be passed to the individual's offspring. In this case, the mutation will be present in every cell in the offspring's body—and may well have an effect on the functioning of the offspring.

Common sense tells us to avoid radiation when possible. Common sense also tells us that all radiation cannot be avoided. Most of it is simply a part of nature.

The average coal-burning power plant is a greater source of airborne radioactive material than is a nuclear power plant.

CONCEPT CHECK
1. Is the human body radioactive?
2. How can radioactivity produce a genetic mutation?

Check Your Answers
1. Yes. The human body is radioactive because of all the radioactive isotopes it contains.
2. Radioactivity produces genetic mutations by damaging a cell's DNA—its genetic blueprint—without killing the cell.

✔ READING CHECK

Where does most of the radioactivity we encounter originate?

16.4 Transmutation of Elements— Changing Identities

When a radioactive nucleus emits an alpha or beta particle, the atomic number of the nucleus is changed. It becomes another element. ✓ The changing of one element to another is called **transmutation.** Consider a uranium-238 nucleus, which contains 92 protons and 146 neutrons. When an alpha particle is ejected, the nucleus loses 2 protons and 2 neutrons. Since an element is defined by the number of protons in its nucleus, the 90 protons and 144 neutrons remaining are no longer uranium. What we have now is the nucleus of a different element— thorium.

This transmutation can be depicted as follows:

Uranium-238 → thorium-234 + helium-4

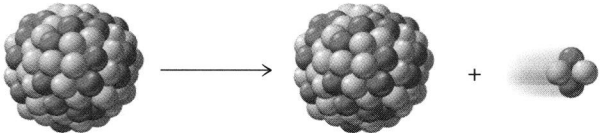

When this transmutation occurs, energy is released. Part of the released energy is in the form of gamma radiation. Most is the kinetic energy of the alpha particle, and some is the kinetic energy of the recoiling thorium atom. Notice that in this and all other nuclear equations, the mass numbers balance (238 = 234 + 4).

Thorium-234 is radioactive. When it decays, it emits a beta particle. As mentioned previously, a beta particle is an electron emitted by a neutron as the neutron transforms to a proton. Thorium has 90 protons, so beta emission leaves the nucleus with one fewer neutron and one more proton. The new nucleus has 91 protons and is no longer thorium. It becomes the element protactinium. Although the atomic number has increased by 1 in this process, the mass number (protons + neutrons) remains the same. This transmutation can be depicted as follows:

Thorium-234 → protactinium-234 + electron

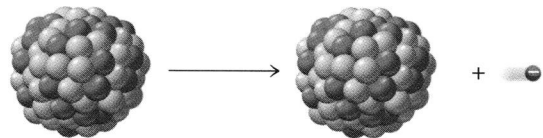

So we see that when an element ejects an alpha particle from its nucleus, the mass number of the remaining atom is decreased by 4 and its atomic number is decreased by 2. The resulting atom is an atom of the element two spaces back in the periodic table because this atom has two fewer protons. When an element ejects a beta particle from its nucleus, the mass of the atom is practically unaffected, meaning there is no change in mass number, but its atomic number increases by 1. The resulting atom becomes the element one space forward in the periodic table because it has one more proton.

The decay of uranium-238 to lead-206 is shown in Figure 16.9. Each green arrow shows an alpha decay, and each red arrow shows a beta decay.

✔ **READING CHECK**

What is the term for the changing of one element into another?

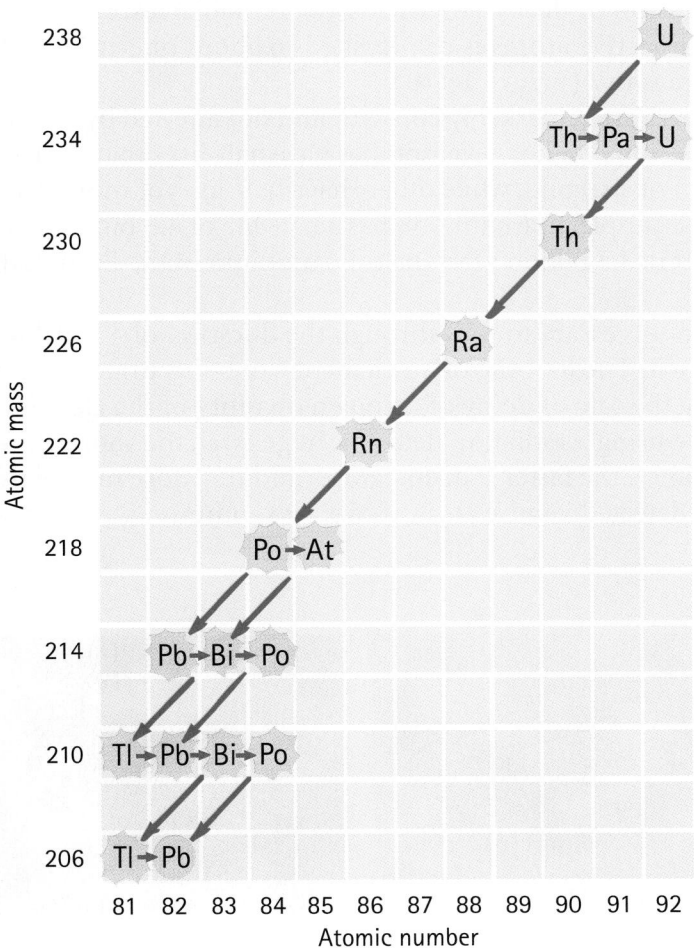

◀ **FIGURE 16.9**
U-238 decays to Pb-206 through a series of alpha and beta decays.

CONCEPT CHECK
What finally becomes of all the uranium that undergoes radioactive decay?

Check Your Answer
All uranium ultimately becomes lead. Along the way, it exists as the elements indicated in Figure 16.9.

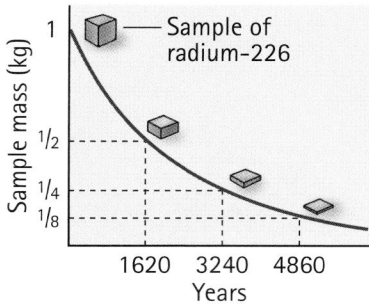

FIGURE 16.10 ▲

Radium-226 has a half-life of 1620 years, meaning that every 1620 years the amount of radium decreases by half as it transmutes to other elements.

✓ **READING CHECK**

How is radioactive half-life defined?

16.5 Half-Life Is a Measure of Radioactive Decay Rate

Radioactive isotopes decay at different rates. The radioactive decay rate is measured in terms of a characteristic time, the **half-life.**

✓ The half-life of a radioactive material is the time needed for half of the radioactive atoms to decay. Radium-226, for example, has a half-life of 1620 years. This means that half of any given specimen of Ra-226 will decay by the end of 1620 years. In the next 1620 years, half of the remaining radium decays, leaving only one-fourth the original number of radium atoms. The other three-fourths convert, by a succession of decays, to lead. After 20 half-lives, an initial quantity of radioactive atoms is diminished to about one-millionth of the original quantity (Figure 16.10).

Half-lives are remarkably constant and not affected by external conditions. Some radioactive isotopes have half-lives that are less than a millionth of a second, while others have half-lives of more than a billion years. For example, uranium-238 has a half-life of 4.5 billion years. This means that in 4.5 billion years, half the uranium in the Earth today will be lead.

It is not necessary to wait through the duration of a half-life in order to measure it. The half-life of an element can be accurately estimated by measuring the rate of decay of a known quantity of the element. This is easily done using a radiation detector. In general, the shorter the half-life of a substance, the faster it disintegrates and the more radioactivity per minute is detected. Figure 16.11 shows two common types of radiation detectors.

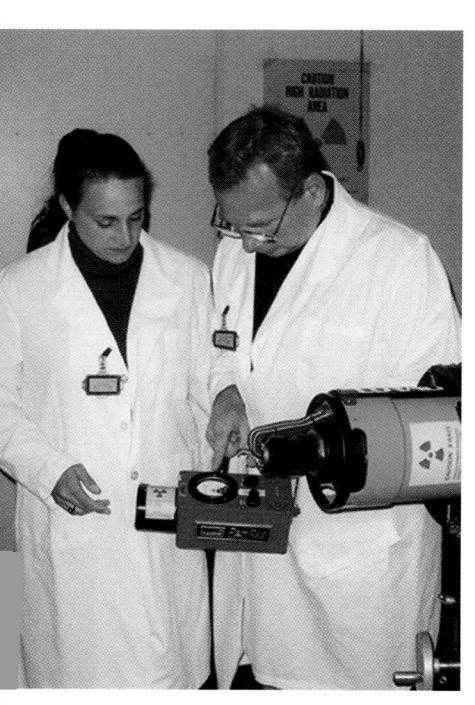

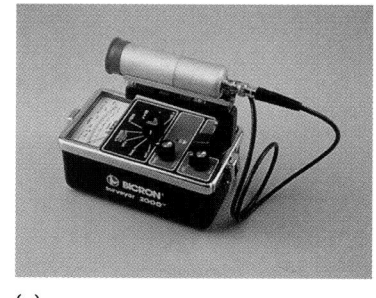
(a)

(b)

FIGURE 16.11 ▲

(a) A Geiger counter detects incoming radiation by the way the radiation affects a gas enclosed in the tube. (b) A scintillation counter detects incoming radiation by flashes of light that are produced when charged particles or gamma rays pass through it.

◀ **FIGURE 16.12**

The film badges worn by Tammy and Larry contain audible alerts for both radiation surge and accumulated exposure. Information from the individualized badges is periodically downloaded to a database for analysis and storage.

CONCEPT CHECK

1. If a sample of radioactive isotopes has a half-life of one day, how much of the original sample will remain at the end of the second day? The third day?
2. Which will give a higher counting rate on a radiation detector—radioactive material that has a short half-life or a long half-life?

Check Your Answers

1. In the second day, one-fourth of the original sample will be left. The three-fourths that underwent decay becomes a different element altogether. At the end of three days, one-eighth of the original sample will remain.
2. The material with the shorter half-life is more active and will show a higher counting rate on a radiation detector.

16.6 Isotopic Dating Measures the Ages of Materials

Cosmic rays continually bombard the Earth's atmosphere. This bombardment causes many atoms in the upper atmosphere to transmute. These transmutations result in many protons and neutrons being "sprayed out" into the environment. Most of the protons are stopped as they collide with the atoms of the upper atmosphere. These protons strip electrons from the atoms they collide with and thus become hydrogen atoms. The neutrons, however, continue for longer distances because they have no electric charge and therefore do not interact electrically with matter. Eventually, many of them collide with atomic nuclei in the lower atmosphere. A nitrogen nucleus that captures a neutron, for instance, may emit a proton and become a nucleus of carbon:

Neutron + nitrogen-14 → carbon-14 + proton

This carbon-14 isotope, which makes up less than one-millionth of 1% of the carbon in the atmosphere, has eight neutrons and is radioactive. (The most common isotope, carbon-12, has six neutrons and is not radioactive.) Because both carbon-12 and carbon-14 are forms of carbon, they have the same chemical properties. Both of these isotopes, for example, chemically react with oxygen to form carbon dioxide, which is consumed by plants. This means that all plants contain a tiny quantity of radioactive carbon-14. ✓ All animals eat either plants or plant-eating animals, and therefore all animals, including us, have a little carbon-14 in them. So we see why all living things on Earth contain some carbon-14.

When carbon-14 emits a beta particle, it is transformed back to nitrogen:

Carbon-14 $\rightarrow$ nitrogen-14 + electron

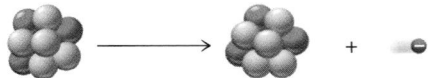

Because plants absorb carbon dioxide only while alive, any carbon-14 that decays is immediately replenished with fresh carbon-14 from the atmosphere. In this way, a radioactive equilibrium is reached where there is a constant ratio of about 1 carbon-14 atom to every 100 billion carbon-12 atoms. When a plant dies, replenishment of carbon-14 ends. Then the percentage of carbon-14 decreases at a rate determined by its half-life. The amount of carbon-12, however, does not change because this isotope does not undergo radioactive decay. The longer a plant or other organism is dead, therefore, the less carbon-14 it contains relative to the constant amount of carbon-12.

The half-life of carbon-14 is about 5730 years. This means that half of the carbon-14 atoms now present in a plant or animal that dies today will decay in the next 5730 years. Half of the remaining carbon-14 atoms will then decay in the following 5730 years, and so on.

With this knowledge, scientists are able to calculate the age of carbon-containing artifacts, such as wooden tools or the skeleton shown in Figure 16.13. Age is measured by their current level of radioactivity. This process, known as **carbon-14 dating,** enables investigators to probe as far as 50,000 years into the past. Beyond this time span, there is too little carbon-14 remaining for accurate analysis.

Mainly due to fluctuations in cosmic ray bombardment over the centuries, carbon dating has an uncertainty of about 15%. This means, for example, that the straw of an old adobe brick dated to be 500 years old may really be only 425 years old on the low side or 575 years old on the high side. For many purposes, this is an acceptable level of uncertainty.

The carbon-dating technique, useful to physicists and biologists, was invented by William Libby, a chemist.

22,920 years ago   17,190 years ago   11,460 years ago   5730 years ago   Present

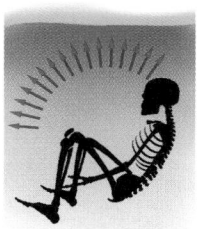 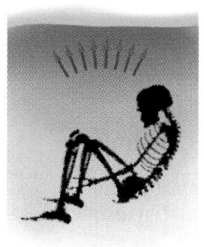 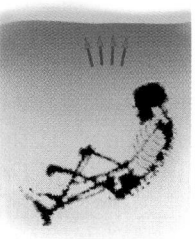 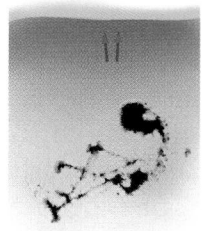 

FIGURE 16.13 ▲
The amount of radioactive carbon-14 in the skeleton is reduced to one-half every 5730 years. The result is that the same skeleton today contains only a trace amount of the original carbon-14. The red arrows represent the relative amounts of carbon-14.

CONCEPT CHECK

1. Suppose that an archaeologist extracts 1 gram of carbon from an ancient ax handle and finds that it is one-fourth as radioactive as 1 gram of carbon extracted from a freshly cut tree branch. About how old is the ax handle?
2. The age of the Dead Sea Scrolls was found by carbon dating. Could this technique have worked if they were carved in stone tablets? Explain.

Check Your Answers

1. Assuming the ratio of C-14/C-12 was the same when the ax was made, the ax handle is as old as two half-lives of C-14, or about 11,460 years old.
2. Stone tablets are not organic and therefore cannot be dated by the carbon-dating technique. Nonliving stone does not ingest carbon and transform it. Carbon dating is useful only for organic material.

Nonliving things can also be dated by the radioactive minerals they contain. The naturally occurring mineral isotopes uranium-238 and uranium-235, for example, decay very slowly and ultimately become lead—but not the common isotope lead-208. Instead, as was shown in Figure 16.9, uranium-238 decays to lead-206. Uranium-235, on the other hand, decays to lead-207. Thus, the lead-206 and lead-207 that now exist in a uranium-bearing rock were at one time uranium. The older the rock, the higher the percentage of these lead isotopes.

If you know the half-lives of uranium isotopes and the percentage of lead isotopes in some uranium-bearing rock, you can calculate the date of rock formation. Rocks dated in this manner have been found to be as much as 3.7 *billion* years old. Samples from the Moon have been dated at 4.2 billion years, which is close to the estimated age of our solar system: 4.6 billion years.

✔ **READING CHECK**

Plants contain a little carbon-14, but why do humans also contain carbon-14?

16.7 Nuclear Fission—The Breaking Apart of Atomic Nuclei

Biology students know that living tissue grows by the division of cells. The splitting in half of living cells is called *fission.* In a similar way, the splitting of atomic nuclei is called **nuclear fission.**

Nuclear fission involves the delicate balance between two forces within the nucleus. One force is the *strong nuclear force,* which is a force that holds all the nucleons together. The second force is the repulsive *electric force* occurring among all the like-sign protons. In all known nuclei, the nuclear strong force dominates. In uranium, however, this domination is weak. If the uranium nucleus is stretched into an elongated shape (Figure 16.14), the electrical forces may push it into an even more elongated shape. If the elongation passes a critical point, electrical forces overwhelm nuclear strong forces, and the nucleus splits. This is nuclear fission.

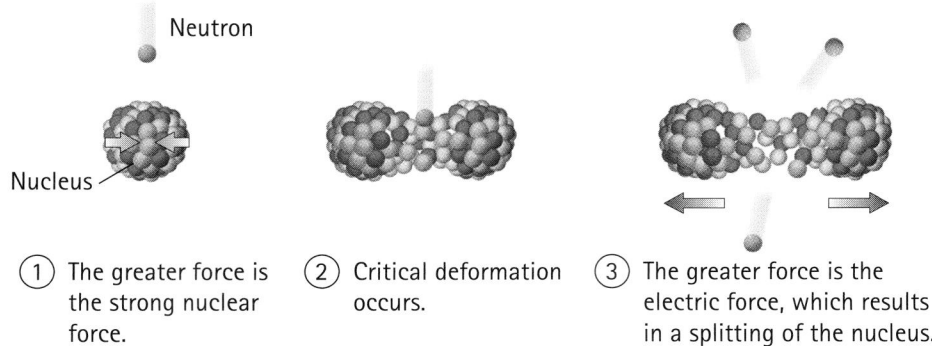

① The greater force is the strong nuclear force.

② Critical deformation occurs.

③ The greater force is the electric force, which results in a splitting of the nucleus.

Neutron

Nucleus

FIGURE 16.14 ▲
An elongation of the nucleus may result in repulsive electric forces overcoming attractive strong nuclear forces, in which case fission occurs.

The absorption of a neutron by a uranium nucleus supplies enough energy to cause such an elongation. The resulting fission process may produce many different combinations of smaller nuclei. More significant, energy released by fission is enormus—about seven million times that of a TNT molecule explosion. This energy is mainly in the form of kinetic energy of the fission fragments, which fly apart from one another. A much smaller amount of energy is released as gamma radiation.

Here is the equation for a typical uranium fission reaction:

Neutron + uranium-235 → fission fragments

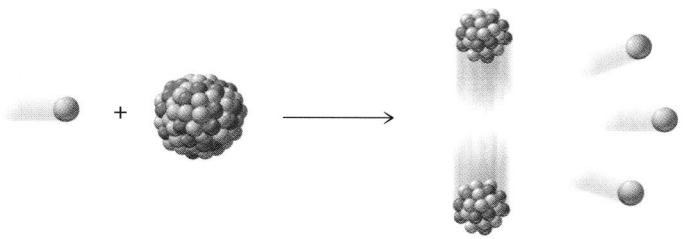

Note in this reaction that 1 neutron starts the fission of the uranium nucleus, which in turn produces 3 neutrons. These new neutrons can cause the fissioning of three other uranium atoms, releasing 9 more neutrons. If each of these 9 neutrons succeeds in splitting a uranium atom, the next step in the reaction produces 27 neutrons, and so on. Such a sequence, illustrated in Figure 16.15, is called a **chain reaction.** ✓ A chain reaction is a self-sustaining reaction in which the products of one reaction event stimulate further reaction events.

Chain reactions are more effective in large chunks of uranium than in smaller chunks. In smaller chunks, neutrons easily find the surface and escape, as shown in Figure 16.16. As the neutrons escape, the chain reaction no longer builds up.

Chain reactions do not occur to any great extent in naturally occurring uranium ore because not all uranium atoms fission so easily. Fission occurs mainly in the rare isotope uranium-235. Uranium-235 makes up only 0.7% of the uranium in pure uranium metal.

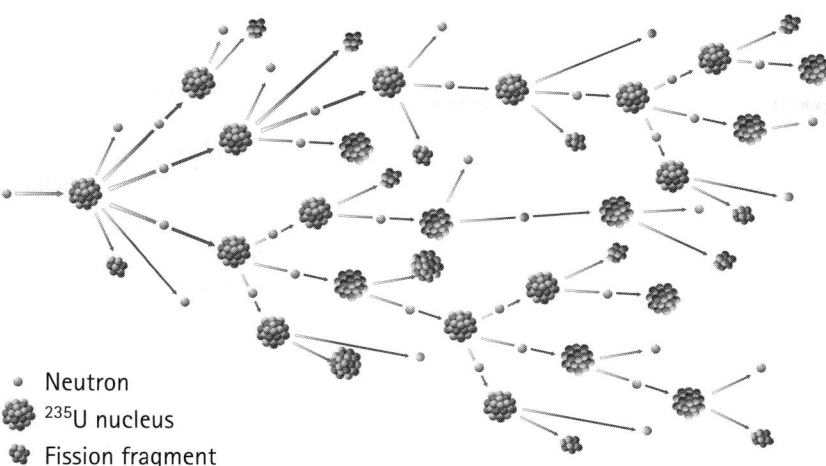

◀ **FIGURE 16.15**
A chain reaction.

- Neutron
- ^{235}U nucleus
- Fission fragment

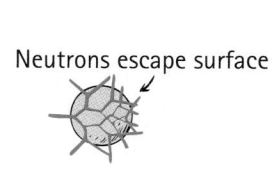
Neutrons escape surface

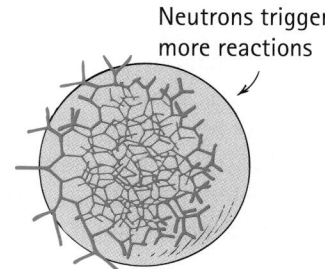
Neutrons trigger more reactions

FIGURE 16.16 ▶
The exaggerated view shows that a chain reaction in a small chunk of pure U-235 is prevented from building up because neutrons are able to escape from the surface. In a larger chunk, however, more uranium and less surface is presented to the neutrons.

If two small chunks are suddenly pushed together, they make a larger chunk. Within this larger chunk, neutrons are no longer able to escape so easily. Instead, they continue the chain reaction, which becomes sustainable. The minimum-mass chunk needed for a sustainable chain reaction is called a **critical mass.** Any chunk at or above the critical mass produces energy. With the correct engineering, this energy can be controlled, which is what happens within a nuclear power plant. A different sort of engineering, as shown in Figure 16.17, produces an explosion of energy, which is the nuclear fission bomb.

Constructing a fission bomb is a formidable task. The difficulty is in separating enough uranium-235 from the more abundant uranium-238. Whereas uranium-235 will fission, uranium-238 will not. Scientists took more than two years to extract enough of the 235 isotope from uranium ore to make the bomb that was detonated at Hiroshima in 1945. To this day, uranium isotope separation remains a difficult process.

✔ **READING CHECK**

What exactly is a chain reaction?

High explosive to drive uranium "shell"   Radioactive neutron source

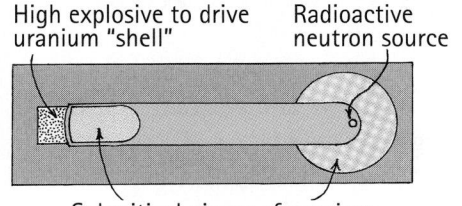

Subcritical pieces of uranium

◀ **FIGURE 16.17**
Simplified diagram of a uranium-fission bomb. Two smaller masses of uranium are initially apart from each other. The bomb is ignited when high explosives (gun powder) force the two masses to come together into a single lump of critical mass.

The awesome energy of nuclear fission was introduced to the world in the form of nuclear bombs, and this violent image still colors our thinking about nuclear power. Therefore, it is difficult for many people to recognize its potential usefulness. Currently, about 20% of electric energy in the United States is generated by nuclear fission reactors (whereas most electric power is nuclear in some other countries—more than 70% in France). These reactors are simply furnaces. They, like fossil-fuel furnaces, merely boil water to produce steam for a turbine. The greatest practical difference is in the amount of fuel required: A single kilogram of uranium fuel, less than the size of a baseball, yields more energy than 30 freight car loads of coal.

A diagram of the most common type of fission reactor is shown below. The nuclear fuel is primarily U-238 blended with about 3% U-235. ✓ Because the U-235 isotopes are so highly diluted with U-238 (which doesn't fission), an explosion like that of a nuclear bomb is not possible. The reaction rate, which depends on the number of neutrons that initiate the fission of other U-235 nuclei, is controlled by rods inserted into the reactor. The control

rods are made of a neutron-absorbing material, such as the metals cadmium or boron.

Heated water around the nuclear fuel is kept under high pressure to prevent boiling. It transfers heat to a second water system with lower pressure, which operates the turbine and electric generator in a conventional fashion. In this design, two separate water systems are used so that no radioactivity reaches the turbine or the outside environment.

One disadvantage of fission power is the generation of radioactive waste products. Safely disposing of these waste products requires special procedures. Although fission has been successfully producing electricity for a half century, the search for satisfactory ways of disposing of radioactive wastes in the United States has remained unsuccessful.

The potential benefits of fission power include plentiful electricity and the conservation of massive amounts of fossil fuels that every year are literally turned to heat and smoke. These fuels, in the long run, may be far more precious as sources of organic molecules than as sources of heat. Also, fission power doesn't generate atmospheric pollutants, such as sulfur oxides or the greenhouse gas carbon dioxide.

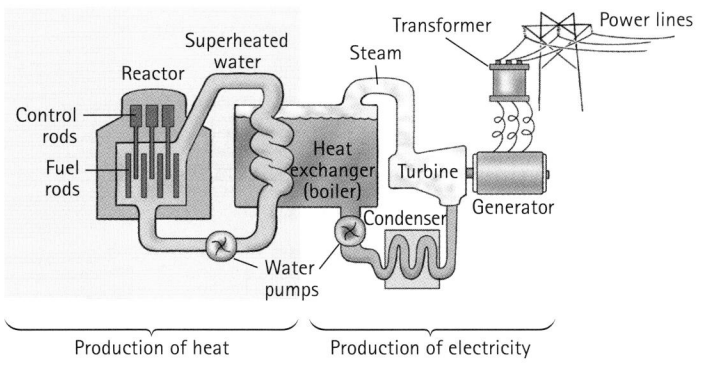

Production of heat   Production of electricity

✓ **READING CHECK**

Why are nuclear explosions not possible in today's nuclear power plants?

16.8 The Mass–Energy Relationship: $E = mc^2$

Clearly, a lot of energy comes from every gram of nuclear fuel that is fissioned. What is the source of this energy? As we will see, it comes from a loss of mass in the process.

In the early 1900s, Albert Einstein discovered that mass is actually "congealed" energy. Mass and energy are two sides of the same coin, as

stated in his celebrated equation $E = mc^2$. ✓ In this equation, *E* stands for the energy contained in any mass when at rest, *m* stands for mass, and *c* is the speed of light. This relationship between energy and mass is the key to understanding why and how energy is released in nuclear reactions.

How easy might it be to pull a nucleon out of a nucleus? To do this, you would have to fight against the strong nuclear force, which holds the nucleon to the nucleus. (Remember, "nucleon" is the generic name for either a proton or a neutron.) So a lot of energy would be required to pull this nucleon out of the nucleus (Figure 16.18). What we learn from Einstein's equation is that the energy you put in to pull the nucleon out is not lost. Instead this energy is absorbed by the nucleon, which thus becomes more massive as you pull it out. For example, if the mass of the nucleon were 1.00000 while in the nucleus, its mass might be a slightly greater 1.00728 after it has been pulled out of the nucleus. The energy you put into pulling the nucleon out converts into mass. Energy and mass are two sides of the same coin. In other words, energy equals mass and mass equals energy. How so? By the equation $E = mc^2$.

So the effective mass of a nucleon depends upon where it is. In general, a nucleon's mass is greatest when it is free by itself outside of the nucleus. The average mass of a nucleon is smallest when it is tightly bound within the nucleus. Not all nuclei, however, are the same. In one nucleus, for example, a nucleon might find itself more tightly bound than in another. The mass of the nucleon, therefore, also depends upon which nucleus it is in. As illustrated in the graph of Figure 16.19, a nucleon has its greatest mass when in the hydrogen nucleus and its smallest mass when in the iron nucleus. The mass of a nucleon then gradually increases as it enters heavier nuclei, such as uranium.

FIGURE 16.18 ▲
Much energy is required to pull a nucleon from an atomic nucleus.

We can't actually measure the mass of a single nucleon while it is inside the nucleus and compare its mass when outside. But as Figure 16.18 suggests, the work done in pulling a nucleon from the nucleus produces a more massive nucleon.

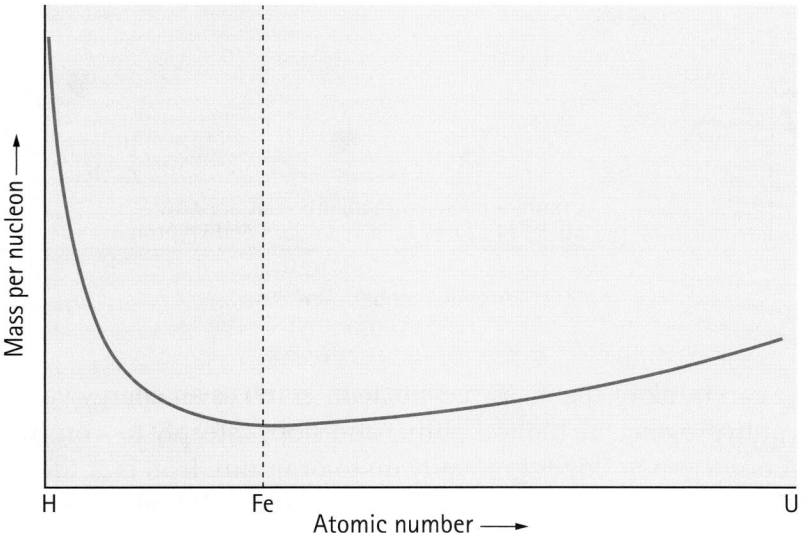

FIGURE 16.19 ▲
This graph shows that the average mass of a nucleon depends on which nucleus it is in. A nucleon has the most mass when in the lightest (hydrogen) nucleus. The nucleon has its least mass when in an iron nucleus, and an intermediate mass when in the heaviest (uranium) nucleus.

Mass is like a super storage battery that stores energy—vast quantities of energy—which can be released if and when the mass decreases.

CONCEPT CHECK
A nucleon has its least mass when in an iron nucleus. Does this mean that it is relatively easy to pull a nucleon out of an iron nucleus?

Check Your Answer
No! Just the opposite. The nucleon has its least mass when in an iron nucleus because that is where it is most tightly bound. To pull the nucleon from the iron nucleus would be much harder than pulling it from any other nucleus.

The graphs shown here reveal the energy of the atomic nucleus, the primary source of energy in stars—which is why they can be considered the most important graphs in this book.

The graph of Figure 16.19 is key to understanding the energy released in nuclear processes. Uranium, being towards the right-hand side of the graph, is shown to have a relatively large amount of mass per nucleon. When the uranium nucleus splits in half, however, smaller nuclei of lower atomic numbers are formed. As shown in Figure 16.20, these nuclei are lower on the graph than uranium, which means that they have a smaller amount of mass per nucleon. Thus, the combined mass of the fission fragments is less than the mass of the uranium nucleus. When this decrease in mass is multiplied by the speed of light squared (c^2 in Einstein's equation), the product is equal to the energy yielded by each uranium nucleus as it undergoes fission.

FIGURE 16.20 ▶
A nucleon's average mass in a uranium nucleus is greater than in any of its nuclear fission fragments. This lost mass is mass that has been transformed into energy, which is why nuclear fission is an energy-releasing process.

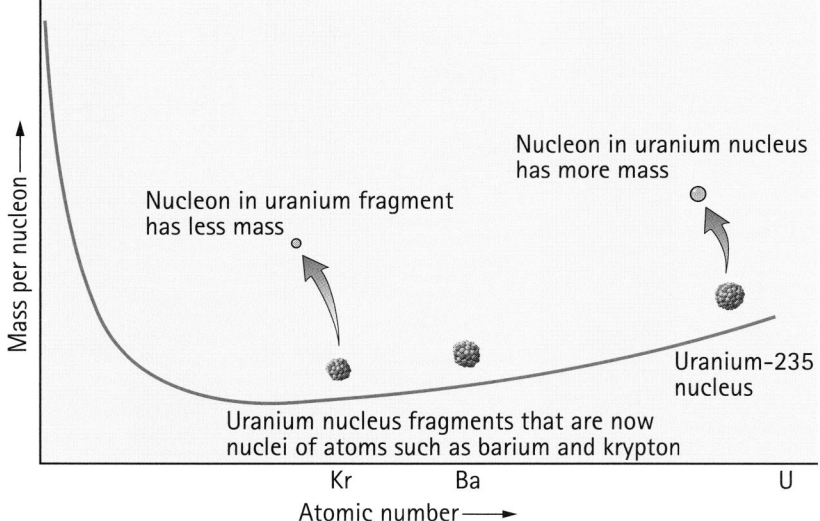

We can think of the mass-per-nucleon graph as an energy valley that starts at hydrogen (the highest point) and slopes steeply to iron (the lowest point), then slopes gradually up to uranium. Iron is at the bottom of the energy valley and is the most stable nucleus. It is also the most tightly bound nucleus; more energy per nucleon is required to separate nucleons from its nucleus than from any other nucleus.

All nuclear power today is produced by nuclear fission. A more promising long-range source of energy is to be found on the left side of the energy valley in a process known as nuclear fusion.

What do the symbols represent in the equation $E = mc^2$?

CONCEPT CHECK

Correct the following incorrect statement: When a heavy element such as uranium undergoes fission, there are fewer nucleons after the reaction than before.

Check Your Answer

When a heavy element such as uranium undergoes fission, there aren't fewer nucleons after the reaction. Instead, there's less mass in the same number of nucleons.

16.9 Nuclear Fusion—The Combining of Atomic Nuclei

Notice that in the graphs of Figures 16.19 and 16.20 the steepest part of the curve is between hydrogen and iron. Energy is gained as light nuclei combine. This combining of nuclei is **nuclear fusion**—the opposite of nuclear fission. We can see from Figure 16.21 that as we move along the list of elements from hydrogen to iron, the average mass per nucleon decreases. ✓ Thus, when two small nuclei fuse—say, a pair of hydrogen isotopes— the mass of the resulting nucleus is less than the mass of the two small nuclei before fusion. Energy is released as smaller nuclei fuse.

For a fusion reaction to occur, the nuclei must collide at a very high speed in order to overcome their mutual electric repulsion. The required speeds correspond to the extremely high temperatures found in the Sun and other stars. Fusion brought about by high temperatures is called **thermonuclear fusion.** In the high temperatures of the Sun approximately 657 million tons of hydrogen is converted into 653 million tons of helium each second. The missing 4 million tons of mass is converted to energy—a tiny bit of which reaches our planet as sunshine.

Such reactions are, quite literally, nuclear burning. Thermonuclear fusion is analogous to ordinary chemical combustion. In both chemical and nuclear burning, a high temperature starts the reaction; the release of energy

With each passing day, the Sun is less massive, and we benefit by receiving its radiant warmth and welcome light.

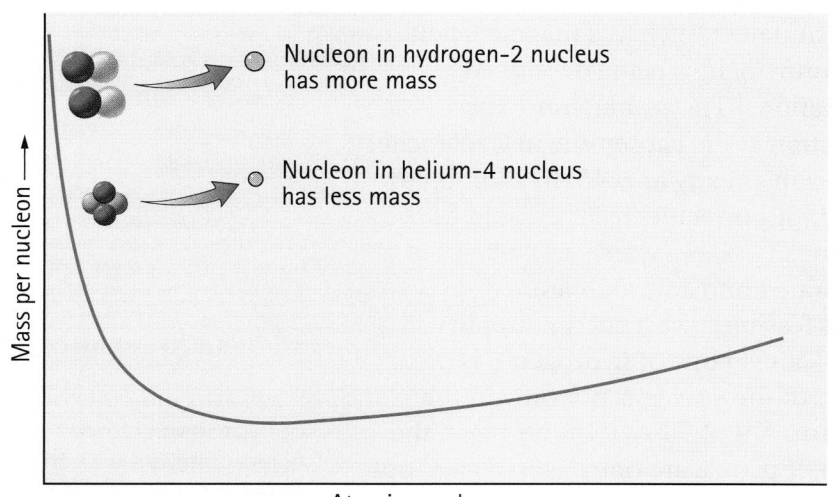

Nucleon in hydrogen-2 nucleus has more mass

Nucleon in helium-4 nucleus has less mass

Mass per nucleon →

Atomic number →

◀ **FIGURE 16.21**

The average mass per nucleon is greater in a hydrogen-2 nucleus than in a helium-4 nucleus. This lost mass is mass that has been converted to energy, which is why nuclear fusion is a process that releases energy.

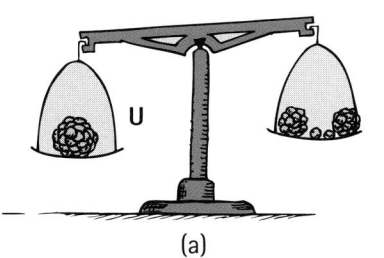 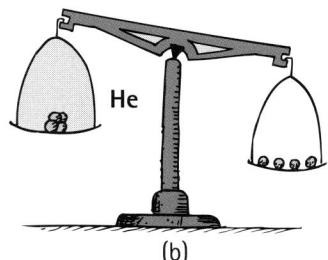

The mass of a nucleus is not equal to the sum of the mass of its parts. (a) The fission fragments of a uranium nucleus are less massive than the uranium nucleus. (b) Two protons and two neutrons are more massive in their free states than when they are combined to form a helium nucleus.

by the reaction maintains a high enough temperature to spread the fire. The net result of the chemical reaction is a combination of atoms into more tightly bound molecules. In nuclear fusion reactions, the net result is more tightly bound nuclei. In both cases, mass decreases as energy is released.

CONCEPT CHECK

1. Fission and fusion are opposite processes, yet each releases energy. Isn't this contradictory?
2. To get a release of nuclear energy from the element iron, should iron undergo fission or fusion?

Check Your Answers

1. No, no, no! This is contradictory only if the same element is said to release energy by both fission and fusion. Only the fusion of light elements and the fission of heavy elements result in a decrease in mass per nucleon and a release of energy.
2. Neither, because iron is at the very bottom of the "energy valley." Fusing a pair of iron nuclei produces an element to the right of iron on the curve, where mass per nucleon is higher. If you split an iron nucleus, the products lie to the left of iron on the curve and also have a higher mass per nucleon. So no energy is released. For energy release, "decrease mass" is the name of the game—any game, chemical or nuclear.

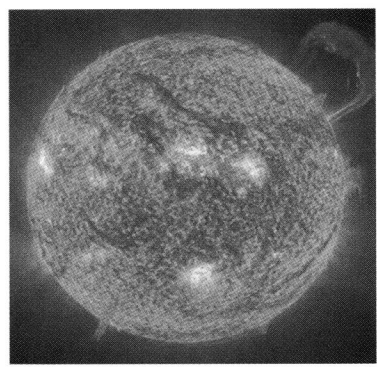

FIGURE 16.23 ▲
Thermonuclear fusion takes place in stars, such as the Sun. You will learn more about this in Chapter 33. Some day, humans may produce vast quantities of energy through thermonuclear fusion as the stars have always done.

Carrying out thermonuclear fusion reactions under controlled conditions requires temperatures of millions of degrees. There are a variety of techniques for attaining high temperatures. But no matter how the temperature is produced, a problem is that all materials melt and vaporize at the temperatures required for fusion. One solution to this problem is to confine the reaction in a nonmaterial container, such as a magnetic field. At this writing, fusion by magnetic confinement is being developed through a project known as the International Thermonuclear Experimental Reactor (ITER). After construction at the chosen site in Cadarache, France, the first fusion reaction may begin as early as 2015. In addition to producing about 500 MW (megawatts) of power, the reactor could be the energy source for the creation of hydrogen gas, H_2, which could be used to power fuel cells, such as those incorporated into automobiles.

If and when nuclear fusion proves feasible, a vast energy supply will become available. The fuel for fusion—an isotope of hydrogen—is plentiful on Earth. It is found in every part of the universe, not only in the stars but also in the space between them. About 90% of the atoms in the universe are estimated to be hydrogen. If people are one day to dart about the universe in the same way we jet about Earth today, their supply of fuel is assured!

 **READING CHECK**

How does the mass of a pair of hydrogen isotopes when fused compare with the mass of the resulting helium nucleus?

Chemistry is the physics of atoms and molecules.

PART THREE

CHEMISTRY

Wait a second! Are you telling me that this is what diamond looks like at the level of its atoms? And that diamond is so hard because its atoms are arranged in a rigid structure, like I see here? And if combined with a few other atoms, such as oxygen and nitrogen, these atoms of diamond could make the molecules of living organisms, such as myself? Are you saying that chemistry is the study of our immediate universe at the level of the very small? Wow. Tell me more!

17

ELEMENTS OF CHEMISTRY

THE MAIN
IDEA

Atoms are the fundamental building blocks of everything around us.

All the stuff around us is made of atoms. This includes the ground we walk on, the food we eat, and the air we breathe. An element is a material made of only one type of atom. Examples include diamond, which is made only of carbon, and pure gold, which is made of only gold atoms. A group of atoms connected together is what we call a *molecule*. A water molecule, for example, consists of two hydrogen atoms together with one oxygen atom, which you may know as H_2O. Add another oxygen atom and you no longer have water. You instead have hydrogen peroxide, H_2O_2, which is a common disinfectant. There are only about 100 different kinds of atoms, but there are a limitless number of different ways in which they can connect together. This is why we have so many different kinds of materials around us. Chemistry is the study of how atoms connect together to form new materials.

Explore!

Can a Gas Extinguish a Flame?
Observe and Record

1. Wear your safety goggles, Add about a teaspoon of baking soda to a tall glass.
2. Slowly add white distilled vinegar so that bubbles don't overflow the edge of the glass. Wait until the bubbles settle.
3. Remove all flammable materials, especially paper towels.
4. With permission from your teacher, light a wooden match and dip the flame into the mouth of the glass. At some point the flame should extinguish. Drop the match into the glass if it does not extinguish.

Analyze and Conclude

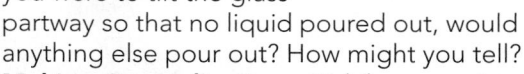

1. **Predicting** Is there a limit to the number of times a flame can be extinguished by the gas in the glass? If you were to tilt the glass partway so that no liquid poured out, would anything else pour out? How might you tell?
2. **Making Generalizations** Did the gas exist before you added the vinegar to the baking soda? Is this gas heavier or lighter than air? How else is this gas different from the air we breathe?

17.1 Chemistry Is Known as the Central Science

When you wonder what the land, sky, or ocean is made of, you are thinking about chemistry. When you wonder how a rain puddle dries up, how a car acquires energy from gasoline, or how your body finds energy from the food you eat, you are again thinking about chemistry. By definition, ✓ chemistry is the study of matter and the transformations it can undergo. Matter is anything that occupies space. It is the stuff that makes up all material things—anything you can touch, taste, smell, see, or hear is matter. The scope of chemistry, therefore, is very broad.

Chemistry is often described as a central science because it touches all the other sciences. It springs from the principles of physics, and it serves as the foundation for the most complex science of all—biology. Indeed, many of the great advances in the life sciences today, such as genetic engineering, are applications of some very exotic chemistry. Chemistry is also the foundation for Earth science. It is also an important component of space science, as described in Figure 17.1. Just as we learned about the origin of the Moon from the chemical analysis of Moon rocks in the early 1970s, we are now learning about the history of Mars and other planets from the chemical information gathered by space probes.

Progress in science is made as scientists conduct research. Research is any activity whose purpose is the discovery of new knowledge. Many scientists focus on **basic research,** which leads us to a greater understanding of how the natural world operates. The foundation of knowledge laid down by basic research frequently leads to useful applications. Research that focuses on developing these applications is known as **applied research.** The majority of chemists choose applied research as their major focus. Applied research in chemistry has provided us with medicine, food, water, shelter, and so many of the material goods that characterize modern life. Just a few of a myriad of examples are shown in Figure 17.2.

Over the course of the past century, we became very good at manipulating atoms and molecules to create materials to match our needs.

Industries within the United States employ about 900,000 chemists.

✓ READING CHECK

What is the definition of chemistry?

There is much better living with chemistry.

◀ **FIGURE 17.1**
Special materials of chemistry, such as rocket fuels, metals for the spaceships, and fabrics for the space suits, were required to allow astronauts to reach and explore the surface of the Moon.

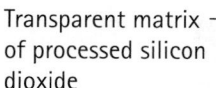

Transparent matrix of processed silicon dioxide

Chemically disinfected drinking water

Caffeine solution

Thermoset polymer

Prescription medicines stored in refrigerator

Chlorofluorocarbon-free refrigerating fluids

Electrical energy from a fossil-fuel or nuclear power plant

Metal alloy

Roasting carbohydrates, fats, proteins, and vitamins

Natural gas laced with odoriferous sulfur compounds

Fertilizer-grown vegetables

FIGURE 17.2 ▲
Most of the material items in any modern house are shaped by some human-devised chemical process.

At the same time, however, mistakes were made when it came to caring for the environment. Waste products were dumped into rivers, buried in the ground, or vented into the air without regard for possible long-term consequences. Many people believed that the Earth was so large that its resources were virtually unlimited and that it could absorb wastes without being significantly harmed.

Most nations now recognize this as a dangerous attitude. As a result, government agencies, industries, and concerned citizens are involved in extensive efforts to take care of the environment. For example, members of the American Chemistry Council, who, as a group, produce 90% of the chemicals manufactured in the United States, have adopted a program called Responsible Care. Through this program, members of this organization have pledged to manufacture without causing environmental damage. The Responsible Care program emblem is shown in Figure 17.3. By using chemistry wisely, most waste products can be minimized, recycled, engineered into useful products, or rendered environmentally safe.

Chemistry has influenced our lives in many important ways, and it will continue to do so in the future. For this reason, it is in everyone's interest to become familiar with the basic concepts of chemistry.

FIGURE 17.3 ▲
The Responsible Care symbol of the American Chemistry Council.

CONCEPT CHECK
Chemists have learned how to produce aspirin using petroleum as a starting material. Is this an example of basic or applied research?

Check Your Answer
This is an example of applied research, because the primary goal was to develop a useful commodity. However, the ability to produce aspirin from petroleum depended on an understanding of atoms and molecules developed from many years of basic research.

Our understanding of the chemistry of life will boost the life sciences to fantastic advances in this 21st century.

17.2 The Submicroscopic World Is Super-Small

From afar, a sand dune appears to be a smooth, continuous material. Up close, however, the dune reveals itself to be made of tiny grains of sand. In a similar fashion, everything around us—no matter how smooth it may appear—is made of the basic units you know as *atoms*. Atoms are so small, however, that a single grain of sand contains on the order of 125 million trillion of them. There are roughly 250,000 times more atoms in a single grain of sand than there are grains of sand in the dunes shown in Figure 17.4.

As small as atoms are, there is much we have learned about them. We know, for example, that there are more than 100 different types of atoms, and they are listed in the widely recognized periodic table. Some atoms link together to form larger but still incredibly small basic units of matter called **molecules.** As shown in Figure 17.4, for example, two hydrogen atoms and one oxygen atom link together to form a single molecule of water, which you know as H_2O. Water molecules are so small that an 8-ounce glass of water contains about a trillion trillion of them.

Our world can be studied at different levels of magnification. At the *macroscopic* level, matter is large enough to be seen, measured, and

FIGURE 17.4 ▲
There are far more atoms in a glass of water than there are grains of sand within this towering sand dune.

Oxygen atom

Hydrogen atoms

Water molecule, H_2O

FIGURE 17.5 ▶
The familiar bulk properties of a solid, a liquid, and a gas. (a) The submicroscopic particles of the solid phase vibrate about fixed positions. (b) The submicroscopic particles of the liquid phase slip past one another. (c) The fast-moving submicroscopic particles of the gaseous phase are separated by large average distances.

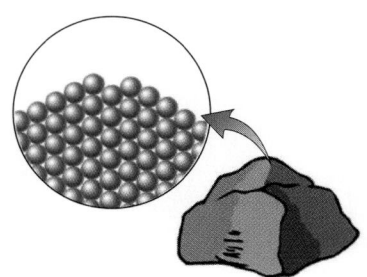

(a) Solid

handled. A handful of sand and a glass of water are macroscopic samples of matter. At the *microscopic* level, physical structure is so fine that it can be seen only with a microscope. A biological cell is microscopic, as is the detail on a dragonfly's wing. Beyond the microscopic level is the **submicroscopic**—the realm of atoms and molecules and an important focus of chemistry.

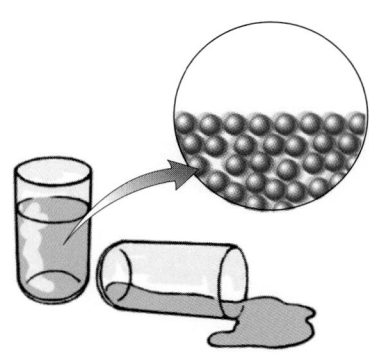

(b) Liquid

Phases of Matter

Matter exists in *phases*, which may be either solid, liquid, or gas. ✔ At the submicroscopic level, solid, liquid, and gaseous phases are distinguished by how the submicroscopic particles hold together. This is illustrated in Figure 17.5. In **solid** matter, such as rock, the attractions between particles are strong enough to hold all the particles together in some fixed three-dimensional arrangement. The particles are able to vibrate back and forth, but they cannot move past one another.

The addition of heat causes these vibrations to increase until, at a certain temperature, the vibrations are rapid enough to break up the fixed arrangements. Rock will melt into magma (a topic of much discussion in Chapter 25). Likewise, ice will melt into water. The particles can then slip past one another and tumble around much like a bunch of marbles tumbling within a bag. This is the **liquid** phase of matter, and it is the mobility of the submicroscopic particles that gives rise to the liquid's fluid character—its ability to flow and to assume the shape of its container.

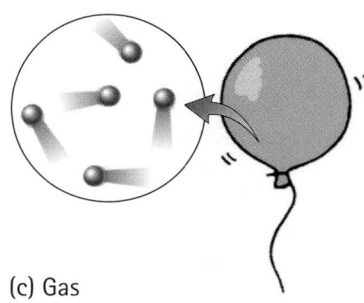

(c) Gas

Further heating causes the submicroscopic particles in a liquid to move so fast that the attractions they have for one another are unable to hold them together. They then separate from one another, forming a **gas.** For magma, this doesn't easily happen, because the particles are strongly attracted to one another. For water, molecules will separate into a gas at 100°C. For a substance like helium, the submicroscopic particles are already in the gaseous phase at room temperature.

Moving at an average speed of 500 meters per second (1100 miles per hour), the particles of a gas are widely separated from one another. Matter in the gaseous phase therefore occupies much more volume than it does in the solid or liquid phase. Applying pressure to a gas squeezes the gas particles closer together, which decreases the volume. The amount of air an underwater diver needs to breathe for many minutes, for example, can be squeezed (compressed) into a tank small enough to be carried on the diver's back.

There are other phases of matter besides solid, liquid, and gas. This includes *plasma*, which is an electrically charged gas. The Sun is made of plasma, as is the inside of a glowing fluorescent lightbulb.

READING CHECK

What determines the phase of a material?

17.3 The Phase of Matter Can Change

✓ In order to change the phase of a substance, you must either add heat or remove heat. This is illustrated in Figure 17.6.

The process of a solid transforming to a liquid is called **melting.** To visualize what happens when heat begins to melt a solid, imagine you are holding hands with a group of people and each of you starts jumping around randomly. The more violently you jump, the more difficult it is to hold onto one another. If everyone jumps violently enough, keeping hold is impossible. Something like this happens to the submicroscopic particles of a solid when it is heated. As heat is added to the solid, the particles vibrate more and more violently. If enough heat is added, the attractive forces between the particles are no longer able to hold them together. The solid melts.

A liquid can be changed to a solid by the removal of heat. This process is called **freezing,** and it is the reverse of melting. As heat is withdrawn from the liquid, particle motion diminishes until the particles, on average, are moving slowly enough for attractive forces between them to take permanent hold. The only motion the particles are capable of then is vibration about fixed positions, which means the liquid has solidified, or frozen.

A liquid can be heated so that it becomes a gas—a process called **evaporation.** As heat is added, the particles move faster. Particles at the liquid surface eventually gain enough energy to jump out of the liquid and enter the air. In other words, they enter the gaseous phase. As more

If you can adequately explain an idea to others, then you have passed the ultimate test of understanding that idea.

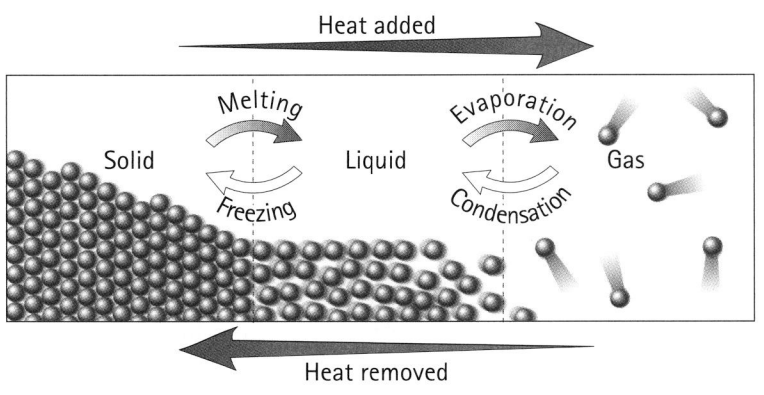

FIGURE 17.6 ▶
Melting and evaporation involve the addition of heat; condensation and freezing involve the removal of heat.

and more particles absorb the heat being added, they too acquire enough energy to escape from the liquid surface and become gas particles. Because a gas results from evaporation, this phase is also sometimes referred to as *vapor*. Water in the gaseous phase, for example, may be referred to as *water vapor*.

The rate at which a liquid evaporates increases with temperature. A puddle of water, for example, evaporates from hot pavement more quickly than it does from your cool kitchen floor. When the temperature is hot enough, evaporation occurs beneath the surface of the liquid. As a result, bubbles form and are buoyed up to the surface. We say that the liquid is **boiling.** A substance is often characterized by its *boiling point,* which is the temperature at which it boils. At sea level, the boiling point of fresh water is 100°C.

The transformation from gas to liquid—the reverse of evaporation—is called **condensation.** This process can occur when the temperature of a gas decreases. The water vapor held in the warm daylight air, for example, may condense to form a wet dew in the cool of the night.

Note that the underlying cause of phase changes is always the transfer of energy. Energy must be added to melt ice into water or to vaporize water. Energy must be removed to condense water vapor back into water or freeze water into ice. The amount of energy needed to change any substance from solid to liquid (and vice versa) is called the **heat of fusion** for the substance. For water, this is 334 joules per gram. And the amount of energy required to change any substance from liquid to gas (and vice versa) is called the **heat of vaporization** for the substance. For water, this is a whopping 2256 joules per gram, which is much larger than its heat of fusion. Water's high heat of vaporization allows you to briefly touch your wetted finger to a hot skillet on a hot stove without harm. As long as your finger remains wet, energy that ordinarily would burn your finger goes instead into changing the phase of the moisture on your finger.

Water vapor is invisible. Upon cooling, however, it condenses into microscopic droplets of liquid, which we see as a cloud. Your breath contains lots of water vapor, which explains why your breath becomes visible when you breathe outside on a cold day.

Once liquid water starts boiling, its temperature remains constant no matter how strong the flame heating the water. The flame's heat is being used to vaporize the water. So, can you add heat to water without changing its temperature? Yes, when the heat is being used to change water's phase.

CONCEPT CHECK
Can you add heat to ice without melting it?

Check Your Answer
Ice can have any temperature below 0°C, down to absolute zero, −273°C. Adding heat to −200°C ice, for example, may raise its temperature to a greater −100°C. As long as its temperature stays below 0°C, the ice does not melt.

✔ **READING CHECK**

What must be added or taken away in order to change the phase of a substance?

17.4 Matter Has Physical and Chemical Properties

Properties that describe the look or feel of a substance, such as color, hardness, density, texture, and phase, are called **physical properties.** Every substance has its own set of characteristic physical properties that we can use to identify that substance (Figure 17.7).

Gold
Opacity: opaque
Color: yellowish
Phase at 25˚C: solid
Density: 19.3 g/mL

Diamond
Opacity: transparent
Color: colorless
Phase at 25˚C: solid
Density: 3.5 g/mL

Water
Opacity: transparent
Color: colorless
Phase at 25˚C: liquid
Density: 1.0 g/mL

◀ **FIGURE 17.7**
Gold, diamond, and water can be identified by their physical properties. If a substance has all the physical properties listed under gold, for example, it must be gold.

The physical properties of a substance can change when conditions change, but that does not mean a different substance is created. Cooling liquid water to below 0°C causes the water to change to solid ice, but the substance is still water, no matter what the phase. The only difference is how the H_2O molecules are arranged. In the liquid phase, the water molecules tumble around one another, whereas in the ice phase, they vibrate about fixed positions. The freezing of water is an example of what chemists call a physical change. During a **physical change,** a substance changes its phase or some other physical property, but not its chemical identity, as Figure 17.8 shows.

CONCEPT CHECK
The melting of gold is a physical change. Why?

Check Your Answer
During a physical change, a substance changes only one or more of its physical properties; its chemical identity does not change. Because melted gold is still gold but in a different form, its melting is only a physical change.

A **chemical property** characterizes the ability of a substance to react with other substances or to transform from one substance to another. Figure 17.9 shows three examples. The methane of natural gas has the chemical property of reacting with oxygen to produce carbon dioxide and water, along with appreciable heat energy. Similarly, it is a chemical property of baking soda to react with vinegar to produce carbon dioxide and water while absorbing a small amount of heat energy. Copper has the chemical property of reacting with carbon dioxide and water to form a greenish-blue solid known as patina. Copper statues exposed to the carbon dioxide and water in the air become coated with patina. The patina is not copper, it is not carbon dioxide, and it is not water. It is a new substance formed by the reaction of these chemicals with one another.

✔ During a chemical change, there is a change in the way the atoms are *chemically bonded* to one another. A **chemical bond** is the force of attraction between two atoms that holds them together. A methane molecule, for example, is made of a single carbon atom bonded to four hydrogen atoms, and an oxygen molecule is made of two oxygen atoms bonded to each other. Figure 17.10 shows the chemical change in which the atoms in a methane molecule and those in two oxygen molecules first pull apart and then form new bonds with different partners, resulting in the formation of molecules of carbon dioxide and water.

Any change in a substance that involves a rearrangement of the way atoms are bonded is called a **chemical change.** Thus the transformation

A chemical property of a substance is its tendency to change into another substance. For example, it is a chemical property of iron to transform into rust.

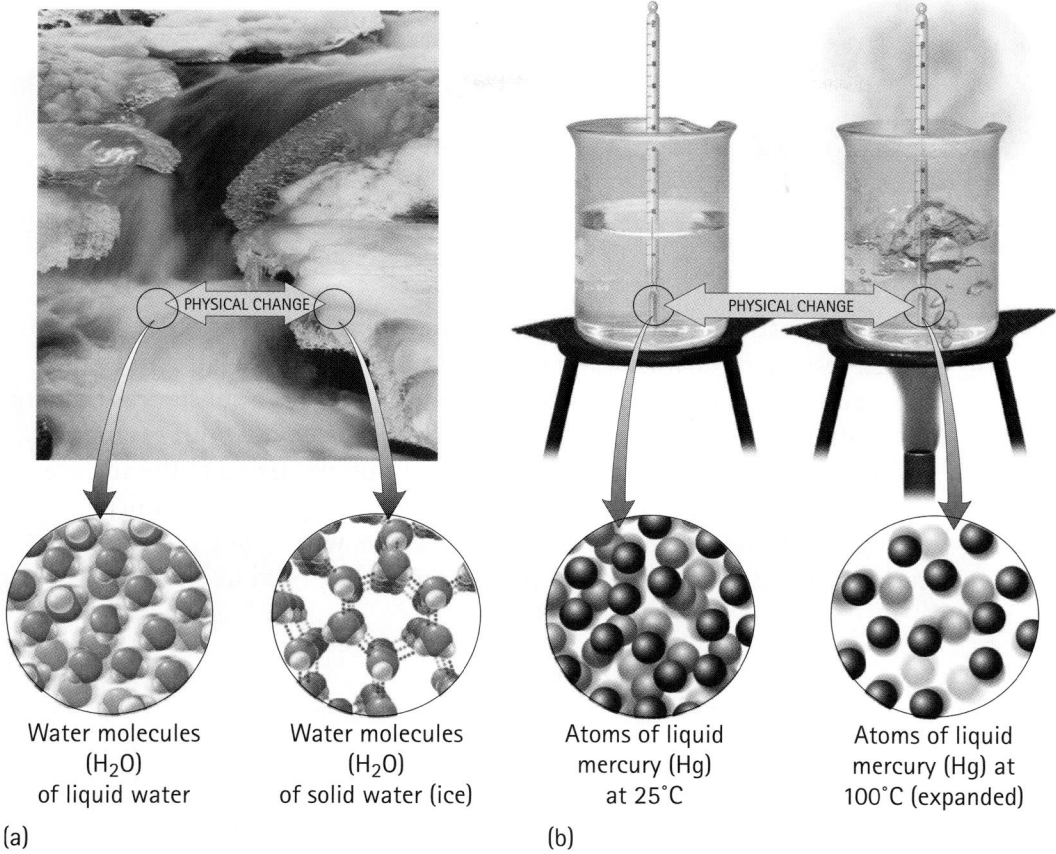

Water molecules
(H₂O)
of liquid water

Water molecules
(H₂O)
of solid water (ice)

Atoms of liquid
mercury (Hg)
at 25°C

Atoms of liquid
mercury (Hg) at
100°C (expanded)

(a)   (b)

FIGURE 17.8 ▲
Two physical changes. (a) Liquid water and ice might appear to be different substances, but, at the submicroscopic level, it is evident that both consist of water molecules. (b) At 25°C, the atoms in a sample of mercury are a certain distance apart, yielding a density of 13.5 grams per milliliter (g/mL). At 100°C, the atoms are farther apart, meaning that each milliliter now contains fewer atoms than at 25°C, and the density is now 13.4 g/mL. The physical property we call density has changed with temperature, but the identity of the substance remains unchanged: Mercury is mercury.

of methane to carbon dioxide and water is a chemical change, as are the other two transformations shown in Figure 17.9. The chemical change shown in Figure 17.11 occurs when an electric current is passed through water. The energy of the current causes the bonds holding atoms together

There are many beautiful chemical changes going on in a campfire.

Methane
Reacts with oxygen to form carbon dioxide and water, giving off lots of heat during the reaction.

Baking soda
Reacts with vinegar to form carbon dioxide and water, absorbing heat during the reaction.

Copper
Reacts with carbon dioxide and water to form the greenish-blue substance called patina.

◄ **FIGURE 17.9**
The chemical properties of substances allow them to transform to new substances. Natural gas and baking soda transform to carbon dioxide and water. Copper transforms to patina.

FIGURE 17.10 ▶
The chemical change in which molecules of methane and oxygen transform to molecules of carbon dioxide and water, as atoms break old bonds and form new ones.

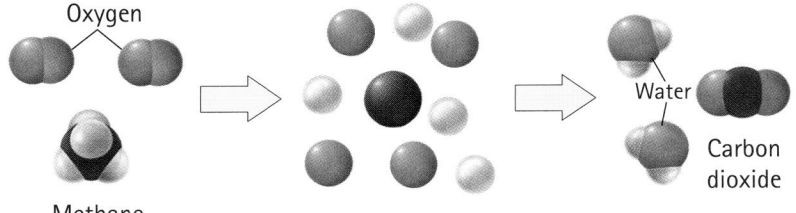

Oxygen

Methane

Water

Carbon dioxide

to break apart. Loose atoms then form new bonds with different atoms, which results in the formation of new molecules. Thus, water molecules are changed to molecules of hydrogen and oxygen, two substances that are very different from water. The hydrogen and oxygen are both gases at room temperature, and they can be seen as bubbles rising to the surface.

In the language of chemistry, materials undergoing a chemical change are said to be *reacting*. Methane reacts with oxygen to form carbon dioxide and water. Water reacts when exposed to electricity

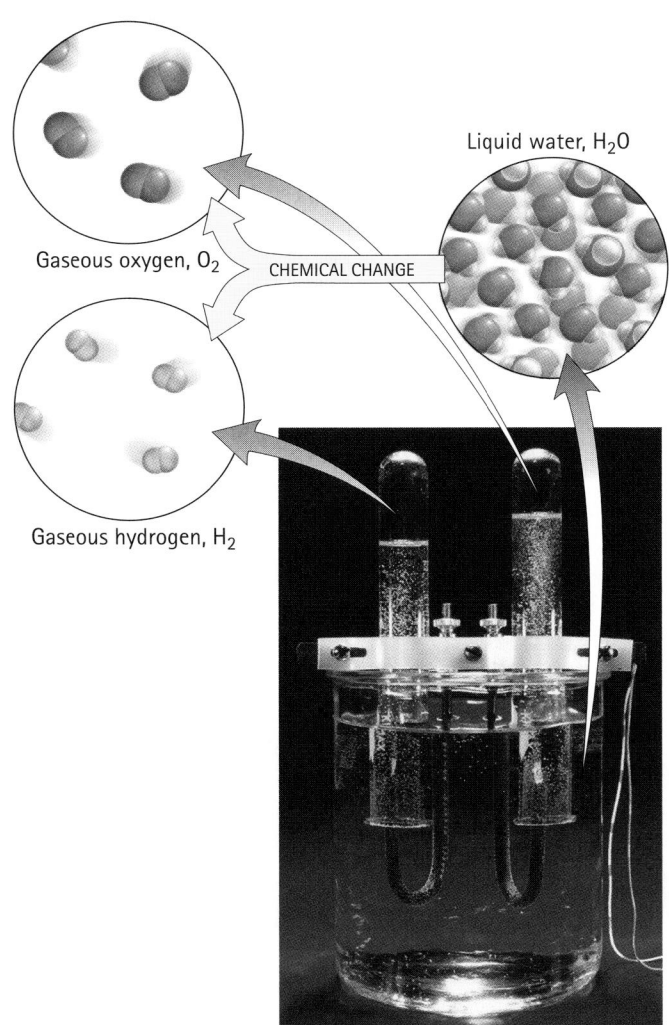

Liquid water, H_2O

Gaseous oxygen, O_2

CHEMICAL CHANGE

Gaseous hydrogen, H_2

FIGURE 17.11 ▶
Water can be transformed to hydrogen gas and oxygen gas by applying the energy of an electric current. This is a chemical change, because new materials (the two gases) are formed as the atoms originally found in the water molecules are rearranged.

to form hydrogen gas and oxygen gas. Thus, the term *chemical change* means the same thing as *chemical reaction*. During a **chemical reaction,** new materials are formed by a change in the way atoms are bonded together. We will explore chemical bonds and their role in chemical reactions when we get to chapters Chapters 18 and 20.

✔ **READING CHECK**

What happens to the atoms within a molecule under-going a chemical reaction?

CONCEPT CHECK

Each sphere in the following diagrams represents an atom. Joined spheres represent molecules. One set of diagrams shows a physical change, and the other shows a chemical change. Which is which?

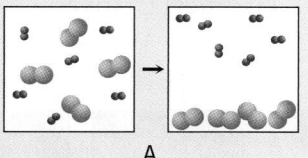 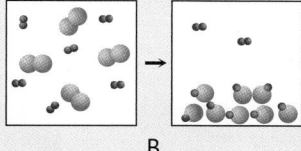

A   B

Check Your Answer

In set A, the molecules before and after the change are the same. They differ only in their positions. Set A, therefore, represents only a physical change. In set B, new molecules, consisting of bonded red and blue spheres, appear after the change. These molecules represent a new material, and so set B represents a chemical change.

Wow, lotta terms: macroscopic, microscopic, submicroscopic, physical properties, physical changes, chemical properties, chemical changes, chemical bonds, chemical reactions . . . hang in there!

17.5 Determining Physical and Chemical Changes Can Be Difficult

✔ After a physical change, the molecules are the same as the ones you started with. After a chemical change, the original molecules have been destroyed and new ones are in their place. In both cases, however, there is a change in physical appearance. Frozen water and melted water, for example, look very different. Likewise, iron and rust look very different.

◀ **FIGURE 17.12**
The transformation of water to ice and the transformation of iron to rust both involve changes in physical appearance. The formation of ice is a physical change, whereas the formation of rust is a chemical change.

Physical change? Chemical change? It's not always easy to tell the difference between the two.

So how can you quickly determine whether an observed change is physical or chemical? After all, we can't see the individual molecules.

There are two powerful guidelines that can help you distinguish between physical and chemical changes. First, in a physical change, a change in appearance is the result of a new set of conditions imposed on the same material. Restoring the original conditions restores the original appearance: Frozen water melts upon warming. Second, in a chemical change, a change in appearance is the result of the formation of a new material that has its own unique set of physical properties. The more evidence you have suggesting that a different material has been formed, the greater the likelihood that the change is a chemical change. Iron is a material that can be used to build cars. Rust is not. This suggests that the rusting of iron is a chemical change.

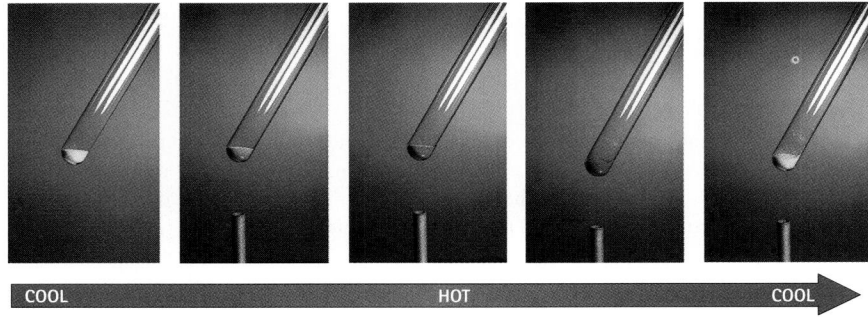

FIGURE 17.13 ▶
Potassium chromate changes color as its temperature changes. This change in color is a physical change. A return to the original temperature restores the original bright yellow color.

Figure 17.13 shows potassium chromate, a material whose color depends on its temperature. At room temperature, potassium chromate is a bright canary yellow. At higher temperatures, it is a deep reddish orange. Upon cooling, the canary color returns, suggesting that the change is physical. With a chemical change, reverting to the original conditions does not restore the original appearance. Ammonium dichromate, shown in Figure 17.14, is an orange material that, when heated, explodes into ammonia, water vapor, and green chromium oxide. When the test tube is returned to the original temperature, there is no trace of orange ammonium dichromate. In its place are new substances having completely different physical properties.

How does a physical change differ from a chemical change?

FIGURE 17.14 ▶
When heated, orange ammonium dichromate undergoes a chemical change to ammonia, water vapor, and chromium(III) oxide. A return to the original temperature does not restore the orange color, because the ammonium dichromate is no longer there.

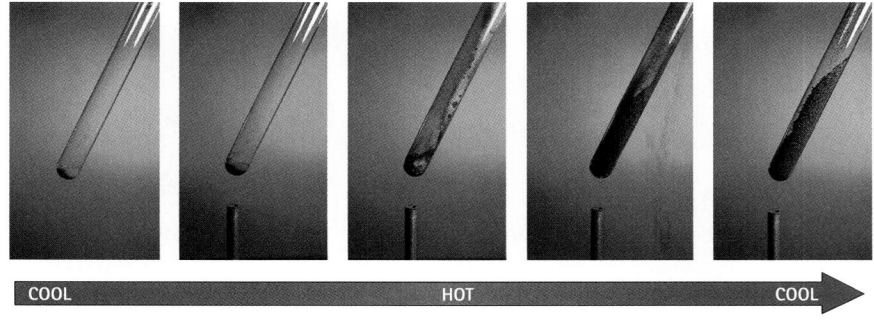

CONCEPT CHECK

Evan has grown an inch in height over the past year. Is this best described as a physical or a chemical change?

Check Your Answer

Are new materials being formed as Evan grows? Absolutely—created out of the food he eats. His body is very different from, say, the spaghetti he ate yesterday. Yet, through some very advanced chemistry, his body is able to absorb the atoms of that spaghetti and rearrange them into new materials. Biological growth, therefore, is best described as a chemical change.

It's okay to learn a little now, and to entrust a lot that remains to others who chose to specialize within this field.

17.6 The Periodic Table Helps Us to Understand the Elements

As was discussed in Chapter 15, the periodic table is a listing of all the known elements with their atomic masses, atomic numbers, and atomic symbols. But there is much more information in this table. The way the elements are organized tells us a lot about the elements' properties. Let's look at how the elements are grouped as metals, nonmetals, and metalloids.

As shown in Figure 17.15, most of the known elements are metals, which are defined as those elements that are shiny, opaque, and good conductors of electricity and heat. ✓ Metals are *malleable,* which means they can be hammered into different shapes or bent without breaking. They are also *ductile,* which means they can be drawn into wires. All but a few metals are solid at room temperature. The exceptions include mercury, Hg; gallium, Ga; cesium, Cs; and francium, Fr, which are all liquids at a warm room temperature of 30°C (86°F). Another interesting exception is hydrogen, H, which takes on the properties of a liquid metal only at very high pressures (Figure 17.16). Under normal conditions, hydrogen behaves as a nonmetallic gas.

The nonmetallic elements, with the exception of hydrogen, are on the right-hand side of the periodic table. Nonmetals are very poor conductors of electricity and heat, and may also be transparent. Solid nonmetals are neither malleable nor ductile. Rather, they are brittle and shatter when hammered. At 30°C (86°F), some nonmetals are solid (carbon, C), others are liquid (bromine, Br), and still others are gaseous (helium, He).

Six elements are classified as metalloids: boron, B; silicon, Si; germanium, Ge; arsenic, As; antimony, Sb; and tellurium, Te. Situated between the metals and the nonmetals in the periodic table, the metalloids have both metallic and nonmetallic characteristics. For example, these elements are weak conductors of electricity, which makes them useful as semiconductors in the integrated circuits of computers. Note

Please put to rest any fear you may have about needing to memorize the periodic table, or even parts of it—better to focus on the many great concepts behind its organization.

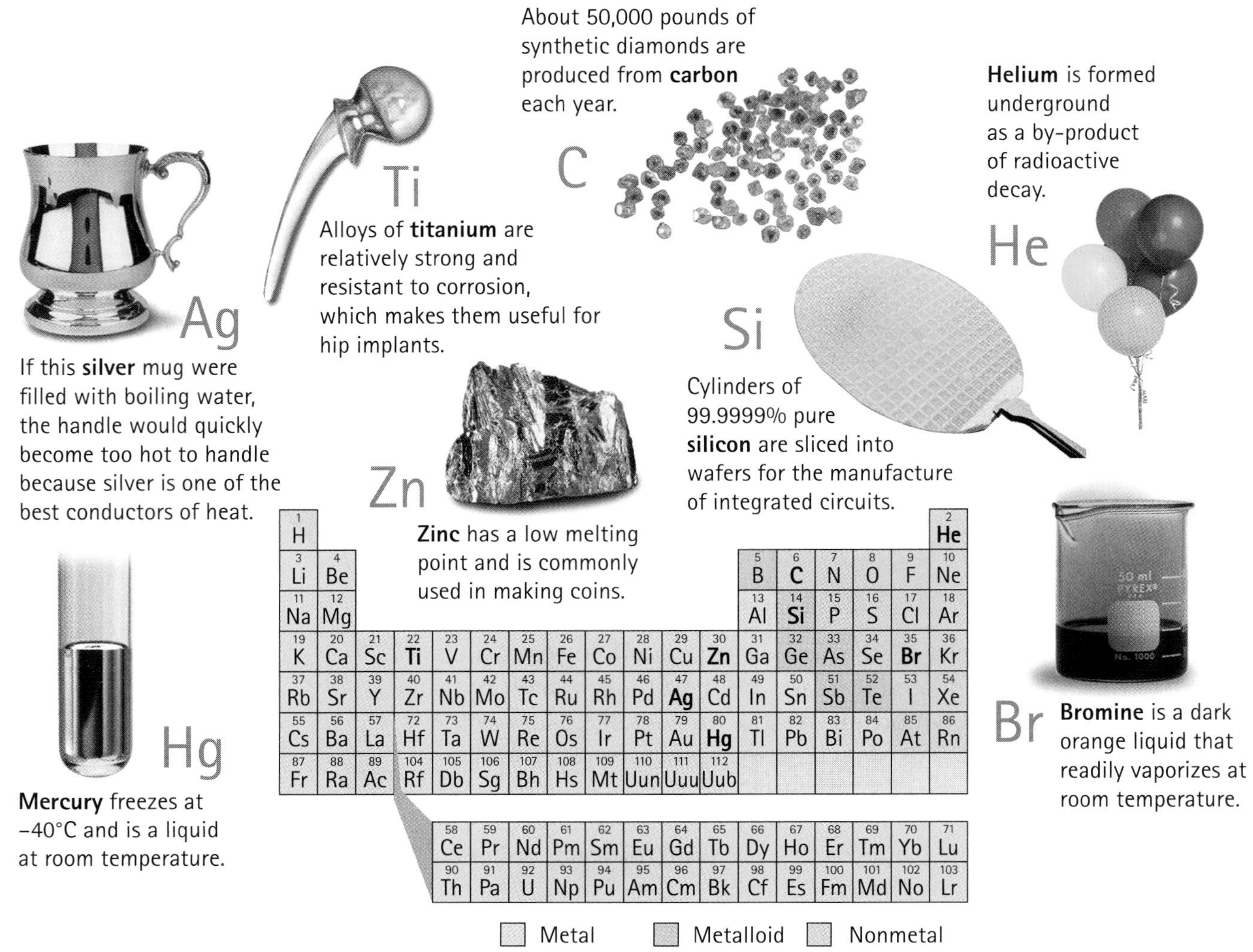

About 50,000 pounds of synthetic diamonds are produced from **carbon** each year.

C

Helium is formed underground as a by-product of radioactive decay.

He

Ti

Alloys of **titanium** are relatively strong and resistant to corrosion, which makes them useful for hip implants.

Ag

If this **silver** mug were filled with boiling water, the handle would quickly become too hot to handle because silver is one of the best conductors of heat.

Si

Cylinders of 99.9999% pure **silicon** are sliced into wafers for the manufacture of integrated circuits.

Zn

Zinc has a low melting point and is commonly used in making coins.

Hg

Mercury freezes at −40°C and is a liquid at room temperature.

Br

Bromine is a dark orange liquid that readily vaporizes at room temperature.

1 H																	2 He
3 Li	4 Be											5 B	6 C	7 N	8 O	9 F	10 Ne
11 Na	12 Mg											13 Al	14 Si	15 P	16 S	17 Cl	18 Ar
19 K	20 Ca	21 Sc	22 Ti	23 V	24 Cr	25 Mn	26 Fe	27 Co	28 Ni	29 Cu	30 Zn	31 Ga	32 Ge	33 As	34 Se	35 Br	36 Kr
37 Rb	38 Sr	39 Y	40 Zr	41 Nb	42 Mo	43 Tc	44 Ru	45 Rh	46 Pd	47 Ag	48 Cd	49 In	50 Sn	51 Sb	52 Te	53 I	54 Xe
55 Cs	56 Ba	57 La	72 Hf	73 Ta	74 W	75 Re	76 Os	77 Ir	78 Pt	79 Au	80 Hg	81 Tl	82 Pb	83 Bi	84 Po	85 At	86 Rn
87 Fr	88 Ra	89 Ac	104 Rf	105 Db	106 Sg	107 Bh	108 Hs	109 Mt	110 Uun	111 Uuu	112 Uub						

58 Ce	59 Pr	60 Nd	61 Pm	62 Sm	63 Eu	64 Gd	65 Tb	66 Dy	67 Ho	68 Er	69 Tm	70 Yb	71 Lu
90 Th	91 Pa	92 U	93 Np	94 Pu	95 Am	96 Cm	97 Bk	98 Cf	99 Es	100 Fm	101 Md	102 No	103 Lr

☐ Metal   ☐ Metalloid   ☐ Nonmetal

FIGURE 17.15 ▲
The periodic table color-coded to show metals, nonmetals, and metalloids.

✔ READING CHECK

Name two physical properties of metals.

from the periodic table how germanium, Ge (number 32), is closer to the metals than to the nonmetals. Because of this positioning, we can deduce that germanium has more metallic properties than silicon, Si (number 14), and is a slightly better conductor of electricity. So we find that integrated circuits fabricated with germanium operate faster than those fabricated with silicon. Because silicon is much more abundant and less expensive to obtain, however, silicon computer chips remain the industry standard.

◀ **FIGURE 17.16**
Geoplanetary models suggest that hydrogen exists as a liquid metal deep beneath the surfaces of Jupiter (shown here) and Saturn. These planets are composed mostly of hydrogen. Inside them, the pressure exceeds 3 million times the Earth's atmospheric pressure. At this tremendously high pressure, hydrogen is pressed to a liquid-metal phase. Back here on the Earth at our relatively low atmospheric pressure, hydrogen exists as a nonmetallic gas.

Periods and Groups

Two other important ways in which the elements are organized in the periodic table are by horizontal rows and vertical columns. As discussed in Section 15.2, each horizontal row is called a *period,* and each vertical column is called a *group* (or sometimes a *family*). As was shown in Figure 15.9, there are 7 periods and 18 groups.

Across any period, the properties of elements gradually change. This gradual change is called a *periodic trend*. Figure 15.10 in Section 15.2 shows how the size of atoms increases as you move from the lower left to the upper right corners of the periodic table. Another important periodic trend is the ease with which an atom loses an electron. Some atoms lose an electron very easily, while others hold onto their electrons much more tightly. As shown in Figure 17.17, atoms of the lower left-hand side of the periodic table lose an electron quite easily. For atoms to the upper right, however, losing an electron is most difficult.

The *carat* is the common unit used to describe the mass of a gem. A 1-carat diamond, for example, has a mass of 0.20 grams. The *karat* is the common unit used to describe the purity of a precious metal, such as gold. A 24-karat gold ring is as pure as can be. A gold ring that is 50% pure is 12 karat.

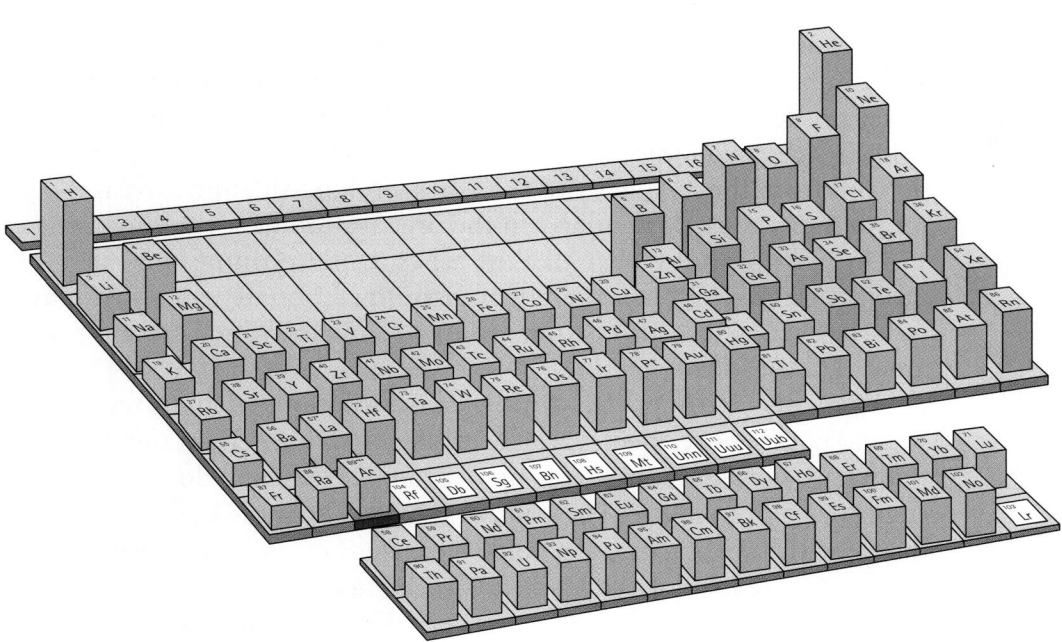

FIGURE 17.17 ▲
The lower the height of the element shown above, the easier it is for it to lose an electron. Atoms to the lower left have only a weak hold on their electrons, which are easily lost. Atoms to the upper right, however, hold onto their electrons much more strongly.

CONCEPT CHECK

Which has a stronger hold on its electrons: a francium atom, Fr, or a helium atom, He?

Check Your Answer

According to Figure 17.17, the francium atom loses an electron much more easily than does a helium atom. The helium atom, therefore, must have a much stronger hold on its electrons.

There's more silicon than germanium in your iPod.

FIGURE 17.18 ▶
The common names for various groups of elements.

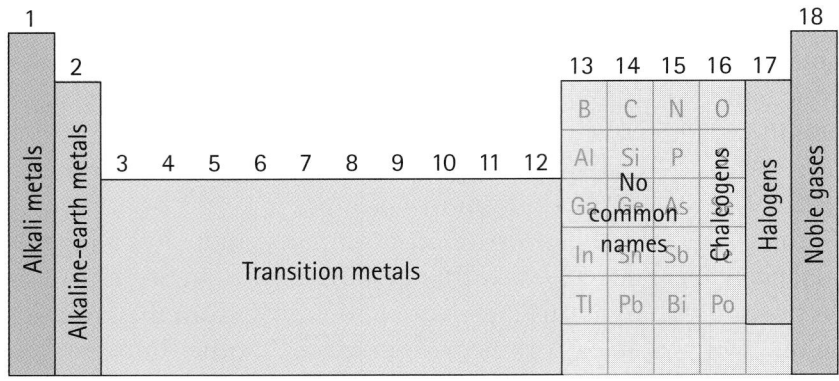

Down any group (vertical column), the properties of elements tend to be remarkably similar, which is why these elements are said to be "grouped" or "in a family." As Figure 17.18 shows, several groups have traditional names that describe the properties of their elements. Early in human history, people discovered that ashes mixed with water produce a slippery solution useful for removing grease. By the Middle Ages, such mixtures were described as being alkaline, a term derived from the Arabic word for ashes, *al-qali*. Alkaline mixtures found many uses, particularly in the preparation of soaps (Figure 17.19). We now know that alkaline ashes contain compounds of group 1 elements, most notably potassium carbonate, also known as potash. Because of this history, group 1 elements, which are metals, are called the *alkali metals.*

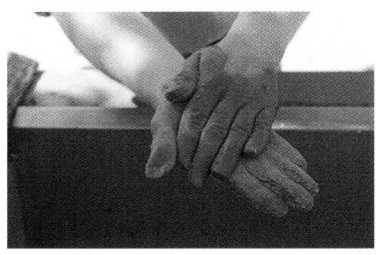

FIGURE 17.19 ▲
Ashes and water make a slippery alkaline solution once used to clean hands.

Elements of group 2 also form alkaline solutions when mixed with water. Furthermore, medieval alchemists noted that certain minerals (which we now know are made up of group 2 elements) do not melt or change when put in fire. These fire-resistant substances were known to the alchemists as "earth." As a holdover from these ancient times, group 2 elements are known as the *alkaline-earth metals.*

Over toward the right side of the periodic table elements of group 16 are known as the *chalcogens* ("ore-forming" in Greek) because the top two elements of this group, oxygen and sulfur, are so commonly found in ores. Elements of group 17 are known as the *halogens* ("salt-forming" in Greek) because of their tendency to form various salts. Group 18 elements are all unreactive gases that tend not to combine with other elements. For this reason, they are called the *noble gases*, presumably because the nobility of earlier times were above interacting with common folk.

The elements of groups 3 through 12 are all metals that do not form alkaline solutions with water. These metals tend to be harder than the alkali metals and less reactive with water; hence they are used for structural purposes. Collectively, they are known as the *transition metals,* a name that denotes their central position in the periodic table. ✔ The transition metals include some of the most familiar and important elements—iron, Fe; copper, Cu; nickel, Ni; chromium, Cr; silver, Ag; and gold, Au. They also include many lesser-known elements that are nonetheless important in

A uranium atom is 40 times heavier than a lithium atom, but only slightly larger in size because its nucleus pulls harder on its electrons.

modern technology. Persons with hip implants appreciate the transition metals titanium, Ti; molybdenum, Mo; and manganese, Mn, because these noncorrosive metals are used in implant devices.

CONCEPT CHECK

The elements copper, Cu, silver, Ag, and gold, Au, are three of the few metals that can be found naturally in their elemental state. These three metals have found great use as currency and jewelry for a number of reasons, including their resistance to corrosion and their remarkable colors. How is the fact that these metals have similar properties reflected in the periodic table?

Check Your Answers

Copper (number 29), silver (number 47), and gold (number 79) are all in the same group in the periodic table (group 11), which suggests they should have similar—though not identical—properties.

Atoms of elements 112 and higher have been created by colliding lighter atoms together at extremely high energies. These ultramassive atoms are unstable and exist for only fractions of a second.

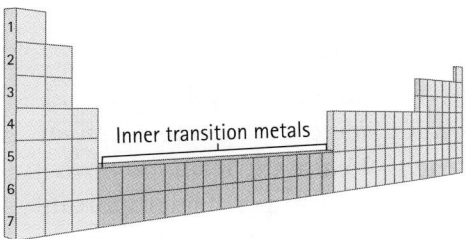

◀ **FIGURE 17.20**
Inserting the inner transition metals between atomic groups 3 and 4 results in a periodic table that is not easy to fit on a standard sheet of paper.

Within the sixth period is a subset of 14 metallic elements (numbers 58 to 71) that are quite unlike any of the other transition metals. A similar subset (numbers 90 to 103) is found within the seventh period. These two subsets are the *inner transition metals*. Inserting the inner transition metals into the main body of the periodic table, as in Figure 17.20, results in a long and cumbersome table. So that the table can fit nicely on a standard paper size, these elements are commonly placed below the main body of the table, as shown in Figure 17.21.

 **READING CHECK**

List several common transition metals.

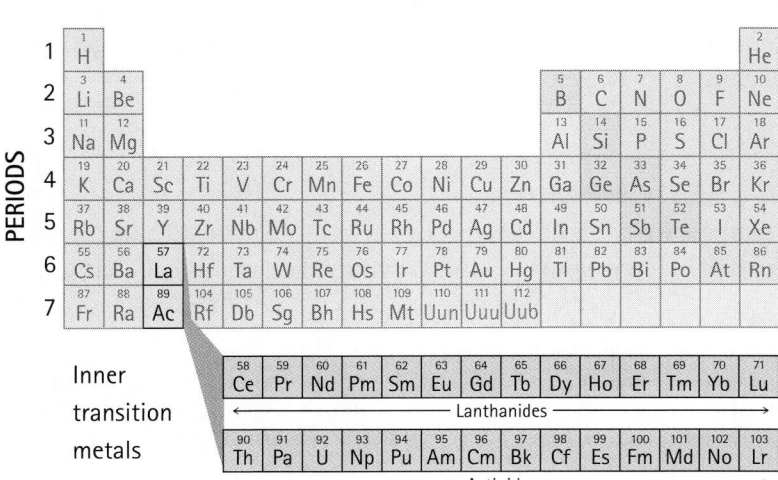

◀ **FIGURE 17.21**
The typical display of the inner transition metals. The count of elements in the sixth period goes from lanthanum (La, 57) to cerium (Ce, 58) on through to lutetium (Lu, 71) and then back to hafnium (Hf, 72). A similar jump is made in the seventh period.

Keep hanging in there! We extend our vocabulary to period, group, periodic trend, chalcogens, halogens, noble gases, transition metals, inner transition metals, lanthanides, and actinides. Now do you see why the fewer terms of physics are taught first in this book?

The sixth-period inner transition metals are called the lanthanides because they fall after lanthanum, La. Because of their similar physical and chemical properties, they tend to occur mixed together in the same locations on Earth. Also because of their similarities, lanthanides are unusually difficult to purify. Recently, the commercial use of lanthanides has increased. Several lanthanide elements, for example, are used in the fabrication of the light-emitting diodes (LEDs) of computer monitors and flat-screen televisions.

The seventh-period inner transition metals are called the actinides because they fall after actinium, Ac. They, too, all have similar properties and hence are not easily purified. The nuclear power industry faces this obstacle because it requires purified samples of two of the most publicized actinides: uranium, U, and plutonium, Pu. Actinides heavier than uranium are not found in nature but are synthesized in the laboratory.

17.7 Elements Can Combine to Form Compounds

As briefly described in Chapter 15, the terms *element* and *atom* are often used in a similar context. You might hear, for example, that gold is an element made of gold atoms. Generally, *element* is used to mean an entire macroscopic or microscopic sample. We use the term *atom* when speaking of the submicroscopic particles in the sample. The important distinction is that elements are made of atoms and not the other way around.

The fundamental unit of an element is indicated by its **elemental formula.** For elements in which the fundamental units are individual atoms, the elemental formula is simply the chemical symbol: Au is the elemental formula for gold, and Li is the elemental formula for lithium, to name just two examples. For some elements, atoms are bonded into molecules. In such cases, the elemental formula is the chemical symbol followed by a subscript indicating the number of atoms in each molecule. For example, elemental nitrogen consists of molecules containing two nitrogen atoms per molecule. Thus, N_2 is the elemental formula for nitrogen. Similarly, O_2 is the elemental formula for the oxygen we breathe, and S_8 is the elemental formula for sulfur.

CONCEPT CHECK

The oxygen we breathe, O_2, is converted to ozone, O_3, in the presence of an electric spark. Is this a physical or chemical change?

Check Your Answer

When atoms regroup, the result is an entirely new substance, and that is what happens here. The oxygen we breathe, O_2, is odorless and life-giving. Ozone, O_3, is toxic, and it has a pungent smell commonly associated with electric motors. The conversion of O_2 to O_3 is therefore a chemical change. However, both O_2 and O_3 are elemental forms of oxygen.

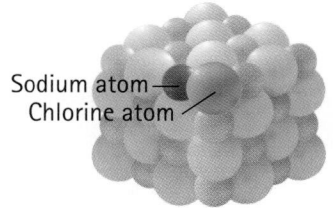

✔ When atoms of *different* elements bond to one another, they make a **compound.** Sodium atoms and chlorine atoms, for example, bond to make the compound sodium chloride, commonly known as table salt. Nitrogen atoms and hydrogen atoms join to make the compound ammonia, which is a common household cleaner.

A compound is represented by its **chemical formula,** in which the symbols for the elements are written together. The chemical formula for sodium chloride is NaCl, and that for ammonia is NH_3. Numerical subscripts indicate the ratio in which the atoms combine. By convention, the subscript 1 is understood and omitted. So the chemical formula NaCl tells us that in the compound sodium chloride, there is one sodium atom for every chlorine atom. The chemical formula NH_3 tells us that in the compound ammonia, there is one nitrogen atom for every three hydrogen atoms, as Figure 17.22 shows.

Compounds have physical and chemical properties that are completely different from the properties of their elemental parts. The sodium chloride, NaCl, shown in Figure 17.23 is very different from the elemental sodium and the elemental chlorine used in its formation. Elemental sodium, Na, consists of nothing but sodium atoms, which form a soft, silvery metal that can be cut easily with a knife. Its melting point is 97.5°C, and it reacts violently with water. Elemental chlorine, Cl_2, consists of chlorine molecules. This material, a yellow-green gas at room temperature, is very toxic, and it was used as a chemical warfare agent during World War I. Its boiling point is –34°C. The compound sodium chloride, NaCl, is a see-through, brittle, colorless crystal having a melting point of 800°C. Sodium chloride does not react chemically with water the way sodium does. It is not toxic like chlorine—in fact, sodium chloride is an essential nutrient for all living organisms. Sodium chloride is not sodium, nor is it chlorine; it is uniquely sodium chloride, a tasty chemical when sprinkled lightly over popcorn.

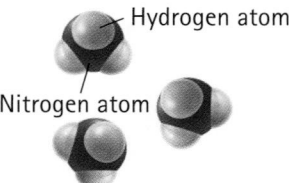

Sodium atom
Chlorine atom

Sodium chloride, NaCl

Hydrogen atom

Nitrogen atom

Ammonia, NH_3

FIGURE 17.22 ▲
The compounds sodium chloride and ammonia are represented by their chemical formulas, NaCl and NH_3. A chemical formula shows the ratio of atoms that constitute the compound.

 **READING CHECK**

Is a compound made from atoms of the same element or of different elements?

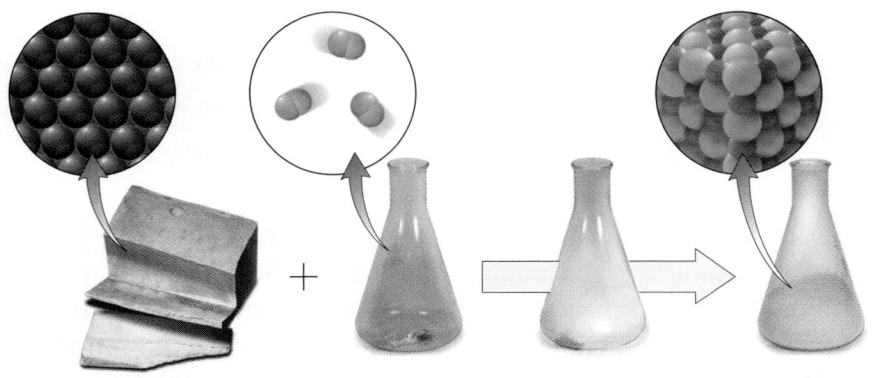

Sodium metal   and   chlorine gas   react to form   sodium chloride

FIGURE 17.23 ▲
Sodium metal and chlorine gas react together to form sodium chloride. Although the compound sodium chloride is made of sodium and chlorine, the physical and chemical properties of sodium chloride are very different from the physical and chemical properties of either sodium metal or chlorine gas.

I find it amazing that something as harmless as salt is made from very dangerous chemicals.

CONCEPT CHECK

Hydrogen sulfide, H_2S, is one of the smelliest compounds. Rotten eggs get their characteristic bad smell from the hydrogen sulfide they release. Can you conclude from this information that elemental sulfur, S_8, is just as smelly?

Check Your Answer

No, you cannot. In fact, the odor of elemental sulfur is negligible compared with that of hydrogen sulfide. Compounds are truly different from the elements from which they are formed. Hydrogen sulfide, H_2S, is as different from elemental sulfur, S_8, as water, H_2O, is from elemental oxygen, O_2.

17.8 There Is a System for Naming Compounds

A system for naming the countless number of possible compounds has been developed by the International Union for Pure and Applied Chemistry (IUPAC). ✔ This system is designed so that a compound's name reflects the elements it contains and how those elements are joined. Anyone familiar with the system, therefore, can deduce the chemical identity of a compound from its systematic name.

As you might imagine, this system is very complex. There is no need for you to learn all its rules. Instead, learning some guidelines will prove most helpful. These guidelines alone will not enable you to name every compound. However, they will acquaint you with how the system works for many simple compounds consisting of only two elements.

Guideline 1　The name of the element farther to the left in the periodic table is followed by the name of the element farther to the right, with the suffix "-ide" added to the name of the latter:

NaCl	sodium chloride	HCl	hydrogen chloride
Li_2O	lithium oxide	MgO	magnesium oxide
CaF_2	calcium fluoride	Sr_3P_2	strontium phosphide

Guideline 2　When two or more compounds have different numbers of the same elements, prefixes are added to remove the ambiguity. The first four prefixes are "mono-" (one), "di-" (two), "tri-" (three), and "tetra-" (four). The prefix mono-, however, is commonly omitted from the beginning of the first word of the name:

Carbon and oxygen

CO	Carbon monoxide
CO_2	Carbon dioxide

READING CHECK

Is it possible to know the elements within a compound simply from the compound's name?

Nitrogen and oxygen

NO_2 Nitrogen dioxide
N_2O_4 Dinitrogen tetroxide

Sulfur and oxygen

SO_2 Sulfur dioxide
SO_3 Sulfur trioxide

Guideline 3   Many compounds are not usually referred to by their systematic names. Instead, they are assigned common names that are more convenient or have been used traditionally for many years. Some common names are water for H_2O, ammonia for NH_3, and methane for CH_4.

CONCEPT CHECK

1. What is the systematic name for NaF?
2. What is the chemical formula for dihydrogen dioxide?
3. What is the common name for dihydrogen monoxide?

Check Your Answers

1. This compound is a cavity-fighting substance added to some toothpastes—sodium fluoride.
2. The chemical formula for dihydrogen dioxide is H_2O_2, which has the common name hydrogen peroxide, sometimes used a skin disinfectant.
3. Water, H_2O, is the common name for dihydrogen monoxide.

17 CHAPTER REVIEW

KEY TERMS

Applied research A branch of scientific research that focuses on developing applications built upon the principles discovered through basic research.

Basic research A branch of scientific research that focuses on a greater understanding of how the natural world operates.

Boiling Evaporation in which bubbles form beneath the liquid surface.

Chemical bond The force of attraction between two atoms that holds them together.

Chemical change A change in which the atoms of one or more substances are rearranged into one or more new substances.

Chemical formula A notation used to indicate the composition of a compound, consisting of the atomic symbols for the different elements of the compound and numerical subscripts indicating the ratio in which the atoms combine.

Chemical property A property that characterizes the ability of a substance to undergo a change that transforms it into a different substance.

Chemical reaction Synonymous with chemical change.

Compound A material in which atoms of different elements are bonded to one another.

Condensation A transformation from a gas to a liquid.

Elemental formula A notation that uses the atomic symbol and (sometimes) a numerical subscript to denote how atoms of the element are bonded together.

Evaporation A transformation from a liquid to a gas.

Freezing A transformation from a liquid to a solid.

Gas Matter that has neither a definite volume nor a definite shape, always filling any space available to it.

Heat of fusion The amount of energy needed to change any substance from solid to liquid, which is the same amount of energy released when the substance transforms from a liquid to a solid.

Heat of vaporization The amount of energy required to change any substance from liquid to gas, which is the same amount of energy released when the substance transforms from a gas to a liquid.

Liquid Matter that has a definite volume but no definite shape, assuming the shape of its container.

Melting A transformation from a solid to a liquid.

Molecule A submicroscopic particle consisting of a group of atoms.

Physical change A change in which a substance changes one or more of its physical properties without transforming it into a new substance.

Physical property Any physical attribute of a substance, such as color, density, or hardness.

Solid Matter that has a definite volume and a definite shape.

Submicroscopic Refers to the realm of atoms and molecules, which is a realm so small that we are unable to observe it directly with optical microscopes.

REVIEW QUESTIONS

Chemistry Is Known as the Central Science

1. Why is chemistry often called the central science?
2. What is the difference between basic research and applied research?
3. What do members of the Chemical Manufacturers Association pledge in the Responsible Care program?

The Submicroscopic World Is Super-Small

4. Are atoms made of molecules or a molecule made of atoms?

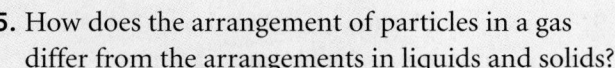

5. How does the arrangement of particles in a gas differ from the arrangements in liquids and solids?

The Phase of Matter Can Change

6. Which requires the removal of heat: melting or freezing?
7. What is it called when evaporation occurs beneath the surface of a liquid?
8. Which is greater: water's heat of fusion or its heat of vaporization?

Matter Has Physical and Chemical Properties

9. What happens to the chemical identity of a substance during a physical change?
10. What is a chemical bond?
11. What changes during a chemical reaction?

Determining Physical and Chemical Changes Can Be Difficult

12. Why is it sometimes difficult to decide whether an observed change is physical or chemical?
13. Why is the rusting of iron considered to be a chemical change?

The Periodic Table Helps Us to Understand the Elements

14. How many periods are there in the periodic table? How many groups?
15. Do properties change or remain the same for elements across any period of the periodic table?

Elements Can Combine to Form Compounds

16. Distinguish between an atom and an element.
17. What is the difference between an element and a compound?
18. What does the chemical formula of a substance tell us about that substance?

There Is a System for Naming Compounds

19. What is the chemical formula for the compound titanium dioxide?
20. Why are common names often used for chemical compounds instead of systematic names?

THINK AND DO

1. When looked at macroscopically, matter appears continuous. On the submicroscopic level, however, we find that matter is made of extremely small particles, such as atoms or molecules. Similarly, a TV screen looked at from a distance appears as a smooth continuous flow of images. Up close, however, we see this is an illusion. What really exists are a series of tiny dots (pixels) that change color in a coordinated way to produce images. Use a magnifying glass to examine closely the screen of a computer monitor or television set.

2. Pour 5 mL of water into a large balloon. Inflate the balloon to full size and then tie the balloon so it remains inflated. Next, add a couple drops of cinnamon oil to 5 mL of water. Pour this cinnamon solution into a second large balloon of a different color. Inflate this second balloon to the same size as the first balloon and then tie it shut. Swish both balloons and then present them to some one who did not see you preparing these balloons. Ask the person which balloon contains the cinnamon. The balloons are sealed. How, then, is the person able to tell one from the other? (*Hint:* Are molecules large or small or very, very small?)

3. The molecules of a gas at room temperature move around violently at speeds of about 500 meters per second. When stuck inside a balloon, these molecules are always colliding with the inner surface of the balloon. Each collision provides a little push outwards on the balloon. All the many collisions working together is what keeps the balloon inflated. To get a "feel" for what's happening here, add about a tablespoon of tiny beads to a large balloon. (Pellets, beans, BB's, grains of rice, etc., also work.) Inflate the balloon to its full size and tie it shut. Hold the balloon in the palms of both hands and shake rapidly. Can you feel the collisions? As you shake the balloon wildly, the flying beads represent the gaseous phase. How should you move the balloon so that the beads represent the liquid phase? The solid phase?

THINK AND COMPARE

1. Water molecules vibrate and jostle about at different speeds depending upon where they are located. Rank the speed of water molecules found in a steam-hot geyser, a frozen glacier, and a flowing river.

2. Rank the following elements in order of the size of their atoms from smallest to largest: cesium, Cs, atomic number 55; silver, Ag, atomic number 47; oxygen, O, atomic number 8.

3. Rank the following elements according to how difficult it is for them to lose an electron: cesium, Cs, atomic number 55; silver, Ag, atomic number 47; oxygen, O, atomic number 8. List the element that has the most difficult time losing an electron last.

THINK AND EXPLAIN

1. Is chemistry the study of the submicroscopic, the microscopic, the macroscopic, or all three? Defend your answer.

2. Of the three sciences physics, chemistry, and biology, which is the most complex?

3. In what sense is a color computer monitor or television screen similar to our view of matter? (*Hint:* Look closely to the pixels on the screen with a magnifying glass.)

4. A cotton ball is dipped in alcohol and wiped across a table top. Explain what happens to the alcohol molecules deposited on the table top. Is this a physical or chemical change?

5. A cotton ball dipped in alcohol is wiped across a table top. Would the resulting smell of the alcohol be more or less noticeable if the table top were much warmer? Explain.

6. Red-colored Kool-aid crystals are added to a still glass of water. The crystals sink to the bottom. Twenty-four hours later, the entire solution is red even though no one stirred the water. Explain.

7. Red-colored Kool-aid crystals are added to a still glass of hot water. The same amount of crystals are added to a second still glass filled with the same amount of cold water. With no stirring, which would you expect to become uniform in color first: the hot water or the cold water? Why?

8. Which has stronger attractions among its submicroscopic particles: a solid at 25°C or a gas of a different substance at 25°C? Explain.

9. Which occupies the greatest volume: 1 g of ice, 1 g of liquid water, or 1 g of water vapor?

10. You take 50 mL of small BB's and combine them with 50 mL of large BB's and you get a total of 90 mL of BB's of mixed size. Explain.

11. You take 50 mL of water and combine it with 50 mL of purified alcohol and you get a total of 98 mL of mixture. Explain.

12. Classify the following changes as physical or chemical. Even if you are incorrect in your assessment, you should be able to defend why you chose as you did.

   a. grape juice turns to wine   _____
   b. wood burns to ashes   _____
   c. water begins to boil   _____
   d. a broken leg mends itself   _____
   e. grass grows   _____
   f. an infant gains 10 pounds   _____
   g. a rock is crushed to powder   _____

13. What physical and chemical changes occur when a wax candle burns?

14. Name 10 elements you have access to macroscopic samples of in your everyday environment.

15. Why not memorize the periodic table?

16. Why is water not classified as an element?

17. Oxygen atoms are used to make water molecules. Does this mean that oxygen, O_2, and water, H_2O, have similar properties?

18. What is the chemical formula for the compound dihydrogen sulfide?

19. What is the chemical name for a compound with the formula Ba_3N_2?

20. What is the common name for dioxygen oxide?

READINESS ASSURANCE TEST (RAT)

If you have a good handle on this chapter—if you really do—then you should be able to score 7 out of 10 on this RAT. Check your answers with your teacher. If you score less than 7, study further before moving on.

Choose the best *answer to the following.*

1. Chemistry is the study of
   (a) matter.
   (b) transformations of matter.
   (c) the submicrosopic realm.
   (d) All of the above.

2. Imagine that you can see individual molecules. You watch a small collection of molecules that are moving around slowly while vibrating and bumping against each other. The slower-moving molecules then start to line up, but as they do so their vibrations increase. Soon all the molecules are aligned and vibrating about fixed positions. What is happening?
   (a) The sample is being cooled and the material is freezing.
   (b) The sample is being heated and the material is melting.
   (c) The sample is being cooled and the material is condensing.
   (d) The sample is being heated and the material is boiling.

3. What chemical change occurs when a wax candle burns?
   (a) The wax near the flame melts.
   (b) The molten wax is pulled upwards through the wick.
   (c) The wax within the wick is heated to about 600°C.
   (d) The heated wax molecules combine with oxygen molecules.

4. The phase in which atoms and molecules no longer move is the
   (a) solid phase.   (b) liquid phase.
   (c) gas phase.   (d) None of the above.

5. Is the following transformation representative of a physical change or a chemical change?

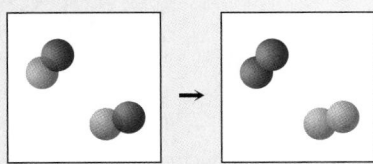

   (a) Chemical because of the formation of elements
   (b) Physical because a new material has been formed
   (c) Chemical because the atoms have changed partners
   (d) Physical because of a change in phase

6. Which elements are some of the oldest known? What is your evidence? The oldest known elements are the ones with

   (a) the lowest atomic numbers.
   (b) the highest atomic numbers.
   (c) atomic symbols that do NOT match their modern names.
   (d) atomic symbols that match their modern names.

7. Strontium, Sr (number 38), is especially dangerous to humans because it tends to accumulate in calcium-dependent bone tissues (calcium, Ca, number 20). This fact relates to the organization of the periodic table in that strontium and calcium are both
   (a) metals.
   (b) in group 2 of the periodic table.
   (c) made of relatively large atoms.
   (d) soluble in water.

8. If you have one molecule of TiO_2, how many molecules of O_2 does it contain?
   (a) One; TiO_2 is a mixture of Ti and O_2.
   (b) None; O_2 is a different molecule from TiO_2.
   (c) Two; TiO_2 is a mixture of Ti and 2 O.
   (d) Three; TiO_2 contains three molecules.

9. The systematic names for water, ammonia, and methane are dihydrogen monoxide, H_2O; trihydrogen nitride, NH_3; and tetrahydrogen carbide, CH_4. Why do most people, including chemists, prefer to use the common names for these compounds?
   (a) The common names are shorter and easier to pronounce.
   (b) These compounds are encountered frequently.
   (c) The common names are more widely known.
   (d) All of the above.

10. What is the name of the following compound: $CaCl_2$?
   (a) Carbon chloride   (b) Dichlorocalcium
   (c) Calc two   (d) Calcium chloride

18

HOW ATOMS BOND AND MOLECULES ATTRACT

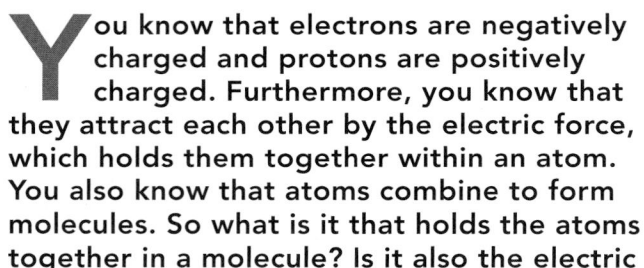 **THE MAIN IDEA**

Atoms and molecules are sticky.

You know that electrons are negatively charged and protons are positively charged. Furthermore, you know that they attract each other by the electric force, which holds them together within an atom. You also know that atoms combine to form molecules. So what is it that holds the atoms together in a molecule? Is it also the electric force? Once you have a molecule, are there any forces of attraction that occur between molecules? Are these once again electrical forces? How do all these workings at the level of atoms and molecules affect the large-scale behavior of matter, such as the formation of the cubic salt crystals shown above?

Explore!

Can Molecular Models Be Built Using Gumdrops?

1. Use different colored gumdrops (or jellybeans) to represent atoms and toothpicks to represent chemical bonds.
2. Each atom has a specific number of bonds it is able to form as follows: carbon (4), hydrogen (1), fluorine (1), oxygen (2).
3. When atoms are placed together within a single molecule, the atoms need to be as far apart from each other as possible while still also connected.
4. Using the above information, build plausible structures for the following compounds: methane, CH_4; difluoromethane, CH_2F_2; ethane, C_2H_6; hydrogen peroxide, H_2O_2; acetylene, C_2H_2.

Analyze and Conclude

1. **Observing** For methane, is it possible to have all five atoms connected laying flat on the table? What is the angle between your hydrogen–carbon–hydrogen bonds? Is it possible to make all these angles greater than 90°? For hydrogen peroxide, would it be preferable to have the two hydrogens on the same side or opposite sides of the oxygen atoms?
2. **Predicting** How many structures are possible for C_2H_6O?
3. **Making Generalizations** Why do the atoms of molecules prefer to be as far apart from each other as possible?

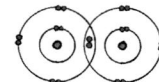

18.1 Electron-Dot Structures Help Us to Understand Bonding

Atoms are held together by an electrical force known as the chemical bond. In this chapter, we will be focusing on three types of bonds: the *ionic bond,* which holds ions together in a crystal; the *metallic bond,* which holds atoms together in a piece of metal; and the *covalent bond,* which holds atoms together in a molecule. We will then be in a position to learn about the forces of attractions occurring among molecules. This, in turn, sets the stage for understanding why matter behaves as it does, which is what chemistry is all about!

For ionic and covalent bonds, we turn to electron-dot structures, which are a shorthand notation of the electron shells presented in Section 15.5. Recall from Chapter 15 how electrons behave as though they are arranged around the atomic nucleus in a series of shells. Electrons in the outermost shell play a significant role in the atom's chemical properties, including its ability to form chemical bonds. To indicate their importance, we call these electrons *valence electrons* (as was described in Section 15.5).

Valence electrons can be conveniently represented as a series of dots surrounding an atomic symbol. This notation is called an **electron-dot structure** or, sometimes, a Lewis dot symbol (in honor of the American chemist G. N. Lewis, who first proposed the concepts of shells and valence electrons). ✔ The electron-dot structures shown in Figure 18.1 help us to understand ionic and covalent bonds. For metallic bonds, electron-dot structures are not important, which is why the metallic groups 3–12 are not included in this illustration.

When you look at the electron-dot structure of an atom, you immediately know how many valence electrons it has and how many of

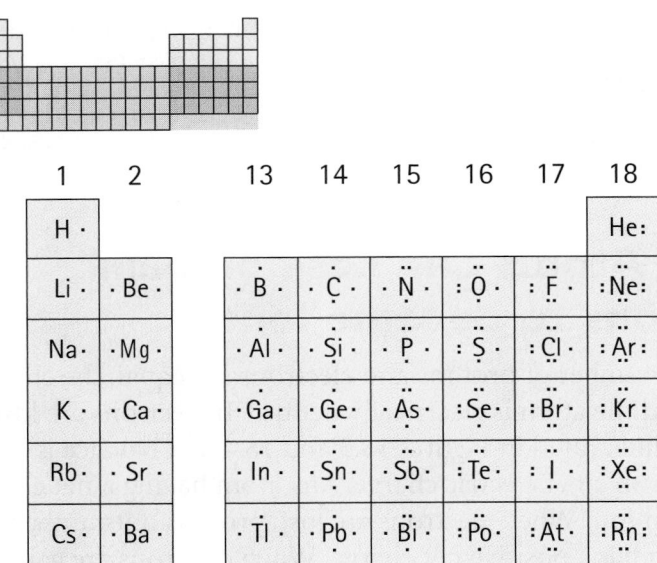

FIGURE 18.1 ▲
The valence electrons of an atom are shown in its electron-dot structure. Note that the first three periods here parallel Figure 15.14.

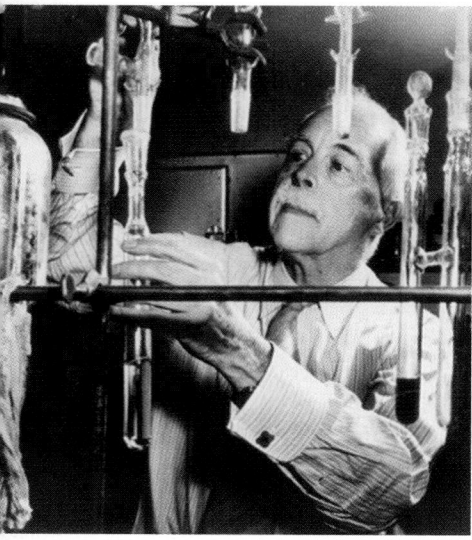

those electrons are *paired*. Chlorine, for example, has three sets of paired electrons and one unpaired electron, and carbon has four unpaired electrons:

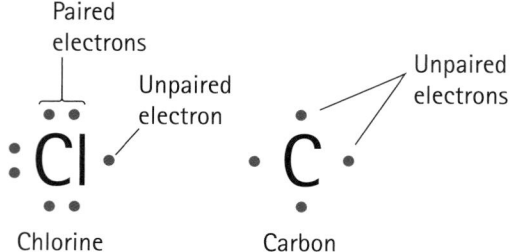

Paired valence electrons are relatively stable—they usually do not form chemical bonds with other atoms. For this reason, electron pairs in an electron-dot structure are called **nonbonding pairs.**

Valence electrons that are *unpaired*, by contrast, have a strong tendency to participate in chemical bonding. By doing so, they become paired with an electron from another atom. The ionic and covalent bonds discussed in this chapter all result from either a transfer or a sharing of unpaired valence electrons.

✔ **READING CHECK**

Electron-dot structures are needed to help us understand what kinds of chemical bonds?

CONCEPT CHECK

Where are valence electrons located, and why are they important?

Check Your Answer
Valence electrons are located in the outermost occupied shell of an atom. They are important because they play a leading role in determining the chemical properties of the atom.

18.2 Atoms Can Lose or Gain Electrons to Become Ions

When the number of protons and electrons are equal, the charges balance and the atom is electrically neutral. If electrons are lost or gained, as illustrated in Figures 18.3 and 18.4, the balance is lost and the atom takes on a net electric charge. Any atom having a net electric charge is an **ion.** When electrons are lost, protons outnumber electrons and the ion has a positive net charge. When electrons are gained, electrons outnumber protons and the ion has a negative net charge.

Chemists use a superscript to the right of the atomic symbol to indicate the strength and sign of an ion's charge. Thus, as shown in

The key to the shell game is ending up as close as you can to a filled shell.

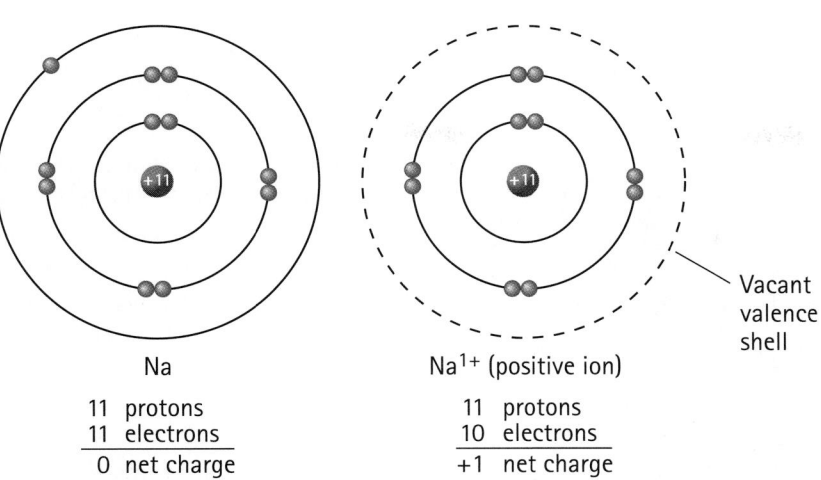

Na
11 protons
11 electrons
0 net charge

Na^{1+} (positive ion)
11 protons
10 electrons
+1 net charge

Vacant valence shell

◀ **FIGURE 18.3**
An electrically neutral sodium atom contains 11 negatively charged electrons surrounding the 11 positively charged protons of the nucleus. When this atom loses an electron, the result is a positive ion.

Figures 18.3 and 18.4, the positive ion formed from the sodium atom is written Na^{1+} and the negative ion formed from the fluorine atom is written F^{1-}. Usually, the numeral 1 is omitted when indicating either a 1+ or 1− charge. Hence, these two ions are most frequently written Na$^+$ and F$^-$. Two more examples: A calcium atom that loses two electrons is written Ca^{2+}, and an oxygen atom that gains two electrons is written O^{2-}.

We can use the shell model to deduce the type of ion an atom tends to form. According to this model, *atoms tend to lose or gain electrons to create a filled outermost shell.* Let's take a moment to consider this point, looking to Figures 18.3 and 18.4 as visual guides.

If an atom has only one or only a few electrons in its outermost shell, it tends to give up (lose) these electrons so that the next shell inward, which is already filled, becomes the outermost shell. The sodium atom of Figure 18.3, for example, has one electron in its third shell. In forming an ion, the sodium atom loses this electron, thereby making the second shell, which is already filled to capacity, the outermost occupied shell. Because the sodium atom has only one valence electron to lose, it tends to form only the 1+ ion.

If the outermost shell of an atom is almost filled, that atom attracts electrons from another atom and so forms a negative ion. The fluorine atom of Figure 18.4, for example, has one space available for an additional electron. After this additional electron is gained, the fluorine atom achieves a filled shell. Fluorine therefore tends to form the 1− ion.

Electrons are negatively charged. So *gaining* an electron results in a negative ion, and *losing* an electron results in a positive ion.

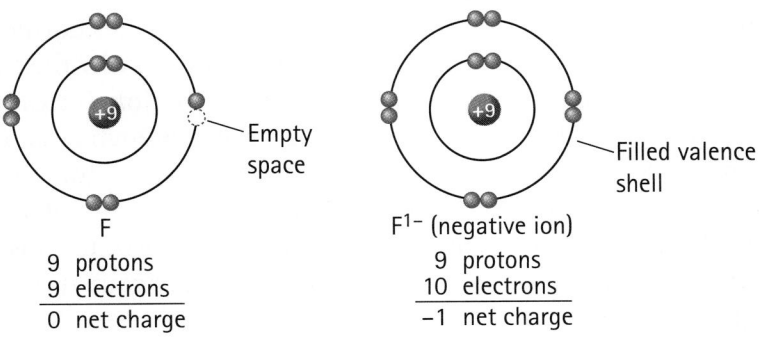

F
9 protons
9 electrons
0 net charge

Empty space

F^{1-} (negative ion)
9 protons
10 electrons
−1 net charge

Filled valence shell

◀ **FIGURE 18.4**
An electrically neutral fluorine atom contains 9 protons and 9 electrons. When this atom gains an electron, the result is a negative ion.

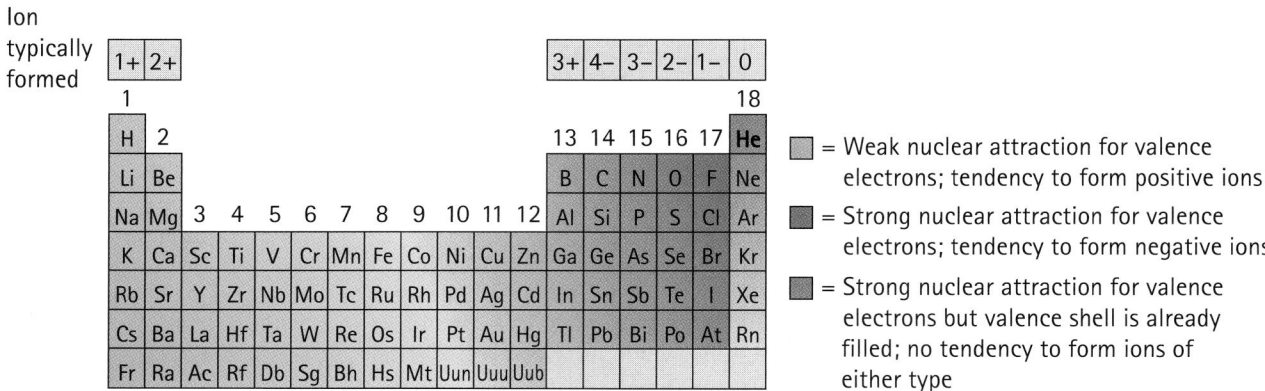

FIGURE 18.5 ▲
The periodic table is your guide to the types of ions that atoms tend to form.

✓ **The periodic table tells us the type of ion each atom tends to form.** As Figure 18.5 shows, each atom of any group 1 element, for example, has only one valence electron and so tends to form the 1+ ion. Each atom of any group 17 element has room for one additional electron in its valence shell and therefore tends to form the 1− ion. Atoms of the noble gas elements tend not to form ions of any type because their valence shells are already filled to capacity.

> **CONCEPT CHECK**
> What type of ion does the magnesium atom, Mg, tend to form?
>
> **Check Your Answer**
> The magnesium atom (atomic number 12) is found in group 2 and has two valence electrons to lose (see Figure 18.2). Therefore, it tends to form the 2+ ion.

As is indicated in Figure 18.5, the attraction between an atom's nucleus and its valence electrons is weakest for elements on the left in the periodic table and strongest for elements on the right. From sodium's position in the table, we can see that a sodium atom's single valence electron is not held very strongly, which explains why it is so easily lost. The attraction the sodium nucleus has for its second-shell electrons, however, is much stronger, which is why the sodium atom rarely loses more than one electron.

At the other side of the periodic table, the nucleus of a fluorine atom holds strongly onto its valence electrons, which explains why the fluorine atom tends not to lose any electrons to form a positive ion. Instead, fluorine's nuclear pull on the valence electrons is strong enough to accommodate even an additional electron taken from some other atom.

The nucleus of a noble gas atom pulls so strongly on its valence electrons that they are very difficult to remove. Because the outermost shell is already filled to capacity, no additional electrons are gained. Thus, a noble gas atom tends not to form ions of any sort.

What do the ions of the following elements have in common: calcium, Ca; chlorine, Cl; chromium, Cr; cobalt, Co; copper, Cu; fluorine, F; iodine, I; iron, Fe; magnesium, Mg; manganese, Mn; molybdenum, Mo; nickel, Ni; phosphorus, P; potassium, K; selenium, Se; sodium, Na; sulfur, S; zinc, Zn? They are all dietary minerals essential for good health, but that can be harmful, even lethal, when they are consumed in excessive amounts.

CONCEPT CHECK

Why does the magnesium atom tend to form the 2+ ion?

Check Your Answer

Magnesium is on the left in the periodic table, and so atoms of this element do not hold onto the two valence electrons very strongly. Because these electrons are not held very tightly, they are easily lost, which is why the magnesium atom tends to form the 2+ ion.

Using our shell model to explain the formation of ions works well for groups 1 and 2 and 13 through 18. This model is too simplified to work well for the transition metals of groups 3 through 12, however, or for the inner transition metals. In general, these metal atoms tend to form positive ions, but the number of electrons lost varies. For example, depending on conditions, an iron atom may lose two electrons to form the Fe^{2+} ion, or it may lose three electrons to form the Fe^{3+} ion.

18.3 Ionic Bonds Result from a Transfer of Electrons

Some atoms tend to lose electrons and others tend to gain them. What do you suppose happens when these two types of atoms are placed together? The answer is that electrons jump from one to the other. The end result is the formation of two oppositely charged ions. This occurs when sodium and chlorine are combined. As shown in Figure 18.6, the sodium atom loses one of its electrons to the chlorine atom, resulting in the formation of a positive sodium ion and a negative chloride ion. The two oppositely charged ions are attracted to each other by the electric force, which holds them close together. ✔ This electric force of attraction between two oppositely charged ions is called an **ionic bond.**

A sodium ion and a chloride ion together make the chemical compound sodium chloride, commonly known as table salt. This and all other chemical compounds containing ions are referred to as **ionic compounds.** All ionic compounds are completely different from the elements from which they are made. As discussed in Section 17.6, sodium chloride is not sodium, nor is it chlorine. Rather, it is a collection of sodium and chloride ions that form a unique material having its own physical and chemical properties.

READING CHECK

Why do you NOT need to memorize the type of ion each atom tends to form?

The ionic bond is merely the electrical force of attraction that holds ions of opposite charge together, in accord with Coulomb's law (Chapter 10).

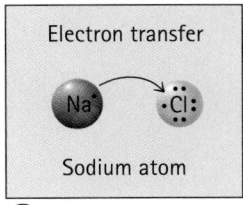

Electron transfer

Sodium atom

①

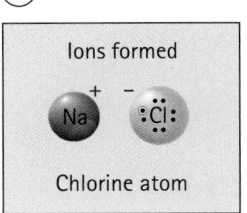

Ions formed

Chlorine atom

②

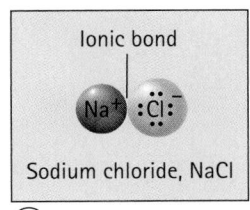

Ionic bond

Sodium chloride, NaCl

③

FIGURE 18.6 ▶

(1) The electrically neutral sodium atom loses an electron to the electrically neutral chlorine atom. (2) This electron transfer results in two oppositely charged ions. (3) The ions are then held together by an ionic bond. The spheres drawn around these and subsequent illustrations of electron-dot structures indicate the relative sizes of the atoms and ions. Note that the sodium ion is smaller than the sodium atom because it lost an electron. The chloride ion is larger than the chlorine atom because it has an added electron.

What type of force gives rise to an ionic bond?

CONCEPT CHECK

Is the transfer of an electron from a sodium atom to a chlorine atom a physical change or a chemical change?

Check Your Answer

Recall, from Chapter 17, that only a chemical change involves the formation of new material. Thus, this or any other electron transfer, because it results in the formation of a new substance, is a chemical change.

As Figure 18.7 shows, ionic compounds typically consist of elements found on opposite sides of the periodic table. Also, because of how the metals and nonmetals are organized in the periodic table, positive ions are generally derived from metallic elements and negative ions are generally derived from nonmetallic elements.

Minerals

What makes a rock? As we explore further in Chapter 25, a rock is a collection of *minerals*. What, then, is a mineral? Geologists describe a mineral as any naturally formed crystalline solid. Examples include calcite, $CaCO_3$; halite, NaCl; and pyrite, FeS_2. Most minerals are examples of ionic compounds.

✓ For all ionic compounds, positive and negative charges must balance. In sodium chloride, for example, there is one sodium 1+ ion for every chloride 1− ion. Charges must also balance in compounds containing ions that carry multiple charges. The calcium ion, for example, carries a charge of 2+, but the fluoride ion carries a charge of only 1−. Because two fluoride ions are needed to balance each calcium ion, the formula for calcium fluoride is CaF_2, as Figure 18.8 illustrates. Calcium fluoride occurs naturally in the drinking water of some communities, where it is a good source of the tooth-strengthening fluoride ion, F−.

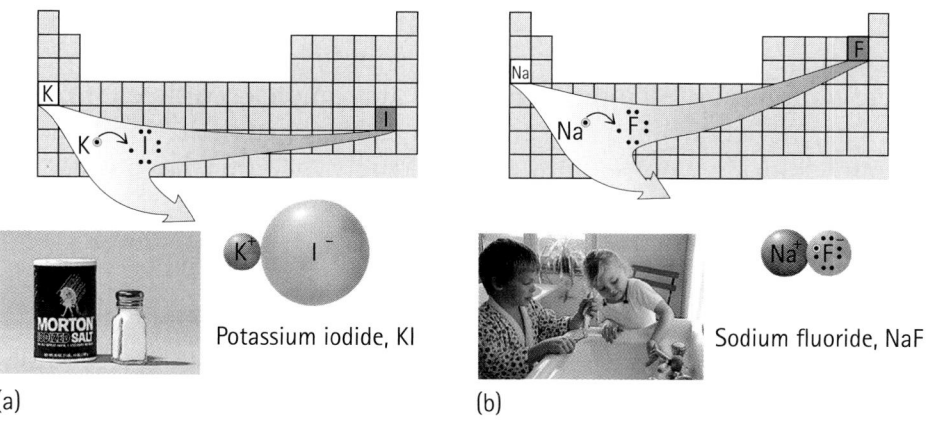

Potassium iodide, KI

(a)

Sodium fluoride, NaF

(b)

FIGURE 18.7 ▲

(a) The ionic compound potassium iodide, KI, is added in minute quantities to commercial salt because the iodide ion, I−, it contains is an essential dietary mineral. (b) The ionic compound sodium fluoride, NaF, is often added to municipal water supplies and toothpastes because it is a good source of the tooth-strengthening fluoride ion, F−.

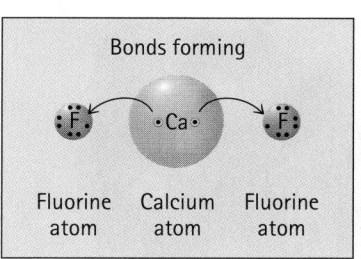

Bonds forming

Fluorine atom Calcium atom Fluorine atom

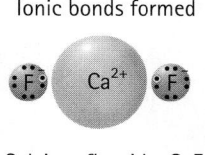

Ionic bonds formed

Calcium fluoride, CaF$_2$

Fluorite

FIGURE 18.8 ▲

A calcium atom loses two electrons to form a calcium ion, Ca^{2+}. These two electrons may be picked up by two fluorine atoms, transforming the atoms to two fluoride ions. Calcium ions and fluoride ions then join to form the ionic compound calcium fluoride, CaF$_2$, which occurs naturally as the mineral fluorite.

An aluminum ion carries a 3+ charge, and an oxide ion carries a 2− charge. Together, these ions make the ionic compound aluminum oxide, Al$_2$O$_3$, the main component of such gemstones as rubies and sapphires. Figure 18.9 illustrates the formation of aluminum oxide. The three oxide ions in Al$_2$O$_3$ carry a total charge of 6−, which balances the total 6+ charge of the two aluminum ions. Interestingly, rubies and sapphires differ in color because of the impurities they contain. Rubies are red because of minor amounts of chromium ions, and sapphires are blue because of minor amounts of iron and titanium ions.

✔ **READING CHECK**

What must be balanced within an ionic compound?

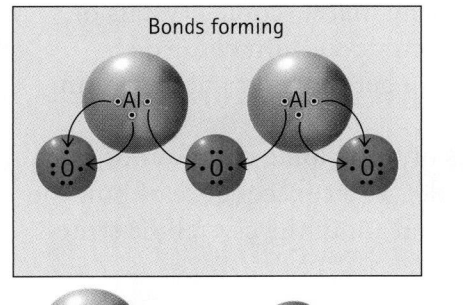

Bonds forming

Aluminum atom   Oxygen atom

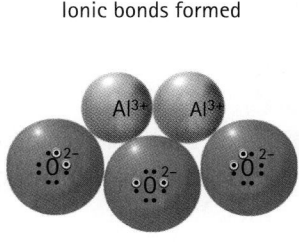

Ionic bonds formed

Aluminum oxide, Al$_2$O$_3$

Ruby

Sapphire

FIGURE 18.9 ▲

Two aluminum atoms lose a total of six electrons to form two aluminum ions, Al^{3+}. These six electrons may be picked up by three oxygen atoms, transforming the atoms to three oxide ions, O^{2-}. The aluminum and oxide ions then join to form the ionic compound aluminum oxide, Al$_2$O$_3$.

CONCEPT CHECK

What is the chemical formula for the ionic compound magnesium oxide?

Check Your Answer

Because magnesium is a group 2 element, you know a magnesium atom must lose two electrons to form a Mg^{2+} ion. Because oxygen is a group 16 element, an oxygen atom gains two electrons to form an O^{2-} ion. These charges balance in a one-to-one ratio, and so the formula for magnesium oxide is MgO.

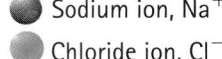

● Sodium ion, Na^+
● Chloride ion, Cl^-

FIGURE 18.10 ▲
(a) Sodium chloride, as well as other ionic compounds, forms ionic crystals in which every internal ion is surrounded by ions of the opposite charge.
(b) A view of crystals of table salt through a microscope shows their cubic structure. The cubic shape is a consequence of the cubic arrangement of sodium and chloride ions.

I appreciate salt in my food even more now that I've come to know more about salt.

An ionic compound typically contains many ions grouped together in a highly ordered three-dimensional structure. In sodium chloride, for example, each sodium ion is surrounded by six chloride ions, and each chloride ion is surrounded by six sodium ions (Figure 18.10). Overall, there is one sodium ion for each chloride ion, but there are no identifiable sodium–chloride pairs. Such an orderly array of ions is known as an *ionic crystal*. The crystalline structure of sodium chloride is cubic, which is why macroscopic crystals of table salt are also cubic. Smash a large cubic sodium chloride crystal with a hammer, and what do you get? Smaller cubic sodium chloride crystals!

Similarly, the crystalline structures of other ionic compounds, such as calcium fluoride and aluminum oxide, are a consequence of how the ions pack together. We go into more detail about the crystalline structures of minerals in Chapter 25.

18.4 Metal Atoms Bond by Losing Their Electrons

In Section 17.6 you learned about the properties of metals. They conduct electricity and heat, are opaque to light, and deform—rather than break—under pressure. Because of these properties, metals are used to build homes, appliances, cars, bridges, airplanes, and sky-scrapers. Metal wires across the landscape transmit communication signals and electric power. We wear metal jewelry, exchange metal currency, and drink from metal cans. Yet, what is it that gives a metal its metallic properties? We can answer this question by looking at the behavior of its atoms.

The outer electrons of most metal atoms tend to be weakly held to the atomic nucleus. Consequently, these electrons are easily lost, leaving behind positively charged metal ions. The many electrons lost from a

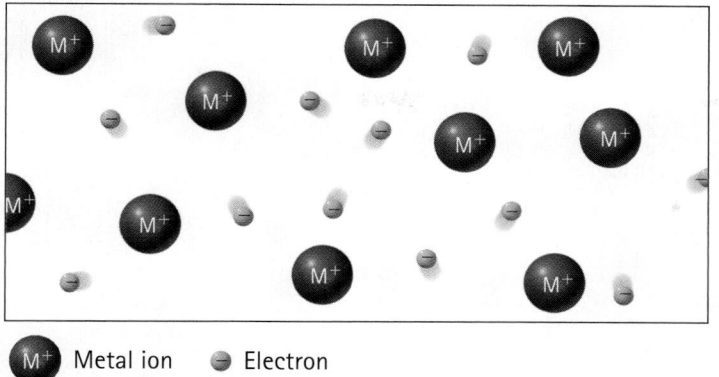

◀ **FIGURE 18.11**
Metal ions are held together by freely flowing electrons. These loose electrons form a kind of "electronic fluid," which flows through the lattice of positively charged ions.

M⁺ Metal ion   ⊖ Electron

large group of metal atoms flow freely through the resulting metal ions, as is depicted in Figure 18.11. ✔ This "fluid" of negatively charged electrons holds the positively charged metal ions together in the type of chemical bond known as a **metallic bond.**

The mobility of electrons in a metal accounts for the metal's significant ability to conduct electricity and heat. Also, metals are opaque and shiny because the free electrons easily vibrate to the oscillations of any light falling on them, reflecting most of it. Furthermore, the metal ions are not rigidly held to fixed positions, as ions are in an ionic crystal. Rather, because the metal ions are held together by a "fluid" of electrons, these ions can move into various orientations relative to one another, which occurs when a metal is pounded, pulled, or molded into a different shape.

Two or more metals can be bonded to each other by metallic bonds. This occurs, for example, when molten gold and molten palladium are blended to form a mixture known as white gold. The quality of the white gold can be modified simply by changing the proportions of gold and palladium. White gold is an example of an **alloy,** which is any mixture composed of two or more metallic elements. By playing around with proportions, metal workers can readily modify the properties of an alloy. For example, in designing the Sacagawea dollar coin, shown in Figure 18.12, the U.S. Mint needed a metal having a gold color—so that it would be popular—and also have the same electrical characteristics as the Susan B. Anthony dollar coin—so that the new coin could substitute for the Anthony coin in vending machines.

Geologic deposits containing high concentrations of metal-containing compounds are called **ores.** The metals industry mines

Steel is an alloy of iron with small percentages of carbon, which adds strength. Stainless steel is steel mixed with noncorroding metals, such as chromium or nickel.

FIGURE 18.12 ▶
The gold color of the Sacagawea U.S. dollar coin is achieved by an outer surface made of an alloy of 77% copper, 12% zinc, 7% manganese, and 4% nickel. The interior of the coin is pure copper.

Metal ores are ionic compounds in which the metal atoms have lost electrons to become positive ions. To convert the ores to metals requires that electrons be given back to the metal ions. That's the main task of a steel mill that generates steel out of iron ore.

FIGURE 18.13 ▲
The world's biggest open-pit mine is the copper mine at Bingham Canyon, Utah.

READING CHECK

What holds the positively charged metal ions together within a metallic bond?

these ores from the ground, as shown in Figure 18.13, and then processes them into metals. Although metal-containing compounds occur just about everywhere, only ores are concentrated enough to make the extraction of the metal economical.

18.5 Covalent Bonds Result from a Sharing of Electrons

Imagine two children playing together and sharing their toys. Perhaps a force that keeps the children together is their mutual attraction to the toys they share. In a similar fashion, two atoms can be held together by their mutual attraction for electrons they share. A fluorine atom, for example, has a strong attraction for one additional electron to fill its outermost shell. As shown in Figure 18.14, a fluorine atom can obtain an additional

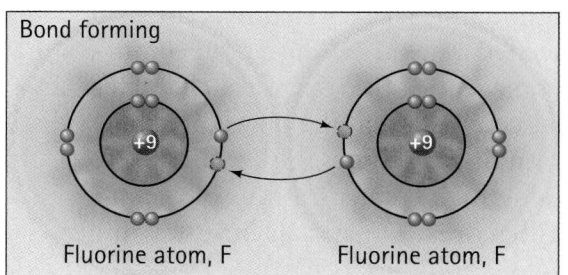
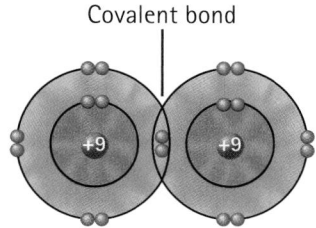

FIGURE 18.14 ▲
The effect of the positive nuclear charge (represented by red shading) of a fluorine atom extends beyond the atom's outermost occupied shell. This positive charge can cause the fluorine atom to become attracted to the unpaired valence electron of a neighboring fluorine atom. Then the two atoms are held together in a fluorine molecule by the attraction they both have for the two shared electrons. Each fluorine atom achieves a filled valence shell.

electron by holding onto the unpaired valence electron of another fluorine atom. This results in a situation in which the two fluorine atoms are mutually attracted to the same two electrons. ✓ This type of electrical attraction in which atoms are held together by their attraction for shared electrons is called a **covalent bond,** where *co-* signifies sharing and *-valent* refers to the fact that it is valence electrons that are being shared.

A substance composed of atoms held together by covalent bonds is a **covalent compound.** The fundamental unit of most covalent compounds is a **molecule,** which we can now formally define as any group of atoms held together by covalent bonds. Figure 18.15 uses the element fluorine to illustrate this principle.

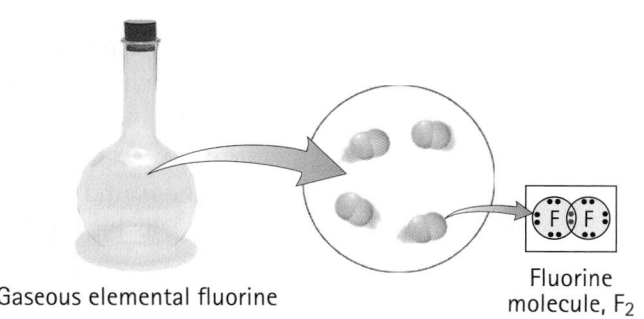

Gaseous elemental fluorine

Fluorine molecule, F_2

When writing electron-dot structures for covalent compounds, chemists often use a straight line to represent the two electrons involved in a covalent bond. In some representations, the nonbonding electron pairs are ignored. This occurs in instances where these electrons play no significant role in the process being illustrated. Here are two frequently used ways of showing the electron-dot structure for a fluorine molecule without using spheres to represent the atoms:

$$:\overset{..}{\underset{..}{F}} - \overset{..}{\underset{..}{F}}: \qquad F - F$$

Remember—the straight line in both versions represents two electrons, one from each atom. Thus, we now have two types of electron pairs to keep track of. The term *nonbonding pair* refers to any pair that exists in the electron-dot structure of an individual atom. The term *bonding pair* refers to any pair that results from formation of a covalent bond.

Recall that an ionic bond is formed when an atom that tends to lose electrons makes contact with an atom that tends to gain them. A covalent bond, by contrast, is formed when two atoms that tend to gain electrons are brought into contact with each other. Atoms that tend to form covalent bonds are therefore primarily atoms of the nonmetallic elements in the upper right corner of the periodic table (with the exception of the noble gas elements, which are very stable and tend not to form bonds).

Hydrogen tends to form covalent bonds because, unlike the other group 1 elements, it has a fairly strong attraction for an additional electron. Two hydrogen atoms, for example, covalently bond to form a hydrogen molecule, H_2, as shown in Figure 18.16.

◀ **FIGURE 18.15**
Molecules are the fundamental units of the gaseous covalent compound fluorine, F_2. Notice that in this model of a fluorine molecule, the spheres overlap, whereas the spheres shown earlier for ionic compounds do not. Now you know that this difference in representation is because of the difference in bond types.

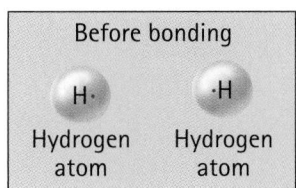

FIGURE 18.16 ▲
Two hydrogen atoms form a covalent bond as they share their unpaired electrons.

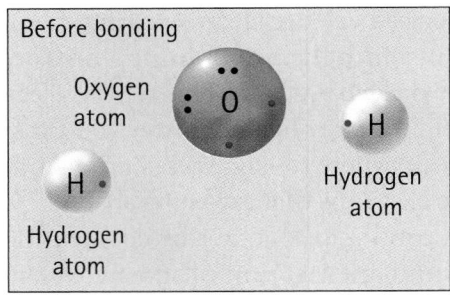
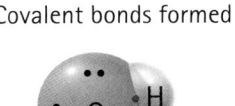

Before bonding

Oxygen atom

Hydrogen atom

Hydrogen atom

Covalent bonds formed

Water molecule, H_2O

FIGURE 18.17 ▲

The two unpaired valence electrons of oxygen pair with the unpaired valence electrons of two hydrogen atoms to form the covalent compound water.

The covalent bond is the electrical force of attraction that two atoms have for a pair of electrons that they share—again, Coulomb's law in action.

The number of covalent bonds an atom can form is equal to the number of additional electrons it can attract, which is the number needed to fill its outermost shell. Hydrogen attracts only one additional electron, and so it forms only one covalent bond. Oxygen, which attracts two additional electrons, finds them when it encounters two hydrogen atoms and reacts with them to form water, H_2O, as Figure 18.17 shows. In water, not only does the oxygen atom have access to two additional electrons by covalently bonding to two hydrogen atoms but each hydrogen atom has access to an additional electron by bonding to the oxygen atom. Each atom thus achieves a filled outermost shell.

Nitrogen attracts three additional electrons and is thus able to form three covalent bonds, as occurs in ammonia, NH_3, shown in Figure 18.18.

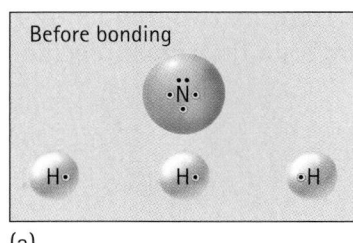
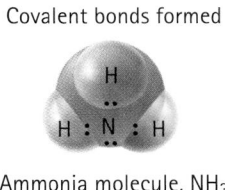

Before bonding

Covalent bonds formed

Ammonia molecule, NH_3

(a)

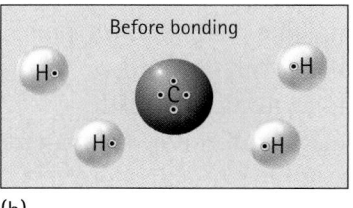

Before bonding

Covalent bonds formed

Methane molecule, CH_4

(b)

FIGURE 18.18 ▲

(a) A nitrogen atom attracts the three electrons in three hydrogen atoms to form ammonia, NH_3, a gas that can dissolve in water to make an effective cleanser. (b) A carbon atom attracts the four electrons in four hydrogen atoms to form methane, CH_4, the primary component of natural gas. In these and most other cases of covalent-bond formation, the result is a filled valence shell for all the atoms involved.

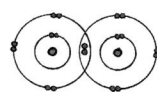

◀ **FIGURE 18.19**
The crystalline structure of diamond is nicely illustrated with sticks to represent the covalent bonds. The molecular nature of a diamond is responsible for its extreme hardness.

Likewise, a carbon atom can attract four additional electrons and is thus able to form four covalent bonds, as occurs in methane, CH_4. Note that the number of covalent bonds formed by these and other nonmetallic elements parallels the type of negative ions they tend to form (see Figure 18.5). This makes sense because covalent-bond formation and negative-ion formation are both applications of the same concept: Nonmetallic atoms tend to gain electrons until their valence shells are filled.

Diamond is a most unusual covalent compound consisting of carbon atoms covalently bonded to one another in four directions. The result is a covalent crystal, which, as shown in Figure 18.19, is a highly ordered, three-dimensional network of covalently bonded atoms. This network of carbon atoms forms a very strong and rigid structure, which is why diamonds are so hard. Also, because a diamond is a group of atoms held together only by covalent bonds, it can be characterized as a single molecule! Unlike most other molecules, a diamond molecule is large enough to be visible to the naked eye, and so it is more appropriately referred to as a macromolecule.

✔ **READING CHECK**

What is a covalent bond?

CONCEPT CHECK
How many electrons make up a covalent bond?

Check Your Answer
Two—one from each participating atom.

It is possible for two atoms to share more than two electrons between them, and Figure 18.20 shows a few examples. Molecular oxygen, O_2, consists of two oxygen atoms connected by four shared electrons.

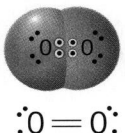

:O═O:

Oxygen, O_2

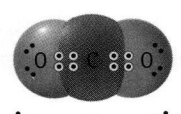

:O═C═O:

Carbon dioxide, CO_2

:N≡N:

Nitrogen, N_2

◀ **FIGURE 18.20**
Double covalent bonds in molecules of oxygen, O_2, and carbon dioxide, CO_2, and a triple covalent bond in a molecule of nitrogen, N_2.

There are always two electrons per covalent bond. A double bond, therefore, consists of four electrons, while a triple bond consists of six electrons.

This arrangement is called a double covalent bond or, for short, a double bond. As another example, the covalent compound carbon dioxide, CO_2, consists of two double bonds connecting two oxygen atoms to a central carbon atom.

Some atoms can form triple covalent bonds, in which six electrons—three from each atom—are shared. One example is molecular nitrogen, N_2. Any double or triple bond is often referred to as a multiple covalent bond. Multiple bonds higher than these, such as the quadruple covalent bond, are not commonly observed.

18.6 Electrons May Be Shared Unevenly in a Covalent Bond

If the two atoms in a covalent bond are identical, their nuclei have the same positive charge, and therefore the electrons are shared evenly. We can represent these electrons as being centrally located by using an electron-dot structure with the electrons situated exactly halfway between the two atomic symbols. Alternatively, we can draw a cloud in which the positions of the two bonding electrons over time are shown as a series of dots. Where the dots are most concentrated is where the electrons have the greatest probability of being located:

$$H : H \qquad H \quad H$$

In a covalent bond between nonidentical atoms, the nuclear charges are different, and consequently the bonding electrons may be shared unevenly. This occurs in a hydrogen–fluorine bond, where electrons are more attracted to fluorine's greater nuclear charge:

$$H : F \qquad H \quad F$$

One side of the hydrogen–fluorine bond has a greater density of electrons and is slightly negative, while the opposite side is slightly positive. This makes up a *dipole*, an extension of electric-charge polarization, as discussed in Chapter 10.

The bonding electrons spend more time around the fluorine atom. For this reason, the fluorine side of the bond is slightly negative and, because the bonding electrons have been drawn away from the hydrogen atom, the hydrogen side of the bond is slightly positive. This separation of charge is called a **dipole** (pronounced *die*-pole) and is represented either by the characters $\delta-$ and $\delta+$ (read "slightly negative" and "slightly positive," respectively) or by a crossed arrow pointing to the negative side of the bond:

$$\begin{array}{cc} \delta+ \quad \delta- \\ H - F \end{array} \qquad \begin{array}{c} \longrightarrow \\ H - F \end{array}$$

H 2.2																	He –
Li 0.98	Be 1.57											B 2.04	C 2.55	N 3.04	O 3.44	F 3.98	Ne –
Na 0.93	Mg 1.31											Al 1.61	Si 1.9	P 2.19	S 2.58	Cl 3.16	Ar –
K 0.82	Ca 1.0	Sc 1.36	Ti 1.54	V 1.63	Cr 1.66	Mn 1.55	Fe 1.83	Co 1.88	Ni 1.91	Cu 1.90	Zn 1.65	Ga 1.81	Ge 2.01	As 2.18	Se 2.55	Br 2.96	Kr –
Rb 0.82	Sr 0.95	Y 1.22	Zr 1.33	Nb 1.6	Mo 2.16	Tc 1.9	Ru 2.2	Rh 2.28	Pd 2.20	Ag 1.93	Cd 1.69	In 1.78	Sn 1.96	Sb 2.05	Te 2.1	I 2.66	Xe –
Cs 0.79	Ba 0.89	La 1.10	Hf 1.3	Ta 1.5	W 2.36	Re 1.9	Os 2.2	Ir 2.20	Pt 2.8	Au 2.54	Hg 2.00	Tl 2.04	Pb 2.33	Bi 2.02	Po 2.0	At 2.2	Rn –
Fr 0.7	Ra 0.9	Ac 1.1	Rf –	Db –	Sg –	Bh –	Hs –	Mt –	Uun –	Uuu –	Uub –						

◀ FIGURE 18.21
The experimentally measured electronegativities of elements.

So atoms forming a chemical bond engage in a tug-of-war for electrons. How strongly an atom is able to tug on bonding electrons has been measured experimentally and quantified as the atom's **electronegativity.** The range of electronegativities runs from 0.7 to 3.98, as Figure 18.21 shows.
✔ The greater an atom's electronegativity, the greater its ability to pull electrons toward itself when bonded. Thus, in hydrogen fluoride, fluorine has a greater electronegativity, or pulling power, than hydrogen.

Electronegativity is greatest for elements at the upper right of the periodic table and lowest for elements at the lower left. Noble gases are not considered in electronegativity discussions because, as previously mentioned, they rarely participate in chemical bonding.

When the two atoms in a covalent bond have the same electronegativity, no dipole is formed (as is the case with H_2) and the bond is classified as a **nonpolar** bond. When the electronegativities of the atoms differ, a dipole may form (as with HF) and the bond is classified as a **polar** bond. Just how polar a bond is depends on the difference between the electronegativity values of the two atoms—the greater the difference, the more polar the bond.

As can be seen in Figure 18.21, the greater the distance between two atoms in the periodic table, the greater the difference in their electronegativities, and hence the greater the polarity of the bond between them. So a chemist can predict which bonds are more polar than others without reading the electronegativities. Bond polarity can be inferred by looking at the relative positions of the atoms in the periodic table—the farther apart they are, especially when one is at the lower left and one is at the upper right, the greater the polarity of the bond between them.

CONCEPT CHECK
List these bonds in order of increasing polarity: P–F, S–F, Ga–F, Ge–F (F, fluorine, atomic number 9; P, phosphorus, atomic number 15; S, sulfur, atomic number 16; Ga, gallium, atomic number 31; Ge, germanium, atomic number 32):
(least polar) _____, _____, _____ (most polar)

Check Your Answer
Note that this answer can be obtained by looking only at the relative positions of these elements in the periodic table rather than by calculating the differences in their electronegativities. The order of increasing polarity is S–F < P–F < Ge–F < Ga–F.

Nonpolar   Polar

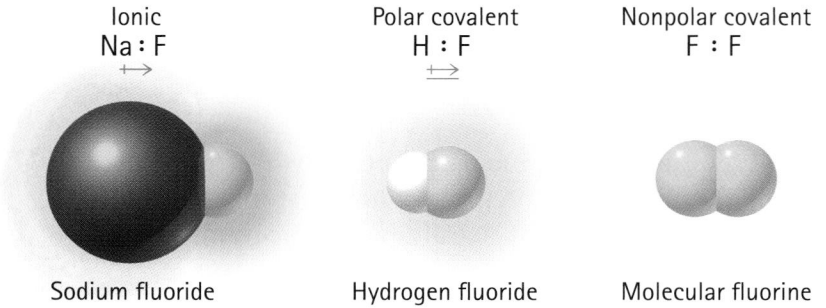

C : C $\delta+$ $\delta-$ C : N $\delta+$ $\delta-$ C : O $\delta+\delta-$ C : F

Electronegativity difference:
0   0.49   0.89   1.43

FIGURE 18.22 ▲
These bonds are in order of increasing polarity from left to right, a trend indicated by the larger and larger crossed arrows and $\delta-$ / $\delta+$ symbols. Which of these pairs of elements are farthest apart in the periodic table?

The magnitude of bond polarity is sometimes indicated by the size of the crossed arrow or the $\delta-$ and $\delta+$ symbols used to depict a dipole, as shown in Figure 18.22.

Note that the electronegativity difference between atoms in an ionic bond can also be calculated. For example, the bond in NaCl has an electronegativity difference of 2.23, far greater than the difference of 1.43 shown for the C–F bond in Figure 18.22.

What is important to understand here is that there is no black-and-white distinction between ionic and covalent bonds. Rather, there is a gradual change from one to the other as the atoms that bond are located farther and farther apart in the periodic table. This is illustrated in Figure 18.23. Atoms on opposite sides of the periodic table have great differences in electronegativity, and hence the bonds between them are highly polar—in other words, ionic. Nonmetallic atoms of the same type have the same electronegativities, and so their bonds are nonpolar covalent. The polar covalent bond with its uneven sharing of electrons and slightly charged atoms is between these two extremes.

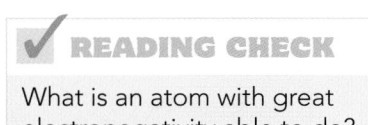

READING CHECK

What is an atom with great electronegativity able to do?

Ionic   Polar covalent   Nonpolar covalent
Na : F   H : F   F : F

Sodium fluoride   Hydrogen fluoride   Molecular fluorine

FIGURE 18.23 ▲
The ionic bond and the nonpolar covalent bond represent the two extremes of chemical bonding. The ionic bond involves a transfer of one or more electrons, and the nonpolar covalent bond involves the equitable sharing of electrons. The character of a polar covalent bond falls between these two extremes.

18.7 Electrons Are Shared Unevenly in a Polar Molecule

When all the bonds in a molecule are nonpolar, the molecule as a whole is also nonpolar—as is the case with H_2, O_2, and N_2. When a molecule consists of only two atoms and the bond between them is polar, the polarity of the molecule is the same as the polarity of the bond—as with HF, HCl, and ClF.

Complexities arise when assessing the polarity of a molecule containing more than two atoms. Consider carbon dioxide, CO_2, shown in Figure 18.24. The cause of the dipole in either one of the carbon–oxygen bonds is oxygen's greater pull on the bonding electrons (because oxygen is more electronegative than carbon). At the same time, however, the oxygen atom on the opposite side of the carbon pulls those electrons back to the carbon. The net result is an even distribution of bonding electrons around the entire molecule. ✓ So, dipoles that are of equal strength but pull in opposite directions in a molecule effectively cancel each other, with the result that the molecule as a whole is nonpolar.

Nonpolar molecules are only weakly attracted to other nonpolar molecules. This weak attraction between nonpolar molecules explains their low boiling points. Recall from Section 17.3 that boiling is a process wherein the molecules of a liquid separate from one another as they go into the gaseous phase. When there are only weak attractions between the molecules of a liquid, less heat energy is required to free the molecules from one another and allow them to enter the gaseous phase. This translates into a relatively low boiling point for the liquid, as, for example, in the nitrogen, N_2, shown in Figure 18.25. The boiling points of hydrogen, H_2; oxygen, O_2; and carbon dioxide, CO_2, are also quite low for the same reason.

There are many instances in which the dipoles of different bonds in a molecule do not cancel each other. Consider the rope analogy of

> A dipole is a vector quantity possessing both magnitude and direction. When two dipoles are equal and opposite, they effectively cancel each other out.

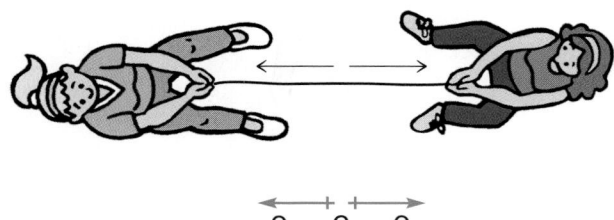

FIGURE 18.24 ▲

There is no net dipole in a carbon dioxide molecule, and so the molecule is nonpolar. This is analogous to two people in a tug-of-war. As long as they pull with equal forces but in opposite directions, the rope remains stationary.

FIGURE 18.25 ▶
Nitrogen is a liquid at temperatures below its chilly boiling point of −196°C. Nitrogen molecules are not very attracted to one another because they are nonpolar. As a result, the small amount of heat energy available at −196°C is enough to separate them and allow them to enter the gaseous phase.

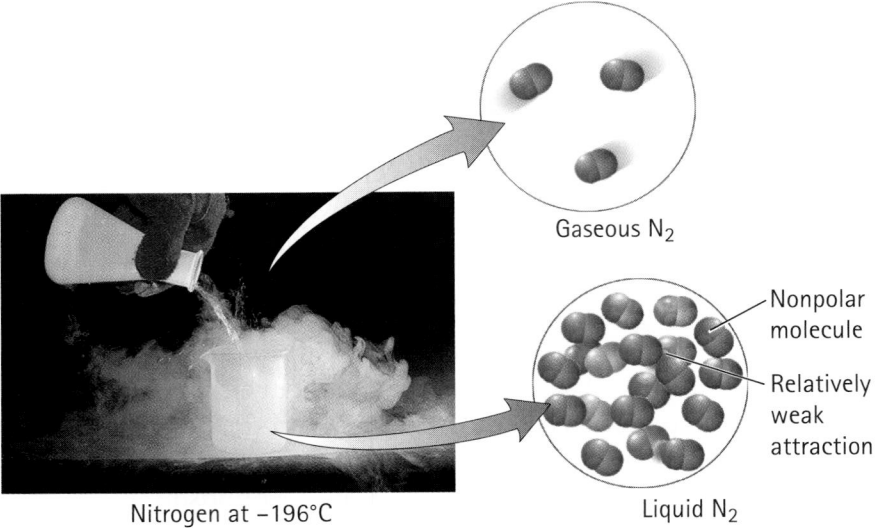

Gaseous N_2

Nonpolar molecule

Relatively weak attraction

Nitrogen at −196°C

Liquid N_2

Figure 18.26. If everyone were to pull equally, the center ring would stay put. Imagine, however, that one person begins to ease off on the rope. Now the pulls are no longer balanced, and the ring begins to move away from the person who is slacking off, as Figure 18.26 shows.

Such a situation occurs in molecules where polar covalent bonds are not equal and opposite each other. Perhaps the most relevant example is water, H_2O. Each hydrogen–oxygen covalent bond has a relatively large dipole because of the great electronegativity difference. Because of the bent shape of the molecule, however, the two dipoles, shown in blue in Figure 18.27, do not cancel each other the way the C–O dipoles in Figure 18.24 do. Instead, the dipoles in the water molecule work together to give an overall dipole, shown in purple, for the molecule.

FIGURE 18.26 ▶
If one person eases off in a three-way tug-of-war but the other two continue to pull, the ring moves in the direction of the purple arrow.

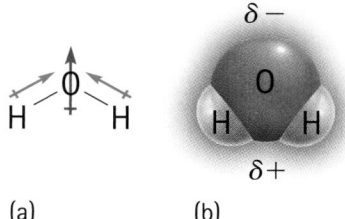

CONCEPT CHECK

Which of these molecules is polar and which is nonpolar?

$$F-C=C-F \qquad H-C=C-F$$

Check Your Answer

Symmetry is often the greatest clue for determining polarity. Because the molecule on the left is symmetrical, the dipoles on the two sides cancel each other. This molecule is therefore nonpolar:

Because the molecule on the right is less symmetrical (more "lopsided"), it is the polar molecule.

FIGURE 18.27 ▲

(a) The individual dipoles in a water molecule add together to give a large overall dipole for the whole molecule, shown in purple. (b) The region around the oxygen atom is therefore slightly negative, and the region around the two hydrogens is slightly positive.

Figure 18.28 illustrates how polar molecules electrically attract one another and, as a result, are relatively difficult to separate. In other words, polar molecules can be thought of as being "sticky," which is why it takes a lot of energy to separate them—to change phase. For this reason, substances composed of polar molecules typically have higher boiling points than substances composed of nonpolar molecules, as Table 18.1 shows.

Water boils at 100°C, whereas carbon dioxide boils at −79°C. This 179°C difference is quite dramatic when you consider that a carbon dioxide molecule is more than twice as massive as a water molecule.

Because molecular "stickiness" can play a lead role in determining a substance's macroscopic properties, molecular polarity is a central concept of chemistry. Figure 18.29 describes an interesting example.

✔ **READING CHECK**

Can the dipoles within a molecule cancel each other out?

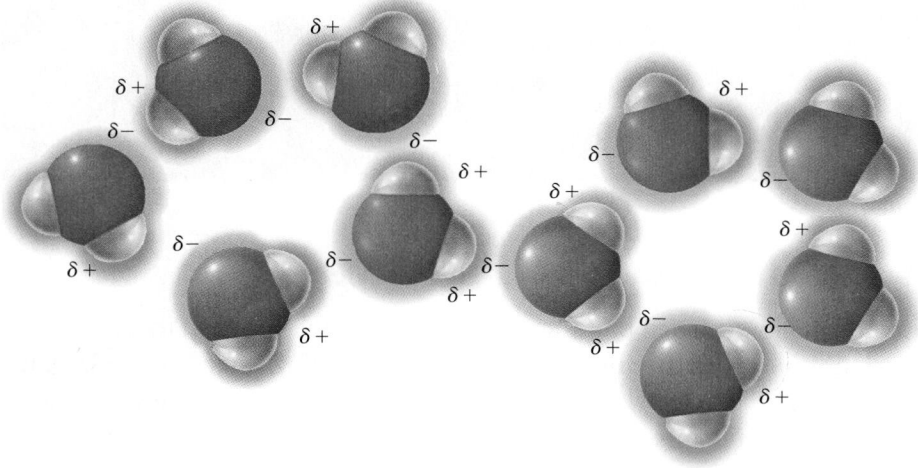

FIGURE 18.28 ▲

Water molecules attract one another because each contains a slightly positive side and a slightly negative side. The molecules position themselves such that the positive side of one faces the negative side of a neighbor.

Water molecules are like little magnets. Rather than having north and south poles, however, their opposite sides have different electric charges.

TABLE 18.1 Boiling Points of Some Polar and Nonpolar Substances

Substance	Boiling Point (°C)
Polar	
Hydrogen fluoride, HF	20
Water, H_2O	100
Ammonia, NH_3	−33
Nonpolar	
Hydrogen, H_2	−253
Oxygen, O_2	−183
Nitrogen, N_2	−196
Boron trifluoride, BF_3	−100
Carbon dioxide, CO_2	−79

FIGURE 18.29 ▶
Oil and water are difficult to mix, as is evident from the oil spill that resulted when this oil tanker broke in half. It's not, however, that oil and water repel each other. Rather, water molecules are so attracted to themselves because of their polarity that they pull themselves together. The nonpolar oil molecules are thus excluded and left to themselves. Being less dense than water, oil floats on the surface, where it poses great danger to birds and other wildlife.

In the first part of this chapter, we talked about how molecules form. Now we see how molecules mix together.

18.8 Molecules Are Attractive

So far you have learned that the atoms of a molecule are held together by covalent bonds. Furthermore, the molecule, behaving itself as a fundamental unit, may have electrical attractions to neighboring molecules. As discussed in the previous section, the greater the polarity of the molecule, the greater its attraction to neighboring molecules. This explains how water has such a high boiling point—the water molecules, being quite polar, are so attracted to one another that a lot of energy is required to separate them from one another into the gaseous phase. If you understand these concepts,

TABLE 18.2 Electrical Attractions Between a Molecule and Its Neighbor

Attraction	Relative Strength
Ion–dipole	Stronger
Dipole–dipole	
Dipole–induced dipole	Weaker

then you are ready for this last section, in which we explore the attractions that occur between different substances, such as water and salt.

Table 18.2 shows three types of electrical attractions involving molecules. ✔ The strength of even the strongest of these attractions is many times weaker than any chemical bond. The attraction between two adjacent water molecules, for example, is only about one-twentieth as strong as the chemical bonds holding the hydrogen and oxygen atoms together in the water molecule. Although molecule-to-molecule attractions are relatively weak, they play a major role in determining the macroscopic behavior of a substance. For example, why does salt dissolve so easily in water? Also, why does the oxygen we breathe NOT dissolve so easily in water?

Ions and Dipoles

Recall that a *polar* molecule is one in which the bonding electrons are unevenly shared. One side of the molecule carries a slight negative charge, and the opposite side carries a slight positive charge. This separation of charge makes up a *dipole*.

So what happens to polar molecules, such as water molecules, when they are near an ionic compound, such as sodium chloride? The opposite charges electrically attract one another. A positive sodium ion attracts the negative side of a water molecule, and a negative chloride ion attracts the positive side of a water molecule. This phenomenon is illustrated in Figure 18.30. Such an attraction between an ion and the dipole of a polar molecule is called an *ion–dipole* attraction.

Ion–dipole attractions are much weaker than ionic bonds. However, a large number of ion–dipole attractions can act collectively to disrupt ionic bonds. This is what happens to sodium chloride (table salt) in water. Attractions exerted by the water molecules break the ionic bonds and pull the ions away from one another. The result, represented in Figure 18.31, is a solution of sodium chloride in water.

✔ **READING CHECK**

How does the strength of attraction between molecules compare to the strength of the chemical bonds occurring within a molecule?

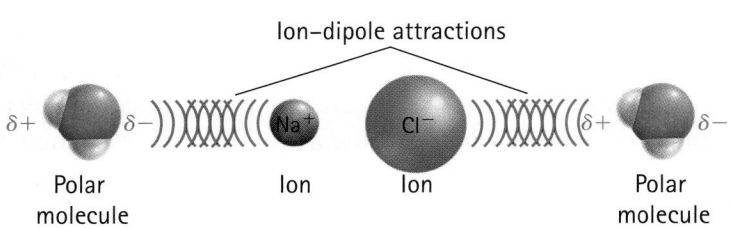

Ion–dipole attractions

$\delta+$ $\delta-$ Na$^+$ Cl$^-$ $\delta+$ $\delta-$

Polar molecule   Ion   Ion   Polar molecule

◀ **FIGURE 18.30**
Electrical attractions are shown as a series of overlapping arcs. The blue arcs indicate negative charge, and the red arcs indicate positive charge.

FIGURE 18.31 ▶
Sodium and chloride ions tightly bound in a crystal lattice are separated from one another by the collective attraction exerted by many water molecules to form an aqueous solution of sodium chloride.

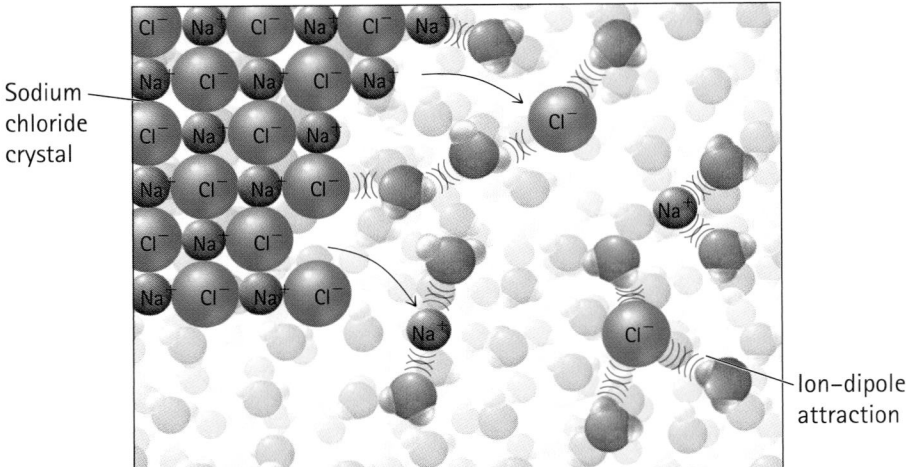

Sodium chloride crystal

Ion–dipole attraction

Aqueous solution of sodium chloride

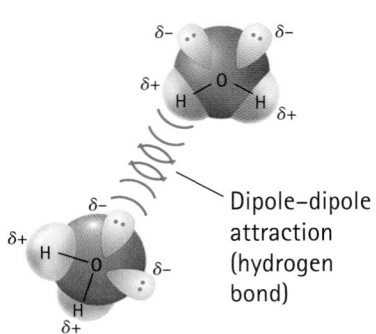

Dipole–dipole attraction (hydrogen bond)

FIGURE 18.32 ▲
The dipole–dipole attraction between two water molecules is a hydrogen bond because it involves hydrogen atoms bonded to highly electronegative oxygen atoms.

Dipoles and Dipoles

An attraction between two polar molecules is called a *dipole–dipole* attraction. This is the type of attraction that holds two or more water molecules together, as was shown in Figure 18.28. An unusually strong dipole–dipole attraction is the **hydrogen bond.** This attraction occurs between molecules that have a hydrogen atom covalently bonded to a highly electronegative atom, usually nitrogen, oxygen, or fluorine. The dipole–dipole attraction between two water molecules is an example of a hydrogen bond, as shown in Figure 18.32. Note how the hydrogen of one molecule is attracted to a pair of nonbonding electrons on the second molecule.

Even though the hydrogen bond is much weaker than any covalent or ionic bond, the effects of hydrogen bonding can be very pronounced. For example, water owes many of its properties, such as its high boiling point, to hydrogen bonds. As will be discussed in later chapters, the hydrogen bond is also of great importance in the chemistry of the molecules of life, such as DNA and proteins, that are found in living organisms.

Dipoles and Induced Dipoles

In many molecules, the electrons are distributed evenly, and so there is no dipole. The oxygen molecule, O_2, is an example. Such a nonpolar molecule can be induced to become a temporary dipole, however, when it is brought close to a water molecule (or to any other polar molecule), as Figure 18.33 illustrates. The slightly negative side of the water molecule pushes the electrons in the oxygen molecule away. Thus, the oxygen molecule's electrons are pushed to the side that is farthest from the water molecule. The result is a temporarily uneven distribution of electrons called an **induced dipole.** The resulting attraction between the permanent dipole (water) and the induced dipole (oxygen) is a *dipole–induced dipole* attraction.

A water molecule is a natural dipole—a bit positive on one end and negative on the other. What's the net charge of a dipole?

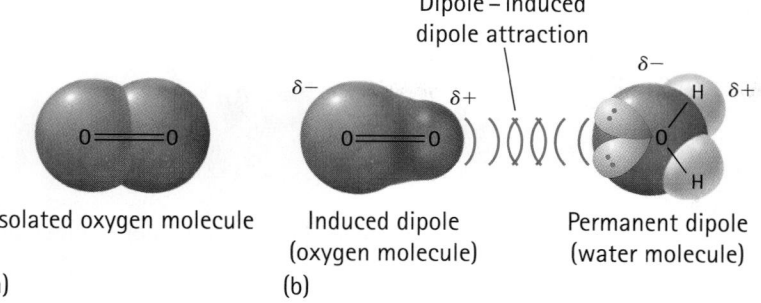

Dipole – induced
dipole attraction

δ− δ+ δ− δ+

Isolated oxygen molecule | Induced dipole (oxygen molecule) | Permanent dipole (water molecule)

(a) (b)

FIGURE 18.33 ▲

(a) An isolated oxygen molecule has no dipole; its electrons are distributed evenly. (b) An adjacent water molecule induces a redistribution of electrons in the oxygen molecule. (The slightly negative side of the oxygen molecule is shown larger than the slightly positive side because the slightly negative side contains more electrons.)

CONCEPT CHECK

How does the electron distribution in an oxygen molecule change when the hydrogen side of a water molecule is nearby?

Check Your Answer

Because the hydrogen side of the water molecule is slightly positive, the electrons in the oxygen molecule are pulled toward the water molecule, inducing in the oxygen molecule a temporary dipole in which the larger side is nearest the water molecule (rather than as far away as possible as it was in Figure 18.33).

You know that two pieces of tape hold together quite firmly when their sticky sides make contact. In a similar manner, the dipole–dipole attraction is rather strong because it involves the coming together of two sticky molecules. The dipole–induced dipole attraction, however, involves the coming together of a sticky and a nonsticky molecule. The dipole–induced dipole attraction, therefore, is much weaker. Also, induced dipoles are only temporary. If the water molecule in Figure 18.33b were removed, the oxygen molecule would return to its normal, nonpolar state.

Although they are weak, dipole–induced dipole attractions are strong enough to hold small quantities of oxygen dissolved in water, as depicted in Figure 18.34. This is the oxygen that fish absorb through their gills in order to survive underwater. So now you know that fish don't breathe water. Instead, they extract the small amounts of oxygen that are dissolved in the water. Remove this dissolved oxygen and the fish would drown.

FIGURE 18.34 ▶

The electrical attraction between water and oxygen molecules is relatively weak, which explains why not much oxygen is able to dissolve in water. For example, water fully aerated at room temperature contains only about 1 oxygen molecule for every 200,000 water molecules. The gills of a fish, therefore, must be highly efficient at extracting molecular oxygen from water.

18 CHAPTER REVIEW

KEY TERMS

Alloy A mixture of two or more metallic elements.

Covalent bond A chemical bond in which atoms are held together by their mutual attraction for two or more electrons they share.

Covalent compound An element or chemical compound in which atoms are held together by covalent bonds.

Dipole A separation of charge that occurs in a chemical bond because of differences in the electronegativities of the bonded atoms.

Electron-dot structure A shorthand notation of the shell model of the atom, in which valence electrons are shown around an atomic symbol.

Electronegativity The ability of an atom to attract a bonding pair of electrons to itself when bonded to another atom.

Hydrogen bond A strong dipole–dipole attraction between a slightly positive hydrogen atom on one molecule and a pair of nonbonding electrons on another molecule.

Induced dipole A dipole temporarily created in an otherwise nonpolar molecule, induced by a neighboring charge.

Ion An electrically charged particle created when an atom either loses or gains one or more electrons.

Ionic bond A chemical bond in which an attractive electric force holds ions of opposite charge together.

Ionic compound Any chemical compound containing ions.

Metallic bond A chemical bond in which positively chargd metal ions are held together within a "fluid."

Molecule A group of atoms held tightly together by covalent bonds.

Nonbonding pairs Two paired valence electrons that tend not to participate in a chemical bond.

Nonpolar Said of a chemical bond that has no dipole.

Ore A geologic deposit containing relatively high concentrations of one or more metal-containing compounds.

Polar Said of a chemical bond that has a dipole.

REVIEW QUESTIONS

Electron-Dot Structures Help Us to Understand Bonding

1. What do the electron-dot structures of elements in the same group in the periodic table have in common with each other?
2. How many unpaired valence electrons are there in the carbon atom?

Atoms Can Lose or Gain Electrons to Become Ions

3. To become a negative ion, does an atom lose or gain electrons?
4. Why does the fluorine atom tend to gain only one electron?

Ionic Bonds Result from a Transfer of Electrons

5. What kind of force holds two atoms together within an ionic bond?
6. Which elements tend to form ionic bonds?
7. Why do all minerals have such high melting points?
8. Why are all crystals of the mineral halite, NaCl, cubic?
9. What is an ionic crystal?
10. What is the electric charge on the calcium ion in calcium chloride, $CaCl_2$?

Metal Atoms Bond by Losing Their Electrons

11. Do metals more readily gain or lose electrons?
12. What is an alloy?

Covalent Bonds Result from a Sharing of Electrons

13. Which elements tend to form covalent bonds?

14. How many electrons are shared in a double covalent bond?

Electrons May Be Shared Unevenly in a Covalent Bond

15. What is a dipole?

16. Which element in the periodic table has the greatest electronegativity? Which has the least electronegativity?

Electrons Are Shared Unevenly in a Polar Molecule

17. How can a molecule be nonpolar when it consists of atoms that have different electronegativities?

18. Which would you describe as "stickier": a polar molecule or a nonpolar one?

Molecules Are Attractive

19. What is the primary difference between a chemical bond and an attraction between two molecules?

20. Are induced dipoles permanent?

CHEMICALS, CHEMICALS, CHEMICALS!

THINK AND DO

1. View crystals of table salt with a magnifying glass or, better yet, a microscope if one is available. If you do have a microscope, crush the crystals with a spoon and examine the resulting powder. Purchase some sodium-free salt, which is potassium chloride, KCl, and examine these ionic crystals, both intact and crushed. Sodium chloride and potassium chloride both form cubic crystals, but there are significant differences. What are they?

2. The following are the structures we hope you came up with for the chemical compounds given in the Explore! at the beginning of this chapter.

Difluoromethane, CH_2F_2, tetrahedron

Ethane, C_2H_6, two tetrahedrons

Hydrogen peroxide, H_2O_2, two bent shapes stuck together

$H — C \equiv C — H$
Acetylene, C_2H_2, linear

Looking for more challenges? Try building the structures for carbon dioxide, CO_2; water, H_2O; and ammonia, NH_3. All of these structures were shown to you within this chapter. For an ultimate challenge, try benzene, C_6H_6, or acetic acid, $C_2H_4O_2$.

THINK AND COMPARE

1. Rank the following bonds in order of increasing polarity:
   (a) $C — H$
   (b) $O — H$
   (c) $N — H$

2. Rank the following compounds in order of increasing boiling point:
   (a) Fluorine, F_2
   (b) Hydrogen fluoride, HF
   (c) Hydrogen chloride, HCl

THINK AND EXPLAIN

1. How is the number of unpaired valence electrons in an atom related to the number of bonds that the atom can form?

2. Magnesium ions carry a 2+ charge, and chloride ions carry a 1− charge. What is the chemical formula for the ionic compound magnesium chloride?

3. Why does the potassium atom tend to lose only one electron?

4. Which should be larger, the potassium atom, K, or the potassium ion, K^+?

5. Two fluorine atoms join together to form a covalent bond. Why don't two potassium atoms do the same thing?

6. Classify the following bonds as ionic, covalent, or neither (O, atomic number 8; F, atomic number 9; Na, atomic number 11; Cl, atomic number 17; U, atomic number 92).

   O with F _____
   Ca with Cl _____
   Na with Na _____
   U with Cl _____

7. Phosphine is a covalent compound of phosphorus, P, and hydrogen, H. What is its chemical formula?

8. Which bond is most polar:
   (a) H—N (b) N—C
   (c) C—O (d) C—C
   (e) O—H (f) C—H

9. Why don't the dipoles of the two hydrogen–oxygen bonds in a water molecule cancel each other out?

10. How many nonbonding pairs of electrons are there in the oxygen atom of a water molecule? How many bonding pairs?

11. The oxygen atom of a water molecule has four pairs of electrons in its outermost shell. Do you suppose all these pairs tend to bunch together on the same side of the atom, or do these pairs tend to spread out as far apart from each other as possible. Why?

12. If water were linear like carbon dioxide, would it be polar or nonpolar? Would its boiling point be higher or lower than 100°C?

13. Why don't oil and water mix?

14. Water, H_2O, and methane, CH_4, have about the same mass and differ by only one type of atom. Why is the boiling point of water so much higher than that of methane?

15. Two kids are sitting across from each other at a table trading their jellybeans. They both start out with the same number of jellybeans, but one of the kids is in a generous mood while the other is in a greedy mood. If each jellybean represents an electron, which kid ends up being slightly negative: the generous kid or the greedy kid? Who ends up being slightly positive? Is the generous kid just as negative as the greedy kid is positive? Would you describe this as a polar or nonpolar situation? What if both kids were equally greedy?

16. Which is stronger: the covalent bond that holds atoms together within a molecule or the electrical attraction between two neighboring molecules?

17. Why is a water molecule more attracted to a calcium ion than a sodium ion?

18. The charges with sodium chloride are all balanced—for every positive sodium ion there is a corresponding negative chloride ion. Since its charges are balanced, how can sodium chloride be attracted to water, and vice versa?

19. How are oxygen molecules attracted to water molecules?

20. Some bottled water is now advertised as containing extra quantities of "Vitamin O," which is a marketing gimmick for selling oxygen, O_2. Might this bottled water actually contain extra quantities of oxygen? How much more than one might find in regular bottled water? How might the amount of oxygen we absorb through our lungs compare to that we might absorb through our stomach—after burping?

READINESS ASSURANCE TEST (RAT)

If you have a good handle on this chapter—if you really do—then you should be able to score 7 out of 10 on this RAT. Check your answers with your teacher. If you score less than 7, study further before moving on.

Choose the best answer to the following.

1. An atom loses an electron to another atom. Is this an example of a physical or chemical change?
   (a) Chemical change involving the formation of ions
   (b) Physical change involving the formation of ions
   (c) Chemical change involving the formation of covalent bonds
   (d) Physical change involving the formation of covalent bonds

2. Aluminum ions carry a 3+ charge, and chloride ions carry a 1− charge. What would be the chemical formula for the ionic compound aluminum chloride?
   (a) Al_3Cl
   (b) $AlCl_3$
   (c) Al_3Cl_3
   (d) AlCl

3. Which would you expect to have a higher melting point: sodium chloride, NaCl, or aluminum oxide, Al_2O_3?
   (a) The aluminum oxide has a higher melting point because it is a larger molecule and has a greater number of molecular interactions.
   (b) NaCl has a higher melting point because it is a solid at room temperature.
   (c) The aluminum oxide has a higher melting point because of the greater charges of the ions, and hence the greater force of attractions between them.
   (d) The aluminum oxide has a higher melting point because of the covalent bonds within the molecule.

4. Atoms of metallic elements can form ionic bonds, but they are not very good at forming covalent bonds. Why?
   (a) These atoms are too large to be able to come in close contact with other atoms.
   (b) They have a great tendency to lose electrons.
   (c) Their valence shells are already filled with electrons.
   (d) They are on the wrong side of the periodic table.

5. In terms of the periodic table, is there an abrupt or gradual change between ionic and covalent bonds?
   (a) An abrupt change that occurs across the metalloids.
   (b) Actually, any element of the periodic table can form a covalent bond.
   (c) There is a gradual change: the farther apart, the more ionic.
   (d) Whether an element forms one or the other depends on nuclear charge and not the relative positions in the periodic table.

6. A hydrogen atom does not form more than one covalent bond because it
   (a) has only one shell of electrons.
   (b) has only one electron to share.
   (c) loses its valence electron so readily.
   (d) has such a strong electronegativity.

7. When nitrogen and fluorine combine to form a molecule, the most likely chemical formula is
   (a) N_3F
   (b) N_2F
   (c) NF_4
   (d) NF
   (e) NF_3

8. A substance consisting of which molecule shown below should have a higher boiling point?

   $$S=C=O \quad O=C=O$$

   (a) The molecule on the left, SCO, because it comes later in the periodic table.
   (b) The molecule on the left, SCO, because it has less symmetry.
   (c) The molecule on the right, OCO, because it has more symmetry.
   (d) The molecule on the right, OCO, because it has more mass.

9. Why are ion–dipole attractions stronger than dipole–dipole attractions?
   (a) The chemical bond in an ion–dipole molecule is similar to a covalent bond.
   (b) Like charge (dipole) does not attract like charge (another dipole).
   (c) Dipole areas are subject to changing from positive to negative regions on the molecule.
   (d) The magnitude of the electric charge associated with an ion is much greater.

10. What is a hydrogen bond?
    (a) The covalent bond between two hydrogen atoms
    (b) A strong glue commonly found in hardware stores
    (c) An unusually strong dipole–dipole attraction
    (d) A type of nuclear explosive

19
HOW CHEMICALS MIX

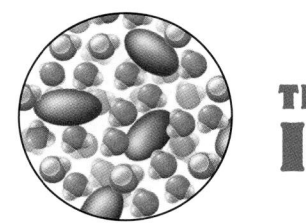

THE MAIN IDEA

Most everything is a mixture.

What's in a glass of tap water? Why is this tap water anything but pure? When tap water is left boiling on the stove too long, it evaporates completely but leaves a chalky residue in the pot. What is this residue and from where did it come? When you stir sugar into water, the sugar crystals disappear, but where do they go? How do chemists classify matter? What do clouds have in common with the blood that runs through our veins? Do fish breathe water or do they take in the oxygen that is dissolved in the water? The answers to these sorts of questions involve an understanding of mixtures.

Explore!

Does Sugar Disappear When It Dissolves in Water?

1. Fill a glass almost to the top with warm water. Mark the water level with tape or a marker and then carefully pour all this warm water into a larger container without spilling a drop.
2. Add about 3 tablespoons of sugar to the emptied glass.
3. Return about half of the warm water to the glass and stir to dissolve all the sugar.
4. Return the remaining water, and as you get close to the mark, predict whether the water level will be less than before, about the same as before, or more than before.

Analyze and Conclude

1. **Observing** Did the water return to its original level? If not, how does the new level relate to the volume of the sugar?

2. **Predicting** Would you get the same results if you instead added the sugar directly to the water in the glass? What if you added sand instead of sugar?
3. **Making Generalizations** Does a solid dissolved in a liquid occupy any volume of space? Is this volume of space more than, less than, or the same as the volume it occupies as an undissolved solid? Which do you suppose displaces more water: a rock or the same rock crushed into powder? (Assume no rock material is lost during the crushing process.) Which dissolves faster in water: a large crystal of sugar or the same crystal crushed into a powder?

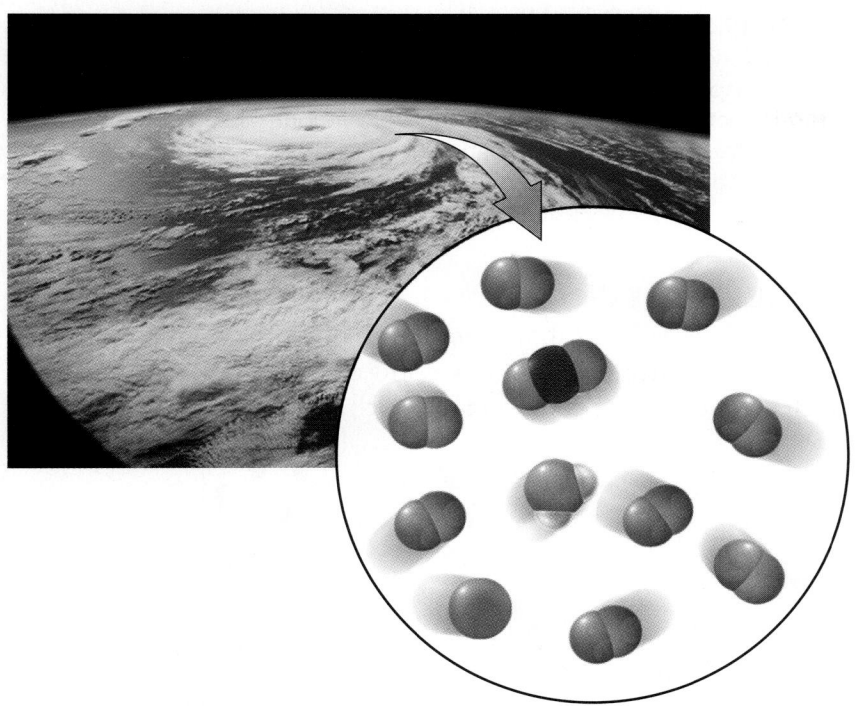

◄ **FIGURE 19.1**
The Earth's atmosphere is a mixture of gaseous elements and compounds. Some of them are shown here.

19.1 Most Materials Are Mixtures

A **mixture** is a combination of two or more substances in which each substance retains its own properties. Most materials we encounter are mixtures: mixtures of elements, mixtures of compounds, or mixtures of elements and compounds. Stainless steel, for example, is a mixture of the elements iron, chromium, nickel, and carbon. Seltzer water is a mixture of a liquid compound, water, and a gaseous compound, carbon dioxide. Our atmosphere, as Figure 19.1 illustrates, is a mixture of the elements nitrogen, oxygen, and argon, plus small amounts of such compounds as carbon dioxide and water vapor.

Tap water is a mixture containing mostly water but also many other compounds. Depending on your location, your water may contain compounds of calcium, magnesium, chlorine, fluorine, iron, and potassium; trace amounts of compounds of lead, mercury, and cadmium; organic compounds; and dissolved oxygen, nitrogen, and carbon dioxide. While it is surely important to minimize any toxic components in your drinking water, it is unnecessary, undesirable, and impossible to remove all other substances from it. Some of the dissolved solids and gases give water its characteristic taste, and many of them promote human health: fluoride compounds protect teeth, chlorine destroys harmful bacteria, and as much as 10 of our daily requirements for iron, potassium, calcium, and magnesium are obtained from drinking water (Figures 19.2 and 19.3).

FIGURE 19.2 ▲
Tap water provides us with H_2O as well as a large number of other compounds, many of which are flavorful and help us grow, as Graham demonstrates at ages 7 and 21. Bottoms up!

◀ **FIGURE 19.3**
Most of the oxygen in the air bubbles produced by an aquarium aerator escapes into the atmosphere. Some of the oxygen, however, mixes with the water. It is this oxygen that fish depend upon to survive. Without this dissolved oxygen, which fish extract from the water with their gills, the fish would promptly drown. So fish don't "breathe" water. They breathe the O_2 that is dissolved in the water.

CONCEPT CHECK

So far, you have learned about three kinds of matter: elements, compounds, and mixtures. Which box below contains only an element? Which contains only a compound? Which contains a mixture?

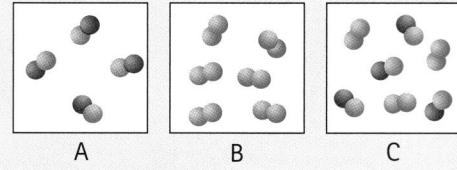

A   B   C

Check Your Answers
The molecules in box A each contain two different types of atoms and so are representative of a compound. The molecules in box B each consist of the same atoms and so are representative of an element. Box C is a mixture of the compound and the element.

Note how the molecules remain intact in the mixture. That is, upon the formation of the mixture, there is no exchange of atoms between the components.

There is a difference between the way materials combine to form mixtures and the way elements combine to form compounds. Each substance in a mixture retains its chemical identity. The sugar molecules in the teaspoon of sugar in Figure 19.4, for example, are identical to the sugar molecules already in the tea. The only difference is that the sugar molecules in the tea are mixed with other substances, mostly water. The

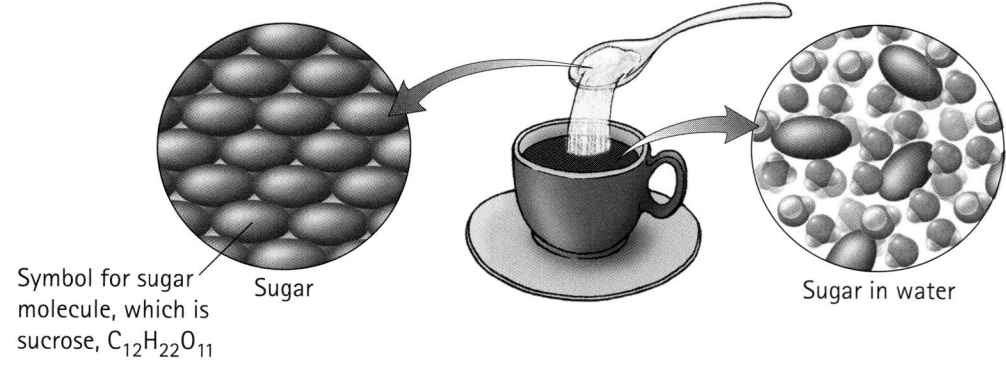

Symbol for sugar molecule, which is sucrose, $C_{12}H_{22}O_{11}$ Sugar   Sugar in water

FIGURE 19.4 ▲
Table sugar is a compound consisting only of sucrose molecules. Once these molecules are mixed into hot tea, they become interspersed among the water and tea molecules and form a sugar-tea-water mixture. No new compounds are formed, and so this is an example of a physical change.

formation of a mixture, therefore, is a physical change. In contrast, as discussed in Section 17.5, there is a change in chemical identity when elements join to form compounds. Recall that sodium chloride is not a mixture of sodium and chlorine atoms but is instead a compound, which means it is entirely different from the elements that compose it. The formation of a compound is therefore a chemical change.

Chemists have devised many ways of separating the components of a mixture. Most of these techniques use the simple principle of separating the components by differences in their physical properties.

Mixtures Can Be Separated by Physical Means

 The components of mixtures can be separated from one another by taking advantage of differences in the components' physical properties. A mixture of solids and liquids, for example, can be separated using filter paper through which the liquids pass but the solids do not. This is how coffee is often made: The caffeine and flavor molecules in the hot water pass through the filter and into the coffee pot, while the solid coffee grounds remain behind. This method of separating a solid-liquid mixture is called *filtration*, and it is a common technique used by chemists.

Mixtures can also be separated by taking advantage of a difference in boiling or melting points. Seawater is a mixture of water and a variety of compounds, mostly sodium chloride. Whereas water boils at 100°C, sodium chloride doesn't even melt until 800°C. One way to separate water from the mixture we call seawater, therefore, is to heat the seawater to about 100°C. At this temperature, the liquid water readily transforms to water vapor, but the sodium chloride stays behind dissolved in the remaining water. As the water vapor rises, it can be channeled into a cooler container, where it condenses into a liquid without the dissolved solids. This process of collecting a vaporized substance, called **distillation,** is illustrated in Figure 19.5. Distillation is a very effective,

✓ READING CHECK

How do we separate the components of a mixture?

(a)

(b)

FIGURE 19.5 ▲

(a) The mixture is boiled in the flask on the left. The rising water vapor is channeled into a downward-slanting tube kept cool by cold water flowing across its outer surface. The water vapor inside the cool tube condenses and collects in the flask on the right. (b) A whiskey still functions by the same principle. A mixture containing alcohol is heated to the point where the alcohol, some flavoring molecules, and some water are vaporized. These vapors travel through the copper coils, where they then condense to a liquid.

◀ **FIGURE 19.6**

At the southern end of San Francisco Bay, there are areas where the seawater has been partitioned off by earthen dikes. These are evaporation ponds, where the water is allowed to evaporate, leaving behind the solids that were dissolved in the seawater. These solids are further refined for commercial sale. The remarkable color of the ponds results from suspended particles of iron oxide and other minerals, which are easily removed during refining.

though costly, way of isolating fresh water from seawater. After all the water has been distilled from seawater, what remains are dry solids. These solids, also a mixture of compounds, contain a variety of commercially valuable compounds, such as sodium chloride and potassium bromide (Figure 19.6).

19.2 The Chemist's Classification of Matter

✓ If a material is **pure,** it consists of only a single element or a single compound. In pure gold, for example, there is nothing but the element gold. In pure table salt, there is nothing but the compound sodium chloride. If a material is **impure,** it is a mixture and contains two or more elements or compounds. This classification scheme is shown in Figure 19.7.

Because atoms and molecules are so small, it is impractical to prepare a sample that is truly pure—that is, truly 100% of a single material. For example, if just one atom or molecule out of a trillion trillion were different, then the 100% pure status would be lost. Samples, however, can be "purified" by various methods, such as distillation. When we say *pure,* it is understood to be a relative term. Comparing the purity of two samples, the purer one contains fewer impurities. A sample of water that

FIGURE 19.7 ▶

A chemical classification of matter.

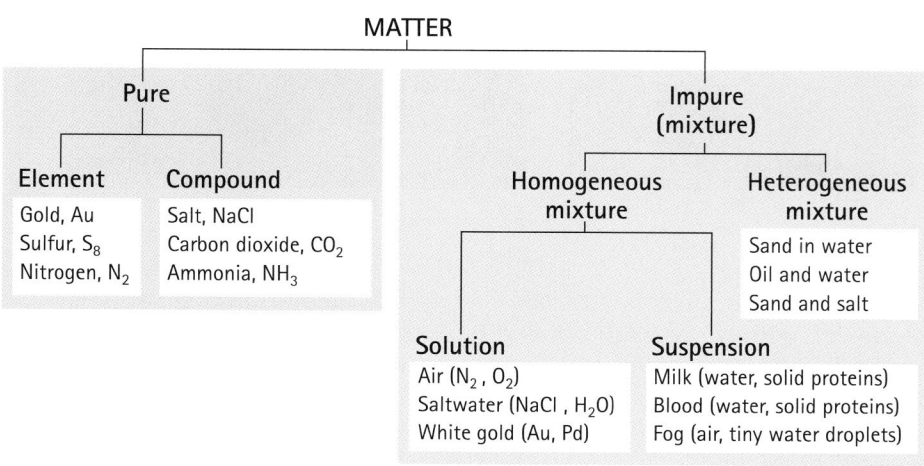

is 99.9% pure has a greater proportion of impurities than does a purer sample of water that is 99.9999% pure.

Sometimes naturally occurring mixtures are labeled as being pure, as in "pure orange juice." Such a statement merely means that nothing artificial has been added. According to a chemist's definition, however, orange juice is anything but pure, as it contains a wide variety of materials, including water, pulp, flavorings, vitamins, and sugars.

Mixtures may be heterogeneous or homogeneous. In a **heterogeneous mixture,** the different components can be seen as individual substances, such as pulp in orange juice, sand in water, or oil globules dispersed in vinegar. The different components are visible. A **homogeneous mixture** has the same composition throughout. Any one region of the mixture has the same ratio of substances as does any other region, and the components cannot be seen as individual identifiable entities. The distinction is shown in Figure 19.8.

A homogeneous mixture may be either a solution or a suspension. In a **solution,** all components are in the same phase. The atmosphere we breathe is a gaseous solution consisting of the gaseous elements nitrogen and oxygen as well as minor amounts of other gaseous materials. Saltwater is a liquid solution because both the water and the dissolved sodium chloride are found in a single liquid phase. An example of a solid solution is white gold, which is a homogeneous mixture of the elements

Orange juice may be 100% natural, but it is never 100% pure.

✔ READING CHECK

How does a chemist define a "pure" material?

Granite

"Snow" in snow globe

Pizza

(a) Heterogeneous mixtures

Air

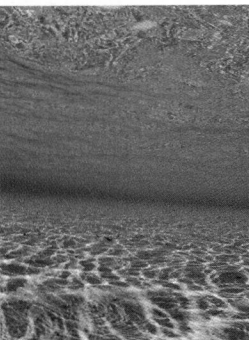
Clear seawater

White gold

(b) Homogeneous mixtures

◄ FIGURE 19.8
(a) In heterogeneous mixtures, the different components can be seen with the naked eye. (b) In homogeneous mixtures, the different components are mixed at a much finer level and so are not readily distinguished.

gold and palladium. We shall be discussing solutions in more detail in the next section.

A **suspension** is a homogeneous mixture in which the different components are in different phases, such as solids in liquids or liquids in gases. In a suspension, the mixing is so thorough that the different phases cannot be readily distinguished. Milk is a suspension because it is a homogeneous mixture of proteins and fats finely dispersed in water. Blood is a suspension composed of finely dispersed blood cells in water. Another example of a suspension is clouds, which are homogeneous mixtures of tiny water droplets suspended in air. Shining a light through a suspension, as is done in Figure 19.9, results in a visible cone as the light is reflected by the suspended components.

The easiest way to distinguish a suspension from a solution in the laboratory is to spin a sample in a centrifuge. This device, spinning at thousands of revolutions per minute, separates the components of suspensions but not those of solutions, as Figure 19.10 shows.

FIGURE 19.9 ▲
The path of light becomes visible when the light passes through a suspension.

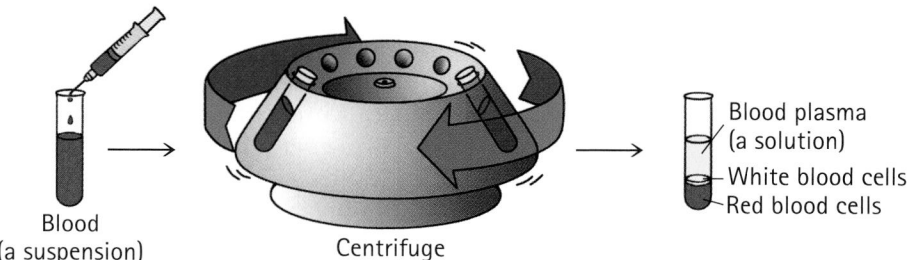

Blood
(a suspension)   Centrifuge

Blood plasma
(a solution)
White blood cells
Red blood cells

FIGURE 19.10 ▲
Blood, because it is a suspension, can be centrifuged into its components, which include the blood plasma (a yellowish solution) and white and red blood cells. The components of the plasma, however, cannot be separated from one another here because a centrifuge has no effect on solutions.

CONCEPT CHECK

Impure water can be purified by which of these?
a. Removing the impure water molecules
b. Removing everything that is not water
c. Breaking down the water to its simplest components
d. Adding some disinfectant such as chlorine

Check Your Answer

The answer is (b): Impure water can be purified by removing everything that isn't water. H_2O is a compound made of the elements hydrogen and oxygen in a 2-to-1 ratio. Every H_2O molecule is exactly the same as every other, and there's no such thing as an impure H_2O molecule. Just about anything, including you, beach balls, rubber ducks, dust particles, and bacteria, can be found in water. When something other than water is found in water, we say that the water is impure. It is important to see that the impurities are in the water and not part of the water, which means that it is possible to remove them by a variety of physical means, such as filtration or distillation.

19.3 A Solution Is a Single-Phase Homogeneous Mixture

What happens when table sugar, known chemically as sucrose, is stirred into water? Is the sucrose destroyed? We know it isn't, because it sweetens the water. Does the sucrose disappear because it because it somehow ceases to occupy space or because it fits within the nooks and crannies of the water? Not so, for the addition of sucrose changes the volume. This may not be noticeable at first, but if you continue adding sucrose to a glass of water, you'll see that the water level rises, just as it would if you were adding sand.

Sucrose stirred into water loses its crystalline form. Each sucrose crystal consists of billions upon billions of sucrose molecules packed neatly together. When the crystal is exposed to water (as was first shown in Figure 19.4 and is shown again here in Figure 19.11), an even greater number of water molecules pull on the sucrose molecules via hydrogen bonds formed between the sucrose molecules and the water molecules. With a little stirring, the sucrose molecules soon mix throughout the water. In place of sucrose crystals and water, we have a homogeneous mixture of sucrose molecules in water. This means that the sweetness of the first sip of the solution is the same as the sweetness of the last sip.

Recall that a homogeneous mixture consisting of a single phase is called a *solution*. Sugar in water is a solution in the liquid phase. Solutions aren't always liquids, however. They can also be solid or gaseous, as Figure 19.12 shows. Gemstones are solid solutions. A ruby, for example, is a solid solution of trace quantities of red chromium compounds in transparent aluminum oxide. Another important example of solid solutions is metal alloys, which are mixtures of different metallic elements. The alloy known as brass is a solid solution of

> To most people, solutions mean finding the answers. To chemists, however, solutions are things that are still all mixed up.

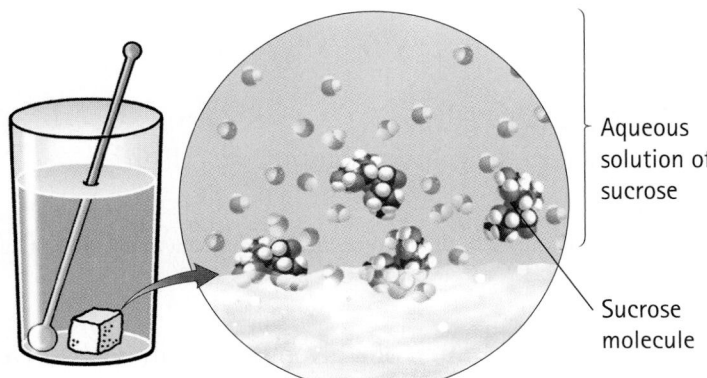

Aqueous solution of sucrose

Sucrose molecule

FIGURE 19.11 ▲
Water molecules pull the sucrose molecules in a sucrose crystal away from one another. This pulling away does not, however, affect the covalent bonds within each sucrose molecule, which is why each dissolved sucrose molecule remains intact as a single molecule.

FIGURE 19.12 ▶
Solutions may occur in (a) the solid phase, (b) the liquid phase, or (c) the gaseous phase.

(a)

(b)

(c)

copper and zinc, for instance, and the alloy stainless steel is a solid solution of iron, chromium, nickel, and carbon.

An example of a gaseous solution is the air we breathe. By volume, this solution is 78% nitrogen gas, 21% oxygen gas, and 1% other gaseous materials, including water vapor and carbon dioxide. The air we exhale is a gaseous solution of 75% nitrogen, 14% oxygen, 5% carbon dioxide, and around 6% water vapor. So we see the air we breathe undergoes a change in composition as it passes through our lungs.

✓ In describing solutions, the component present in the largest amount is the **solvent,** and any other components are **solutes.** For example, when a teaspoon of table sugar is mixed with 1 liter of water, we identify the sugar as the solute and the water as the solvent.

The process of a solute mixing with a solvent is called dissolution, or **dissolving.** To make a solution, a solute must dissolve in a solvent; that is, the solute and solvent must form a homogeneous mixture. Whether or not one material dissolves in another is a function of their electrical attractions for each other. The stronger the attractions, the greater the ability to dissolve.

CONCEPT CHECK
What is the solvent in the gaseous solution we call air?

Check Your Answer
Nitrogen is the solvent, because it is the component that is present in the greatest quantity.

There is a limit to how much of a given solute can be dissolved in a given solvent, as Figure 19.13 illustrates. We know that when you add table sugar to a glass of water, for example, the sugar rapidly dissolves. As you continue to add sugar, however, there comes a point when it no longer dissolves. Instead, it collects at the bottom of the glass, even after stirring. At this point, the water is saturated with sugar, meaning that the water cannot accept any more sugar. When this happens, we have what is

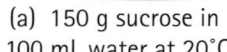

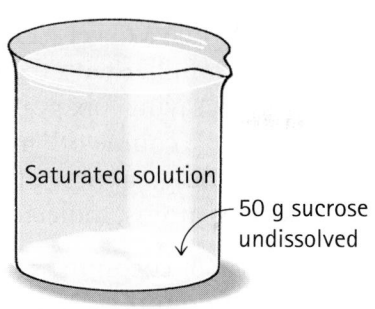

(a) 150 g sucrose in 100 mL water at 20°C

(b) 200 g sucrose in 100 mL water at 20°C

(c) 250 g sucrose in 100 mL water at 20°C

50 g sucrose undissolved

FIGURE 19.13 ▲
A maximum of 200 g of sucrose dissolves in 100 mL of water at 20°C. (a) Mixing 150 g of sucrose in 100 mL of water at 20°C produces an unsaturated solution. (b) Mixing 200 g of sucrose in 100 mL of water at 20°C produces a saturated solution. (c) If 250 g of sucrose is mixed with 100 mL of water at 20°C, 50 g of sucrose remains undissolved.

called a **saturated solution,** defined as one in which no more solute can be dissolved. A solution that has not reached the limit of solute that will dissolve is called an **unsaturated solution.**

✔ **READING CHECK**

What is the difference between the solvent and the solute?

19.4 Concentration Is Given as Moles per Liter

✔ The quantity of solute dissolved in a solution is described in mathematical terms by the solution's **concentration,** which is the amount of solute dissolved per amount of solution:

Concentration = amount of solute/amount of solution.

For example, a sucrose-water solution may have a concentration of 1 gram of sucrose for every liter of solution. This can be compared with concentrations of other solutions. A sucrose-water solution containing 2 grams of sucrose per liter of solution, for example, is more *concentrated,* and one containing only 0.5 gram of sucrose per liter of solution is less concentrated, or more *dilute.*

Chemists are often more interested in the number of solute particles in a solution than in the number of grams of solute. Submicroscopic particles, however, are so very small that the number of them in any observable sample is incredibly large. To avoid awkwardly large numbers, scientists use a unit called the mole. One **mole** of any type of particle is, by definition, 6.02×10^{23} particles. (This superlarge number is about 602 billion trillion, or 602,000,000,000,000,000,000,000 particles. Interestingly, the term mole is derived from the Latin word *moles,* meaning "heap," "mass," or "pile.") One mole of gold atoms, for example, is 6.02×10^{23} gold atoms, and 1 mole of sucrose molecules is 6.02×10^{23} sucrose molecules.

Whew! This espresso is really concentrated!

Ahh. My regular coffee is more dilute.

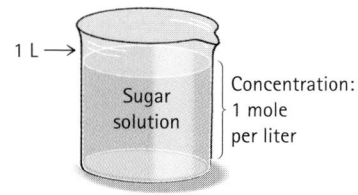

1 mole of sucrose
equals
342 g of sucrose
equals
6.02×10^{23} molecules of sucrose

FIGURE 19.14 ▲

An aqueous solution of sucrose that has a concentration of 1 mole of sucrose per liter of solution contains 6.02×10^{23} sucrose molecules (342 g) in every liter of solution.

Even if you've never heard the term *mole* in your life before now, you are already familiar with the basic idea. Saying "one mole" is just a shorthand way of saying "six point oh two times ten to the twenty-third particles." Just as "a couple of" means 2 of something and "a dozen of" means 12 of something, "a mole of" means 6.02×10^{23} of some elementary unit, such as atoms, molecules, or ions. It's as simple as that:

- a couple of coconuts = 2 coconuts
- a dozen donuts = 12 donuts
- a mole of molecules = 6.02×10^{23} molecules

One mole is an unbelievably large number. A stack containing "1 mole" of pennies would reach a height of about 860 quadrillion kilometers, which is roughly equal to the diameter of our galaxy, the Milky Way. And "1 mole" of marbles would be enough to cover the entire land area of the 50 United States to a depth greater than 1.1 kilometers.

But sucrose molecules are so small that there are 6.02×10^{23} of them in only 342 grams of sucrose, which is about a cupful. Thus, because 342 grams of sucrose contain 6.02×10^{23} molecules of sucrose, we can use our shorthand wording and say that 342 grams of sucrose contain 1 mole of sucrose. As Figure 19.14 shows, therefore, an aqueous solution that has a concentration of 342 grams of sucrose per liter of solution also has a concentration of 6.02×10^{23} sucrose msolecules per liter of solution or, by definition, a concentration of 1 mole of sucrose per liter of solution. The number of grams tells you the mass of solute in a given solution, and the number of moles indicates the actual number of molecules.

A common unit of concentration used by chemists is **molarity,** which is the solution's concentration expressed in moles of solute per liter of solution:

Molarity = number of moles of solute/liters of solution.

A solution that contains 1 mole of solute per liter of solution is a 1 molar solution, which is often abbreviated 1 M. A 2 molar (2 M) solution contains 2 moles of solute per liter of solution.

The difference between referring to the number of molecules of solute and referring to the number of grams of solute can be illustrated by the following question. A saturated solution of sucrose contains 200 grams of sucrose and 100 grams of water. Which is the solvent: sucrose or water?

As shown in Figure 19.15, there are 3.5×10^{23} molecules of sucrose in 200 grams of sucrose, but there are almost 10 times as many molecules of water in 100 grams of water—3.3×10^{24} molecules. As defined

Saturated solution of sucrose in water at 20°C

Component	Mass	Number of molecules
Sucrose	200 g	3.5×10^{23}
Water	100 g	3.3×10^{24}

◄ **FIGURE 19.15**

Although 200 g of sucrose is twice as massive as 100 g of water, there are about 10 times as many water molecules in 100 g of water as there are sucrose molecules in 200 g of sucrose. How can this be? Each water molecule is about 1/20th as massive (and much smaller) than each sucrose molecule, which means that about 10 times as many water molecules can fit within half the mass.

earlier, the solvent is the component present in the largest amount, but what do we mean by amount? If amount means number of molecules, then water is the solvent. If amount means mass, then sucrose is the solvent. So, the answer depends on how you look at it. From a chemist's point of view, amount typically means the number of molecules, and so water is the solvent in this case.

✔ READING CHECK

What does concentration measure?

CONCEPT CHECK

1. How many moles of sucrose are there in 0.5 L of a 2 molar solution? How many molecules of sucrose is this?
2. Does 1 L of a 1 molar solution of sucrose in water contain 1 L of water, less than 1 L of water, or more than 1 L of water?

Check Your Answers

1. First you need to understand that 2 molar means 2 moles of sucrose per liter of solution. To obtain the amount of solute, you should multiply solution concentration by amount of solution:

$$(2 \text{ moles/L})(0.5 \text{ L}) = 1 \text{ mole},$$

which is the same as 6.02×10^{23} molecules.

2. The definition of molarity refers to the number of liters of solution, not to the number of liters of solvent. When sucrose is added to a given volume of water, the volume of the solution increases. So, if 1 mole of sucrose is added to 1 L of water, the result is more than 1 L of solution. Therefore, 1 L of a 1 molar solution requires less than 1 L of water.

19.5 Solubility Measures How Well a Solute Dissolves

The **solubility** of a solute is its *ability* to dissolve in a solvent. ✔ As can be expected, solubility depends on the submicroscopic attractions between solute particles and solvent particles. In general, the stronger these attractions, the greater the solubility. If a solute has any appreciable solubility in a solvent, then that solute is said to be **soluble** in that solvent.

Solubility also depends on the attractions of solute particles for one another. As shown in Figure 19.16, a sucrose molecule has many polar hydrogen–oxygen bonds. Sucrose molecules, therefore, can form multiple hydrogen bonds with one another. These hydrogen bonds are strong enough to make sucrose a solid at room temperature and to give it the relatively high melting point of 185°C. In order for sucrose to dissolve in water, the water molecules must first pull sucrose molecules away from one another. This puts a limit on the amount of sucrose that can dissolve in water—eventually, a point is reached at which there are not enough water molecules to separate the sucrose molecules from one another. As we discussed in Section 19.3, this is the point of saturation, and any additional sucrose added to the solution does not dissolve.

Grease is soluble in paint thinner, which is why paint thinner can be used to clean one's hands of grease. But body oils are also soluble in paint thinner, which is why hands cleaned with paint thinner feel dry and chapped.

FIGURE 19.16 ▶
A sucrose molecule is shown here, using a "space-filling" model (above) and the more traditional "stick" structure (below). Sucrose is a very polar molecule because of its many O–H bonds. Neighboring sucrose molecules, therefore, are strongly attracted to each other.

Sucrose

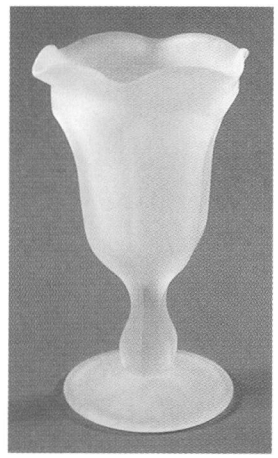

FIGURE 19.17 ▲
Glass is frosted by dissolving its outer surface in hydrofluoric acid.

Let's now look at a solute that has very little solubility, namely, oxygen, O_2, in water. In contrast to sucrose, which has a solubility of 200 grams per 100 milliliters of water, only 0.004 gram of oxygen can dissolve in 100 milliliters of water. We can account for oxygen's low solubility in water by noting that the only electrical attractions that occur between oxygen molecules and water molecules are relatively weak dipole–induced dipole attractions. These weak attractions must compete with the stronger attraction of water molecules for water molecules, which effectively excludes oxygen molecules from intermingling.

A material that does not dissolve in a solvent to any appreciable extent is said to be **insoluble** in that solvent. There are many substances we consider to be insoluble in water, including sand and glass. Just because a material is not soluble in one solvent, however, does not mean it won't dissolve in another. Sand and glass, for example, are soluble in hydrofluoric acid, HF, which is used to give glass the decorative frosted look shown in Figure 19.17. Also, although Styrofoam is insoluble in water, it is soluble in acetone, a solvent once commonly used in fingernail-polish remover. Pour a little acetone into a Styrofoam cup, and the acetone soon causes the Styrofoam to deform, as demonstrated in Figure 19.18.

FIGURE 19.18 ▲
Is this cup melting or dissolving?

CONCEPT CHECK
Why is there are limit to the amount of sucrose that will dissolve in water?

Check Your Answer
The attraction between two sucrose molecules is much stronger than the attraction between a sucrose molecule and a water molecule. Because of this, sucrose dissolves in water only so long as the number of water molecules far exceeds the number of sucrose molecules. When there are too few water molecules to dissolve any additional sucrose, the solution is saturated.

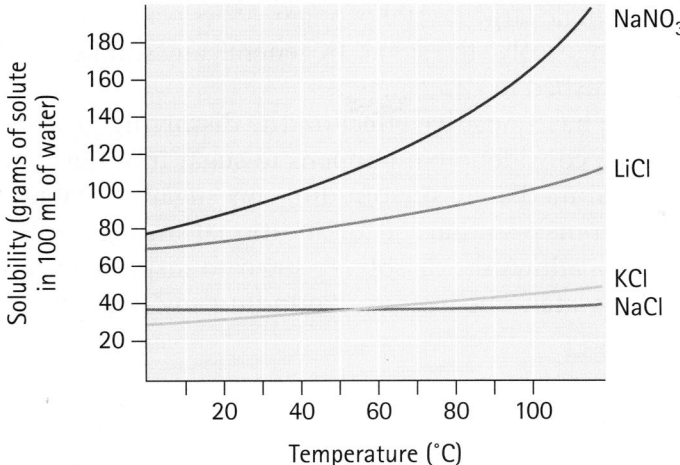

FIGURE 19.19 ▲
The solubility of many water-soluble solids increases with temperature, while the solubility of others is only very slightly affected by temperature.

Solubility Changes with Temperature

You probably know from experience that water-soluble solids usually dissolve better in hot water than in cold water. A highly concentrated solution of sucrose in water, for example, can be made by heating the solution almost to the boiling point. This is how syrups and hard candy are made.

Solubility increases with increasing temperature because hot water molecules have greater kinetic energy and therefore are able to collide with the solid solute more vigorously. The vigorous collisions facilitate the disruption of electrical particle-to-particle attractions in the solid.

Although the solubilities of many solid solutes—sucrose, to name just one example—are greatly increased by temperature increases, the solubilities of other solid solutes, such as sodium chloride, are only mildly affected, as Figure 19.19 shows. This difference involves a number of factors, including the strength of the chemical bonds in the solute molecules and the way those molecules are packed together. Some chemicals, such as calcium carbonate, $CaCO_3$, actually become *less* soluble as the water temperature increases. This explains why the inner surface of tea kettles is often coated with calcium carbonate residues.

When a sugar solution saturated at a high temperature is allowed to cool, some of the sugar usually comes out of solution and forms what is called a **precipitate.** When this occurs, the solute, sugar in this case, is said to have *precipitated* from the solution.

Solid Solutes Dissolve Quicker When Crushed and Stirred

Drop a large crystal of sugar into a cup of water and you will find it takes a long while to dissolve. Crush that sugar crystal into a fine powder, however, and the sugar dissolves quite readily. The reason for this is because the crushed sugar has more surface area in contact with the

READING CHECK

What helps to make a solute soluble in a solvent?

Air is a gaseous solution, and one of its minor components is water vapor. The process of this water coming "out of solution" in the form of rain or snow is called "precipitation." The rain or snow is the "precipitate."

water than the single large crystal of sugar. More water molecules, therefore, can work to dissolve the sugar. In general, any solid solute dissolves quicker when crushed to a fine powder.

Stirring also has a positive effect on the dissolving process. To understand this, consider what happens to sugar in water if you *don't* stir. The water immediately surrounding any sugar crystal becomes saturated, which makes it harder for additional sugar molecules to dissolve. Stirring allows the dissolving sugar to disperse evenly throughout the water. The water surrounding any sugar crystal, therefore, is as fresh as possible.

Solubility of Gases

In contrast to the solubilities of most solids, the solubilities of gases in liquids *decrease* with increasing temperature, as Table 19.1 below shows. This effect occurs because, with an increase in temperature, the solvent molecules have more kinetic energy. This makes it more difficult for a gaseous solute to remain in solution because the solute molecules are literally ejected by the high-energy solvent molecules.

Perhaps you have noticed that warm carbonated beverages go flat faster than cold ones. The higher temperature causes the molecules of carbon dioxide gas to leave the liquid solvent at a higher rate.

The solubility of a gas in a liquid also depends on the pressure of the gas immediately above the liquid. In general, a higher gas pressure above the liquid means more of the gas dissolves. A gas at a high pressure has many, many gas particles crammed into a given volume. The "empty" space in an unopened soft-drink bottle, for example, is crammed with carbon dioxide molecules in the gaseous phase. With nowhere else to go, many of these molecules dissolve in the liquid, as shown in Figure 19.20. Alternatively, we might say that the great pressure forces the carbon dioxide molecules into solution. When the bottle is opened, the "head" of highly pressurized carbon dioxide gas escapes. Now the gas pressure above the liquid is lower than before. As a result,

TABLE 19.1 Temperature-Dependent Solubility of Oxygen Gas in Water at a Pressure of 1 Atmosphere

Temperature (°C)	O_2 Solubility (g O_2/L H_2O)
0	0.0141
10	0.0109
20	0.0092
25	0.0083
30	0.0077
35	0.0070
40	0.0065

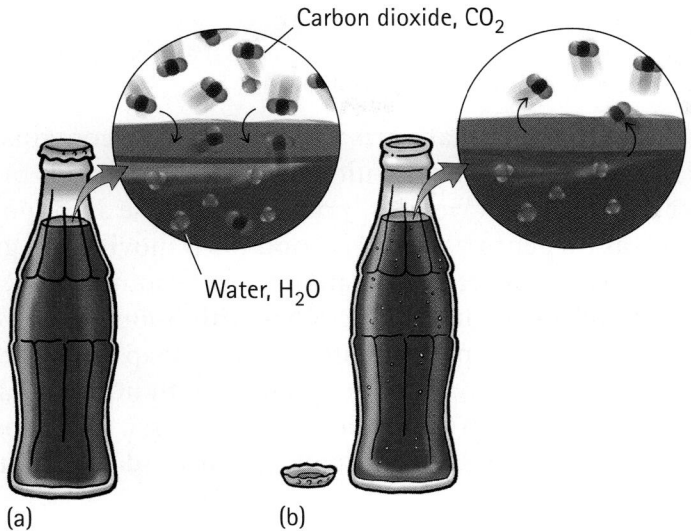

Carbon dioxide, CO_2

Water, H_2O

(a)   (b)

FIGURE 19.20 ▲
(a) The carbon dioxide gas above the liquid in an unopened soft-drink bottle consists of many tightly packed carbon dioxide molecules that are forced by pressure into solution. (b) When the bottle is opened, the pressure is released, and carbon dioxide molecules originally dissolved in the liquid can escape into the air.

the solubility of the carbon dioxide drops, and the carbon dioxide molecules that were once squeezed into the solution begin to escape into the air above the liquid.

The rate at which carbon dioxide molecules leave an opened soft drink is relatively slow. You can increase the rate by pouring in granulated sugar, salt, or sand. The microscopic nooks and crannies on the surfaces of the grains serve as *nucleation sites* where carbon dioxide bubbles are able to form rapidly and then to escape by buoyant forces. Shaking the beverage also increases the surface area of the liquid-to-gas interface, making it easier for the carbon dioxide to escape from the solution. Once the solution is shaken, the rate at which carbon dioxide escapes becomes so great that the beverage froths over. You also increase the rate at which carbon dioxide escapes when you pour the beverage into your mouth, which abounds in nucleation sites. You can feel the resulting tingly sensation. No doubt about it—carbonated beverages are a real treat.

CONCEPT CHECK
You open two cans of soft drink, one from a warm kitchen shelf and the other from the coldest depths of your refrigerator. Which fizzes more in your mouth?

Check Your Answer
The solubility of carbon dioxide in water decreases with increasing temperature. The warm drink, therefore, will fizz in your mouth more than the cold one will.

Polar solvents, such as water, work well to dissolve polar solutes, such as sugar. Likewise, nonpolar solvents, such as paint thinner, work well to dissolve nonpolar solvents, such as oil.

19.6 Soap Works by Being Both Polar and Nonpolar

Dirt and grease together make *grime*. Because grime contains many nonpolar components, it is difficult to remove from hands or clothing with water alone. To remove most grime, we can use a nonpolar solvent, such as turpentine, which is good for removing the grime left on hands after such activities as changing a car's motor oil. Rather than washing our dirty hands and clothes with nonpolar solvents, however, we have a more pleasant alternative—soap and water.

✓ Soap works because soap molecules have both nonpolar and polar properties. A typical soap molecule has two parts: a long, nonpolar tail of carbon and hydrogen atoms and a polar head containing at least one ionic bond.

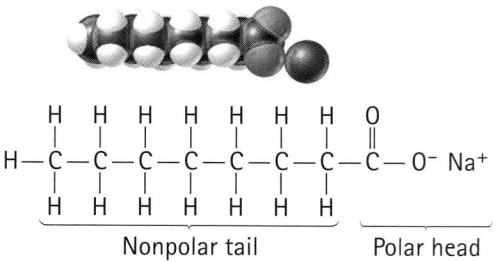

Nonpolar tail   Polar head

Because most of a soap molecule is nonpolar, it attracts nonpolar grime molecules, as Figure 19.21 illustrates. In fact, grime quickly finds itself surrounded in three dimensions by the nonpolar tails of soap molecules. This attraction is usually sufficient to lift the grime away from the surface being cleaned. With the nonpolar tails facing inward toward the grime, the polar heads are all directed outward, where they are attracted to water molecules by relatively strong ion–dipole attractions. If the water is flowing, the whole conglomeration of grime and soap molecules flows with it, away from your hands or clothes and then down the drain.

For the past several centuries, soaps have been prepared by treating animal fats with sodium hydroxide, NaOH, also known as caustic lye. In this reaction, which is still used today, each fat molecule is broken down into three *fatty acid* soap molecules and one glycerol molecule:

FIGURE 19.21 ▲
Nonpolar grime attracts and is surrounded by the nonpolar tails of soap molecules. The polar heads of the soap molecules are attracted by ion–dipole attractions to water molecules, which then carry the soap–grime combination away.

READING CHECK

What properties do soap molecules have?

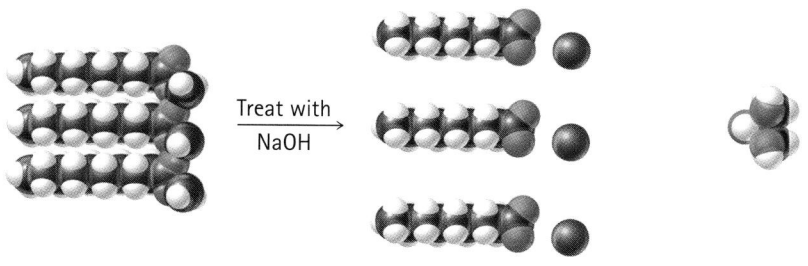

Fat molecule   Three fatty acid soap molecules   Glycerol molecule

In the 1940s, chemists began developing a class of synthetic soaplike compounds, known as *detergents,* that offer several advantages over true soaps, such as stronger grease penetration and lower price. The chemical structure of detergent molecules is similar to that of soap molecules in that both possess a polar head attached to a nonpolar tail. The polar head in a detergent molecule, however, typically consists of either a sulfate group, $-OSO_3^-$, or a sulfonate group, $-SO_3^-$, and the nonpolar tail can have an assortment of structures.

One of the most common sulfate detergents is sodium lauryl sulfate, a main ingredient of many toothpastes. A common sulfonate detergent is sodium dodecyl benzenesulfonate, also known as a linear alkylsulfonate, or LAS, often found in dishwashing liquids. Both of these detergents are biodegradable, which means that microorganisms can break down the molecules once they are released into the environment.

fyi
- "Dry cleaning" is the process of washing clothes without water. The most common dry cleaning solvent is perchloroethylene, C_2Cl_4, which is gentle on the clothing and can clean a full load in under 10 minutes. After a washing cycle, the solvent is centrifuged out of the machine, filtered, distilled, and recycled for the next load. Clothes come out of the machine already dry and ready for folding.

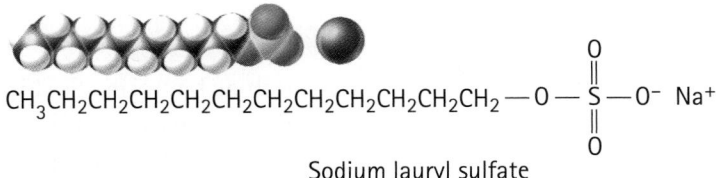

$CH_3CH_2CH_2CH_2CH_2CH_2CH_2CH_2CH_2CH_2CH_2CH_2 - O - \overset{\overset{O}{\|}}{\underset{\underset{O}{\|}}{S}} - O^-\ Na^+$

Sodium lauryl sulfate

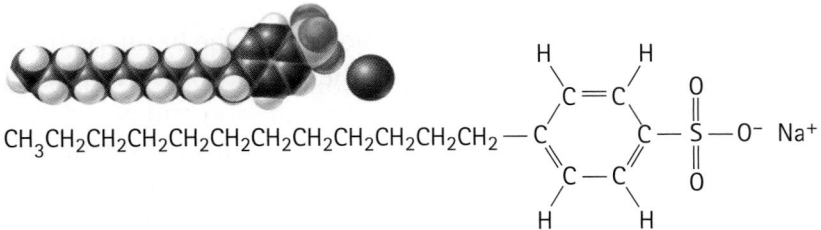

$CH_3CH_2CH_2CH_2CH_2CH_2CH_2CH_2CH_2CH_2CH_2CH_2 - C \cdots C - \overset{\overset{O}{\|}}{\underset{\underset{O}{\|}}{S}} - O^-\ Na^+$

Sodium dodecyl benzenesulfonate

CONCEPT CHECK

Is it possible for a molecule to be both polar and nonpolar at the same time?

Check Your Answers

Yes, and soap molecules are a perfect example.

19.7 Purifying the Water We Drink

As was discussed earlier, it is impossible to obtain 100% pure water. However, we are able to purify water to meet our needs. ✔ We do this by taking advantage of the differences in physical properties of water and the solutes or particulates it contains.

Water that is safe for drinking is said to be *potable.* In the United States, potable water is currently used for everything from cooking to flushing our toilets. The first step most public utilities take to produce potable water from natural sources such as groundwater, lakes, or rivers

Lack of clean drinking water is one of the world's leading causes of death, especially among children of developing nations.

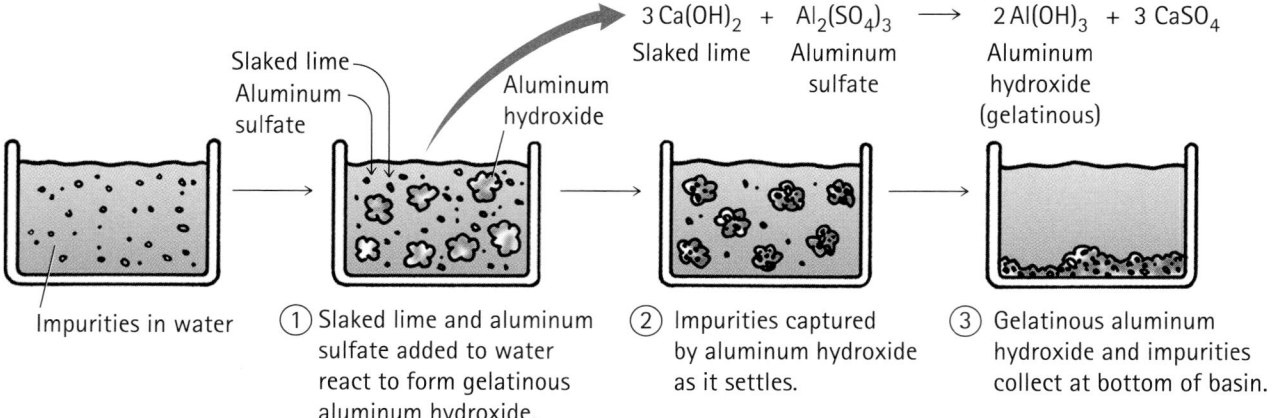

$$3\,Ca(OH)_2 \;+\; Al_2(SO_4)_3 \;\longrightarrow\; 2\,Al(OH)_3 \;+\; 3\,CaSO_4$$

Slaked lime   Aluminum   Aluminum
           sulfate   hydroxide
                         (gelatinous)

Slaked lime
Aluminum
sulfate

Aluminum
hydroxide

Impurities in water

① Slaked lime and aluminum sulfate added to water react to form gelatinous aluminum hydroxide.

② Impurities captured by aluminum hydroxide as it settles.

③ Gelatinous aluminum hydroxide and impurities collect at bottom of basin.

FIGURE 19.22 ▲
Slaked lime, $Ca(OH)_2$, and aluminum sulfate, $Al_2(SO_4)_3$, react to form aluminum hydroxide, $Al(OH)_3$, and calcium sulfate, $CaSO_4$, which together form a gelatinous material.

FIGURE 19.23 ▲
Volatile impurities are removed from drinking water by cascading it through the columns of air within each of these stacks.

In the early 1990s, an anti-water-chlorination campaign in Peru led the government to stop chlorinating the drinking water. Within months there were 1.3 million new cases of cholera resulting in 13,000 deaths.

is to remove any dirt particles or pathogens, such as bacteria. This is done by mixing the water with certain minerals, such as slaked lime and aluminum sulfate, which forms a gelatinous material, aluminum hydroxide, that intersperses throughout the water (Figure 19.22). This is done in a large settling basin. Slow stirring causes the gelatinous material to clump together and settle to the bottom of the basin. As these clumps form and settle, they carry with them many of the dirt particles and bacteria. The water is then filtered through sand and gravel.

To improve the odor and flavor of the water, many treatment facilities also *aerate* the water by cascading it through a column of air, as shown in Figure 19.23. Aeration removes many unpleasant-smelling volatile chemicals, such as sulfur compounds. At the same time, air dissolves into the water giving it a better taste—without dissolved air, the water tastes flat. As a final step, the water is treated with a disinfectant, usually chlorine gas, Cl_2, but sometimes ozone, O_3, and then stored in a holding tank that feeds into the city mains.

Developed countries have the technology and infrastructure to produce vast quantities of water suitable for drinking—as a result, many citizens take their drinking water for granted. The number of public water-treatment facilities in developing nations, however, is relatively small. In these these locations, many people drink their water in the form of a hot beverage, such as tea, which is disinfected through boiling. Alternatively, disinfecting iodine tablets can be used.

Fuel for boiling and tablets for disinfecting, however, are not always available. As a result, more than 400 people in the world (mostly children) die every hour from preventable diseases or infections such as cholera, typhoid fever, dysentery, and hepatitis, which they contract by drinking contaminated water. In response, several American manufacturers have developed tabletop systems that bathe water with pathogen-killing ultraviolet light. One prototype model, shown in Figure 19.24, disinfects 15 gallons per minute, weighs about 15 pounds, and is powered

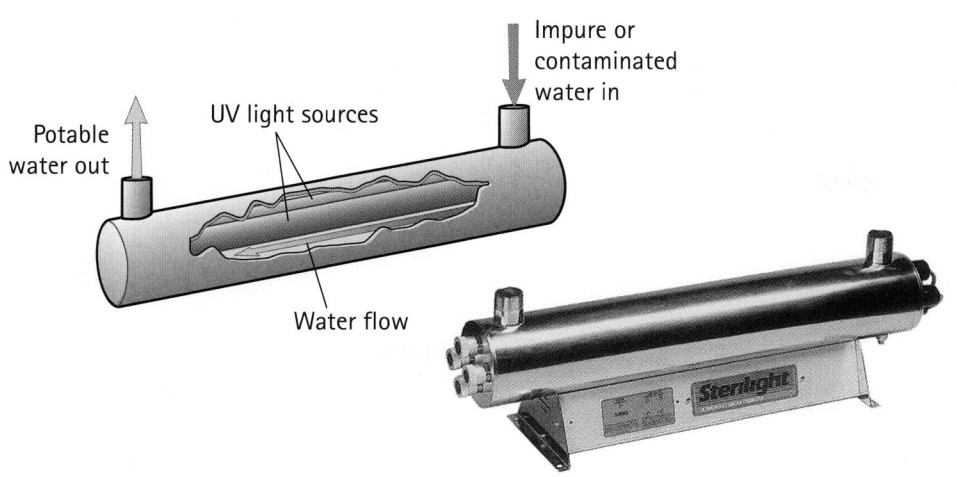

Impure or contaminated water in

Potable water out

UV light sources

Water flow

◀ **FIGURE 19.24**
Small-scale water-disinfecting units, such as the one shown here, have great value in regions of the world where potable water is scarce.

by photovoltaic solar cells, which permit it to run unsupervised in remote locations.

Aside from pathogens, untreated water from wells or rivers may contain toxic metals that seep into the water supply from natural geologic formations. Many of the wells in Bangladesh, for example, are made very deep so as to avoid the pathogens that thrive in the surface waters of this region. The water obtained from these deep wells, however, is highly contaminated with arsenic—a naturally occurring element in the Earth's crust. The arsenic is in the underlying rock, which formed from river sediments carried down from the Himalayas. Because this region is so densely populated, as many as 70 million people may be subject to some level of arsenic poisoning, which manifests itself as skin lesions and a higher susceptibility to cancer. Low-cost methods for removing arsenic from well water are greatly needed, as are worldwide recognition of this problem and the political, economic, and social support to overcome it.

fyi

- In 1908, Jersey City, New Jersey, became the first American city to begin chlorinating its drinking water. By 1910, as disinfecting drinking water with chlorine became more widespread, the death rate from typhoid fever dropped to 20 lives per 100,000. In 1935, the death rate fell to 3 lives per 100,000. By 1960, fewer than 20 persons in the entire United States died from typhoid fever.

CONCEPT CHECK
At a water-treatment facility, how does adding slaked lime and aluminum sulfate to water purify the water?

Check Your Answer
The water entering a water treatment plant is usually a heterogeneous mixture containing suspended solids. Adding slaked lime and aluminum sulfate serves to capture these suspended solids, which then sink to the bottom, where they are easily removed.

✔ **READING CHECK**

How do we purify our drinking water?

19 CHAPTER REVIEW

KEY TERMS

Concentration A quantitative measure of the amount of solute in a solution.

Dissolving The process of mixing a solute in a solvent to produce a homogeneous mixture.

Distillation A purifying process in which a vaporized substance is collected by exposing it to cooler temperatures over a receiving flask, which collects the condensed purified liquid.

Heterogeneous mixture A mixture in which the various components can be seen as individual substances.

Homogeneous mixture A mixture in which the components are so finely mixed that the composition is the same throughout.

Impure In chemistry, this term refers to a material that is a mixture of more than one element or compound.

Insoluble Not capable of dissolving to any appreciable extent in a given solvent.

Mixture A combination of two or more substances in which each substance retains its properties.

Molarity A unit of concentration equal to the number of moles of a solute per liter of solution.

Mole The amount of any pure substance that contains as many atoms, molecules, ions, or other elementary units as the number of atoms in 12 grams of carbon-12. This is equal to 6.02×10^{23} particles.

Precipitate A solute that has come out of solution.

Pure Having a uniform composition, or being without impurities. In chemistry, the term is used to denote a material that consists of a single element or compound.

Saturated solution A solution containing the maximum amount of solute that will dissolve in its solvent.

Solubility The ability of a solute to dissolve in a given solvent.

Soluble Capable of dissolving to an appreciable extent in a given solvent.

Solute Any component in a solution that is not the solvent.

Solution A homogeneous mixture in which all components are in the same phase.

Solvent The component in a solution that is present in the largest amount.

Suspension A homogeneous mixture in which the various components are in different phases.

Unsaturated solution A solution that is capable of dissolving additional solute.

REVIEW QUESTIONS

Most Materials Are Mixtures

1. What defines a material as being a mixture?
2. How can the components of a mixture be separated from one another?
3. How does distillation separate the components of a mixture?

The Chemist's Classification of Matter

4. Why is it not practical to have a macroscopic sample that is 100% pure?
5. How is a solution different from a suspension?
6. How can a solution be distinguished from a suspension?

A Solution Is a Single-Phase Homogeneous Mixture

7. What happens to the volume of a sugar solution as more sugar is dissolved in it?
8. Why is a ruby considered to be a solution?
9. Distinguish between a solute and a solvent.

Concentration Is Given As Moles Per Liter

10. What does it mean to say that a solution is concentrated?

11. Is 1 mole of particles a very large number of particles or a very small number?

12. Is concentration typically given with the volume of solvent or the volume of solution?

Solubility Measures How Well a Solute Dissolves

13. Why does the solubility of a gas solute in a liquid solvent decrease with increasing temperature?

14. Why do sugar crystals dissolve faster when crushed?

15. Is sugar a polar or nonpolar substance?

Soap Works by Being Both Polar and Nonpolar

16. Water and soap are attracted to each other by what type of electrical attraction?

17. What is the difference between a soap and a detergent?

Purifying the Water We Drink

18. Why is treated water sprayed into the air prior to being piped to users?

19. What are two ways in which people disinfect water in areas where municipal treatment facilities are not available?

20. What naturally occurring element has been contaminating the water supply of Bangladesh?

THINK AND DO

1. To see the gases dissolved in your water, fill a clean cooking pot with water and let it stand at room temperature for several hours. Note the bubbles that adhere to the inner sides of the pot. Where do the bubbles originate? What do you suppose they contain? For further experimentation, place warm water from the kitchen faucet in one pot. In a second pot, place boiled water that has cooled down to the same temperature. You'll find that boiling *deaerates* the water—that is, it removes the atmospheric gases. Why don't fish live very long in deaerated water?

2. Place a concentrated dot of ink from a black felt-tip pen at the center of the piece of porous paper. Carefully place one drop of water on top of the dot, and watch the ink spread radially with the solvent. Because the different components of the ink have different affinities for the water (and the paper) they travel with the water at different rates. Just after the drop of water is completely absorbed, add a second drop, then a third, and so on until the ink components have separated to your satisfaction. Try other types of inks.

This technique, called *paper chromatography*, was originally developed to separate plant pigments from one another. The separated pigments had different colors, which is how this technique got its name—*chroma* is Latin for "color." Try the above procedure using spinach extract!

THINK AND COMPARE

1. Rank the following solutions in order of increasing concentration: Solution A, 0.5 moles of sucrose in 2.0 liters of solution; Solution B, 1.0 moles of sucrose in 3.0 liters of solution; Solution C, 1.5 moles of sucrose in 4.0 liters of solution.

2. Rank the following compounds in order of increasing solubility in water:

$$CH_3CH_2{-}OH$$

Ethanol

$$CH_3CH_2CH_2CH_2{-}OH$$

Butanol

$$CH_3CH_2CH_2CH_2CH_2CH_2{-}OH$$

Hexanol

THINK AND EXPLAIN

1. Which of the following boxes contains an element? A compound? A mixture? How many different types of molecules are shown altogether in all three boxes?

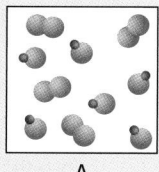 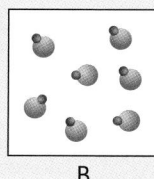 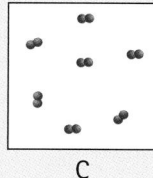

    A   B   C

2. Why can't the elements of a compound be separated from one another by physical means?

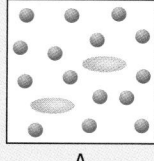 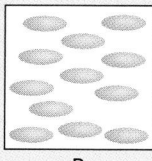 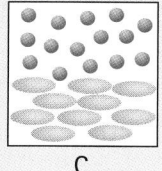

    A   B   C

3. Which of the above best represents a suspension?
4. Which of the above best represents a solution?
5. Which of the above best represents a compound?
6. Many dry cereals are fortified with iron, which is added to the cereal in the form of small iron particles. How might these particles be separated from the cereal?
7. The boiling point of 1,4-butanediol is 230°C. Would you expect this compound to be soluble or insoluble in room-temperature water? Explain.

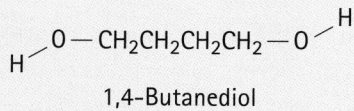

1,4-Butanediol

8. Why does oxygen have such a low solubility in water?
9. Why does the solubility of a gas solute in a liquid solvent decrease with increasing temperature?
10. Hydrogen chloride, HCl, is a gas at room temperature. Would you expect this material to be very soluble or not very soluble in water?
11. Distinguish between a saturated solution and an unsaturated solution.
12. The volume of many liquid solvents expands with increasing temperature. What happens to the concentration of a solution made with such a solvent as the temperature of the solution is increased?

13. Why is it not possible to calculate the concentration of a solution that contains 2 moles of sugar in 10 L of water?
14. If 342 g of sugar contains 1 mole of sugar molecules, then how many moles of sugar molecules are there in 171 g of sugar?
15. What is the count of sugar molecules in 0.5 mole of sugar?
16. How necessary is soap for removing salt from your hands? Why?
17. When you set a pot of tap water on the stove to boil, you'll often see bubbles start to form well before boiling temperature is reached. Explain this observation.
18. Some people fear drinking distilled water because they have heard it leaches minerals from the body. Using your knowledge of chemistry, explain how these fears have no basis and how distilled water is in fact very good for drinking.
19. Many homeowners get their drinking water piped up from wells dug on their property. Sometime this well water smells bad because of trace quantities of the gaseous compound hydrogen sulfide, H_2S. How might this odor be removed from water already taken from the tap?
20. Why is flushing a toilet with clean water from a municipal supply about as wasteful as flushing it with bottled water?

THINK AND SOLVE

1. Assume the total number of molecules in a sample of liquid is about 3 million trillion. One million trillion of these are molecules of some poison, while 2 million trillion of these are water molecules. Show that the purity of this water is about 67%.
2. Assume the total number of molecules in a glass of liquid is about 1,000,000 million trillion. One million trillion of these are molecules of some poison, while 999,999 million trillion of these are water molecules. What is the purity of the water? In other words, what percentage of all the molecules in the glass are water?
3. You drink a small glass of water that is 99.9999% pure water and 0.0001% some poison. Assume the glass contains about a million million trillion molecules, which is about 30 mL. Show that there were about 1 million trillion poison molecules in this drink. Should you be concerned?

4. How much sodium chloride, in grams, is needed to make 15 L of a solution that has a concentration of 3.0 g of sodium chloride per liter of solution?

5. If water is added to 1 mole of sodium chloride in a flask until the volume of the solution is 1 L, what is the molarity of the solution? What is the molarity when water is added to 2 moles of sodium chloride to make 0.5 L of solution?

READINESS ASSURANCE TEST (RAT)

If you have a good handle on this chapter—if you really do—then you should be able to score 7 out of 10 on this RAT. Check your answers with your teacher. If you score less than 7, study further before moving on.

Choose the best *answer to the following.*

1. Someone argues that he or she doesn't drink tap water because it contains thousands of molecules of some impurity in each glass. How would you respond in defense of the water's purity, if it indeed does contain thousands of molecules of some impurity per glass?
   (a) Impurities aren't necessarily bad, in fact, they may be good for you.
   (b) The water contains water molecules and each water molecule is pure.
   (c) There's no defense. If the water contains impurities, it should not be drunk.
   (d) Compared to the billions and billions of water molecules, a thousand molecules of something else is practically nothing.

2. What is the difference between a compound and a mixture?
   (a) They both consist of atoms from different elements.

   (b) The way in which their atoms are bonded together.
   (c) One is a solid and the other is a liquid.
   (d) The components of a mixture are not chemically bonded together.

3. Is the air in your house a homogeneous or heterogeneous mixture?
   (a) Homogeneous because it is mixed very well
   (b) Heterogeneous because of the dust particles it contains
   (c) Homogeneous because it is all at the same temperature
   (d) Heterogeneous because it consists of different types of molecules

4. Why is half-frozen fruit punch always sweeter than the same fruit punch completely melted?
   (a) Because the sugar sinks to the bottom
   (b) Because sugar molecules don't crystallize with the water
   (c) Because the half-frozen fruit punch is warmer
   (d) Because sugar molecules precipitate as crystals

5. How many grams of sugar (sucrose) are there in 5 L of sugar water that has a concentration of 0.5 g per liter of solution?
   (a) 50 g
   (b) 25 g
   (c) 2.5 g
   (d) 1.5 g

6. Suggest why sodium chloride, NaCl, is insoluble in gasoline. Consider the electrical attractions.
   (a) Since this molecule is so small, there is not much opportunity for the gasoline to interact with it through any electrical attractions.
   (b) Since gasoline is a very polar molecule, the salt can only form dipole–induced dipole bonds, which are very weak, giving it a low solubility in gasoline.
   (c) Since gasoline is so strongly attracted to itself, the salt, NaCl, is excluded.
   (d) Salt is composed of ions that are too attracted to themselves. Gasoline is nonpolar so salt and gasoline will not interact very well.

7. Fish don't live very long in water that has just been boiled and brought back to room temperature. Suggest why.
   (a) There is now a higher concentration of dissolved CO_2 in the water.
   (b) The nutrients in the water have been destroyed.
   (c) Since some of the water was evaporated while boiling, the salts in the water are now more concentrated. This has a negative effect on the fish.
   (d) The boiling process removes the air that was dissolved in the water. Upon cooling, the water is void of its usual air content, hence, the fish drown.

8. Would you expect to find more dissolved oxygen in polar or tropical ocean waters? Why?
   (a) There would be more dissolved oxygen in the tropical oceans because intense tropical storms mix up the atmospheric oxygen into the ocean water.
   (b) There would be more dissolved oxygen in the polar oceans because the colder oxygen would "sink" and dissolve into the water.
   (c) There would be more dissolved oxygen in the tropical oceans because the heated oxygen molecules in the air would collide with and mix into the water.
   (d) There would be more dissolved oxygen in the polar oceans because the solubility of oxygen in water *decreases* with increasing temperature.

9. What is the boiling temperature of a single water molecule? Does this question make sense?
   (a) Boiling involves the separation of many molecules (plural). With only one molecule, the concept of boiling is meaningless.
   (b) Yes, this question does make sense because temperature measures the average kinetic energy of a molecule, which is 100°C for water.
   (c) 100°C indicates when the covalent bonds of the water molecule have been broken to give rise to hydrogen and oxygen atoms, which are released into atmosphere.
   (d) No, this question does not make sense because you need at least two molecules to get the average kinetic energy.

10. What is an advantage of using chlorine gas to disinfect drinking water supplies?
    (a) It provides residual protection against pathogens.
    (b) It gives the water a fresh taste.
    (c) Residual chlorine in water helps to whiten teeth.
    (d) Excess chlorine is absorbed into our bodies as a mineral supplement.
    (e) All of the above.

20

HOW CHEMICALS REACT

THE MAIN IDEA

Atoms change partners during a chemical reaction.

What happens in a chemical reaction? Why are new materials produced? Why must chemical equations always be balanced? If two molecules meet, will they always react to form new molecules? Why or why not? What is a catalyst and how are they used to improve our standards of living? What role do catalysts play in our environment? Why do some chemical reactions, such as the burning of wood, produce energy while other reactions, such as those occurring in the cooking of food, require the input of energy? What is the ultimate driving force for all chemical reactions? The goal of this chapter is to provide you with a stronger handle on the basics of chemical reactions, which were introduced in Chapter 17.

Explore!

Is Energy Released or Absorbed When Substances Mix Together?

1. Add lukewarm water to two plastic cups. (Do *not* use insulating Styrofoam cups.)
2. Transfer the water back and forth between cups to ensure equal temperatures.
3. Stir several tablespoons of table salt into only one of the cups. Compare the temperatures of the two cups by holding them up to your temperature-sensitive cheeks.

Analyze and Conclude

1. **Predicting** Hold some room-temperature water in the cupped palm of your hand over a sink. Will this water become warmer or cooler after you pour in a small amount (about 1 mL) of room-temperature rubbing alcohol?
2. **Making Generalizations** If two attracting molecules were to float toward each other, would they accelerate to faster speeds just before they collided? What happens to the temperature of a material when its molecules are accelerated to faster speeds?

20.1 Chemical Reactions Are Represented by Chemical Equations

As was discussed in Chapter 17, during a chemical reaction, atoms rearrange to create one or more new compounds. This activity is neatly summed up in written form as a **chemical equation.** A chemical equation shows the reacting substances, called **reactants,** to the left of an arrow that points to the newly formed substances, called **products:**

Reactants → products

Typically, reactants and products are represented by their elemental or chemical formulas. Sometimes molecular models or, simply, names may be used instead. Phases are also often shown: (*s*) for solid, (*l*) for liquid, and (*g*) for gas. Compounds dissolved in water are designated (*aq*) for aqueous solution. Lastly, numbers are placed in front of the reactants or products to show the ratio in which they either combine or form. These numbers are called *coefficients,* and they represent numbers of individual atoms and molecules. For instance, to represent the chemical reaction in which coal burns in the presence of oxygen to form gaseous carbon dioxide, we write the chemical equation using coefficients of 1:

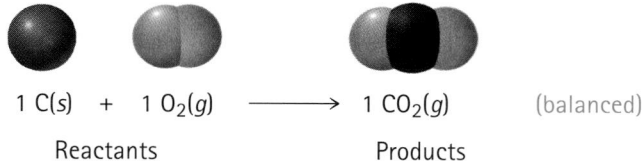

$$1\ C(s)\ +\ 1\ O_2(g)\ \longrightarrow\ 1\ CO_2(g) \qquad \text{(balanced)}$$

Reactants　　　　　　　　Products

One of the most important principles of chemistry is the **law of mass conservation.** The law of mass conservation states that matter is neither created nor destroyed during a chemical reaction. The atoms present at the beginning of a reaction merely rearrange to form new molecules. This means that no atoms are lost or gained during any reaction. The chemical equation must therefore be *balanced.* In a balanced equation, each atom must appear on both sides of the arrow the same number of times. The equation for the formation of carbon dioxide is balanced because each side shows one carbon atom and two oxygen atoms. You can count the number of atoms in the models to see this for yourself.

In another chemical reaction, two hydrogen gas molecules, H_2, react with one oxygen gas molecule, O_2, to produce two molecules of water, H_2O, in the gaseous phase:

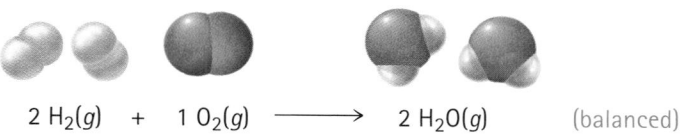

$$2\ H_2(g)\ +\ 1\ O_2(g)\ \longrightarrow\ 2\ H_2O(g) \qquad \text{(balanced)}$$

This equation for the formation of water is also balanced—there are four hydrogen and two oxygen atoms before and after the arrow.

A coefficient in front of a chemical formula tells us the number of times that element or compound must be counted. For example, 2 H_2O indicates two water molecules, which contain a total of four hydrogen atoms and two oxygen atoms. By convention, the coefficient 1 is omitted so that the above chemical equations are typically written

$$C(s) + O_2(g) \rightarrow CO_2(g) \qquad \text{(balanced)}$$

$$2\,H_2(g) + O_2(g) \rightarrow 2\,H_2O(g) \qquad \text{(balanced)}$$

CONCEPT CHECK

How many oxygen atoms are indicated by the following balanced equation?

$$3\,O_2(g) \rightarrow 2\,O_3(g)$$

Check Your Answer

There are six oxygen atoms. Before the reaction, these six oxygen atoms are found in three O_2 molecules. After the reaction, these same six atoms are found in two O_3 molecules.

An unbalanced chemical equation shows the reactants and products without the correct coefficients. For example, the equation

$$NO(g) \rightarrow N_2O(g) + NO_2(g) \qquad \text{(not balanced)}$$

is not balanced because there is one nitrogen atom and one oxygen atom before the arrow, but three nitrogen atoms and three oxygen atoms after the arrow.

You can balance unbalanced equations by adding or changing coefficients to produce correct ratios. (It's important not to change subscripts, however, because to do so changes the compound's identity—H_2O is water, but H_2O_2 is hydrogen peroxide!) For example, to balance the equation above, place a 3 before the NO:

$$3\,NO(g) \rightarrow N_2O(g) + NO_2(g) \qquad \text{(balanced)}$$

Now there are three nitrogen atoms and three oxygen atoms on each side of the arrow, and the law of mass conservation is not violated.

CONCEPT CHECK

Write a balanced equation for the reaction showing hydrogen gas, H_2, and nitrogen gas, N_2, forming ammonia gas, NH_3:

$$\underline{\qquad}\ H_2(g) + \underline{\qquad}\ N_2(g) \rightarrow \underline{\qquad}\ NH_3(g)$$

Check Your Answer

Initially, we see two hydrogen atoms before the reaction arrow and three on the right. This can be remedied by placing a coefficient of 3 by the hydrogen, H_2, and a coefficient of 2 by the ammonia, NH_3. This makes for six hydrogen atoms both before and after the reaction arrow. Meanwhile, the coefficient of 2 by the ammonia also makes for two nitrogen atoms after the arrow, which balances out the two nitrogen atoms appearing before the arrow. The full balanced equation, therefore, is

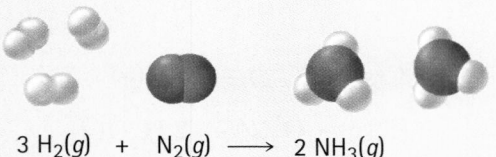

$$3\ H_2(g)\ +\ N_2(g)\ \longrightarrow\ 2\ NH_3(g)$$

Examine the *Conceptual Physical Science Exploration Practice Book* for some methods on how to balance chemical equations. Your instructor may also share with you his or her favorite method. For more practice balancing equations, see the questions at the end of this chapter.

Atoms are neither created nor destroyed in a chemical reaction. Instead, they simply change partners.

✔ **READING CHECK**

Why must the chemical equation be balanced?

Practicing chemists develop a skill for balancing equations. This skill involves creative energy and, like other skills, improves with experience. More important than being an expert at balancing equations is knowing why they need to be balanced. ✔ And the reason is the **law of mass conservation,** which tells us that atoms are neither created nor destroyed in a chemical reaction—they are simply rearranged. So every atom present before the reaction must be present after the reaction, even though the groupings of atoms are different.

20.2 Chemical Reactions Can Be Slow or Fast

A balanced chemical equation tells us what reactants react to form products. The equation, however, tells us little about how this process actually occurs. For the remainder of this chapter, we explore a few of the details of what exactly happens when reactants react to form products. We begin by showing how the speed of a reaction depends on the concentration and temperature of reacting molecules.

Some chemical reactions, such as the rusting of iron, are slow; others, such as the burning of gasoline, are fast. The speed of any reaction is indicated by its *reaction rate,* which is an indicator of how quickly the reactants transform to products. As shown in Figure 20.1, initially a flask may contain only reactant molecules. Over time, these reactants form product molecules, and, as a result, the concentration of product molecules increases. The **reaction rate,** therefore, can be defined either as how quickly the concentration of products increases or as how quickly the concentration of reactants decreases.

What determines the rate of a chemical reaction? ✔ The answer is complex, but one important factor is that reactant molecules must physically come together. Because molecules move rapidly, this physical

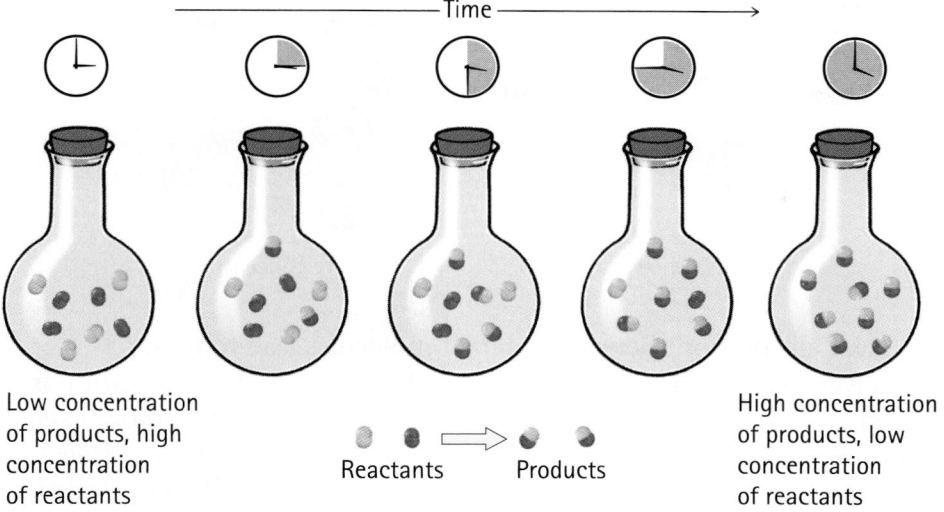

Time

Low concentration of products, high concentration of reactants

Reactants   Products

High concentration of products, low concentration of reactants

FIGURE 20.1 ▲

Over time, the reactants in this reaction flask may transform to products. If this happens quickly, the reaction rate is fast. If this happens slowly, the reaction rate is slow.

contact is appropriately described as a collision. We can illustrate the relationship between molecular collisions and reaction rate by considering the reaction of gaseous nitrogen and gaseous oxygen to form gaseous nitrogen monoxide, as shown in Figure 20.2.

Because reactant molecules must collide in order for a reaction to occur, the rate of a reaction can be increased by increasing the number of collisions. An effective way to increase the number of collisions is to increase the concentration of the reactants. Figure 20.3 shows that, with higher concentrations, there are more molecules in a given volume, which makes collisions between molecules more probable. As an analogy, consider a group of people on a dance floor—as the number of people increases, so does the rate at which they bump into one another. An increase in the concentration of nitrogen and oxygen molecules, therefore, leads to a greater number of number of collisions between these molecules—hence, a greater number of nitrogen monoxide molecules will form in a given period of time.

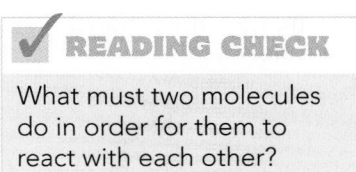

✓ **READING CHECK**

What must two molecules do in order for them to react with each other?

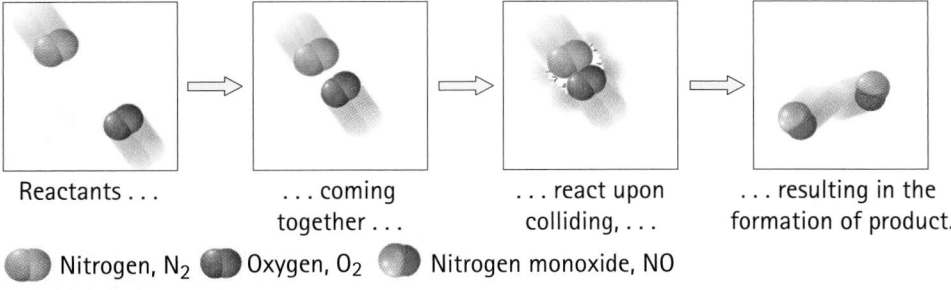

Reactants . . .   . . . coming together . . .   . . . react upon colliding, ... resulting in the formation of product.

Nitrogen, N_2 Oxygen, O_2 Nitrogen monoxide, NO

FIGURE 20.2 ▲

During a reaction, reactant molecules collide with one another.

FIGURE 20.3 ▶
The more concentrated a sample of nitrogen and oxygen, the greater the probability that N_2 and O_2 molecules will collide and form nitrogen monoxide.

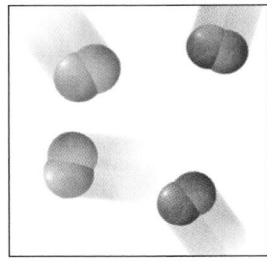
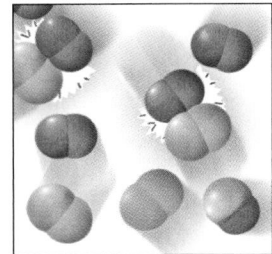

Less concentrated   More concentrated

Not all collisions between reactant molecules lead to products, however, because the molecules must collide in a certain orientation in order to react. Nitrogen and oxygen, for example, are much more likely to form nitrogen monoxide when the molecules collide in the parallel orientation shown in Figure 20.2. When they collide in the perpendicular orientation shown in Figure 20.4, nitrogen monoxide does not form. For larger molecules, which can have numerous orientations, this orientation requirement is even more restrictive.

FIGURE 20.4 ▶
The orientation of reactant molecules in a collision can determine whether or not a reaction occurs. A perpendicular collision between N_2 and O_2 tends not to result in formation of a product molecule.

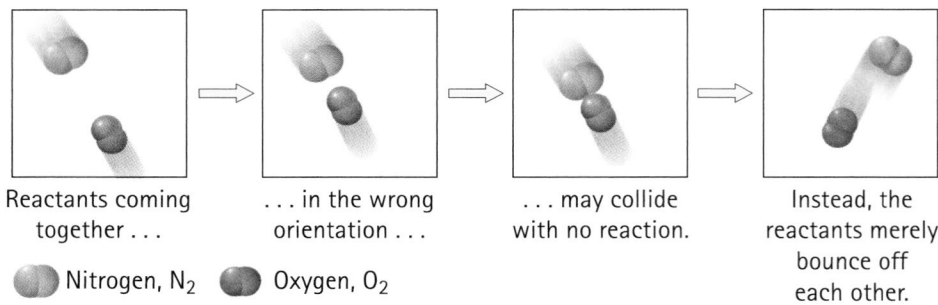

Reactants coming together . . .   . . . in the wrong orientation . . .   . . . may collide with no reaction.   Instead, the reactants merely bounce off each other.

Nitrogen, N_2 Oxygen, O_2

A second reason that not all collisions lead to product formation is that the reactant molecules must also collide with enough kinetic energy to break their bonds. Only then is it possible for the atoms in the reactant molecules to change bonding partners and form product molecules. The bonds in N_2 and O_2 molecules, for example, are quite strong. In order for these bonds to be broken, collisions between the molecules must contain enough energy to break the bonds. As a result, collisions between slow-moving N_2 and O_2 molecules, even those that collide in the proper orientation, may not form NO, as is shown in Figure 20.5.

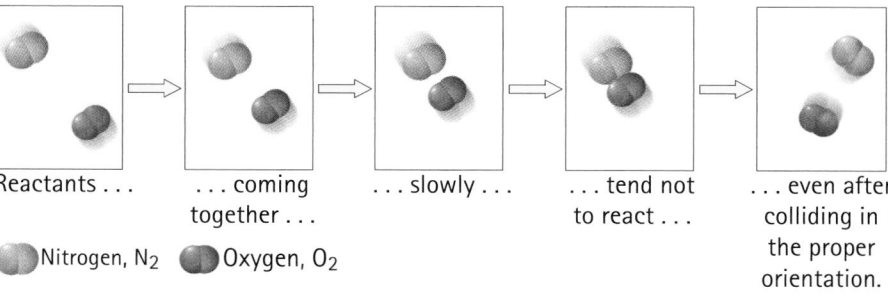

FIGURE 20.5 ▶
Slow-moving molecules may collide with insufficient force to break their bonds. As a result, they cannot react to form product molecules.

Reactants . . .   . . . coming together . . .   . . . slowly . . .   . . . tend not to react . . .   . . . even after colliding in the proper orientation.

Nitrogen, N_2 Oxygen, O_2

The higher the temperature of a material, the faster its molecules move and the more forceful the collisions between them. Higher temperatures, therefore, increase reaction rates. The nitrogen and oxygen molecules that make up our atmosphere, for example, are continually colliding with one another. At the normal temperatures of our atmosphere, however, these molecules do not generally have sufficient kinetic energy for the formation of nitrogen monoxide. The heat of a lightning bolt, however, dramatically increases the kinetic energy of these molecules to the point that a large portion of the collisions in the vicinity of the bolt result in the formation of nitrogen monoxide. The nitrogen monoxide formed in this manner undergoes further atmospheric reactions to form chemicals known as nitrates that plants depend on for survival. This is an example of *nitrogen fixation,* which you may have learned about in a biology class.

CONCEPT CHECK

An internal-combustion engine works by drawing a mixture of air and gasoline vapors into a chamber. The action of a piston then compresses these gases into a smaller volume prior to ignition by the spark of a spark plug. What is the advantage of squeezing the vapors to a smaller volume?

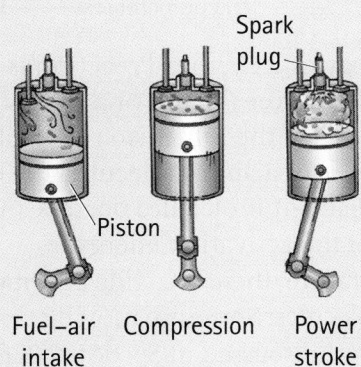

Fuel–air   Compression   Power
intake   stroke

Check Your Answer

Squeezing the vapors to a smaller volume effectively increases their concentration and, hence, the number of collisions between molecules. This, in turn, promotes the chemical reaction. As discussed in Section 8.5, compression also increases the temperature, which further favors the chemical reaction.

The life sciences involve fantastic applications of chemistry, nitrogen fixation being just one example. Others include photosynthesis, cellular respiration, and molecular genetics. So there are distinct advantages to learning about chemistry *before* advancing to the life sciences.

The energy required to break bonds can also come from the absorption of electromagnetic radiation. As the radiation is absorbed by reactant molecules, the atoms in the molecules may start to vibrate so rapidly that the bonds between them are easily broken. In many instances, the direct absorption of electromagnetic radiation is sufficient to break chemical bonds and to initiate a chemical reaction. The common atmospheric pollutant nitrogen dioxide, NO_2, for example, may transform to nitrogen monoxide and atomic oxygen merely upon exposure to sunlight:

$$NO_2 + \text{sunlight} \rightarrow NO + O$$

Activation Energy Is the Energy Needed for Reactants to React

Whether they result from collisions, or from the absorption of electromagnetic radiation, or both, broken bonds are a necessary first step in most chemical reactions. The energy required for this initial breaking of bonds can be viewed as an *energy barrier*. The minimum energy required to overcome this energy barrier is known as the **activation energy** (E_a). In the reaction between nitrogen and oxygen to form nitrogen monoxide, the activation energy is so high (because the bonds in N_2 and O_2 are strong) that only the fastest-moving nitrogen and oxygen molecules possess sufficient energy to react. Figure 20.6 shows the activation energy in this chemical reaction as a vertical hump.

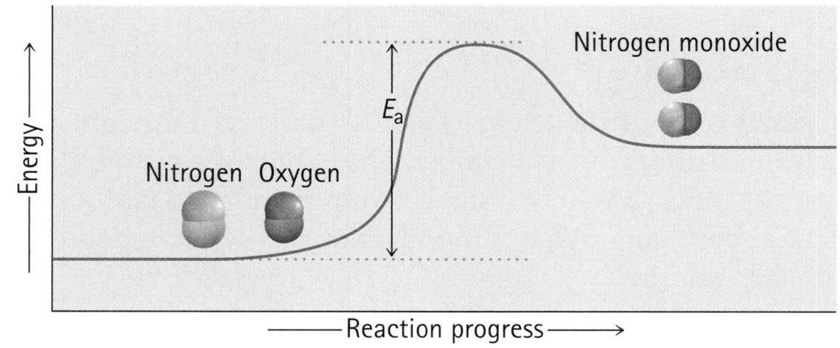

FIGURE 20.6 ▶
Reactant molecules must gain a minimum amount of energy, called the activation energy, E_a, before they can transform to product molecules.

The activation energy of a chemical reaction is analogous to the energy a car needs to drive over the top of a hill. Without sufficient energy to climb to the top of the hill, it isn't possible for the car to get to the other side. Likewise, reactant molecules can transform to product molecules only if the reactant molecules possess an amount of energy equal to or greater than the activation energy.

At any given temperature, there is a wide distribution of kinetic energies in reactant molecules. Some are moving slowly, and others are moving quickly. As we discussed in Section 9.2, the temperature of a material is related to the average of all these kinetic energies. The few fast-moving reactant molecules in Figure 20.7 have enough energy to pass over the energy barrier and are the first to transform to product molecules.

When the temperature of the reactants is increased, the number of reactant molecules possessing sufficient energy to pass over the barrier also increases, which is why reactions are generally faster at higher temperatures. Conversely, at lower temperatures, there are fewer molecules having sufficient energy to pass over the barrier. Hence, reactions are generally slower at lower temperatures.

In order for two chemicals to be able to react, they must first collide in the proper orientation. Second, they must have sufficient kinetic energy to initiate the breaking of chemical bonds so that new bonds can form.

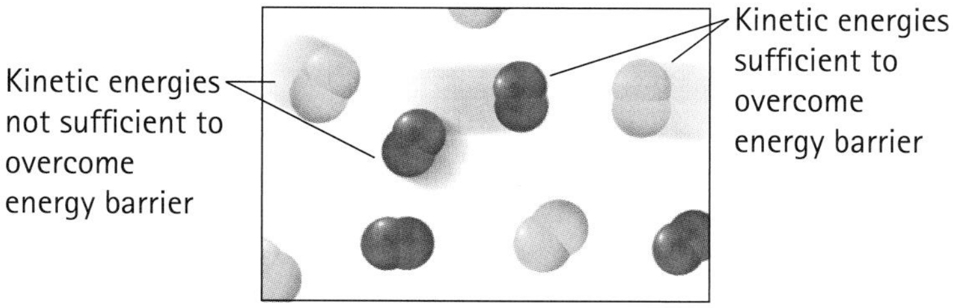

FIGURE 20.7 ▶
Because fast-moving reactant molecules possess sufficient energy to pass over the energy barrier, they are the first ones to transform to product molecules.

Kinetic energies not sufficient to overcome energy barrier

Kinetic energies sufficient to overcome energy barrier

20.4 Chemical Reactions Can Be Either Exothermic or Endothermic

Once a reaction is complete there may be either a net release or a net absorption of energy. Reactions in which there is a net release of energy are called **exothermic.** Rocket ships lift off into space and campfires glow red hot as a result of exothermic reactions. Reactions in which there is a net absorption of energy are called **endothermic.** Photosynthesis, for example, involves a series of endothermic reactions that are driven by the energy of sunlight. Both exothermic and endothermic reactions, illustrated in Figure 20.14, can be understood through the concept of bond energy.

During a chemical reaction, chemical bonds are broken and atoms rearrange to form new chemical bonds. ✓ Such breaking and forming of chemical bonds involves changes in energy. As an analogy, consider a pair of magnets. To separate them requires an input of "muscle energy." Conversely, when the two separated magnets collide, they become slightly warmer than they were, and this warmth is evidence of energy released. Energy must be absorbed by the magnets if they are to move apart, and energy is released as they come together. The same principle applies to atoms. To pull bonded atoms apart requires an energy input. When atoms combine, there is an energy output, usually in the form of faster-moving atoms and molecules, electromagnetic radiation, or both.

The amount of energy required to pull two bonded atoms apart is the same as the amount released when they are brought together. This energy is called **bond energy.** Each chemical bond has its own characteristic bond energy. The hydrogen–hydrogen bond energy, for example, is 436 kilojoules (kJ) per mole. This means that 436 kJ of energy is absorbed as 1 mole of hydrogen–hydrogen bonds break apart,

◀ **FIGURE 20.14**
For the chemical reactions that occur when wood is burning, there is a net release of energy. For the chemical reactions that occur in a photosynthetic plant, there is a net absorption of energy.

I must supply energy to these magnets in order to pull them apart.

Energy is released when they come together!

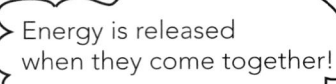

TABLE 20.1 Selected Bond Energies

Bond	Bond Energy (kJ/mole)	Bond	Bond Energy (kJ/mole)
H—H	436	N—N	159
H—C	414	O—O	138
H—N	389	Cl—Cl	243
H—O	464	C=O	803
H—F	569	N=O	631
H—S	339	O=O	498
H—Cl	431	C≡C	837
C—C	347	N≡N	946

and 436 kJ of energy is released upon the formation of 1 mole of hydrogen–hydrogen bonds. Different bonds involving different elements have different bond energies, as Table 20.1 shows. You can refer to the table as you study this section, but please do not memorize these bond energies. Instead, focus on understanding what they mean.

CONCEPT CHECK
Do all covalent single bonds have the same bond energy?

Check Your Answer
No. Bond energy depends on the types of atoms bonding. The H—H single bond, for example, has a bond energy of 436 kJ per mole, but the H—O single bond has a bond energy of 464 kJ per mole. All covalent single bonds do not have the same bond energy.

An Exothermic Reaction Involves a Net Release of Energy

For any chemical reaction, the total amount of energy absorbed in breaking bonds in reactants is always different from the total amount of the energy released as bonds form in the products. Consider the reaction in which hydrogen and oxygen react to form water:

$$H—H + H—H + O=O \longrightarrow H—O\backslash_H + H\backslash O/H$$

In the reactants, hydrogen atoms are bonded to hydrogen atoms, and oxygen atoms are double-bonded to oxygen atoms. The total amount of energy absorbed as these bonds break is +1370 kJ. Note, we use the plus sign to indicate the amount of energy *absorbed* to break bonds.

 **READING CHECK**

What do the breaking and forming of chemical bonds involve?

Type of Bond	Number of Moles	Bond Energy	Total Energy
H—H	2	+436 kJ/mole	+872 kJ
O=O	1	+498 kJ/mole	+498 kJ
		Total energy absorbed:	+1370 kJ

In the products there are four hydrogen–oxygen bonds. The total amount of energy released as these bonds form is −1856 kilojoules. Note, we use the minus sign to indicate the amount of energy *released* as bonds are formed.

Type of Bond	Number of Moles	Bond Energy	Total Energy
H—O	4	−464 kJ/mole	−1856 kJ
		Total energy released:	−1856 kJ

The amount of energy released in this reaction exceeds the amount of energy absorbed. The net energy of the reaction is found by adding the two quantities:

Net energy of reaction = energy absorbed + energy released

$$= +1370 \text{ kJ} + (-1856 \text{ kJ})$$

$$= -486 \text{ kJ}$$

The negative sign on the net energy indicates that there is a net release of energy, and so the reaction is exothermic. For any exothermic reaction, energy can be considered a product and is thus sometimes included after the arrow of the chemical equation:

$$2 \, H_2 + O_2 \rightarrow 2 \, H_2O + \text{energy}$$

In an exothermic reaction, the potential energy of atoms in the product molecules is lower than their potential energy in the reactant molecules. This is illustrated in the reaction profile shown in Figure 20.15. The lowered potential energy of the atoms in the product molecules is due to their being more tightly held together. This is analogous to two attracting magnets, whose potential energy decreases as they come closer together. The loss of potential energy is balanced by a gain in kinetic energy. Like two free-floating magnets coming together and accelerating to higher speeds, the potential energy of the reactants is converted to faster-moving atoms and molecules, electromagnetic radiation, or both. This kinetic energy released by the reaction is equal to the difference between the potential energy of the reactants and the potential energy of the products, as is indicated in Figure 20.15.

Remember, in a chemical reaction, the bonds being formed are different from the bonds that were broken. The bond energies of the bonds being formed, therefore, are also different from those of the bonds that were broken.

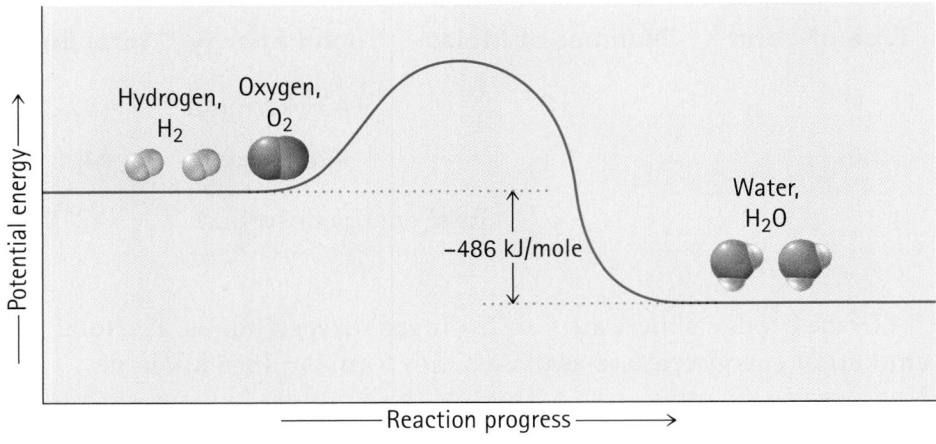

FIGURE 20.15 ▲

In an exothermic reaction, the product molecules are at a lower potential energy than the reactant molecules. The net amount of energy released by the reaction is equal to the difference in potential energies of the reactants and products.

FIGURE 20.16 ▲

A space shuttle uses exothermic chemical reactions to lift off from Earth's surface.

> Recall, from Chapter 4, that for every action there is an opposite and equal reaction. A rocket is thrust upwards, for example, only as its exhaust chemicals are thrust downwards.

It is important to understand that the energy released by an exothermic reaction is not created by the reaction. This is in accord with the *law of conservation of energy*, which tells us that energy is neither created nor destroyed in a chemical reaction. Instead, energy is merely converted from one form to another. During an exothermic reaction, energy that was once in the form of the potential energy of chemical bonds is released as the kinetic energy of fast-moving molecules and/or as electromagnetic radiation.

The amount of energy released in an exothermic reaction depends on the amounts of the reactants. The reaction of large amounts of hydrogen and oxygen, for example, provides the energy to lift the space shuttle, shown in Figure 20.16, into orbit. There are two compartments in the large central tank, to which the orbiter is attached—one filled with liquid hydrogen and the other filled with liquid oxygen. Upon ignition, these two liquids mix and react chemically to form water vapor, which produces the needed thrust as it is expelled out the rocket cones. Additional thrust is obtained by a pair of solid-fuel rocket boosters containing a mixture of ammonium perchlorate, NH_4ClO_4, and powdered aluminum. Upon ignition, these chemicals react to form products that are expelled at the rear of the rocket. The balanced equation representing this reaction is

$$3\ NH_4ClO_4 + 3\ Al \rightarrow Al_2O_3 + AlCl_3 + 3\ NO + 6\ H_2O + \textbf{energy}$$

CONCEPT CHECK

Where does the net energy released in an exothermic reaction go?

Check Your Answer

This energy goes into increasing the speeds of reactant atoms and molecules and often into electromagnetic radiation.

An Endothermic Reaction Involves a Net Absorption of Energy

When the amount of energy released in product formation is *less* than the amount of energy absorbed when reactant bonds break, the reaction is endothermic. An example is the reaction of atmospheric nitrogen and oxygen to form nitrogen monoxide, which is the same reaction used for many of the discussions earlier in this chapter:

$$N{\equiv}N + O{=}O \rightarrow N{=}O + N{=}O$$

The amount of energy absorbed as the chemical bonds in the reactants break is

Type of Bond	Number of Moles	Bond Energy	Total Energy
N≡N	+1	+946 kJ/mole	+946 kJ
O=O	+1	+498 kJ/mole	+498 kJ
		Total energy absorbed:	+1444 kJ

The amount of energy released upon the formation of bonds in the products is

Type of Bond	Number of Moles	Bond Energy	Total Energy
N=O	2	−631 kJ/mole	−1262 kJ
		Total energy released:	−1262 kJ

As before, the net energy of the reaction is found by adding the two quantities:

$$\textbf{Net energy of reaction} = \textbf{energy absorbed} + \textbf{energy released}$$

$$= \textbf{+1444 kJ} + \textbf{(−1262 kJ)}$$

$$= \textbf{+182 kJ}$$

The positive sign indicates that there is a net *absorption* of energy, meaning the reaction is endothermic. For any endothermic reaction, energy can be considered a reactant and is thus sometimes included before the arrow of the chemical equation:

$$\textbf{Energy} + \textbf{N}_2 + \textbf{O}_2 \rightarrow \textbf{2 NO}$$

In an endothermic reaction, the potential energy of atoms in the product molecules is higher than their potential energy in the reactant molecules. This is illustrated in the reaction profile shown in Figure 20.17. Raising the potential energy of the atoms in the product molecules requires a net input of energy, which must come from some

FIGURE 20.17 ▶
In an endothermic reaction, the product molecules are at a higher potential energy than the reactant molecules. The net amount of energy absorbed by the reaction is equal to the difference in potential energies of the reactants and products.

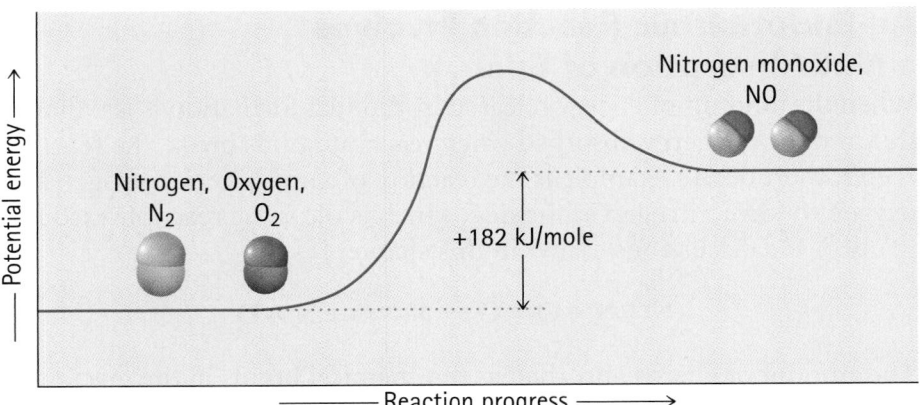

FIGURE 20.17 ▶
In an endothermic reaction, the product molecules are at a higher potential energy than the reactant molecules. The net amount of energy absorbed by the reaction is equal to the difference in potential energies of the reactants and products.

external source, such as electromagnetic radiation, electricity, or heat. Nitrogen and oxygen react to form nitrogen monoxide with the application of much heat, as occurs adjacent to a lightning bolt or in an internal-combustion engine.

CONCEPT CHECK

Should the following reaction be endothermic or exothermic?

$$O_2N-NO_2 \rightarrow O_2N + NO_2$$

Check Your Answer

No calculations are necessary. This reaction is endothermic because it involves only the breaking of a chemical bond. For practice with examples requiring calculations, see the questions at the end of this chapter.

20.5 Chemical Reactions Are Driven by Entropy

Energy tends to disperse. It flows from where it is concentrated to where it is spread out. The energy of a hot pan, for example, does not stay concentrated in the pan once the pan is taken off the stove. Instead, the energy disperses away from the pan and into the cooler surroundings. Similarly, the concentrated chemical energy found in gasoline disperses into the heat of many smaller, lower-energy molecules upon combustion. Some of this heat is used by the engine to get the car moving. The rest spreads into the engine block, radiator fluid, or out the exhaust pipe.

✔ Scientists consider this tendency of energy to disperse as one of the central reasons for the occurrence of any physical or chemical process. In other words, processes that result in the dispersion of energy tend to occur on their own—they are favored. This includes the cooling down of a hot pan or the burning of gasoline. In both cases, there is a dispersal of energy to the environment.

The opposite also holds true. Processes that result in the concentration of energy tend *not* to occur—they are not favored. Heat from the room, for example, will not spontaneously move back into the pan to heat it up.

Likewise, low-energy exhaust molecules coming out of a car's tailpipe will not spontaneously come back together to form higher-energy gasoline molecules. The natural flow of energy is always a one-way trip from where it is concentrated to where it is less concentrated, or "spread out."

Entropy is the term we use to describe this natural spreading of energy. Applied to chemistry, entropy helps us to answer a most fundamental question: If you take two materials and put them together, will they react to form new materials? If the reaction results in an overall increase in entropy (a dispersal of energy), then the answer is yes. Conversely, if the reaction results in an overall decrease in entropy (a concentration of energy), then the reaction will *not* occur by itself.

A quick way to determine whether a reaction might be favorable is to assess whether the reaction leads to an overall dispersal of energy, which is the same thing as an increase in *entropy*.

Because it is the natural tendency for energy to disperse, a reaction that leads to an increase in entropy will likely occur, while a reaction that leads to a decrease in entropy will *not* likely occur.

Using this concept of entropy, you are now in a position to understand why exothermic reactions are self-sustaining while most endothermic reactions need a continual prodding. Exothermic reactions spread energy out to the surroundings, much like a cooling hot pan. This is an increase in entropy, hence, exothermic reactions are favored to occur. An endothermic reaction, by contrast, requires that energy from the surroundings be absorbed by the reactants. This is a concentration of energy, which is counter to energy's natural tendency to disperse. Endothermic reactions, therefore, can only be sustained with the continual input of some external source of energy.* For photosynthesis, the source of this energy is the Sun, which is a hothouse of greater entropy-producing exothermic nuclear reactions.

✔ **READING CHECK**

What is one of the central reasons for the occurrence of any physical and chemical process?

CONCEPT CHECK
Sugar crystals form naturally within a supersaturated solution of sugar water. Does the formation of these crystals, in which the sugar molecules are aligned in an orderly fashion, result in an increase or decrease in entropy?

Check Your Answer
The formation of these sugar crystals results in an *increase* in entropy. Your clue to an increase in entropy here is that the crystals form "on their own" without the input of an external source of energy. Interestingly, energy is released when molecules come together to form a solid. For example, when water freezes, heat is actually released. This release of heat is the dispersal of energy, which is, by definition, an increase in entropy.

FIGURE 20.18 ▲
Some of the Sun's dispersed energy is used to drive endothermic reactions that allow the living organisms to function.

---

* Interestingly, there are examples of endothermic reactions that proceed spontaneously, absorbing heat from the environment. A classic example is the mixing of a salt in water (see the Explore! activity at the beginning of this chapter). In such cases, heat is absorbed (cooling occurs) as energy-containing atoms and molecules disperse into solution.

20 CHAPTER REVIEW

KEY TERMS

Activation energy The minimum energy required in order for a chemical reaction to proceed.

Bond energy The amount of energy that is either absorbed as a chemical bond breaks or is released as a chemical bond forms.

Catalyst Any substance that increases the rate of a chemical reaction without itself being consumed by the reaction.

Chemical equation A representation of a chemical reaction in which reactants are drawn before an arrow that points to the products.

Endothermic A term that describes a chemical reaction in which there is a net absorption of energy.

Exothermic A term that describes a chemical reaction in which there is a net release of energy.

Law of mass conservation Matter is neither created nor destroyed during a chemical reaction—atoms merely rearrange, without any apparent loss or gain of mass, to form new molecules.

Products The new materials formed in a chemical reaction.

Reactants The reacting substances in a chemical reaction.

Reaction rate A measure of how quickly the concentration of products in a chemical reaction increases or the concentration of reactants decreases.

REVIEW QUESTIONS

Chemical Reactions Are Represented by Chemical Equations

1. What is the purpose of coefficients in a chemical equation?

2. How many chromium atoms and how many oxygen atoms are indicated on the right side of this balanced chemical equation:

$$4\,Cr(s) + 3\,O_2(g) \rightarrow 2\,Cr_2O_3(g)$$

3. Why is it important that a chemical equation be balanced?

4. Why is it important never to change a subscript in a chemical formula when balancing a chemical equation?

Chemical Reactions Can Be Slow or Fast

5. Why don't all collisions between reactant molecules lead to product formation?

6. What generally happens to the rate of a chemical reaction with increasing temperature?

7. Which reactant molecules are the first to pass over the energy barrier?

8. What term is used to describe the minimum amount of energy required in order for a reaction to proceed?

Catalysts Speed Up Chemical Reactions

9. What catalyst is effective in the destruction of atmospheric ozone, O_3?

10. What is the purpose of a catalytic converter?

11. What does a catalyst do to the energy of activation for a reaction?

12. What net effect does a chemical reaction have on a catalyst?

13. Why are catalysts so important to our economy?

Chemical Reactions Can Be Either Exothermic or Endothermic

14. If it takes 436 kJ to break a bond, how many kilojoules are released when the same bond is formed?

15. Is there any energy consumed at any time during an exothermic reaction?

16. What is released by an exothermic reaction?

17. What is absorbed by an endothermic reaction?

Chemical Reactions Are Driven by Entropy

18. As energy disperses, where does it go?

19. What is always increasing?

20. Why are exothermic reactions self-sustaining?

THINK AND DO

1. If you ever have the opportunity to play with an electric train set, be sure to smell the engine car after it has been operating. You will note a slight "electric smell." This is the smell of ozone gas, which is created as the oxygen in the air is zapped with electrical sparks. Why is this smell also sometimes apparent during a lightning storm? Is the formation of ozone from oxygen an endothermic or exothermic reaction?

2. An Alka-Seltzer antacid tablet reacts vigorously with water. But how does this tablet react to a a solution of half water and half corn syrup? Propose an explanation involving the relationship between reaction speed and the frequency of molecular collisions.

3. Baker's yeast contains a biological catalyst known as catalase, which catalyzes the transformation of hydrogen peroxide, H_2O_2, into oxygen, O_2, and water, H_2O. Write a balanced equation for this reaction. Add a couple milliliters of 3% hydrogen peroxide to a glass containing a small amount of baker's yeast. What happens? Why?

THINK AND COMPARE

1. Rank the following reaction profiles in order of increasing reaction speed:

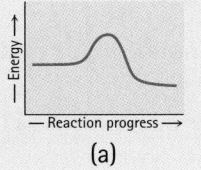   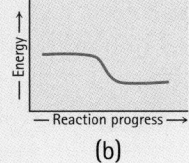   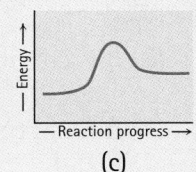

(a)   (b)   (c)

2. Rank the following covalent bonds in order of increasing bond strength:

(a) C≡C   **(b)** C=C   **(c)** C—C

THINK AND EXPLAIN

1. Balance these equations:

(a) ____ Fe(s) + ____ O_2(g) → ____ Fe_2O_3(s)

(b) ____ H_2(g) + ____ N_2(g) → ____ NH_3(g)

(c) ____ Cl_2(g) + ____ KBr(aq) → ____ Br_2(l) + ____ KCl(aq)

(d) ____ CH_4(g) + ____ O_2(g) → ____ CO_2(g) + ____ H_2O(l)

Use the following illustration to answer Questions 2–4:

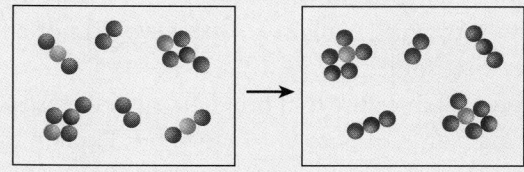

2. Is this reaction balanced?

3. There is an excess of at least one of the reactant molecules. Which one?

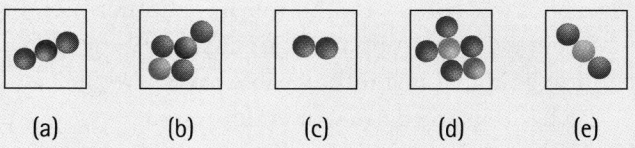

(a)   (b)   (c)   (d)   (e)

4. Which equation best describes this reaction?

(a) $2 AB_2 + 2 DCB_3 + B_2 → 2 DBA_4 + 2 CA_2$

(b) $2 AB_2 + 2 CDA_3 + B_2 → 2 C_2A_4 + 2 DBA$

(c) $2 AB_2 + 2 CDA_3 + A_2 → 2 DBA_4 + 2 CA_2$

(d) $2 BA_2 + 2 DCA_3 + A_2 → 2 DBA_4 + 2 CA_2$

5. The reactants shown schematically on the left represent iron oxide, Fe_2O_3, and carbon monoxide, CO. Write out the full balanced chemical equation that is depicted.

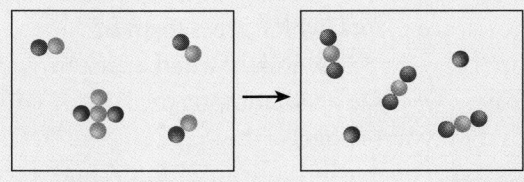

6. What two aspects of a collision between two reactant molecules determine whether or not the collision results in the formation of product molecules?

7. Why does a glowing splint of wood burn only slowly in air but burst into flames when placed in pure oxygen?

8. Why is heat often added to chemical reactions performed in the laboratory?

9. Does a refrigerator prevent or delay the spoilage of food? Explain.

10. What can you deduce about the activation energy of a reaction that takes billions of years to go to completion? How about a reaction that takes only fractions of a second?

11. Many people hear about atmospheric ozone depletion and wonder why we don't simply replace that which has been destroyed. Knowing about chlorofluorocarbons and knowing how catalysts work, explain how this would not be a lasting solution.

12. Note in Table 20.1 that bond energy increases going from C—N to C—O to C—F. Explain this trend based on the sizes of these atoms as deduced from their positions in the periodic table.

13. In an endothermic reaction, which has greater potential energy: the reactants or the products?

14. Are the chemical reactions that take place in a disposable battery exothermic or endothermic? What evidence supports your answer? Is the reaction going on in a rechargeable battery while it is recharging exothermic or endothermic?

15. Under what conditions will a hot pie not lose heat to its surroundings?

16. Why do exothermic reactions typically favor the formation of products?

17. Exothermic reactions are favored because they release heat to the environment. Would an exothermic reaction be more favored or less favored if it were carried out within a superheated chamber?

18. What role does entropy play in chemical reactions?

19. A gardener finds weeds growing "all by themselves" within her garden. Yet these weeds, she knows, are plants, which are a form of concentrated energy. She wonders whether the growing of these weeds is a violation of the laws of entropy. What do you tell her?

20. How is it possible to cause an endothermic reaction, such as the one shown below, to proceed when the reaction causes energy to become less dispersed?

$$6\,CO_2(g) + 6\,H_2O(l) \rightarrow C_6H_{12}O_6(s) + 6\,O_2(g)$$

THINK AND SOLVE

1. Use the bond energies in Table 20.1 and the accounting format shown in Section 20.3 to determine whether these reactions are exothermic or endothermic:
(a) $H_2 + Cl_2 \rightarrow 2\,HCl$
(b) $2\,HC\equiv CH + 5\,O_2 \rightarrow 4\,CO_2 + 2\,H_2O$

2. Use the bond energies in Table 20.1 and the accounting format shown in Section 20.3 to determine whether these reactions are exothermic or endothermic:
(a) $H_2N—NH_2 \rightarrow H_2 + H_2 + N_2$
(b) $2\,H_2O_2 \rightarrow O_2 + 2\,H_2O$

READINESS ASSURANCE TEST (RAT)

If you have a good handle on this chapter—if you really do—then you should be able to score 7 out of 10 on this RAT. Check your answers with your teacher. If you score less than 7, study further before moving on.

Choose the best answer to the following.

1. What coefficients balance the following equation:

$$\_\_\_\_\,P_4(s) + \_\_\_\_\,H_2(g) \rightarrow \_\_\_\_\,PH_3(g)$$

(a) 4, 2, 3
(b) 1, 6, 4
(c) 1, 4, 4
(d) 2, 10, 8

2. Which equation is balanced?
(a) $Mg(s) + 2\,HCl(aq) \rightarrow MgCl_2(aq) + H_2(g)$
(b) $3\,Al(s) + 3\,Br_2(l) \rightarrow Al_2Br_3(s)$
(c) $2\,HgO(s) \rightarrow 2\,Hg(l) + 2\,O_2(g)$
(d) All of them.

3. Is the following chemical equation balanced?

$$2\,C_4H_{10}(g) + 13\,O_2(g) \rightarrow 8\,CO_2(g) + 10\,H_2O(l)$$

(a) Yes, it is balanced.
(b) No, it is not balanced.
(c) It is balanced, but liquids cannot be made from gases.
(d) It is not balanced because one side contains a liquid.

4. How is it possible for a jet airplane carrying 110 tons of jet fuel to emit 340 tons of carbon dioxide?
   (a) Not possible! This would be a violation of the principle of mass conservation.
   (b) The fuel combines with oxygen to produce carbon dioxide, which is more massive.
   (c) As it combusts, the fuel expands to a gaseous phase, which occupies more volume.
   (d) The fuel exhaust also contain soot, which adds to the mass.

5. The yeast in bread dough feeds on sugar to produce carbon dioxide. Why does the dough rise faster in a warmer area?
   (a) There is a greater number of effective collisions among reacting molecules.
   (b) Atmospheric pressure decreases with increasing temperature.
   (c) The yeast tends to "wake up" with warmer temperatures, which is why baker's yeast is best stored in the refrigerator.
   (d) The rate of evaporation increases with increasing temperature.

6. What can you deduce about the activation energy of a reaction that takes billions of years to go to completion? How about a reaction that takes only fractions of a second?
   (a) The activation energy of both these reactions must be very low.
   (b) The activation energy of both these reactions must be very high.
   (c) The slow reaction must have a high activation energy while the fast reaction must have a low activation energy.
   (d) The slow reaction must have a low activation energy while the fast reaction must have a high activation energy.

7. What role do CFCs play in the catalytic destruction of ozone?
   (a) Ozone is destroyed upon binding to a CFC molecule that has been energized by ultraviolet light.
   (b) There is no strong scientific evidence that CFCs play a significant role in the catalytic destruction of ozone.

   (c) CFC molecules activate chlorine atoms into their catalytic action.
   (d) CFC molecules migrate to the stratosphere where they generate chlorine atoms upon being destroyed by ultraviolet light.

8. Is the synthesis of ozone, O_3, from oxygen, O_2, an example of an exothermic or endothermic reaction?
   (a) Exothermic because ultraviolet light is emitted during its formation
   (b) Endothermic because ultraviolet light is emitted during its formation
   (c) Exothermic because ultraviolet light is absorbed during its formation
   (d) Endothermic because ultraviolet light is absorbed during its formation

9. How much energy, in kilojoules, is released or absorbed from the reaction of 1 mole of nitrogen, N_2, with three moles of molecular hydrogen, H_2, to form 2 moles of ammonia, NH_3? Consult Table 20.1 for bond energies.
   (a) +899 kJ/mol
   (b) −993 kJ/mol
   (c) +80 kJ/mol
   (d) −80 kj/mol

10. How is it possible to cause an endothermic reaction to proceed when the reaction causes energy to become less dispersed?
    (a) The reaction should be placed in a vacuum.
    (b) The reaction should be cooled down.
    (c) The concentration of the reactants should be increased.
    (d) The reaction should be heated.

21
TWO TYPES OF CHEMICAL REACTIONS

THE MAIN IDEA

Protons are exchanged in acid–base reactions; electrons are exchanged in oxidation–reduction reactions.

In this chapter, we explore two main types of chemical reactions, acid–base reactions and oxidation–reduction reactions. Acid–base reactions involve the transfer of *protons* from one reactant to another. These sorts of reactions within your stomach help you to digest your food. Most consumer goods can trace their origins to acid–base chemical reactions, which also play an important role in the environment.

Oxidation–reductions reactions involve the transfer of one or more *electrons* from one reactant to another. The burning of wood is an oxidation–reduction reaction, as are the reactions your body uses to transform the food you eat into biochemical energy. Oxidation–reduction reactions are responsible for the rusting of a car. They are also the source of a battery's electrical energy.

Explore!

Can You Make Your Own pH Indicator?

1. Boil a couple leaves of red cabbage in about two cups of water for five minutes. Strain the broth and allow to cool.
2. Add the purple broth to at least three clear plastic cups.
3. Add a teaspoon of white vinegar to the first cup, a teaspoon of baking soda to the second cup.

Analyze and Conclude

1. **Observing** What color changes did you see?
2. **Predicting** What would happen to the baking soda solution if you were to slowly add vinegar to it?
3. How do chemists often visualize the pH of a solution?

21.1 Acids Donate and Bases Accept Hydrogen Ions

The term *acid* comes from the Latin *acidus,* which means "sour." The sour taste of vinegar and citrus fruits is due to the presence of acids. Figure 21.1 shows only a very few of the acids we commonly encounter. Acids are essential in the chemical industry. For example, sulfuric acid, H_2SO_4, is used to make fertilizers, detergents, paint dyes, plastics, pharmaceuticals, and storage batteries, as well as to produce iron and steel.

Bases are characterized by their bitter taste and slippery feel. Interestingly, bases themselves are not slippery. Rather, they cause skin oils to transform into slippery solutions of soap. Most commercial preparations for unclogging drains contain sodium hydroxide, NaOH (also known as lye), which is extremely basic and hazardous when concentrated. Bases are also heavily used in industry. Solutions containing bases are often called *alkaline,* a term derived from the Arabic *al-qali* ("the ashes"). Ashes are slippery when wet because of the presence of the base, potassium carbonate, K_2CO_3. Figure 21.2 shows some familiar bases.

Acids and bases may be defined in several ways. ✔ For our purposes, an appropriate definition is as follows: an **acid** is any chemical that donates a hydrogen ion, H^+, and a **base** is any chemical that accepts a hydrogen ion. Recall that a hydrogen atom consists of one electron surrounding a one-proton nucleus. A hydrogen ion, H^+, formed from the loss of an electron, therefore, is nothing more than a lone proton. Thus, it is also sometimes said that an acid is a chemical that donates a proton and a base is a chemical that accepts a proton.

(a)   (b)   (c)   (d)

FIGURE 21.1 ▲

Examples of acids. (a) Citrus fruits contain many types of acids, including ascorbic acid, $C_6H_8O_6$, which is vitamin C. (b) Vinegar contains acetic acid, $C_2H_4O_2$, and can be used to preserve foods. (c) Many toilet-bowl cleaners are formulated with hydrochloric acid, HCl. (d) All carbonated beverages contain carbonic acid, H_2CO_3, while many also contain phosphoric acid, H_3PO_4.

> The hydrogen ion, H^+, does not readily exist in water because any hydrogen ion formed is quickly picked up by a water molecule and transformed to the hydronium ion, H_3O^+.

(a)

(b)

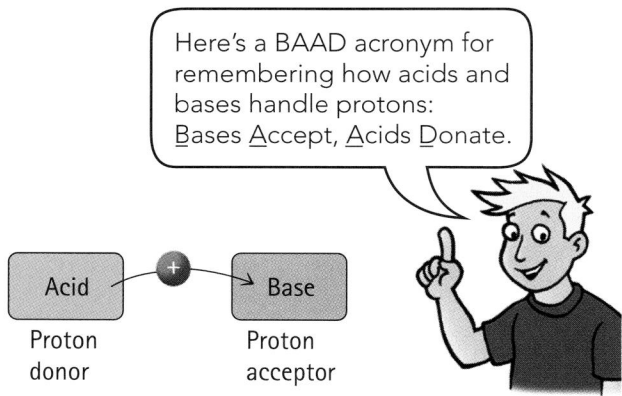
(c)

(d)

FIGURE 21.2 ▲

Examples of bases. (a) Reactions involving sodium bicarbonate, NaHCO₃, cause baked goods to rise. (b) Ashes contain potassium carbonate, K₂CO₃. (c) Soap is made by reacting bases with animal or vegetable oils. The soap itself, then, is slightly alkaline. (d) Powerful bases, such as sodium hydroxide, NaOH, are used in drain cleaners.

Here's a BAAD acronym for remembering how acids and bases handle protons: Bases Accept, Acids Donate.

Acid ⊕→ Base
Proton donor   Proton acceptor

Consider what happens when hydrogen chloride is mixed into water:

Recall that a hydrogen ion with a positive charge is simply a lone proton.

Hydrogen atom   Positive hydrogen ion (lone proton)

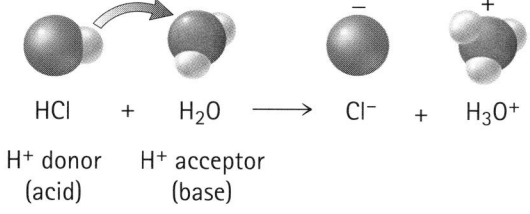

$$HCl \quad + \quad H_2O \quad \longrightarrow \quad Cl^- \quad + \quad H_3O^+$$

H⁺ donor   H⁺ acceptor
(acid)   (base)

Hydrogen chloride donates a hydrogen ion to a water molecule, resulting in a third hydrogen bonded to the oxygen. In this case, hydrogen chloride behaves as an acid (proton donor) and water behaves as a base

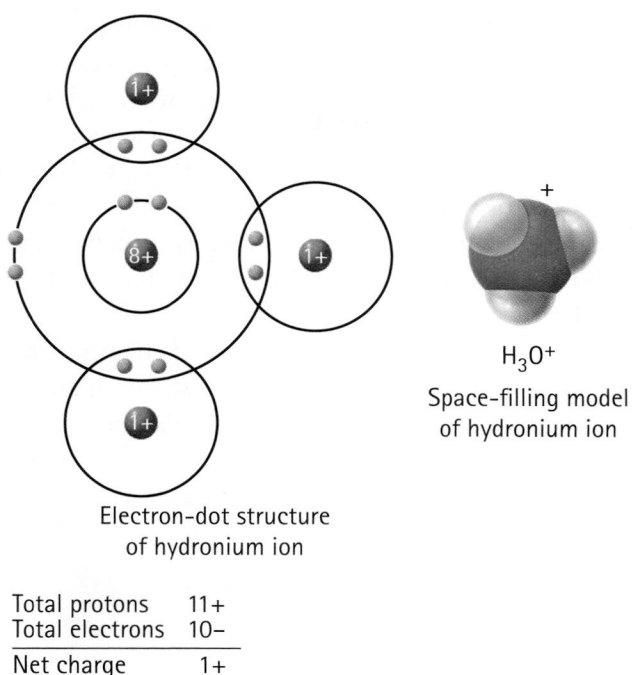

Electron-dot structure
of hydronium ion

Space-filling model
of hydronium ion

H_3O^+

Total protons   11+
Total electrons   10–
_____
Net charge   1+

◀ **FIGURE 21.3**
The hydronium ion's positive
charge is a consequence of
the extra proton this molecule
has acquired.

(proton acceptor). The products of this reaction are a chloride ion and a
hydronium ion, H_3O^+, which, as Figure 21.3 shows, is a water molecule
with an extra proton.

When added to water, ammonia behaves as a base as it accepts a
hydrogen ion from water, which, in this case, behaves as an acid:

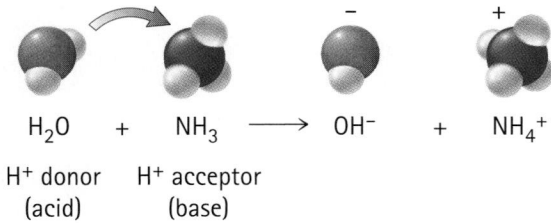

H_2O + NH_3 $\longrightarrow$ OH^- + NH_4^+

H⁺ donor   H⁺ acceptor
(acid)   (base)

This reaction results in the formation of an ammonium ion and a
hydroxide ion, which, as shown in Figure 21.4, is a water molecule without the nucleus of one of the hydrogen atoms.

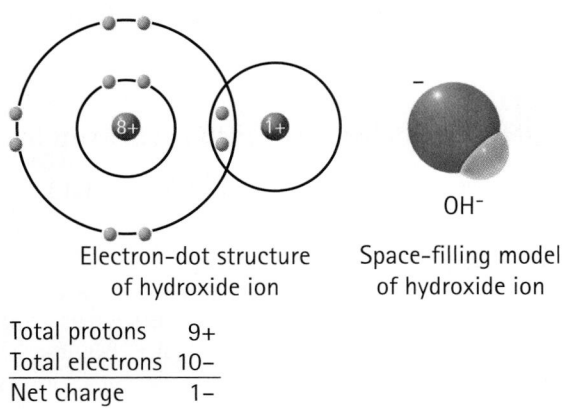

Electron-dot structure
of hydroxide ion

Space-filling model
of hydroxide ion

OH^-

Total protons   9+
Total electrons   10–
_____
Net charge   1–

◀ **FIGURE 21.4**
Hydroxide ions have a net
negative charge, which is a
consequence of having lost a
proton. Like hydronium ions,
they play a part in many
acid–base reactions.

How we behave depends on who we're with. Likewise for chemicals.

An important aspect of this acid–base definition is that it uses *behavior* to define a substance as an acid or a base. We say, for example, that hydrogen chloride *behaves* as an acid when mixed with water, which *behaves* as a base. Similarly, ammonia *behaves* as a base when mixed with water, which under this circumstance *behaves* as an acid. Because acid–base is seen as a behavior, there is no contradiction when a chemical like water behaves as a base in one instance but as an acid in another instance. By analogy, consider yourself. You are who you are, but your behavior changes depending on whom you are with. Likewise, it is a chemical property of water to behave as a base (to accept H^+) when mixed with hydrogen chloride and as an acid (to donate H^+) when mixed with ammonia.

The products of an acid–base reaction can also behave as acids or as bases. An ammonium ion, for example, may donate a hydrogen ion back to a hydroxide ion to re-form ammonia and water:

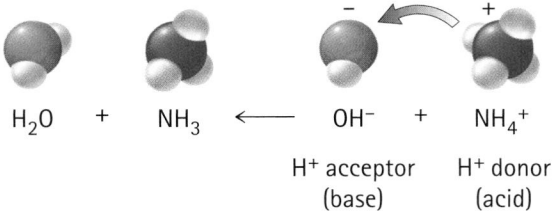

$$H_2O \quad + \quad NH_3 \quad \longleftarrow \quad OH^- \quad + \quad NH_4^+$$

$$\qquad\qquad\qquad\qquad\qquad\quad \text{H}^+ \text{ acceptor} \quad \text{H}^+ \text{ donor}$$
$$\qquad\qquad\qquad\qquad\qquad\quad \text{(base)} \qquad\quad \text{(acid)}$$

Forward and reverse acid–base reactions proceed simultaneously and can therefore be represented as occurring at the same time by using two oppositely facing arrows:

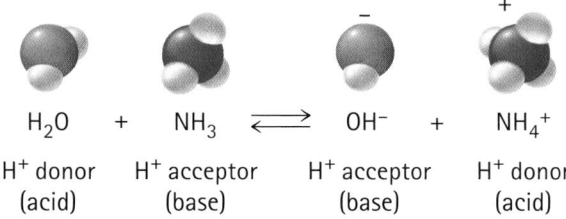

$$H_2O \quad + \quad NH_3 \quad \rightleftharpoons \quad OH^- \quad + \quad NH_4^+$$

$$\text{H}^+ \text{ donor} \quad \text{H}^+ \text{ acceptor} \qquad \text{H}^+ \text{ acceptor} \quad \text{H}^+ \text{ donor}$$
$$\text{(acid)} \qquad\quad \text{(base)} \qquad\qquad \text{(base)} \qquad\quad \text{(acid)}$$

✔ **READING CHECK**

What is the definition of an acid and a base?

When the equation is viewed from left to right, the ammonia behaves as a base because it accepts a hydrogen ion from the water, which therefore acts as an acid. Viewed in the reverse direction, the equation shows that the ammonium ion behaves as an acid because it donates a hydrogen ion to the hydroxide ion, which therefore behaves as a base.

CONCEPT CHECK

Identify the acid or base behavior of each participant in the reaction

$$H_2PO_4^- + H_3O^+ \leftrightarrow H_3PO_4 + H_2O$$

Check Your Answer

In the forward reaction (left to right), $H_2PO_4^-$ gains a hydrogen ion to become H_3PO_4. In accepting the hydrogen ion, $H_2PO_4^-$ is behaving as a base. It gets the hydrogen ion from the H_3O^+, which is behaving as an acid. In the reverse direction, H_3PO_4 loses a hydrogen ion to become $H_2PO_4^-$ and is thus behaving as an acid. The recipient of the hydrogen ion is the H_2O, which is behaving as a base as it transforms to H_3O^+.

A Salt Is the Ionic Product of an Acid–Base Reaction

In everyday language, the word *salt* implies sodium chloride, NaCl, table salt. In the language of chemistry, however, **salt** is a general term meaning any ionic compound formed from the reaction between an acid and a base. Hydrogen chloride and sodium hydroxide, for example, react to produce the salt sodium chloride and water:

$$\text{HCl} \quad + \quad \text{NaOH} \quad \rightarrow \quad \text{NaCl} \quad + \quad \text{H}_2\text{O}$$

| Hydrogen chloride (acid) | Sodium hydroxide (base) | Sodium chloride (salt) | Water |

Similarly, the reaction between hydrogen chloride and potassium hydroxide yields the salt potassium chloride and water:

$$\text{HCl} \quad + \quad \text{KOH} \quad \rightarrow \quad \text{KCl} \quad + \quad \text{H}_2\text{O}$$

| Hydrogen chloride (acid) | Potassium hydroxide (base) | Potassium chloride (salt) | Water |

Potassium chloride is the main ingredient in "salt-free" table salt, as noted in Figure 21.5.

Salts are generally far less corrosive than the acids and bases from which they are formed. A corrosive chemical has the power to disintegrate a material or wear away its surface. Hydrogen chloride is a remarkably corrosive acid, which makes it useful for cleaning toilet bowls and etching metal surfaces. Sodium hydroxide is a very corrosive base used for unclogging drains. Mixing hydrogen chloride and sodium hydroxide together in equal portions, however, produces a solution of sodium chloride—saltwater, which is not nearly as destructive as either starting material.

There are as many salts as there are acids and bases. Sodium cyanide, NaCN, is a deadly poison. "Saltpeter," which is potassium nitrate, KNO_3, is useful as a fertilizer and in the formulation of gunpowder. Calcium chloride, $CaCl_2$, is commonly used to de-ice walkways, and sodium fluoride, NaF, helps to prevent tooth decay. The acid–base reactions forming these salts are shown in Table 21.1.

The reaction between an acid and a base is called a **neutralization** reaction. As can be seen in the color-coding of the neutralization reactions in Table 21.1, the positive ion of a salt comes from the base and the negative ion comes from the acid. The remaining hydrogen and hydroxide ions join to form water.

FIGURE 21.5 ▶

"Salt-free" table salt substitutes contain potassium chloride in place of sodium chloride. Caution is advised in using these products, however, because excessive quantities of potassium salts can lead to serious illness. Furthermore, sodium ions are a vital component of our diet and should never be totally excluded. For a good balance of these two important ions, you might inquire about commercially available half-and-half mixtures of sodium chloride and potassium chloride, such as the one shown here.

TABLE 21.1 Acid–Base Reactions and the Salts Formed

Acid		Base		Salt		Water
HCN Hydrogen cyanide	+	NaOH Sodium hydroxide	→	NaCN Sodium cyanide	+	H_2O
HNO_3 Nitric acid	+	KOH Potassium hydroxide	→	KNO_3 Potassium nitrate	+	H_2O
2 HCl Hydrogen chloride	+	$Ca(OH)_2$ Calcium hydroxide	→	$CaCl_2$ Calcium chloride	+	$2 H_2O$
HF Hydrogen fluoride	+	NaOH Sodium hydroxide	→	NaF Sodium fluoride	+	H_2O

Not all neutralization reactions result in the formation of water. In the presence of hydrogen chloride, for example, the drug pseudoephedrine behaves as a base by accepting H^+ from a hydrogen chloride. The negative Cl^- then joins the pseudoephedrine–H^+ ion to form the salt pseudoephedrine hydrochloride, which is a common nasal decongestant, shown in Figure 21.6. This salt is soluble in water and can be absorbed through the digestive system. In fact, most drugs taken by mouth are bases that have been reacted with an acid to form a water-soluble salt. Drugs that can't form salts are usually best administered by syringe.

FIGURE 21.6 ▶
Hydrogen chloride and pseudoephedrine react to form the salt *pseudoephedrine hydrochloride*, which, because of its solubility in water, is readily absorbed into the body.

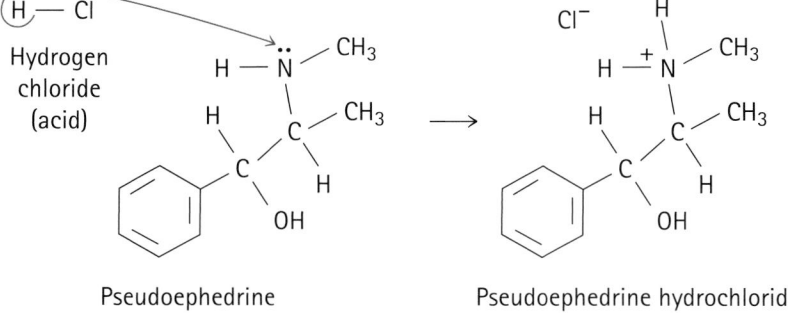

Pseudoephedrine (base) Pseudoephedrine hydrochloride (salt)

CONCEPT CHECK
Is a neutralization reaction best described as a physical change or a chemical change?

Check Your Answer
New chemicals are formed during a neutralization reaction, meaning the reaction is a chemical change.

21.2 Some Acids and Bases Are Stronger Than Others

In general, the stronger an acid, the more readily it donates hydrogen ions. Likewise, the stronger a base, the more readily it accepts hydrogen ions. An example of a strong acid is hydrogen chloride, HCl, and an example of a strong base is sodium hydroxide, NaOH. The corrosiveness of these materials is a result of their strength. ✓ Strong acids and bases tend to generate more ions than weak acids or bases. Therefore, a strong acid (or strong base) solution is a better conductor of electricity than is a weak acid (or base) solution of the same concentration. This idea is illustrated in Figure 21.7.

Weak acids and weak bases have only a small tendency to donate or accept hydronium ions. An example of a weak acid is acetic acid, $C_2H_4O_2$, which is the active ingredient of vinegar. Baking soda, $NaHCO_3$, is an example of a weak base. Both of these substances are only mildly corrosive. The mild corrosive action of vinegar, however, is sufficient to clean the gunk off older pennies. Drop an old penny into some vinegar and see for yourself.

Just because an acid or base is strong doesn't mean a solution of that acid or base is corrosive. The corrosive action of an acidic solution is caused by the hydronium ions rather than by the acid that generated those hydronium ions. Similarly, the corrosive action of a basic solution results from the hydroxide ions it contains, regardless of the base that generated those hydroxide ions. A *very* dilute solution of a strong acid or a strong base will have little corrosive action because in such solutions there are only a few hydronium or hydroxide ions. You shouldn't be too alarmed, therefore, when you discover that some toothpastes are formulated with small amounts of sodium hydroxide, one of the strongest bases known.

A concentrated solution of acetic acid, a weak acid, may be just as corrosive or even more corrosive than a dilute solution of hydrochloric acid, which is a strong acid. The strengths of two acids (or bases), therefore, can only be compared when they have the same concentration.

✓ **READING CHECK**

What do strong acids generate more of than do weak acids?

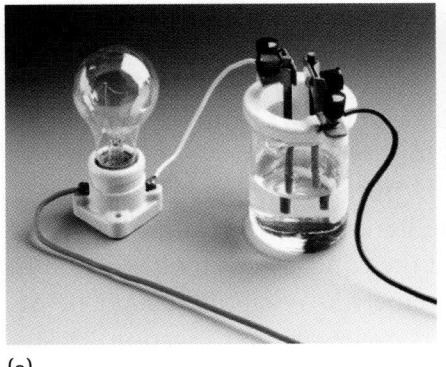  

(a)　　　　(b)　　　　(c)

FIGURE 21.7 ▲
(a) The pure water in this circuit is unable to conduct electricity because it contains practically no ions. The lightbulb in the circuit therefore remains unlit. (b) Because HCl is a strong acid, nearly all of its molecules break apart in water, giving a high concentration of ions, which are able to conduct an electric current that lights the bulb. (c) Acetic acid, $C_2H_4O_2$, is a weak acid; in water, only a small portion of its molecules break up into ions. Because fewer ions are generated, only a weak current exists, and the bulb is therefore dimmer.

CONCEPT CHECK
According to the aqueous solutions illustrated here, which is the stronger base, NH_3 or NaOH?

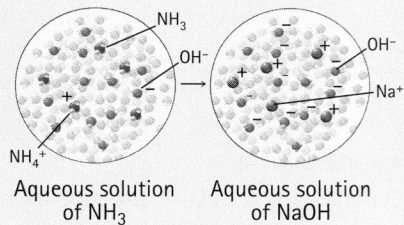

Aqueous solution of NH_3 · Aqueous solution of NaOH

Check Your Answer
The solution on the right contains the greater number of ions, meaning sodium hydroxide, NaOH, is the stronger base. Ammonia, NH_3, is the weaker base, indicated by the relatively few ions in the solution on the left.

21.3 Solutions Can Be Acidic, Basic, or Neutral

Water has an interesting property of being able to behave both as an acid and as a base. Because of this, it can react with itself. In behaving as an acid, a water molecule donates a hydrogen ion to a neighboring water molecule, which, in accepting the hydrogen ion, is behaving as a base. This reaction produces a hydroxide ion and a hydronium ion, which react to re-form the water molecules:

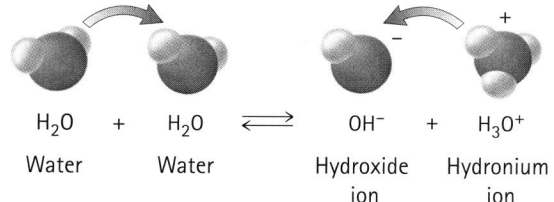

H_2O + H_2O ⇌ OH^- + H_3O^+
Water · Water · Hydroxide ion · Hydronium ion

When a water molecule gains a hydrogen ion, a second water molecule must lose a hydrogen ion. So for every one hydronium ion formed, one hydroxide ion also forms. In pure water, therefore, the total number of hydronium ions must be the same as the total number of hydroxide ions. Experiments reveal that the concentration of hydronium and hydroxide ions in pure water is extremely low—about 0.0000001 M for each, where M stands for molarity, or moles per liter (Section 19.4). Water by itself, therefore, is a very weak acid as well as a very weak base, as evidenced by the unlit lightbulb in Figure 21.7a.

CONCEPT CHECK
Do water molecules react with one another?

Check Your Answer
Yes, but not to any large extent. When they do react, they form hydronium and hydroxide ions.

Any solution containing an equal number of hydronium and hydroxide ions is said to be **neutral.** ✔ Pure water is an example of a neutral solution—not because it contains so few hydronium or hydroxide ions, but because it contains equal numbers of these ions. A neutral solution is also obtained when equal quantities of acid and base are combined, which explains why acids and bases are said to *neutralize* each other.

The balance of hydronium and hydroxide ions in a neutral solution is upset by adding either an acid or a base. Add an acid and the water will react with that acid to produce more hydronium ions. Many of these additional hydronium ions neutralize hydroxide ions, which then become fewer. The final result is that the hydronium ion concentration is greater than the hydroxide ion concentration. Such a solution is said to be **acidic.**

Add a base to water and the reverse happens. The water will react with that base to produce more hydroxide ions. Many of these additional hydroxide ions will neutralize hydronium ions, which then become fewer. The final result is that the hydronium ion concentration is less than the hydroxide ion concentration. Such a solution is said to be **basic.** This is all summarized in Figure 21.8.

fyi
- A beautician can control how long hair retains artificial coloring by modifying the pH of the hair-coloring solution. With an acidic solution, microscopic openings, called cuticles, on the surface of each hair close shut so that the dye binds only to the outside. This results is a temporary hair coloring, which may come off with the next hair washing. With an alkaline solution, the cuticles open up, which allows the dye to penetrate into the hair for a more permanent effect.

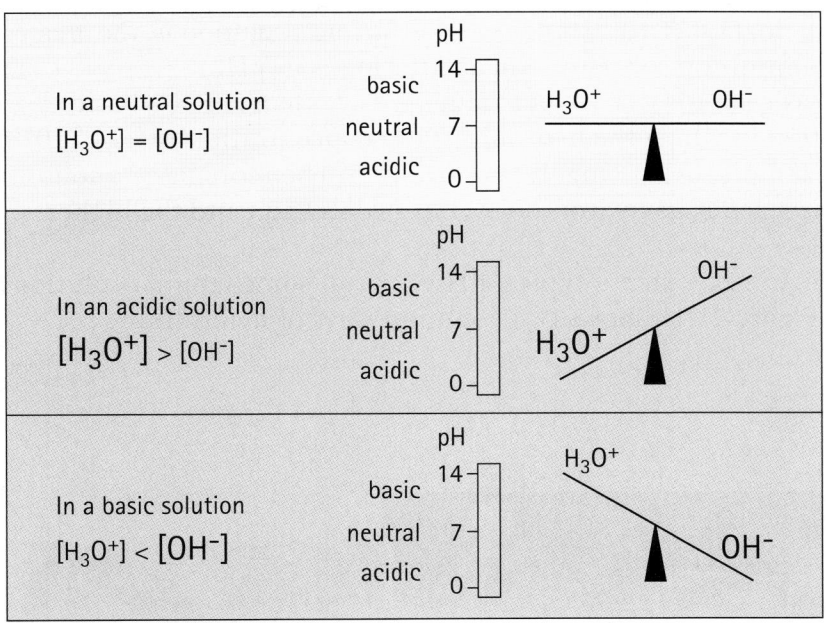

◀ **FIGURE 21.8**
The relative concentrations of hydronium and hydroxide ions determine whether a solution is acidic, basic, or neutral.

CONCEPT CHECK

How does adding ammonia, NH_3, to water make a basic solution when there are no hydroxide ions in the formula for ammonia?

Check Your Answer

Ammonia indirectly increases the hydroxide ion concentration by reacting with water:

$$NH_3 + H_2O \rightarrow NH_4^+ + OH^-$$

This reaction raises the hydroxide-ion concentration, which has the effect of lowering the hydronium-ion concentration. With the hydroxide-ion concentration now higher than the hydronium-ion concentration, the solution is basic.

✔ **READING CHECK**

Why is pure water a neutral solution?

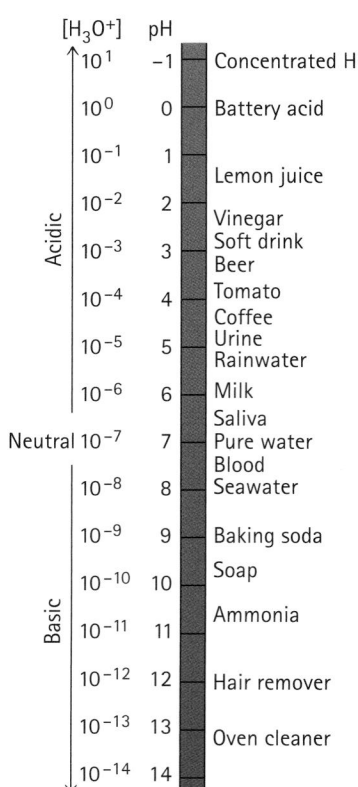

[H₃O⁺] and pH scale:

$[H_3O^+]$	pH	
10^1	−1	Concentrated HCl
10^0	0	Battery acid
10^{-1}	1	
		Lemon juice
10^{-2}	2	Vinegar
10^{-3}	3	Soft drink / Beer
10^{-4}	4	Tomato
		Coffee
10^{-5}	5	Urine / Rainwater
10^{-6}	6	Milk
		Saliva
10^{-7}	7	Pure water / Blood
10^{-8}	8	Seawater
10^{-9}	9	Baking soda
10^{-10}	10	Soap
10^{-11}	11	Ammonia
10^{-12}	12	Hair remover
10^{-13}	13	Oven cleaner
10^{-14}	14	

Acidic (top), Neutral at 10^{-7} / pH 7, Basic (bottom)

FIGURE 21.9 ▲
The pH values of some common solutions.

The pH Scale Is Used to Describe Acidity

The *pH scale* is a numeric scale used to express the acidity of a solution. Mathematically, **pH** is equal to the negative logarithm of the hydronium-ion concentration:

$$pH = -\log[H_3O^+]$$

Note that brackets are used to represent molar concentrations, meaning $[H_3O^+]$ is read "the molar concentration of hydronium ions." For an understanding of the logarithm function, see the Math Connection.

Consider a neutral solution that has a hydronium-ion concentration of $1.0 \times 10^{-7}\ M$. To find the pH of this solution, we first take the logarithm of this value, which is −7 (see the Math Connection on logarithms). The pH is by definition the negative of this value, which means $-(-7) = 7$. Hence, in a neutral solution, where the hydronium-ion concentration equals $1.0 \times 10^{-7}\ M$, the pH is 7.

Acidic solutions have pH values less than 7. For an acidic solution in which the hydronium-ion concentration is $1.0 \times 10^{-4}\ M$, for example, $pH = -\log(1.0 \times 10^{-4}) = 4$. The more acidic a solution is, the greater its hydronium-ion concentration and the lower its pH.

Basic solutions have pH values greater than 7. For a basic solution in which the hydronium-ion concentration is $1.0 \times 10^{-8}\ M$, for example, $pH = -\log(1.0 \times 10^{-8}) = 8$. The more basic a solution is, the smaller its hydronium-ion concentration and the higher its pH.

Figure 21.9 shows typical pH values of some familiar solutions, and Figure 21.10 shows two common ways of determining pH values.

(a)

(b)

FIGURE 21.10 ▲
(a) The pH of a solution can be measured electronically using a pH meter. (b) A rough estimate of the pH of a solution can be obtained with litmus paper, which is coated with a dye that changes color with pH.

LOGARITHMS AND pH

The logarithm is simply the power to which 10 is raised. For example, the logarithm of 10^2 is 2, which is the power to which 10 is raised. Likewise, the logarithm of 10^3 is 3. Logarithms are that simple! To take it one step farther, if you know that 10^2 is equal to 100, then you'll understand that the logarithm of 100 also is 2. Check this out on your calculator. Similarly, the logarithm of 1000 is 3 because 10 raised to the third power, 10^3, equals 1000.

Any positive number, including a very small one, has a logarithm. The logarithm of 0.0001, which equals 10^{-4}, for example, is −4 (the power to which 10 is raised to equal this number).

Example
What is the logarithm of 0.01?

Answer
The number 0.01 is 10^{-2}, the logarithm of which is −2 (the power to which 10 is raised).

The concentration of hydronium ions in most solutions is typically much less than 1 M. Recall, for example, that in neutral water the hydronium ion concentration is 0.0000001 M (10^{-7} M). The logarithm of any number smaller than 1 (but greater than zero) is a negative number. The definition of pH includes the minus sign so as to transform the logarithm of the hydronium ion concentration to a positive number.

When a solution has a hydronium ion concentration of 1 M, the pH is 0 because 1 $M = 10^0$ M. A 10 M solution has a pH of −1 because 10 $M = 10^1$ M.

Example
What is the pH of a solution that has a hydronium-ion concentration of 0.001 M?

Answer
The number 0.001 is 10^{-3}, and so

$$\begin{aligned} pH &= -\log[H_3O^+] \\ &= -\log 10^{-3} \\ &= -(-3) \\ &= 3 \end{aligned}$$

21.4 Rainwater Is Acidic and Ocean Water Is Basic

Rainwater is naturally acidic. One source of this acidity is carbon dioxide, the same gas that gives fizz to soda drinks. There are 810 billion tons of CO_2 in the atmosphere, most of it from such natural sources as volcanoes and decaying organic matter but a growing amount (about 135 billion tons) from human activities.

Water in the atmosphere reacts with carbon dioxide to form *carbonic acid*:

$$CO_2(g) + H_2O(l) \rightarrow H_2CO_3(aq)$$

Carbon   Water   Carbonic
dioxide   acid

Carbonic acid, as its name implies, behaves as an acid and lowers the pH of water. The CO_2 in the atmosphere brings the pH of rainwater to about 5.6—noticeably below the neutral pH value of 7. Because of local fluctuations, the normal pH of rainwater varies between 5 and 7. This natural acidity of rainwater may accelerate the erosion of land and, under certain circumstances, can lead to the formation of underground caves, as we explore in Section 28.3.

By convention, *acid rain* is a term used for rain having a pH lower than 5.6. Acid rain is created when airborne pollutants, such as sulfur

fyi

- Significant progress has been made towards fixing the problem of acid rain. In the United States, sulfur dioxide and nitrogen oxide emissions have been reduced by nearly half since 1980.

(a)

(b)

FIGURE 21.11 ▲
(a) These two photographs show the same obelisk in New York City's Central Park before and after the effects of acid rain. (b) Many forests downwind from heavily industrialized areas, such as in the northeastern United States and in Europe, have been noticeably hard-hit by acid rain.

dioxide, are absorbed by atmospheric moisture. Sulfur dioxide is readily converted to sulfur trioxide, which reacts with water to form *sulfuric acid:*

$$2\,SO_2(g) + O_2(g) \rightarrow SO_3(g)$$

Sulfur   Oxygen   Sulfur
dioxide   trioxide

$$SO_3(g) + H_2O(l) \rightarrow H_2SO_4(aq)$$

Sulfur   Water   Sulfuric
trioxide   acid

Each year about 20 million tons of SO_2 is released into the atmosphere by the combustion of sulfur-containing coal and oil. Sulfuric acid is much stronger than carbonic acid, and, as a result, rain laced with sulfuric acid eventually corrodes metal, paint, and other exposed substances. Each year, the damage costs billions of dollars. The cost to the environment is also high (Figure 21.11). Many rivers and lakes receiving acid rain become less capable of sustaining life. Much vegetation that receives acid rain doesn't survive. This is particularly evident in heavily industrialized regions.

CONCEPT CHECK
When sulfuric acid, H_2SO_4, is added to water, what makes the resulting aqueous solution corrosive?

Check Your Answer
Because H_2SO_4 is a strong acid, it readily forms hydronium ions when dissolved in water. Hydronium ions are responsible for the corrosive action.

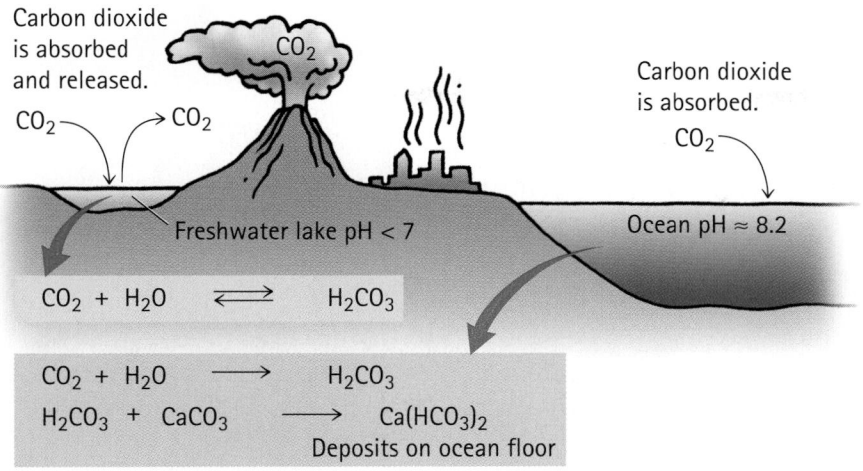

FIGURE 21.12 ▲
Carbon dioxide forms carbonic acid upon entering any body of water. In fresh water, this reaction is reversible, and the carbon dioxide is released back into the atmosphere. In the alkaline ocean, the carbonic acid is neutralized to such compounds as calcium bicarbonate, $Ca(HCO_3)_2$, which precipitate to the ocean floor. As a result, most of the atmospheric carbon dioxide that enters our oceans remains there.

A long-term solution to acid rain is to prevent most of the generated sulfur dioxide and other pollutants from entering the atmosphere in the first place. Toward this end, smokestacks have been designed or retrofitted to minimize the quantities of pollutants released. Though costly, the positive effects of these adjustments have been demonstrated. An ultimate long-term solution, however, would be a shift from fossil fuels to cleaner energy sources, such as nuclear and solar energy.

It should come as no surprise that the amount of carbon dioxide put into the atmosphere by human activities is growing. What is surprising, however, is that the atmospheric concentration of CO_2 is not increasing as fast as we put it there. A likely explanation has to do with the oceans, as illustrated in Figure 21.12. ☑ When atmospheric CO_2 dissolves in any body of water—a raindrop, a lake, or the ocean—it forms carbonic acid. Carbonic acid in the ocean, however, is quickly neutralized (the ocean is alkaline, pH ≈ 8.2). Thus, carbonic acid neutralization in the ocean prevents CO_2 from being released back into the atmosphere. The ocean, therefore, is a carbon dioxide *sink*—most of the CO_2 that goes in doesn't come out. So, pushing more CO_2 into our atmosphere means pushing more of it into our vast oceans. This is another of the many ways in which the oceans regulate our global environment.

But the ocean doesn't absorb carbon dioxide without cost. Shelled marine organisms create their shells using dissolved carbonate ions, CO_3^{2-}. These ions are neutralized (destroyed) by carbonic acid. So as the oceans continue to absorb atmospheric carbon dioxide, the concentration of carbonate ions will decrease. This is bad news for all those carbonate ion-dependent shelled marine organisms, many of which form the bottom of the marine food chain.

The pollution humans release knows no political boundaries. Iron smelters operating in China, for example, release pollutants that are readily detected in Seattle, Washington.

FIGURE 21.13 ▶

Researchers at the Mauna Loa Weather Observatory in Hawaii have recorded increasing concentrations of atmospheric carbon dioxide since they began collecting data in the 1950s. This famous graph is known as the Keeling curve, after the scientist, Charles Keeling, who initiated this project and first noted the trends. Interestingly, the oscillations within the Keeling curve reflect seasonal changes in CO_2 levels.

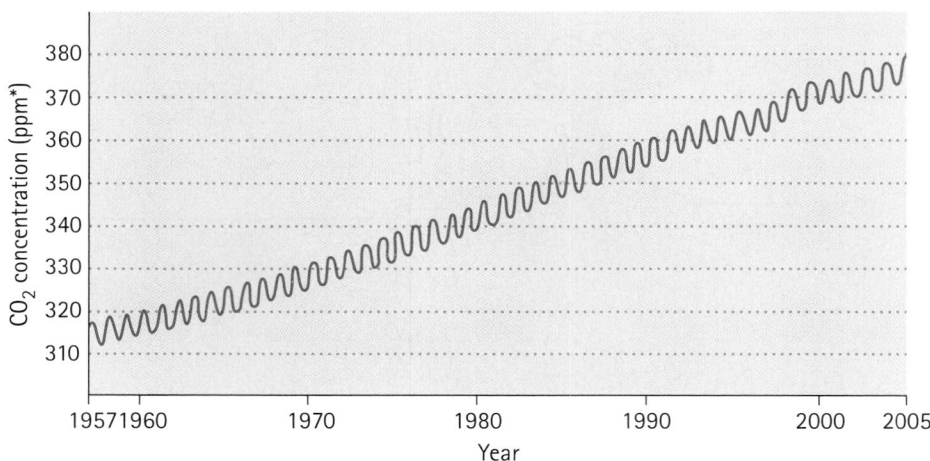

*ppm = parts per million, which tells us the number of carbon dioxide molecules for every million molecules of air.

✔ **READING CHECK**

What happens to carbon dioxide as it dissolves in water?

As Figure 21.13 shows, the concentration of atmospheric CO_2 is increasing. Carbon dioxide is being produced faster than the ocean can absorb it, and this may alter Earth's environment. Carbon dioxide is a *greenhouse gas*, which means it helps keep the surface of the Earth warm by preventing infrared radiation from escaping into outer space. Without greenhouse gases in the atmosphere, Earth's surface would average a frigid −18°C. However, with increasing concentration of CO_2 in the atmosphere, we might experience higher average temperatures. Higher temperatures may significantly alter global weather patterns as well as raise the average sea level, as the polar ice caps melt and the volume of seawater increases because of thermal expansion. Global warming is explored in greater detail in Chapter 30. No matter how you look at it, the increasing concentration of atmospheric carbon dioxide is a serious concern.

So the pH of rain depends in great part on the concentration of atmospheric CO_2, which depends on the pH of the oceans. These systems are interconnected with global temperatures, which naturally connect to the countless living systems on Earth. How true it is: All the parts are intricately connected, down to the level of atoms and molecules!

21.5 Oxidation Is the Loss of Electrons and Reduction Is the Gain of Electrons

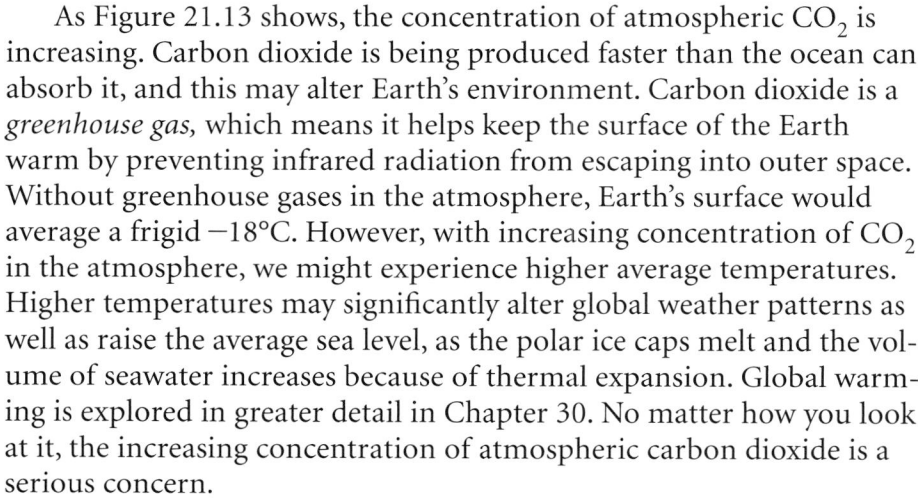

FIGURE 21.14 ▲

In the exothermic formation of sodium chloride, sodium metal is oxidized by chlorine gas, and chlorine gas is reduced by sodium metal.

During an acid–base reaction, protons are transferred from one reactant (the acid) to another (the base). ✔ A related class of reactions is the oxidation–reduction reactions in which electrons are transferred between reactants. An oxidation–reduction reaction occurs when sodium and chlorine react to form sodium chloride, as shown in Figure 21.14. The equation for this reaction is

$$2\,Na(s) + Cl_2(g) \longrightarrow 2\,NaCl(s)$$

To see how electrons are transferred in this reaction, we can look at each reactant individually. Each electrically neutral sodium atom loses an electron and becomes a positively charged ion. This process of losing an electron is called **oxidation.**

$$2\,\text{Na}(s) \rightarrow 2\,\text{Na}^+ + 2e^- \qquad \text{Oxidation}$$

Each electrically neutral chlorine atom captures an electron to form chloride ions. This process of gaining electrons is called **reduction.**

$$\text{Cl}_2 + 2e^- \rightarrow 2\text{Cl}^- \qquad \text{Reduction}$$

The final result is that the two electrons lost by the sodium atoms are transferred to the chlorine atoms. Note that oxidation and reduction always occur together; you cannot have one without the other. The electrons lost by one chemical in an oxidation reaction are always gained by another chemical in a reduction reaction.

Because the sodium causes reduction of the chlorine, the sodium is acting as a *reducing agent*. A reducing agent is any reactant that causes another reactant to be reduced. Note that sodium is oxidized when it behaves as a reducing agent—it loses electrons. Conversely, the chlorine causes oxidation of the sodium and so is acting as an *oxidizing agent*. Because it gains electrons in the process, an oxidizing agent is reduced. Just remember that **l**oss of **e**lectrons is **o**xidation, and **g**ain of **e**lectrons is **r**eduction. Here is a helpful mnemonic adapted from a once-popular children's story: **Leo** the lion went **"ger."**

Different elements have different oxidation and reduction tendencies—some lose electrons more readily, while others gain electrons more readily, as Figure 21.15 illustrates. This periodic trend follows the same periodic trend of *electronegativity* discussed in Section 18.6. Recall that electronegativity is the ability of an atom to pull electrons towards itself. The stronger the electronegativity, therefore, the stronger the ability to behave as an oxidizing agent.

> When we say a substance was oxidized, we're saying that it lost electrons. When we say a substance was reduced, we're saying that it gained electrons.

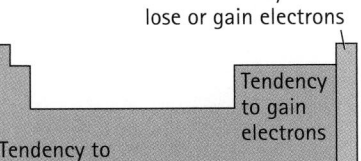

FIGURE 21.15 ▲
The ability of an atom to gain or lose electrons is indicated by its position in the periodic table. Those at the upper right tend to gain electrons, and those at the lower left tend to lose them.

CONCEPT CHECK
True or false:
1. Reducing agents are oxidized in oxidation–reduction reactions.
2. Oxidizing agents are reduced in oxidation–reduction reactions.

Check Your Answer
Both statements are true.

READING CHECK

What gets transferred in an oxidation–reduction reaction?

21.6 The Energy of Flowing Electrons Can Be Harnessed

Electrochemistry is the study of the relationship between electrical energy and chemical change. ✓ It involves either the use of an oxidation–reduction reaction to produce an electric current or the use of an electric current to produce an oxidation–reduction reaction. To understand how

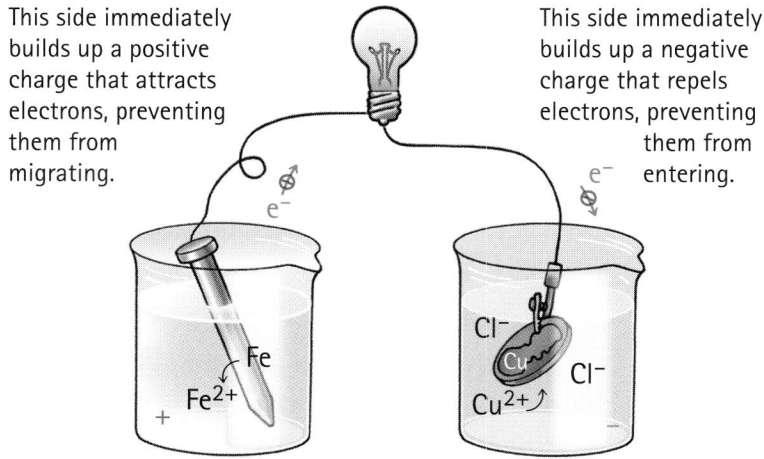

This side immediately builds up a positive charge that attracts electrons, preventing them from migrating.

This side immediately builds up a negative charge that repels electrons, preventing them from entering.

FIGURE 21.16 ▲

An iron nail (reducing agent) is placed in water and connected by a conducting wire to a solution of copper ions (oxidizing agent). Electrons, however, won't flow from the iron to the copper ions because of a buildup of charge in each container.

What does electrochemistry involve?

A salt bridge may be as simple as a paper towel soaked in saltwater.

an oxidation–reduction reaction can generate an electric current, consider what happens when a reducing agent is placed in direct contact with an oxidizing agent: Electrons flow from the reducing agent to the oxidizing agent. This flow of electrons is an electric current, which is a form of kinetic energy that can be harnessed for useful purposes.

Simply placing a reducing agent in contact with an oxidizing agent, however, does not allow for a continuous flow of electrons. Why? Because as the reducing agent loses electrons it builds up a positive charge. Meanwhile, as the oxidizing agent gains electrons it builds up a negative charge. Remember that electrons are negative, so the last thing they want to do is flow from where it is positive to where it is negative. Any electric current between the reducing and oxidizing agents quickly stops, as is illustrated in Figure 21.16.

The solution to this problem is to build what is called a *salt bridge*, which allows allow ions to migrate back and forth between the separated reducing and oxidizing agents. As shown in Figure 21.17 on page 477, a salt bridge may be a U-shaped tube filled with a salt, such as sodium nitrate, $NaNO_3$, and closed with semiporous plugs. The flow of ions between the two containers prevents a buildup of charge, which permits electrons to flow freely.

Batteries

So we can see that, with the proper setup, it is possible to harness electrical energy from an oxidation–reduction reaction. The device shown in Figure 21.17 illustrates the basic principles, but it is not well suited to lighting up your flashlight. Using the same principles, but different chemicals and different designs, it's possible to create an all-in-one, self-contained unit that you know as a *battery*. There are many

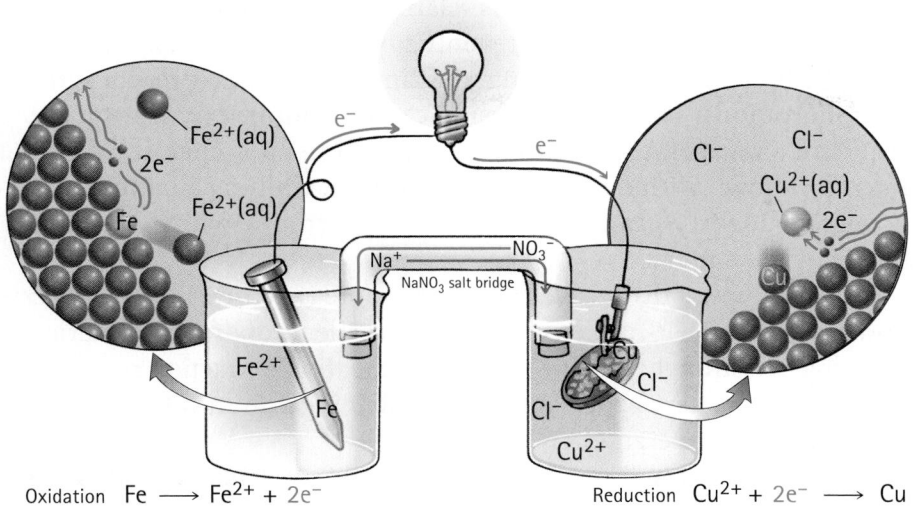

Oxidation $Fe \longrightarrow Fe^{2+} + 2e^-$ Reduction $Cu^{2+} + 2e^- \longrightarrow Cu$

◀ **FIGURE 21.17**
The salt bridge completes the electric circuit. Electrons freed as the iron is oxidized pass through the wire to the container on the right. Nitrate ions, NO_3^-, from the salt bridge flow into the left container to balance the positive charges of the Fe^{2+} ions that form, thereby preventing any buildup of positive charge. Meanwhile, Na^+ ions from the salt bridge enter the right container to balance the Cl^- ions "abandoned" by the Cu^{2+} ions as the Cu^{2+} ions pick up electrons to become metallic copper.

different types of batteries, but they all function by the same principle: Two materials that oxidize and reduce each other are connected by a medium through which ions travel to balance an external flow of electrons.

The common *dry-cell battery*, which was invented in the 1860s, is still used today, and it is probably the cheapest disposable energy source for flashlights, toys, and the like. The basic design consists of a zinc cup filled with a thick paste of ammonium chloride, NH_4Cl, zinc chloride, $ZnCl_2$, and manganese dioxide, MnO_2. Immersed in this paste is a porous stick of graphite that projects to the top of the battery, as shown in Figure 21.18.

Reduction $2\,NH_4^+ + 2e^- \longrightarrow 2\,NH_3 + H_2$

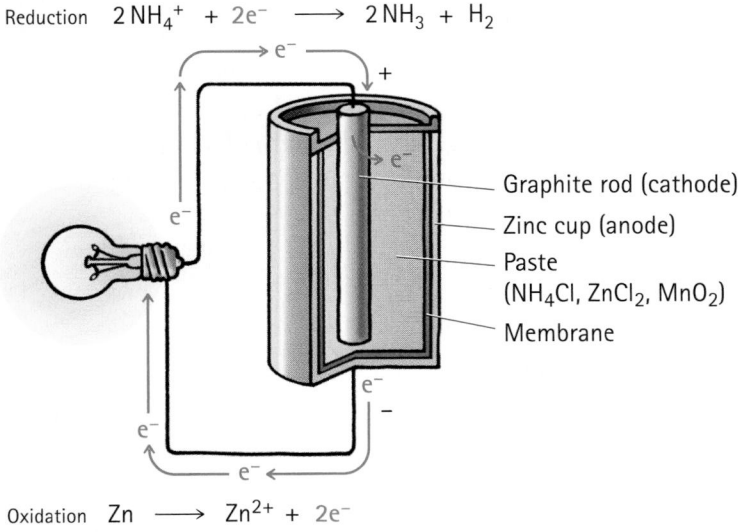

Graphite rod (cathode)
Zinc cup (anode)
Paste
(NH_4Cl, $ZnCl_2$, MnO_2)
Membrane

Oxidation $Zn \longrightarrow Zn^{2+} + 2e^-$

FIGURE 21.18 ▲
A common dry-cell battery with a graphite rod immersed in a paste of ammonium chloride, manganese dioxide, and zinc chloride.

If you store your extra flashlight batteries in the refrigerator, they'll last longer.

FIGURE 21.19 ▲
Alkaline batteries last a lot longer than dry-cell batteries and give a steadier voltage, but they are more expensive.

Aside from the initial charge of a brand-new battery, the energy in a car battery ultimately comes from fuel in the gas tank through the process of recharging.

FIGURE 21.20 ▲
As of 2008, over 1 million Prius hybrids have been sold worldwide, about 600,000 of them in the United States. Look now for hybrid vehicles that can be plugged into your home electrical outlet, charged at night, and driven the next day using no gasoline for up to 60 miles.

The life of a dry-cell battery is relatively short. Oxidation causes the zinc cup to deteriorate, and eventually the contents leak out. Even while the battery is not operating, the zinc corrodes as it reacts with ammonium ions. This zinc corrosion can be inhibited by storing the battery in a refrigerator. As discussed in Section 20.2, chemical reactions slow down with decreasing temperature. Chilling a battery, therefore, slows down the rate at which the zinc corrodes, which increases the life of the battery.

Another type of disposable battery, the more expensive *alkaline battery,* shown in Figure 21.19, avoids many of the problems of dry-cell batteries by operating in a strongly alkaline paste. The oxidation–reduction reactions in an alkaline battery are better suited to maintaining a given voltage during longer periods of operation.

The small lithium disposable batteries used for calculators and cameras are variations of the alkaline battery. In the lithium battery, lithium metal is used as the source of electrons rather than zinc. Not only is lithium able to maintain a higher voltage than zinc but it is also about one-thirteenth as dense, which allows for a lighter battery.

Disposable batteries have relatively short lives because electron-producing chemicals are consumed. The main feature of *rechargeable* batteries is the reversibility of the oxidation and reduction reactions. So recharging a rechargeable battery simply means regenerating the chemicals. For a traditional car battery, this chemical is simply lead, Pb, which transforms into lead sulfate, $PbSO_4$, as it releases electrons. As the car battery is recharged, the $PbSO_4$ is transformed back into lead, Pb.

Rechargeable lithium ion batteries have found a wide range of applications, from powering computer laptops to cell phones. Safer lithium phosphate iron batteries are used for hybrid cars, such as the popular Toyota Prius shown in Figure 21.20. Hybrids have improved gas mileage because as the car slows down, its kinetic energy is transformed into the electric potential energy of the battery rather than being wasted as heat from the car's brake pads. The captured electrical energy of the battery is subsequently used to assist the gas-powered engine to get the car moving. Also, the hybrid's battery system allows for the engine to shut off when the car is merely idling or moving slowly, as occurs in heavy traffic.

CONCEPT CHECK
What happens to the chemicals in a rechargeable battery that doesn't happen to the chemicals in a disposable battery?

Check Your Answer
The chemicals of any battery are consumed as it generates electricity. Within a rechargeable battery, however, these chemicals are regenerated as the battery is recharged. The types of chemical reactions taking place in a disposable battery are not reversible.

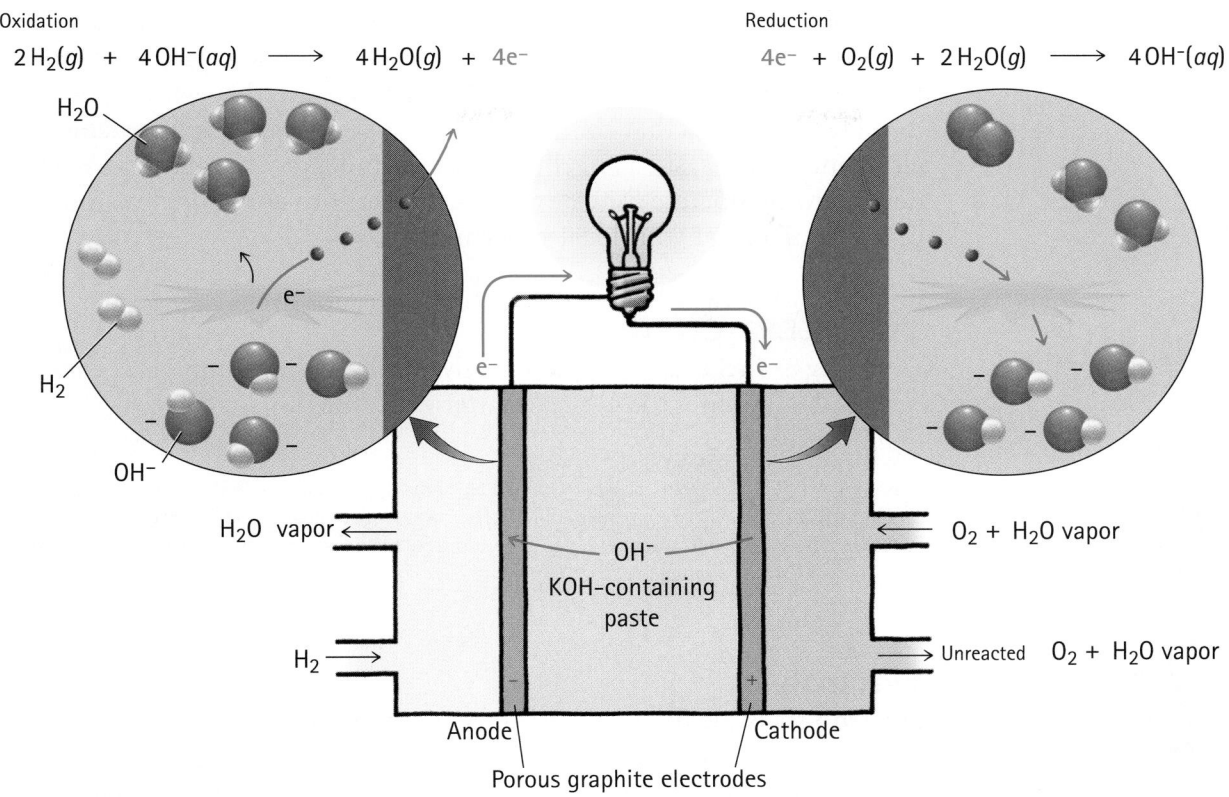

Oxidation

$$2H_2(g) + 4OH^-(aq) \longrightarrow 4H_2O(g) + 4e^-$$

Reduction

$$4e^- + O_2(g) + 2H_2O(g) \longrightarrow 4OH^-(aq)$$

H$_2$O

e$^-$

H$_2$

OH$^-$

H$_2$O vapor ←

O$_2$ + H$_2$O vapor

e$^-$ e$^-$

OH$^-$
KOH-containing
paste

H$_2$ →

→ Unreacted O$_2$ + H$_2$O vapor

Anode   Cathode
Porous graphite electrodes

FIGURE 21.21 ▲
The hydrogen–oxygen fuel cell.

Fuel Cells

A *fuel cell* is a device that converts the chemical energy of a fuel to electrical energy. Fuel cells are by far the most efficient means of generating electricity. A hydrogen–oxygen fuel cell is shown in Figure 21.21. It has two compartments, one for entering hydrogen fuel and the other for entering oxygen fuel. Hydrogen is oxidized upon contact with hydroxide ions at the negative electrode. The electrons from this oxidation flow through an external circuit and provide electric power before meeting up with oxygen at the positive electrode. The oxygen readily picks up the electrons (in other words, the oxygen is reduced) and reacts with water to form hydroxide ions. These hydroxide ions migrate across the ionic paste of potassium hydroxide, KOH, to join with hydrogen, thus completing the circuit.

Although fuel cells are similar to dry-cell batteries, they don't run down as long as fuel is supplied. The space shuttle uses hydrogen–oxygen fuel cells to meet its electrical needs. The cells also produce more than 100 gallons of drinking water for the astronauts during a typical week-long mission. Back on Earth, researchers are developing fuel cells for buses and automobiles. As shown in Figure 21.22, experimental fuel-cell buses are already operating in several cities, such as Vancouver, British Columbia, and Chicago, Illinois. These vehicles produce very few pollutants and can run much more efficiently than vehicles that burn fossil fuels.

FIGURE 21.22 ▲
Because this bus is powered by a fuel cell, its tailpipe emits mostly water vapor.

In the future, commercial buildings as well as individual homes may be outfitted with fuel cells as an alternative to receiving electricity (and heat) from regional power stations. Researchers are also working on miniature fuel cells that could replace the batteries used for portable electronic devices, such as cellular phones and laptop computers. Such devices could operate for extended periods of time on a single "ampule" of fuel available at your local supermarket.

> **CONCEPT CHECK**
>
> As long as fuel is available to it, a given fuel cell can supply electrical energy indefinitely. Why can't batteries do the same?
>
> **Check Your Answer**
> Batteries generate electricity as the chemical reactants they contain are reduced and oxidized. Once these reactants are consumed, the battery can no longer generate electricity. A rechargeable battery can be made to operate again, but only after the energy flow is interrupted so that the reactants can be replenished.

Electrolysis

Electrolysis is the use of electrical energy to produce chemical change. The recharging of a car battery is an example of electrolysis. Another, shown in Figure 21.23, is passing an electric current through water, a process that breaks the water down into its elemental components:

$$\text{Electrical energy} + 2\,H_2O(l) \rightarrow 2\,H_2(g) + O_2(g)$$

Electrolysis is used to purify metals from metal ores. An example is aluminum, the third most abundant element in Earth's crust. Aluminum occurs naturally bonded to oxygen in an ore called bauxite. Through electrolysis, the bauxite can be converted into aluminum metal. Today, worldwide production of aluminum is about 16 million tons annually. For each ton produced from ore, about 16,000 kilowatt-hours of electrical energy is required, as much as a typical American household consumes in 18 months. Processing recycled aluminum, on the other hand, consumes only about 700 kilowatt-hours for every ton. Thus, recycling aluminum not only reduces litter but also helps to reduce the load on power companies, which, in turn, reduces air pollution. Furthermore, reserves of high-quality aluminum oxide ores are already depleted in the United States. Recycling aluminum, therefore, also helps to minimize the need for developing new bauxite mines in foreign countries.

> Chemical change can produce electricity, so it makes sense that electricity can produce chemical change. Science is symmetrical.

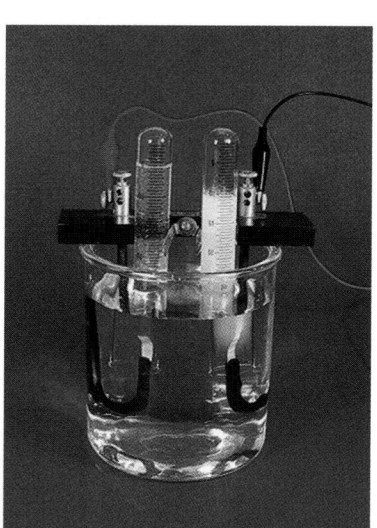

FIGURE 21.23 ▲
The electrolysis of water produces hydrogen gas and oxygen gas in a 2:1 ratio by volume, in accord with the chemical formula for water: H_2O. In order for this process to work, ions must be dissolved in the water so that electric charge can be conducted between the electrodes.

> **CONCEPT CHECK**
>
> Is the exothermic reaction in a hydrogen–oxygen fuel cell an example of electrolysis?
>
> **Check Your Answer**
> No. During electrolysis, electrical energy is used to produce chemical change. In the hydrogen–oxygen fuel cell, chemical change is used to produce electrical energy.

21.7 Oxygen Is Responsible for Corrosion and Combustion

✓ If you look to the upper right of the periodic table, you will find one of the most common oxidizing agents—oxygen. In fact, if you haven't guessed already, the term *oxidation* is derived from the name of this element. Oxygen is able to pluck electrons from many other elements, especially those that lie at the lower left of the periodic table. Two common oxidation–reduction reactions involving oxygen as the oxidizing agent are *corrosion* and *combustion*.

CONCEPT CHECK

Oxygen is a good oxidizing agent, but so is chlorine. What does this indicate about their relative positions in the periodic table?

Check Your Answer
Chlorine and oxygen must lie in the same area of the periodic table. Both have strong effective nuclear charges and are strong oxidizing agents.

Corrosion is the process whereby a metal deteriorates. Corrosion caused by atmospheric oxygen is a widespread and costly problem. About one-quarter of the steel produced in the United States, for example, goes into replacing corroded iron at a cost of billions of dollars annually. Iron corrodes when it reacts with atmospheric oxygen and water to form iron oxide trihydrate, which is the naturally occurring reddish-brown substance you know as rust, shown in Figure 21.24:

$$4\,Fe + 3\,O_2 + 3\,H_2O \rightarrow 2\,Fe_2O_3 \cdot 3\,H_2O$$
Iron   Oxygen   Water   Rust

Another common metal oxidized by oxygen is aluminum. The product of aluminum oxidation is aluminum oxide, Al_2O_3, which is not water soluble. Because of its insolubility, aluminum oxide forms a protective coat that shields the metal from further oxidation. This coat is so thin that it's transparent, which is why aluminum maintains its metallic shine.

A protective, water-insoluble oxidized coat is the principle underlying a process called *galvanization*. Zinc has a slightly greater tendency to oxidize than does iron. For this reason, many iron objects, such as the nails pictured in Figure 21.25, are *galvanized* by coating them with a thin layer of zinc. The zinc oxidizes to zinc oxide, an inert, insoluble substance that protects the iron underneath it from rusting.

Combustion is an oxidation–reduction reaction between a nonmetallic material and molecular oxygen. Combustion reactions are characteristically exothermic (energy releasing). A violent combustion reaction is the formation of water from hydrogen and oxygen. As discussed in Section 20.4, the energy from this reaction is used to

FIGURE 21.24 ▲
Rust itself is not harmful to the iron structures on which it forms. It is the loss of metallic iron that ruins the structural integrity of these objects.

FIGURE 21.25 ▲
The galvanized nail (*bottom*) is protected from rusting by the sacrificial oxidation of zinc.

The air above a campfire is always moist. Why? Because of all the water vapor it contains. This water vapor is one of the products of the combustion of the firewood.

✔ READING CHECK

What is one of the most common oxidizing agents?

power rockets into space. More common examples of combustion include the burning of wood and fossil fuels. The combustion of these and other carbon-based chemicals forms carbon dioxide and water. Consider, for example, the combustion of methane, the major component of natural gas:

$$CH_4 + 2\,O_2 \rightarrow CO_2 + 2\,H_2O + \textbf{energy}$$

Methane Oxygen   Carbon   Water
dioxide

In combustion, electrons are transferred in the creation of polar covalent bonds. This concept is illustrated in Figure 21.26, which compares the electronic structures of the combustion starting material, molecular oxygen, and the combustion product, water. Molecular oxygen is a nonpolar covalent compound. After combustion, however, the oxygens are bonded with hydrogen atoms. The electrons shared between the oxygen and hydrogen atoms are pulled to the oxygen. This gives the oxygen a negative charge, which is another way of saying it has gained electrons and has thus been reduced. At the same time, the hydrogen atoms develop a positive charge, which is another way of saying they have lost electrons and have thus been oxidized. This gain of electrons by oxygen and loss of electrons by hydrogen is an energy-releasing process. Typically, the energy is released either as molecular kinetic energy (heat) or as light (the flame).

FIGURE 21.26 ▶
(a) Neither atom in an oxygen molecule is able to preferentially attract the bonding electrons. (b) The oxygen atom of a water molecule pulls the bonding electrons away from the hydrogen atoms on the water molecule, making the oxygen slightly negative and the two hydrogens slightly positive.

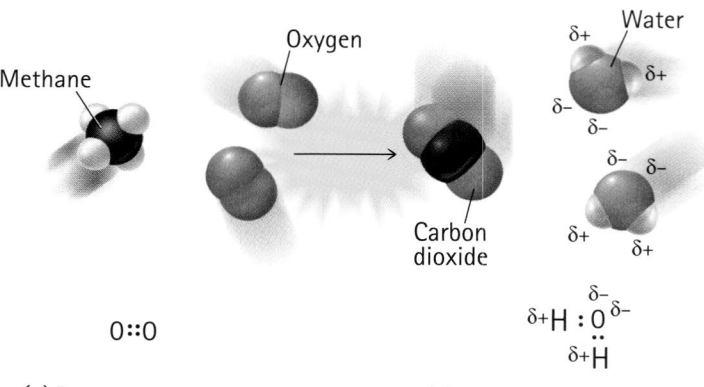

(a) Reactant oxygen atoms share electrons equally in O_2 molecules.

(b) Product oxygen atoms pull electrons away from H atoms in H_2O molecules and are reduced.

CONCEPT CHECK
What is the difference between corrosion and combustion?

Check Your Answer
Corrosion is the oxidation of metals whereas combustion is the oxidation of nonmetallic compounds, such as those found in wood.

Interestingly, combustion oxidation–reduction reactions occur throughout your body. You can visualize a simplified model of your metabolism by reviewing Figure 21.26 and substituting a food molecule for the methane. Food molecules relinquish their electrons to the oxygen molecules you inhale. The products are carbon dioxide, water vapor, and energy. You exhale the carbon dioxide and water vapor, but much of the energy from the reaction is used to keep your body warm and to drive the many other biochemical reactions necessary for life.

21.8 Hydrogen Sulfide Can Induce Suspended Animation

Our bodies require lots of energy for living. We get this energy by oxidizing food molecules with oxygen, which we get from the air we breathe. When you stop breathing, as tragically occurs with choking or drowning, the amount of oxygen available to your cells quickly drops. After minutes of not breathing, your body begins to die.

But instead of dying, why doesn't the body simply turn off until oxygen becomes available again? Interestingly, deathly damage occurs not so much because your body no longer has oxygen available to it. The problem is that small amounts of oxygen remain in the cells of the body for a long time after breathing has stopped. Many cellular processes are able to continue even at these low oxygen levels. ✓ With some parts working and others not working, the cell is thrown so far off balance that it dies, which ultimately leads to the death of the individual.

Might there be a way around this? One trick is to make sure that all cellular processes shut down together at the same time. This explains how some people who drown in frozen waters can sometimes be revived even though they have not been breathing for several hours! Their cells were shut down uniformly due to the rapid onslaught of the extreme cold. They entered a state known as *suspended animation* where their bodies are completely shut off, like a light, with the potential to reawaken once the conditions are favorable.

Experimenting with mice, researchers have discovered an alternative way of inducing suspended animation. As the mouse breathes certain concentrations of the gas hydrogen sulfide, H_2S, its cellular functions slow to a halt and its body temperature drops to only a few degrees above the surrounding temperature. In effect, the animal becomes cold-blooded, which is what happens to bears and ground squirrels when they hibernate. The reason they think this works is because the hydrogen sulfide has many properties similar to molecular oxygen, O_2. Cells absorb and try to use the H_2S as though it were O_2. But without the oxidative powers of O_2, the cell's machinery simply shuts down. That the cells shut down uniformly is key to the subsequent revival of the organism.

> Experimenting with what? Hydrogen sulfide? Isn't that the molecule responsible for the stinky smell of rotten eggs? The answer begins with a "Y."

READING CHECK

What happens when only some parts of your cells stop working while others keep on going?

FIGURE 21.27 ▶
Science-fictional "warp drives" allow astronauts to travel vast interstellar distances in a matter of days. With our current technology, however, it will take over a year for our astronauts just to reach our next-door neighbor planet Mars.

Billions of years ago, hydrogen sulfide was once an abundant gas in our atmosphere. Primitive life forms likely utilized this hydrogen sulfide until molecular oxygen, O_2, became more abundant. Our susceptibility to hydrogen sulfide may be a trait we inherited from these ancient times.

If applicable to humans, hydrogen sulfide-induced suspended animation holds many possibilities. For example, suspended animation could protect people who suffer strokes, heart attacks, or other critical injuries in which blood flow or blood supply is severely limited. This technology may also help transplant organs remain healthy for longer periods of time prior to transplantation. This is important because of the long time it often takes to transport the organ from the donor to the recipient, who are often many miles apart. Perhaps people with incurable diseases might one day put themselves on H_2S suspended animation to be awakened decades later when a cure has been found. Perhaps astronauts one day may rely on H_2S-induced suspended animation for long space voyages (see Figure 21.27 above).

CONCEPT CHECK

Cells functioning with only small amounts of available oxygen continue to produce carbon dioxide. This carbon dioxide enters the blood but doesn't leave when breathing has stopped. What then happens to the pH of the blood?

Check Your Answer

The carbon dioxide reacts with water to form carbonic acid, which changes the blood's pH to lethally low levels. So artificial respiration not only helps get needed oxygen into the unconscious victim, it also helps to remove waste carbon dioxide from the victim's blood.

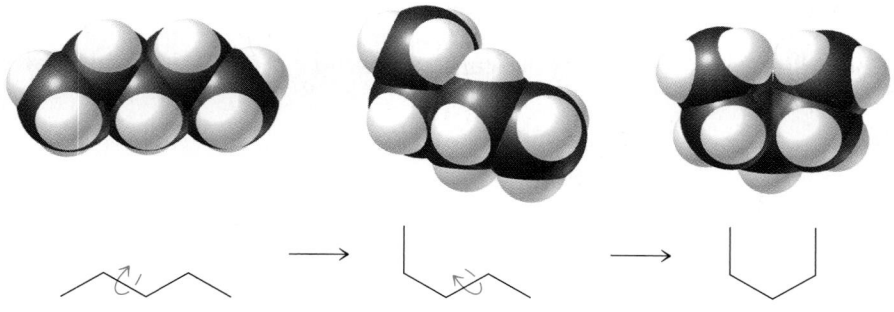

◀ **FIGURE 22.2**
Three conformations for a molecule of *n*-pentane. The molecule looks different in each conformation, but the five-carbon framework is the same in all three conformations. In a sample of liquid *n*-pentane, the molecules are found in all conformations—not unlike a bucket of worms.

CONCEPT CHECK

Which carbon–carbon bond was rotated to go from the "before" conformation of *iso*-pentane to the "after" conformation:

Check Your Answer

Bond "c." This rotation is similar to that of the arm of an arm wrestler who, with the arm just above the table while on the brink of losing, suddenly gets a surge of strength and swings the opponent's arm through a half-circle arc and wins. The best way to understand the geometrical shapes of organic molecules is to get your hands on a set of molecular models.

Before   After

In looking at the stick structures, remember that each corner or end represents a carbon atom and that each carbon atom must be bonded four times. Hydrogen atoms are not usually shown.

The number of carbon atoms within a hydrocarbon is indicated by the hydrocarbon's name, as shown in Table 22.1.

When the hydrocarbon is branched, the name of the hydrocarbon is based upon the longest carbon chain. Smaller branches off of this longest chain are written as a prefix ending in -yl. As shown in Table 22.1, a one-carbon branch would be indicated by the prefix methyl, where "meth" means a single carbon. Also, the longest chain is numbered to indicate where the branching takes place. For example, the following compound is 3-methylhexane because it has a methyl group branching off of the third carbon of hexane:

longest chain

branch off of 3rd carbon

The name of an organic molecule tells you the structure of that molecule.

TABLE 22.1 Name of Simple Hydrocarbons

Formula	Hydrocarbon Name	-yl Prefix
CH_4	Methane	methyl
C_2H_6	Ethane	ethyl
C_3H_8	Propane	propyl
C_4H_{10}	Butane	butyl
C_5H_{12}	Pentane	pentyl
C_6H_{14}	Hexane	hexyl
C_7H_{16}	Heptane	heptyl
C_8H_{18}	Octane	octyl
C_9H_{20}	Nonane	nonyl
$C_{10}H_{22}$	Decane	decyl

Below is the structure for 2,3-dimethylhexane, which tells us that there are two methyl groups—one located at the second carbon and another located at the third carbon. Number the longest chain backwards and you would have 4,5-dimethylhexane, which is the identical structure. The convention, however, is to always use the lowest numerals possible when naming an organic compound.

CONCEPT CHECK
As shown in Figure 22.1, *neo*-pentane has an alternate name based upon the naming methodology described above. What is this alternate name for *neo*-pentane?

Check Your Answer
Neo-pentane also goes by the name 2,2-dimethylpropane. The "di" is a prefix that means *two*, which in this case means two methyls. The numerals tell us to what carbon these two methyl groups are attached.

Hydrocarbons are obtained primarily from coal and petroleum. Most of the coal and petroleum that exists today was formed between 290 million and 354 million years ago when plant and animal matter decayed in the absence of oxygen. At that time, the Earth was covered with extensive swamps that, because they were close to sea level, periodically became submerged. The organic matter of the swamps was buried beneath layers of marine sediments and was eventually transformed into either coal or petroleum.

Coal is a solid material containing many large, complex hydrocarbon molecules. Most of the coal mined today is used for the production of steel and for generating electricity at coal-burning power plants.

READING CHECK

What types of atoms are found in hydrocarbons?

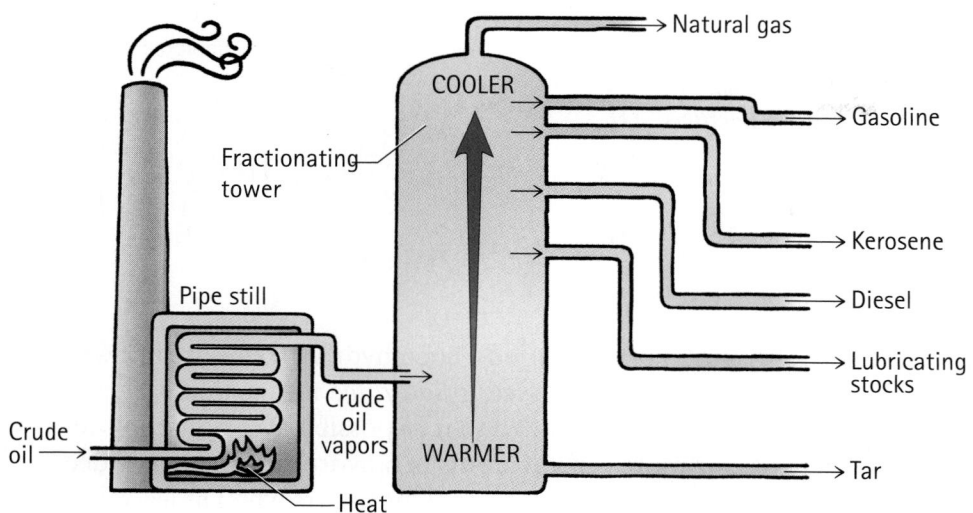

◀ **FIGURE 22.3**
A schematic for the fractional distillation of petroleum into its useful hydrocarbon components.

Petroleum, also called crude oil, is a liquid readily separated into its hydrocarbon components through a process known as *fractional distillation,* shown in Figure 22.3. The crude oil is heated to a temperature high enough to vaporize most of the components. The hot vapor flows into the bottom of a fractionating tower, which is warmer at the bottom than at the top. As the vapor rises in the tower and cools, the various components begin to condense. Hydrocarbons that have high boiling points, such as tar, condense first at warmer temperatures. Hydrocarbons that have low boiling points, such as gasoline, travel to the cooler regions at the top of the tower before condensing. Pipes drain the various liquid hydrocarbon fractions from the tower. Natural gas, which is primarily methane, does not condense. It remains a gas and is collected at the top of the tower.

The lower the boiling point of the hydrocarbon, the higher it travels up the fractionation tower.

CONCEPT CHECK
Are the molecules of crude oil that rise to the top of the fractionation tower the heavier or lighter ones?

Check Your Answer
The lighter molecules found in crude oil are the ones that rise highest within the fractionation tower. These are the molecules that have the lower boiling points. So, the lower the boiling point, the higher the molecules rise.

22.2 Unsaturated Hydrocarbons

Recall, from Section 18.1, that carbon has four unpaired valence electrons. As shown in Figure 22.4, each of these electrons is available for pairing with an electron from another atom, such as hydrogen, to form a covalent bond.

In all the hydrocarbons discussed so far, including the methane shown in Figure 22.4, each carbon atom is bonded to four neighboring

FIGURE 22.4 ▶
Carbon has four valence electrons. Each electron pairs with an electron from a hydrogen atom in the four covalent bonds of methane.

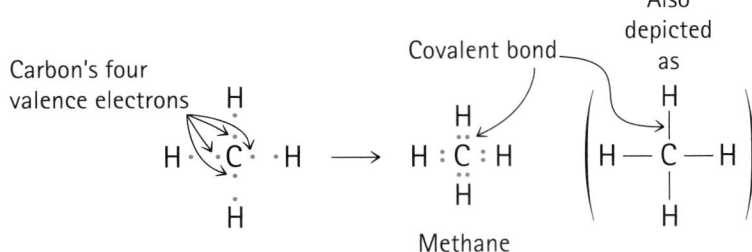

Carbon's four valence electrons

Covalent bond

Also depicted as

Methane

Unsaturated hydrocarbons have at least one double or triple bond. Saturated hydrocarbons have only single bonds.

atoms by four single covalent bonds. Such hydrocarbons are known as **saturated hydrocarbons.** The term *saturated* means that each carbon has as many atoms bonded to it as possible. We now explore cases where one or more carbon atoms in a hydrocarbon are bonded to fewer than four neighboring atoms. This occurs when at least one of the bonds between a carbon and a neighboring atom is a multiple bond. (See Section 18.5 for a review of multiple bonds.) ✓ A hydrocarbon that has a multiple bond—either double or triple—is known as an **unsaturated hydrocarbon.** Because of the multiple bond, two of the carbons are bonded to fewer than four other atoms. These carbons are thus said to be *unsaturated.*

Figure 22.5 compares the saturated hydrocarbon *n*-butane with the unsaturated hydrocarbon 2-butene. The number of atoms that are bonded to each of the two middle carbons of *n*-butane is four, whereas each of the two middle carbons of 2-butene is bonded to only three other atoms—a hydrogen and two carbons. By convention, the name of a saturated hydrocarbon uses the suffix –ane, as in *butane.* An unsaturated hydrocarbon containing a double bond is indicated using the suffix –ene, as in *butene.*

An important unsaturated hydrocarbon is benzene, C_6H_6, which may be drawn as three double bonds contained within a flat hexagonal ring, as is shown in Figure 22.6a. Unlike the double-bond electrons in most other unsaturated hydrocarbons, the electrons of the double bonds in benzene are not fixed between any two carbon atoms. Instead, these

Saturated hydrocarbon

Unsaturated hydrocarbon

n-Butane, C_4H_{10}

2-Butene, C_4H_8

FIGURE 22.5 ▲
The carbons of the hydrocarbon *n*-butane are *saturated,* each being bonded to four other atoms. Because of the double bond, two of the carbons of the unsaturated hydrocarbon 2-butene are bonded to only three other atoms, which makes the molecule an unsaturated hydrocarbon.

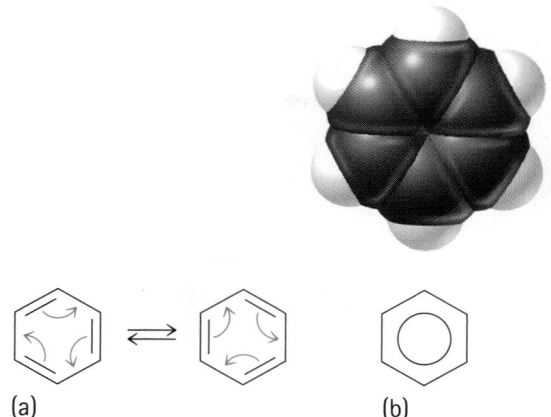

◀ **FIGURE 22.6**
(a) The double bonds of benzene, C_6H_6, are able to migrate around the ring. (b) For this reason, they are often represented by a circle within the ring.

(a)   (b)

electrons are able to move freely around the ring. This is commonly represented by drawing a circle within the ring, as shown in Figure 22.6b, rather than by individual double bonds.

Many organic compounds contain one or more benzene rings in their structure. Because many of these compounds are fragrant, any organic molecule containing a benzene ring is classified as an **aromatic compound** (even if it is not particularly fragrant). Figure 22.7 shows a few examples. Toluene, a common solvent used as a paint thinner, is toxic and gives airplane glue its distinctive odor. Some aromatic compounds, such as naphthalene, contain two or more benzene rings fused together. At one time, mothballs were made of naphthalene. Most mothballs sold today, however, are made of the less toxic 1,4-dichlorobenzene.

An example of an unsaturated hydrocarbon containing a triple bond is acetylene, C_2H_2. A confined flame of acetylene burning in oxygen is hot enough to melt iron, which makes acetylene a choice fuel for welding (Figure 22.8).

READING CHECK

What type of hydrocarbon has double or triple bonds?

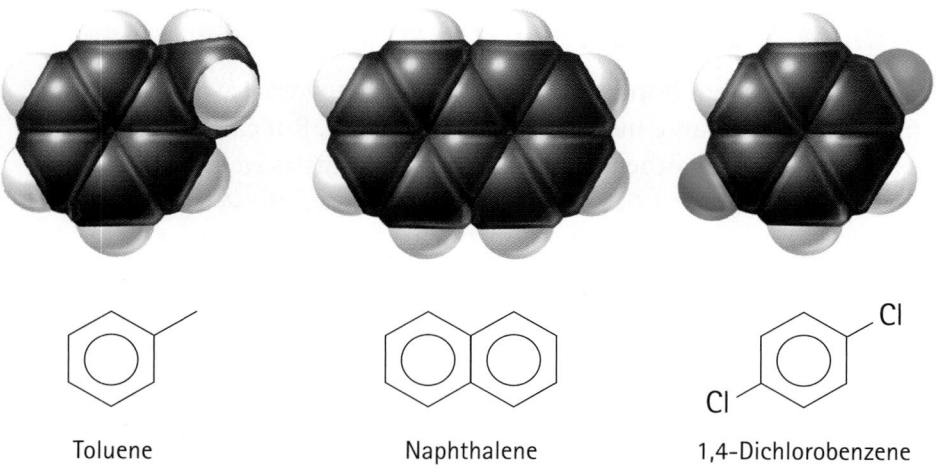

Toluene   Naphthalene   1,4-Dichlorobenzene

FIGURE 22.7 ▲
The structures for three odoriferous organic compounds containing one or more benzene rings: toluene, naphthalene, and 1,4-dichlorobenzene.

FIGURE 22.8 ▶
The unsaturated hydrocarbon acetylene, C_2H_2, when burned in this torch, produces a flame that is hot enough to melt iron.

H—C≡C—H
Acetylene

CONCEPT CHECK

Prolonged exposure to benzene increases the risk of developing certain cancers. The structure of aspirin contains a benzene ring. Does this indicate that prolonged exposure to aspirin will increase a person's risk of developing cancer?

Benzene ring

Aspirin

Check Your Answer

No. Although benzene and aspirin both contain a benzene ring, these two molecules have different overall structures and quite different chemical properties. Each carbon-containing organic compound has its own set of unique physical, chemical, and biological properties. While benzene may cause cancer, aspirin works as a safe remedy for headaches.

22.3 Functional Groups

Carbon atoms can bond to one another in many ways, which results in an incredibly large number of hydrocarbons. But carbon atoms can bond to many other different types of atoms as well. This further increases the number of possible organic molecules. ✓ In organic chemistry, any atom other than carbon or hydrogen in an organic molecule is called a **heteroatom,** where *hetero-* means "different from either carbon or hydrogen."

A hydrocarbon structure can serve as a framework for the attachment of various heteroatoms. This is analogous to a tree serving as the structure on which ornaments are hung. Just as the ornaments give character to the tree, so do heteroatoms give character to an organic molecule. Heteroatoms have profound effects on the properties of an organic molecule.

Consider ethane, C_2H_6, and ethanol, C_2H_6O, which differ from each other by only a single oxygen atom. Ethane has a boiling point of $-88°C$,

The chemistry of hydrocarbons is surely interesting, but start adding heteroatoms to these organic molecules and the chemistry becomes extraordinarily interesting. The organic chemicals of living organisms, for example, all contain heteroatoms.

making it a gas at room temperature, and it does not dissolve in water very well. Ethanol, by contrast, has a boiling point of +78°C, making it a liquid at room temperature. It is infinitely soluble in water, and it is the active ingredient of alcoholic beverages. Consider further ethylamine, C_2H_7N, which has a nitrogen atom on the same basic two-carbon framework. This compound is a corrosive, pungent, highly toxic gas—most unlike either ethane or ethanol.

Organic molecules are classified according to the functional groups they contain, where a **functional group** is defined as a combination of atoms that behave as a unit. Most functional groups are distinguished by the heteroatoms they contain, and some common groups are listed in Table 22.2. As you study these functional groups in the following sections, do *not* focus on memorizing these functional groups. Instead, focus on understanding how these functional groups determine the

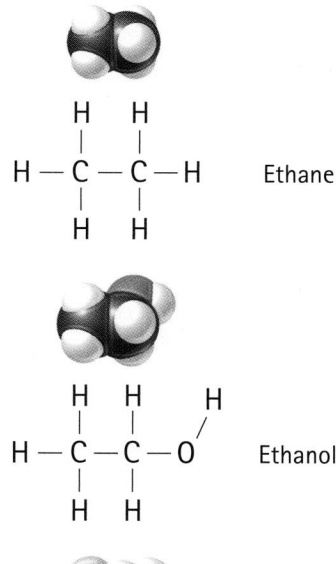

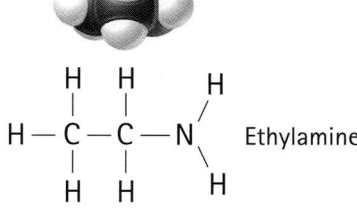

TABLE 22.2 Functional Groups in Organic Molecules

General Structure	Name	Class
Hydroxyl group	Hydroxyl group	Alcohols
Ether group	Ether group	Ethers
Amine group	Amine group	Amines
Ketone group	Ketone group	Ketones
Aldehyde group	Aldehyde group	Aldehydes
Amide group	Amide group	Amides
Carboxyl group	Carboxyl group	Carboxylic acids
Ester group	Ester group	Esters

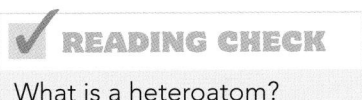

✔ READING CHECK

What is a heteroatom?

chemical and physical properties of the organic molecule. Doing so will give you a greater appreciation of the remarkable diversity of organic molecules and their many applications.

CONCEPT CHECK

What is the significance of heteroatoms in an organic molecule?

Check Your Answer

Heteroatoms largely determine an organic molecule's "personality."

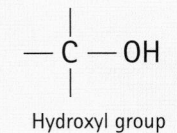

Hydroxyl group

22.4 Alcohols and Ethers

Alcohols are organic molecules in which a *hydroxyl group* is bonded to a saturated carbon. The hydroxyl group consists of an oxygen bonded to a hydrogen. Because of the polarity of the oxygen–hydrogen bond, low-formula-mass alcohols are often soluble in water. Some common alcohols and their melting and boiling points are listed in Table 22.3.

More than 11 billion pounds of methanol, CH_3OH, is produced annually in the United States. Most of it is used for making formaldehyde and acetic acid, important starting materials in the production of plastics. In addition, methanol is used as a solvent and an anti-icing agent in gasoline. Sometimes called wood alcohol because it can be obtained from wood, methanol is very toxic and should never be ingested.

TABLE 22.3 Some Simple Alcohols

Structure	Scientific Name	Common Name	Melting Point (°C)	Boiling Point (°C)
	Methanol	Methyl alcohol	−97	65
	Ethanol	Ethyl alcohol	−115	78
	2-Propanol	Isopropyl alcohol	−126	97

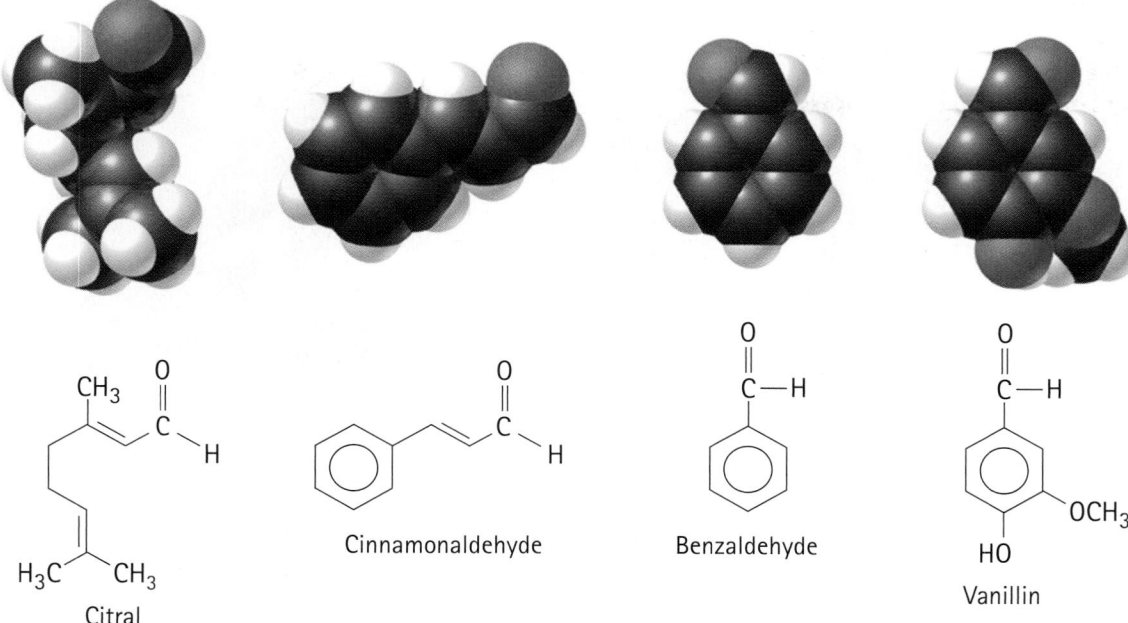

FIGURE 22.16 ▲
Aldehydes are responsible for many familiar fragrances.

aldehydes citral, cinnamaldehyde, and benzaldehyde, respectively. The structures of these three aldehydes are shown in Figure 22.16. Another aldehyde, vanillin, is the key flavoring molecule derived from seed pods of the vanilla orchid. Imitation vanilla flavoring is less expensive because it is economically synthesized from waste chemicals from the wood-pulp industry. Many books manufactured in the days before "acid-free" paper smell of vanilla because of the vanillin formed and released as the paper ages, a process that is accelerated by the acids the paper contains.

An **amide** is a carbonyl-containing organic molecule in which the carbonyl carbon is bonded to a nitrogen atom. The active ingredient of most mosquito repellents is an amide whose chemical name is *N, N*-diethyl-*m*-toluamide but is commercially known as DEET, shown in Figure 22.17. This compound causes certain insects, especially mosquitoes, to lose their sense of direction, which effectively protects DEET wearers from being bitten.

A **carboxylic acid** is a carbonyl-containing organic molecule in which the carbonyl carbon is bonded to a hydroxyl group. As its name implies, this functional group is able to donate hydrogen ions. Organic molecules that contain it are therefore acidic. An example is acetic acid, $C_2H_4O_2$. Vinegar is a solution of acetic acid in water. You may recall that this organic compound was used as an example of a weak acid back in Chapter 21.

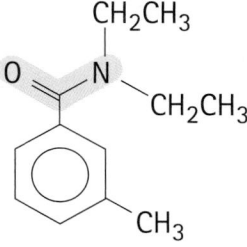

N,N-Diethyl-*m*-toluamide
(DEET)

FIGURE 22.17 ▲
N,N-diethyl-*m*-toluamide is an example of an amide. Amides contain the amide group, shown highlighted in blue.

Amide group

Carboxyl group

FIGURE 22.18 ▶
Aspirin, acetylsalicylic acid, contains both the carboxylic acid and ester functional groups.

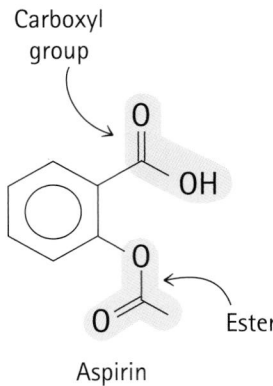

Carboxyl group

OH

Ester

Aspirin
(acetylsalicylic acid)

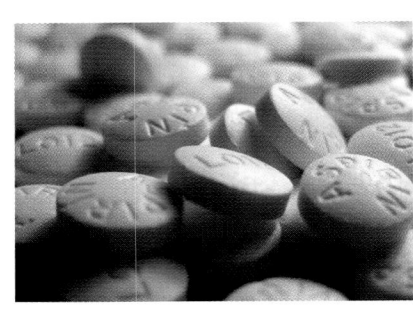

Ester group

An **ester** is an organic molecule similar to a carboxylic acid except that in the ester, the hydroxyl hydrogen is replaced by a carbon. Unlike carboxylic acids, esters are not acidic because they lack the hydrogen of the hydroxyl group. An interesting example of an organic compound containing both a carboxylic acid group and an ester group is acetylsalicylic acid, more commonly known as aspirin, as shown in Figure 22.18.

CONCEPT CHECK
Identify all the functional groups in these two molecules (ignore the sulfur group in penicillin G):

Testosterone

Penicillin G

Check Your Answers
Testosterone: alcohol and ketone. Penicillin G: amide (two amide groups), carboxylic acid.

✔ **READING CHECK**

What type of bond is found between the carbon and oxygen of a carbonyl group?

Like aldehydes, many simple esters have notable fragrances and are often used as flavorings. Some familiar ones are listed in Table 22.5.

TABLE 22.5 Some Esters and Their Flavors and Odors

Structure	Name	Flavor/Odor
H–C(=O)–O–CH₂CH₃	Ethyl formate	Rum
H₃C–C(=O)–O–CH₂CH₂–C(CH₃)(CH₃)–H	Isopentyl acetate	Banana
H₃C–C(=O)–O–CH₂(CH₂)₆CH₃	Octyl acetate	Orange
CH₃CH₂CH₂–C(=O)–O–CH₃	Methyl butyrate	Apple
(benzene ring with OH)–C(=O)–O–CH₃	Methyl salicylate	Wintergreen

We eat organic chemicals daily. In fact, organic chemicals are the *only* things we eat, except for some important minerals, such as the ions of sodium and calcium.

22.7 Polymers

Polymers are exceedingly long molecules that consist of repeating molecular units called **monomers,** as Figure 22.19 illustrates. Monomers have relatively simple structures consisting of anywhere from 4 to 100 atoms per molecule. When monomers are chained together, they can form polymers consisting of hundreds of thousands of atoms per molecule. These large molecules are still too small to be seen with the unaided eye. They are, however, giants in the submicroscopic world—if a typical polymer molecule were as thick as a kite string, it would be 1 kilometer long.

Human-made polymers, also known as synthetic polymers, make up the class of materials commonly known as plastics. There are two major

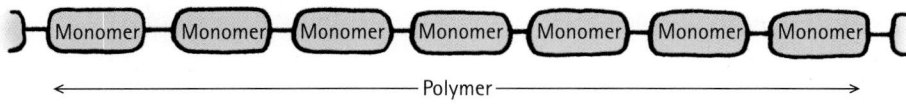

FIGURE 22.19 ▲
A polymer is a long molecule consisting of many smaller monomer molecules linked together.

TABLE 22.6 Addition and Condensation Polymers

Polymers	Repeating Unit	Common Uses	Recycling Code
Polyethylene (PE)	H H | | ···C—C··· | | H H	Plastic bags, bottles	2 HDPE 4 LDPE
Polypropylene (PP)	H H | | ···C—C··· | | H CH$_3$	Indoor–outdoor carpets	5 PP
Polystyrene (PS)	H H | | ···C—C··· | H ⬡	Plastic utensils, insulation	6 PS
Polyvinyl chloride (PVC)	H H | | ···C—C··· | | H Cl	Shower curtains, tubing	3 V
Polyethylene terephthalate	O O ‖ ‖ ···C—⬡—C O—CH$_2$CH$_2$—O···	Clothing, plastic bottles	1 PET

types of synthetic polymers used today—*addition polymers* and *condensation polymers*. As shown in Table 22.6, addition and condensation polymers have a wide variety of uses. Solely the product of human design, these polymers pervade modern living. In the United States, for example, synthetic polymers have surpassed steel as the most widely used material.

Addition Polymers

Addition polymers form simply by the joining together of monomer units. For this to happen, each monomer must contain at least one double bond. As shown in Figure 22.20, polymerization occurs when two of the electrons from each double bond split away from each other to form new covalent bonds with neighboring monomer molecules. ✔ During this process, no atoms are lost, so the total mass of an addition polymer is equal to the sum of the masses of all the monomers.

Nearly 12 million tons of polyethylene is produced annually in the United States; that's about 90 pounds per U.S. citizen. The monomer from which it is synthesized, ethylene, is an unsaturated hydrocarbon produced in large quantities from petroleum. High-density polyethylene

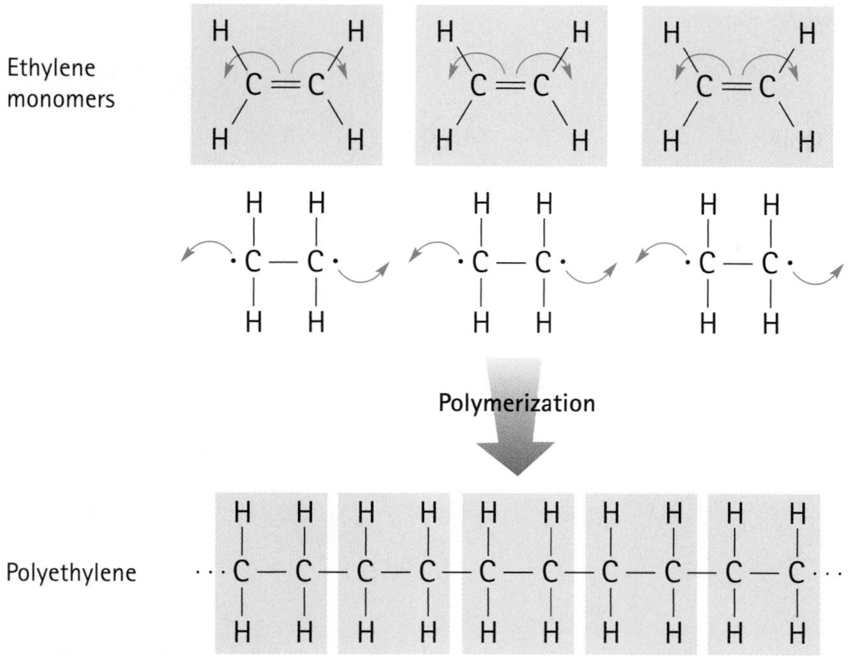

Ethylene monomers

Polymerization

Polyethylene

FIGURE 22.20 ▲
The addition polymer polyethylene is formed as electrons from the double bonds of ethylene monomer molecules split away and become unpaired valence electrons. Each unpaired electron then joins with an unpaired electron of a neighboring carbon atom to form a new covalent bond that links two monomer units together.

(HDPE), shown schematically in Figure 22.21a, consists of long strands of straight-chain molecules packed closely together. The tight alignment of neighboring strands makes HDPE a relatively rigid, tough plastic useful for such things as bottles and milk jugs. Low-density polyethylene (LDPE), shown in Figure 22.21b, is made of strands of highly branched chains, a pattern that prevents the strands from packing closely together. This makes LDPE more bendable than HDPE and gives it a lower melting point. While HDPE holds its shape in boiling water, LDPE deforms. It is most useful for such items as plastic bags, photographic film, and electrical-wire insulation.

Other addition polymers are created by using different monomers. The only requirement is that the monomer must contain a double bond.

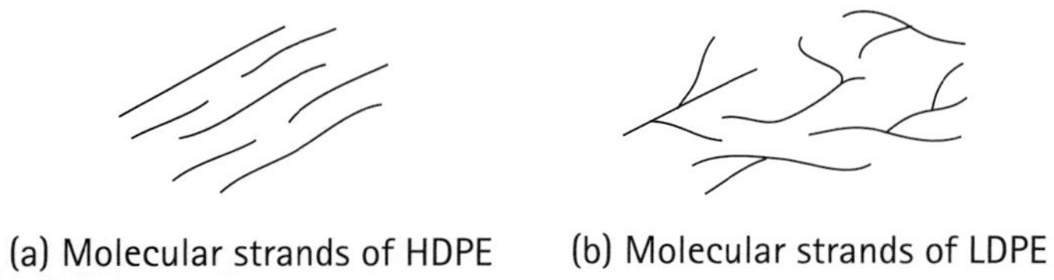

(a) Molecular strands of HDPE   (b) Molecular strands of LDPE

FIGURE 22.21 ▲
(a) The polyethylene strands of HDPE are able to pack closely together, much like strands of uncooked spaghetti. (b) The polyethylene strands of LDPE are branched, which prevents the strands from packing well.

FIGURE 22.22 ▶
Propylene monomers polymerize to form polypropylene.

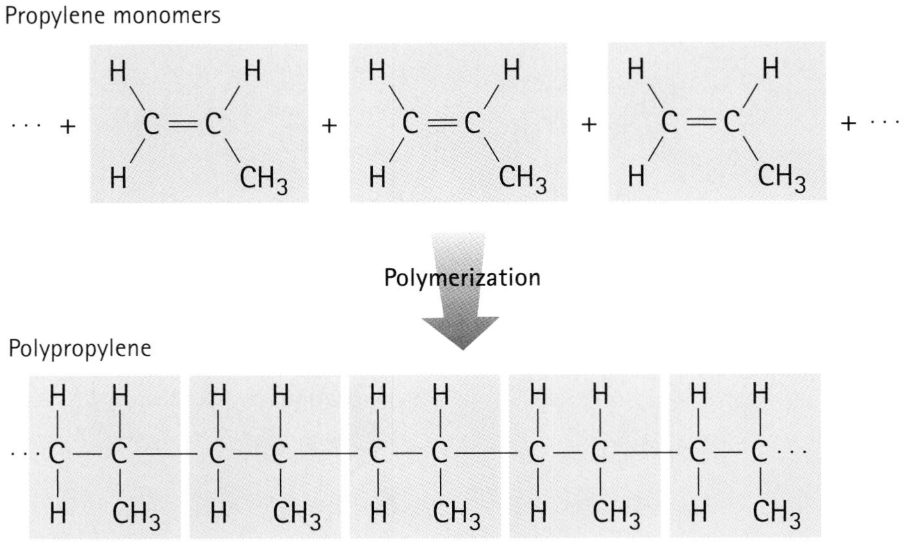

Propylene monomers

Polymerization

Polypropylene

The monomer propylene, for example, yields polypropylene, as shown in Figure 22.22. Polypropylene is a tough plastic material useful for pipes, hard-shell suitcases, and appliance parts. Fibers of polypropylene are used for upholstery, indoor–outdoor carpets, and even thermal underwear.

Figure 22.23 shows that using styrene as the monomer yields polystyrene. Transparent plastic cups are made of polystyrene, as are thousands of other household items. Blowing air into liquid polystyrene generates Styrofoam, which is widely used for coffee cups, packing material, and insulation.

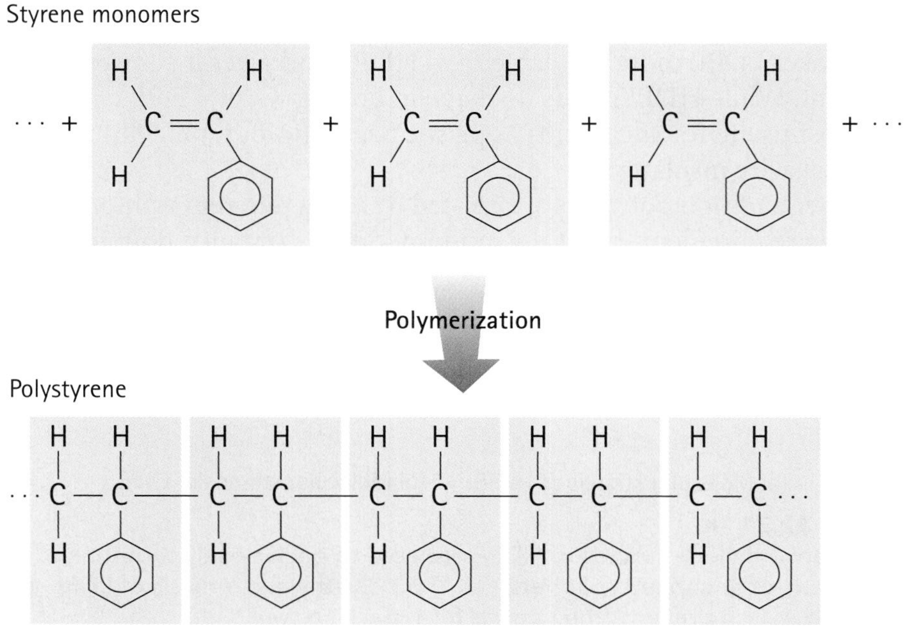

Styrene monomers

Polymerization

Polystyrene

FIGURE 22.23 ▶
Styrene monomers polymerize to form polystyrene.

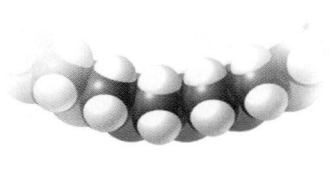

◀ **FIGURE 22.24**
PVC is tough and easily
molded, which is why it often
is used to fabricate many
household items.

Polyvinyl chloride (PVC)

Another important addition polymer is polyvinylchloride (PVC), which is tough and easily molded. Floor tiles, shower curtains, and pipes are most often made of PVC, shown in Figure 22.24.

Rigid polymers such as PVC can be made soft by incorporating small molecules called plasticizers. Pure PVC, for example, is a tough material great for making pipes. Mixed with a plasticizer, the PVC becomes soft and flexible and thus useful for making shower curtains, toys, and many other products now found in most households. One of the more commonly used plasticizers are the phthalates, some of which have been shown to disrupt the development of reproductive organs, especially in the fetus and in growing children. Governments and manufacturers are now working to phase out these plasticizers. But some phthalates, such as DINP, have been shown to be much less dangerous.

The addition polymer polyvinylidene chloride (trade name Saran), shown in Figure 22.25, is used to make plastic wrap for food. The large

Should all phthalates be banned or just the ones proven to be harmful? This is a social and political question that has yet to be resolved.

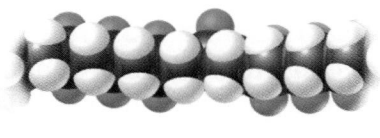

Polyvinylidene chloride (Saran)

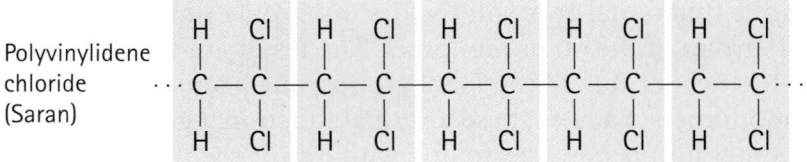

◀ **FIGURE 22.25**
The large chlorine atoms in
polyvinylidene chloride make
this addition polymer sticky.

FIGURE 22.26 ▶
The fluorine atoms in poly-tetrafluoroethylene tend not to experience molecular attractions, which is why this addition polymer is used as a nonstick coating and lubricant.

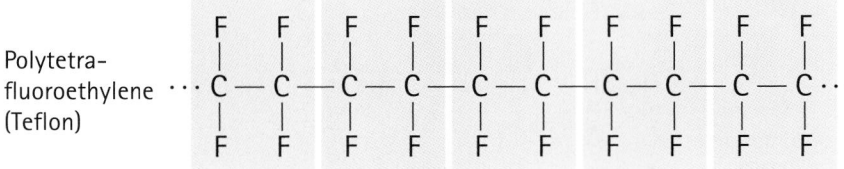

Polytetra-fluoroethylene (Teflon)

READING CHECK

Which has more mass: an addition polymer or the monomers that combined to make that addition polymer?

chlorine atoms in this polymer help it to stick to such surfaces as glass by dipole–induced-dipole attractions, as discussed in Section 18.8.

The addition polymer polytetrafluoroethylene, shown in Figure 22.26, is what you know as Teflon. In contrast to the chlorine-containing Saran, fluorine-containing Teflon has a nonstick surface because the fluorine atoms tend not to experience any molecular attractions. In addition, because carbon–fluorine bonds are unusually strong, Teflon can be heated to high temperatures before decomposing. These properties make Teflon an ideal coating for cooking surfaces. It is also relatively inert, which is why many corrosive chemicals are shipped or stored in Teflon containers.

CONCEPT CHECK
What do all monomers that are used to make addition polymers have in common?

Check Your Answer
A double covalent bond between two carbon atoms.

Condensation Polymers

A **condensation polymer** is one that is formed when the joining of monomer units is accompanied by the loss of a small molecule, such as water or hydrochloric acid. Any monomer capable of becoming part of a condensation polymer must have a functional group on each end. When two such monomers come together to form a condensation polymer, one functional group of the first monomer links up with one functional group of the other monomer. The result is a two-monomer unit that has two terminal functional groups, one from each of the two original monomers. Each of these terminal functional groups in the two-monomer unit is now free to link with one of the functional

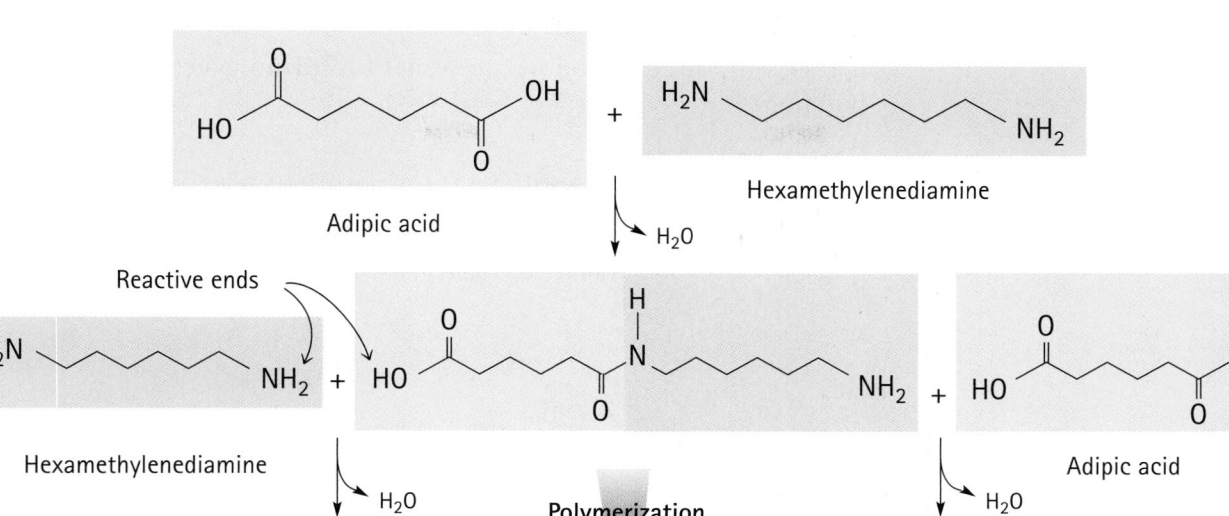

FIGURE 22.27 ▲
Adipic acid and hexamethylenediamine polymerize to form the condensation copolymer nylon.

groups of a third monomer, and then a fourth, and so on. In this way a polymer chain is built.

Figure 22.27 shows this process for the condensation polymer called nylon, which is composed of two different monomers. One monomer is adipic acid, which contains two reactive end groups, both carboxyl groups. The second monomer is hexamethylenediamine, in which two amine groups are the reactive end groups. One end of an adipic acid molecule and one end of a hexamethylamine molecule can be made to react with each other, splitting off a water molecule in the process. After two monomers have joined, reactive ends still remain for further reactions, which leads to a growing polymer chain. Aside from its use in hosiery, nylon also finds important uses in the manufacture of ropes, parachutes, clothing, and carpets.

Nylon was invented just prior to World War II, during which it was widely used and perceived as a miracle material.

CONCEPT CHECK
The structure of 6-aminohexanoic acid is as follows:

Is this compound a suitable monomer for forming a condensation polymer?

Check Your Answer
Yes, because the molecule has two reactive ends. You know both ends are reactive because they are the ends shown in Figure 22.27. The only difference here is that both types of reactive ends are on the same molecule. Monomers of 6-aminohexanoic acid combine by splitting off water molecules to form the polymer known as nylon-6:

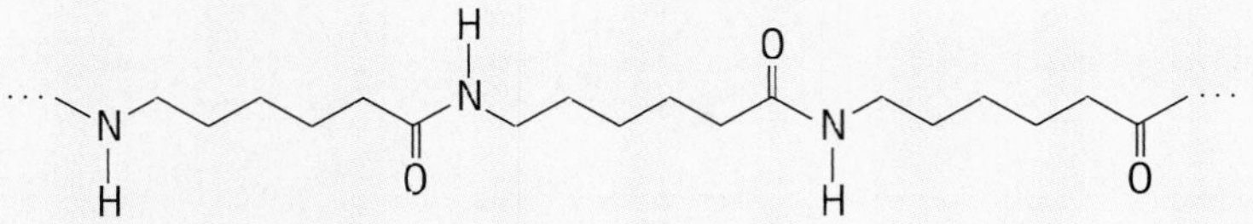

Another widely used condensation polymer is polyethylene terephthalate (PET), which is formed from the polymerization of ethylene glycol and terephthalic acid, as shown in Figure 22.28. Plastic soda bottles are made from this polymer. Also, PET fibers are sold as Dacron polyester, a product used in clothing and stuffing for pillows and sleeping bags. Thin films of PET, which are called Mylar, can be coated with metal particles to make magnetic recording tape or those metallic-looking balloons you see for sale at most grocery store checkout counters.

Monomers that contain three reactive functional groups can also form polymer chains. These chains become interlocked in a rigid 3-dimensional network that lends considerable strength and durability to the polymer. Once formed, these condensation polymers cannot be

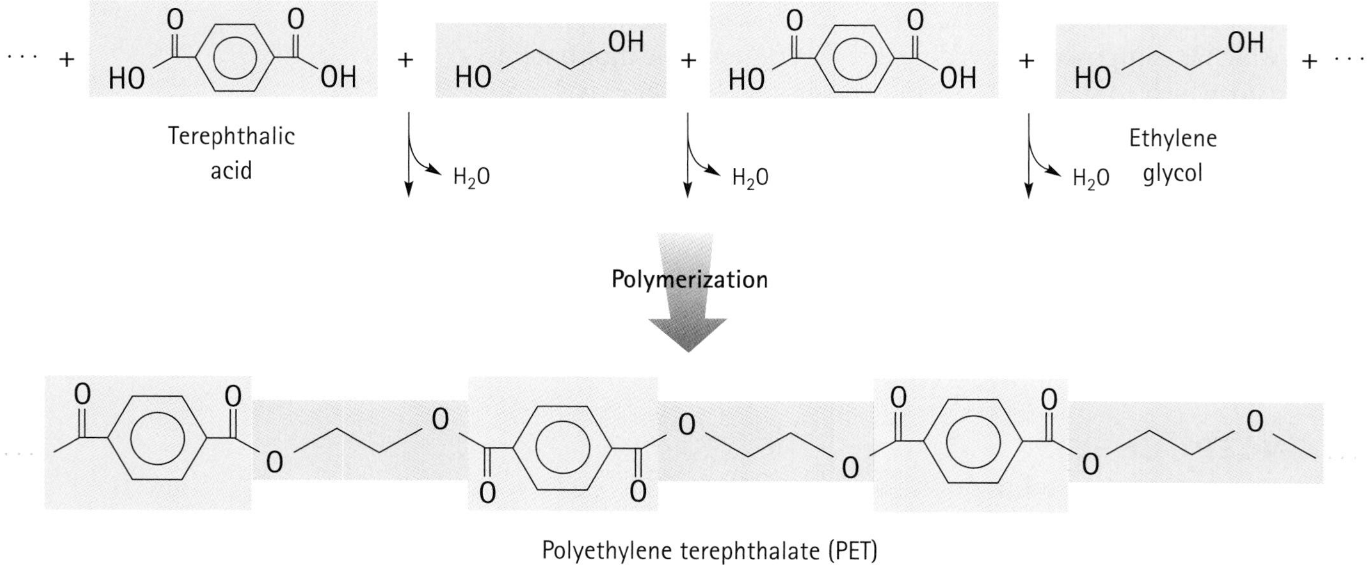

Polyethylene terephthalate (PET)

FIGURE 22.28 ▲
Terephthalic acid and ethylene glycol polymerize to form the condensation copolymer polyethylene terephthalate.

remelted or reshaped, which makes them hard-set, or *thermoset*, polymers. Hard plastic dishes (Melmac) and countertops (Formica) are made of this material. A similar polymer, Bakelite, made from formaldehyde and phenols containing multiple oxygen atoms, is used to bind plywood and particle board. Bakelite was synthesized in the early 1900s, and it was the first widely used polymer.

The synthetic-polymers industry has grown remarkably over the past half century. Annual production of polymers in the United States alone has grown from 3 billion pounds in 1950 to more than 100 billion pounds in 2008. Today, it is a challenge to find any consumer item that does *not* contain a plastic of one sort or another. In the future, watch for new kinds of polymers having a wide range of remarkable properties. One interesting application is shown in Figure 22.29. We already have polymers that conduct electricity, others that emit light, others that replace body parts, and still others that are stronger but much lighter than steel. Imagine synthetic polymers that mimic photosynthesis by transforming solar energy to chemical energy, or that efficiently separate freshwater from ocean water. These are not dreams. They are realities that chemists have already been demonstrating in the laboratory. Polymers hold a clear promise for the future.

◀ **FIGURE 22.29**
Flexible and flat video displays can now be fabricated from polymers.

22 CHAPTER REVIEW

KEY TERMS

Addition polymer A polymer formed by the joining together of monomer units with no atoms being lost as the polymer forms.

Alcohol An organic molecule that contains a hydroxyl group bonded to a saturated carbon.

Aldehyde An organic molecule containing a carbonyl group, the carbon of which is bonded either to one carbon atom and one hydrogen atom or to two hydrogen atoms.

Amide An organic molecule containing a carbonyl group, the carbon of which is bonded to a nitrogen atom.

Amine An organic molecule containing a nitrogen atom bonded to one or more saturated carbon atoms.

Aromatic compound Any organic molecule containing a benzene ring.

Carbonyl group A carbon atom double-bonded to an oxygen atom; it is found in ketones, aldehydes, amides, carboxylic acids, and esters.

Carboxylic acid An organic molecule containing a carbonyl group, the carbon of which is bonded to a hydroxyl group.

Condensation polymer A polymer formed by the joining together of monomer units accompanied by the loss of small molecules, such as water.

Configuration A term used to describe how the atoms within a molecule are connected. For example, two structural isomers will consist of the same number and same kinds of atoms, but in different configurations.

Conformation One of a wide range of possible spatial orientations of a particular configuration.

Ester An organic molecule containing a carbonyl group, the carbon of which is bonded to one carbon atom and one oxygen atom bonded to another carbon atom.

Ether An organic molecule containing an oxygen atom bonded to two carbon atoms.

Functional group A specific combination of atoms that behaves as a unit in an organic molecule.

Heteroatom Any atom other than carbon or hydrogen in an organic molecule.

Hydrocarbon A chemical compound containing only carbon and hydrogen atoms.

Ketone An organic molecule containing a carbonyl group, the carbon of which is bonded to two carbon atoms.

Monomers The small molecular units from which a polymer is formed.

Organic chemistry The study of carbon-containing compounds.

Polymer A long organic molecule made of many repeating units.

Saturated hydrocarbon A hydrocarbon containing no multiple covalent bonds, with each carbon atom bonded to four other atoms.

Structural isomers Molecules that have the same molecular formula but different chemical structures.

Unsaturated hydrocarbon A hydrocarbon containing at least one multiple covalent bond.

REVIEW QUESTIONS

Hydrocarbons

1. What are some uses of hydrocarbons?
2. How do two structural isomers differ from each other?
3. How are two structural isomers similar to each other?
4. What physical property of hydrocarbons is used in fractional distillation?
5. To how many atoms is a saturated carbon atom bonded?

Unsaturated Hydrocarbons

6. What is the difference between a saturated hydrocarbon and an unsaturated hydrocarbon?

7. How many multiple bonds must a hydrocarbon have in order to be classified as unsaturated?

8. Aromatic compounds contain what kind of ring?

Functional Groups

9. What is a heteroatom?

Alcohols and Ethers

10. Which heteroatom is characteristic of an alcohol?

11. Why are low-formula-mass alcohols soluble in water?

12. What distinguishes an alcohol from an ether?

Amines and Alkaloids

13. Which heteroatom is characteristic of an amine?

14. Do amines tend to be acidic, neutral, or basic?

15. Are alkaloids found in nature?

Carbonyl Compounds

16. Which elements make up the carbonyl group?

17. How are ketones and aldehydes related to each other? How are they different from each other?

18. How are amides and carboxylic acids related to each other? How are they different from each other?

Polymers

19. What happens to the double bond of a monomer participating in the formation of an addition polymer?

20. What is released in the formation of a condensation polymer?

THINK AND DO

1. Two carbon atoms connected by a single bond can rotate relative to each other. As we discussed in Section 22.1, this ability to rotate can give rise to numerous conformations (spatial orientations) of an organic molecule. Is it also possible for two carbon atoms connected by a double bond to rotate relative to each other? Perform this quick activity to see for yourself.

Attach one jellybean to each end of a single toothpick. Hold one of the jellybeans firmly with one hand while rotating the second jellybean with your other hand. Observe how there is no restriction on the different orientations of the two jellybeans relative to each other.

Hold two toothpicks side by side and attach one jellybean to each end such that each jellybean has both toothpicks poked into it. As before, hold one jellybean while rotating the other. What kind of rotations are possible now? Relate what you observe to the carbon–carbon double bond. Which structure of Figure 22.5 do you suppose has more possible conformations: *n*-butane or 2-butene? What do you suppose is true about the ability of atoms connected by a carbon–carbon triple bond to twist relative to each other?

2. The chemical composition of a polymer has a significant effect on its macroscopic properties. To see this for yourself, place a drop of water on a new plastic sandwich bag, and then tilt the bag vertically so that the drop races off. Observe the behavior of the water carefully. Now race a drop of water off a freshly pulled strip of plastic food wrap. How does the behavior of the drop on the wrap compare with the behavior of the drop on the sandwich bag?

Most brands of sandwich bags are made of polyethylene terephthalate, and most brands of food wrap are made of polyvinylidene chloride. Look carefully at the chemical composition of these polymers, shown in Section 22.7. Which contains larger atoms? Interestingly, the larger the atom, the easier it is to induce a dipole (Review Section 18.8). So, water sticks better to the plastic containing the larger atoms. Which plastic is this?

THINK AND COMPARE

1. Rank the following hydrocarbons in order of increasing number of hydrogens:

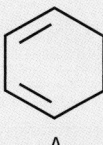
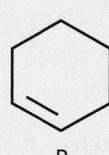

A   B   C

2. Rank the following hydrocarbons in order of smallest to largest carbon chain:

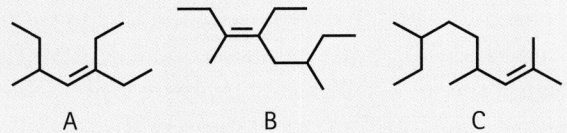

A   B   C

3. Rank the following organic molecules in order of increasing solubility in water:

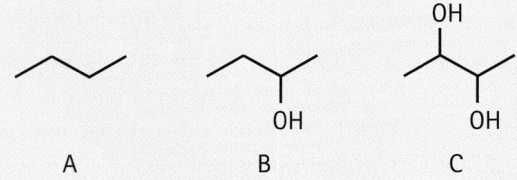

A   B   C

4. Rank the following organic molecules in order of increasing solubility in water:

A   B   C

THINK AND EXPLAIN

1. What property of carbon allows for the formation of so many different organic molecules?

2. According to Figure 22.3, which has a higher boiling point: gasoline or kerosene?

3. Hydrocarbons release a lot of energy when ignited. Where does this energy come from?

4. There are five atoms in the methane molecule, CH_4. One out of these five is a carbon atom, which is $1/5 \times 100 = 20\%$ carbon. What is the percentage of carbon in ethane, C_2H_6? Propane, C_3H_8? Butane, C_4H_{10}? Do heavier hydrocarbons tend to produce more or less carbon dioxide upon combustion compared to lighter hydrocarbons? Why?

5. What is the chemical formula for the following structure?

6. What is the chemical formula for the following structure?

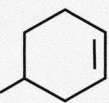

7. Shown below is the structure of 2-methyl-pentane. What is the structure of 3-methyl-pentane?

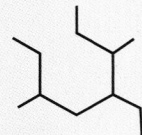

2-methyl-pentane

8. What is the name of the following structure?

9. How many possible chemical structures (configurations) can be made from the formula C_3H_8?

10. How many possible chemical structures (configurations) can be made from the formula C_2H_2BrF?

11. How many possible chemical structures (configurations) can be made from C_3H_6?

12. Why do heteroatoms make such a difference in the physical and chemical properties of an organic molecule?

13. Why do ethers typically have lower boiling points than alcohols?

14. A common inactive ingredient in products such as sunscreen lotions and shampoo is triethyl amine, also known as TEA. What is the chemical structure for this compound?

15. Suggest an explanation for why aspirin has a sour taste.

16. An amino acid is an organic molecule that contains both an amine group and a carboxyl group. At an acidic pH, which structure is most likely:

Explain your answer.

17. Identify the following functional groups in this organic molecule—amide, ester, ketone, ether, alcohol, aldehyde, amine:

18. The chemical compound lysine is shown below. What functional group must be removed in order to produce cadaverine, as shown in Figure 22.11?

Lysine

19. The copolymer styrene-butadiene rubber (SBR), shown below, is used for making tires as well as bubble gum. Is it an addition polymer or a condensation polymer?

SBR

20. Why is plastic food wrap a stickier plastic than polyethylene?

READINESS ASSURANCE TEST (RAT)

If you have a good handle on this chapter—if you really do—then you should be able to score 7 out of 10 on this RAT. Check your answers with your teacher. If you score less than 7, study further before moving on.

Choose the best answer to the following.

1. Why does the melting point of hydrocarbons increase as the number of carbon atoms per molecule increases?
   (a) An increase in the number of carbon atoms per molecules also means an increase in the density of the hydrocarbon.
   (b) Because of greater induced dipole–induced dipole molecular attractions.
   (c) Larger hydrocarbon chains tend to be branched.
   (d) Because the molecular mass also increases.

2. How many structural isomers are there for hydrocarbons having the molecular formula C_4H_{10}?
   (a) None
   (b) One
   (c) Two
   (d) Three

3. What is the name for the structure shown below?

   (a) 3-methyl-4-ethyl-6-methyloctane
   (b) 3,6-dimethyl-4-ethyloctane
   (c) 2-ethyl-4-ethyl-5-methylheptane
   (d) 2,4-diethyl-5-methylheptane

4. Heteroatoms make a difference in the physical and chemical properties of an organic molecule because
   (a) they add extra mass to the hydrocarbon structure.
   (b) each heteroatom has its own characteristic chemistry.
   (c) they can enhance the polarity of the organic molecule.
   (d) All of the above.

5. Why might a high-formula-mass alcohol be insoluble in water?

(a) A high-formula-mass alcohol would be too attracted to itself to be soluble in water.

(b) The bulk of a high-formula-mass alcohol likely consists of nonpolar hydrocarbons.

(c) Such an alcohol would likely be in a solid phase.

(d) In order for two substances to be soluble in each other, their molecules need to be of comparable mass.

6. Alkaloid salts are not very soluble in the organic solvent diethyl ether. What might happen to the free-base form of caffeine (an alkaloid) dissolved in diethyl ether if gaseous hydrogen chloride, HCl, were bubbled into the solution?

(a) A second layer of water would form.

(b) Nothing and the HCl gas would merely bubble out of solution.

(c) The diethyl-ether-insoluble caffeine salt would form as a white precipitate.

(d) The acid–base reaction would release heat, which would cause the diethyl ether to start evaporating.

7. Explain why caprylic acid, $CH_3(CH_2)_6COOH$, dissolves in a 5% aqueous solution of sodium hydroxide but caprylaldehyde, $CH_3(CH_2)_6CHO$, does not.

(a) With two oxygens, the caprylic acid is about twice as polar as the caprylaldehyde.

(b) The caprylaldehyde is a gas at room temperature.

(c) The caprylaldehyde behaves as a reducing agent, which neutralizes the sodium hydroxide.

(d) The caprylic acid reacts to form the water-soluble salt.

8. How many oxygen atoms are bonded to the carbon of the carbonyl of an ester functional group?

(a) None

(b) One

(c) Two

(d) Three

9. One solution to the problem of our overflowing landfills is to burn plastic objects instead of burying them. What would be some of the advantages and disadvantages of this practice?

(a) Disadvantage: toxic air pollutants; advantage: reduced landfill volume

(b) Disadvantage: loss of vital petroleum-based resource; advantage: generation of electricity

(c) Disadvantage: discourages recycling; advantage: provides new jobs

(d) All of the above.

10. Which would you expect to be more viscous, a polymer made of long molecular strands or one made of short molecular stands? Why?

(a) Long molecular strands because they tend to tangle among themselves

(b) Short molecular strands because of a greater density

(c) Long molecular strands because of a greater molecular mass

(d) Short molecular strands because their ends are typically polar

23

THE NUTRIENTS OF LIFE

THE MAIN IDEA

There are only four types of biomolecules

We eat because of the steady demand our bodies have for new molecules, which we need for energy as well as for building tissues. Interestingly, within about seven years most of the molecules in a human body have been replaced by new ones—the body you have today is not the one you had seven years ago! What are the four types of biomolecules? How are they used by our bodies? How does knowing about the chemicals of life help us make good decisions about the food we eat?

Explore!

Can You Experience the Action of Enzymes?

1. Look at a soda cracker and think about eating it. This should help to stimulate your salivary glands to produce extra enzyme-containing saliva. Use wheat-free crackers if you have any allergies to wheat-based products.
2. Chew on the cracker without swallowing for as long as possible. Pay careful attention to the flavor.

Analyze and Conclude

1. **Observing** Enzymes (biochemical catalysts) in your saliva work to break down starch molecules, which are bland, into simple carbohydrate molecules, which are sweet. Did you notice that the cracker tasted sweeter the longer it was in your mouth?
2. **Predicting** Rice candies are popular within Asian cultures. What do you suppose is done to the rice to convert it to a "candy"?
3. **Making Generalizations** Why is it important to chew your food thoroughly before swallowing it? Are enzymes found only in your saliva or throughout your body?

23.1 Biomolecules Are Molecules Produced and Used by Organisms

The fundamental unit of almost all organisms is the *cell*. Cells are typically so small that you need a microscope to see them individually. About 10 average-sized human cells, for example, could fit within the period at the end of this sentence. Figure 23.1 shows a typical animal cell and a typical plant cell.

Housed within each cell is the *cell nucleus*, which contains the genetic code. Just outside the nucleus is the *cytoplasm*, which is a solution containing biomolecules as well as organelles. The organelles

FIGURE 23.1 ▶
Macrosopic, microscopic, and submicroscopic views of an animal and a plant.

are tiny structures, each of which has its own purpose, such as producing energy or creating biomolecules.

The great majority of the biomolecules used by cells are either carbohydrates, lipids, proteins, or nucleic acids. In addition, most cells need small amounts of vitamins and minerals in order to function properly.

There are many different molecules in the food you eat, but the bulk of these molecules are either carbohydrates, lipids, proteins, or nucleic acids. That's it!

23.2 Carbohydrates Give Structure and Energy

Carbohydrates are molecules of carbon, hydrogen, and oxygen produced by plants through photosynthesis. The term *carbohydrate* is derived from the fact that plants make these molecules from carbon (from atmospheric carbon dioxide) and water. The term **saccharide** is a synonym for *carbohydrate*, and a *monosaccharide* ("one saccharide") is the fundamental carbohydrate unit. There are many kinds of monosaccharides. The structures of the two most common ones, glucose and fructose, are shown in Figure 23.2.

Monosaccharides are the building blocks of *disaccharides*, which are carbohydrate molecules containing two monosaccharide units. Figure 23.3 shows table sugar—sucrose—the most well-known disaccharide. In the digestive tract, sucrose is readily cleaved into its monosaccharide units, glucose and fructose.

Monosaccharides and disaccharides are classified as *simple carbohydrates*, where the word *simple* is used because these food molecules consist of only one or two monosaccharide units. Most simple carbohydrates have some degree of sweetness and are also known as sugars.

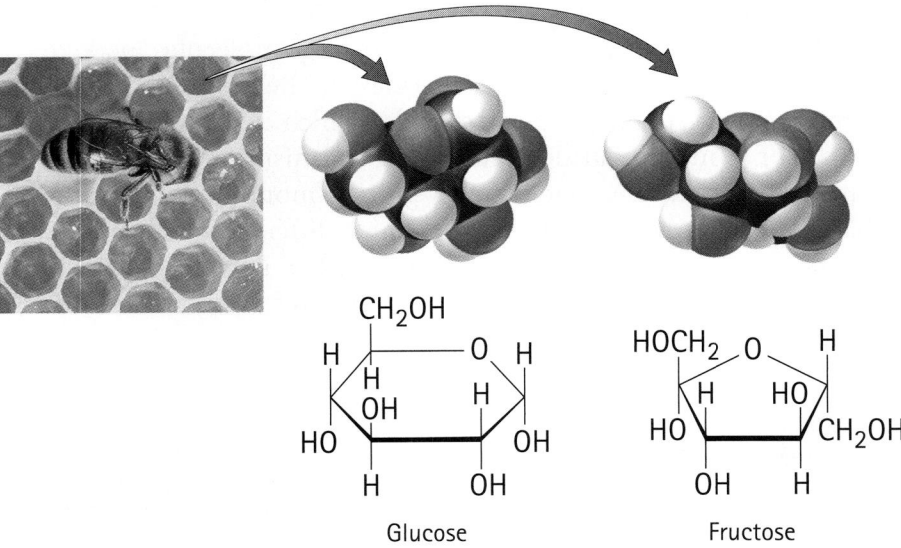

Glucose   Fructose

FIGURE 23.2 ▲
Honey is a mixture of the monosaccharides glucose and fructose. Glucose is a six-membered ring, and fructose is a five-membered ring. For simplicity, the stick structures introduced in Chapter 22 are shown below each molecular model.

FIGURE 23.3 ▲
Disaccharides, such as sucrose, consist of two chemically bonded monosac-charide units, which are cleaved during digestion.

Polysaccharides Are Complex Carbohydrates

✓ Recall from Chapter 22 that polymers are large molecules made of repeating monomer units. Monosaccharides are the monomers that link to form the polymers called *polysaccharides*, which contain hundreds to thousands of monosaccharide units. Living organisms produce and consume polysaccharides made of only glucose monomers. These poly-saccharides include *starch* and *cellulose*, which differ from one another only in how the glucose units are chained together. Polysaccharides are known as *complex carbohydrates*, where *complex* refers to the multitude of monosaccharide units linked together.

Plants store glucose as starch. But we animals store glucose as a third type of polysaccharide called *glycogen*. Your athletic endurance is related to how much glycogen you have stored away.

CONCEPT CHECK

What makes a simple carbohydrate simple and a complex carbohy-drate complex?

Check Your Answer

Simple carbohydrates are simple in the sense that they consist of only one or two monosac-charide units per molecule. Complex carbohydrates are complex in the sense of consisting of high numbers of monosaccharide units per molecule.

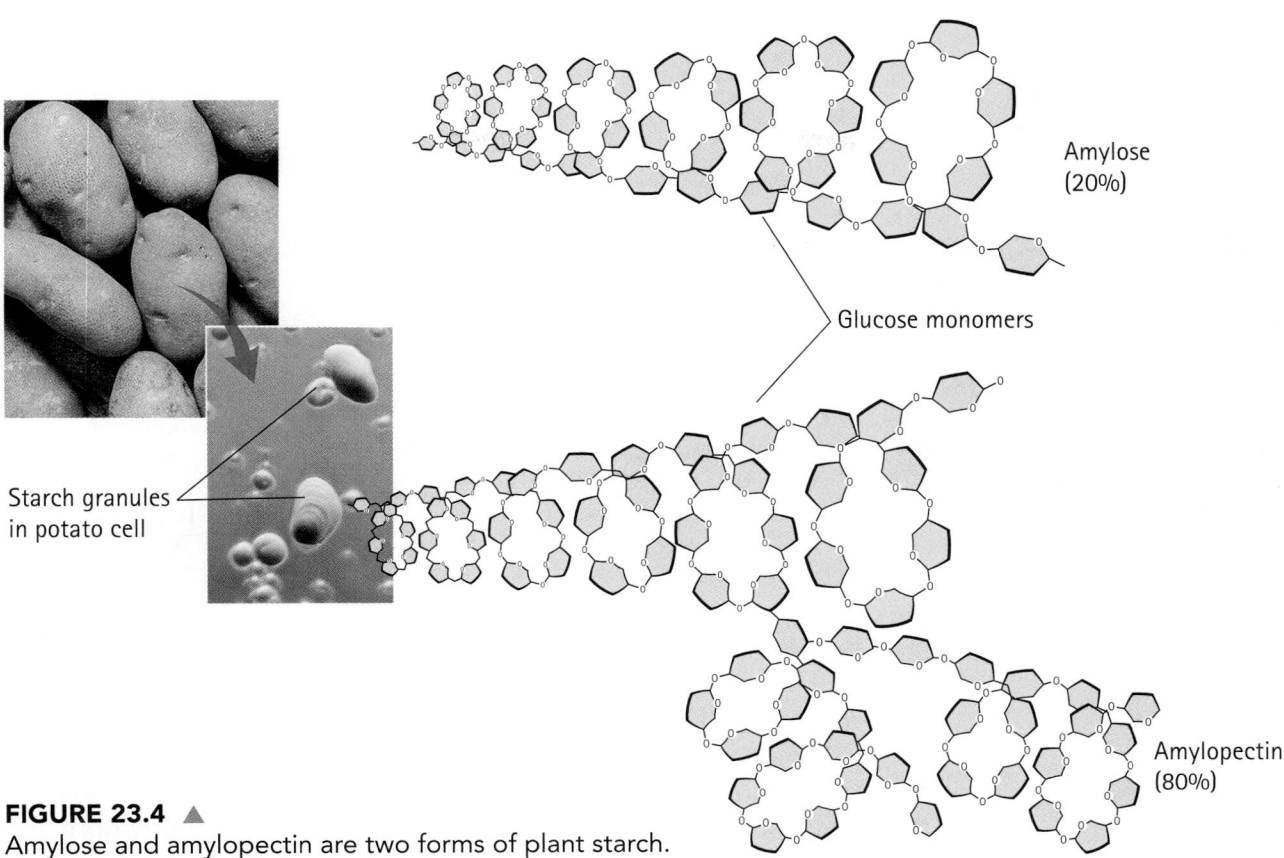

Amylose (20%)

Glucose monomers

Amylopectin (80%)

FIGURE 23.4 ▲
Amylose and amylopectin are two forms of plant starch.

Starch is a polysaccharide produced by plants to store the abundance of glucose formed during photosynthesis. The starch of a potato plant is shown in Figure 23.4. Animals can also obtain glucose from plant starch, which makes plant starch an all-important food source.

Cellulose, a structural component of plant cell walls, is also a polysaccharide of glucose. Because of the way the glucose units are put together, however, the cellulose polysaccharide does not coil. Also, there is no branching. These two attributes of cellulose allow the polysaccharide strands to align much like strands of uncooked spaghetti. This alignment maximizes the number of hydrogen bonds between strands, which makes cellulose a tough material. Added strength is given as the plant lays down microscopic fibers of cellulose in a criss-cross fashion, as shown in Figure 23.5.

Cellulose serves as the primary structural component of all plants. Cotton is nearly pure cellulose. Wood, made largely of cellulose, can support trees that are as much as 30 meters tall. Cellulose is by far the most abundant organic compound on Earth.

Most animals, including humans, are not able to break cellulose down to glucose. Instead, the cellulose in the food we eat serves as dietary fiber that helps in regulating bowel movements. In the large intestine, cellulose-based fiber absorbs water and has a laxative effect. Waste products are therefore moved along faster. Microorganisms that live in the digestive tracts of wood-eating termites and grass-eating ruminants (cows, sheep, and goats) can break cellulose down to glucose.

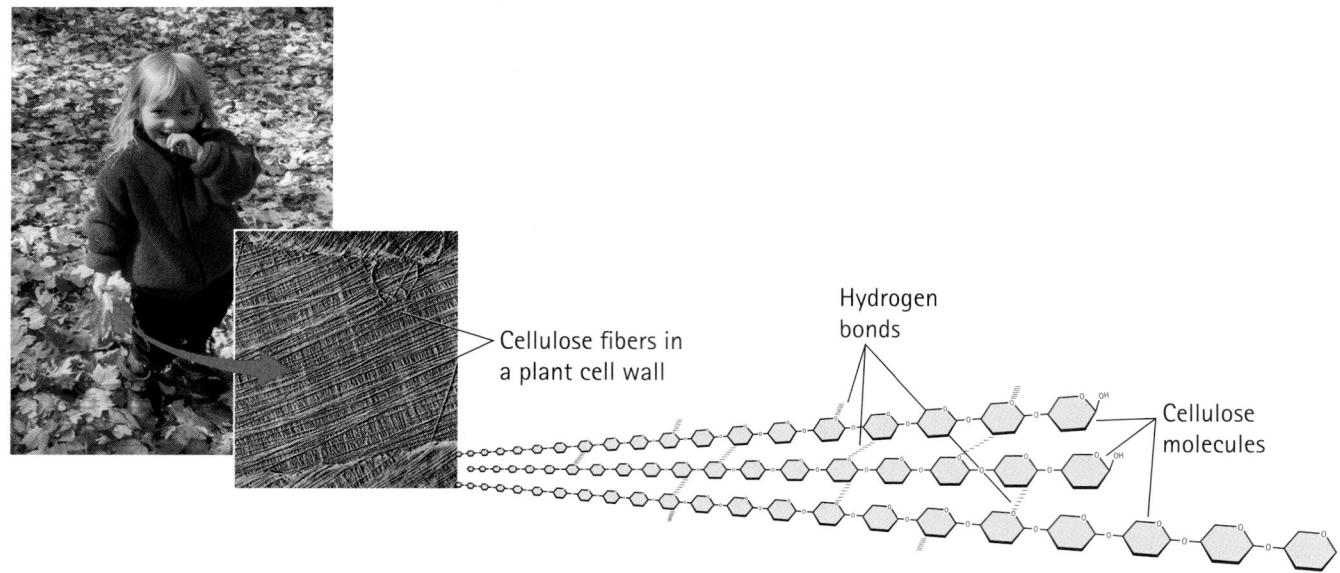

FIGURE 23.5 ▲
Strands of cellulose in a plant, including the plant leaves held by Maitreya, are joined by hydrogen bonds. These microscopic fibers are laid down in a criss-cross pattern to give strength in many directions.

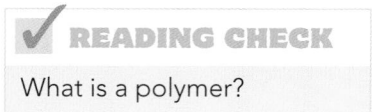

READING CHECK

What is a polymer?

Strictly speaking, termites and ruminants do not digest cellulose. Rather, they digest the glucose produced by the cellulose-digesting microorganisms that live inside them.

23.3 Lipids Are Insoluble in Water

✓ All **lipids** are insoluble in water. There are two main types of lipids: *fats* and *steroids*. Fats are stored in the body in *fat deposits*. These deposits serve as important reservoirs of energy. Fat deposits directly under the skin help to insulate us from the cold—good news for the walrus shown in Figure 23.6. In addition, vital organs, such as the heart and kidneys, are cushioned against injury by fat deposits.

A **fat** molecule is made from the combination of a glycerol molecule, $C_3H_8O_3$, and three fatty acid molecules, as shown in Figure 23.7. A *fatty acid* is a long-chain hydrocarbon terminating in a carboxylic acid group. The chain is typically

FIGURE 23.6 ▶
The walrus and other polar species are insulated from the cold by a thick layer of fat beneath their skin.

Glycerol   Fatty acid molecules   Fat molecule, a triglyceride

FIGURE 23.7 ▲
A typical fat molecule, also known as a triglyceride, is the combination of one glycerol unit and three fatty acid molecules. Note that this reaction involves the formation of three ester functional groups.

between 12 and 18 carbon atoms long and may be either saturated or unsaturated. Recall from Chapter 22 that a *saturated* chain contains no carbon–carbon double bonds; an *unsaturated* chain contains one, two, or three carbon–carbon double bonds. Note that, like carbohydrates, fats contain only carbon, hydrogen, and oxygen. Because fats are made from three fatty acids and glycerol, they are also known as *triglycerides*.

The digestion of fat is accompanied by the release of considerably more energy than is produced by the digestion of an equivalent amount of either carbohydrate or protein. There are about 38 kilojoules (9 Calories) of energy in 1 gram of fat but only about 17 kilojoules (4 Calories) of energy in 1 gram of carbohydrate or protein. (Recall that the energy content of food is often reported in Calories, with an uppercase C, where 1 Calorie = 1 kilocalorie = 1000 calories.)

✔ **READING CHECK**

Are lipids soluble in water?

CONCEPT CHECK
Give two reasons animals living in cold climates tend to form a thick layer of fat just prior to the onset of winter.

Check Your Answer
Fat provides a source of energy during the winter, when food is generally scarce, and it provides insulation from cold winter temperatures.

An unsaturated fat can be chemically transformed into a saturated fat by reacting it with hydrogen, H_2. This is called *hydrogenation*. An oily fat can be made more solid by partial hydrogenation in which only some of the double bonds are saturated. This allows for candy bars that don't melt in your hand.

As shown in Figure 23.8a, saturated fat molecules are able to pack tightly together because their saturated fatty acid chains point straight out and align with one another. This gives saturated fats, such as those found within beef, high melting points, and as a result they tend to be solid at room temperature. Unsaturated fats—those made from unsaturated fatty acids—have fatty acid chains that are "kinked" wherever double bonds occur, as shown in Figure 23.8b. The kinks inhibit alignment, and as a result unsaturated fats tend to have low melting points. These fats are liquid at room temperature and are commonly referred to as *oils*. Most vegetable oils are liquid at room temperature because of the high proportion of unsaturated fats they contain.

The fat from an animal or plant is a mixture of different fat molecules having various degrees of unsaturation. Fat molecules containing

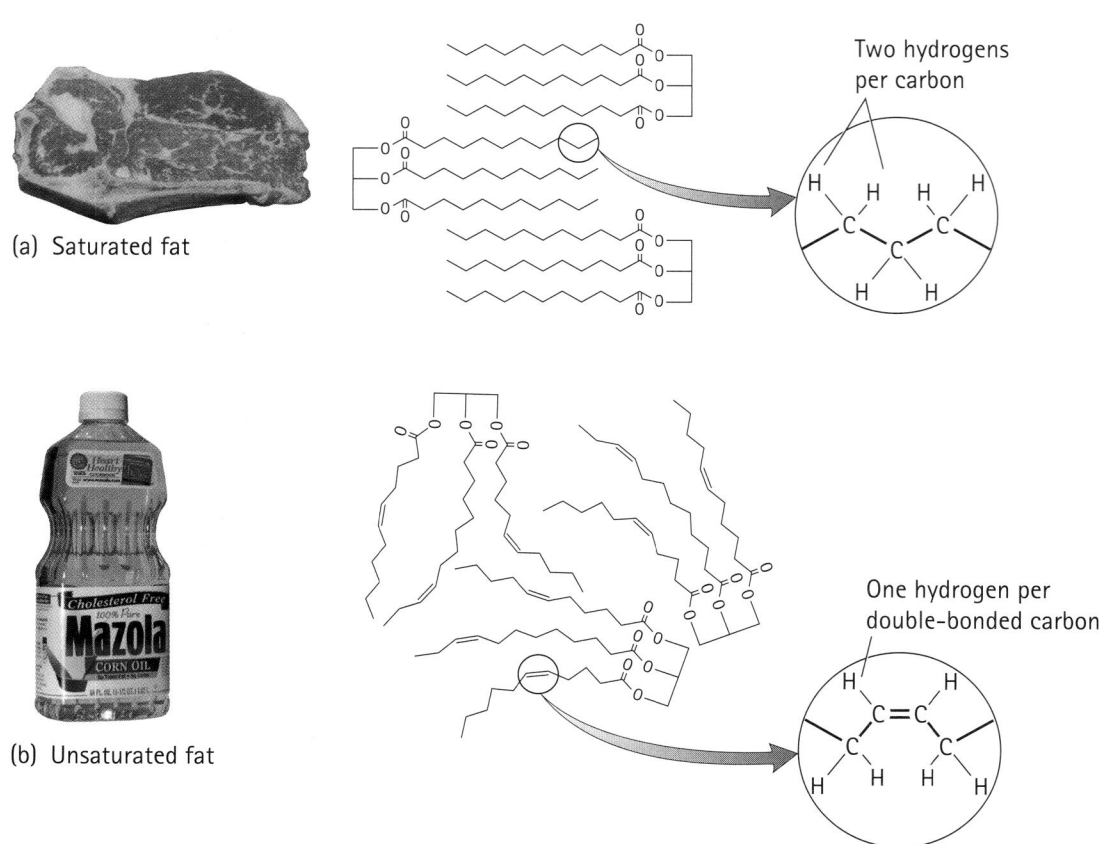

FIGURE 23.8 ▲
(a) Saturated fats are typically solid at room temperature because of molecular attractions between fatty acid chains. (b) Unsaturated fats are typically liquid at room temperature because molecular attractions are inhibited by the kinked nature of the fatty acid chains.

NH$_2$

Adenine

Ribose sugar

O

OH   OH

Adenosine triphosphate (ATP)

H$_2$C — O — P — O — P — O — P — O$^-$

Phosphate ions

◀ **FIGURE 23.19**
Phosphate ions are an important part of the ATP molecule.

molecules into and out of cells. The human body goes through lots of ATP—about 8 grams per minute during strenuous exercise. It is a short-lived molecule and so must be produced continuously.

23.7 Metabolism Is the Cycling of Biomolecules Through the Body

Your body takes in biomolecules in the food you eat and breaks them down to their molecular components. Then one of two things happens: either your body "burns" these molecular components for their energy content through a process known as *cellular respiration,* or these components are used as the building blocks for your body's own versions of carbohydrates, lipids, proteins, and nucleic acids. The sum total of all these biochemical activities is what we call **metabolism.** ✔ Two forms of metabolism are *catabolism* and *anabolism*. Figure 23.20 shows the major catabolic and anabolic pathways of living organisms.

All metabolic reactions that involve the tearing down of biomolecules are grouped under the heading of **catabolism.** Digestion and cellular respiration are examples of catabolic reactions. Digestion begins with the hydrolysis of food molecules, a reaction in which water is used to sever bonds in the molecules. The small molecules formed in digestion, such as the glucose units of complex carbohydrates, then migrate to all the various cells of the body and take part in cellular respiration. There the small food molecules lose electrons to the oxygen that was inhaled through our lungs, and as a result break down to even smaller molecules as carbon dioxide, water, and ammonia, which are excreted. Through this process, high-energy molecules, such as ATP, are created. These high-energy molecules are able to drive reactions that produce body heat, muscle movement, and nerve impulses. They also are responsible for allowing **anabolism,** which is the general term for all the chemical reactions that produce large biomolecules from smaller molecules.

The types of biomolecules produced by anabolism are the same as the types found in food—carbohydrates, lipids, proteins, and nucleic acids. These products of anabolism are, if you will, the host's own

ATP is also used by the body to allow muscles to relax after contraction. When the body dies, no matter what the cause, ATP synthesis comes to a halt, and all body muscles become stiff—a condition known as rigor mortis.

Anabolic steroids are biomolecules that promote the building up of body tissues. Athletes have found that these steroids improve performance, but they also have many negative side effects. Athlete beware!

FIGURE 23.20 ▶
Metabolic pathways for the food we ingest. Catabolic pathways are indicated by the purple arrows, anabolic pathways by the blue arrow.

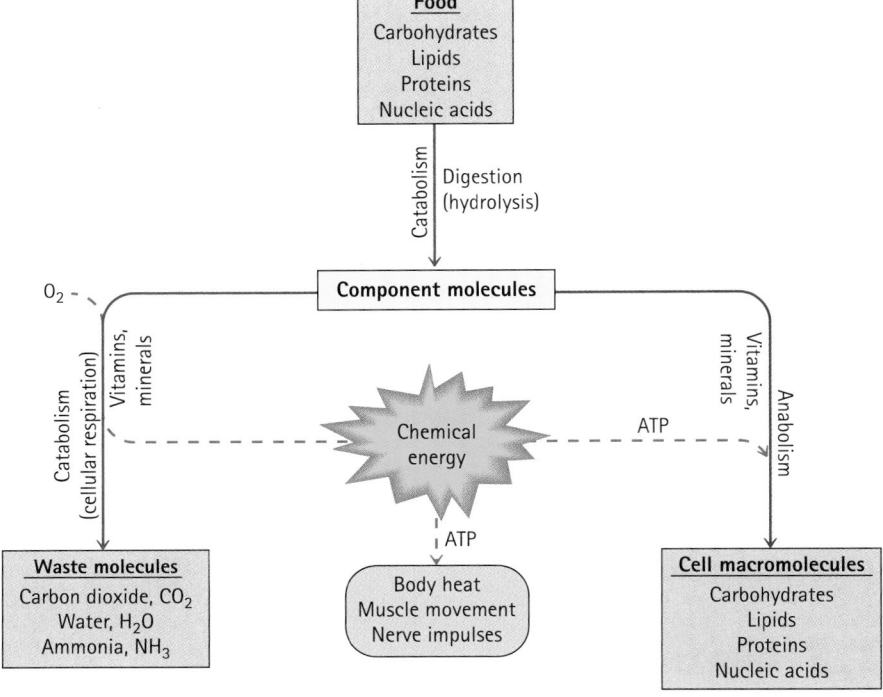

READING CHECK

What are the two forms of metabolism?

version of what the food once was. And if the host ever becomes food, anabolic reactions in the subsequent host will result in different versions of the molecules. Thus, organisms in a food chain live off one another by absorbing one another's energy via catabolic reactions and then rearranging the remaining atoms and molecules via anabolic reactions into the biomolecules they need to survive.

Catabolism and anabolism work together. In healthy muscle tissue, for example, the rate of muscle degradation (catabolism) is matched by the rate of muscle building (anabolism). If you increase your food supply and exercise vigorously, it is possible to favor the muscle-building anabolic reactions over the muscle-destroying catabolic reactions. The result is an increase in muscle mass. Stop eating and exercising, however, and these anabolic reactions lose out to the catabolic reactions. The result is a decrease in muscle mass—you begin to waste away.

CONCEPT CHECK
Anabolic steroids help people gain muscle mass. If there were such a thing as a catabolic steroid, what would be its effect?

Check Your Answer
Anabolic? Catabolic? Which is which? Many students recognize the term *anabolic steroids* from the sports news media, which are quick to report on famous athletes caught using these steroids for improved performance. Anabolism, therefore, is muscle building, and so catabolism must be muscle degrading. A catabolic steroid would cause a loss of muscle mass.

23.8 The Food Pyramid Summarizes a Healthful Diet

The food pyramid, shown in Figure 23.21, summarizes the food intake recommendations of the United States Department of Agriculture (USDA). According to this pyramid, an individual's daily diet should consist mostly of bread, cereals, grains, pastas, fruits, and vegetables, with the amount of dairy products and meats fairly limited and foods high in sugars or fats consumed only sparingly.

We can get some insight into the reasons behind these recommendations by looking at how the body handles the biomolecules contained in these foods.

Carbohydrates Predominate in Most Foods

The breads, cereals, grains, pastas, fruits, and vegetables that take up the majority of the pyramid are important sources of food primarily because they contain a good balance of all nutrients—carbohydrates, fats, proteins, nucleic acids, vitamins, and minerals. The predominant component of these foods, however, is carbohydrates. There are two types of carbohydrates—nondigestible, called dietary *fiber,* and digestible, mainly starches and sugars.

As discussed in Section 23.2, dietary fiber helps keep things moving in the bowels, especially in the large intestine. There are two kinds of fiber—water-insoluble and water-soluble. Insoluble fiber consists mainly of cellulose, which is found in all food derived from plants. In general, the less processed the food, the higher the insoluble-fiber content. Brown rice, for example, has a greater proportion of insoluble fiber than does white rice, which is made by milling away the rice seed's outer coating (along with numerous vitamins and minerals).

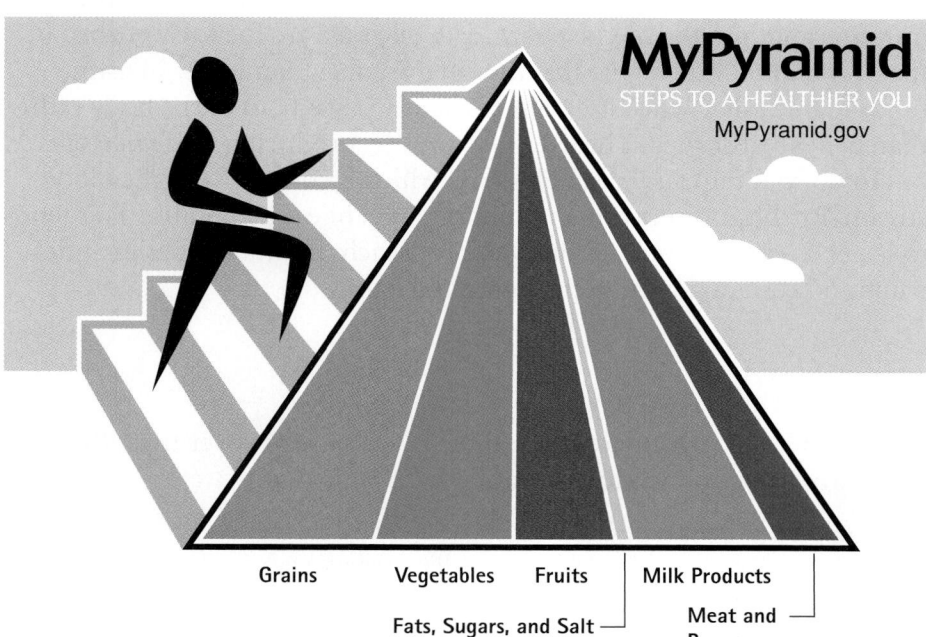

◀ **FIGURE 23.21**
The food pyramid.

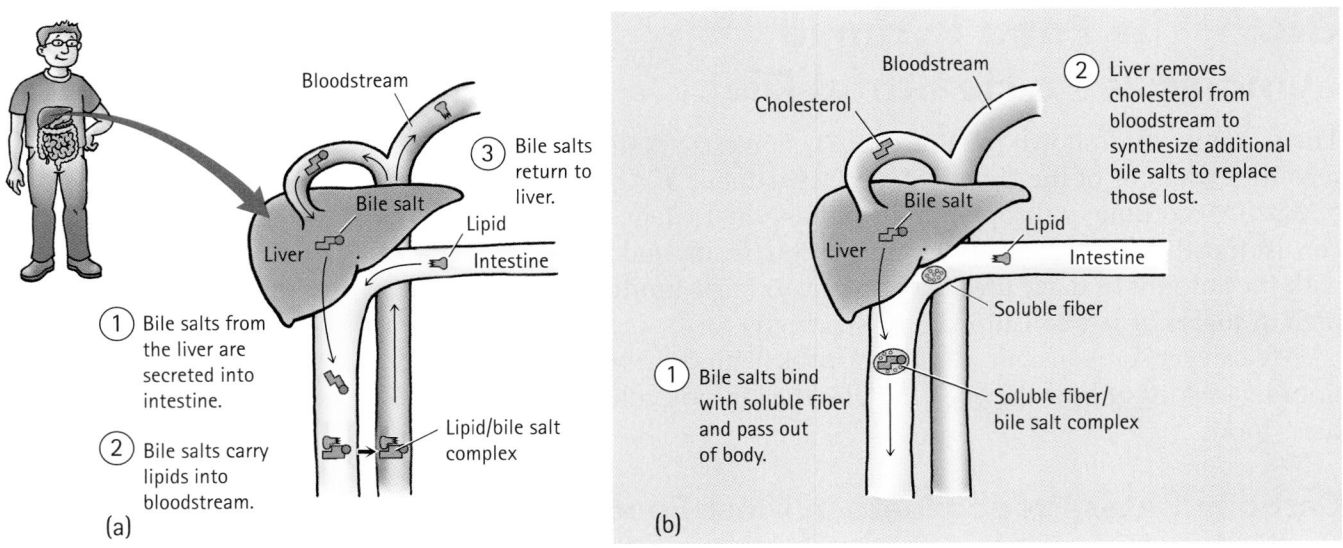

FIGURE 23.22 ▲

(a) With no soluble fiber present, bile salts recycle to the liver and no new ones need to be made. (b) In the presence of soluble fiber, bile salts are removed from the body. The liver then must use cholesterol from the blood to make new supplies of bile salts. Thus by binding with bile salts, soluble fiber indirectly decreases the amount of cholesterol in the blood.

Soluble fiber is made of certain types of starches that are resistant to digestion in the small intestine. An example is pectin, which is added to jams and jellies because it acts as a thickening agent, becoming a gel when dissolved in a limited amount of water. Soluble fiber tends to lower cholesterol levels in the blood because of how it interacts with *bile salts,* which are cholesterol-derived substances produced in the liver and then secreted into the intestine. As shown in Figure 23.22, one of the functions of bile salts is to carry ingested lipids through the membranes of the intestine and into the bloodstream. The bile salts are then reabsorbed by the liver and cycled back to the intestine. Soluble fiber in the intestine binds to bile salts, which are then efficiently passed out of the body rather than being reabsorbed. The liver responds by producing more bile salts, but to do so it must utilize cholesterol, which it collects from the bloodstream. By this indirect route of binding with bile salts, soluble fiber tends to lower a person's cholesterol level. Foods rich in soluble fiber include fruits and certain grains, such as oats and barley.

CONCEPT CHECK

How do lipids, which are not soluble in water, cross over from the digestive tract to the bloodstream, which is primarily water?

Check Your Answer

The nonpolar lipids are made soluble in water when they are coupled with bile salts from the liver. Soluble fibers help to remove bile salts, which makes it harder for the lipids to be absorbed into the bloodstream.

During digestion, the digestible carbohydrates—both starches and sugars—are transformed to glucose, which is absorbed into the blood-stream through the walls of the small intestine. The body then utilizes this glucose to build energy molecules, such as ATP. Carbohydrate-containing foods are rated for how quickly they cause an increase in blood glucose levels. This rating is done with what is known as the *glycemic index*. The index compares how much a given food increases a person's blood glucose level relative to the increase seen when pure glucose is ingested, with the latter increase assigned a standard value of 100. In general, foods that are high in starch or sugar but low in dietary fiber are high on the glycemic index, a baked potato being a prime example.

The glycemic index for a particular food can vary greatly from one person to the next. How the food was prepared can also make a big dif-ference. Thus, index values, such as the ones shown in Table 23.4, are to be taken lightly—merely as ballpark figures. Given this qualification,

TABLE 23.4 Glycemic Index for Select Foods

Food	Glycemic Index	Food	Glycemic Index
Glucose	100	Honey	58
Baked potato	85	Sweet corn	55
Cornflakes	83	Brown rice	55
Microwaved potato	82	Popcorn	55
Jelly beans	80	Oatmeal cookies	55
Vanilla wafers	77	Sweet potato	54
French fries	75	Banana	54
Cheerios	74	Milk chocolate	49
White bread	71	Orange	44
Mashed potato	70	Snickers candy bar	40
Life-Savers candy	70	Pinto beans	39
Shredded Wheat	69	Apple	38
Wheat bread	68	Spaghetti, boiled 5 minutes	36
Sucrose	64		
Raisins	64	Skim milk	32
Mars candy bar	64	Whole milk	27
High-fructose corn syrup	62	Grapefruit	25
		Soy beans	18
White rice	58	Peanuts	15

Source: Jennie Brand Miller et al., *The Glucose Revolution: The Authoritative Guide to the Glycemic Index.* Sydney: Marlowe & Company, 1999.

Diabetes is a disease in which the body does not produce enough insulin or the insulin it produces is not effective. Glucose, therefore, is not readily absorbed by cells where it is needed. Instead, the glucose builds up to high concentrations in the bloodstream.

however, the index provides valuable information for people, such as those with diabetes, who need to pay close attention to their blood sugar levels.

There are a number of problems associated with eating carbohydrate foods that have a high glycemic index. For example, the rapid spike in blood glucose level causes the body to produce extra *insulin*, a blood-soluble protein that causes glucose to be moved out of the blood and into cells to be metabolized. Insulin is very effective at what it does, however, and soon the extra insulin in the blood leads to a depletion of blood glucose. The body responds by releasing glucose from its glucose stores, but also by triggering a sense of hunger, even if the person just ate. A meal rich in foods high on the glycemic index therefore promotes overeating, which usually leads to obesity.

CONCEPT CHECK
Why does your blood sugar level drop after eating a lot of sugar?

Check Your Answer
Eating a lot of sugar initially raises your blood sugar level, but it also causes the release of insulin, which can go overboard in removing that sugar from your blood.

✓ Many professional organizations, such as the American Diabetes Association, caution that priority should be given to the quantities of carbohydrates ingested rather than to the glycemic index of the food containing those carbohydrates. What really counts is the total number of calories absorbed, not whether these calories came from foods high or low on the index. For most people, however, ingesting foods low on the index makes maintaining a healthful caloric intake more manageable.

Another advantage of eating carbohydrates from foods that are low on the index is that these foods provide energy to the body over an extended period of time. They do this because the glucose molecules they contain are released slowly. Furthermore, maintaining moderate glucose levels in the blood allows the body to continue using fats for its energy needs. As was discussed in Section 23.3, fats provide much more energy per gram (and thus more ATP) than do carbohydrates.

For athletes, a diet rich in foods low on the index, such as spaghetti, translates to greater endurance. Interestingly, this greater endurance is just as useful for bodybuilders as it is for marathon runners. The energy required for building muscles is far more critical than the supply of raw materials needed. Furthermore, the body's metabolism is versatile enough to generate proteins out of glucose (just as it is able to generate glucose out of proteins). Thus, the body-builder's supply of proteins is assured. A diet rich in carbohydrates is therefore more effective at allowing a bodybuilder to build muscles than is a diet rich in proteins.

A candy bar is good for a quick energy fix, but chow down on a spaghetti feast the night before a strenuous workout for long-run energy.

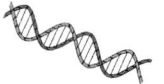

Despite the many advantages of eating carbohydrates low on the glycemic index, foods rich in carbohydrates high on the index, such as sucrose, are now more popular than ever. Many of these foods are highly processed. Although they are good at providing energy, the USDA recommends that they be consumed only sparingly because they lack many of the essential nutrients present in grains, vegetables, and fruits.

✓ READING CHECK

What does the American Diabetes Association recommend about the glycemic index?

Unsaturated Fats Are Generally More Healthful Than Saturated Fats

Because your body uses saturated fats to synthesize cholesterol, the more saturated fats you ingest, the more cholesterol your body is able to synthesize. Unsaturated fats, by contrast, are not ideal starting materials for cholesterol synthesis.

Another reason unsaturated fats are more healthful has to do with how fats associate with cholesterol. Fats and cholesterol are both nonpolar lipids, which, on their own, are insoluble in blood. In order to move through the bloodstream, these compounds are packaged with bile salts, as was discussed earlier. Most lipids, however, are made water-soluble by being packaged with water-soluble proteins in complexes called *lipoproteins*. Lipoproteins are classified according to density, as noted in Table 23.5. Very-low-density lipoproteins (VLDL) serve primarily in the transport of fats throughout the body. Low-density lipoproteins (LDL) transport cholesterol to the cells, where it is used to build cell walls. High-density lipoproteins (HDL) bring cholesterol to the liver, where it is transformed to a variety of useful biomolecules.

A diet high in saturated fats leads to elevated VLDL and LDL levels in the bloodstream. This is undesirable because these lipoproteins tend to form fatty deposits called *plaque* in the artery walls. Plaque deposits can become inflamed to the point where they rupture, releasing blood-clotting factors into the bloodstream. A blood clot formed around the rupture site is let loose into the bloodstream, where it can become lodged and block the flow of blood to a particular region of the body. When that region is in the heart, the result is a heart attack. When that region is in the brain, the result is a stroke.

Cholesterol is produced mainly by us animals. Plants also produce cholesterol, but only in very small amounts.

VLDL, LDL, HDL?? Which should be higher? Which should be lower? Confused? Ask your doctor. He or she will be very impressed you know these terms.

TABLE 23.5 The Classification of Lipoproteins

Lipoprotein	Percent Protein	Density (g/mL)	Primary Function
Very-low-density (VLDL)	5	1.006–1.019	Fat transport
Low-density (LDL)	25	1.019–1.063	Cholesterol transport (to cells to build cell walls)
High-density (HDL)	50	1.063–1.210	Cholesterol transport (to liver for processing)

In contrast to saturated fats, unsaturated fats tend to increase blood HDL levels, which is desirable because these lipoproteins are effective at *removing* plaque from artery walls.

CONCEPT CHECK

For what two reasons are unsaturated fats better for you than saturated fats?

Check Your Answer

Unsaturated fats are not so readily used by your body to synthesize cholesterol. They also tend to increase the proportion of high-density lipoproteins, which lower the level of cholesterol in your blood and help relieve the buildup of arterial plaque.

Mix a partially hydrogenated vegetable oil with yellow food coloring, a little salt, and the organic compound butyric acid for flavor, and you have margarine, which became popular around the time of World War II as an alternative to butter.

Unsaturated fats, as noted in Section 23.3, tend to be liquids at room temperature. They can be transformed to a more solid consistency, however, by *hydrogenation*, a chemical process in which hydrogen atoms are added to carbon–carbon double bonds. Many food products, such as chocolate bars, contain partially hydrogenated vegetable oils so that they are of a consistency that sells well in the marketplace. Hydrogenation increases the percentage of saturated fats, however, and therefore makes these fats less healthful. Furthermore, as Figure 23.23 shows, some of the double bonds that remain are transformed to the *trans* structural isomer. Because carbon chains containing *trans* bonds tend to be less kinked than chains containing *cis* bonds, the partially hydrogenated fat has straighter chains. This means the fat is more likely to mimic the action of saturated fats in the body.

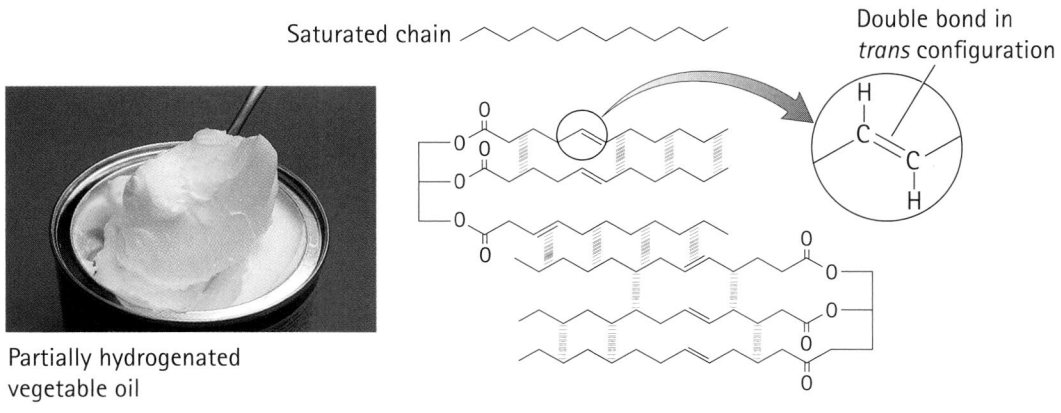

Partially hydrogenated vegetable oil

FIGURE 23.23 ▲

Hydrogenation can lead to *trans* double bonds in the fatty acid chain, which as a result points straight out, much as the chain of a saturated fatty acid does.

Our Intake of Essential Amino Acids Should Be Carefully Monitored

Proteins are useful for their energy content, just as starches, sugars, and fats are, but perhaps the greatest importance of proteins lies in how our bodies use them for building such structures as enzymes, bones, muscles, and skin. Of the 20 amino acids the human body uses to build proteins, the adult body is able to produce 12 of them in amounts sufficient for its needs—it produces these amino acids from carbohydrates and fatty acids. The remaining eight, listed in Table 23.6, must be obtained from food. Because the body needs these eight amino acids but cannot synthesize them, they are called *essential amino acids,* in the sense that it is essential we get adequate amounts of them from our food. To support rapid growth, infants and children require, in addition to the eight amino acids listed for adults in Table 23.6, large amounts of arginine and histidine, which can be obtained only from the diet. For infants and juveniles, therefore, there are a total of 10 essential amino acids. (The term *essential* is unfortunate because, in truth, all 20 amino acids are vital to our good health.)

Why our bodies produce ample amounts of some amino acids and not others can be explained by looking at the chemical structures of the amino acid side groups, some shown in Figure 23.11. The nonessential amino acids have side groups that tend to be simple and therefore can be produced by the body without much effort. The essential amino acids, however, tend to be biochemically more difficult to make. The body therefore can save energy by obtaining these amino acids from outside sources. Over the course of evolution, our capacity to build these amino acids diminished. In a similar manner, we lost the capacity to build vitamins, which are also complex molecules more efficiently obtained through our diet. In other words, we let other living organisms go

TABLE 23.6 The Essential Amino Acids

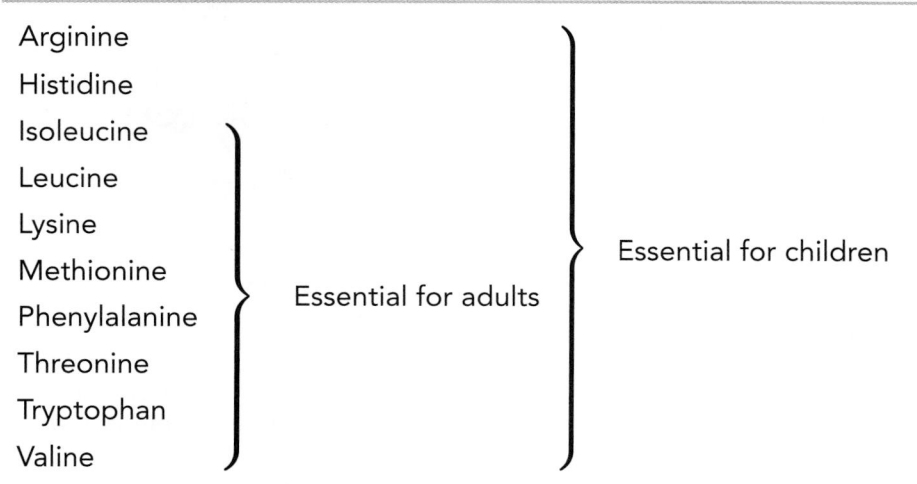

CONCEPT CHECK
What is an essential amino acid?

Check Your Answer
An essential amino acid is an amino acid that we must obtain from our diet because our bodies have lost the ability to make this amino acid.

through the metabolic expense of building these biomolecules, and then we eat those organisms.

In general, the more closely the amino acid composition of ingested protein resembles the amino acid composition of the animal eating the protein, the higher the nutritional quality of that protein. For humans, mammalian protein is of the highest nutritional quality, followed by fish and poultry, then by fruits and vegetables. Plant proteins in particular are often deficient in lysine, methionine, or tryptophan. A vegetarian diet provides adequate protein only if it contains a variety of protein sources, with a deficiency in one source being compensated for by an excess in another source, as shown in Figure 23.24.

The old adage "you are what you eat" has a literal foundation. With the exception of the oxygen you obtain through your lungs, nearly every atom or molecule in your body got there by first passing through your mouth and into your stomach. All the biomolecules needed for the energy and growth of a fetus growing in the womb,

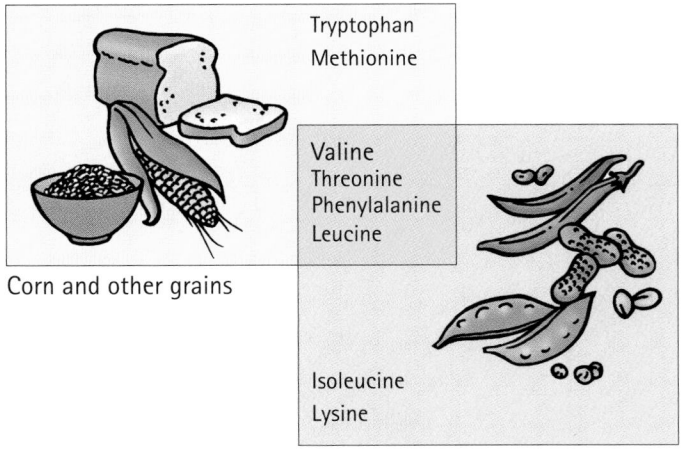

Corn and other grains

Tryptophan
Methionine

Valine
Threonine
Phenylalanine
Leucine

Isoleucine
Lysine

Beans and other legumes

FIGURE 23.24 ▲
Sufficient protein can generally be obtained in a vegetarian diet by combining a legume, such as peas or beans, with a grain, such as wheat or corn. Familiar meals containing such a combination include a peanut butter sandwich, corn tortillas and refried beans, and rice and tofu.

23 REVIEW

KEY TERMS

Amino acid The monomers of polypeptides, each monomer consisting of an amine group and a carboxylic acid group bonded to the same carbon atom.

Anabolism Chemical reactions that synthesize biomolecules in the body.

Carbohydrate A biomolecule that contains only carbon, hydrogen, and oxygen atoms and is produced by plants through photosynthesis.

Catabolism Chemical reactions that break down biomolecules in the body.

Deoxyribonucleic acid A nucleic acid containing a deoxygenated ribose sugar, having a double helical structure, and carrying genetic code in the nucleotide sequence.

Fat A biomolecule that packs a lot of energy per gram and consists of a glycerol unit attached to three fatty acid molecules.

Lipid A broad class of biomolecules that are not soluble in water.

Metabolism The general term describing all chemical reactions in the body.

Mineral Inorganic chemicals that play a wide variety of roles in the body.

Nucleic acid A long polymeric chain of nucleotide monomers.

Nucleotide A nucleic acid monomer consisting of three parts: a nitrogenous base, a ribose sugar, and an ionic phosphate group.

Protein A polymer of amino acids, also known as a polypeptide.

Ribonucleic acid A nucleic acid containing a fully oxygenated ribose sugar.

Saccharide Another term for carbohydrate. The prefixes *mono-*, *di-*, and *poly-* are used before this term to indicate the length of the carbohydrate.

Vitamin Organic chemicals that assist in various biochemical reactions in the body and can be obtained only from food.

REVIEW QUESTIONS

Biomolecules Are Molecules Produced and Used by Organisms

1. Is the cell nucleus within the cytoplasm or is the cytoplasm within the cell nucleus?
2. What are the four major categories of biomolecules discussed in this chapter?

Carbohydrates Give Structure and Energy

3. Are all carbohydrates digestible by humans?
4. Which monosaccharide do starches and cellulose have in common?
5. What is the most abundant organic compound on Earth?

Lipids Are Insoluble in Water

6. What are the structural components of a triglyceride?
7. What makes a saturated fat saturated?
8. What do all steroids have in common?

Proteins Are Polymers of Amino Acids

9. How do various amino acids differ from one another?
10. What do a peptide, polypeptide, and protein all have in common?

Nucleic Acids Code for Proteins

11. What is the difference between a nucleic acid and a nucleotide?
12. Where in the cell are deoxyribonucleic acids found?
13. Which four nitrogenous bases are found in DNA? In RNA?

Vitamins Are Organic, Minerals Are Inorganic

14. What are two classes of vitamins?
15. Why is it often more healthful to eat vegetables that have been steamed rather than boiled?

Metabolism Is the Cycling of Biomolecules Through the Body

16. What is the general outcome of catabolism?
17. What is the general outcome of anabolism?

The Food Pyramid Summarizes a Healthful Diet

18. Which type of biomolecule does the food pyramid recommend we eat the most of?
19. Are all dietary fibers made of cellulose?
20. Which type of lipoproteins have a greater association with the formation of plaque on artery wall: LDLs or HDLs?

THINK AND DO

1. To each of two drinking glasses add a teaspoon of potato broth, which you can make by boiling a few slices of potato to mushiness in about a cup of water. Also add a tablespoon of fresh water to each glass to dilute the broth. Collect a good wad of saliva in your mouth, and then gracefully spit into one of the glasses. Swirl to mix. Wait for a few minutes, and then add a drop of iodine solution (available from stores as a disinfectant) to each glass. The iodine reacts with the starch to form a blue substance. The more starch, the darker the color. Did one solution turn darker than the other? Why?
2. Obtain several different brands of "light" butter or margarine spreads. Place each sample in a test tube or narrow glass. Add enough so that, when melted,

the liquid will be at least 2 cm deep. Label each test tube with the brand it contains. Melt all samples in a microwave oven. (Watch carefully because this doesn't take long.) How many layers do you see form in each test tube? Which is more dense: water or fat? Which brand contains the most water? What happens when you cool the test tube in a refrigerator? Which sample do you suppose contains the greatest proportion of saturated or trans fats?

THINK AND COMPARE

1. Rank the following molecules in order of increasing sweetness: glucose, cellulose, starch.
2. Rank the following molecules in order of increasing molecular mass: cholesterol, glycine, deoxyribonucleic acid.
3. Rank the following mineral ions in order of how much we need each day: sodium, potassium, chromium.

THINK AND EXPLAIN

1. Does a carbohydrate contain water?
2. What is another biological use for carbohydrates besides energy?
3. Why do plants produce starch?
4. In what ways are cellulose and starch similar to each other? In what ways are they different from each other?
5. Why are lipids insoluble in water?
6. Why is it important to have cholesterol in your body?
7. Could a food product containing glycerol and fatty acids but no triglycerides be advertised as being fat-free? If so, how might such advertising be misleading?
8. What are the building blocks of a protein molecule?
9. Why doesn't the human body synthesize the essential amino acids?
10. In what way is the chemical structure of DNA different from that of RNA?
11. Both water-soluble and water-insoluble vitamins can be toxic in large quantities. Our bodies are much more tolerant of the water-soluble ones, however. Why?
12. The dietary minerals must be in ionic form in order for the body to make use of them. Why?

13. A friend of yours loads up on vitamin C once a week instead of spacing it out over time. She argues the convenience of not having to take pills every day. What advice do you have for her?

14. Which statement is more accurate?
    (a) Vitamins are needed by the body to avoid vitamin-deficiency diseases, such as scurvy.
    (b) Vitamins are needed by the body so that many of its catabolic and anabolic reactions can proceed efficiently.

15. Suggest why the glycemic index for sucrose is only about 64% that of glucose.

16. Is it possible to eat a food low on the glycemic index and still experience a significant increase in blood glucose?

17. Mammals cannot produce polyunsaturated fatty acids. How is it then that the lard obtained from pigs contains up to 10% polyunsaturated fatty acids?

18. Peanut butter has more protein per gram than a hard-boiled egg, and yet the egg represents a better source of protein. Why?

19. The human body stores excess glucose as glycogen and excess fat as fatty tissue that can accumulate beneath the skin. How does the body store any excess amino acids?

20. Cold cereal is often fortified with all sorts of vitamins and minerals but is deficient in the amino acid lysine. How might this deficiency be compensated for in a breakfast meal?

READINESS ASSURANCE TEST (RAT)

If you have a good handle on this chapter—if you really do—then you should be able to score 7 out of 10 on this RAT. Check your answers with your teacher. If you score less than 7, study further before moving on.

Choose the best answer to the following.

1. Which of the following statements is most accurate?
   (a) You are undergoing continuous chemical change within your body from moment to moment.
   (b) You have the exact same molecules in your body as when you were born.
   (c) You have the exact same types of chemicals in your body as when you were younger.
   (d) You are chemically the same as you were when you were born.
   (e) None of the above.

2. Which of the following statements best describes a monosaccharide?
   (a) It is the fundamental unit of genetic material.
   (b) It is the fundamental unit of life.
   (c) It is the monomer of the sucrose polymer.
   (d) It is the basic repeating unit of a polymer.
   (e) It is the fundamental unit of a carbohydrate.

3. Which of the following functional groups play the biggest role in the properties of carbohydrates?
   (a) alcohols
   (b) amides
   (c) esters
   (d) ethers
   (e) phenols

4. Which of the following is not a function of fat within the body?
   (a) It acts as a cushion to prevent injury.
   (b) It acts as an energy reserve.
   (c) It acts as a source of glucose.
   (d) It acts as insulation.
   (e) All of the above are functions of fat.

5. Why do most steroids have chemical structures very similar to cholesterol?
   (a) Most steroids are synthesized from cholesterol.
   (b) Cholesterol is synthesized from saturated fats and steroids are made from unsaturated fats.
   (c) Cholesterol is a fat, which is a lipid like steroids.
   (d) Steroids are made from saturated fats and cholesterol is made from unsaturated fats.
   (e) None of the above.

6. How are proteins similar to starch?
   (a) Both are polymers.
   (b) Both can be consumed and used as an energy source.
   (c) Both are made of glucose monomers.
   (d) Both are lipids.
   (e) (a) and (b).

7. The structure of ATP is most closely related to that of a
   (a) carbohydrate.
   (b) lipid.
   (c) amino acid.
   (d) nucleic acid.
   (e) nucleotide.

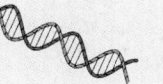

8. Vitamins such as vitamin B and vitamin C are often lost by boiling vegetables. Why?
   (a) They are water-soluble and are poured down the drain with the water.
   (b) The heat of the boiling water causes the vitamin to evaporate.
   (c) They are fat-soluble compounds and vegetables do not have much fat.
   (d) The heat of the boiling water destroys the vitamin.
   (e) None of the above.

9. Which of the following statements is true about metabolism?
   (a) Metabolism involves the use of energy to make new biomolecules.
   (b) Metabolism involves the rate of cellular respiration.
   (c) Metabolism involves the destruction of biomolecules and the production of energy.
   (d) All of the above.
   (e) None of the above.

10. Which of the following statements about proteins is true?
    (a) The source of a protein (mammal, fish, or poultry) has little effect on nutritional value.
    (b) The simpler an amino acid is, the more likely it is to be generated within the body.
    (c) All of the amino acids in the human body can be synthesized by our cells.
    (d) All of the above are true.
    (e) None of the above is true.

24

MEDICINAL CHEMISTRY

Medicines are like keys that unlock various biological responses.

Where do most medicines come from? How do chemists design medicines in the laboratory? How do medicines cure us of infections? Why do some keep us awake, while others make us drowsy? And how do medicines stop us from experiencing pain? What is the difference between a medicine and a drug? How is caffeine removed from coffee and tea? Why do cigarette smokers usually smoke after eating a meal? For answers, we turn to medicinal chemistry, which is the study of how certain organic molecules are able to exert special effects on our bodies.

Explore!

Can You Build a Drug-Receptor Site Model with Clay and a Sheet of Paper?

1. Soften up a teaspoon of craft clay by kneading it with your hands.
2. Press the softened clay into one of the nooks of a tightly crumpled paper ball.

Analyze and Conclude

1. **Observing** Most drugs work by binding to tiny nooks within our bodies called receptor sites. For this activity, the clay represents a drug, which has a receptor site found within the paper ball. Once you pull your clay out of the paper, how easily does it fit back into its receptor site? Without deforming, how well does the clay fit within other nooks and crannies?

2. **Predicting** If two drugs have the same biological effect, what might be true about their shapes? Some drugs have a rigid chemical structure that does not change shape easily. Others have a loose chemical structure that can contort. Which of these should have a greater variety of biological effects? Which should have a more specific biological effect?

3. **Making Generalizations** The drug-receptor site model is often called the "lock-and-key" model. Which is the lock? Which is the key? Why is the 3-dimensional shape of a drug important?

24.1 Medicines Are Drugs That Benefit the Body

What is a medicine? A *medicine* is a drug taken for the purpose of improving a person's health. So what, then, is a drug? Loosely defined, a drug is any substance other than food or water that affects how the body functions. The word *drug* refers to a wide range of chemical substances. All medicines are drugs, but not all drugs are medicines. Many drugs are used for nonmedical purposes, some legal and others illegal. Legal nonmedical drugs include alcohol, caffeine, and nicotine. Illegal nonmedical drugs include LSD and cocaine.

There are a variety of ways to classify drugs. For example, drugs can be classified according to how they are derived, as is shown in Table 24.1. Drugs that are natural products come directly from terrestrial or marine plants or animals. Drugs that are chemical derivatives are natural products that have been chemically modified to increase potency or decrease side effects. Synthetic drugs are those that originate in the laboratory.

Perhaps the most common way to classify drugs is according to their primary biological effect. ✓ It must be noted, however, that most drugs exhibit a broad spectrum of activity, which means they have multiple effects on the body. Therefore, they may fall under several classifications. Aspirin, for example, relieves pain, but it also reduces fever and inflammation, thins the blood, and causes ringing in the ears. Morphine relieves pain, but it also constipates and suppresses the urge to cough.

TABLE 24.1 The Origin of Some Common Drugs

Origin	Drug	Biological Effect
Natural product	Caffeine	Nerve stimulant
	Reserpine	Hypertension reducer
	Vincristine	Anticancer agent
	Penicillin	Antibiotic
	Morphine	Analgesic
Chemical derivative of natural product	Prednisone	Antirheumatic
	Ampicillin	Antibiotic
	LSD	Hallucinogenic
	Chloroquinine	Antimalarial
	Ethynodiol diacetate	Contraceptive
Synthetic	Valium	Antidepressant
	Benadryl	Antihistamine
	Allobarbital	Sedative-hypnotic
	Phencyclidine	Veterinary anesthetic
	Methadone	Analgesic

Most all drugs have a multitude of effects on the body.

At times, the multiple effects of a drug are desirable. Aspirin's pain-reducing and fever-reducing properties work well together in treating flu symptoms in adults, for instance. Additionally, aspirin's blood-thinning ability helps prevent heart disease. Morphine was widely used during the American Civil War both for relieving the pain of battle wounds and for controlling diarrhea. Often, however, the side effects of a drug are less desirable. Ringing in the ears and upset stomach are a few of the negative side effects of aspirin, and a major side effect of morphine is its addictiveness. A main goal of drug research, therefore, is to find drugs that are specific in their action and have minimal side effects.

Although two drugs that are taken together may have different primary activities, they may share a common secondary activity. The effect that both drugs share can be amplified when the two drugs are taken together. One drug enhancing the action of another is called the **synergistic effect.** A synergistic effect is often more powerful than the sum of the activities of the two drugs taken separately. One of the great challenges of physicians and pharmacists is keeping track of all the possible combinations of drugs and potential synergistic effects.

The synergism that results from mixing drugs that have the same primary effect is particularly hazardous. For example, a moderate dose of a sedative that makes you sleepy combined with a moderate amount of alcohol may be lethal. In fact, most drug overdoses are the result of a combination of drugs rather than the abuse of a single drug.

✔ **READING CHECK**

Do most drugs have a singular effect on the body or a multitude of effects?

CONCEPT CHECK
Distinguish between a drug and a medicine.

Check Your Answer
A drug is any substance administered so as to affect body function. A medicine is any drug administered for its therapeutic effect. All medicines are drugs, but not all drugs are medicines.

24.2 The Lock-and-Key Model Guides Chemists in Creating New Medicines

To find new and more effective medicines, chemists use various models that describe how drugs work. By far, the most useful model of drug action is the **lock-and-key model.** ✔ The basis of this model is that there is a connection between a drug's chemical structure and its biological effect. For example, morphine and all related pain-relieving opioids, such as codeine and heroin, have the T-shaped structure shown in Figure 24.1.

T-shaped 3-dimensional
structure found in all opioids

Morphine

Codeine

Heroin

FIGURE 24.1 ▲
All drugs that act like morphine have the same basic 3-dimensional shape as morphine.

According to the lock-and-key model, illustrated in Figure 24.2, biologically active molecules function by fitting into *receptor sites* in the body, where they are held by molecular attractions, such as hydrogen bonding. When a drug molecule fits into a receptor site the way a key fits into a lock, a particular biological event is triggered, such as a nerve impulse or even a chemical reaction. In order for a molecule to fit into a particular receptor site, however, it must have the proper shape, just as a key must have properly shaped notches in order to fit a lock.

Another facet of this model is that the molecular attractions holding a drug to a receptor site are easily broken. (Recall from Section 18.8 that most molecular attractions are many times weaker than chemical bonds.) A drug is therefore held to a receptor site only temporarily. Once the drug is removed from the receptor site, body chemistry destroys the drug's chemical structure and the effects of the drug are said to have worn off.

> Many drugs, though not all, are thought to work by the lock-and-key mechanism.

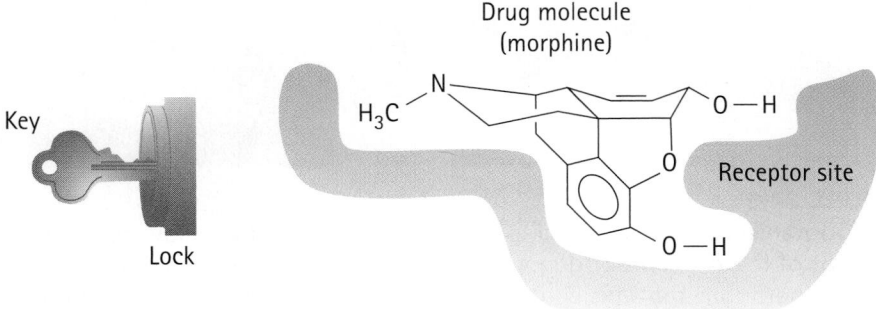

FIGURE 24.2 ▲
Many drugs act by fitting into receptor sites on molecules found in the body, much as a key fits in a lock.

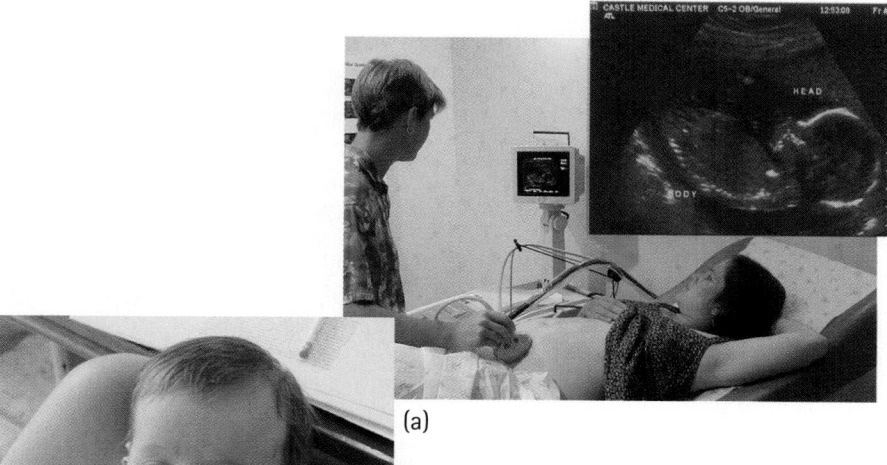

(a)

(b)

◀ **FIGURE 23.25**
(a) As a fetus, Maitreya is undergoing the most rapid growth rate of her life, and thus her dependence on a healthful diet is as great as it will ever be. (b) As a baby, Maitreya's nutritional needs are still great.

like Maitreya Suchocki in Figure 23.25, must first pass through the lungs and mouth of her mother, which is why it is so vital that her mother eat right and maintain a healthful lifestyle while pregnant. And then, a mere 40 weeks later, her mother's food has been transformed by the actions of Maitreya's DNA into a whole new body ripe for exploring the world around her.

✔ READING CHECK

What is the basis of the lock-and-key model?

Using this model, we can understand why some drugs are more potent than others. Heroin, for example, is a more potent painkiller than is morphine because the chemical structure of heroin allows for tighter binding to its receptor sites.

The lock-and-key model has developed into one of the most important tools of pharmaceutical study. Knowing the precise shape of a target receptor site allows chemists to design molecules that have an optimum fit and a specific biological effect.

Biochemical systems are so complex, however, that our knowledge is still limited, as is our capacity to design effective medicinal drugs. For this reason, most new medicinal drugs are still discovered instead of designed. One important avenue for drug discovery is *ethnobotany*. An *ethnobotanist* is a researcher who learns about the medicinal plants used in indigenous cultures, such as the root of the Bobgunnua tree, shown in Figure 24.3. Today, there are hundreds of clinically useful prescription drugs derived from plants. About three-quarters of these came to the attention of the pharmaceutical industry as a result of their use in folk medicine.

Another important method of drug discovery is the random screening of vast numbers of compounds. Each year, for example, the National Cancer Institute screens some 20,000 compounds for anticancer activity. One successful hit was the compound Taxol, shown in Figure 24.4. This compound has significant activity against several forms of cancer, especially ovarian cancer.

A drug isolated from a natural source is not necessarily better or more gentle than one produced in the laboratory. Aspirin, for example, is a human-made chemical derivative, and it is certainly more

FIGURE 24.3 ▲
Ethnobotanists directed natural-products chemists to the yellow coating on the root of the African Bobgunnua tree. Indigenous people have known for many generations that this coating has medicinal properties. From extracts of the coating, the chemists isolated a compound that is highly effective in treating fungal infections. This compound, produced by the tree to protect itself from root rot, shows much promise in the treatment of the opportunistic fungal infections that plague those suffering from AIDS.

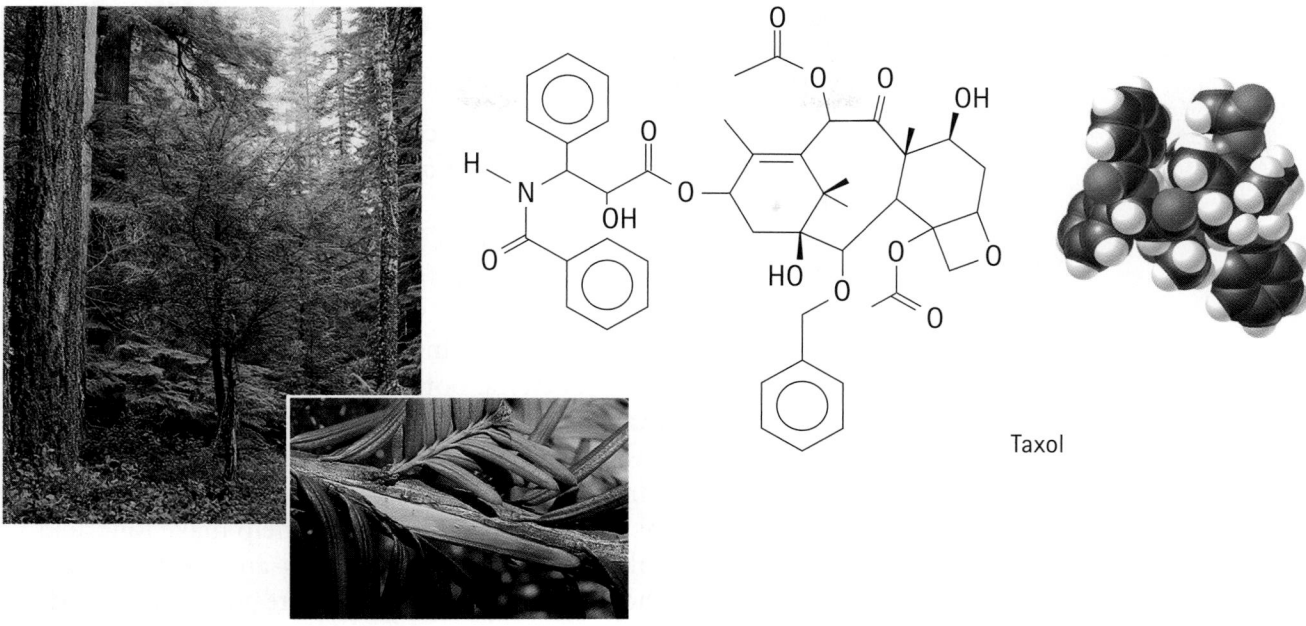

Taxol

FIGURE 24.4 ▲
Originally isolated from the bark of the Pacific yew tree, Taxol is a complex natural product useful in the treatment of various forms of cancer.

gentle than cocaine, which is 100% natural. The main advantage of natural products is their great *diversity*. Each year, more than 3000 new chemical compounds are discovered from plants. Many of these compounds are biologically active, serving the plant as a chemical defense against disease or predators. Nicotine, for example, is a naturally occurring insecticide produced by the tobacco plant to protect itself from insects.

It has been estimated that only 5000 plant species have been studied exhaustively for possible medical applications. This is a minor fraction of the estimated 250,000 to 300,000 plant species on our planet, most of which are located in tropical rain forests. That we know little or nothing about much of the plant kingdom has raised justified and well-publicized concern. For as rain forests are being destroyed, also being destroyed are plant species that might yield useful medicines.

Medicines are just one of many reasons that our planet's rain forests need to be preserved. Ah . . . Home sweet home!

CONCEPT CHECK

Why are chemicals from natural sources so suitable for making drugs?

Check Your Answer

Because there are so many naturally derived chemicals with biological effects. Their vast diversity permits the manufacture of the many different types of medicines needed to combat the many different types of illnesses humans are susceptible to.

24.3 Chemotherapy Cures the Host by Killing the Disease

The use of drugs that destroy disease-causing agents without destroying the animal host is known as **chemotherapy**. This approach is effective in the treatment of many diseases, including bacterial infections. ✓ It works by taking advantage of the ways a disease-causing agent, also known as a *pathogen,* is different from a host.

Developed in the 1930s, *sulfa drugs* were the first drugs used to treat bacterial infections. They work by taking advantage of a striking difference between humans and bacteria. Both humans and bacteria must have the nutrient *folic acid* in order to remain healthy. We humans can obtain folic acid from what we eat. But bacteria cannot absorb folic acid from outside sources. Instead, bacteria must make their own supply of folic acid. For this, they possess receptor sites that help make folic acid from a simpler molecule found in all bacteria, *para*-aminobenzoic acid (PABA). The PABA attaches to the specific receptor site and is converted to folic acid, as shown in Figure 24.5.

Sulfa drugs have a close structural resemblance to PABA. When taken by a person suffering from a bacterial infection, a sulfa drug is transformed by the body into the compound *sulfanilamide,* which attaches to the bacterial receptor sites designed for PABA, as shown in Figure 24.6. This prevents the bacteria from synthesizing folic acid. Without folic acid, the bacteria soon die. The patient, however, lives on because he or she receives folic acid from the diet.

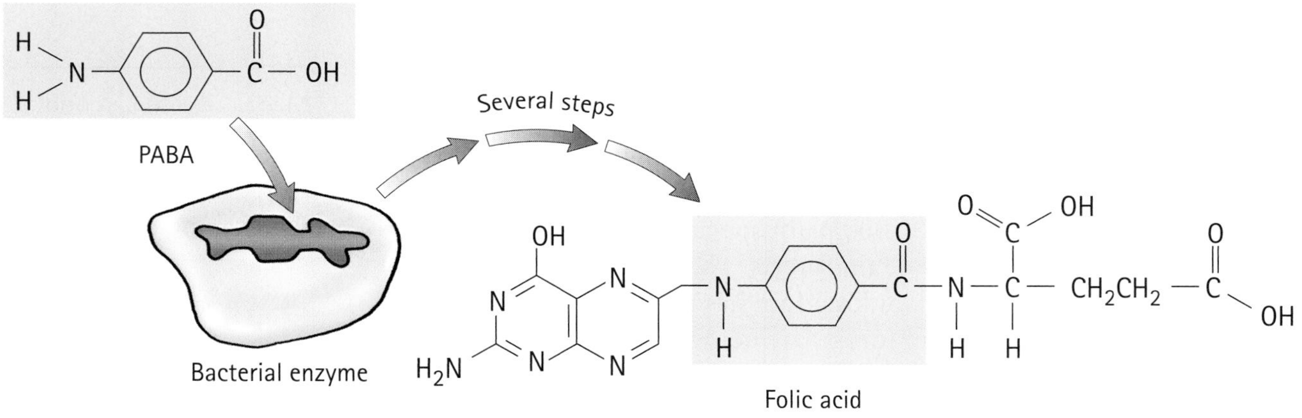

FIGURE 24.5 ▲
Bacterial enzymes use *para*-aminobenzoic acid (PABA) to synthesize folic acid.

CONCEPT CHECK
How is sulfanilamide poisonous to bacteria but not to humans?

Check Your Answer
Sulfanilamide is poisonous to bacteria because it prevents them from synthesizing the folic acid they need to survive. Humans utilize folic acid from their diet and so they are not bothered by sulfanilamide's ability to disrupt the synthesis of folic acid.

FIGURE 24.6
In the body, sulfa drugs are transformed into sulfanilamide, which binds to the bacterial receptor sites and keeps them from doing their job.

Sulfanilamide

PABA

No folic acid; bacteria die

Bacterial enzyme

Antibiotics are chemicals that prevent the growth of bacteria. They are produced by such microorganisms as molds, other fungi, and even bacteria. The first antibiotic discovered was penicillin. Many derivatives of penicillin, such as the penicillin G shown in Figure 24.7, have since been isolated from microorganisms as well as prepared in the laboratory. Penicillins and the closely related compounds known as cephalosporins, also shown in Figure 24.7, kill bacteria by inactivating receptor sites responsible for strengthening the bacterial cell wall. With this receptor site inactivated, bacterial cell walls grow weak and eventually burst.

✔ READING CHECK

How does chemotherapy work?

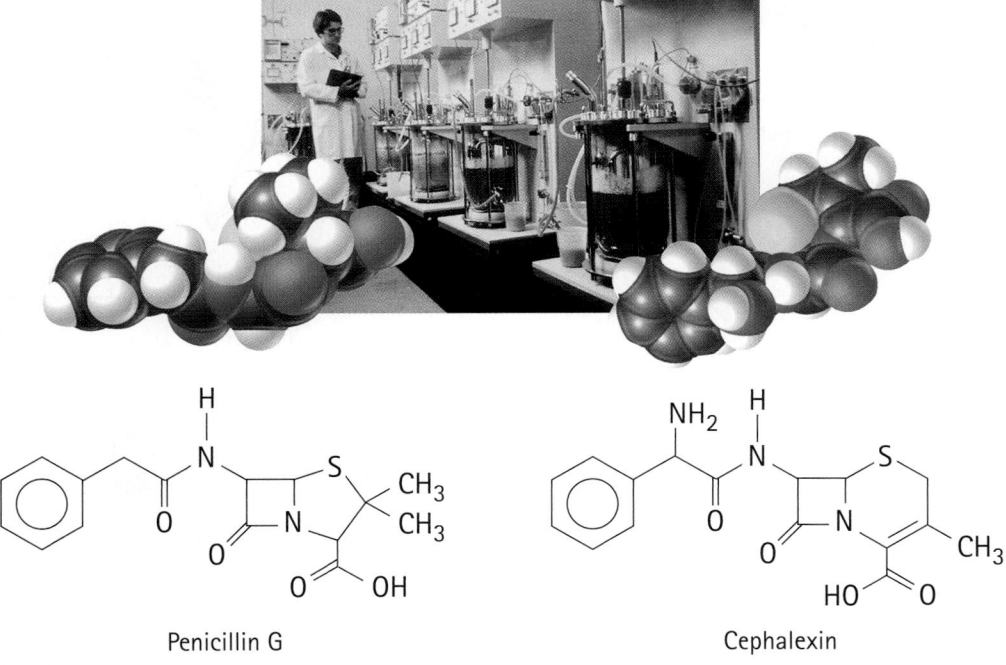

Penicillin G

Cephalexin

FIGURE 24.7 ▲
Penicillins, such as penicillin G, and cephalosporins, such as cephalexin, as well as most other antibiotics, are produced by microorganisms that can be mass-produced in large vats. The antibiotics are then harvested and purified.

As I pump iron, my nerves are pumping sodium ions!

24.4 The Nervous System Is a Network of Neurons

Many drugs function by affecting the nervous system. To understand how these drugs work, it is important to know the basic structure and functions of the nervous system.

Thoughts, physical actions, and sensory input all involve the transmission of electrical signals through the body. The path for these signals is a network of nerve cells, or *neurons.* **Neurons** are specialized cells capable of sending electrical impulses. First, in what is called the *resting phase,* a nerve cell primes itself for an impulse by ejecting sodium ions, as shown in Figure 24.8a. More sodium ions outside the neuron than inside creates a separation of charge. And the separation of charge gives rise to an electric potential of around −70 millivolts (mV) across the cell membrane. ✔ As shown in Figure 24.8b, a nerve impulse is a reversal in this electric potential that travels down the length of the neuron to the *synaptic terminals.* The reversal of the electric potential within an impulse occurs as sodium ions flush back into the neuron.

After the impulse passes a given point along the neuron, the cell again ejects sodium ions at that point to reestablish the original distribution of ions and the −70 mV potential.

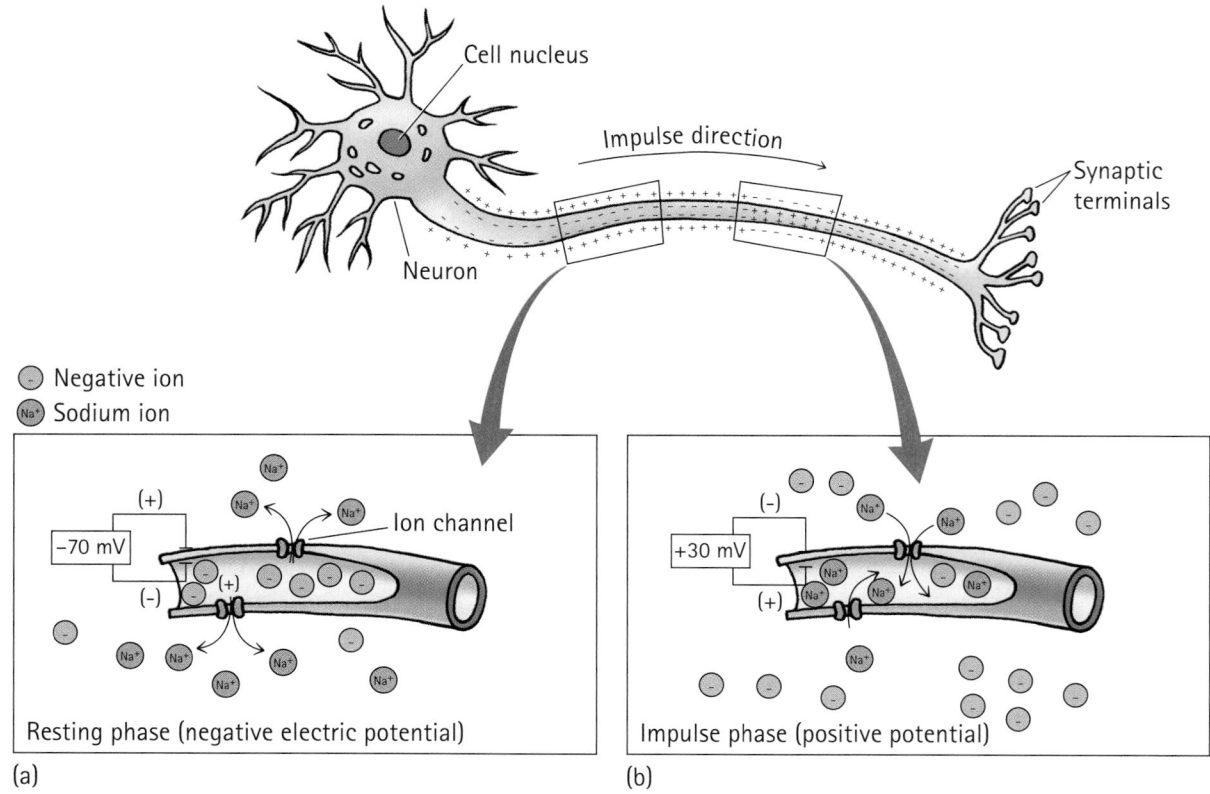

FIGURE 24.8 ▲

(a) The resting phase of a neuron maintains a greater concentration of sodium ions outside the cell. This results in a voltage of about −70 mV. (b) In the impulse phase, sodium ions flush back into the cell to give a voltage of about +30 mV.

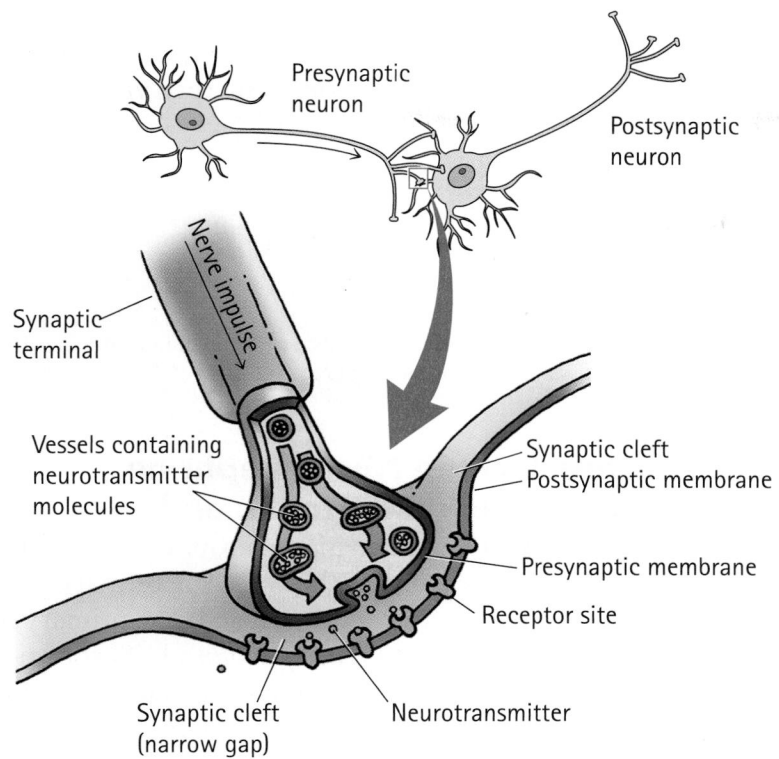

Unlike the wires in an electric circuit, most neurons are not physically connected to one another. Nor are they connected to the muscles or glands on which they act. Rather, as Figure 24.9 shows, they are separated from one another or from a muscle or gland by a narrow gap known as the **synaptic cleft.**

A nerve impulse reaching a synaptic cleft causes bubble-like compartments in the terminal, called *vessels,* to release neurotransmitters into the cleft. **Neurotransmitters** are organic compounds released by a neuron and capable of activating receptor sites.

A neurotransmitter, once released into the synaptic cleft, migrates across the cleft to receptor sites on the opposite side. If the receptor sites are located on a *postsynaptic neuron,* as shown in Figure 24.9, the binding of the neurotransmitter may start a nerve impulse in that neuron. If the receptor sites are located on a muscle or organ, then binding of the neurotransmitter may start a bodily response, such as muscle contraction or the release of hormones.

Two important classes of neurons include the *stress* and *maintenance* neurons. Both types are always firing. But in times of stress, as when facing an angry bear or giving a speech, the stress neurons are more active than the maintenance neurons. This condition is the *fight-or-flight* response, during which fear causes stress neurons to trigger rapid bodily changes to help defend against impending danger: The mind becomes alert, air passages in the nose and lungs open to bring in more oxygen, the heart beats faster to spread the oxygenated blood throughout the body, and nonessential activities such as digestion are temporarily stopped. In times of relaxation, such as sitting down in front of the television with a bowl of potato

✔ **READING CHECK**

What is a nerve impulse?

chips, the maintenance neurons are more active than the stress neurons. Under these conditions, digestive juices are secreted, intestinal muscles push food through the gut, the pupils constrict to sharpen vision, and the heart pulses at a minimal rate.

CONCEPT CHECK
What is a neurotransmitter?

Check Your Answer
A neurotransmitter is a small organic molecule released by a neuron. It influences neighboring tissues, such as nerve membranes, by binding to receptor sites.

Norepinephrine

Acetylcholine

FIGURE 24.10 ▲
The chemical structures of the stress neurotransmitter norepinephrine and the maintenance neurotransmitter acetylcholine.

Neurotransmitters Include Norepinephrine, Acetylcholine, Dopamine, Serotonin, and GABA

On the chemical level, stress and maintenance neurons can be distinguished by the types of neurotransmitters they use. The primary neurotransmitter for stress neurons is *norepinephrine*. The primary neurotransmitter for maintenance neurons is *acetylcholine*, both shown in Figure 24.10. As we shall see in the following sections, many drugs function by altering the balance of stress and maintenance neuron activity.

In addition to norepinephrine and acetylcholine, a host of other neurotransmitters contribute to a broad range of effects. Three examples are the neurotransmitters dopamine, serotonin, and gamma aminobutyric acid, shown in Figure 24.11.

Dopamine plays a significant role in activating the brain's reward center. The brain's reward center is located in the hypothalamus. This is an area at the lower middle of the brain, as illustrated in Figure 24.12. The hypothalamus is the main control center for the involuntary part of the peripheral nervous system and for emotional response and behavior. Stimulation of the reward center by dopamine results in a pleasurable sense of *euphoria*, which is an exaggerated sense of well-being.

Serotonin is the neurotransmitter used by the brain to block unneeded nerve impulses. To make sense of the world, the frontal lobes of the brain selectively block out a multitude of signals coming

Dopamine

Serotonin

Gamma aminobutyric acid (GABA)

FIGURE 24.11 ▲
The chemical structures of three neurotransmitters important to the central nervous system.

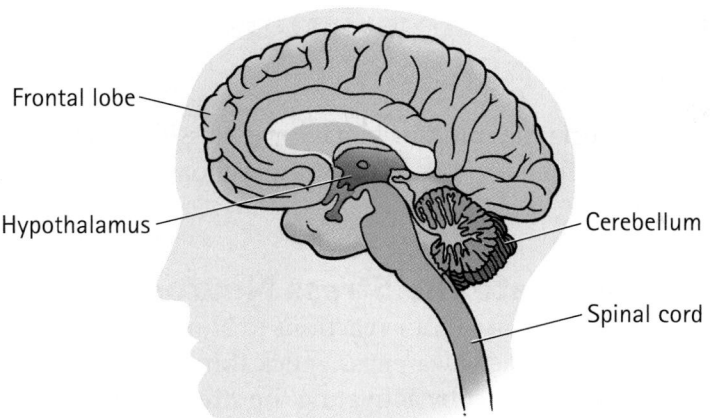

Frontal lobe

Hypothalamus

Cerebellum

Spinal cord

from the lower brain and from the peripheral nervous system. We are not born with this ability to selectively block out information. In order to have an appropriate focus on the world, newborns must learn from experience which lights, sounds, smells, and feelings outside and inside their bodies must be dampened. A healthy, mature brain is one in which serotonin successfully suppresses lower-brain nerve signals. Information that does make it to the higher brain can then be sorted efficiently.

Drugs such as LSD that modify the action of serotonin alter the brain's ability to sort information, and this alters perception. While hallucinating, for example, an LSD user rarely sees something that isn't there. Rather, the user has an altered perception of something that does exist.

The control of physical responses ultimately allows us to perform such complex tasks as driving a car or playing the piano. The control of emotional responses allows us to refine our behavior, such as overcoming anxiety in tense social interactions or remaining calm in an emergency. The brain controls both physical and emotional responses by inhibiting the transmission of nerve impulses. The neurotransmitter responsible for this inhibition—*gamma aminobutyric acid* (GABA)—is *the* major inhibitory neurotransmitter of the brain. Without it, coordinated movements and emotional skills would not be possible.

All these bizarre neurotransmitters are what help me to be who I am!

CONCEPT CHECK

Match the neurotransmitter to its primary function:

_____ norepinephrine
_____ acetylcholine
_____ dopamine
_____ serotonin
_____ GABA

a. inhibits nerve transmission
b. stimulates reward center
c. selectively blocks nerve impulses
d. maintains stressed state
e. maintains relaxed state

Check Your Answers
d, e, b, c, a

24.5 Psychoactive Drugs Alter the Mind or Behavior

Any drug that affects the mind or behavior is classified as **psychoactive.** In this chapter, we focus on two classes of psychoactive drugs: stimulants and depressants.

Stimulants Activate the Stress Neurons

By enhancing the intensity of our reactions to stimuli, *stimulants* cause brief periods of heightened awareness, quick thinking, and elevated mood. They exert this effect by activating the stress neurons. Four widely recognized stimulants are amphetamines, cocaine, caffeine, and nicotine.

Amphetamines are a family of stimulants that include the parent compound *amphetamine* (also known as "speed") and such derivatives as methamphetamine and pseudoephedrine. As you can see by comparing Figure 24.13 with Figures 24.10 and 24.11, these drugs are structurally similar to the neurotransmitters norepinephrine and dopamine. Amphetamines bind to receptor sites for these neurotransmitters. So amphetamines mimic many of the effects of norepinephrine and dopamine on the stress neurons, including the fight-or-flight response and the ability to give a person a sense of euphoria.

The stimulating and mood-altering effects of amphetamines give them a high abuse potential. Side effects include insomnia, irritability, loss of appetite, and paranoia. Amphetamines take a particularly hard toll on the heart. Hyperactive heart muscles are prone to tearing. Subsequent scarring of tissue ultimately leads to a weaker heart. Furthermore, amphetamines cause blood vessels to constrict and blood pressure to rise. And these conditions increase the likelihood of heart attack or stroke, especially for the one person out of four whose blood pressure is already high.

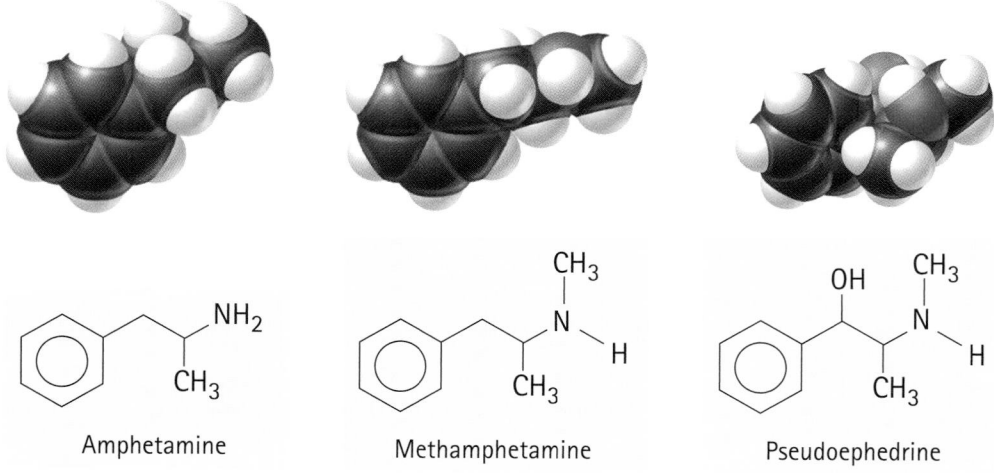

FIGURE 24.13 ▲
Amphetamines are a family of compounds structurally related to the neurotransmitters norepinephrine and dopamine.

One of the more notorious and abused stimulants is *cocaine,* a natural product isolated from the South American coca plant, shown in Figure 24.14. Once in the bloodstream, cocaine produces a sense of euphoria and increased stamina. It is also a powerful local anesthetic when applied topically. Within a few decades of its first isolation from plant material in 1860, cocaine was used as a local anesthetic for eye surgery and dentistry. This practice was stopped once safer local anesthetics were discovered in the early 1900s.

Cocaine and amphetamines share a similar profile of addictiveness, though cocaine's addictive properties are more intense. The cocaine that is inhaled nasally is the hydrochloride salt. The free-base form of cocaine, called *crack cocaine,* is also abused. As with the street drug "ice," which is the free-base form of methamphetamine, crack cocaine is volatile. It may be smoked for what is an intense but profoundly dangerous and addictive high.

✓ Drug addiction is not completely understood, but scientists do know that it involves both physical and psychological dependence. **Physical dependence** is the need to continue taking the drug to avoid withdrawal symptoms. For amphetamines, drug withdrawal symptoms include depression, fatigue, and a strong desire to eat. **Psychological dependence** is the *craving* to continue drug use. This craving may be the most serious and deep-rooted aspect of addiction. It can persist even after withdrawal from physical dependence, frequently leading to renewed drug-seeking behavior.

Long-term cocaine or amphetamine abuse leads to a deterioration of the nervous system. The body recognizes the excessive stimulatory actions produced by these drugs. To deal with the overstimulation, the body creates more depressant receptor sites for neurotransmitters that inhibit nerve transmission. A tolerance for the drugs therefore develops. Then, to receive the same stimulatory effect, the abuser is forced to increase the dose. And this induces the body to create even more depressant receptor sites. The end result over the long term is that the abuser's natural levels of dopamine and norepinephrine are insufficient to compensate for the excessive number of depressant sites. Lasting personality changes are thus often observed. Addicts, even when recovered, often report feelings of psychological depression.

Everything changes, including your body's response to drugs that are used for a long time.

FIGURE 24.14 ▶
The South American coca plant has been used by indigenous cultures for many years in religious ceremonies and as an aid to staying awake on long hunting trips. Leaves are either chewed or ground to a powder that is inhaled nasally.

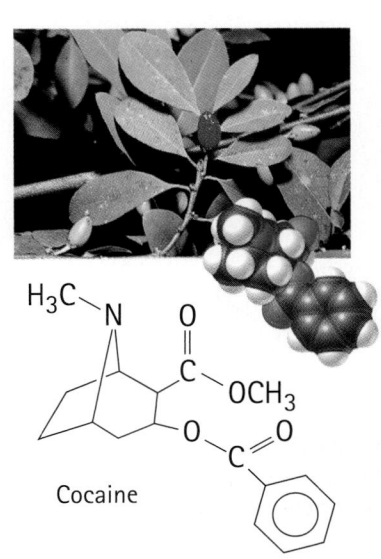

Cocaine

CONCEPT CHECK
How do amphetamines and cocaine exert their effects?

Check Your Answer
Amphetamines and cocaine in the synaptic cleft both cause an overstimulation of receptor sites for norepinephrine and dopamine.

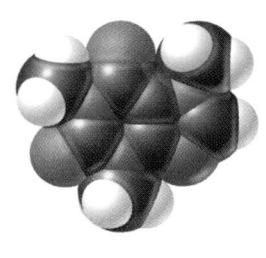

Caffeine

FIGURE 24.15 ▲
A coffee plant with its ripening caffeine-containing beans.

A much milder and legal stimulant is *caffeine*, depicted in Figure 24.15. A number of mechanisms have been proposed for caffeine's stimulatory effects. Perhaps the most straightforward mechanism is that caffeine facilitates the release of norepinephrine into synaptic clefts. Caffeine also exerts many other effects on the body, such as dilation of arteries, relaxation of bronchial and gastrointestinal muscles, and stimulation of stomach-acid secretion.

The caffeine people ingest comes from various natural sources, including coffee beans, teas, kolanuts, and cocoa beans. Kolanut extracts are used for making cola drinks, and cocoa beans (not to be confused with the cocaine-producing coca plant) are roasted and then ground to a paste used for making chocolate. Caffeine is relatively easy to remove from these natural products by using high-pressure carbon dioxide, which selectively dissolves the caffeine. This allows for the economical production of "decaffeinated" beverages, many of which, however, still contain small amounts of caffeine. Interestingly, cola drink manufacturers use decaffeinated kolanut extract in their beverages. The caffeine is added in a separate step to guarantee a particular caffeine concentration. In the United States, about 2 million pounds of caffeine is added to soft drinks each year. Table 24.2 shows the caffeine content of various commercial products. For comparison, the maximum daily dose of caffeine tolerable by most adults is about 1500 milligrams.

Another legal, but far more toxic, stimulant is *nicotine.* As noted earlier, tobacco plants produce nicotine as a chemical defense against insects. This compound is so potent that a lethal dose in humans is only about 60 milligrams. A single cigarette may contain up to 5 milligrams of nicotine. Most of it is destroyed by the heat of the burning embers, however, so that less than 1 milligram is typically inhaled by the smoker.

Nicotine and the neurotransmitter acetylcholine, which acts on maintenance neurons, have similar structures, as Figure 24.16 illustrates. Nicotine molecules are therefore able to bind to acetylcholine receptor sites and trigger many of acetylcholine's effects, including relaxation and increased digestion. This explains the tendency of smokers to smoke after eating meals. In addition, acetylcholine is used for muscle

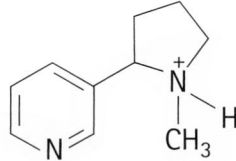

Nicotine

Acetylcholine

FIGURE 24.16 ▲
Nicotine is able to bind to receptor sites for acetylcholine because of structural similarities.

TABLE 24.2 Approximate Caffeine Content of Various Products

Product	Caffeine Content
Brewed coffee	100–150 mg/cup
Instant coffee	50–100 mg/cup
Decaffeinated coffee	2–10 mg/cup
Black tea	50–150 mg/cup
Cola drink	35–55 mg/12 oz
Chocolate bar	1–2 mg/oz
Over-the-counter stimulant	100 mg/dose
Over-the-counter analgesic	30–60 mg/dose

contraction, and so the smoker may also experience some muscle stimulation immediately after smoking. After these initial responses, however, nicotine molecules remain bound to the acetylcholine receptor sites. This blocks acetylcholine molecules from binding. The result is that the activity of these neurons is depressed.

Recall that maintenance neurons and stress neurons are both always working. Thus inhibiting the activity of one type makes the other type more effective. So, as nicotine depresses the maintenance neurons, it favors the stress neurons. This raises the smoker's blood pressure and stresses the heart.

Animal studies show inhaled nicotine to be about six times more addictive than injected heroin. Because nicotine leaves the body quickly, withdrawal symptoms begin about 1 hour after a cigarette is smoked, which means the smoker is inclined to light up frequently.

Figure 24.17 shows what a smoker's lungs look like. In the United States, about 450,000 individuals die each year from such tobacco-related health problems as emphysema, heart failure, and various forms of cancer, especially lung cancer, which is brought on primarily by tobacco's tar component. Some relief from the addiction can be obtained with nicotine chewing gum and nicotine skin patches. In order for any method to be effective, however, the smoker must first genuinely desire to quit smoking.

READING CHECK

What are the two types of dependencies involved with drug addiction?

CONCEPT CHECK

Caffeine and nicotine both add stress to the nervous system, but they do so by different means. Briefly describe the difference.

Check Your Answer

Caffeine stimulates the release of the stress neurotransmitter norepinephrine, and nicotine both depresses the action of the maintenance neurotransmitter acetylcholine and enhances the release of norepinephrine.

Tobacco field

Tobacco curing on racks

Cigarette manufacture

User

Blackened lungs

FIGURE 24.17 ▲
The path of tobacco from the field to a smoker's lungs. According to the World Health Organization, of the more than 1.3 billion smokers alive today, about 650 million will eventually die from their use of tobacco.

Depressants Inhibit the Ability of Neurons to Conduct Impulses

Depressants are a class of drugs that inhibit the ability of neurons to conduct impulses. Two commonly used depressants are ethanol and benzodiazepines.

Ethanol, also known simply as *alcohol,* is by far the most widely used depressant. Its structure is shown in Figure 24.18. In the United States,

FIGURE 24.18 ▶
One of the initial effects of alcohol is a depression of social inhibitions, which can serve to bolster mood. Alcohol is not a stimulant, however. From the first sip to the last, body systems are being depressed.

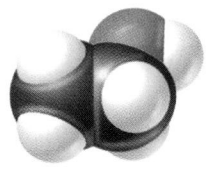

$CH_3CH_2 — OH$
Ethanol

about a third of the population, or about 100 million people, drink alcohol. It is well established that alcohol consumption leads to about 150,000 deaths each year in the United States. The causes of these deaths are overdoses of alcohol alone, overdoses of alcohol combined with other depressants, alcohol-induced violent crime, cirrhosis of the liver, and alcohol-related traffic accidents.

Benzodiazepines are a potent class of antianxiety agents. Compared to many other types of depressants, benzodiazepines are relatively safe and rarely produce cardiovascular and respiratory depression. Their antianxiety effects were identified in 1957 by chance. During a routine laboratory clean-up, a synthesized compound that had been sitting on the shelf for two years was submitted for routine testing despite the fact that compounds thought to have similar structures had shown no promising pharmacologic activity. This particular compound, however, shown in Figure 24.19 and now known as chlordiazepoxide, contained an unexpected seven-membered ring. Chlordiazepoxide showed a significant calming effect in humans, and by 1960 was marketed under the trade name Librium® as an antianxiety agent. Shortly thereafter, a derivative, diazepam, was found to be five to ten times more potent than Librium. In 1963, diazepam hit the market under the trade name Valium®.

A primary way in which alcohol and benzodiazepines exert their depressant effect is by enhancing the action of GABA. As shown in Figure 24.20, GABA keeps electrical impulses from passing through a neuron by binding to a receptor site on a channel that penetrates the cell membrane of the neuron. Figure 24.20a shows that when GABA binds to the receptor site, the channel opens, allowing chloride ions to

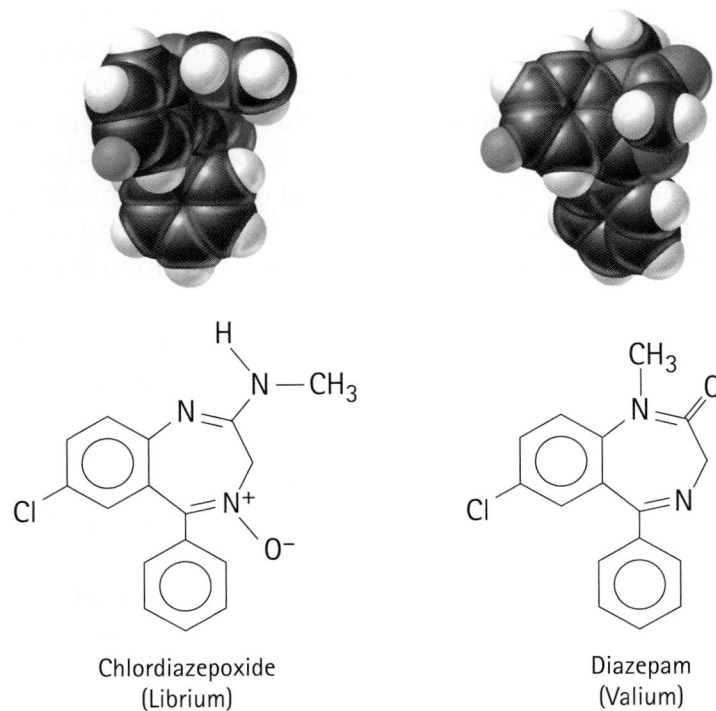

Chlordiazepoxide
(Librium)

Diazepam
(Valium)

◀ **FIGURE 24.19**
The benzodiazepines Librium and Valium.

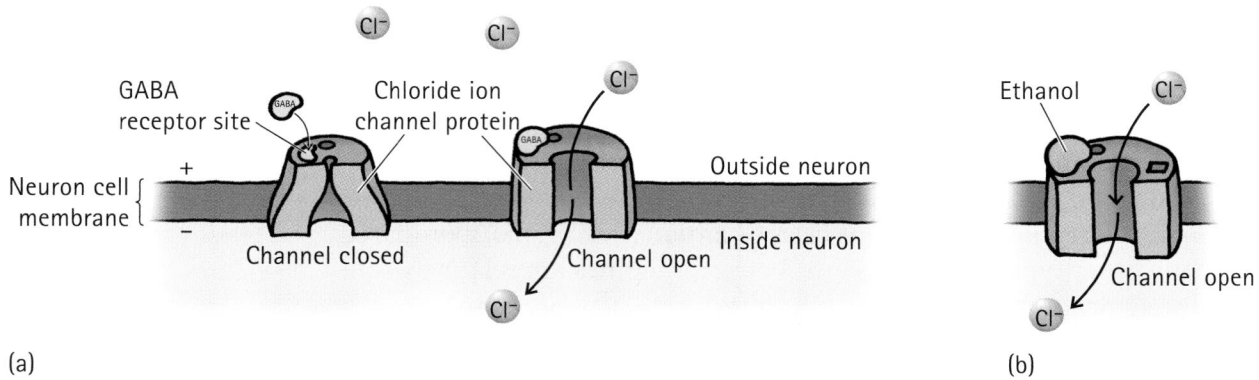

FIGURE 24.20 ▲
(a) When GABA binds to its receptor site, a channel opens to allow negatively charged chloride ions into the neuron. The high concentration of negative ions inside the neuron prevents the electric potential from reversing from negative to positive. Because that reversal is necessary if an impulse is to travel through a neuron, no impulse can move through the neuron.
(b) Ethanol mimics GABA by binding to GABA receptor sites.

migrate into the neuron. The resulting negative charge buildup in the neuron maintains the negative electric potential across the cell membrane. This inhibits a reversal to a positive potential and prevents an impulse from traveling along the neuron. (If you are confused, go back and review Figure 24.8 and the text describing it. Perhaps you now know why doctors spend many years hitting the books in their medical training.)

Ethanol mimics the effect of GABA by binding to GABA receptor sites. This allows chloride ions to enter the neuron, as shown in Figure 24.20b. The effect of alcohol is dose-dependent, which means that the greater the amount drunk, the greater the effect. At small concentrations, few chloride ions are permitted into the neuron; these low concentrations of ions decrease inhibitions, alter judgment, and impair muscle control. As the person continues to drink and the chloride ion concentration inside the neuron rises, both reflexes and consciousness diminish, eventually to the point of coma and then death.

Figure 24.21 illustrates how benzodiazepines exert their depressant effects by binding to receptor sites located adjacent to GABA receptor sites. Benzodiazepine binding merely helps GABA bind. Because benzodiazepine doesn't directly open chloride-ion channels, overdoses of this compound are less hazardous than alternatives. This makes the benzodiazepines the drugs of choice for treating symptoms of anxiety.

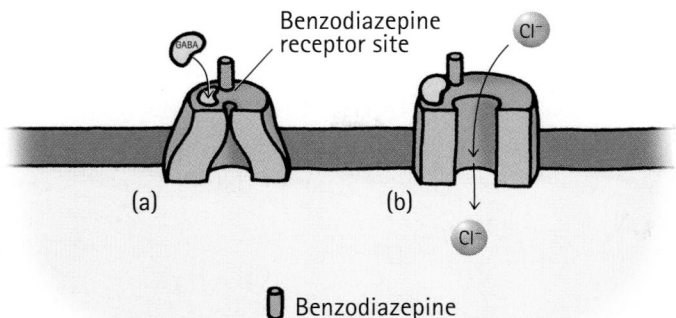

FIGURE 24.21 ▲
The receptor sites for benzodiazepines are adjacent to GABA receptor sites.
(a) Benzodiazepines cannot open up the chloride channel on their own.
(b) Rather, benzodiazepines help GABA in its channel-opening task.

> **CONCEPT CHECK**
>
> Does the activity of a neuron increase or decrease as chloride ions are allowed to pass into it?
>
> **Check Your Answer**
>
> Chloride ions inside the neuron help to maintain the negative electric potential. This inhibits the neuron from being able to conduct an impulse (see Figure 24.8). Chloride ions, therefore, decrease the activity of a neuron.

24.6 Pain Relievers Inhibit the Transmission or Perception of Pain

Physical pain is a complex body response to injury. On the cellular level, pain-inducing biochemicals are rapidly synthesized at the site of injury, where they initiate swelling, inflammation, and other responses that get your body's attention. These pain signals are sent through the nervous system to the brain, where the pain is perceived. To alleviate pain, drugs act at various stages of this process, as shown in Figure 24.22.

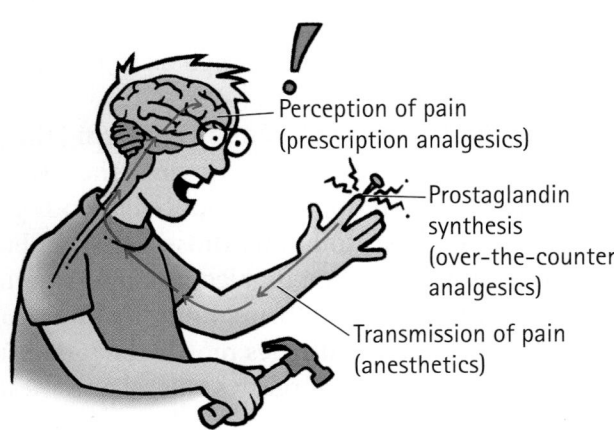

◀ **FIGURE 24.22**
Injury to tissue causes the transmission of pain signals to the brain. Pain relievers prevent this transmission, inhibit the inflammation response, or dampen the brain's ability to perceive the pain.

FIGURE 24.23 ▶
Local anesthetics have similar structural features, including an aromatic ring, an intermediate chain, and an amine group. Ask your dentist which ones he or she uses for your treatment.

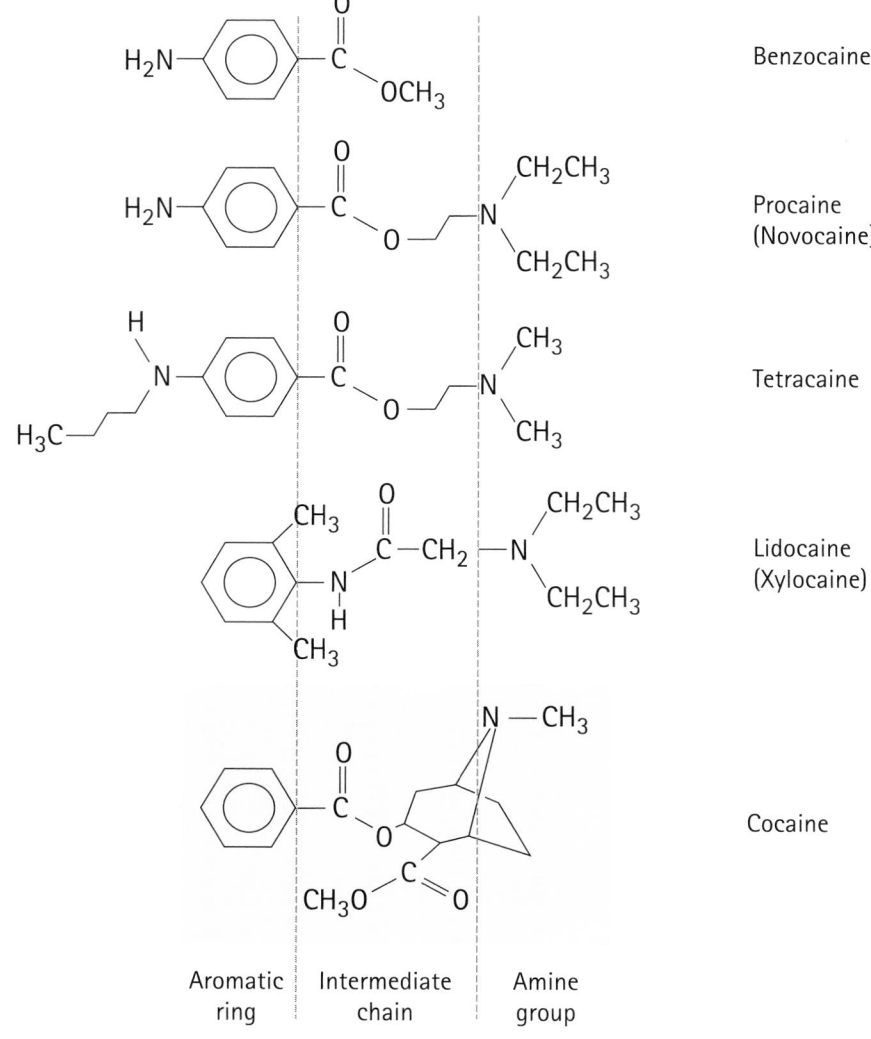

Benzocaine

Procaine (Novocaine)

Tetracaine

Lidocaine (Xylocaine)

Cocaine

| Aromatic ring | Intermediate chain | Amine group |

Sevoflurane

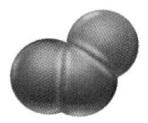

Nitrous oxide

FIGURE 24.24 ▲
The chemical structures of sevoflurane and nitrous oxide.

Anesthetics prevent neurons from transmitting sensations to the brain. *Local anesthetics* are applied either topically to numb the skin or by injection to numb deeper tissues. These mild anesthetics are useful for minor surgical or dental procedures. As described earlier, cocaine was the first medically used local anesthetic. Others having fewer side effects soon followed, such as the ones shown in Figure 24.23.

✔ A *general anesthetic* blocks out pain by rendering the patient unconscious. As discussed in Section 22.4, diethyl ether was one of the first general anesthetics. Two of the more popular gaseous general anesthetics used by anesthesiologists today are those shown in Figure 24.24, sevoflurane and nitrous oxide. When inhaled, these compounds enter the bloodstream and are distributed throughout the body. At certain blood concentrations, general anesthetics render the individual unconscious, which is useful for invasive surgery. General anesthesia must be monitored very carefully, however, so as to avoid a major shutdown of the nervous system and subsequent death.

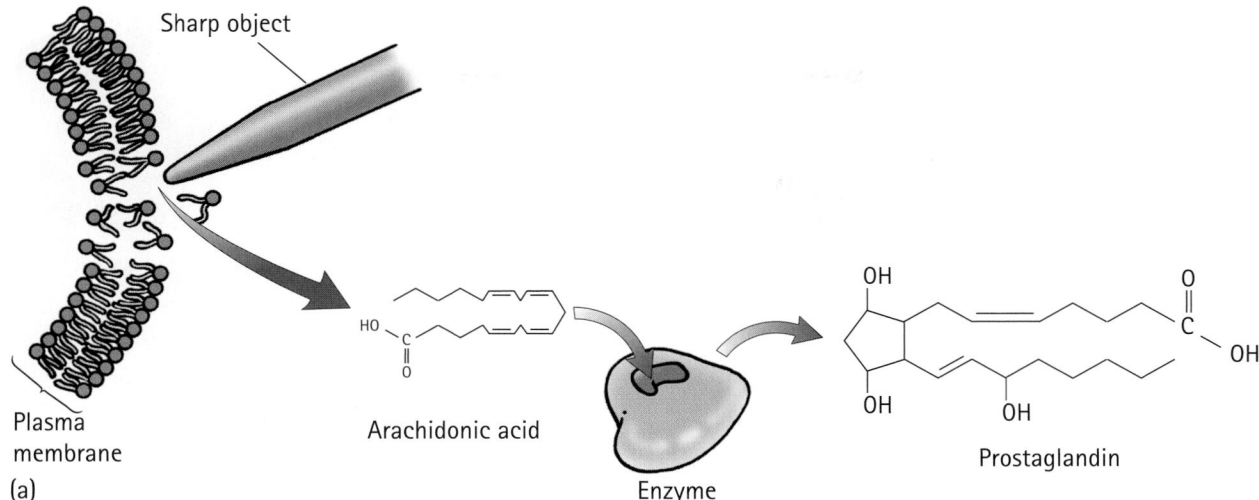

FIGURE 24.25 ▲

(a) Prostaglandins, which cause pain signals to be sent to the brain, are synthesized by the body in response to injury. The starting material for all prostaglandins is arachidonic acid, which is found in the membrane of all cells. Arachidonic acid is transformed to prostaglandins with the help of a receptor site. There are a variety of prostaglandins, each having its own effect, but all have a chemical structure resembling the one shown here.
(b) Analgesics inhibit the synthesis of prostaglandins by binding to the arachidonic acid receptor site. With no prostaglandins, no pain signals are generated.

Analgesics are a class of drugs that enhance our ability to tolerate pain without abolishing nerve sensations. Over-the-counter analgesics, such as aspirin, ibuprofen, and acetaminophen, inhibit the formation of *prostaglandins*. As Figure 24.25 illustrates, prostaglandins are biochemicals the body quickly synthesizes to generate pain signals. These analgesics also reduce fever because of the role prostaglandins play in raising body temperature. In addition to reducing pain and fever, aspirin and ibuprofen act as anti-inflammatory agents because they block the formation of a certain type of prostaglandin responsible for inflammation. Acetaminophen does not act on inflammation. These three analgesics are shown in Figure 24.26.

The more potent opioid analgesics—morphine, codeine, and heroin (Figure 24.1)—moderate the brain's perception of pain by binding to

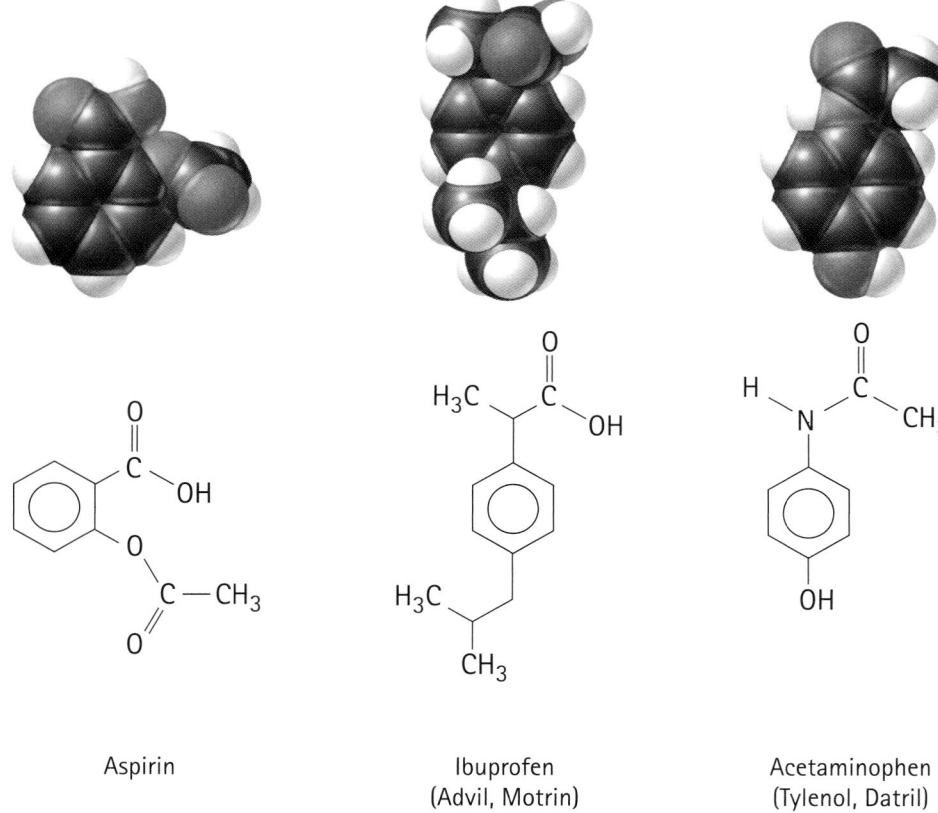

Aspirin

Ibuprofen
(Advil, Motrin)

Acetaminophen
(Tylenol, Datril)

FIGURE 24.26 ▲
Aspirin and ibuprofen block the formation of prostaglandins responsible for pain, fever, and inflammation. Acetaminophen blocks the formation only of prostaglandins responsible for pain and fever.

receptor sites on neurons in the central nervous system, which includes the brain and spinal column. Initial discovery of these receptor sites raised the question of why they exist. Perhaps, it was hypothesized, opioids mimic the action of a naturally occurring brain chemical. *Endorphins,* a group of large biomolecules that have strong opioid activity, were subsequently isolated from brain tissue. It has been suggested that endorphins evolved as a means of suppressing awareness of pain that would otherwise be incapacitating in life-threatening situations. The "runner's high" experienced by many athletes after a vigorous workout is caused by endorphins.

Endorphins are also implicated in the *placebo effect,* in which patients experience a reduction in pain after taking what they believe is a drug but is actually a sugar pill. (A *placebo* is any inactive substance used as a control in a scientific experiment.) Through the placebo effect, it is the patients' belief in the effectiveness of a medicine rather than the medicine itself that leads to pain relief. The involvement of endorphins in the placebo effect has been demonstrated by replacing the sugar pills with drugs that block opioids or endorphins from binding to their receptor sites. Under these circumstances, the placebo effect vanishes.

In addition to acting as analgesics, opioids can induce euphoria, which is why they are so frequently abused. With repeated use, the body develops a tolerance to these drugs: More and more must be administered to achieve the same effect. Abusers also become physically dependent on opioids, which means they must continue to take the opioids to avoid severe withdrawal symptoms, such as chills, sweating, stiffness, abdominal cramps, vomiting, weight loss, and anxiety. Interestingly, when opioids are used primarily for pain relief rather than for pleasure, the withdrawal symptoms are much less dramatic—especially when the patient does not know he or she has been on these drugs.

✔ READING CHECK

How does a general anesthetic knock out pain?

CONCEPT CHECK
Distinguish between an anesthetic and an analgesic.

Check Your Answer
An anesthetic blocks pain signals from reaching the brain. An analgesic facilitates the ability to manage pain signals once they are received by the brain.

Perhaps nowhere is the impact of chemistry on society more evident than in the development of drugs. On the whole, they have increased our lifespan and improved our quality of living. They have also presented us with a number of ethical and social questions. How do we care for an expanding elderly population? What drugs, if any, should be permissible for recreational use? How do we deal with drug addiction—as a crime, as a disease, or both? As we continue to learn more about ourselves and our ills, we can be sure that more powerful drugs will become available. All drugs, however, carry certain risks that we should be aware of. As most physicians would point out, drugs offer many benefits, but they are no substitute for a healthy lifestyle and preventative approaches to medicine.

Take good care of your body and your body takes good care of you. That's good chemistry!

24 CHAPTER REVIEW

KEY TERMS

Analgesic A drug that enhances the ability to tolerate pain without abolishing nerve sensations.

Anesthetic A drug that prevents neurons from transmitting sensations to the brain.

Chemotherapy The use of drugs to destroy pathogens without destroying the animal host.

Lock-and-key model A model that explains how drugs interact with receptor sites.

Neuron A specialized cell capable of receiving and sending electrical impulses.

Neurotransmitter An organic compound capable of activating receptor sites on proteins embedded in the membrane of a neuron.

Physical dependence A dependence characterized by the need to continue taking a drug to avoid withdrawal symptoms.

Psychoactive Said of a drug that affects the mind or behavior.

Psychological dependence A deep-rooted craving for a drug.

Synaptic cleft A narrow gap across which neurotransmitters pass either from one neuron to the next or from a neuron to a muscle or gland.

Synergistic effect One drug enhancing the effect of another.

REVIEW QUESTIONS

Medicines Are Drugs That Benefit the Body

1. What are the three origins of drugs?
2. Are a drug's side effects necessarily bad?
3. What is the synergistic effect?

The Lock-and-Key Model Guides Chemists in Creating New Medicines

4. In the lock-and-key model, is a drug viewed as the lock or the key?
5. What holds a drug to its receptor site?

Chemotherapy Cures the Host by Killing the Disease

6. Why do bacteria need PABA but humans can do without it?
7. How does penicillin G cure bacterial infections?

The Nervous System Is a Network of Neurons

8. How does a neuron maintain an electric potential difference across its membrane?
9. What ion is pumped out of the neuron to create a resting phase?
10. What are some of the things going on in the body when maintenance neurons are more active than stress neurons?
11. What neurotransmitter functions most in the brain's reward center?
12. What is the role of GABA in the nervous system?

Psychoactive Drugs Alter the Mind or Behavior

13. How is psychological dependence distinguished from physical dependence?
14. What is one mechanism for how caffeine stimulates the nervous system?
15. What neurotransmitter does nicotine mimic?
16. What drugs enhance the action of GABA?

Pain Relievers Inhibit the Transmission or Perception of Pain

17. What is an anesthetic?

18. What is an analgesic?

19. Where are the major opioid receptor sites located?

20. What biochemical is thought to be responsible for the placebo effect?

THINK AND DISCUSS

1. Medicines, such as pain relievers and antidepressants, are being found in the drinking water supplies of many municipalities. How did these medicines get there? Does it matter that they are there? Should something be done about it? If so, what?

2. Alcohol-free and caffeine-free beverages have been quite successful in the marketplace, while nicotine-free tobacco products have yet to be introduced. Speculate about possible reasons.

3. Would making tobacco illegal help or hurt people trying to kick the habit of using tobacco products such as cigarettes? What unintended consequences might arise from the prohibition of tobacco? How can society best convince people NOT to pick up the tobacco habit?

4. Why might someone find it more challenging to take a drug for a mental illness versus a physical illness?

5. Should you be permitted access to any medicine that might save your life? What if you had no health insurance and could not afford the medicine? What if you were an impoverished citizen of Bangladesh? What should be the moral, social, and economic responsibilities of a company that develops and produces medicines?

THINK AND EXPLAIN

1. Aspirin can cure a headache, but when you pop an aspirin tablet, how does the aspirin know to go to your head rather than your big toe?

2. Which is better for you: a drug that is a natural product or one that is synthetic?

3. When is a drug overdose most likely to happen?

4. How does chemotherapy work to fight a disease or infection?

5. How do some molds or fungi protect themselves from bacterial infections?

6. Would formulating a sulfa drug with PABA be likely to increase or decrease its antibacterial properties?

7. What is the main difference between stress neurons and maintenance neurons?

8. What is an advantage of synaptic clefts between neurons rather than direct connections?

9. How is psychological dependence distinguished from physical dependence?

10. How is a drug addict's addiction similar to our need for food? How is it different?

11. When a neuron is at rest there is a tiny electric potential across the neuron's membrane. How was this electric potential created?

12. What flushes back into a neuron as a nerve impulse passes through the neuron?

13. What are the symptoms that a person's stress neurons have been activated?

14. Why do heavy drinkers have a greater tolerance for alcohol?

15. An excess of which of the neurotransmitters discussed in this chapter would mostly readily hinder your ability to perform a complicated physical performance, such as a ballet dance?

16. Nicotine solutions are available from lawn and garden stores as an insecticide. Why must gardeners handle this product with extreme care?

17. A variety of gaseous compounds behave as general anesthetics even though their structures have very little in common. Does this support the role of a receptor site for their mode of action?

18. How might the structure of benzocaine be modified to create a compound having greater anesthetic properties?

19. Which is the more appropriate statement: Opioids have endorphin activity, or endorphins have opioid activity? Explain your answer.

20. A person may feel more relaxed after smoking a cigarette, but his or her heart is actually stressed. Why?

READINESS ASSURANCE TEST (RAT)

If you have a good handle on this chapter—if you really do—then you should be able to score 7 out of 10 on this RAT. Check your answers with your teacher. If you score less than 7, study further before moving on.

Choose the best answer to the following.

1. Which of the following is true about a medicine?

  (a) It is a drug that provides a euphoric effect.

  (b) It is a drug that kills bacteria.

  (c) It is a drug that is isolated from a plant.

  (d) It is a drug that has a therapeutic properties.

  (e) None of the above.

2. Which of the following would be an example of a synergistic effect?
   (a) Penicillin kills infectious bacteria and beneficial bacteria in the intestine.
   (b) Antidepressants and cold medicine together can lead to seizures.
   (c) Caffeine suppresses the appetite and causes the jitters.
   (d) Aspirin reduces fever and thins the blood.
   (e) None of the above.

3. What are side effects?
   (a) They are a multiplying effect seen in a drug due to other interactions.
   (b) They are a beneficial aspect of a drug that has not been exploited.
   (c) Any behavior of a drug that is opposite of its primary function.
   (d) Any behavior of a drug that is not fulfilling its primary function.
   (e) None of the above.

4. Which of the following would not be part of the drug discovery process?
   (a) Examining folklore for herbs and potions
   (b) Changing an old drug a little and testing the new drug for activity
   (c) Learning the exact shape of various receptor sites
   (d) Random testing of new compounds for drug activity
   (e) All of the above are part of the process.

5. What is the main difference between bacteria and humans that gives sulfa drugs their antibacterial properties?
   (a) Bacteria cells can readily absorb sulfa drugs; human cells cannot.
   (b) Human cells can readily absorb sulfa drugs; bacteria cells cannot.
   (c) Bacteria cells can readily absorb folic acid; human cells cannot.
   (d) Human cells can readily absorb folic acid; bacteria cells cannot.
   (e) None of the above.

6. How is a set of neurons different from a wire conductor?
   (a) The wire can carry more information.
   (b) Neurons do not need to physically touch to conduct.
   (c) One conducts electricity; the other does not.
   (d) Wires are made of metal.
   (e) All of the above.

7. What is taking place in the synaptic cleft?
   (a) Receptor sites are forming due to neuron activity.
   (b) Negative ions are accumulating due to neuron firing.
   (c) Sodium ions are generating charge.
   (d) Neurotransmitters are diffusing between the cells.
   (e) None of the above.

8. Which of the following neurotransmitters would most likely help you avoid a car accident?
   (a) Dopamine   (b) Norepinephrine
   (c) Acetylcholine   (d) GABA
   (e) Serotonin

9. What is the cause of physical pain?
   (a) Rapid damage response synthesis of chemicals in the body
   (b) Nerve impulses
   (c) A sensation only perceived in the brain
   (d) Inflammation
   (e) All of the above.

10. How do over-the-counter analgesics like aspirin work?
    (a) They block nerve impulses.
    (b) They alter the perception of pain in the brain.
    (c) They block the production of pain-producing chemicals.
    (d) All of the above.
    (e) None of the above.

Earth Science is the physics, chemistry, and biology of Earth.

PART FOUR

EARTH SCIENCE

Millions of years ago the sand in this sandstone was sand on a beach.

And millions of years before that it wasn't even sand. It was granite rock in mountains hundreds of miles from here.

25

ROCKS AND MINERALS

THE MAIN IDEA

Elements are the building blocks of minerals, which in turn are the building blocks of rocks.

Earth is composed of layers of different materials. We live on Earth's outer surface layer—the crust. Earth's crust is broken up into a dozen or so **tectonic plates.** Some plates are made of ocean floor only, but most are a combination of continental land joined to ocean floor (Figure 25.1). Whichever type—continental or oceanic—the plates move in response to the cooling of Earth's interior. And as the plates move, Earth's surface changes. Heat escaping from Earth's interior allows molten rock to move upward and create new solid rock—igneous rock. Then at the surface, where the rock is touched by water and air—it begins to break down. Rock weathers and erodes into smaller pieces. In time, these smaller rock pieces form another new rock—sedimentary rock. The continual movement of plates and release of internal heat transforms rocks even more. Old rocks change into new rocks—metamorphic rocks. Each of these three rock types are made of minerals, and all minerals are made of elements.

Explore!

Earth's Elements Are Not Evenly Distributed—How Did They Separate from One Another?

1. Fill an empty large-mouthed jar (for example, a peanut butter jar) with clay, sand, pebbles, and coarse gravel.
2. Add water and cap the jar.
3. Shake.

Analyze and Conclude

1. **Observing** After shaking, the mixture of materials begins to settle into various layers.

2. **Predicting** Denser materials settle to the bottom of the jar; less-dense materials migrate to the top of the jar.
3. **Making Generalizations** Earth's elements long ago also separated according to density. Heavy, iron-rich elements sank to early Earth's center and lighter, silicate elements migrated toward the surface.

FIGURE 25.1 ▶
Earth's surface—the crust—is a mosaic of tectonic plates that move in response to heat flow and convection in Earth's interior. As the plates move, Earth's surface changes.

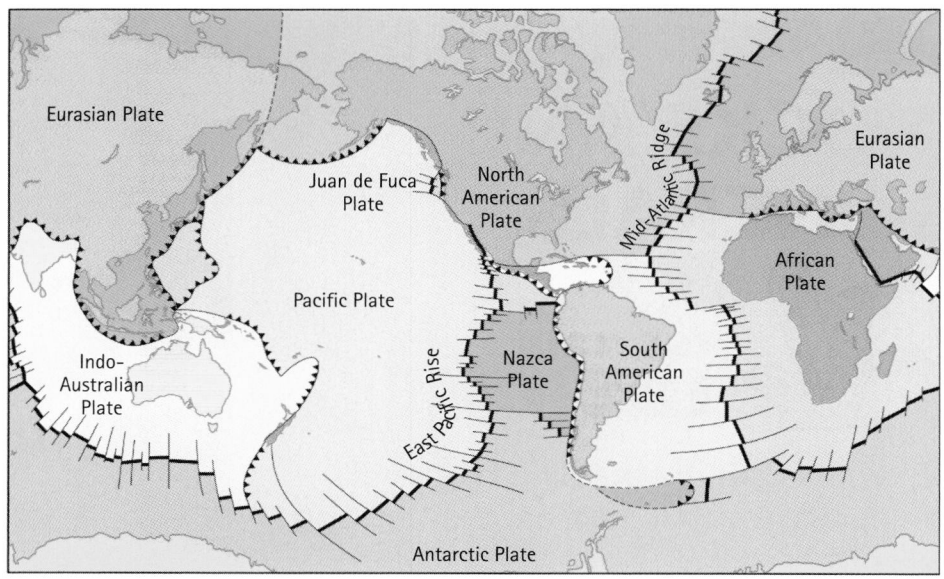

25.1 Our Rocky Planet

You may recall that there are about a hundred different chemical elements. But only eight elements make up 98% of Earth's entire mass (Figure 25.2)! All of the other elements combined make up the remaining 2%.

Elements are distributed unevenly on Earth. For example, denser elements are more concentrated near Earth's center, and less-dense elements are more abundant near Earth's surface. To explain this lopsided distribution, we need to examine the very beginnings of Earth.

Our solar system formed about 4.5 billion years ago. The dust, gases, and rocky and metallic debris orbiting the newly forming Sun collided and joined together to become the planets, asteroids, and comets that we know today. One such rocky mass became our Earth, which formed as chunks of all sizes accumulated. When first formed, Earth's elements *were* distributed evenly throughout the planet. But all that was about to change.

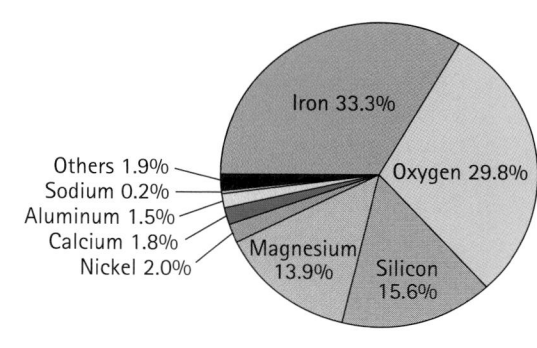

FIGURE 25.2 ▶
Only eight of the chemical elements are found in abundance on Earth.

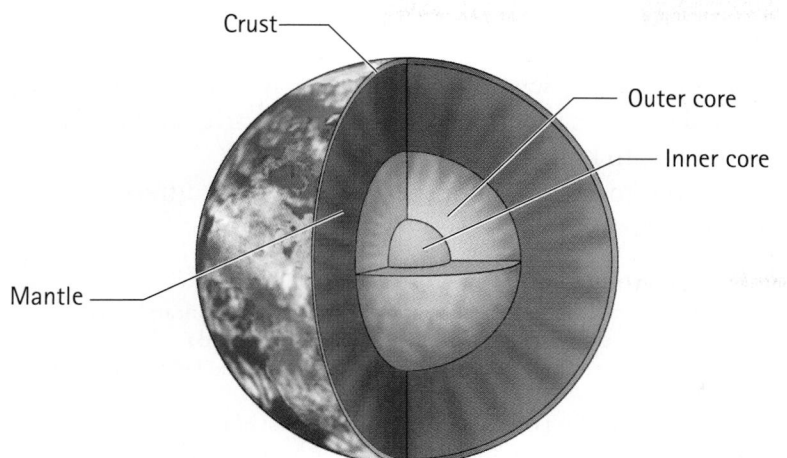

◀ **FIGURE 25.3**
Earth has a layered internal structure. The layers—the crust, mantle, and core—differ in composition and density.

With each meteorite impact, heat was generated. Then as Earth grew, its own gravity attracted even more debris. The attraction became so strong that young Earth actually squeezed itself into a smaller volume. This action produced even more heat. A third source of heat came from the decay of naturally occurring radioactive elements. These three sources of heat—*impact heating*, *gravitational contraction heating*, and *radioactive decay heating*—acted together to bring young Earth to its melting point.

So in a molten, or nearly molten state, and under the influence of gravity, dense material sank to Earth's center and less-dense material rose toward the surface. This *density segregation* led to the formation of a dense, metallic *core*; a less-dense, rocky *mantle*; and an even less-dense, rocky *crust* (Figure 25.3).

In Figure 25.4 we see the current composition of Earth's crust. When you compare the composition of the crust to that of Earth as a whole, you see that the same few elements appear in both. The percentages, however, are quite different. As expected, the crust is composed of mostly lighter elements. In fact, almost half the mass of Earth's crust is the element oxygen (O) and about a fourth is the element silicon (Si)!

In terms of radioactive decay, the amount of heat generated in any cubic meter of rock is actually quite small. But considering the trillions of cubic meters of rock within Earth, the amount of heat adds up to be very significant.

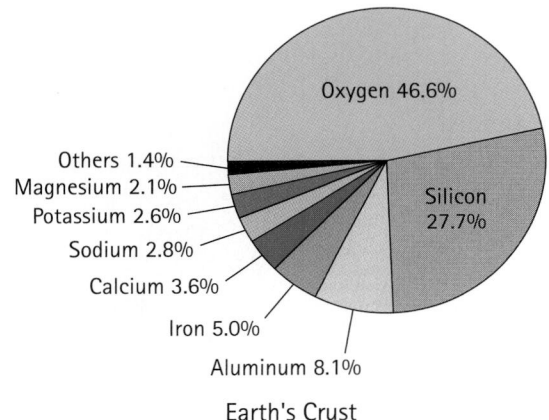

Oxygen 46.6%
Others 1.4%
Magnesium 2.1%
Potassium 2.6%
Sodium 2.8%
Calcium 3.6%
Iron 5.0%
Aluminum 8.1%
Silicon 27.7%
Earth's Crust

◀ **FIGURE 25.4**
Percentage of elements in Earth's crust, by mass. Oxygen and silicon make up more than 75% of Earth's crust.

The same type of density segregation occurs in a mixture of oil and water. The denser water sinks to form a layer at the bottom and the less-dense oil rises to form a layer at the top.

Crystal glassware is not a crystalline solid. All types of glass are *amorphous*—they do not have an ordered atomic structure. The atoms in any type of glass solidified in a random and disorganized manner.

fyi

• Minerals found in rocks and in dietary supplements are similar, yet different. The minerals in rocks are naturally occurring, crystalline solids with a definite chemical composition. Minerals found in dietary supplements are human-made compounds that contain elements necessary for life functions. The elements used to make dietary supplements, however, ultimately come from the naturally occurring minerals of Earth's crust.

✔️ **READING CHECK**

What are the three requirements for a substance to be considered as a mineral?

25.2 What Is a Mineral?

In everyday language, minerals are a part of our diet ("vitamins and minerals"), and minerals provide the raw materials needed for industry (aluminum for cans, iron for steel, etc.). From these two simple examples, it is easy to see the importance of minerals in the geosphere and in our lives. But what exactly is a mineral?

✔️ A **mineral** is a naturally formed, crystalline solid (a solid with an ordered arrangement of atoms), with a specific chemical composition.

- To be *naturally formed* means that it is not manufactured in a laboratory. So cubic zirconia and other synthetic gems are not minerals.
- To be a *crystalline solid* means that the atoms that make up a mineral are always arranged in an orderly geometric pattern. So minerals of the same type always have the same geometric arrangement of atoms.
- To have a *specific chemical composition* means that for two samples to be considered the same type of mineral, they must have the same basic chemical composition.

FIGURE 25.5 ▶
The basic structural form of the mineral halite (table salt) is cubic. This form is repeated over and over in three dimensions. The internal order of halite crystals is revealed in its macroscopic mineral grains.

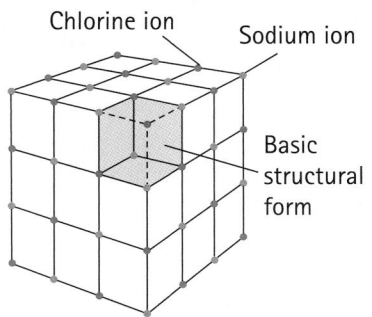

Chlorine ion   Sodium ion

Basic structural form

(a) Crystalline structure form of halite

25.3 Mineral Properties

✓ Minerals are classified by chemical composition (which elements are present) and crystal structure (how the atoms are arranged). Although composition and structure are microscopic, they determine a mineral's observable physical properties. In other words, what we see on the outside depends on what is on the inside. The physical properties of minerals include crystal form, hardness, fracture or cleavage, luster, color, streak, and density.

Crystal Form Expresses the Arrangement of Atoms in a Mineral

Have you ever seen table salt (the mineral *halite*) under a magnifying glass? Each salt grain has the form of a tiny cube (Figure 25.5). A crystal's shape, or its **crystal form,** is an expression of the orderly arrangement of its atoms. When you look at a fully formed crystal, what you see is the actual arrangement of atoms in its structure. Each type of mineral has a unique composition and crystal form (Figure 25.6). Unfortunately, well-shaped crystals are rare in nature because of space constraints—most crystals grow in cramped spaces.

(b) Grains of the mineral halite (table salt)

Wow! When I look at this crystal I'm actually seeing the mineral's arrangement of atoms!

(a)   (b)   (c)   (d)   (e)   (f)

FIGURE 25.6 ▲
Many minerals are easily recognized by their crystal form. (a) Amethyst, the purple variety of quartz, has a hexagonal crystal form with pointed ends. (b) Pyrite, or "fool's gold," typically forms cubic crystals marked with parallel lines called *striations*. (c) Rosasite has radiating bluish-green crystals that group into balls. (d) Rhodochrosite (whose name means "rose-colored") has a rhombohedral crystal form. Some minerals have distinctive growth patterns. (e) The mineral hematite often grows in a grape-clustered form. (f) Asbestos minerals have a fibrous form.

✓ **READING CHECK**

What two factors determine a mineral's classification?

Salt Crystals
Look at some crystals of table salt under a microscope or magnifying glass and observe their generally cubic shapes. There's no machine at the salt factory specifically designed to give salt crystals these cubic shapes, as opposed to round or triangular ones. The cubic shape occurs naturally and reveals how the atoms of salt are organized—cubically. Smash a few of these salt cubes and then look at them again carefully. What you'll see are smaller salt cubes!

CONCEPT CHECK
Many minerals can be identified by their physical properties—crystal form, hardness, fracture, cleavage, luster, color, streak, and density. Why is identifying a mineral by its crystal form usually difficult?

Check Your Answer
Well-shaped crystals are rare in nature because minerals typically grow in cramped spaces.

Hardness Is the Resistance of a Mineral to Scratching

Hardness does not refer to how easily a mineral breaks, but rather it's resistance to scratching. For example, a quartz crystal can scratch a feldspar crystal because quartz is harder than feldspar. The ability of one mineral to scratch another and the resistance of a mineral to being scratched are measures of hardness. We use the **Mohs scale of hardness** (Table 25.1) to compare the hardness of different minerals.

Why are some minerals harder than others? ✔ Hardness depends on the strength of a mineral's chemical bonds—the stronger its bonds, the harder the mineral. The factors that influence bond strength are ionic charge, atom or ion size, and packing (Chapter 18). Strong bonds are generally found between highly charged ions—the greater the attraction, the stronger the bond. Size affects bond strength as well, because small atoms and ions can generally pack closer together than large atoms and ions. Closely packed atoms and ions have a smaller distance between one another, and thus they form stronger bonds because they attract one another with more force. Gold, with its large atoms, is soft. Its atoms are rather loosely packed and loosely bonded. Diamond, with its small carbon atoms and tightly packed structure, is very hard—the hardest mineral known (Figure 25.7).

READING CHECK

What factors determine a mineral's hardness?

TABLE 25.1 Mohs Scale of Hardness

Mineral	Hardness	Object of Similar Hardness
Talc	1	
Gypsum	2	Fingernail (2.5)
Calcite	3	Copper wire or coin (3.5)
Fluorite	4	
Apatite	5	Steel knife blade, glass (5.5)
Feldspar	6	Unglazed porcelain plate (6.5)
Quartz	7	
Topaz	8	
Corundum	9	
Diamond	10	

The stronger the bond, the harder the mineral.

Cleavage and Fracture Are Ways in Which Minerals Break

If you shatter the mineral calcite with a hammer, its break surface is smooth and flat. This "clean" type of breakage occurs parallel to a mineral's *planes of weakness*—planes along which chemical bonds are weak or few in number. **Cleavage** is the tendency for a mineral to break along such planes of weakness. ✔ Cleavage planes are determined by crystal structure and chemical bond strength.

Some minerals cleave more easily than others. In general, minerals that have strong bonds between planar (flat) crystal surfaces show poor cleavage, whereas those with weak bonds along planar surfaces show more distinct cleavage. The minerals muscovite (mica) and calcite both have well-defined cleavage. For example, mica's crystal structure consists of atoms arranged in sheets. The atoms *within* the individual sheets are connected by strong bonds, but *between* the sheets, the bonds are weak. So muscovite cleaves where the bonding is weak—between its planar sheets (Figure 25.8). You can even peel muscovite off in thin layers. Shiny flakes of muscovite are used in glittering body paints, and they

READING CHECK

What factors determine a mineral's cleavage?

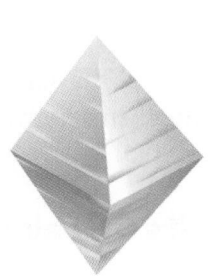

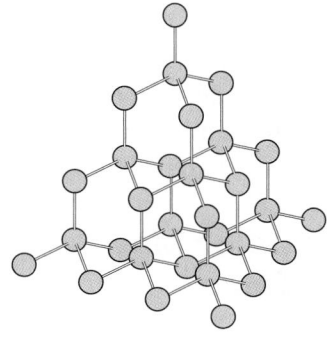

◀ **FIGURE 25.7**
Diamond is pure carbon and has a tightly packed symmetrical structure. It is the hardest mineral known.

FIGURE 25.8 ▶

A mineral's cleavage is very useful in its identification. (a) Muscovite, a mineral of the mica group, has perfect cleavage in one direction. It breaks apart into sheets. (b) Calcite (calcium carbonate) has perfect cleavage in three directions. It breaks apart into smaller rhombohedral shapes.

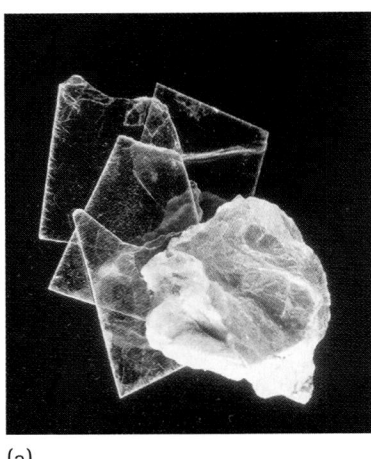
(a)

(b)

add shimmer to autobody paints as well. Minerals that have no planar alignment of bonds, like quartz, do not display cleavage and always **fracture.** The degree and type of cleavage or fracture are useful guides for identifying minerals.

CONCEPT CHECK

1. When pieces of calcite and fluorite are scraped together, which scratches which?
2. The mineral muscovite has very distinct cleavage, yet the mineral quartz fractures. How does this relate to each mineral's crystal structure?

Check Your Answers

1. Looking at Table 25.1, we see that fluorite is harder than calcite. So fluorite scratches calcite.
2. Muscovite forms as a layered sheet-like structure. The bonds between the different layers are weaker than the bonds within the individual layers. Mica minerals cleave between the layers. Quartz has a more complicated structure, with no layering and no distinct planes of weakness. Therefore, quartz fractures.

Luster Is the Appearance of a Mineral's Surface in Reflected Light

The **luster** of a mineral is the way its surface appears when it reflects light. Luster does not depend on color. Minerals of the same color may have different lusters, and minerals of the same luster may have different colors. Mineral lusters are listed in Table 25.2.

A Mineral's Color May Vary, but Its Streak Is Always the Same

Although color is an obvious feature of a mineral, it is not a reliable means of identification. Some minerals—copper and turquoise are two

The physical properties of a mineral all relate back to the mineral's chemistry.

TABLE 25.2 Mineral Lusters

Mineral Luster	Appearance
Metallic	Strong reflection; polished or dull
Vitreous	Bright, glassy
Resinous	Waxy
Greasy	Like oily glass, also may feel greasy
Pearly	Pearly iridescence
Silky	Sheen of silk
Adamantine	Diamond, brilliant

FIGURE 25.9 ▲
The mineral corundum (Al_2O_3) comes in a variety of colors as a result of chemical impurities. The addition of small amounts of chromium in place of aluminum produces the red gemstone *ruby*. With the addition of small amounts of iron and titanium, the result is the blue gemstone *sapphire*.

examples—have a distinctive color. But most minerals either occur in a variety of colors or can be colorless.

Chemical impurities in a mineral affect color. For example, the common mineral quartz, SiO_2, can be found in many colors. It can be clear and colorless if it is pure, or it can be milky white from small impurities. Rose-colored quartz results from small amounts of titanium; purple quartz (amethyst) results from small amounts of iron. The color of the mineral corundum, Al_2O_3, is commonly white or grayish. But impurities in corundum give us rubies and sapphires (Figure 25.9).

Streak, the name given to the color of a mineral in its powdered form, is an important characteristic for identifying minerals that have a metallic or semimetallic luster. When rubbed across an unglazed porcelain plate, all minerals leave behind a thin layer of powder—a streak. Although different samples of the same mineral can have different colors, the color of the mineral's streak is always the same (Figure 25.10). For example, the mineral hematite varies in color (red, brown, or black) but it always makes a reddish-brown streak. Magnetite can be gray or brown to black, but always makes a black streak. Minerals that do not have a metallic luster generally leave behind a white streak. A white streak cannot be used to identify minerals.

Density Is a Ratio of Mass to Volume

Density is a property of all matter, minerals included. In practical terms, the density of a mineral tells us how heavy a mineral feels for its size. More specifically, a mineral's density is the ratio of its mass to its volume. The densities of some minerals are shown in Table 25.3.

Gold's particularly high density of 19.3 g/cm^3 is nicely taken advantage of by miners panning for gold. Fine gold pieces hidden in a mixture of mud and sand settle to the bottom of the pan when the mixture is swirled in water. Water and less-dense materials spill out when the mixture is swirled. After a succession of douses and swirls, only the substance with the highest density remains in the pan—gold!

FIGURE 25.10 ▲
The streak test can be used to identify minerals that have a metallic or semimetallic luster.

TABLE 25.3 Density of Various Minerals (g/cm³)

Borax	1.7	Pyrite	5.0
Quartz	2.65	Hematite	5.26
Talc	2.8	Copper	8.9
Mica	3.0	Silver	10.5
Chromite	4.6	Gold	19.3

The density of a mineral depends on a number of factors—the masses of the mineral's constituent atoms and the packing of these atoms, which, in turn, is a function of the atoms' sizes.

fyi

- All silicate minerals have the same fundamental structure of atoms, the silicon–oxygen tetrahedron—four oxygen atoms joined to one silicon atom $(SiO_4)^{4-}$. The powerful bond that unites the oxygen and silicon ions is akin to the cement that holds Earth's crust together.

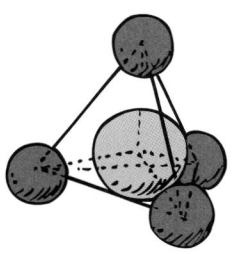

✔ **READING CHECK**

Why are silicate minerals so common? What are the two most common silicate minerals?

25.4 Classification of Rock-Forming Minerals

There are more than 4000 known minerals, and new ones are discovered every year. Despite this large number, most minerals are actually quite rare. In fact, only about a dozen or so minerals make up most of the rocks exposed at Earth's surface. These are the *rock-forming minerals*.

Minerals are classified by their chemical composition. Doing so produces two main categories: the **silicates** and the **nonsilicates** (Figure 25.11). Look back at Figure 25.4 and you will understand how Earth scientists arrived at this simple division. ✔ Oxygen is the most abundant element in Earth's crust, and silicon is the second most abundant. Minerals that contain both silicon (Si) and oxygen (O) as part of their chemical composition are called *silicates*. Minerals that do not contain these two elements are called *nonsilicates*.

Silicates Make Up Nearly 90% of Earth's Crust

Silicon has a great affinity for oxygen. In fact, silicon has such a strong tendency to bond with oxygen that silicon is never found in nature as a pure element; it is *always* chemically combined with oxygen. The combination of silicon and oxygen is simply called *silica* (SiO_2). The silicates are the most common mineral group, making up more than 90% of Earth's crust. Most silicates also contain some of the other eight common elements, which include Fe, Mg, Ca, and Al, but the basic building block of *all* silicates is Si and O.

The silicates are subdivided into two groups—those that contain iron and/or magnesium (*ferromagnesian*) and those that do not (*nonferromagnesian*). Because of the presence of iron and/or magnesium, ferromagnesian silicates tend to be dense and dark in color. Nonferromagnesian silicates do not contain significant amounts of iron or magnesium; therefore, they generally have relatively low densities and are light in color. ✔ The most abundant mineral in the crust is feldspar, a nonferromagnesian silicate that contains aluminum, sodium, potassium, and/or calcium, plus silicon and oxygen. Feldspar makes up more than 50% of the crust. Quartz (SiO_2), the second most common mineral in Earth's crust, is composed of only silicon and oxygen. If you have ever collected rocks and minerals, you probably have some quartz and feldspar specimens in your collection.

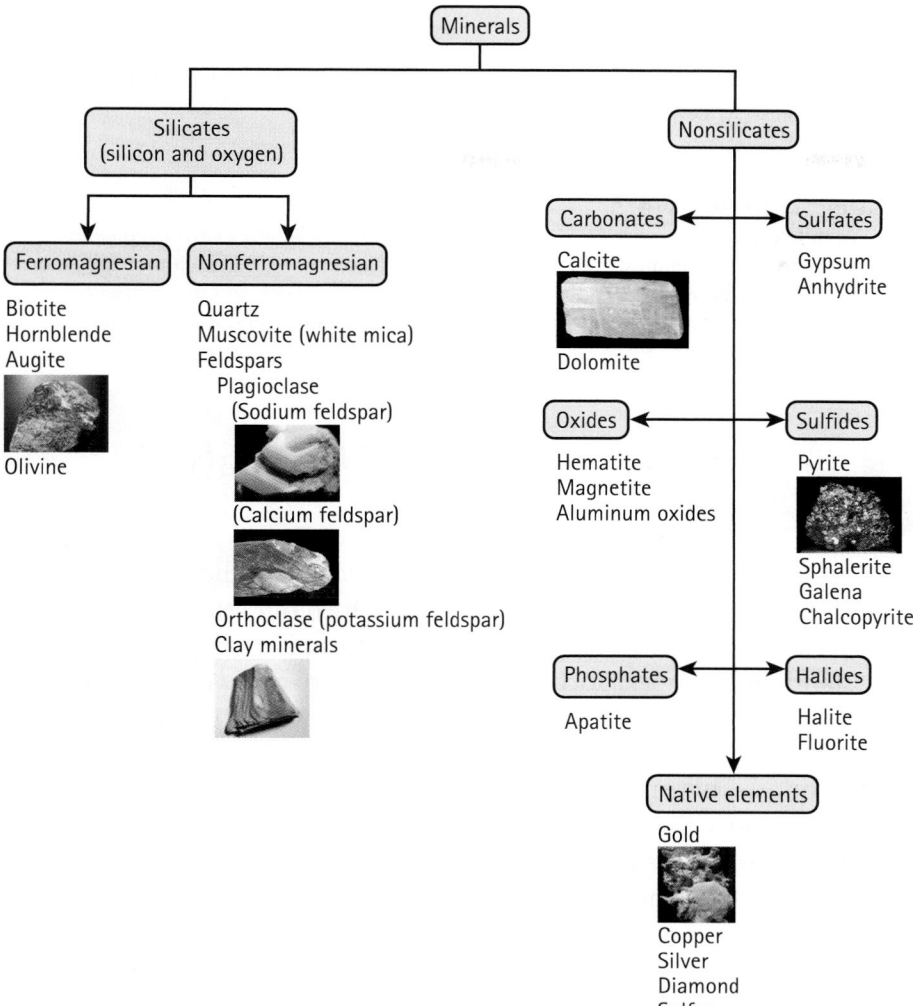

fyi

- The carbonate ion $CO_3{}^{2-}$ has a structure that features a central carbon atom bonded to three oxygen atoms.

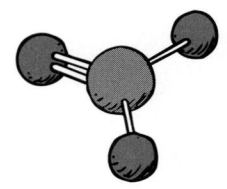

There are lots of definitions in this chapter—do not get bogged down by memorizing so many terms. Instead, let the terms help you with your understanding. The terms *ferromagnesian* and *nonferromagnesian* are long words indeed, but use these terms conceptually: *ferro-* means iron; *nonferro-* means no iron. Keep it simple.

Nonsilicates Include the Carbonates, Oxides, and Sulfide Minerals

Nonsilicate minerals make up just 8% of Earth's crust by mass. ✓ The nonsilicates include the carbonates, oxides, sulfides, the native elements such as gold and silver, and a few others.

The carbonates are the most abundant nonsilicate minerals. Two common carbonate minerals are calcite ($CaCO_3$) and dolomite ($CaMg(CO_3)_2$)—the main minerals found in the group of rocks called *limestone*. Oxide minerals contain oxygen combined with one or more metals. These metals include iron (which forms hematite and magnetite), chromium (which forms chromite), manganese (which forms pyrolusite), tin (which forms cassiterite), and uranium (which forms uraninite). Sulfide minerals also contain many metallic elements. The sulfide and oxide minerals are economically important—they form many of the metal ores necessary for industrial and technological manufacturing. Other important rock-forming mineral groups are the sulfates, halides, and phosphates.

✓ READING CHECK

Aside from the silicates, what are some other rock-forming mineral groups?

25.5 The Formation of Minerals and Rock

So far we have explored the definition of a mineral and the different properties and classifications of minerals. Now we turn our attention to how minerals form. Understanding how minerals form is a stepping-stone to understanding how rocks form. Rocks, after all, are made of minerals.

Minerals are formed by **crystallization**—the growth of a crystalline solid from a liquid, gas, or other solid. Crystallization starts when atoms begin to bond with each other in a particular geometric pattern. As the number of atoms and bonds increases, a single crystal forms. And as more and more atoms bond to the microscopic crystal, repeating the underlying pattern, the crystal grows.

Minerals commonly crystallize from two different sources: from **magma**—molten rock—and from water solutions. Minerals formed from the crystallization of magma make up igneous rock, and minerals formed from the precipitation or evaporation of water make up some types of sedimentary rock. Minerals, in their many forms, are the building blocks of the many different rocks on Earth.

Ice crystals forming in water is a familiar form of crystallization. Just as there is water and ice, there is magma and rock.

25.6 Rocks Are Divided into Three Main Groups

✓ A **rock** is an aggregate of minerals. While minerals are chemical mixtures or compounds, rocks are physical mixtures. In some rocks, the grains are "cemented" together; in others, the grains are tightly interlocked. In many rocks, you can see mineral crystals. Granite, one of the most common rocks in Earth's continental crust, contains visible crystals of the minerals feldspar, quartz, and hornblende (Figure 25.12). On the other hand, in rocks such as basalt, shale, or slate, individual mineral grains are difficult to distinguish—they are too small to be seen with the unaided eye.

Rocks are divided into the following three groups (Figure 25.13):

Igneous rocks are formed by the cooling and crystallization of *magma*. The word *igneous* means "formed by fire." Igneous rock can form at or below Earth's surface. Granite and basalt are common igneous rocks.

◀ **FIGURE 25.12**
A rock is an aggregate of one or more minerals. This granite is an aggregate of the minerals quartz, horn-blende, and feldspar.

Quartz
(Mineral)
+

Hornblende
(Mineral)
+

Feldspar
(Mineral)

=

Granite
(Rock)

Sedimentary rocks are formed from the cementation or compaction of *sediment*. Some sedimentary rocks form when minerals precipitate out of water solutions at or near Earth's surface. Sandstone, shale, and limestone are common sedimentary rocks.

Metamorphic rocks are formed from older, preexisting rocks (igneous, sedimentary, or metamorphic) that were transformed in Earth's interior by high temperature, high pressure, or both—without melting. The word *metamorphic* means "changed in form." For example, marble is metamorphosed limestone, and slate is metamorphosed shale.

(a)   Basalt   Granite

(b)   Sandstone   Limestone

(c)   Marble   Slate

FIGURE 25.13 ▶
The three main types of rock. (a) Basalt and granite are igneous rocks. (b) Sandstone and limestone are sedimentary rocks. (c) Marble and slate are metamorphic rocks.

25.7 Igneous Rocks Form When Magma Cools

All igneous rocks are formed from magma, and magma is made from rocks and minerals that have melted. Magma is classified by the amount of silica it contains. There are three major types of magma: *basaltic* (about 50% silica), *andesitic* (~60% silica), and *granitic* (~70% silica).

The largest region of Earth's interior is the mantle, which is solid. The mantle is composed of low-silica-content igneous rocks. ✔ When these mantle rocks partially melt, basaltic magma is formed. Basaltic magma that solidifies forms the dark igneous rock known as *basalt.* The Hawaiian Islands and the oceanic crust are made from basalt. Andesitic magma solidifies to form the igneous rock *andesite,* which gets its name from the Andes Mountains in South America, where it is very common. When granitic magma cools slowly below Earth's surface it forms *granite* and other similar granitic-type rocks. Of all the igneous rocks in the crust, oceanic and continental crust combined, approximately 80% was formed from basaltic magma, 10% from andesitic magma, and 10% from granitic magma.

Depending on where magma cools, igneous rocks may form at or below Earth's surface. On the continents, the most common igneous rocks are granite and andesite. On the ocean floor, basalt is the most common kind of rock.

READING CHECK

What is the origin of the basaltic rocks we see on Earth's crust?

CONCEPT CHECK

If 80% of all igneous rocks are formed from basaltic magma, why do we see so much granite?

Check Your Answer

Basalt is the most common igneous rock on the ocean floor. Look at a globe to see that the oceans cover nearly three-quarters of Earth's surface. We see so much granite because it is the most common igneous rock on continental land.

Igneous Rocks at Earth's Surface

Igneous rocks that form by the eruption of molten rock at Earth's surface are called **volcanic rocks.** Magma that moves upward from inside Earth and erupts onto the surface is called **lava.** The term *lava* refers both to the molten rock and to the solid rock that forms from it. ✔ Igneous rocks formed at Earth's surface are often referred to as *extrusive* rocks. The lava pushes out of Earth's interior to flow on the surface. The word *extrusive* means "pushed out of."

Lava may be extruded through cracks and *fissures* (long cracks) in Earth's surface or through a central vent—a **volcano.** Eruptions from a volcano are familiar to us because they are very exciting to see. But the outpourings of magma from fissures are much more common than volcanic eruptions.

READING CHECK

Why are volcanic rocks referred to as extrusive igneous rocks?

FIGURE 25.14 ▶
The flood basalts that produced the Columbia Plateau covered more than 200,000 km² of the preexisting land surface.

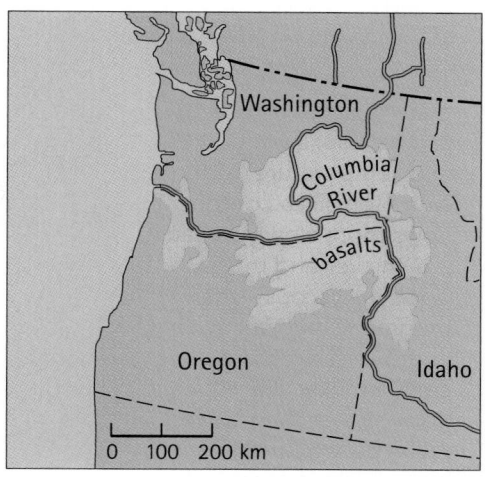

Fissure Eruptions Occur Under Water and on Land

Most fissure eruptions occur when basaltic lava erupts at the bottom of the ocean. Such underwater fissure eruptions form the ocean floors. Fissure eruptions also occur on land. Lava out-pourings known as *flood basalts* have flooded large areas and created extensive lava plains. The Columbia Plateau in the Pacific Northwest is the result of extensive flood basalts (Figure 25.14), as is the Deccan Plateau in India.

Volcanoes Come in a Variety of Shapes and Sizes

Volcanoes are openings where magma rises to Earth's surface and erupts as lava. Three main types of volcanoes exist—shield, cinder cone, and composite (Figure 25.15).

Shield volcanoes are built by a steady supply of easily flowing basaltic lava that flows out in all directions to make a broad, gently sloping cone that resembles a shield. Mauna Loa in Hawaii, the largest volcano on Earth, is a shield volcano standing 4145 m above sea level and more than 9750 m above the ocean floor (Figure 25.16).

Cinder cone volcanoes are very steep but small in comparison to shield volcanoes. Cinder cones are formed from the piling up of ash, cinders, and rocks that have been explosively erupted from a single vent to form a symmetrical, steep-sided cone. Two well-known examples of cinder cones are Sunset Crater in Arizona and Parícutin in Mexico.

When a volcano erupts both lava and ash, a *composite cone* of alternating layers of lava, ash, and mud is produced (the word *composite* means mixture). The layers build up to form a volcano with a steep-sided summit and gently sloping lower flanks. Composite cone volcanoes are bigger than cinder cones because the mixture of lava and ash helps to old the cone together. Mount Fujiyama (Mt. Fuji) is a classic example of a majestic composite cone volcano.

Composite volcanoes tend to erupt explosively because their magmas and lavas usually do not flow easily. This thicker magma traps volcanic gases, which increases the pressure inside the volcano. We can compare the gases in magma to the gases in a bottle of carbonated soda. If we cover the top of the bottle and shake vigorously, the gases separate from the soda and form bubbles. When we remove the cover, pressure

(a)

(b)

(c)

FIGURE 25.15 ▶
The three types of volcanoes. (a) Shield volcanoes, such as Mauna Loa. (b) Cinder cone volcanoes, such as Sunset Crater in Arizona. (c) Composite volcanoes, such as picturesque Mount Fujiyama (Mt. Fuji).

fyi

- The flow behavior of magma is influenced by two factors: silica content and temperature. Magma with high silica content flows more slowly because it is thicker and more gooey—*viscous*—than magma with lower silica content. (This is like a spilled milkshake, which flows slower than spilled milk.) Basaltic magma is an important example of a low-silica, fast-flowing magma. Temperature also affects the ability of magma to flow. Hotter magma flows more easily than cooler magma.

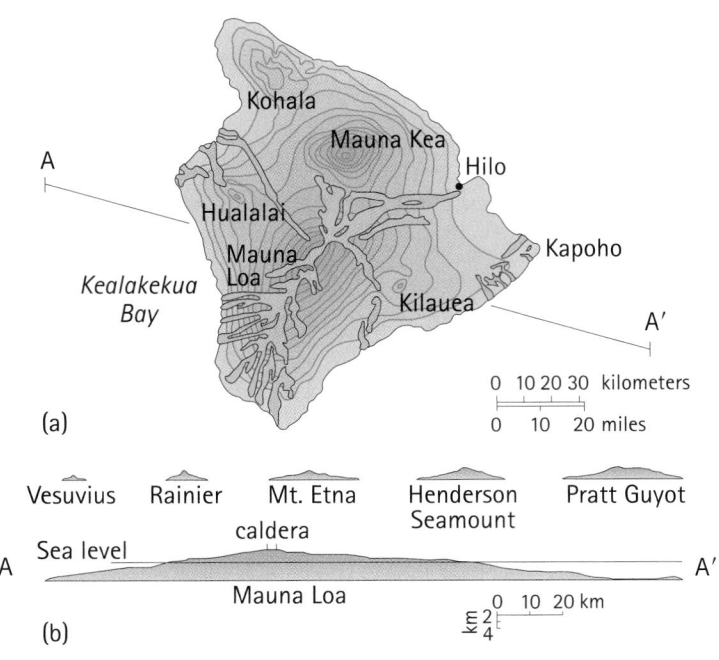

(a)

(b)

FIGURE 25.16

(a) Mauna Loa, a shield volcano on the island of Hawaii, is the largest volcano on Earth. (b) When compared with other large volcanoes, its immense size and volume is dramatic.

is released and gases and liquid explode from the bottle. The gases in magma behave in much the same way. In a volcanic blast, the pressure and temperature increase and the whole mass of gooey magma and overlying rock explodes into dust and rubble. When mixed with volcanic ash, this mixture can expand and destroy everything in its path. Examples of this kind of volcanic activity occurred at Mount Vesuvius in AD 79, at Mount Pelée in 1902, and at Mount St. Helens in 1980.

Igneous Rocks Beneath Earth's Surface

When magma cools beneath Earth's surface, the igneous rock that forms is called **plutonic rock.** The word *plutonic* is derived from Pluto, the mythological god of the underworld. ✔ Formed below Earth's surface, these igneous rocks are referred to as *intrusive* rocks. As magma moves upward, it pushes into existing rock, where it cools and solidifies (the

READING CHECK

Why are plutonic rocks referred to as intrusive igneous rocks?

LINK TO MYTHOLOGY: VOLCANOES

According to Greco-Roman mythology, volcanic activity can be traced to Vulcan, the Roman god of volcanic fire and metalworking. The word *volcano* comes from the island of Vulcano, off the coast of southern Italy. In ancient times, Vulcano was believed to be the metal workshop of Vulcan. The people living near Vulcano believed that the lava fragments and glowing ash that erupted from the island were a result of Vulcan's work as he forged thunderbolts for Jupiter, king of the gods, and weapons for Mars, god of war. The people of Polynesia, who attribute volcanic activity to Pele, goddess of volcanoes, tell a similar story.

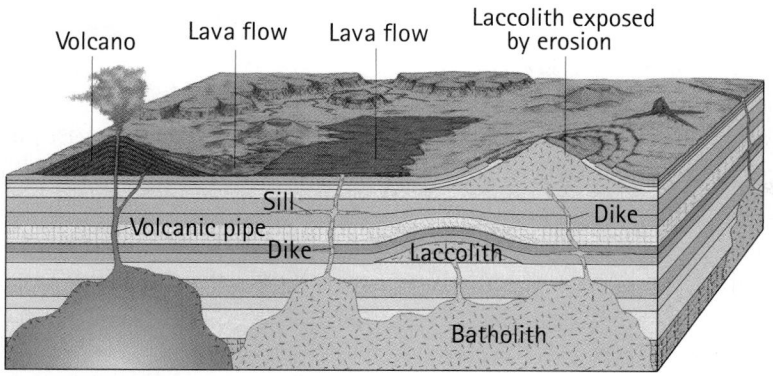

FIGURE 25.17 ▲
Intrusive igneous features in cross-sectional view.

Volcanic rocks (such as basalt) form at Earth's surface where they cool quickly; they tend to have small-to-microscopic crystals. Plutonic rocks (such as granite) form below Earth's surface, where they cool slowly; they tend to have much larger crystals that are easily seen without magnification.

word *intrusive* means "pushed into"). All intrusive igneous rock bodies are called *plutons.* Being intrusive, plutonic rocks can be studied only after they are exposed by uplift and erosion at Earth's surface. The most common plutonic rock is granite.

Plutons occur in a great variety of shapes and sizes, ranging from small pipe-like dikes to large, expansive *batholiths* (Figure 25.17). Batholiths, the largest plutons, are created by numerous intrusive events over millions of years. They form the cores of many major mountain systems around the world. Generally speaking, they are the crystallized magma chambers that fed long-since-eroded volcanoes. Many modern mountains are actually the exposed batholith cores of larger mountains that have long since eroded away. Some of the largest batholiths in North America include the Coast Range batholith and the Sierra Nevada batholith (Figure 25.18).

CONCEPT CHECK

1. Why is it incorrect to say that igneous rocks may form from the intrusion of lava?
2. Is it correct to say that igneous rocks may form from the extrusion of lava?

Check Your Answers

1. The terminology in the statement is used incorrectly. The term *intrusion* refers to solidification that occurs in Earth's interior and therefore has nothing to do with lava. Lava is not a synonym for magma, but is the term for magma that has been extruded at Earth's surface in molten form. By definition, there is no lava beneath Earth's surface. Magma is intruded and lava is extruded to form igneous rocks.
2. Yes, once magma is extruded from Earth it is called lava, which when solidified becomes igneous rock.

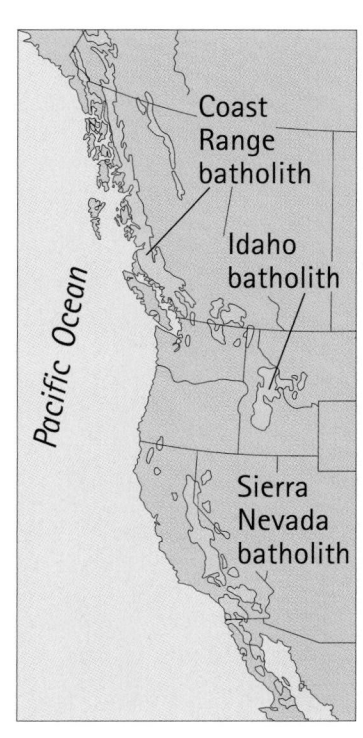

FIGURE 25.18 ▶
Some of the largest batholiths in North America include the Coast Range batholith and the Sierra Nevada batholith.

25.8 Sedimentary Rocks Blanket Most of Earth's Surface

Sedimentary rocks are the most common rocks in the uppermost part of the crust. They cover two-thirds of Earth's surface, forming a thin, extensive blanket over older igneous and metamorphic rocks. Because sedimentary rocks contain the remains of organisms and older rocks they provide information about geological events that occurred over time at Earth's surface.

The Formation of Sedimentary Rock

✓ Sedimentary rock forms in a long process over four stages: *weathering, erosion, deposition*, and *sedimentation.* **Weathering** is the disintegration or decomposition of rock at or near Earth's surface. Agents such as water and reactive chemicals weather the rock—breaking it into smaller pieces, cracking its surface, rounding and smoothing its edges and corners, and sometimes transforming its chemical composition.

There are two types of weathering—*mechanical* and *chemical.* Both produce sediment. *Mechanical weathering* physically breaks rocks into smaller and smaller pieces (Figure 25.19). For example, the freezing and thawing of ice can widen small cracks in a rock. In *chemical weathering*, reactions with water decompose rock (Figure 25.20). Because liquid water and water vapor are just about everywhere, chemical weathering can actually produce more sediment than mechanical weathering.

As rock weathers, it also erodes. **Erosion** is the process by which weathered rock particles are removed and transported away by water,

✓ **READING CHECK**

What are the four stages in the formation of sedimentary rock?

FIGURE 25.19 ▲
The rocks on this mountain peak have been split apart by mechanical weathering. As water freezes and expands in cracks, the rock splits and breaks apart.

25.9 Metamorphic Rocks Are Changed Rocks

What happens when a mass of rock is brought to a location that has a much higher temperature and pressure than the environment in which it originally formed? The rock transforms—old rock changes into new rock.

✓ The change in rock due to new physical and chemical conditions is called **metamorphism.** All rocks, whether igneous, sedimentary, or metamorphic, can undergo metamorphism. An everyday example of metamorphism is potter's clay. Potter's clay is soft at room temperature. But when heated, it becomes a hard ceramic. Similarly, limestone subjected to enough heat and pressure becomes marble. And shale is metamorphosed to slate. Rocks may also be drastically stretched or compressed. It is important to note that during metamorphism, minerals do not melt. Once minerals melt, metamorphism has ended and igneous activity has begun. In metamorphism, change occurs instead by *recrystallization* of preexisting minerals or by *mechanical deformation* of rock.

Recrystallization occurs when the minerals in a rock change because they have been exposed to high temperatures and pressures. Under these new conditions, atoms can actually migrate through the heated solids to form larger crystals, or they can totally recombine to form new minerals. *Mechanical deformation* occurs when rock is subjected to physical stress. It may or may not involve elevated temperatures. For example, surface rocks can become buried and subjected to increased pressure. Such stress may cause the rocks to bend and fold. Or the increased pressure may deform and flatten the rocks. Such physical stress occurs deep in Earth's crust.

Two Kinds of Metamorphism: Contact and Regional

The most common types of metamorphism are *contact metamorphism* and *regional metamorphism.* Each type of metamorphism is characterized by differences in mechanical deformation and recrystallization.

Contact metamorphism occurs when a body of rock is intruded by magma (Figure 25.27). The high temperature of the magma produces

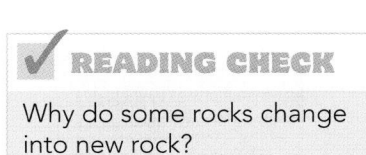

✓ **READING CHECK**

Why do some rocks change into new rock?

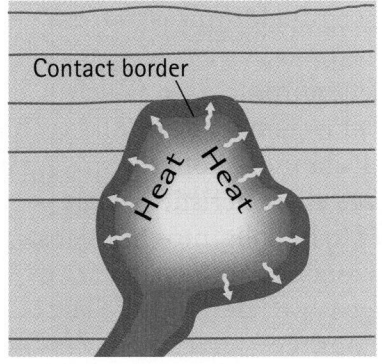

(a)

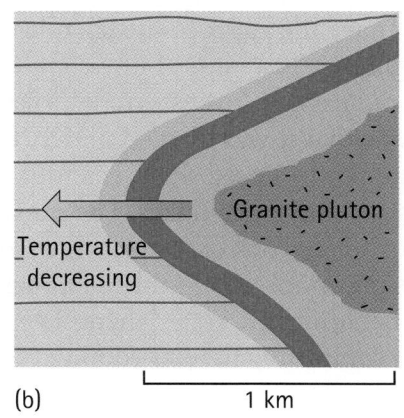

(b) |⎯⎯⎯⎯⎯⎯⎯| 1 km

◀ **FIGURE 25.27**
(a) Contact metamorphism is the result of rising molten magma that intrudes a rock body. (b) Surrounding the solidified intrusive rock is a zone of alteration. Alteration is greatest at the contact area, and it decreases farther away from the contact area.

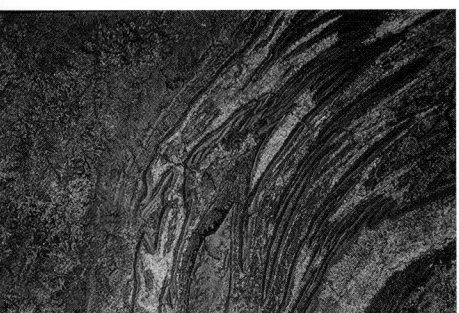

FIGURE 25.28 ▲
This satellite photo reveals regional-scale folding of metamorphic rocks in the Appalachian Mountains of central Pennsylvania.

Areas of regional metamorphism are wonderful hunting grounds for gem prospectors. The heat and pressure that accompanies regional metamorphism can produce beautiful minerals.

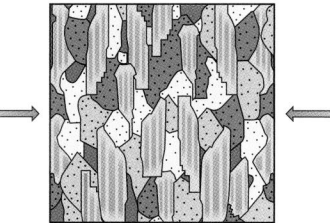

FIGURE 25.29 ▲
As compressive forces squeeze platy and sheet-structured minerals, the grains align themselves perpendicularly to the main direction of force. Arrows indicate the direction of compressive force.

a zone of alteration that surrounds the intrusion. As you would expect, alteration is greatest near the intrusion (at the contact) and lessens away from the contact. For example, one of the most common changes is an increase in crystal size due to recrystallization. Crystal size is greatest at the contact and decreases with increasing distance from that point. The area of the "altered zone" depends on the size of the intrusion. Contact metamorphism is typically associated with lots of chemical activity and little or no mechanical deformation.

Regional metamorphism is the alteration of rock by both heat and pressure over an entire region rather than just near a contact between rock bodies. For example, during the process of mountain building, Earth's crust is severely compressed into a mass of highly deformed rock. This deformation can be seen in the folded and fractured rock layers in many mountain ranges (Figure 25.28). Intense igneous activity also occurs during the process of mountain building. Large bodies of magma intrude the crust, raising the temperature of preexisting rocks in the region. Regionally metamorphosed rocks are found in all the major mountain belts of the world. Regional metamorphism combines recrystallization with mechanical deformation.

Classifying Metamorphic Rocks

☑ Metamorphic rocks are defined by their appearance and the minerals they contain. For classification and identification, metamorphic rocks are divided into two groups: *foliated* and *nonfoliated*.

Foliated Metamorphic Rocks Have a Layered Appearance

When rock is subjected to increased pressure, some of its minerals re-align into parallel planes as they recrystallize. The face of each of these parallel planes is perpendicular to the main direction of the compressive force. This leads to a layered appearance called *foliation*. Sheet-structured minerals, such as micas, orient themselves as they grow so that their sheets are perpendicular to the direction of maximum pressure (Figure 25.29). The new rock, which now has parallel flakes, or plates, of mica, is said to be foliated. The most common foliated metamorphic rocks—slate, schist, and gneiss—are derived from sedimentary rocks that have the appropriate chemical composition for micas to form (Figure 25.30).

Slate is the "lowest-grade" foliated metamorphic rock, which means that it was formed under relatively low temperature. Slate, which is metamorphosed shale, is a foliated rock composed of very small particles and tiny mica flakes. The most obvious characteristic of slate is its excellent rock cleavage, which allows it to be split into thin slabs. The best pool tables and chalkboards are made from slate quarried in metamorphic areas where slaty cleavage is well developed. Slate is also commonly used as roofing tile and floor tile.

(a)

(b)

(c)

FIGURE 25.30 ▲
Common foliated metamorphic rocks: (a) slate, (b) schist, and (c) gneiss.

Schist is one of the most easily recognizable metamorphic rocks because it is very shiny. Schist forms under higher temperature than slate. The extra heat allows the mineral grains to grow large enough to be identified with the naked eye. Schists usually contain about 50% platy minerals—most commonly muscovite and biotite. The larger mica flakes give the rock a highly reflective surface that is quite striking. Schists are named according to the major minerals in the rock (biotite schist, staurolite-garnet schist, and so on).

Gneiss (pronounced "nice") is a foliated metamorphic rock that contains alternating layers of dark platy minerals and lighter granular minerals. The layers give this metamorphic rock its characteristic banded appearance. This appearance results from even greater temperature than those that create schist. The most common granular minerals found in gneiss are quartz and feldspar. These are also the most common granular minerals in granite. In fact, some gneisses are actually metamorphosed granites.

Nonfoliated Metamorphic Rocks Have a Uniform Appearance

Nonfoliated metamorphic rocks can form either in areas of increased temperature and pressure or in areas where only the temperature increased. Even under high pressure, foliation cannot develop if the rock does not have the right chemical composition for platy minerals (such as micas) to form. If the chemical composition is correct, but the pressure is not high enough, such as in contact metamorphism, foliation cannot develop. Two common nonfoliated rocks are marble and quartzite.

Marble (Figure 25.31a) is a crystalline, metamorphosed limestone. Pure marble is white and is virtually 100% calcite, which is neither platy nor elongated. Because of its color and its relative softness (hardness 3), marble is a popular building stone. Often the limestone from which marble formed contained impurities that produce various colors in the marble. Thus, marble can vary in color from pink to gray, green, or even black.

Quartzite (Figure 25.31b) is metamorphosed quartz sandstone, and it is therefore very hard (hardness 7). Quartz is another mineral that is

(a)

(b)

FIGURE 25.31 ▲
Nonfoliated metamorphic rocks: (a) marble and (b) quartzite.

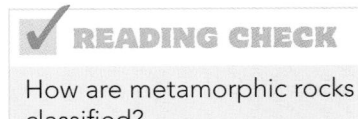
✓ READING CHECK

How are metamorphic rocks classified?

not platy or elongated. The recrystallization of quartzite can be so complete that when struck, the rock can split across the original quartz sand grains, rather than between them. Although pure quartzite is white, it commonly contains impurities that can cause it to be a variety of colors, such as pink, green, or light gray.

CONCEPT CHECK

Under conditions of extreme temperature, when can a rock no longer undergo metamorphism?

Check Your Answer

When it melts. Once a rock melts, it becomes magma. And when magma cools to form rock, the new rock, by definition, is igneous rock.

PhysicsPlace.com

Tutorial
The Rock Cycle Activity
Video
The Rock Cycle

25.10 The Rock Cycle

Earth is a dynamic, ever-changing, and active planet. Made up of atoms and molecules, Earth's elements combine to make minerals, which are formed by the process of crystallization from either magma or water solutions. The minerals formed are determined by the elements present and the conditions that lead to their formation. These factors in turn determine the arrangement of atoms in each mineral and the strength of the bonds that hold the atoms together. More than 90% of Earth's minerals are silicates—composed predominantly of silicon and oxygen plus other elements such as aluminum, iron, calcium, sodium, potassium, and magnesium. Minerals combine to make rocks—the igneous, sedimentary, and metamorphic rocks that we see all around us.

Most minerals (and hence, rocks) are formed by the crystallization of magma. And magma forms when rock melts. Magma comes in three basic compositions—basaltic, andesitic, and granitic. These magma types lead to different types of igneous rocks.

Although most of Earth's crust is composed of igneous and metamorphic rock, the rock we see at the surface is mainly sedimentary. Sedimentary rock forms from the remains of rock that has been weathered and eroded. Sedimentary rock provides a record of environmental and biological changes on Earth's surface. And when sedimentary rock is forced deep within Earth or involved in mountain building, great temperatures and pressures can transform it into metamorphic rock. Under the proper conditions, metamorphic rock can melt and become magma, which eventually solidifies as igneous rock to complete the **rock cycle** (Figure 25.32).

The rock cycle varies in its paths. Igneous rock, for example, may be subjected to heat and pressure far below Earth's surface to become metamorphic rock. Or metamorphic or sedimentary rocks at Earth's surface may decompose to become sediment that becomes new sedimentary rock. There are many possible variations in the cycle. Cycles within cycles even occur. Whatever the path, Earth's crust forms when molten rock rises from its depths, cools, and solidifies to form a crust

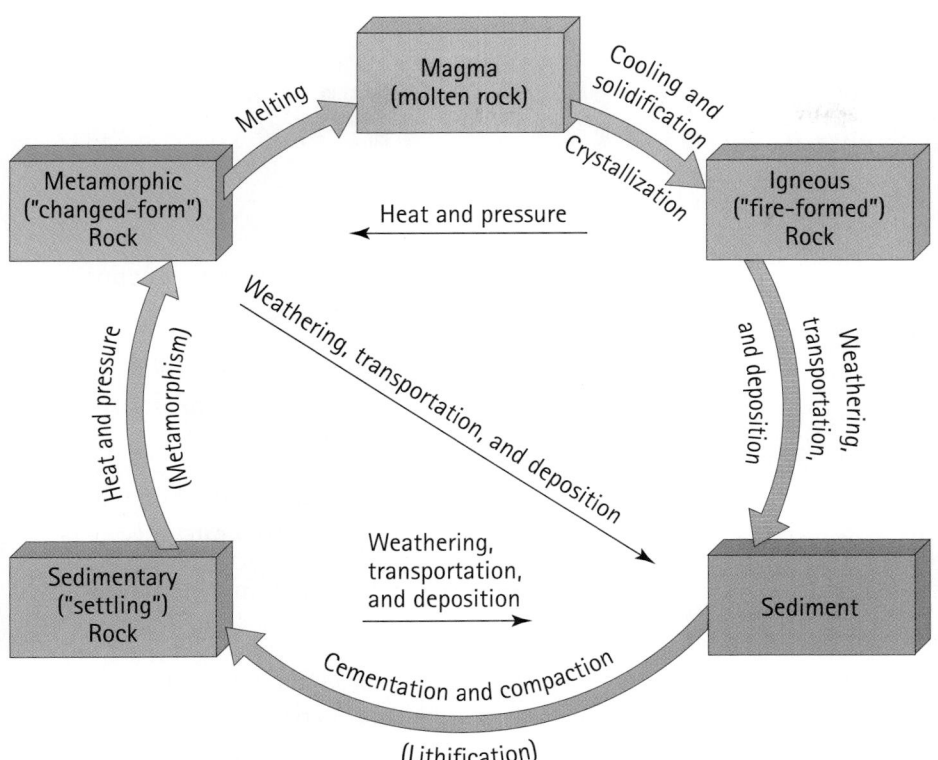

◀ **FIGURE 25.32**
The rock cycle: Igneous rock, subjected to heat and pressure far below Earth's surface, may become metamorphic rock; and igneous, metamorphic, or sedimentary rocks at Earth's surface may decompose to become sediment that in turn becomes new sedimentary rock. Whatever the route, molten rock rises from the depths of Earth, cools, and solidifies to form a crust that, over eons, is reworked by shifting and erosion, only to return eventually to become magma in Earth's interior.

that, over time, is reworked by shifting and erosion and can eventually be returned to the interior. There it may be completely melted and once again become magma.

A great cycle indeed. But what about Earth's interior—what is going on in Earth's innards? Let us now turn our attention to the exploration of Earth's interior.

25 CHAPTER REVIEW

KEY TERMS

Chemical sediments Sediments that form by the precipitation of minerals from water on Earth's surface.

Cleavage The tendency of a mineral to break along planes of weakness.

Crystal form The outward expression of the orderly internal arrangement of atoms in a crystal.

Crystallization The growth of a solid from liquid or gas whose atoms come together in specific chemical proportions and crystalline arrangements.

Density The ratio between the mass of a substance and its volume.

Deposition The stage of sedimentary rock formation in which eroded particles come to rest.

Erosion The wearing away of rocks and the processes by which rock particles are transported by water, wind, or ice.

Fracture A break that does not occur along a plane of weakness.

Igneous rocks Rocks formed by the cooling and crystallization of hot, molten rock material called magma (or lava).

Lava Molten magma that moves upward from inside Earth and flows onto the surface. The term *lava* refers both to the molten rock and to the solid rocks that form from it.

Luster The appearance of a mineral's surface when it reflects light.

Magma Molten rock in Earth's interior.

Metamorphic rocks Rocks formed from preexisting rocks that have been changed or transformed by high temperature, high pressure, or both.

Metamorphism The changes in rock that happen as physical and chemical conditions change.

Mineral A naturally formed crystalline solid, composed of an ordered arrangement of atoms with a specific chemical composition.

Mohs scale of hardness A ranking of a mineral's hardness, which is its resistance to scratching.

Nonsilicate A mineral that does not contain silica (silicon + oxygen).

Plutonic rock Intrusive igneous rock formed from magma that cools beneath Earth's surface. Granite is a plutonic rock.

Rock An aggregate of minerals. Some rocks are aggregates of fossil shell fragments, solid organic matter, or any combination of these components.

Rock cycle A sequence of events involving the formation, destruction, alteration, and reformation of rocks as a result of the generation and movement of magma; the weathering, erosion, transportation, and deposition of sediment; and the metamorphism of preexisting rocks.

Sedimentary rocks Rocks formed from the accumulation of weathered material (sediments) that has been eroded by water, wind, or ice.

Sedimentation The stage of sedimentary rock formation in which deposited sediments accumulate and change into sedimentary rock through the processes of compaction and, usually, cementation.

Silicate A mineral that contains both silicon and oxygen and (usually) other elements in its chemical composition; silicates are the largest and most common rock-forming mineral group.

Streak The name given to the color of a mineral in its powdered form.

Tectonic plates Sections into which Earth's crust is broken up; they move in response to heat flow and convection in Earth's interior.

Volcanic rocks Extrusive igneous rocks formed by the eruption of molten rock at Earth's surface. Basalt is a volcanic rock.

Volcano A central vent through which lava, gases, and ash erupt and flow.

Weathering Disintegration and/or decomposition of rock at or near Earth's surface.

REVIEW QUESTIONS

Our Rocky Planet

1. How did density segregation contribute to Earth's internal layers?
2. What is the most abundant element for Earth as a whole?
3. What is the most abundant element in Earth's crust? What is the second most abundant element?

What Is a Mineral?

4. What is a mineral?

Mineral Properties

5. What physical properties are used to identify minerals?
6. Most mineral samples do not display their crystal forms. Why not?
7. Why is color not always the best way to identify a mineral?

Classification of Rock-Forming Minerals

8. What is the difference between a silicate mineral and a nonsilicate mineral?
9. What is the most abundant mineral in Earth's crust? What is the second most abundant mineral?

The Formation of Minerals and Rock

10. What is the process of crystallization?

Rocks Are Divided into Three Main Groups

11. Name the three major types of rocks and describe the conditions of their origin.

Igneous Rocks Form When Magma Cools

12. What are the most common igneous rocks, and where do they generally occur?
13. Where on Earth's surface are lava flows most common?
14. What are the three major types of volcanoes?
15. What are three common types of plutons?

Sedimentary Rocks Blanket Most of Earth's Surface

16. How does weathering produce sediment? Distinguish between weathering and erosion.
17. What is a clastic sedimentary rock?
18. When water evaporates from a body of water, what type of sediment is left behind?

Metamorphic Rocks Are Changed Rocks

19. What is metamorphism? What causes it?
20. Distinguish between foliated and nonfoliated metamorphic rocks.

THINK AND EXPLAIN

1. What three sources of heat contributed to the melting and density segregation of early Earth?
2. Which type(s) of rock is (are) made from previously existing rock? Which does not require high temperature and pressure for its formation?
3. Which type of volcano produces the most violent eruptions? Which type produces the quietest eruptions?
4. What two mineral groups provide most of the ore that society needs?
5. Would you expect to find any fossils in limestone? Why or why not?

6. Is cleavage the same thing as crystal form? Why or why not?

7. Are the Hawaiian Islands made up primarily of igneous, sedimentary, or metamorphic rock?

8. Where does most magma originate?

9. What patterns of alteration are characteristic of contact metamorphism?

10. In what two ways does sediment turn into sedimentary rock?

11. Name two mica minerals that can give a metamorphic rock its foliation.

12. How is foliation different from sedimentary layering?

13. What are the two processes by which rock is changed during metamorphism?

14. What type of rock is formed when magma rises slowly and solidifies before reaching Earth's surface? Give an example.

15. Each of the following statements describes one or more characteristics of a particular metamorphic rock. For each statement, name the metamorphic rock being described.
    (a) Foliated rock, sometimes derived from granite
    (b) Hard, nonfoliated, single-mineral rock, formed under high-to-moderate pressure
    (c) Foliated rock, possessing excellent rock cleavage; generally used in making blackboards
    (d) Nonfoliated rock composed of carbonate minerals
    (e) Foliated rock containing about 50% platy minerals; named according to the major minerals in the rock

16. How do chemical sediments produce rock? Name two chemical sedimentary rocks.

17. How are most carbonate rocks formed?

18. What is a fossil? How are fossils used in the study of geology?

19. Imagine that we have a liquid with a density of 3.5 g/cm³. Knowing that objects of higher density will sink in the liquid, will a piece of quartz sink or float in the liquid? How about a piece of chromite?

20. The factors that influence bond strength influence mineral hardness. What are these factors?

THINK AND SOLVE

Gold has a density of 19.3 g/cm³. A 5-gal pail of water (density of water = 1.0 g/cm³) has a mass of about 18 kg. What is the mass of a 5-gal pail of gold?

READINESS ASSURANCE TEST (RAT)

If you have a good handle on this chapter—if you really do—then you should be able to score 7 out of 10 on this RAT. Check your answers with your teacher. If you score less than 7, study further before moving on.

Choose the best answer to each of the following

1. The silicates are the largest mineral group because silicon and oxygen are
   (a) the hardest elements on Earth's surface.
   (b) the two most abundant elements in Earth's crust.
   (c) found in the common mineral quartz.
   (d) stable at Earth's surface.

2. Compaction and cementation of sediments leads to
   (a) magma generation.
   (b) sedimentary rocks.
   (c) formation of pore water.
   (d) metamorphism.

3. Why are silicon and oxygen concentrated near Earth's surface while iron is concentrated at the core?
   (a) Earth's materials separated early in its history through the process of density segregation.
   (b) Silicon and oxygen are less dense than iron.
   (c) Both of these.
   (d) Neither of these.

4. All of the following can occur during metamorphism EXCEPT
   (a) melting.
   (b) recrystallization.
   (c) changes in composition.
   (d) changes in crystal size.

5. The most characteristic feature of sedimentary rocks is that they
   (a) are observed in great thicknesses.
   (b) are formed in layered sequences.
   (c) contain fossils.
   (d) are made from unconsolidated sediments.

6. In a sedimentary rock, the degree of particle roundness can indicate
   (a) the duration and/or length of travel.
   (b) where the sediment particles originated.
   (c) where the particles were deposited.
   (d) how the particles were cemented and compacted.

7. The characteristics of regional metamorphism include
   (a) deformed rock layers.
   (b) distinctly foliated rocks.
   (c) zoned sequences of minerals.
   (d) All of these.

8. Coarse-grained plutonic igneous rocks occur because
   (a) lava intrudes deep into Earth's interior.
   (b) minerals cooled and grew quickly.
   (c) minerals cooled and grew over long periods of time.
   (d) larger minerals are more stable than smaller ones.

9. What most strongly influences a mineral's hardness?
   (a) The geometry of a mineral's atomic structure
   (b) The strength of a mineral's chemical bonds
   (c) The silica content
   (d) The number of planes of weakness

10. Sedimentary rocks often contain the remains of ancient life forms—fossils. And fossils in these rocks help us understand
    (a) how humans evolved.
    (b) Earth's early formation.
    (c) the zoned sequences of minerals.
    (d) Earth's geologic and biologic history.

26
THE ARCHITECTURE OF EARTH

THE MAIN IDEA

Seismic waves let us "see" Earth's inner structure.

Did you ever think about digging a hole to China? The idea is intriguing, for why go all the way around the Earth if you can take a shortcut straight through? Digging such a hole, unfortunately, is not a realistic possibility. But if such a hole were possible, what would we find in Earth's interior? Is Earth's interior different from its surface? If so, in what ways? More important, how do we know that it is different?

We learn about Earth's interior by making observations at its surface—not only surface rocks but events that create surface rocks, such as volcanic eruptions. Does the upward movement of magma cause Earth's surface to tremble? The answer is yes. In fact, any movement in Earth's interior can cause the surface to shake. Depending on size, these "earthquakes" might be mild or fearsome and destructive events. What is important is that they provide a vital key to understanding Earth's internal structure. Rocks, volcanic eruptions, and earthquakes are all external expressions of Earth's internal processes.

Explore!

How Do Seismic Waves Travel Through Earth's Interior?

1. Place a slinky on a table. With two people holding the ends of the slinky, stretch it so it is about 5 feet in length.
2. On one end, pull the slinky backward and then push forward rapidly. Stop, and observe.

Analyze and Conclude

1. **Observing** The pull-push force creates vibrations (waves) in the slinky. The waves move from one person to the other along the length of the

slinky. Part of the slinky compresses and part of it stretches as the waves travel back and forth.
2. **Predicting** What would happen if you moved the slinky in a side-to-side motion, similar to the motion of a slithering snake?
3. **Making Generalizations** Push-pull seismic waves compress and expand the rock as they move through it, but side-to-side seismic waves do not. They simply shake the rock up and down or side to side, perpendicular to the direction the waves are traveling.

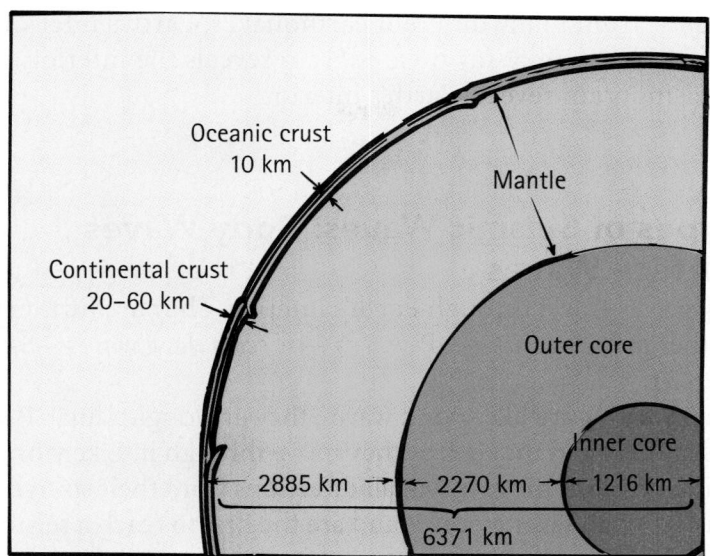

26.1 Earthquakes Make Seismic Waves

An **earthquake** is the shaking or trembling of the ground that occurs when rock beneath Earth's surface breaks. All earthquakes create waves that travel through Earth's interior. Such earthquake-generated waves are called *seismic waves.* The way these waves travel has provided Earth scientists with a view into Earth's interior revealing a planet that is layered. The major layers of Earth are the *crust, mantle,* and *outer* and *inner core* (Figure 26.1).

Recall from Chapter 12 that wave speed depends on the medium through which the waves travel. We learned that sound waves made by two rocks clicking together travel faster through water than through air. And sound waves travel even faster through solids than liquids. Just like sound waves, ✔ the speed of seismic waves depends on the properties of the material through which they travel. Different materials cause different wave speeds. So measuring the speeds of seismic waves tells us a lot about Earth's internal composition.

During an earthquake, energy is released within Earth's interior and radiates outward in all directions. As the energy travels to Earth's surface, an earthquake occurs as the ground shakes and moves. This ground movement is recorded on a machine called a *seismograph* (Figure 26.2), which provides information about the strength and speed of seismic waves. By combining seismograph records (called

Wave speed is dependent on the properties of the rock through which the waves travel. If the rock is denser and more elastic, wave speed increases. If the rock is less dense and less elastic, wave speed decreases. Each change in speed corresponds to a change in the rock's properties.

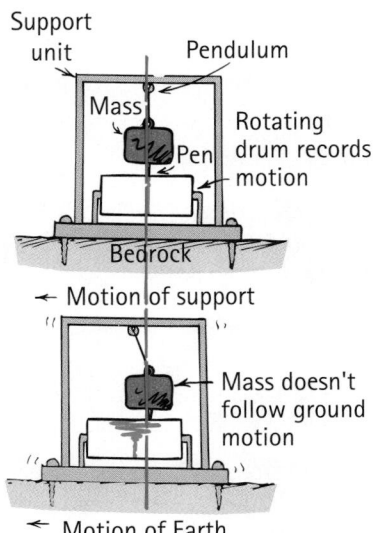

FIGURE 26.2 ▶
Diagram of a seismograph. When the ground shakes, the support unit attached to the ground also moves, but because of inertia, the mass at the end of the pendulum tends to stay in place. A pen attached to the mass marks the relative displacement on the slowly rotating drum beneath. In this way, the seismograph records ground movement.

READING CHECK

How do seismic waves reveal Earth's internal composition?

Waves stopped by a boundary between differing materials bounce back—they reflect. Waves that pass through a boundary change speed. A change in speed causes the wave to bend and change direction—they refract.

READING CHECK

What are the two basic types of seismic waves?

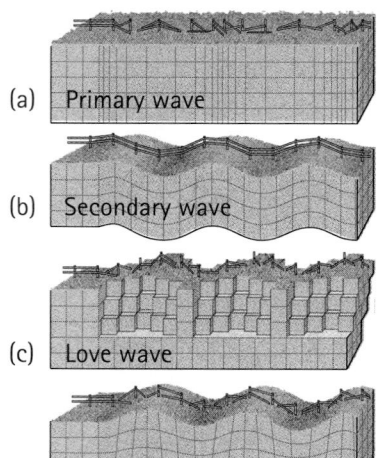

(a) Primary wave

(b) Secondary wave

(c) Love wave

(d) Rayleigh wave

seismograms) from many different earthquakes, Earth's interior is revealed. So just as an X-ray or sonogram reveals the interior of your body, a seismogram reveals Earth's interior.

Two Types of Seismic Waves: Body Waves and Surface Waves

Body waves travel through Earth's interior. They are further classified as either *primary waves*—P-waves—or *secondary waves*—S-waves (Figure 26.3).

Primary waves are like sound waves, they are longitudinal. P-waves compress and expand the rock as they move through it. Like vibrations in a bell, primary waves move out in all directions from their source. They are the fastest of all seismic waves and are the first to reach a seismograph. Because both solids and fluids can compress and expand, P-waves travel through any type of material—solid rock, magma, water, or air.

Secondary waves are transverse, like the waves on a vibrating violin string. S-waves vibrate the particles of their medium up and down and from side to side. Vibrations are perpendicular to the direction of wave travel. Because S-waves travel more slowly than P-waves, they are the second waves to register on a seismograph. Importantly, S-waves cannot move through fluids—they travel only through solids.

Surface waves travel on Earth's surface. There are also two types of surface waves: *Rayleigh waves* and *Love waves*. **Rayleigh waves** are both longitudinal and transverse and move in an up-and-down motion. **Love waves** are transverse and move in a side-to-side, whiplike motion. Surface waves travel more slowly than P- and S-waves, so they are the last to register on a seismograph. Despite their speed, however, because they travel on Earth's surface they do the most damage.

In Earth's interior, seismic waves are reflected by "boundary surfaces" between differing materials. And when seismic waves pass into a different material, their wave speed changes, causing the wave to bend (refract). Geoscientists study the reflection, refraction, and speeds of the various types of seismic waves to piece together a story about Earth's interior. Seismic-wave research has revealed the architecture of Earth's internal layers (Figure 26.4).

◄ **FIGURE 26.3**
Block diagrams show the effects of seismic waves. The yellow portion on the left side of each diagram represents the undisturbed area. (a) Primary body waves alternately compress and expand the material through which they travel—as shown by the different spacing between the vertical lines. This is similar to the action of a spring. (b) Secondary body waves cause the material through which they travel to move up and down and from side to side. (c) Love surface waves whip back and forth like secondary body waves, but only in the horizontal direction, along the ground surface. (d) Rayleigh surface waves have a rolling, up-and-down motion, similar to ocean waves.

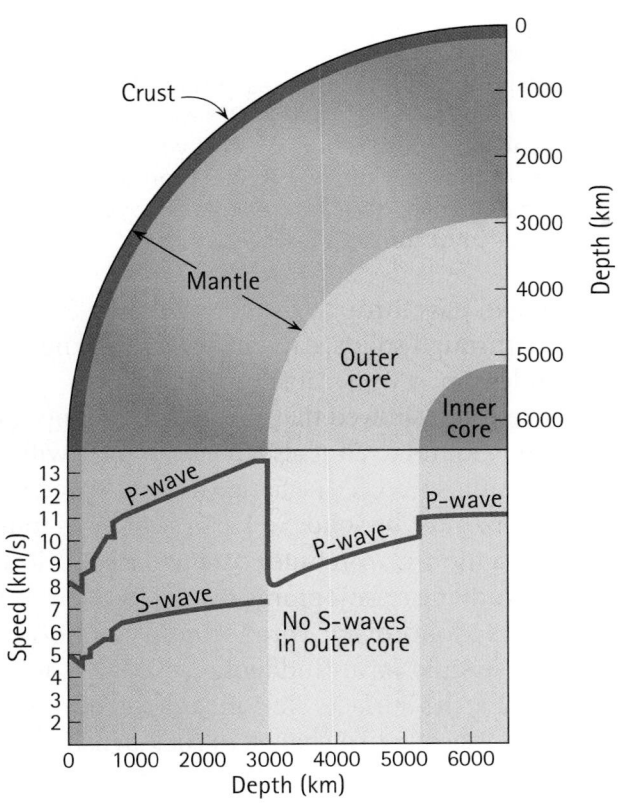

FIGURE 26.4 ▲
Cross section of Earth's internal layers, showing the increases and decreases of P-wave and S-wave velocity in the different layers.

fyi

- Much that we know about Earth's interior was learned as a result of the Cold War between the United States and the former Soviet Union. In the 1960s, when testing of nuclear weapons was very common, underground nuclear explosions were found to produce seismic waves. Both countries installed sensitive seismographic stations to monitor their opponent's activities. It was the seismograms from this network of stations that revealed many of the details of the unseen structure of our planet.

26.2 Seismic Waves Reveal Earth's Internal Layers

The discovery of Earth's internal structure did not happen all at once. In fact, like most scientific discoveries, it took many years to put the entire picture together. ✔ But all the discoveries were made by the scientific analysis of seismograms and commonsense thinking.

In the early 1900s, earthquake seismograms showed that Earth has a central core. S-waves were found to travel some distance through Earth and then stop. P-waves traveled as far into Earth as the S-waves, but instead of stopping they refracted and lost speed. These variations in P-waves and S-waves were found to be caused by a boundary at approximately 2900 km depth—the core–mantle boundary. When P-waves reach this boundary, they reflect and refract so strongly that the core actually casts a P-wave "shadow" over part of Earth (Figure 26.5). The shadow is a region where no waves are detected at the surface. Because the boundary is so distinct, it marks an important change in the density of the materials present in Earth's interior.

Both the overall density of Earth and the speed with which P-waves travel through the core suggest that the core is composed of iron, a metal that is much denser than the silicate rocks that make up the mantle. The S-wave shadow is even more extensive than the P-wave shadow—S-waves

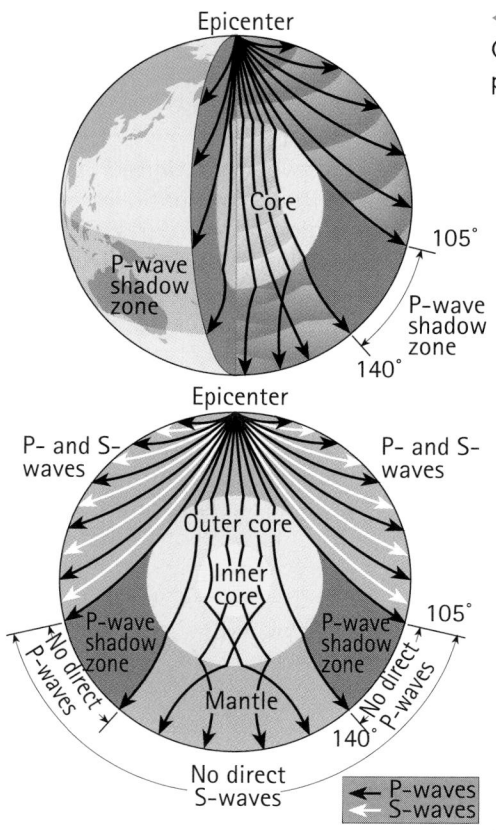

◄ **FIGURE 26.5**
Cutaway and cross-sectional diagrams showing the change in wave paths at the major internal boundaries and the P-wave shadow. The P-wave shadow between 105° and 140° from an earthquake's epicenter is caused by the refraction of P-waves at the core–mantle boundary. Note that any location that is more than 105° from an earthquake's epicenter does not receive S-waves because the liquid outer core does not transmit S-waves.

are not able to pass through the core. Because S-waves travel only through solids, early investigators knew that Earth's core, or at least part of it, must be liquid.

Later research showed that P-waves refract not only at the core–mantle boundary, but also at a boundary within the core, where they gain speed. This change in wave speed tells us that the inner core must be solid. So Earth's core was found to have two parts—a molten-iron outer core and a solid-iron inner core.

While studying seismograms from a local earthquake in 1909, Andrija Mohorovičić (pronounced "moho-rovu-chick") noticed that seismic waves suddenly picked up speed at a certain depth below Earth's surface. He concluded that the increase in wave speed was due to a change in rock density. The seismograms had literally drawn a map of the upper boundary of Earth's *mantle*, a layer of denser rock underlying the less dense crust. This boundary, known as the **Mohorovičić discontinuity** (called the "Moho" for short), separates Earth's crust from rocks of different composition in the mantle below.

Taken together, these discoveries showed that Earth has a layered structure: crust, mantle, and an inner and outer core. Do you suppose these internal layers influence the geologic changes our planet experiences? The answer is yes, as we will now see.

✔ **READING CHECK**

How were Earth's internal layers discovered?

CONCEPT CHECK
What evidence supports the idea that Earth's inner core is solid and the outer core is liquid?

Check Your Answer
The evidence is in the way P- and S-waves move through Earth's interior. As the waves encounter the core–mantle boundary, P-waves are reflected and refracted, but S-waves are only reflected. This results in very pronounced wave shadows. Because S-waves cannot travel through liquids, the S-wave reflection implies a liquid outer core. The P-waves continue to move through the outer core. Then at a depth of nearly 5200 km, they suddenly increase in speed. Knowing that P-waves travel faster in solids than in liquids, we infer that the inner core is solid.

Earth's core is much hotter than Earth's surface. Temperature and pressure increase with depth in Earth's interior.

The Core Has Two Parts: A Solid Inner Core and a Liquid Outer Core

Earth's **core** is composed mainly of iron and smaller amounts of nickel. In the inner core, the iron and nickel are solid. Although the inner core is indeed very hot, intense pressure from the weight of the rest of Earth prevents the inner core from melting (just as increased pressure in a pressure cooker prevents high-temperature water from boiling).

Whereas the inner core is solid, the outer core is a molten liquid. This is because the outer core is under less pressure due to less weight of Earth above. The molten outer core flows at the very slow rate of several kilometers per year. This flow is evident far outside Earth's surface. ✓ The flowing molten outer core produces a flowing electric charge—an electric current. This electric current powers Earth's magnetic field. This magnetic field is not stable but has changed throughout geologic time. Recall from Chapter 11 that there have been times when Earth's magnetic field has diminished to zero, only to build up again with the poles reversed. These magnetic pole reversals probably result from changes in the direction of fluid flow in the molten outer core of Earth.

CONCEPT CHECK
Iron's normal melting point is 1535°C, yet Earth's inner core temperature is at least 5000°C. Why doesn't the solid inner core melt?

Check Your Answer
The intense pressure from the weight of Earth above crushes atoms together so tightly that even the high temperature cannot budge them. Because of the pressure, melting cannot occur and the inner core remains solid (similar to how high-pressure gas in a pressure cooker prevents boiling).

The Mantle Is Dynamic
✓ Surrounding the core of the planet is the **mantle,** a rocky layer some 2900 km thick. The mantle is Earth's thickest layer and makes up about 84% of Earth's volume. From top to bottom, the mantle's composition is relatively uniform—composed of hot, iron-rich silicate rocks. In general, these mantle rocks behave like an elastic solid. And in most parts of the mantle, the rocks can actually flow, even though they are solid. This behavior—the ability of rocks to flow without breaking—is called *plasticity.* So even though it is uniform in composition, the mantle varies in its physical properties. How do we know that it varies? The answer is seismic studies!

Information from seismic studies divides the mantle into two portions—the lower mantle and the upper mantle. The lower mantle extends from the outer core to a depth of about 700 km (the dashed line in Figure 26.4). The lower mantle is completely solid because pressure in this region is too great for melting to occur.

The upper mantle, which extends from the 700-km depth to the crust–mantle boundary, has two zones (Figure 26.6). The lower zone of the upper mantle is called the **asthenosphere.** This zone is solid but contains small amounts of liquid from the partial melting of mantle rocks. The asthenosphere is especially plastic, and it flows more easily than the lower mantle. Plastic flow in the mantle takes the form of very slowly moving *convection currents*—hot material rises, cools, and then

The discovery of magnetic pole reversals helped explain how continents move.

 READING CHECK

In what way does flow in Earth's core influence places far outside of the core?

Have you ever played with Silly Putty? If so, you are familiar with plastic flow. Silly Putty behaves like a solid under sudden stress—it breaks—but behaves like a fluid when stress is applied slowly—it flows.

 READING CHECK

Which of Earth's layers has the most volume?

FIGURE 26.6 ▶
The core, mantle, and crust each have a different composition. The mantle extends from the core–mantle boundary to the base of the crust. The lower mantle is essentially a single unit from a depth of 2900 km up to about 700 km. Although the upper mantle is of a similar composition, it is divided into two distinct units. The lower portion of the upper mantle is the plastic asthenosphere. The top portion of the upper mantle, and the entire crust, together form the rigid lithosphere.

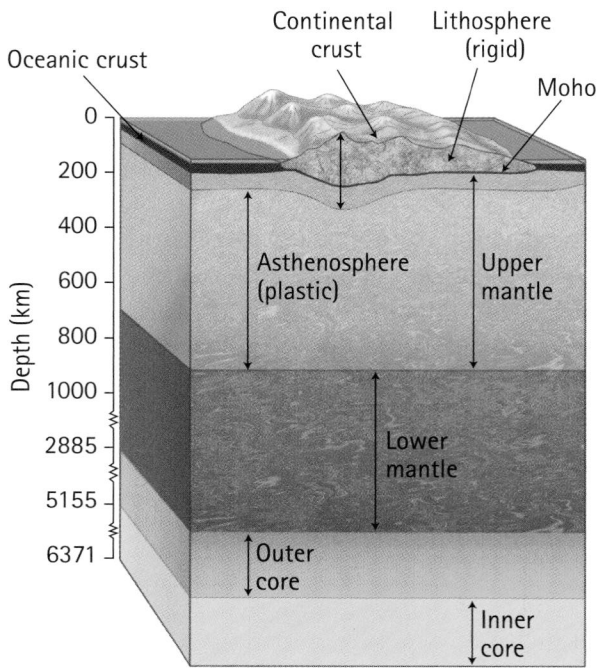

So the upper mantle includes the asthenosphere and part of the lithosphere, and the lithosphere is part mantle and part crust.

(a)

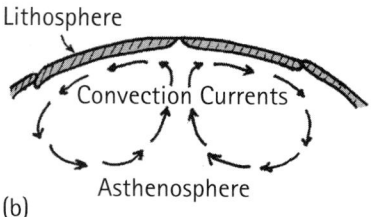

(b)

sinks (Figure 26.7). As we will soon see, the constant flowing movements in the asthenosphere greatly affect the surface features of our planet.

Above the asthenosphere is the **lithosphere.** The lithosphere is about 100 km thick and includes the entire crust and the uppermost part of the mantle. Unlike the asthenosphere, the lithosphere is more rigid and brittle and does not flow. The lithosphere is, in a sense, riding on top of the asthenosphere like a raft on a pond. The lithosphere moves along with the motions of the material beneath it in the asthenosphere. The motions in the mantle, however, are not uniform. Because of this, the brittle lithosphere is broken into many individual pieces called *plates.*

Mantle convection currents move at a leisurely pace, taking hundreds of millions of years to complete one loop. Even so, heat-driven motion in Earth's interior shapes and reshapes many of our surface features. The lithospheric plates are always in motion. Their movement causes earthquakes, volcanic activity, and the deformation of large masses of rock that create mountains.

CONCEPT CHECK

The mantle has a fairly uniform composition, yet its physical properties vary. Why is this?

Check Your Answer

The mantle makes up 84% of Earth's volume. It is huge! Conditions—such as temperature and pressure—nearer to the crust are very different from those closer to the core. From top to bottom, as conditions vary, so do the physical properties of the mantle material.

◀ **FIGURE 26.7**
(a) A familiar example of convection is seen when water is heated in a pan.
(b) A simple model showing convection currents in the asthenosphere.

The Crust Makes up the Ocean Floor and Continental Land

The top part of the lithosphere is the **crust.** We live on the continental crust; the oceanic crust makes up the ocean floor. Oceanic crust is very different from continental crust. The crust of the ocean basins is compact; it's only about 10 km thick and composed of dense basaltic rocks. Continental crust is thicker (20 to 60 km in thickness), and it is composed of granitic rocks. ✔ Because granitic rocks are less dense than basaltic rocks, most of the continental crust is above sea level.

If the continental crust is so much thicker than oceanic crust, why are the ocean basins underwater and the continents high and dry? The answer is found in their density differences and buoyancy (Chapter 8). Remember, the entire lithosphere—uppermost mantle and crust—"floats" on the asthenosphere. Objects float because a *buoyant force* acts on them. In the case of the "floating lithosphere," the buoyant force is produced by the underlying mantle. The upward push of the asthenosphere opposes the downward pull of gravity. When the buoyant and gravitational forces are in balance, the vertical position of the crust is stable. This is the principle of *isostasy.* So, the thin oceanic crust and the thick continental crust are in isostatic balance with respect to the asthenosphere. The upward-acting buoyant force of the asthenosphere equals the weight of the entire lithosphere.

Focusing on the crust, the concept of isostasy can be made clear with the following analogy. Imagine that Earth's crust is a cargo ship and that the mantle is the ocean. The ship establishes its vertical position in the water when the net force on it is zero. This happens when the gravitational force pulling the ship downward (its weight) equals the buoyant force pushing it upward (Figure 26.8). When the ship is loaded, it is denser and floats lower in the water—more of it is submerged than when it is empty. Likewise, the crust's vertical position rises and falls according to variations in density (Figure 26.9). So, thin, dense oceanic crust sits lower than thicker, less-dense continental crust.

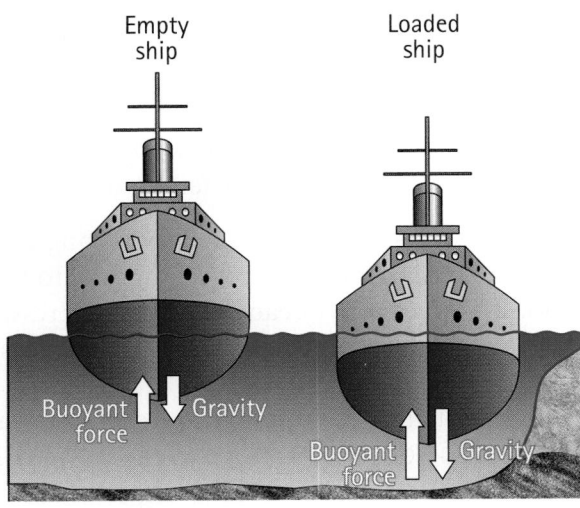

Empty ship   Loaded ship

Buoyant force — Gravity

Buoyant force — Gravity

Although the core conducts heat to the mantle, some of the mantle's heat comes from the decay of radioactive elements within it. So even though concentrations of radioactive materials are greater in Earth's crust, most of Earth's radioactive heat comes from the mantle due to its much greater volume.

If Earth were the size of a cue ball, it would be just as smooth! Its highest mountains are insignificant compared with the Earth's radius.

✔ **READING CHECK**

For what physics reason are the continents above sea level?

◀ **FIGURE 26.8**
The vertical position of the crust is stable when the gravitational and buoyant forces balance. Denser oceanic crust therefore has a lower vertical position than the less-dense continental crust, just as the loaded ship sits lower in the water than the unloaded ship.

FIGURE 26.9 ▶
Continental crust is thicker and less dense than oceanic crust. As such, continental crust floats higher on the mantle than oceanic crust. To achieve balance, the thicker crust has deeper roots.

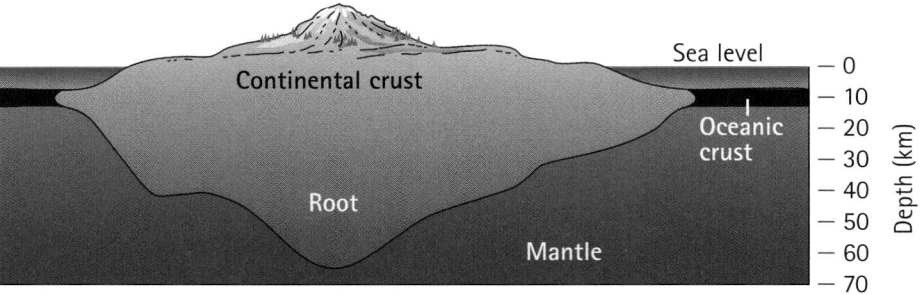

CONCEPT CHECK

If you wished to drill the shortest hole to the mantle, would you drill in western Colorado or in Florida?

Check Your Answer

Put the question another way: If you wanted to drill the shortest hole through ice to the water below, would you drill atop an iceberg or through a slab of ice that hardly extends out of the water? You would drill your hole in the slab, of course, and likewise you should drill through the thinner crust of mountain-free Florida. In mountainous western Colorado, the crust is much thicker. (If you really want the shortest hole, you should drill through the ocean floor—exactly what scientists have done in Project MoHole, in the East Pacific Ocean.)

26.3 Internal Motion Deforms Earth's Surface

Lithospheric plates move slowly but constantly as they float across the asthenosphere. ✔ The motion pushes and pulls on the plates, which puts stress on them. Rocks subjected to stress begin to deform. There are three types of stress, and each stress has a different result:

> *Compressional stress* occurs when slabs of rock are pushed together. The result is shortening of the rock bodies.
>
> *Tensional stress* occurs when slabs of rock are pulled apart. The result is extension of the rock bodies.
>
> *Shear stress* occurs when slabs of rock are both pulled and pushed. The result is the sliding of one slab of rock past the other in opposite directions, without any noticeable shortening or extension.

Rock bodies respond to stress in different ways. Although we generally think of rocks as a hard solid, rocks are actually elastic to varying degrees. *Elastic* doesn't necessarily mean "stretchy." A stressed elastic material simply returns to its original shape when the stress is removed. For example, a stretched rubber band returns to its original shape after it is released. If the rubber band breaks when stretched, it means that the *elastic limit* was exceeded. Rocks also return to their original shape after the stress is removed—unless the elastic limit is exceeded.

✓ When stress exceeds the elastic limit of the rock, the rock permanently loses its original form. The rock either breaks or flows. Rock that breaks undergoes *brittle deformation;* rock that flows has *plastic deformation.* The response of the rock to stress depends on temperature, pressure, and the composition of the rock. Brittle deformation occurs near Earth's surface where temperatures and pressures are low. Brittle deformation produces faults and fractured rocks. Plastic deformation typically occurs deep below Earth's surface where temperature and pressure conditions are high. Plastic deformation causes rock to fold and flow. The geologic structures we see at Earth's surface, such as folds, faults, and related mountains, are all examples of rock deformation from stresses that exceeded the strength of the rock.

Folded Rock: An Expression of Compressive Force

✓ Compressive stresses push rocks together. The rocks then begin to buckle and fold (Figure 26.10). This is similar to the wrinkles you might find in a throw rug when you push one end of the rug toward the other end. Of course, to wrinkle large sections of rock requires strong forces, which come from the movement of lithospheric plates.

Recall from Chapter 25 that sediments are deposited horizontally layer by layer, with the bottom layer deposited first. Thus, the bottom layer is the oldest, and the top layer is the youngest. As these originally flat-lying sedimentary rock layers are subjected to compressive stress, they tilt and become folded. Rock can be up-folded or down-folded. Each fold has an axis, with rock layers on one side of the axis a mirror image of the rock layers on the other side. You can imagine the axis as a plane extending downward into Earth, as Figure 26.11 shows. When the layers tilt toward the fold axis, so that if you put a marble on the rock it would roll toward

✓ **READING CHECK**

What is the cause of stress in rocks? What happens when stress exceeds the rock's strength?

✓ **READING CHECK**

What kind of stresses cause rocks to buckle and fold?

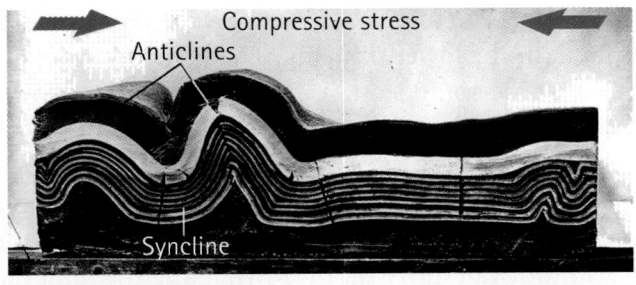
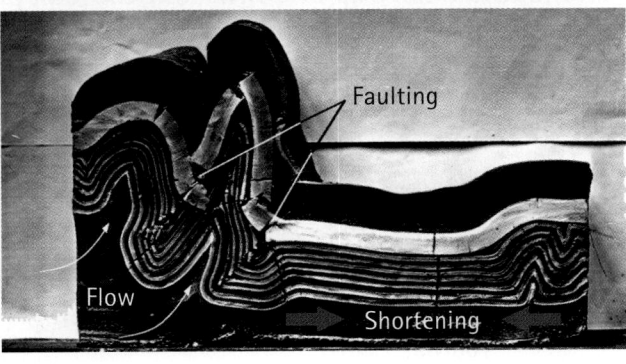

◀ **FIGURE 26.10**
These photos show a cross section of wax layers subjected to compressive stress. The wax layers represent sedimentary layers. When compressed from opposite sides, the wax layers become shortened and deform into folds. With more stress, the wax layers continue to deform until they break—forming faults.

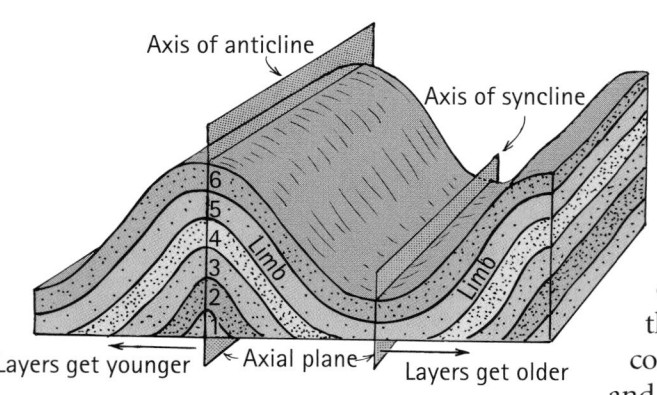

FIGURE 26.11 ▲
Anticline and syncline folds. Layer 1 is the oldest rock, and layer 6 is the youngest. The limbs of an anticline (an up-fold) tilt away from the axis of the fold, and the rock layers are oldest at the core of the fold. The limbs of a syncline (a down-fold) tilt toward the axis of the fold, and the rocks are youngest at the core.

the axis, the fold is called a **syncline.** The rocks at the center, or *core,* of a syncline are the youngest. As you move horizontally away from the axis, the rocks get older and older. If the fold layers tilt away from the axis, so that if you put a marble on the rock it would roll away from the axis, the fold is called an **anticline.** The rocks in an anticline are oldest at the core, and as you move horizontally away from the axis, the rocks get younger. Another way to think about this concept is that anticlines are pushed upward into an arch, and synclines are pushed downward into a sag.

> ## CONCEPT CHECK
> Why are rocks at the core of a syncline younger than those farther out from the core, whereas the opposite is true for an anticline?
>
> ### Check Your Answer
> Think of the rug example. Assume the top surface of the rug is younger than the lower surface. When you push the rug it can (1) fold upward (like the letter "A"), or (2) fold downward (like a sag). In the first case, the bottom surface makes up the core—an anticline. In the second case, the top surface makes up the core—a syncline. Makes sense!

Faults Are Caused by the Forces of Compression, Tension, or Shearing

☑ When rock deformation is brittle, the rock fractures into separate blocks. If one block then moves relative to the other block, the fracture is called a **fault.** Movement along a fault can occur rapidly and produce an earthquake, or slowly over time without significant earthquakes occurring.

Faults are classified by the relative direction of their movement. Look at Figure 26.12 and note the oblique line in the top drawing—this line represents a fault. Imagine that you could pull the block diagram apart at the fault, as shown in the lower drawing. The half containing the fault surface where someone could stand is the *footwall* block. The fault surface of the other half is inclined and would make standing impossible; this is the *hanging wall* block. These terms were coined by miners because one could hang a lamp on a hanging wall, and one could stand on a footwall. Movement on this type of fault is mostly up-and-down motion—the hanging wall and footwall move up or down along the fault plane.

For a fault resulting from compressional forces, the hanging wall is pushed upward along the fault plane relative to the footwall, as Figure 26.13

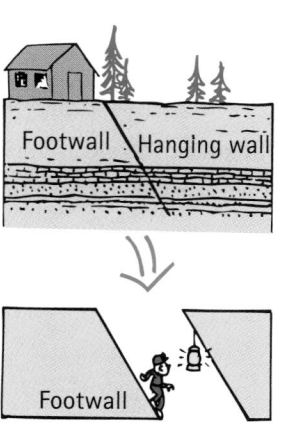

FIGURE 26.12 ▲
The terms *footwall* and *hanging wall* were commonly used by miners because one could hang a lamp on a hanging wall and stand on a footwall.

FIGURE 26.13 ▶
A reverse fault. In a zone of compressional faulting, the hanging wall is pushed up relative to the footwall. (a) A reverse fault before erosion; (b) the same reverse fault after erosion.

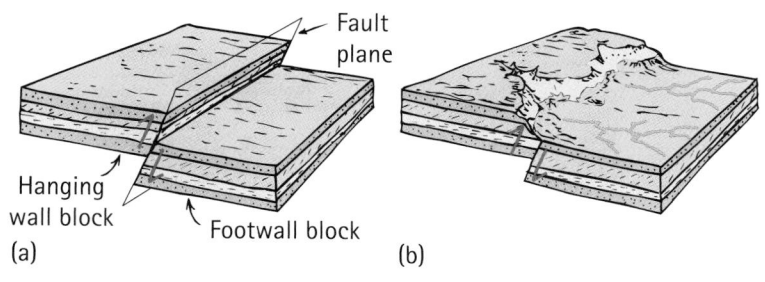

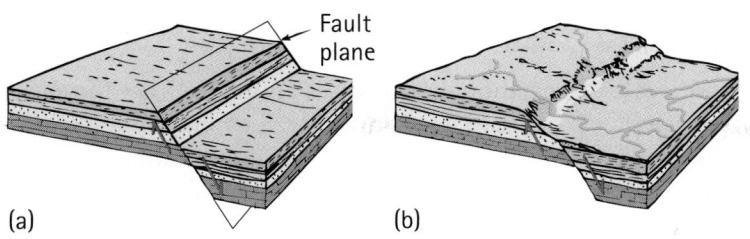

FIGURE 26.14 ▲
A normal fault. In a zone of tensional faulting, the hanging wall drops down relative to the footwall, forming a normal fault. (a) A normal fault before erosion; (b) the same normal fault after erosion.

shows. This type of fault is called a *reverse fault*. The Rocky Mountain foreland (east of the highest peaks), the Canadian Rockies, and the Appalachian Mountains, to name a few, were formed in part by reverse faulting.

Stress in rock also occurs because of tension. Tensional forces, which pull at rocks, are the opposite of compressional forces, which push. Tension causes the hanging wall to drop downward along the fault plane relative to the adjacent footwall. This type of fault is called a *normal fault* (Figure 26.14). Virtually the entire state of Nevada, eastern California, southern Oregon, southern Idaho, and western Utah are greatly affected by normal faulting.

Faults that have horizontal movement, in which blocks of rock slip horizontally past one another with very little vertical movement, are called *strike-slip faults* (Figure 26.15). Some of the world's most famous faults, such as the San Andreas Fault in California, are strike-slip faults. Sticking and slipping of rock blocks along the San Andreas Fault (really a zone of faults) generates many of the earthquakes California is noted for, including the great San Francisco quake of 1906.

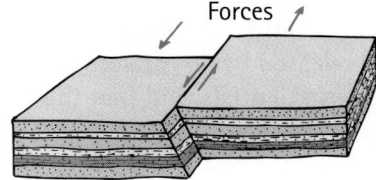

FIGURE 26.15 ▲
The relative movement of a strike-slip fault is horizontal. The rock bodies are neither shortened nor extended.

CONCEPT CHECK

1. Reverse faults are the result of compressional forces. What happens to Earth's crust in a zone of reverse faulting?
2. Normal faults result from tensional stress. What surface feature may we expect to find in a zone of normal faulting?

Check Your Answers

1. Compressional stress pushes rocks together. So in a zone of reverse faulting we would find shortening of the crust. Look at Figure 26.13; can you see how compression shortens the crust?
2. Tensional stress pulls the crust apart. So in a zone of normal faulting we would find extension of the crust. Look at Figure 26.14; can you see how tension extends and elongates the crust?

✔ **READING CHECK**

What causes rock to break apart and then form a fault?

Earthquakes—The Result of Sudden Deformation

Most earthquakes are related to movement in Earth's interior. We have learned how this movement creates stress, which can ultimately create faults. Stress continues to build up in fault zones when rocks stick together. Faults are not smooth planes—they have irregular

EARTHQUAKE MEASUREMENT—MAGNITUDE SCALES

Charles Richter developed the first magnitude scale in 1935. Magnitude scales are designed to measure the amount of ground shaking and energy released by an earthquake and are based on data from seismograms. The scales are logarithmic. This means that each 1-point increase on the scale is equivalent to a 10-fold increase in the amplitude of ground shaking. So a magnitude 6 earthquake shakes the ground 10 times more than a magnitude 5 earthquake but 100 times more than a magnitude 4 earthquake.

But how does magnitude relate to the energy released by an earthquake? Through careful analysis, scientists find that the energy released by an earthquake increases about 30 times for each increase in magnitude. So a magnitude 6 earthquake is 10 times larger but 30 times more energetic when compared to a magnitude 5 earthquake. And a magnitude 6 earthquake is 100 times bigger but 900 times more energetic than a magnitude 4 earthquake!

The Richter magnitude scale was developed for moderate-size, shallow-focus earthquakes. For larger earthquakes, the *moment magnitude scale* is used. This scale is similar to the Richter magnitude scale but also takes into account the length of fault rupture and the area over which the rupture occurred.

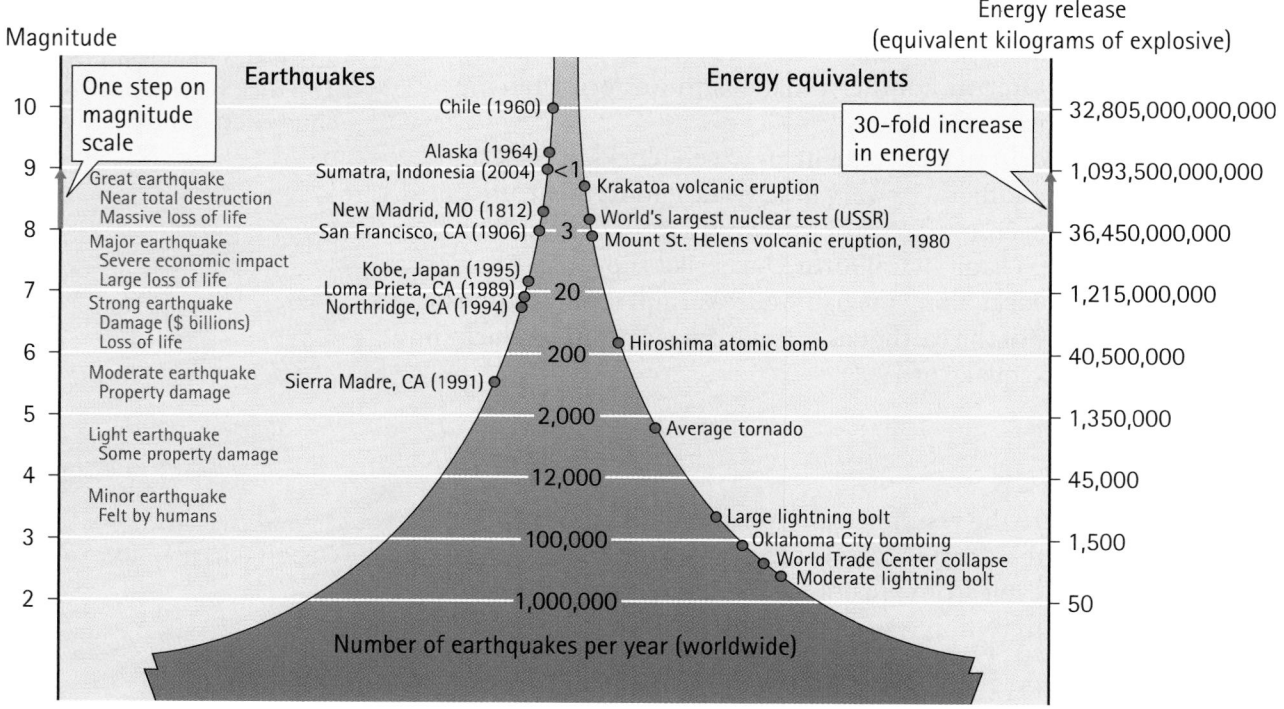

▲ This diagram shows earthquake magnitude compared to the energy released by an earthquake and the energy released by natural and human-caused events. It also shows how many earthquakes of a given magnitude occur each year. For example, the 1906 San Francisco earthquake had a magnitude of 7.8, and about three earthquakes of this size occur on a yearly basis.

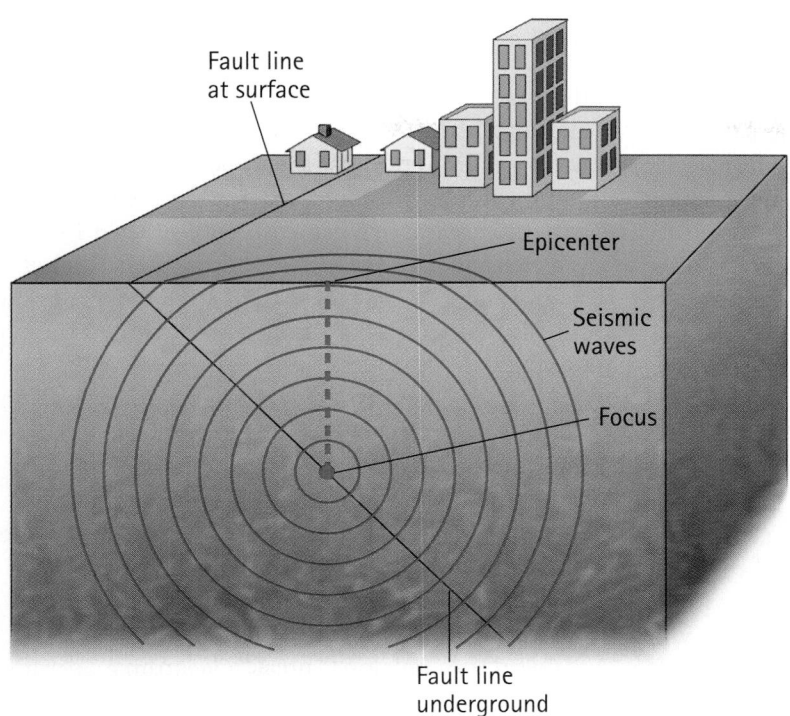

Fault line
at surface

Epicenter

Seismic
waves

Focus

Fault line
underground

◄ FIGURE 26.16
The actual underground
location where fault rupture
occurs is the *focus.* Directly
overhead at the ground surface is the *epicenter.* Seismic
waves radiate out from the
focus in all directions.

The stored-up elastic energy
is a form of potential energy,
which when released is
motion energy.

surfaces that interlock and resist movement. Rocks on opposite sides
of the fault become stuck and locked into position. This locked rock
stores elastic energy. Then when stress builds to the point where
it exceeds the strength of the fault, the rock suddenly snaps—an
earthquake!

The sudden release of stored elastic energy is very similar to the
recoil of a spring. The released energy radiates away from the initial
break site as seismic waves. This break site, the point of origin where the
rock initially slips, is called the *focus* (Figure 26.16). The point on Earth's
surface directly above the focus is called the *epicenter.*

✔️ The stored energy released in an earthquake does not come as
a single large quake. Remember, fault surfaces are irregular and not all
movement is at the same time. Although most of the energy is released
during the main quake, some of the energy comes before—*foreshocks*—
and some comes after—*aftershocks*. You have probably heard these terms
in news reports after an earthquake. You have probably also heard the
term *magnitude*. Magnitude refers to the earthquake's size and the
amount of energy it releases. Large earthquakes have a greater magnitude and release more energy than smaller earthquakes.

Because seismic waves weaken with increasing distance from an
earthquake, the strongest ground shaking is generally at the epicenter. But
ground shaking is not the only hazard people face during earthquakes.
Earthquakes can also trigger landslides. Another hazard is liquefaction,
in which wet sediment behaves like a fluid instead of a solid.

The soil and underlying rock can actually increase the size, or
amplitude, of seismic waves away from the epicenter. Waves carry

■ Earthquakes cause the
ground to shake and rup-
ture. As the ground shakes,
so do buildings on top of
the land. It is often said
that earthquakes don't kill
people, but falling buildings
do. Earthquakes can cause
general property damage,
lack of basic necessities,
collapse of buildings, loss
of life, higher insurance
premiums, disease, land-
slides and avalanches, road
and bridge damage, and
fires generated by broken
gas and electric lines.

✔️ READING CHECK

The release of stored
elastic energy results in an
earthquake. Is this energy
released all at once?

LINK TO PHYSICS: WAVE MOTION—TSUNAMI

A tsunami is a wave, or series of waves, generated in a large body of water by any type of powerful disturbance that vertically displaces the water column. Earthquakes, landslides, and explosions can generate a tsunami. When an earthquake causes the disturbance, a tsunami is sometimes referred to as a seismic sea wave.

Most tsunami arise when an earthquake occurs on an underwater reverse fault. Rapid movement of the seafloor—generally upward—quickly thrusts the water above the uplifted area upward. The huge displaced mass of water then drops back down to sea level and a large wave is generated—a tsunami (Figure 26.17).

✓ For all water waves, the depth over which wave energy travels is equal to one-half of the wave's wavelength. And this makes the behavior of tsunami different from most other ocean waves. For example, wind-driven ocean waves have wavelengths usually less than 150 m. So, most ocean waves affect only the uppermost part of the water column. ✓ Tsunami, however, have very long wavelengths—100 km or more. One-half of a tsunami's wavelength—about 50 km—is far deeper than the average ocean depth—about 4 km. So the energy within a tsunami is spread over the entire water column, not just the uppermost layer. This is an enormous amount of energy! In the open ocean, tsunami are barely visible because of their huge wavelength. The ocean rises perhaps 1 m higher than usual, but that 1-m rise is spread over 100 km! And tsunami are fast—they travel through the open ocean at 800 km/h!

The true power of a tsunami becomes apparent as the wave approaches shore. Wave speed depends on water depth, so as the tsunami enters shallow water, the wave slows down. The energy that was spread over a 4-km depth is now squeezed into the space of a few tens of meters, and then even less. Where does this energy go? Up! Wave height can grow from a mere meter to up to 30 m!

And tsunami don't "break" like ordinary ocean waves. As viewed from shore, a tsunami can appear as a very rapidly rising tide—a fast-moving wall of water. But if a wave trough gets to shore first instead of a crest, it appears as a rapidly receding, very low tide. Unfortunately, this can cause people to investigate

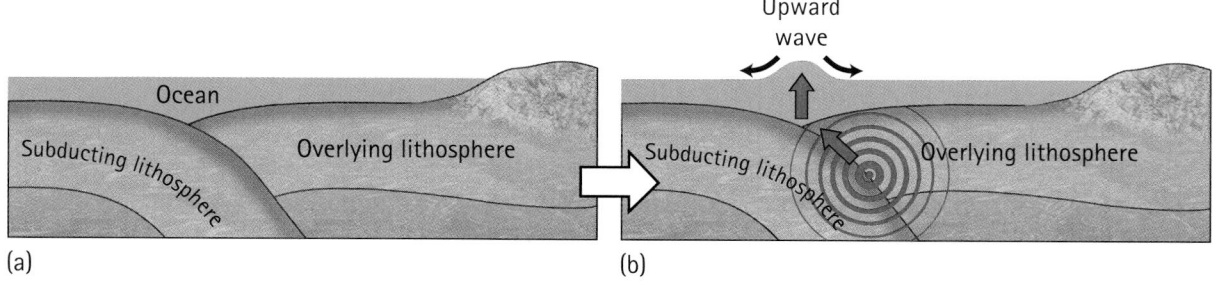

FIGURE 26.17 ▲
Most tsunami are generated by earthquakes in *subduction zones*—where one lithospheric plate moves below another plate. (a) When the edge of the overriding plate is pulled downward, elastic energy builds up. The release of the energy causes the earthquake. (b) When the edge of the bent plate snaps back into place, the entire water column is flung upward and then drops back down, causing the tsunami.

energy, and larger amplitude waves carry more energy than smaller amplitude waves. Seismic waves can be amplified when they are forced to slow down by some soil and rock types. The seismic energy that would have otherwise moved quickly through an area "piles up" instead. This increases the amplitude of the waves, which causes stronger ground shaking.

the odd sudden appearance of exposed seafloor, complete with flopping fish that have been temporarily stranded. The wave crest quickly follows, drowning unsuspecting onlookers. So in many ways, a tsunami on land is like an enormous sledgehammer delivering an extreme blow.

The 2004 Sumatra tsunami was generated in a subduction zone. Because of its size (magnitude 9.2) and huge release of energy, the Sumatran quake is termed a *megathrust* earthquake. The energy released by this quake is comparable to the amount of energy released by a bomb made up of 100 billion tons of TNT (if it were possible to make such a bomb). Although the quake itself caused severe damage and casualties, the tsunami that followed in the earthquake's aftermath was even more devastating. Coastal areas throughout the Indian Ocean basin were destroyed. This Indonesian tsunami killed more than 184,000 people in 12 countries (Figure 26.18).

- In lands bordering the Pacific Ocean, effective tsunami warning stations have been in existence since the 1960s. Unfortunately, the warning system did not extend into the Indian Ocean. After the catastrophe of the 2004 Sumatran tsunami, a warning system for the Indian Ocean was put in place. There is even the possibility of creating a unified global tsunami warning system, to include the Atlantic Ocean and the Caribbean.

✔ **READING CHECK**

How are tsunami waves different from regular ocean waves?

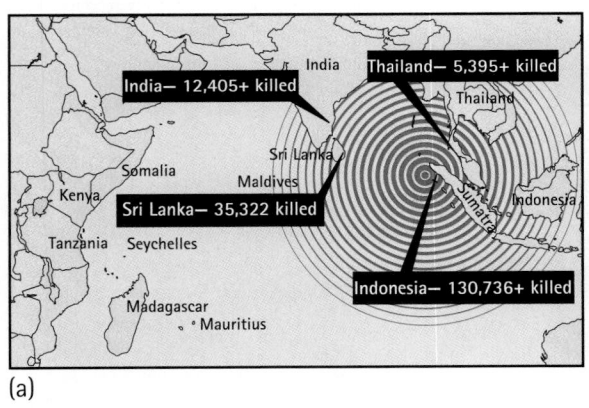

(a)

(b)

FIGURE 26.18 ▲
The Indonesian tsunami of 2004 was generated by a strong subduction-related earthquake off the coast of Sumatra. The tsunami struck the surrounding coastlines with catastrophic force; more than 184,000 people were killed, and many livelihoods were destroyed.

Seismic waves travel faster in hard, rigid rock, and slower in softer, less-rigid rock. And they travel even slower in unconsolidated sediment or artificial fill. Put another way, it takes less time to move seismic waves through a kilometer of hard material than it does to move seismic waves through a kilometer of soft material. Seismic waves linger and grow larger in less-rigid rock material.

- A primary focus of earthquake preparation today is to make sure buildings and other structures are engineered to withstand the maximum amount of shaking without completely collapsing. For example, during the 1989 Loma Prieta quake, the upper floors of the Transamerica Pyramid building swayed more than a foot from side to side, but its earthquake-resistant construction allowed the landmark to survive without damage.

✓ READING CHECK

Where do most earthquakes occur?

Let's use this knowledge as we look back at the 1989 Loma Prieta earthquake in California. Parts of San Francisco on rigid bedrock had the smallest wave amplitudes—so little damage occurred to structures built on bedrock. On the other hand, structures built on soft mud close to the bay suffered severe damage or collapse. For example, the Cypress Freeway structure collapsed, killing 40 people. And many homes in San Francisco's Marina district were severely damaged and caused five related deaths. Both of these damage sites were due to structures built on unconsolidated sediment. Shaken by the quake, the loose, soft sediment behaved like a slushy Jello—unable to support structures. The earthquake lasted approximately 15 seconds, but resulted in 67 deaths and 4000 injuries and left more than 12,000 people homeless. A lot of damage for 15 seconds of shaking!

Where Do Most Earthquakes Occur?

We see the impact of Earth's internal movement in the folds and faults of Earth's surface. ✓ Most earthquakes are associated with areas where lithospheric plates meet. So, not surprisingly, the majority of earthquakes occur in just a few narrow zones (Figure 26.19). For example, the area of the Pacific Rim is a zone of great seismic activity, as is the area of the Mid-Atlantic Ridge. Compare Figure 26.19 to Figure 25.1. Can you see that most earthquakes occur at plate boundaries?

Even so, earthquakes can happen anywhere in the world, and not just where plates meet. Earth is very old, and there have been many

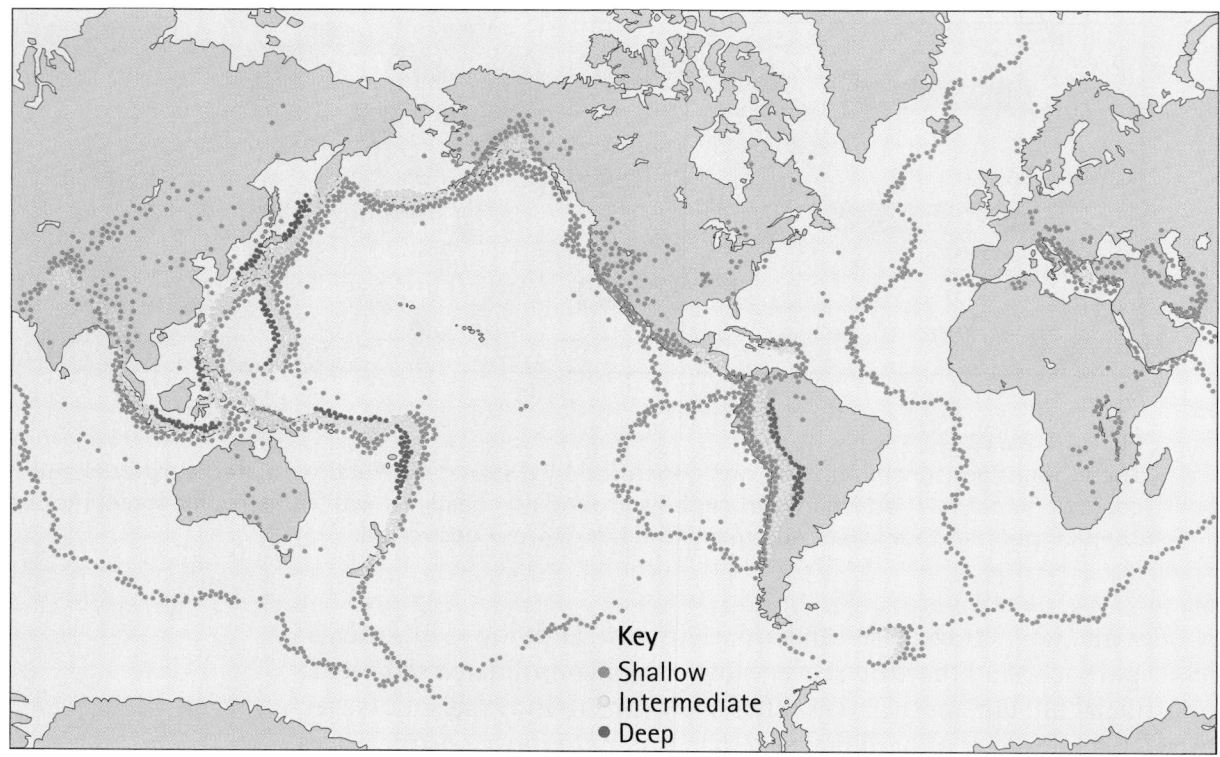

Key
- Shallow
- Intermediate
- Deep

FIGURE 26.19 ▲
Most earthquakes occur in just a few narrow zones.

TABLE 26.1 Some Notable Earthquakes

Year	Location	Magnitude	Estimated Deaths	Comments
1811	New Madrid, MO	8.0	few	
1906	San Francisco, CA	7.8	700	Fires, extensive damage
1923	Tokyo, Japan	8.2	150,000	Fires, extensive destruction
1960	Southern Chile	9.5	5,700	Largest earthquake recorded
1964	Anchorage, AK	9.2	131	>$300 million in damage
1970	Peru	7.9	66,000	Great rockslide
1975	Liaoning, China	7.5	few	First quake predicted
1985	Mexico City, Mexico	8.0	9,000	
1989	Loma Prieta, CA	7.1	62	>$6 billion in damage
1994	Northridge, CA	6.7	57	>$25 billion in damage
1995	Kobe, Japan	7.2	5,500	>$125 billion in damage
1999	Izmit, Turkey	7.6	17,000	
2001	India	7.6	20,000	
2001	El Salvador	7.7	1,000	
2003	Bam, Iran	6.6	>30,000	
2004	Sumatra, Indonesia	9.2	>184,000	Tsunami, >12 countries affected
2007	Sumatra, Indonesia	6.4	70	
2008	Sichuan, China	7.9	50,000	

changes to its surface, so faults and related earthquakes can also be found far from present-day plate boundaries. For example, the New Madrid seismic zone in the central Mississippi Valley is of particular interest. In the winter of 1811–1812, a series of strong earthquakes along this zone permanently changed the course of the Mississippi River! With an estimated magnitude of 8.0, the earthquakes were so strong they caused church bells to ring in Boston, Massachusetts, 1600 km (1000 mi) away! Fortunately, because the region was sparsely settled at the time, the loss to human life and property was minimal.

In terms of the different fault types, devastating earthquakes occur with all three types: reverse, normal, or strike-slip faults. For example, the 1906 San Francisco earthquake had a magnitude of 7.8 and resulted in 700 deaths and extensive fire damage. And the 1989 Loma Prieta earthquake with a magnitude of 7.1 caused

About 80% of the world's large earthquakes occur in the Pacific Rim area encircling the Pacific Ocean. Interestingly, about 75% of the world's volcanoes are located here as well. This area is referred to as the "Ring of Fire." Do you think earthquakes and volcanoes have a common connection?

26 CHAPTER REVIEW

KEY TERMS

Anticline An up-fold in rock with relatively older rocks at the fold core; rock age decreases with horizontal distance from the fold core.

Asthenosphere A subdivision of the upper mantle situated below the lithosphere, a zone of plastic, easily deformed rock.

Body wave A type of seismic wave that travels through Earth's interior.

Core The central layer of Earth's interior, divided into an outer liquid core and an inner solid core.

Crust Earth's outermost layer.

Earthquake The shaking or trembling of the ground that happens when rock under Earth's surface breaks.

Fault A fracture along which movement of rock on one side relative to rock on the other has occurred.

Lithosphere The entire crust plus the rigid portion of the mantle that is situated above the asthenosphere.

Love waves Surface waves that move in a side-to-side, whiplike motion.

Mantle The middle layer in Earth's interior, between the crust and the core.

Mohorovičić discontinuity (Moho) The crust–mantle boundary. This marks one of the depths where the speed of P-waves traveling through Earth increases.

Primary wave (P-wave) A longitudinal body wave that compresses and expands the material through which it moves; it travels through solids, liquids, and gases and is the fastest seismic wave.

Rayleigh waves Waves that move in an up-and-down motion along the Earth's surface.

Secondary wave (S-wave) A transverse body wave that vibrates the material through which it moves side to side or up and down; an S-wave cannot travel through liquids and so does not travel through Earth's outer core.

Surface wave A type of seismic wave that travels along Earth's surface.

Syncline A down-fold in rock with relatively young rocks at the fold core; rock age increases with horizontal distance from the fold core.

REVIEW QUESTIONS

Earthquakes Make Seismic Waves

1. How do P-waves travel through Earth's interior?
2. How do S-waves travel through Earth's interior?
3. Can S-waves travel through liquids?
4. Name the two types of surface waves and describe the motion of each.

Seismic Waves Reveal Earth's Internal Layers

5. What was Andrija Mohorovičić's major contribution to Earth science?
6. How did seismic waves contribute to the discovery of Earth's core?

7. What is the evidence that Earth's inner core is solid?

8. What is the evidence that Earth's outer core is liquid?

9. In what ways are the asthenosphere and the lithosphere different from each other?

10. How does continental crust differ from oceanic crust?

11. Why does continental crust stand higher on the mantle than oceanic crust?

Internal Motion Deforms Earth's Surface

12. What happens to rock when stress exceeds a rock's elastic limit?

13. Are folded rocks primarily the result of compressional or tensional forces?

14. Distinguish between anticlines and synclines.

15. What is the difference between reverse faults and normal faults?

16. Which kind of fault forms primarily from tension in Earth's crust? Primarily from compression?

17. Where does most of an earthquake's damage generally occur?

18. Where do most of the world's earthquakes occur?

19. What device do scientists use to measure an earthquake?

20. What is the source of a tsunami's huge amount of energy?

THINK AND EXPLAIN

1. Compare the relative speeds of primary and secondary seismic waves. Which type of material can each travel through?

2. How can seismic waves indicate whether regions within Earth are solid or liquid?

3. How do seismic waves indicate layering of materials in Earth's interior?

4. What is the evidence that Earth's inner core is solid?

5. Speculate on why the lower part of the lithosphere is rigid and the asthenosphere is plastic, even though they are both part of the mantle.

6. Even though the inner and outer cores are both predominantly composed of iron and nickel, the inner core is solid and the outer core is liquid. Why?

7. What does the P-wave shadow tell us about Earth's composition?

8. If Earth's mantle is composed of rock, how can we say that the crust floats on the mantle?

9. Why is Earth's crust thicker beneath a mountain range?

10. What type of fault is associated with the 1964 earthquake in Alaska?

11. How much more does the ground shake during a 6.6-magnitude earthquake than it does during a 5.6-magnitude earthquake?

12. In 1960, a large tsunami struck the Hawaiian Islands without warning, devastating the coastal town of Hilo. Since that time, a tsunami warning station has been established for the coastal areas of the Pacific. Why do you think these stations are located around the Pacific Rim rather than around the Atlantic?

13. How do faults and folds support the idea that lithospheric plates move?

14. Reverse faults are created by compressional forces. Where in the United States do we find evidence of reverse faults?

15. Normal faults are created by tensional forces. Where in the United States do we find evidence of normal faults?

16. Strike-slip faults show horizontal motion. Where in the United States do we find strike-slip faulting?

17. If you found tilted beds of sedimentary rock that are part of a fold, what detail, other than the direction they tilted, would you need to know in order to tell if the fold was an anticline or a syncline?

18. Does the fact that the mantle is beneath the crust necessarily mean that the mantle is denser than the crust? Explain.

19. In an earthquake, what type of land surface would be more safe—rigid bedrock or sandy soil? Defend your answer.

20. In an earthquake, does the release of energy usually happen all at once? Defend your answer.

READINESS ASSURANCE TEST (RAT)

If you have a good handle on this chapter—if you really do—then you should be able to score 7 out of 10 on this RAT. Check your answers with your teacher. If you score less than 7, study further before moving on.

Choose the best answer to each of the following.

1. Which of the following statements is false?
   - (a) The mantle includes part of the crust.
   - (b) The lithosphere includes the entire crust.
   - (c) The mantle includes part of the lithosphere.
   - (d) The mantle includes the entire asthenosphere.

2. Seismic waves increase in speed when
   - (a) they pass through a liquid.
   - (b) rocks become less dense and less rigid.
   - (c) they form a wave shadow.
   - (d) the elasticity of the rock increases.

3. Which part of Earth's crust is most dense?
   - (a) The oceanic crust
   - (b) The continental crust
   - (c) The asthenosphere
   - (d) The lithosphere

4. The source of a tsunami is usually
   - (a) a reverse fault.
   - (b) an earthquake.
   - (c) unusual motion when approaching shore.
   - (d) an excess in wavelength size.

5. Which of the following types of seismic waves are the fastest?
   - (a) Love waves
   - (b) S-waves
   - (c) P-waves
   - (d) Rayleigh waves

6. Earthquakes are primarily caused by
   - (a) vibrations of sandy sediment.
   - (b) the sudden release of stored elastic energy in stressed rocks.
   - (c) the expansion of Earth's crust.
   - (d) convection motion in Earth's core.

7. Earth's internal composition is revealed by
   - (a) the change in speed of seismic waves.
   - (b) reflection and refraction of seismic waves.
   - (c) seismograms.
   - (d) All of the above.

8. When stress exceeds a rock's strength, the rock body
   - (a) breaks, which forms a fault.
   - (b) permanently changes its form—it breaks or flows.
   - (c) becomes brittle, creating folded rock layers
   - (d) becomes plastic.

9. Rocks buckle and fold when subjected to
   - (a) tensional force.
   - (b) the release of stored elastic energy.
   - (c) stretching of Earth's crust.
   - (d) compressional force.

10. A fracture in a rock body is a fault if
    - (a) stress exceeds the rock's strength.
    - (b) one block of the rock body moves relative to the block on the other side of the fracture.
    - (c) the rock body crumbles.
    - (d) the footwall moves upward.

27

PLATE TECTONICS— A UNIFYING THEORY

 ## THE MAIN IDEA

Movement inside of Earth greatly affects Earth's surface.

Why do the great majority of earthquakes and volcanic eruptions occur in narrow zones? And why do very few earthquakes and volcanic eruptions occur outside these zones? How do the answers to these questions relate to the formation of mountains? Why do we find granitic rocks and basaltic rocks where we do? What kind of motion in Earth's lithosphere produced the Atlantic Ocean? In this chapter we apply what we have learned about rocks—how they form, fold, and break—to the global perspective.

Explore!

How Did Different Continents from Long Ago Fit Together into One Large Supercontinent?

1. From any world map, cut out the different continents.
2. Like working a jigsaw puzzle, look to see where the edges of the different continents can match up.

Analyze and Conclude

1. **Observing** Notice that the pieces do not always match up perfectly. And to make things fit, notice that some pieces need to be rotated.
2. **Predicting** Wouldn't some of the rocks, mountains, and life forms be similar on pieces that fit together?
3. **Making Generalizations** The continents (the lithosphere) are not completely stationary. They "float" on the asthenosphere. So as the asthenosphere circulates, the continents drift.

27.1 Continental Drift—An Idea Before Its Time

Have you ever noticed on a map that Africa and South America fit together like pieces of a jigsaw puzzle (Figure 27.1)? One person who took this observation seriously was Alfred Wegener (Figure 27.2). Wegener saw the Earth as a dynamic planet with the continents in slow, but constant motion. Wegener's hypothesis, known as **continental drift,** stated that the world's continents were once joined together as a single supercontinent that he called **Pangaea** (pronounced Pan-gee-ah), meaning "all land." He further hypothesized that Pangaea had fractured into a number of large pieces, and that South America and Africa had indeed once been joined together as part of a larger land mass.

Wegener supported his hypothesis with impressive evidence. He proposed that the boundary of each continent was not at its shoreline, but at the edge of its *continental shelf.* The continental shelf is the gently sloping platform between the shoreline and the much steeper slope that leads to the deep ocean floor. When Wegener fit Africa and South America together along their continental shelves, the fit was nearly perfect (Figure 27.3). He then investigated rocks that are now separated by the Atlantic Ocean. The rocks were similar in both age and in type. In addition, many mountain chains were found to be continuous across the ocean. Mountains in North America matched up with mountains in Europe, and mountains in South America matched up with those in Africa. Wegener also looked at the fossil record. He found fossils of identical land-dwelling animals in South America and Africa but nowhere else. This was a strange finding because today, animals and plants of these

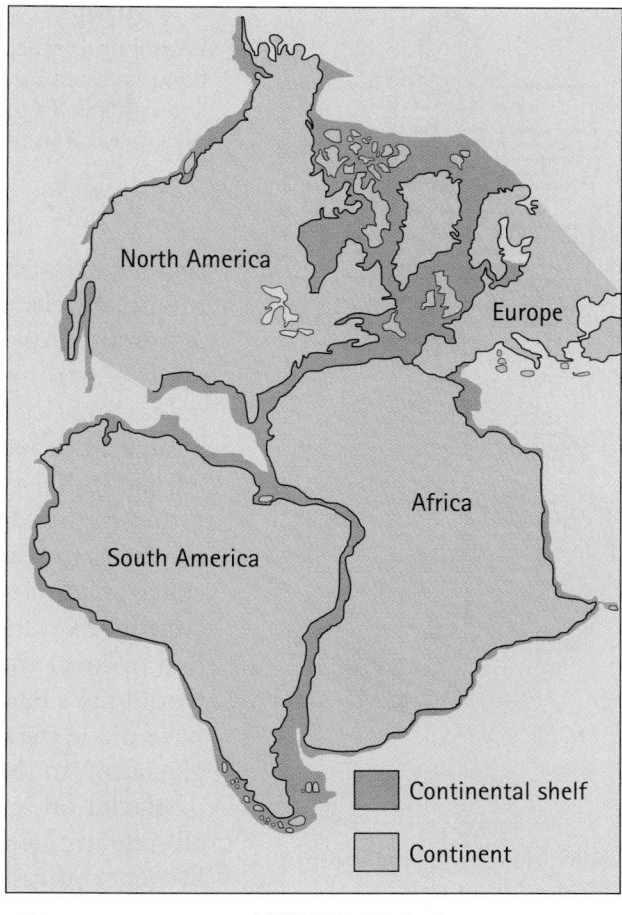

FIGURE 27.1 ▲
The jigsaw-puzzle fit between continents is even better at the continental shelves than at the shorelines of the continents.

FIGURE 27.2 ▶
Alfred Wegener (1880–1930) was a brilliant interdisciplinary scientist. His hypothesis of continental drift in 1915 eventually led to the theory of plate tectonics. His interests included not only meteorology and climatology but also astronomy, geology, geophysics, oceanography, and paleontology. Throughout his life, Wegener had a fascination with exploring the Arctic, and he was fortunate to survive several Arctic adventures. In 1930, at age 50, however, his luck ran out. Wegener died while crossing an ice sheet on an expedition in Greenland. His body still remains as part of the Greenland glacier. The life of Alfred Wegener—a productive life indeed!

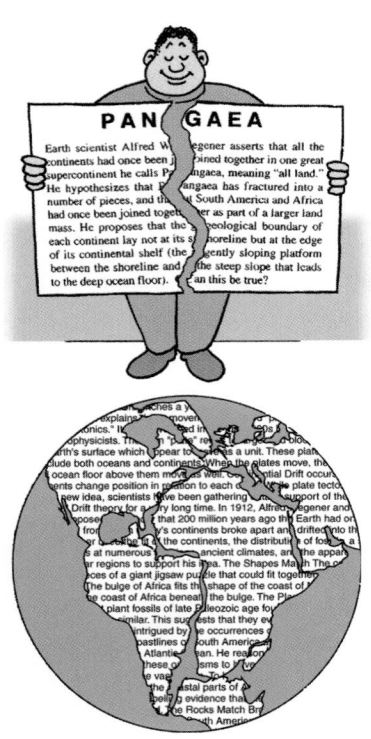

New ideas are more often ignored than welcomed.

Video
The Mantle and Crust

◄ **FIGURE 27.3**
Wegener related finding similar fossils and rocks on separated continents to two pieces of torn newspaper with matching lines of type. If the edges and the lines of type fit together, he said, the two pieces of newspaper must have originally been one.

regions are notable for their striking differences. And fossils of nearly identical trees were found in South America, India, Australia, and Antarctica.

Even stronger evidence for a supercontinent was found by studying paleoclimatic (ancient climate) data. More than 300 million years ago, a huge continental ice sheet covered parts of South America, southern Africa, India, and southern Australia. Evidence of this ice sheet is found in thousands of well-preserved glacial striations. As a glacier, or ice sheet, moves over the land, it gouges the surface. These gouges, called *striations*, reveal the directions of ice flow (Figure 27.4). If these continents were in their present positions, the ice sheet would have had to cover the entire Southern Hemisphere, and, in some places, would have had to cross the equator! An ice sheet that extensive would have made the world climate very cold. But there is no evidence of glaciation in the Northern Hemisphere at that time. In fact, this time of glaciation in the Southern Hemisphere was a time of subtropical climate in the Northern Hemisphere. To account for this inconsistency, Wegener proposed that Pangaea had been in existence 300 million years ago, with South Africa located over the South Pole. This reconstruction would bring all the glaciated regions into close proximity near the South Pole and place the modern northern continents nearer to the tropics.

Wegener described continental drift in his book *The Origin of Continents and Oceans*, published in 1915. Although he used evidence from many different scientific disciplines, his well-founded hypothesis was ridiculed by the scientific community. Skeptics claimed that Wegener failed to provide a reasonable driving force for why the continents moved. (Wegener wrongly proposed that the tidal influence of the Moon could produce the needed force. He also proposed that the continents broke through Earth's crust like icebreakers cutting through ice.) Without a convincing explanation for his hypothesis, it was dismissed. Only later, in the light of newfound discoveries, did the scientific community accept Wegener's concept.

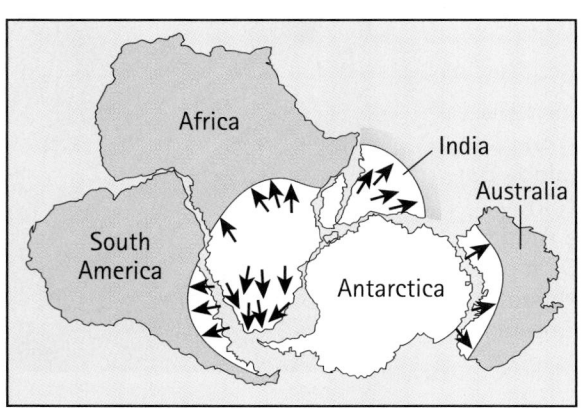

◄ **FIGURE 27.4**
Glacial striations in rock outcrops in South America, Africa, Antarctica, India, and Australia provide paleoclimatic evidence that these continents were once positioned together. The arrows depict the direction of glacial movement.

CONCEPT CHECK

1. What evidence might lead someone with no understanding of science to suspect that the continents were once connected?
2. Scientists of Wegener's day rejected continental drift because they couldn't imagine how massive, rocky continents could grind through the solid rock of the ocean floor. What did these scientists evidently not know about the mantle?
3. What evidence did Wegener use to support his hypothesis?

Check Your Answers

1. The most obvious evidence is the matching of the edges of the African and South American continents, which can be seen on any world map or globe.
2. Wegener's contemporaries assumed that if the continents were to move, they would have to push through solid rock. They did not know that the mantle has a "plastic-like" layer—the asthenosphere, over which "floating" continents can readily glide.
3. Geologic—On both sides of the Atlantic Ocean there are similar rocks and mountain chains. Biologic—Similar animal and plant fossils on separated continents. Paleoclimatologic—Glacial striations in southern continents, but not in northern continents.

✓ READING CHECK

How does continental drift relate to the ancient super-continent of Pangaea?

27.2 Search for the Mechanism to Support Continental Drift

✓ One of the first key discoveries in support of continental drift came about through studies of Earth's magnetic field. We know from Chapter 11 that Earth is a huge magnet. Its magnetic north and south pole are near the geographic poles. Because certain minerals align themselves with the magnetic field when a rock is formed, many rocks have a preserved record of changes in Earth's magnetism over time. These changes include times when the magnetic north and south poles were reversed. This magnetism from the geologic past is known as **paleomagnetism.**

In the 1950s, a plot of the positions of the magnetic north pole through time revealed that over the past 500 million years, the position of the pole had apparently wandered throughout the world (Figure 27.5). It

Paleo- means "old" or "ancient." For example, *paleoclimate* is a way to describe ancient climates, *paleomagnetism* describes ancient magnetic data, and *paleontology* is the study of life in ancient geologic time.

FIGURE 27.5 ▶
The path of the magnetic north pole during the last 500 million years. (The unit m.y.a. stands for "million years ago.") The lower red line is from evidence collected in Europe, and the upper red line is from evidence collected in North America. One would expect that these two lines would overlie each other. Thus, either the magnetic pole wanders erratically, or the continents have moved. But how could the pole be in more than one place at the same time?

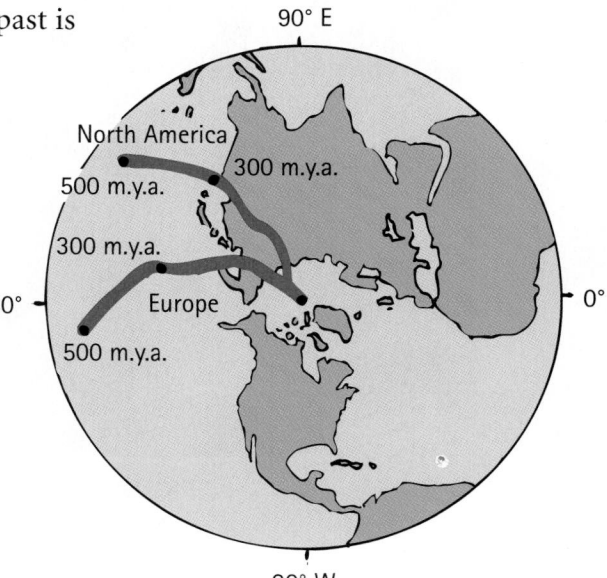

Earthquakes in Indonesia are very common; in 2006 and 2007 this region experienced four earthquakes registering more than a magnitude 6.0. A very shaky area, indeed.

62 deaths and did more than $6 billion in damage. Strike-slip movement on the San Andreas Fault was involved in both of these earthquakes.

Although earthquakes associated with normal faults are quite common, the most catastrophic earthquakes occur along reverse faults. For example, the 1964 Anchorage, Alaska, earthquake (magnitude 9.2) caused 131 deaths and over $300 million in damage.* Some of the greatest tragedies of more recent times include the January 2001 earthquake in El Salvador (7.7 magnitude, 844 deaths), the January 2001 earthquake in India (7.6 magnitude, 20,000 deaths), the December 2004 earthquake in Sumatra (9.2 magnitude, 184,000 deaths), and the May 2008 Sichuan earthquake in China (7.9 magnitude) that resulted in 70,000 deaths and left more than 5 million people homeless.

Understanding earthquakes is obviously of major importance to society. Unfortunately, earthquakes and fault movement occur with little or no immediate warning and thus are very difficult to predict. The best that can be done is to calculate the probability that earthquakes will occur within a given time frame. Such probability estimates are based on the past occurrences of earthquakes in the target area. Table 26.1 lists some of the most notable earthquakes according to their impact on society.

---

* The death toll was largely due to great seismic sea waves, or tsunami. A tsunami is generated from the displacement of water as a result of an earthquake, a submarine landslide, or an underwater volcanic eruption.

FIGURE 27.6 ▲
Harry Hess (1906–1969). Hess's hypothesis of seafloor spreading helped establish a mechanism for Wegener's hypothesis of continental drift.

seemed that either the magnetic poles had migrated through time or the continents had drifted. Because the apparent path of polar movement varied from continent to continent, it was conceivable that the continents had moved. This supported the hypothesis of continental drift. But a mechanism to explain how the movement occurred was still lacking.

Just as Wegener used evidence from different disciplines to support his hypothesis, the mechanism to explain continental drift was assembled from several different disciplines. ✓ Piece by piece, it all came together—and much of it as a result of seafloor exploration.

A key player who helped solve the puzzle of continental drift was Harry Hess (Figure 27.6), a geology professor who also served as a naval captain during World War II. To help his attack ship maneuver near shore during beach landings, Hess used a *fathometer*, an innovative depth sounder, to map the underwater topography. But Hess, a scientist as well as a sailor, continued using the fathometer in the open sea. With the data, Hess constructed a detailed profile of the ocean floor. His findings expanded on and supported other discoveries and emerging ideas.

Extensive and detailed mapping of the ocean floors revealed huge mountain ranges running down the middle of the Atlantic, Pacific, and Indian oceans (Figure 27.7). The Mid-Atlantic Ridge, for example, was found to stretch along the center of the Atlantic Ocean basin parallel to

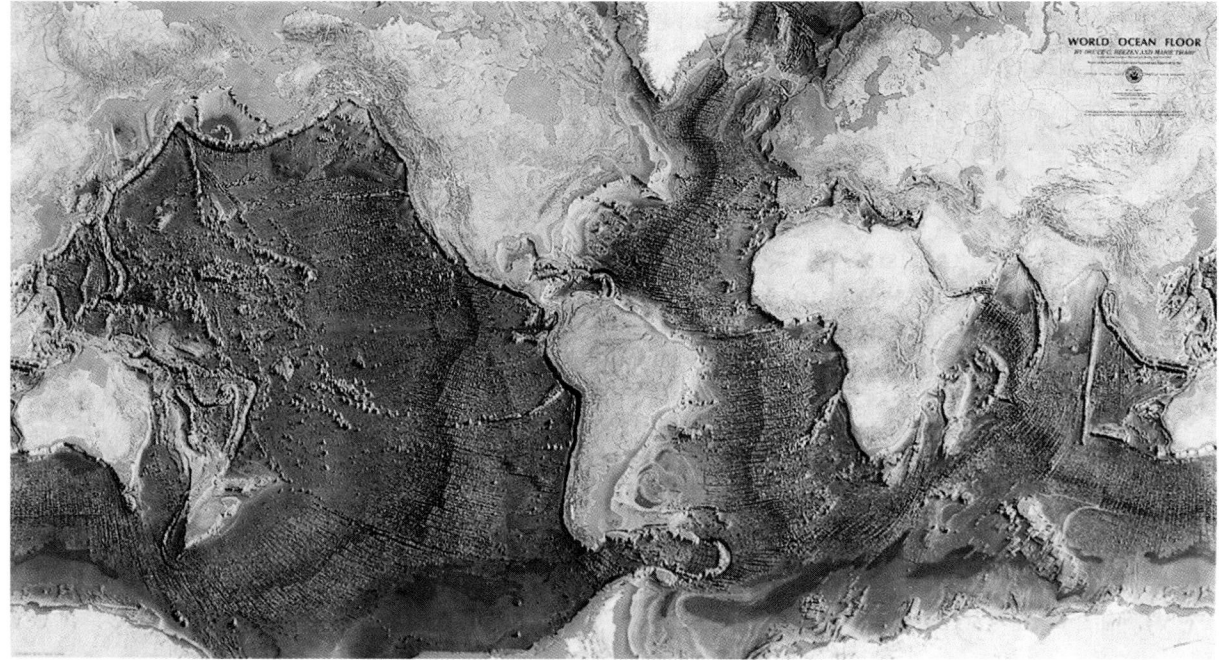

FIGURE 27.7 ▲
The first detailed map of the ocean floors. Based on sonar readings, the map reveals enormous mountain ranges in the middle of the oceans and deep ocean trenches near some continental landmasses.

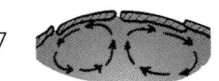

FIGURE 27.8 ▶
The Mid-Atlantic Ridge runs down the center of the Atlantic Ocean. Its highest peaks emerge above the water in several places, creating oceanic islands such as Iceland. This photo shows the exposed rift valley on Iceland.

the American, European, and African coastlines. The ridge stretches 19,312 km, and its highest peaks emerge above sea level to form oceanic islands, such as Iceland and the Azores (Figure 27.8). In the center of the ridge and all along its length is a deep fissure—a volcanic *rift zone*. Another ocean floor feature was the discovery of deep *ocean trenches* (long, deep troughs in the seafloor) near some continental landmasses, particularly around the edges of the Pacific. So it was revealed that some of the deepest parts of the ocean are actually near some of the continents, and some of the shallowest waters are in the middle of the oceans, at the mid-ocean ridges.

CONCEPT CHECK
Why was Wegener's theory of continental drift not taken more seriously when first proposed?

Check Your Answer
Wegener failed to propose a suitable mechanism that could cause the continents to move. Even if he had proposed the role of the convective interior, we can only guess how quickly the scientific community would have accepted his hypothesis. Scientists, like all other human beings, tend to identify with the ideas of their time. Do advances in knowledge, scientific or otherwise, occur because they are accepted by the status quo or because holders of the status quo eventually die off? Knowledge that is radical and unacceptable to the old guard is often easily accepted by newcomers who use it to push the knowledge frontier further. Hooray for the young (and the young-at-heart)!

✔ **READING CHECK**

What two areas of study helped revive the search for a mechanism for continental drift?

Seafloor Spreading—A Scientific Revolution
With the discovery of the mid-ocean rifts, Hess was inspired to look back at his data from years before. In 1960, he proposed that the seafloor is not permanent but is constantly being renewed. Hess hypothesized that the ocean ridges are located above upwelling convection currents in the mantle. As material from the mantle oozes upward, new lithosphere is formed. Old lithosphere is simultaneously destroyed in the deep-ocean trenches near the edges of some continents. Thus, in a conveyor-belt fashion, new lithosphere forms at a spreading center and older lithosphere moves out from the ridge crest, eventually to be recycled back into the mantle at a deep-ocean trench (Figure 27.9). Hess called his hypothesis **seafloor spreading.**

Interestingly, support for this theory came from paleomagnetic studies of the ocean floor. As new basalt is extruded at a mid-ocean ridge, it is magnetized in a direction that aligns with the existing magnetic field. The magnetic surveys obtained from mapping the

Measurements of the ocean floor began during World Wars I and II with echo-sounding devices. These sonar systems measured ocean depth by recording the time it took for a ship's sound signal (commonly called a "ping") to bounce off the ocean floor. These signals revealed that the ocean floor was much more rugged than previously thought.

Evidence of seafloor spreading tells us that continents must move.

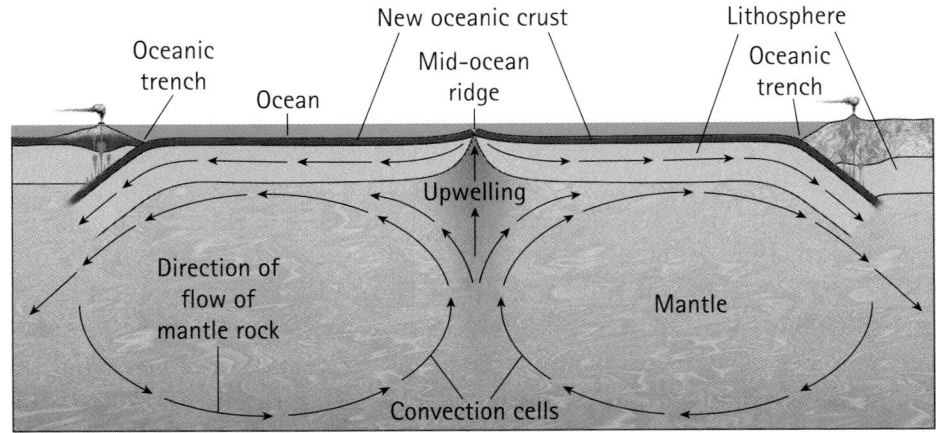

FIGURE 27.9 ▲
In conveyor-belt fashion, new lithosphere is formed at the mid-ocean ridges (also called "spreading centers") as old lithosphere is recycled back into the asthenosphere at a deep ocean trench.

fyi

- East Africa may be the site of our Earth's next major ocean! Spreading processes have already torn Saudi Arabia away from Africa to form the Red Sea. With continued spreading, the edge of the present-day African continent will separate completely, the Indian Ocean will flood into the area, and the easternmost corner of Africa (the Horn of Africa) will become a large island!

ocean's floor showed alternating stripes of normal and reversed polarity, paralleling either side of the ridge areas (Figure 27.10). As in a very slow magnetic tape recording, ✓ the magnetic history of Earth is recorded in the spreading ocean floors. In this way, we have a continuous record of the movement of the seafloors. Since the dates of pole reversals can be determined, the magnetic pattern of the spreading seafloor tells us both the age of the seafloor and the rate at which it spreads.

The hypothesis of seafloor spreading provided the mechanism to explain continental drift. The time was right for the revolutionary concepts to follow. The tide of scientific opinion had indeed switched in favor of a mobile Earth.

FIGURE 27.10 ▶
As new material is extruded at an oceanic ridge (spreading center), it is magnetized according to the existing magnetic field. Magnetic surveys show alternating stripes of normal and reversed polarity paralleling both sides of the rift area. Like a very slow magnetic tape recording, Earth's magnetic history is recorded in the spreading ocean floors.

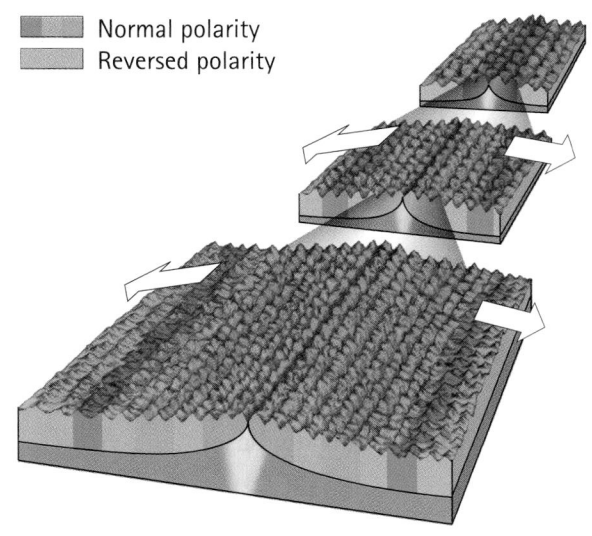

Normal polarity
Reversed polarity

CONCEPT CHECK

What is seafloor spreading, and what is its relationship to the rock basalt?

Check Your Answer

Seafloor spreading is summed up by the conveyor-belt model of convection. Partial melting of mantle rock produces basaltic magma, which oozes upward at the mid-ocean ridges to form new oceanic crust. Old crust moves away from the ridge to be recycled in the deep ocean trenches. And the rock that forms the new crust—basalt.

✔ **READING CHECK**

How is seafloor spreading related to magnetic striping of the ocean floor?

27.3 The Theory of Plate Tectonics

Plate tectonics describes the motions of Earth's lithosphere that create ocean basins, mountain ranges, earthquake belts, and other large-scale features of Earth's surface. ✔ The theory of plate tectonics states that Earth's outer shell, the lithosphere, is divided into eight relatively large plates and a number of smaller ones (Figure 27.11). These lithospheric plates ride atop the relatively plastic asthenosphere below. So Wegener's hypothesis of continental drift was on the right track. The continents really do move—they move because they are embedded in the drifting tectonic plates.

Ultimately, lithospheric plates move in response to convection in Earth's interior. Recall from Chapter 9 that heat naturally moves from warmer regions to cooler regions. Inside Earth, heat moves from the hot core and mantle to the cooler crust. Heat flow from the core to the mantle is mostly due to conduction. In the mantle, however, most of the heat flow is due to

PhysicsPlace.com
Video
Plate Tectonics

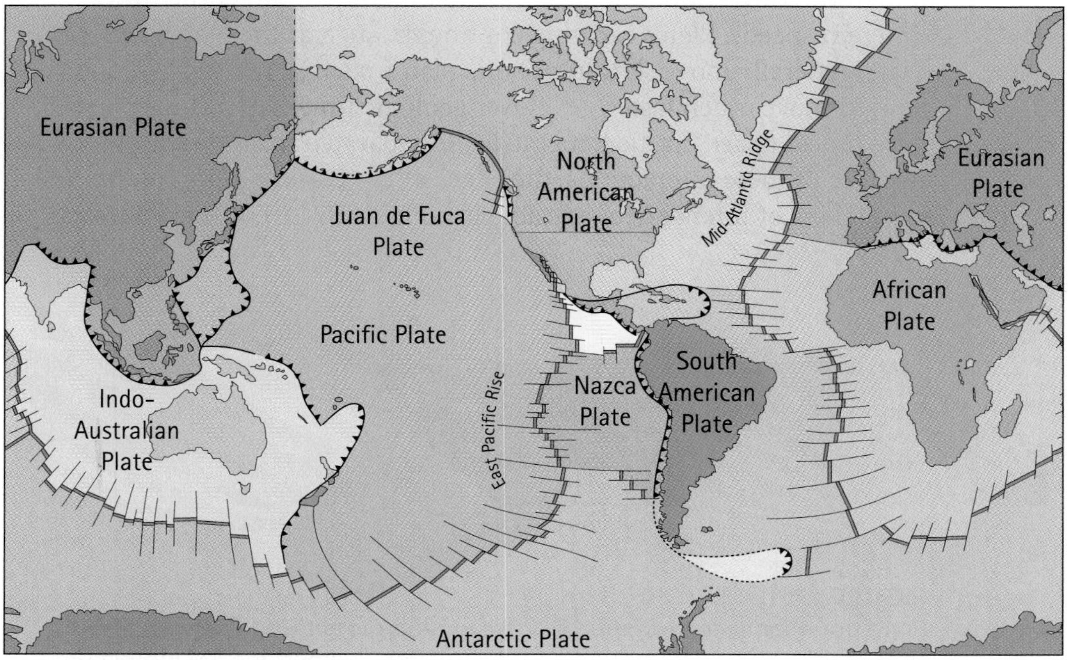

FIGURE 27.11 ▲

The lithosphere is divided into eight large plates and a number of smaller ones.

Unequal heating and gravity produce convection in both Earth's mantle and its atmosphere.

convection. As hot mantle rock rises, it expands. Then, closer to Earth's surface, the rock cools, contracts, and sinks. From this example, can you see how temperature differences are a key part of convection?

Interestingly, however, gravity plays an even greater part. When rock expands it becomes less dense, and when rock contracts it becomes more dense. Convection in the hot mantle occurs because gravity pulls the denser rock downward relative to the less-dense rock, which continues to rise upward. As the less-dense rock rises it takes the place of the sinking dense rock pulled downward by gravity—convection! So gravity and heat are what cause mantle convection. And the lithospheric plates move because they are the upper part of the mantle convection cells.

✔ **READING CHECK**

What is the theory of plate tectonics and what is the relationship between lithospheric plates and the underlying asthenosphere?

CONCEPT CHECK
What role does gravity play in mantle convection?

Check Your Answer
Gravity acts on everything. Earth is round because gravity pulls everything inward toward Earth's center. Gravity's pull on denser material generates the convection cycle. So, strictly speaking, it is the density difference more than the temperature difference that causes the convective motion.

27.4 Three Types of Plate Boundaries

PhysicsPlace.com

Tutorial
Plate Boundaries and Plate Tectonics
Video
Folds and Faults in Earth's Crust

Earth's lithospheric plates move in a conveyor-belt manner in response to mantle convection. The plates move in different directions and at different speeds. Plates carrying continents, such as the North American Plate, generally move slower; oceanic plates, such as the Pacific Plate, tend to move much faster. ✔ Over geologic time, the various plates have pulled apart, crashed, merged, and separated from one another. Because of these interactions, the edges of plates—the *plate boundaries*— are regions of intense geologic activity (Figure 27.12). While interiors

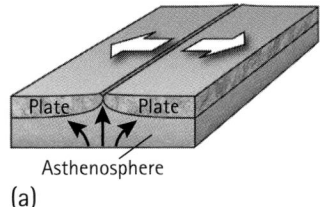
(a)

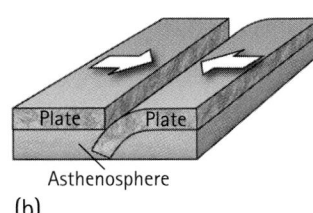
(b)

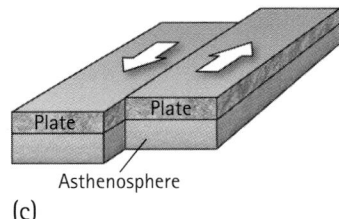
(c)

FIGURE 27.12 ▲
Plate boundaries are regions of intense geologic activity. They are also the sites of lithospheric formation and destruction. Named for the movement they accommodate, the three types of plate boundaries are (a) divergent, (b) convergent, and (c) transform boundaries.

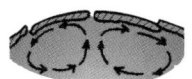

HOT SPOTS AND GPS—A MEASUREMENT OF TECTONIC PLATE MOTION

Motion is relative. Whenever we discuss the motion of something, all we can do is describe its motion relative to something else. We call the place from which motion is observed and measured a *reference frame.* Living in a world where everything is in motion, how can we measure rates of plate motion? What do we choose for our reference frame?

One reference frame comes from volcanic activity far removed from plate boundaries. For example, the Hawaiian Islands in the middle of the Pacific Plate. The Hawaiian Islands are the tips of huge volcanoes that formed as the Pacific seafloor moved over a fixed *hot spot*—a magma source rising from Earth's interior. Because we can date the age of the islands, we can measure the rate of plate movement over the hot spot. The concept of a concentrated or fixed hot spot provides a stationary reference point on Earth's surface from which plate motion can be determined. There are about 50 such reference hot spots around the world, and they are used together to determine rates of plate movements.

There is, however, some debate as to whether the hot spots are truly stationary. Fortunately, a more precise way to measure Earth movement is by GPS— Global Positioning System. GPS uses radio signals from special satellites orbiting Earth. Although not stationary, satellites are an ideal reference frame for two reasons: Their location is precisely known and they do not move with Earth's lithospheric plates.

The time it takes for a signal to travel from a satellite to a GPS receiver on the ground determines the distance between the receiver and the satellite. For accuracy, the position of each ground station is determined by simultaneous signals from several satellites. Once this data is known, repeat measurements are periodically done to monitor change and thus track continental movement. More than one hundred GPS stations from around the world monitor tectonic plate motion. Results from the data show that Hawaii is moving in a northwesterly direction toward Japan at a rate of 8.3 cm per year. And Maryland is moving away from England at a rate of 1.7 cm per year. Amazing!

▲ GPS satellites are used to measure plate motion.

of plates are relatively quiet, most earthquakes, volcanic eruptions, and mountain building events occur at plate boundaries. There are three types of plate boundaries:

Divergent boundaries—*where plates move away from each other*

Convergent boundaries—*where plates move toward each other*

Transform boundaries—*where plates slide past each other*

Divergent Plate Boundaries

Heat-driven convection cells in the mantle operate in symmetric loops. When two upward-moving convection currents meet at the surface, the lithospheric plates spread apart. Tension is the dominant force where

✔ **READING CHECK**

How have various plates interacted over geologic time?

Don't forget gravity! The elevation of the mid-ocean ridge is high compared to the surrounding seafloor. So, gravity causes the plates to slide down and away from the mid-ocean ridge, like cookies sliding off a tilted cookie sheet. But why is the ridge higher? Because of the rising, hot mantle rock.

✔ **READING CHECK**

How does new lithosphere form at divergent boundaries?

plates move away from each other. These spreading centers are **divergent plate boundaries** (Figure 27.12a).

✔ Magma generated at a divergent boundary is from the partial melting of mantle rock brought upward with the rising convection currents. Basaltic lava erupts where the plates diverge and fills the gap between the spreading plates. When cooled, the basalt becomes new oceanic crust. The crust is spread equally on both sides of the ridge. It is thin at the ridge, and thickens away from the ridge.

Mid-ocean ridges mark the locations of most divergent plate boundaries. For example, the Mid-Atlantic Ridge is the divergent boundary between the North American and Eurasian Plates in the North Atlantic, and the South American and African Plates in the South Atlantic. The rate of spreading at the Mid-Atlantic Ridge ranges between 1 cm and 6 cm per year. Although this spreading may seem slow, through geologic time the effect has been tremendous. Over the past 190 million years, seafloor spreading has transformed a tiny waterway through Africa, Europe, and the Americas into the vast Atlantic Ocean of today!

Spreading centers are not restricted to the ocean floors. They also develop on land. Hot, molten material in Earth's interior rising beneath continental landmasses generates tension. This causes the crust to stretch and bend upward (this is called *upwarping*). Gaps in the crust are produced, and large slabs of rock slide and sink down into these

---

CALCULATION CORNER

Calculating the Age of the Atlantic Ocean

If you can estimate the rate of seafloor spreading and you know the present width of an ocean, you can calculate the ocean basin's age. Between the United States and Africa, the Atlantic Ocean is currently about 4,830 km or 4.8×10^8 cm wide. Let's assume that the rate of seafloor spreading in the Atlantic has been a constant 2.5 cm/yr over geologic time. We can then apply the familiar equation that relates speed, time, and distance:

$$\textbf{Time} = \textbf{distance/speed}$$
$$= (4.8 \times 10^8 \text{ cm})/(2.5 \text{ cm/yr})$$
$$= 1.92 \times 10^8 \text{ yr}$$
$$\sim \textbf{190 million years}$$

Based on these estimates, the age of the Atlantic Ocean is about 190 million years.

Sample Problem

The Red Sea is presently a narrow body of water located over a divergent plate boundary. The

plates began to diverge from one another about 30 million years ago. Knowing the age and the width, what is the average rate of spreading?

Solution:

If we take the current width of the Red Sea to be 300 km, or 3.0×10^7 cm, and the time of spreading to be 30 million years, the speed at which the seafloor spreads is then

$$\textbf{Speed} = \textbf{distance/time}$$
$$= (3.0 \times 10^7 \text{ cm})/(3.0 \times 10^7 \text{ yr})$$
$$= \textbf{1 cm/yr}$$

At this rate, it will take about 400 million years for the Red Sea to be as wide as the Atlantic Ocean. But this may not actually occur because Earth is dynamic and ever changing. For example, in 2006 the Ethiopian rift zone spread by about 8 m! This rate of change is dynamic indeed!

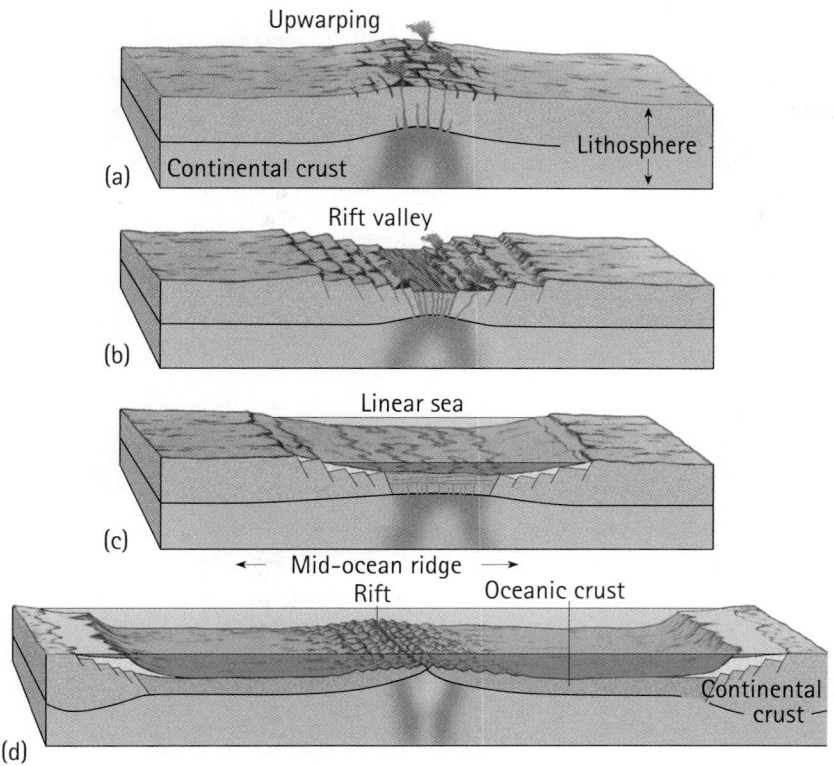

Upwarping

Lithosphere

(a) Continental crust

Rift valley

(b)

Linear sea

(c)

← Mid-ocean ridge →
Rift Oceanic crust

Continental crust

(d)

◀ **FIGURE 27.13**
Formation of a rift valley and its transformation into an ocean basin. (a) Rising magma uplifts continental crust, causing the surface to crack. (b) Rift valley forms as crust is pulled apart. Africa's Great Rift Valley is in this stage today. (The two sides of the valley move away from each other because they are located above mantle convection cells that have the same circulation pattern as the cells in Figure 27.9.) (c) Water from the ocean drains in as the rift drops below sea level, forming a linear sea, so called because it is usually long and narrow. (d) Over millions of years, the rift continues to widen and becomes an ocean basin.

gaps. The large down-dropped valleys generated by this process are called either **rifts** or *rift valleys* (Figure 27.13). The Great Rift Valley of East Africa is an excellent example of such a feature. If the spreading continues, it may be the beginning of a new ocean basin.

Convergent Plate Boundaries

Convergent plate boundaries, as the name implies, are where plates come together, or converge. Convergent boundaries are areas of compressive stress and, depending on the nature of the plates, the recycling, or destruction, of lithosphere. These regions of plate collisions are also regions of great mountain building. The type of convergence—or "slow collision"—that takes place depends on the type of lithosphere that is involved. The three kinds of convergent plate boundaries are

- Oceanic–oceanic convergence (Figure 27.14a)
- Oceanic–continental convergence (Figure 27.14b)
- Continental–continental convergence (Figure 27.14c)

Oceanic–Oceanic Convergence When two oceanic plates meet, the older (and therefore cooler and denser) plate slides beneath the younger, less-dense plate. ✔ The process in which one plate bends and descends beneath the other is called **subduction,** and the area where this occurs is called a *subduction zone.* At Earth's surface, subduction zones are marked by deep ocean trenches that run parallel to the edges of convergent

The rate of seafloor spreading at the Mid-Atlantic Ridge is about equal to the rate fingernails grow.

Imagine a "fly on a star" seeing in quick time how Earth's surface has changed over geologic time. What a show!

17. What are the three types of plate collisions that occur at convergent boundaries?
18. What is a transform boundary?
19. What kind of plate boundary separates the North American Plate from the Pacific Plate?

The Theory That Explains Much

20. Cite at least three geologic features that are explained by plate tectonics.

THINK AND EXPLAIN

1. What kind of boundary separates the South American Plate from the African Plate?
2. What is the driving force for mountain building in the Andes?
3. What is the explanation for the generation of earthquakes near plate boundaries?
4. What is the explanation for the movements of the lithospheric plates?
5. Is it likely that present-day ocean basins are a permanent feature on our planet? Defend your answer.
6. Are the present-day continents a permanent feature on our planet? Defend your answer.
7. Are metamorphic rocks found at all three types of plate boundaries?
8. Provide an explanation for why granite frequently forms at oceanic–continental convergent boundaries but infrequently at oceanic–oceanic convergent boundaries.
9. Make a clear distinction between continental drift and plate tectonics.
10. Why are the most ancient rocks found on the continents, rather than on the ocean floor?
11. Upon crystallization, magnetic minerals align themselves in the direction of the surrounding magnetic field. This provides a magnetic fossil imprint. How does the seafloor's magnetic record support the theory of continental drift?

12. How are the theories of seafloor spreading and continental drift supported by paleomagnetic data?
13. What types of boundaries are associated with seafloor spreading centers?
14. What does Earth's crust have in common with a conveyor belt?
15. Subduction is the process of one lithospheric plate descending beneath another. Why does the oceanic portion of the lithosphere undergo subduction while the continental portion does not?
16. What is the cause of earthquakes in southern California?
17. What type of magma do we find at each of the different plate boundaries?
18. How old is the Atlantic Ocean thought to be?
19. What type of lava erupts at divergent boundaries? What types erupt at convergent boundaries?
20. What clues can we use to recognize the boundaries between ancient plates that are no longer in existence?

THINK AND SOLVE

1. If the mid-Atlantic ocean is spreading at 2.5 cm per year, how many years has it taken for it to reach its present width of about 5000 km?
2. If you know the rate of movement along a fault, the amount of offset over a period of time can be calculated. The basic relationship is

$$\text{Rate} = \frac{\text{distance}}{\text{time}}$$

Movement along the San Andreas Fault is about 3.5 cm/yr. If a fence were built across the fault in 2000, how far apart will the two sections of the now-broken fence be in 2010?

FIGURE 27.14 ▶
The three types of
convergent boundaries:
(a) oceanic–oceanic,
(b) oceanic–continental, and
(c) continental–continental.

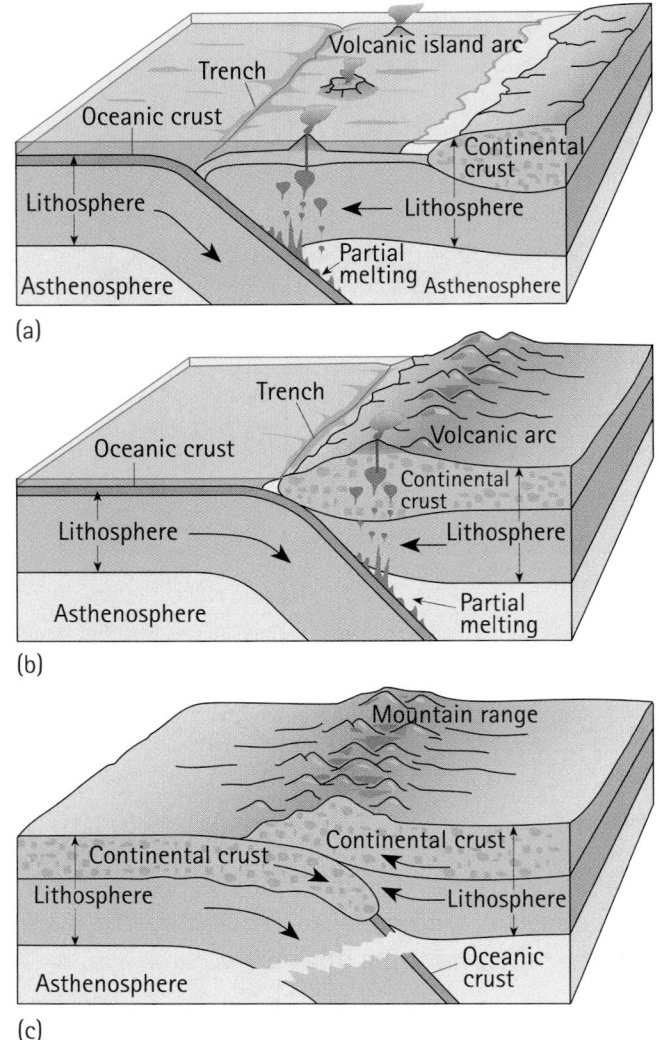

(a)

(b)

(c)

boundaries. The Marianas Trench, for example, is where the Pacific
Plate and the slower-moving Philippine Plate collide. Dropping 11,000 m
(7 mi) below sea level, the Marianas Trench is the deepest location on
Earth's crust. If the world's highest mountain, Mt. Everest, were sunk to
the bottom of the Marianas Trench, there would still be more than a mile
of ocean above it!

✓ Subduction is an important part of mantle convection. Each
downgoing plate, or *slab,* controls the downward part of a convection
cell. So, once again, gravity plays a large role. As gravity pulls the
oldest edge of the subducting slab into Earth's interior, the rest of the
subducting plate is also pulled trenchward (a process called *slab-pull*).
The longer the subduction zone, the more weight there is to pull on
the rest of the downgoing plate. So, longer subduction zones mean
faster plate movement. You can visualize a plate being pulled into
the asthenosphere as being like a tablecloth being pulled slowly off
a table (Figure 27.15).

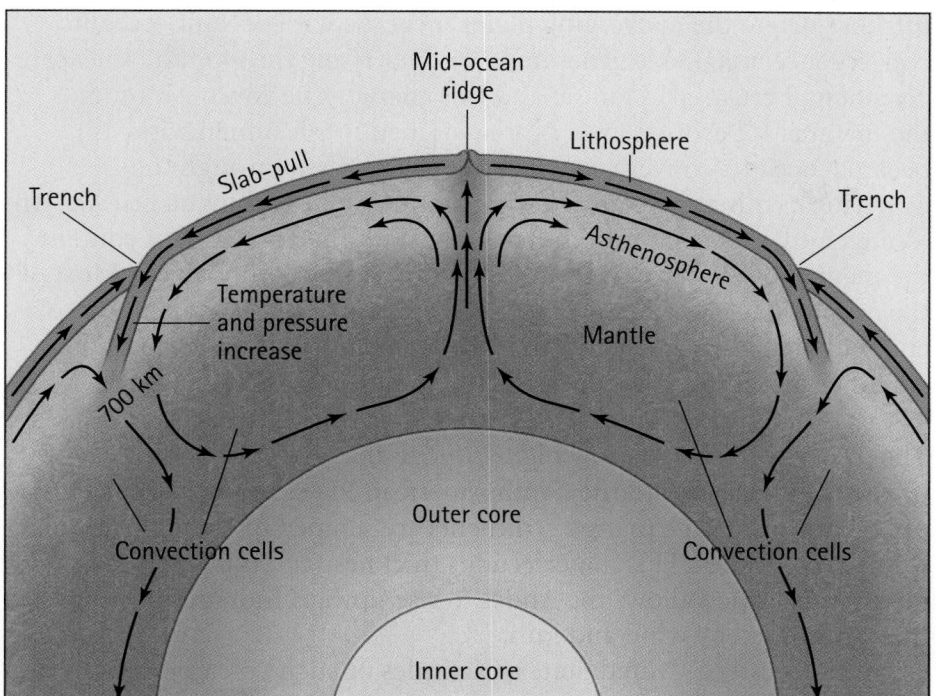

◀ **FIGURE 27.15**
Simplified view of convection cells within the mantle, showing slab-pull. The downgoing part of the plate—the "slab"—heats up but generally does not melt. Because temperature and pressure are high, this is a location of intense metamorphism.

The process of subduction induces the generation of magma and the formation of volcanoes on the seafloor. As the descending, seawater-saturated plate is pulled downward, it heats up. Fluids released by the wet rock interact with the wedge of mantle rock between the two plates (Figure 27.14a). These fluids lower the melting point of the mantle rock, causing it to partially melt and form basaltic magma.

As the magma rises, reservoirs of basaltic magma form in the crust. When magma reaches the ocean floor, undersea volcanoes form, which often start out with basaltic lava. Much of the basaltic magma erupts, but some undergoes crystallization. Partial melting of oceanic crust also occurs. Both crystallization and partial melting act to increase the magma's silica content. The volcanoes then begin erupting andesitic magma, which allows them to grow significantly higher. The volcanoes eventually break the surface of the ocean as a series of islands called an *island arc*. The size and elevation of the islands in an island arc increase over time because of continued volcanic activity. Such island arcs have formed the Aleutian Islands, the Marianas Islands, and the Tonga island group in the South Pacific Ocean, as well as the island-arc systems of the Alaskan Peninsula, the Philippines, and Japan.

Oceanic–Continental Convergence
When an oceanic plate collides with a continental plate, the denser oceanic plate subducts beneath the less-dense continental plate. A deep-ocean trench forms

READING CHECK

What is subduction, and how is it related to the plate tectonic model?

The main difference between magma generation at convergent and divergent boundaries is distance and time. Magma generation at divergent boundaries is very near the surface, so basaltic magma moves upward unimpeded. At convergent boundaries, magma generation occurs deeper in the mantle. This means more distance and more time. Then, when the rising magma encounters the overlying lithosphere, travel takes even more time. But no worries—more time and distance means a greater variety of igneous rocks!

✓ **READING CHECK**

Why does magma composition change in areas of oceanic–continental convergence?

offshore where the converging plates meet. As with oceanic–oceanic convergence, magma is generated. ✓ But rising through the thicker continental crust takes longer and this changes the composition of the magma. It becomes even more enriched in silica than with oceanic–oceanic convergence. As the magma rises through the overlying continental plate, varying amounts of continental crust begin to melt and are mixed into the magma. With the higher silica content, the magma develops a more andesitic or granitic composition. Most of the erupted lava is andesitic, and such eruptions can be quite intense and violent. The majority of the granitic magma does not erupt, but solidifies underground to form intrusive plutonic rock—granite.

The Andes Mountains of western South America formed in this way. The Andes continue to grow higher due to the ongoing subduction of the Nazca Plate beneath the South American Plate. As the Nazca Plate is pulled downward, its marine sediments are scraped off onto the granitic roots of the Andes. This material adds thickness and buoyancy to the mountains, which allows the Andes to rise upward more rapidly than they are eroded by wind and rain.

In the western United States, examples of such volcanic activity are found in the Sierra Nevada, an ancient volcanic range, and the Cascade Range, which is currently active. The Sierra Nevada were produced by subduction of the ancient Farallon Plate beneath the North American Plate. The Sierra Nevada batholith is a remnant of the original volcanic range, and the California Coast Range has remnants of the sediments that accumulated in the trench. The Cascade Range, produced from the subduction of the Juan de Fuca Plate (part of the Farallon Plate) beneath the North American Plate, includes the volcanoes Mount Rainier, Mount Shasta, and Mount St. Helens. The 1980 eruption of Mount St. Helens is proof that the Cascade Range is still active.

As you may expect, active subduction zones are areas of intense earthquakes. Earthquakes occur along the subduction zones as the subducted plate grinds against the overriding plate. Earthquakes become steadily deeper and deeper in the direction of subduction (Figure 27.16).

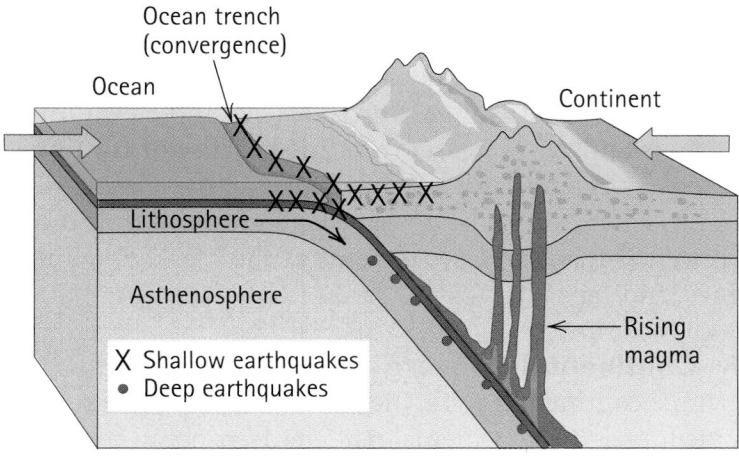

FIGURE 27.16 ▶
Earthquakes at a subduction zone get deeper and deeper in the direction of subduction.

CONCEPT CHECK

Erosion wears mountains down, and yet the Andes Mountains grow taller each year. Why?

Check Your Answer

Subduction is still occurring, which causes the uplift of the Andes Mountains. Because the rate of uplift is greater than the erosion rate, the mountains continue to grow. The Andes are not alone—the Sierra Nevada Mountains are growing as well.

Continental–Continental Convergence When two plates of continental crust collide there is no subduction. This happens because ✓ both plates are made of buoyant granitic-type rocks. The colliding rocks have similar densities, so neither sinks below the other—no subduction. Instead, convergence between two continental plates is more like a head-on collision (Figure 27.14c). Compression causes the plates to break and fold up on each other, making the crust very thick. Intensely compressed and metamorphic rocks define the zones where continental plates meet. In contrast to convergence involving two oceanic plates or one continental and one oceanic plate, volcanic activity is not a characteristic feature of continental–continental collisions—but earthquakes are!

The collision between continental plates has produced some of the most famous mountain ranges. One majestic example is the snow-capped Himalayas, the highest mountain range in the world. This chain of towering peaks is still being thrust upward as India continues crunching up against Asia (Figure 27.17). The European Alps were formed in a similar fashion when part of the African Plate collided with the Eurasian Plate some 40 million years ago. Relentless pressure between the two plates continues, and it is slowly closing up the Mediterranean Sea. In America, the Appalachian Mountains were produced from a continental–continental collision that ultimately resulted in the formation of the supercontinent Pangaea.

READING CHECK

Why is there no subduction where two continental plates collide?

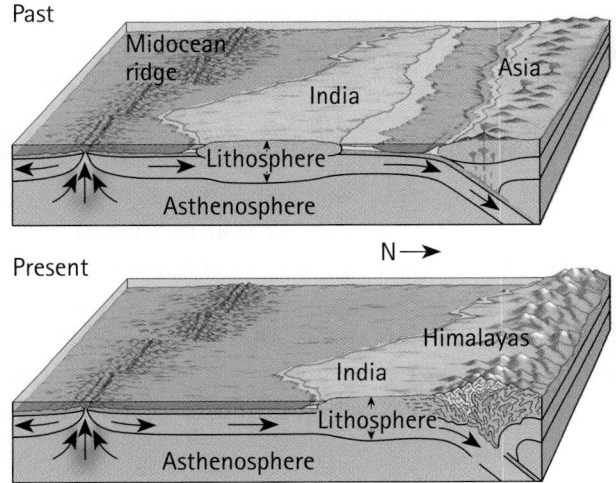

FIGURE 27.17 ▲
The continent–continent collision of India with Asia produced—and is still producing—the Himalayas.

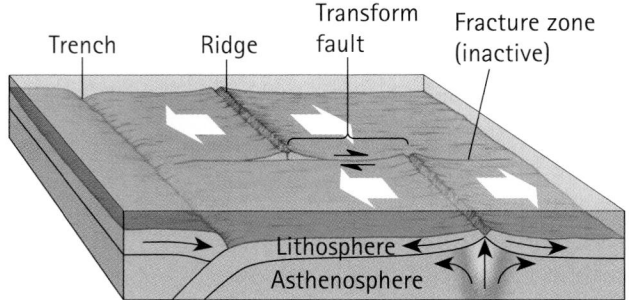

FIGURE 27.18 ▲
Transform faults allow two plates to slide past one another where two ridge segments are offset.

Transform Plate Boundaries

A **transform plate boundary** occurs where two plates are neither colliding nor pulling apart. Rather, they are sliding horizontally past each other (Figure 27.18). Most transform boundaries are found in ocean basins and connect offsets in the mid-ocean ridges. Look at the Mid-Atlantic Ridge in Figure 27.19 and notice how it is broken up into segments. The offset ridge segments are connected by strike-slip faults that "transform" the motion from one ridge segment to another. Between ridge segments, lithosphere coming from one ridge moves in the opposite direction of lithosphere coming from the other ridge. Now look at Figure 27.18 in the area of the inactive fracture zone. The sections of lithosphere on opposite sides of the fracture zone are part of the same plate; both sides are moving in the same direction. But along the transform boundary, lithosphere is moving in opposite directions. ✔ Because there is no tension or compression between the plates as they slide by each other, there is no creation or destruction of the lithosphere. A transform boundary simply accommodates horizontal plate movement.

Although most transform boundaries are relatively short and located within ocean basins, a few are quite long, such as the San Andreas Fault in California (Figure 27.20). The San Andreas Fault stretches for 1500 km from Cape Mendocino in northern California to the East Pacific Rise in the Gulf of California. The Pacific Plate is moving northwest at a rate of about 5.0 cm per year relative to the North American Plate. The San Andreas Fault accommodates about 70% of this motion, or about 3.5 cm per year. The rest of the motion occurs along other faults (such as the Hayward Fault). Grinding and crushing take place as the two plates move past each other. When sections of the

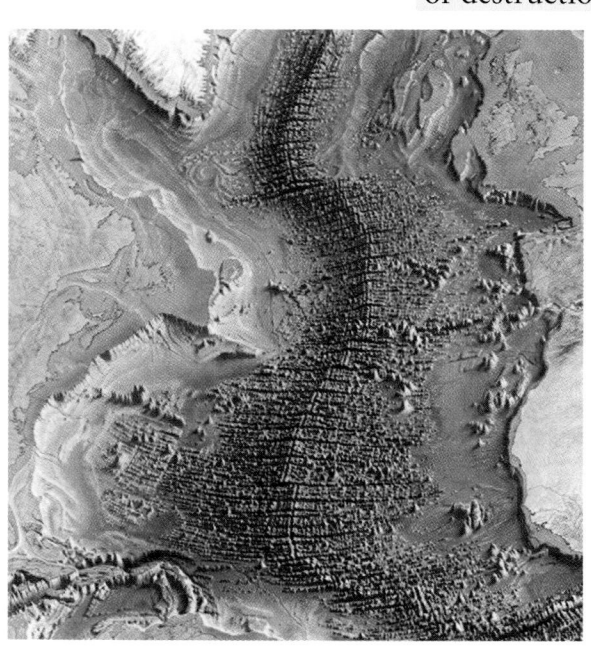

◄ **FIGURE 27.19**
Most transform boundaries occur in ocean basins where they offset oceanic ridges—for example, the Mid-Atlantic Ridge.

CALCULATION CORNER

Moving Faults

Knowing the rate of movement along a fault, it's easy to calculate the amount of *offset* over a period of time. Movement along the San Andreas Fault is about 3.5 cm per year. If a fence was built across the fault in 1990, we can figure out how far apart the two sections of the now-broken fence will be in 2010. The period of time is $2010 - 1990 = 20$ years. Thus, the two parts of the fence will be separated by

$$\frac{3.5 \text{ cm}}{\text{yr}} \times 20 \text{ yr} = 70.0 \text{ cm}$$

If you want to know how many feet that is, we can convert by knowing that there are 2.54 cm in an inch and 12 in in a foot:

$$70.0 \text{ cm} \times \frac{1 \text{ in}}{2.54 \text{ cm}} \times \frac{1 \text{ ft}}{12 \text{ in}} = 2.296 \text{ ft}$$

Which rounds off to 2.3 feet.

Sample Problem

Two trees were planted on opposite sides of the San Andreas Fault in 1907. How far apart were they in 2007?

Solution:

The period of time is $2007 - 1907 = 100$ years.

$$100 \text{ yr} \times \frac{3.5 \text{ cm}}{\text{yr}} = 350 \text{ cm}$$

---

plates become locked together, stress builds up until it is relieved in the form of an earthquake. On April 18, 1906, the Pacific Plate lurched about 6 m northward over a 434-km stretch of the fault, releasing the built-up stress and resulting in the catastrophic San Francisco earthquake.

> ✔ **READING CHECK**
>
> Why is no lithosphere created or destroyed at a transform boundary?

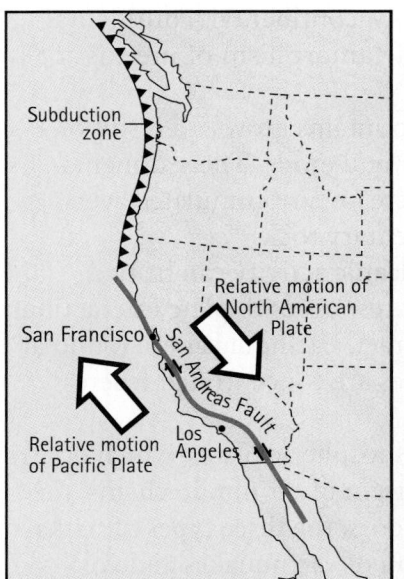

(a)

(b)

> The 1906 San Francisco earthquake did quite a bit of damage. Interestingly, most of the damage was due to the fires that burned the city afterward. The reason—all the water mains had burst and broken during the earthquake, so there was no water to put out the fires!

FIGURE 27.20 ▲
(a) The San Andreas Fault is a transform boundary famous for its earthquakes. The slice of California moving northwesterly lies on the Pacific Plate, while the rest of California sits on the North American Plate. (b) In this photo of the San Andreas Fault, notice the long valley created by many years of rock grinding along the fault.

27.5 The Theory That Explains Much

Before the theory of plate tectonics, processes such as mountain building, folding, and faulting were poorly understood. Plate tectonics explains the where and why of many geologic phenomena. Indeed, it can be thought of as a unifying theory because it links causes and effects.

Why are the Appalachian Mountains located where they are? What about the Sierra Nevada? The Rocky Mountains? The Alps? The plate tectonics model gives an answer: mountain-building events take place near convergent plate boundaries.

We can also relate the formation of the three different rock types to the plate tectonics theory. All rocks are tied to plate interaction in one manner or another. First, consider the most common igneous volcanic rock. The formation of huge volumes of basalt is linked to divergent plate boundaries, where mantle rock partially melts to form new basaltic oceanic crust.

Magma is also generated at subduction zones. Plates that are pulled downward become heated and release fluids. These fluids interact with the overlying mantle rock and lithosphere, causing the material to partially melt. Some of this magma crystallizes. The processes of partial melting and crystallization enrich the magma's silica content. Andesitic magma forms first followed by granitic magma. So the formation of andesite and granite is linked to subduction processes at convergent boundaries. Lastly, wherever plates collide there is intense heat and pressure. And what kind of rock is associated with heat and pressure? Metamorphic rock. Most notably, continental–continental convergence zones are where extensive metamorphism of preexisting rock occurs.

What about sedimentary rocks? As mountains grow as a result of plate collisions, they also begin to weather and erode. The sediments produced are transported downslope. There, they accumulate, layer upon layer, and eventually become sedimentary rock.

Lastly, virtually all earthquake and volcanic activity can be tied directly to plate tectonics. These energetic responses to plate interactions are almost always found where plates interact. Earthquakes are found at all types of plate boundaries, and volcanoes are concentrated where plates either collide or pull apart.

So, the tectonic interaction between lithospheric plates, which occurs mostly at their boundaries, explains the origin of mountain chains, the formation and destruction of the ocean floors, the three types of rocks found on Earth, and the global distribution of earthquakes and volcanoes. The internal motions that change Earth's surface do so in a cycle. The study of geology uses observable processes that occur today to understand what may have occurred in the past. This concept is commonly stated as "the present is the key to the past." So, what has happened in the past provides clues as to what may happen in the future. The Earth is indeed a dynamic planet.

27 CHAPTER REVIEW

KEY TERMS

Continental drift A hypothesis by Alfred Wegener that the world's continents are mobile and have moved to their present positions as the ancient supercontinent Pangaea broke apart.

Convergent plate boundary A plate boundary where tectonic plates move toward one another; an area of compressive stress where lithosphere is recycled into the mantle, or shortened by folding and faulting.

Divergent plate boundary A plate boundary where lithospheric plates move away from one another (a spreading center); an area of tensional stress where new lithospheric crust is formed.

Paleomagnetism The natural, ancient magnetization in a rock that can be used to determine the polarity of Earth's magnetic field and the rock's location of formation.

Pangaea A single large landmass—a supercontinent—that existed in the geologic past and was composed of all the present-day continents.

Plate tectonics The theory that Earth's lithosphere is broken into pieces (plates) that move over the asthenosphere; boundaries between plates are where most earthquakes and volcanoes occur and where lithosphere is created and recycled.

Rift (rift valley) A long, narrow gap that forms as a result of two plates diverging.

Seafloor spreading The moving apart of two oceanic plates at a rift in the seafloor.

Subduction The process in which one tectonic plate bends and descends beneath another plate at a convergent boundary.

Transform plate boundary A plate boundary where two plates are sliding horizontally past each other, without appreciable vertical movement.

REVIEW QUESTIONS

Continental Drift—An Idea Before Its Time

1. What key evidence did Alfred Wegener use to support his idea of continental drift?
2. Wegener proposed that the world's continents had at one time all been joined together into one supercontinent. What was the name of this supercontinent?
3. What are glacial striations?
4. Why did the scientific community first reject Wegener's hypothesis of continental drift?

Search for the Mechanism to Support Continental Drift

5. Describe how apparent polar wandering helped revive the idea that continents move over geologic time.
6. What role did paleomagnetism play in supporting the hypothesis of continental drift?
7. What did the detailed mapping of the oceans reveal about their topography?
8. What was the major discovery of H. H. Hess?
9. How is the ocean floor similar to a gigantic slow-moving tape recorder?
10. In what way does seafloor spreading support the theory of continental drift?

The Theory of Plate Tectonics

11. The lithosphere moves because of convection currents in the mantle. What causes the convection currents?
12. Describe plate tectonics in one simple statement.

Three Types of Plate Boundaries

13. Name and describe the three types of plate boundaries.
14. What is a rift? Give an example.
15. Where is the world's longest mountain chain located?
16. In what locations are the deepest parts of the ocean?

3. The San Andreas Fault separates the northwest-moving Pacific Plate, on which Los Angeles sits, from the North American Plate, on which San Francisco sits. If the plates slide past one another at a rate of 3.5 cm/yr, how long will it take the two cities to form one large city? (The present distance between Los Angeles and San Francisco is 600 km.)

READINESS ASSURANCE TEST (RAT)

If you have a good handle on this chapter—if you really do—then you should be able to score 7 out of 10 on this RAT. Check your answers with your teacher. If you score less than 7, study further before moving on.

Choose the best *answer to each of the following.*

1. At divergent boundaries, basaltic magma is generated by the
   (a) melting of coral reefs.
   (b) partial melting of continental crust.
   (c) partial melting of mantle rock.
   (d) subduction of mantle rock.

2. The theory of plate tectonics is *NOT* supported by
   (a) seafloor spreading.
   (b) paleomagnetism.
   (c) fathometers.
   (d) glacial striations.

3. Seafloor spreading provided a driving force for continental drift because
   (a) the youngest seafloor is found near the continents.
   (b) seafloor spreading pushes continents apart.
   (c) mantle convection causes slippage.
   (d) subduction creates the youngest seafloor.

4. Lithosphere is created at _____ boundaries and destroyed at _____ boundaries.
   (a) convergent; divergent
   (b) divergent; convergent
   (c) divergent; transform
   (d) convergent; transform

5. The hypothesis of continental drift is *NOT* supported by
   (a) similar fossils.
   (b) glacial striations.
   (c) similar rock types on different continents.
   (d) the tidal influence of the Moon.

6. Subduction occurs as a result of
   (a) gravity pulling the older and denser lithosphere downward.
   (b) horizontal plate accommodation.
   (c) upwelling of hot mantle material along the trench.
   (d) lubrication from the generation of andesitic magma.

7. The dominant force at divergent boundaries is
   (a) compression.
   (b) tension.
   (c) shearing.
   (d) similar to that in reverse faulting.

8. The dominant force at convergent boundaries is
   (a) compression.
   (b) tension.
   (c) shearing.
   (d) similar to that in normal faulting.

9. Magma is generated at all of these plate boundaries EXCEPT
   (a) divergent boundaries.
   (b) oceanic–oceanic convergent boundaries.
   (c) oceanic–continental convergent boundaries.
   (d) continental–continental convergent boundaries.

10. Convection in the mantle is caused primarily by
    (a) heat moving from the core to the crust.
    (b) conduction.
    (c) gravity and temperature differences.
    (d) friction of overlying lithosphere.

28

SHAPING EARTH'S SURFACE

THE MAIN IDEA

Water on Earth operates in a continuous cycle. As water moves through this cycle, it shapes our planet's surface.

A view of Earth from space shows our planet to be a vast expanse of water interrupted here and there by island-like continents. Indeed, about 70% of Earth's surface is covered with water. Continental land makes up only 30% of the surface. But on this land we find stark white glaciers, dry arid deserts, towering mountains, snaking rivers, deeply incised canyons, wide-open plains, and fan-shaped deltas where rivers meet the sea. What is the common factor in all of these varied landscapes? Water.

Water has many forms—it is a liquid, a solid, and a gas. And water, in each of these forms, shapes Earth's surface. Although water is the primary agent of change, gravity plays a major role. As we have seen throughout our exploration of physical science, the force of gravity is always at work. Wind is also an agent of change. So with gravity, the agents of water, wind, and ice are at work as they change and shape Earth's surface. Let us investigate how these agents flow and their effect on Earth's surface.

Explore!

Water Shapes Earth's Surface

1. Fill a rectangular aluminum baking pan partway with wet sand—the sand should be wet to the point where it holds its shape.
2. Elevate one end of the cake pan to create a slope.
3. Using a spouted beaker, slowly pour water onto the elevated side of the pan. Change the level of the pan as you continue to pour the water.

Analyze and Conclude

1. **Observing** Does the water form into a small river? Does slope influence the formation of the river?

2. **Predicting** At first, water simply soaks into the sand. Then when the sand is saturated, the water flows into the form of a small river.

3. **Making Generalizations** A small river will form more quickly and flow faster when the slope is steepened.

28.1 The Hydrologic Cycle

✓ Water on Earth is constantly circulating, driven by the heat of the Sun and the force of gravity. As the Sun's energy evaporates ocean water, a cycle begins (Figure 28.2). Evaporation moves water molecules from Earth's surface to become part of the atmosphere. The resulting moist air may be transported great distances by wind. Some of the water molecules condense to form clouds and then precipitate as rain or snow. The total amount of water vapor in the atmosphere remains relatively constant—evaporation and precipitation are in balance.

If precipitation falls on the ocean, the cycle is complete—from ocean back to ocean. The cycle is longer when precipitation falls on land. In this case, water may drain to small streams, then to rivers, and then journey back into the ocean. Or it may percolate into the ground or evaporate back into the atmosphere before reaching the ocean. Also, water falling on land may become part of a snow pack or glacier. Although snow or ice may lock water up for many years, such water eventually melts or evaporates and returns to the cycle. This natural circulation of water—from the oceans to the air, to the ground, to the oceans, and then back to the atmosphere—is called the **hydrologic cycle.**

The rain or snow that falls on the continents is Earth's only natural supply of freshwater. More than three-quarters of Earth's freshwater is locked up in polar ice caps and glaciers. Surprisingly, most of the freely flowing freshwater is not in lakes and rivers, but rather beneath Earth's surface. As rain falls and soaks into the ground, it percolates downward to fill the open pore spaces between sediment grains. This water is **groundwater.**

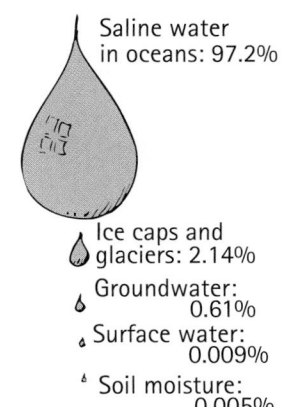

Saline water in oceans: 97.2%

Ice caps and glaciers: 2.14%
Groundwater: 0.61%
Surface water: 0.009%
Soil moisture: 0.005%

FIGURE 28.1 ▲
Distribution of Earth's water supply.

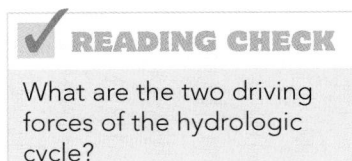

✓ **READING CHECK**

What are the two driving forces of the hydrologic cycle?

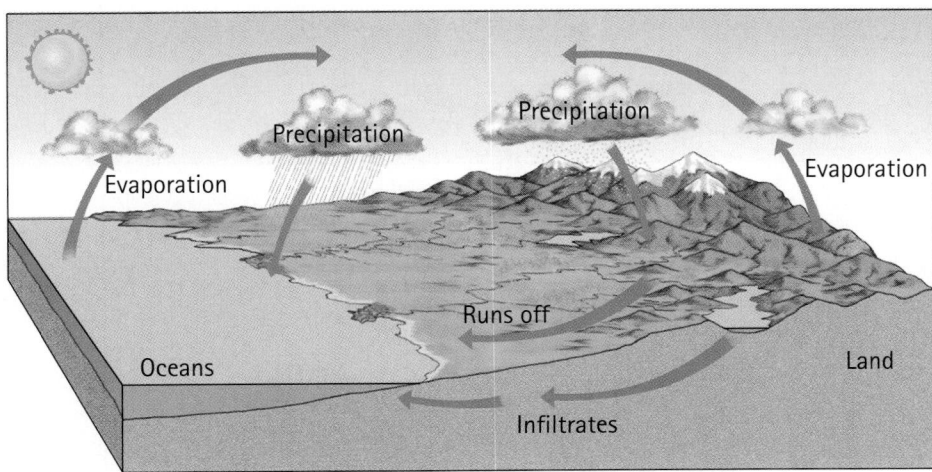

Precipitation

Precipitation

Evaporation

Evaporation

Runs off

Oceans

Land

Infiltrates

FIGURE 28.2 ▲
The hydrologic cycle. Water evaporated at Earth's surface enters the atmosphere as water vapor, condenses into clouds, precipitates as rain or snow, and falls back to the surface, only to evaporate again and go through the cycle yet another time.

CONCEPT CHECK
What percentage of Earth's water supply is freshwater?

Check Your Answer
Less than 3%, as you can see by adding the freshwater values in Figure 28.1: 2.14% (glaciers) + 0.61% (groundwater) + 0.009% (surface water) + 0.005% (soil moisture) = 2.764%.

28.2 Groundwater—Water Below the Surface

The liquid water in lakes, ponds, rivers, streams, springs, and puddles is the only freshwater that meets our eye, but all these water sources together hold only about 1.5% of Earth's non-ice freshwater. The other 98.5% resides in porous regions beneath Earth's surface—groundwater.

Have you ever noticed how during a rainstorm sandy ground soaks up rain like a sponge? Water literally disappears into the ground. The type of surface material influences the ease with which water goes into the ground. Some soils, like sand, readily soak up the water. Other soils, like clay, do not. Rocky surfaces with little or no soil are the poorest absorbers of water, with water penetration occurring mostly through cracks in the rock. Rain that does not soak into the ground becomes *runoff*, which then finds its way to bodies of water such as lakes and rivers or evaporates.

✓ Water beneath the ground exists as groundwater and *soil moisture.* The region where water has completely filled all the open pore spaces is called the *saturated zone.* (Figure 28.3). Above the saturated zone is the *unsaturated zone,* where soil moisture resides. Pore spaces in

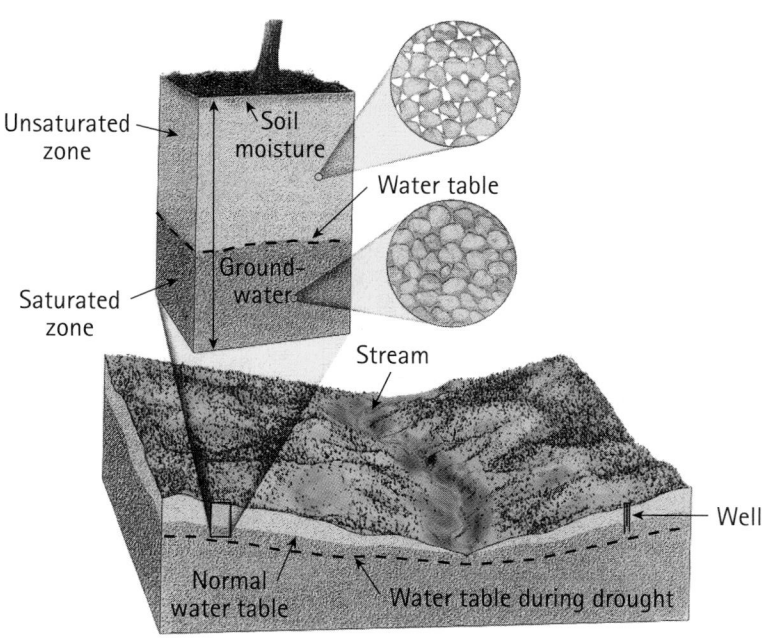

FIGURE 28.3 ▶
The unsaturated zone is above the saturated zone. In the unsaturated zone, pore spaces are filled with both air and water. This water is soil moisture. In the saturated zone, pore spaces are completely filled with water. This water is groundwater.

(a)

(b)

FIGURE 28.4 ▲
Porosity and permeability. (a) The sediment particles in clay are small, flat, and tightly packed, with small, poorly connected pore spaces. Thus, clays have high porosity but low permeability. (b) Sediment particles in sand or gravel are relatively uniform in size and shape, with large and well-connected pore spaces. This allows water to flow freely. So sands and gravels can have high porosity and high permeability.

the unsaturated zone are not completely filled with water—they contain a significant amount of air. As with water in a swimming pool, the pressure in groundwater increases with depth. Just as we can pump water from a swimming pool, we can pump groundwater from the ground. However, the presence of air in pore spaces prevents us from withdrawing water from the unsaturated zone.

The amount of water that the ground can hold depends on the porosity of the soil or rock. **Porosity** is the volume of open pore space in soil, sediment, or rock compared to the total volume (the solids plus the open pore spaces). Porosity depends on the size, shape, and packing of the soil or sediment particles. For example, a soil made up of rounded particles of the same size will have a higher porosity than a soil made up of rounded particles of many different sizes. This is because the smaller particles fill up the spaces between the larger particles, thereby reducing the soil's overall porosity.

Porosity is a measure of the maximum *amount* of groundwater at a given location. But porosity does not tell us how groundwater *moves*. The degree to which a porous material permits fluid to flow is called **permeability.** When pore spaces are well connected, water flows easily. If the pore spaces are extremely small and poorly connected (as is the case with flattened clay particles), water may barely move. Think of it this way: It's a lot easier to sip soda through a large straw than through one of those very small straws intended for stirring coffee. Likewise, it is difficult for water to move through the pores of clay. The permeability of clay is almost zero, even though the porosity of most clay can be very high. In contrast, sand and gravel have large, well-connected pore spaces, and water moves freely from one pore space to the next. Thus, sand and gravel are highly porous and highly permeable (Figure 28.4).

 **READING CHECK**

What is water below the ground surface called?

CALCULATION CORNER

Porosity

Porosity tells us the ratio of open space to total volume of soil, sediment, or rock sample:

$$\text{Porosity} = \frac{\text{volume of open space}}{\text{volume of open space } + \text{ volume of solids}}$$

Sample Problem
The volume of solids in a sediment sample is 975 cm³ and the volume of open space is 325 cm³. What is the porosity?

Solution:

$$\text{Porosity} = \frac{325 \text{ cm}^3}{325 \text{ cm}^3 + 975 \text{ cm}^3} = 0.25$$

So the volume of open space is only one-fourth of the total volume.

Want to see the water table? Most ponds and lakes are simply a place where the land surface dips below the water table.

The Water Table

If we were to dig a hole into the ground, we'd find that the wetness of the soil varies with depth. Just below the surface, in the unsaturated zone, pore spaces are partially filled with water. Then as we dig farther, we enter the saturated zone, where pore spaces are completely filled with water. If our hole is entirely within the unsaturated zone, the hole does not fill with water. But if we dig our hole deeper into the saturated zone, it does fill with water. The upper boundary of the saturated zone is called the **water table** (Figure 28.3). The level of the water in our hole is at the same level as the water table. In fact, the level of water in our hole *is* the water table at that location.

The depth of the water table beneath the surface varies with precipitation and climate. It ranges from zero in marshes and swamps to hundreds of meters in some parts of the deserts. The water table also tends to rise and fall with the surface topography (Figure 28.5). At lakes and perennial streams (streams that flow all year), the water table is above the land surface.

Aquifers and Springs

✔ Any water-bearing underground region that groundwater can flow through is called an *aquifer*. These reservoirs of groundwater are beneath the land surface in many places and contain an enormous amount of water. Aquifers can be either confined or unconfined. In an unconfined aquifer, the soil or sediment above the water table is permeable. So water

FIGURE 28.5 ▶

The water table roughly parallels the ground surface. In times of drought, the water table falls, reducing stream flow and drying up wells. The water table can also fall if water pumped out of a well exceeds the amount of groundwater recharge.

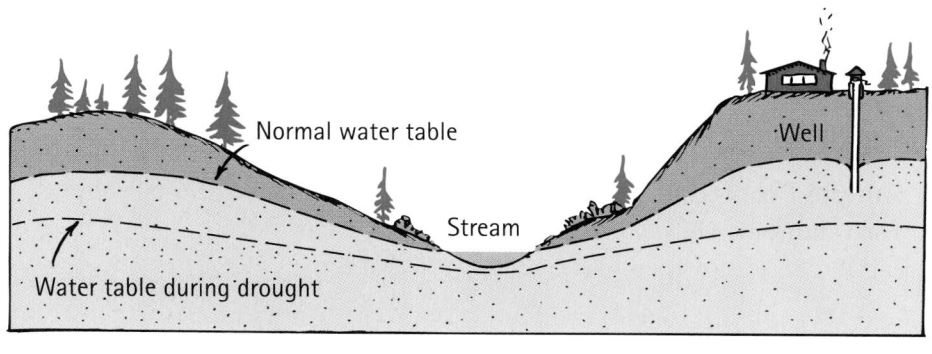

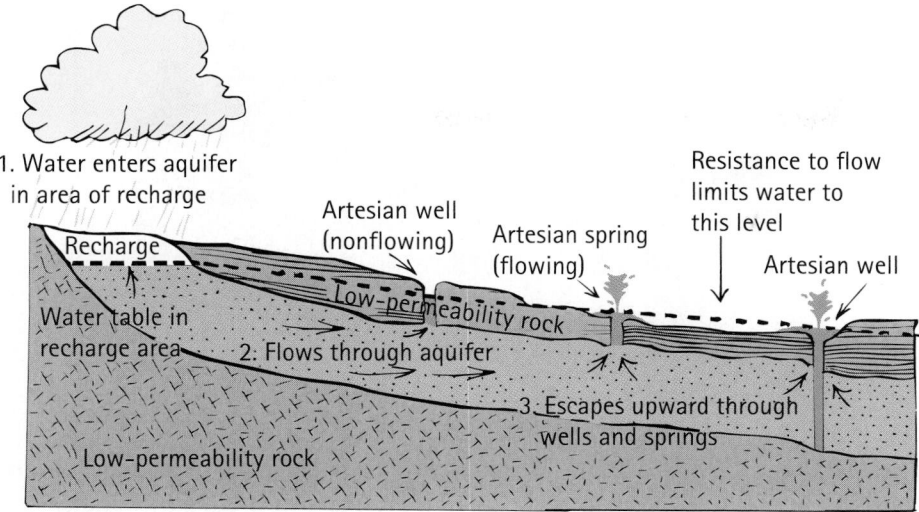

FIGURE 28.6 ▲
An artesian system is formed when groundwater in a confined aquifer rises to the surface through any opening that taps the aquifer. Water flows freely if the height of the water table in the recharge area is greater than the height of the opening (flowing artesian well and artesian spring). If the height of the opening is greater than the height of the water table in the recharge area, the water does not flow (nonflowing artesian well).

Water withdrawn from the saturated zone is groundwater discharge. Water added to the saturated zone is groundwater recharge.

that soaks into the ground directly flows into the aquifer. Water added to an aquifer is called *recharge.* All aquifers are at least partially unconfined, as illustrated in Figure 28.6.

In a *confined* aquifer, the soil or sediment above the aquifer has a low permeability. So water that soaks into the ground cannot flow directly into the aquifer. An aquifer is considered confined if it is sandwiched between continuous low-permeability layers (Figure 28.6). Confining layers are generally found in sedimentary deposits that have alternating sand and clay layers or alternating sandstone and shale layers. So water recharge to a confined aquifer comes from water infiltration at the unconfined portions of the aquifer at higher elevation.

As we learned in Chapter 8, pressure in water depends on the height of water above. Water anywhere in the confined portion of an aquifer is below the level of the water table in the recharge area. So groundwater in a confined aquifer is under pressure from the water above. This pressure causes the water to flow out through openings at lower elevations. This is called an **artesian system.** If the opening is natural and water flows out of the ground, it is an *artesian spring.* If the opening has been drilled, it is an *artesian well.*

Because rock layers are not always continuous, sometimes a low-permeability layer can stop and hold the downward-percolating water above the water table. When this happens, a *perched* water table is created. Wherever the water table meets the land surface, groundwater can emerge as a spring, stream, or lake (Figure 28.7). Springs can generally be found where the water table (or a perched water table) meets the

The Ogallala aquifer stretches from South Dakota to Texas and from Colorado to Arkansas!

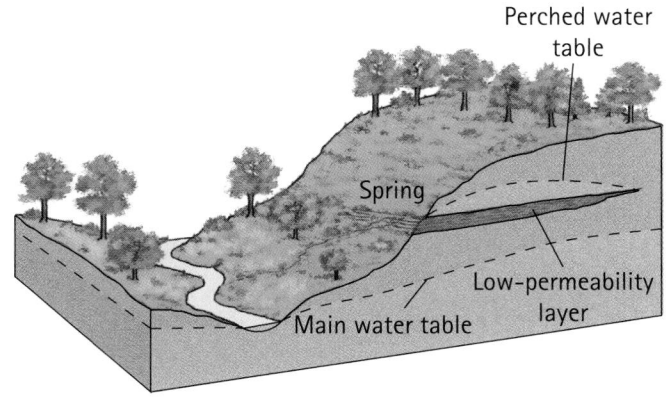

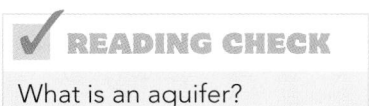

What is an aquifer?

surface along a slope, such as on a hillside or coastal cliff. Because water
tends to leak out of the ground through cracks and breaks in a rock,
springs are often associated with faults. In fact, field geologists can often
locate faults by looking for springs.

CONCEPT CHECK

1. If you dig a hole in the ground and pour water into it, the water
   seeps out. Why doesn't water seep out of ponds and lakes?
2. What principal condition is required for an artesian system to
   occur?

Check Your Answers

1. It does! Water does seep in and out of ponds, but because the surrounding ground (to
   the sides and below the pond) is already soaking wet, seepage in equals seepage out.
2. For an artesian system to occur, confining layers are needed. Confining layers allow
   groundwater in confined aquifers to be under higher pressure than groundwater in
   unconfined aquifers. This allows water to rise above the top of the confined aquifer at
   natural or human-made openings.

Groundwater Movement

How does water move through the ground? We have learned that a
material's permeability affects groundwater flow. The higher the permeability, the easier the flow. But there is another factor that affects the flow
of groundwater—gravity. All water flows downhill because of gravity.
For example, in a stream at Earth's surface, water flows from areas of
high elevation to areas of low elevation. Likewise for groundwater, but
beneath the ground it is the elevation of the water table that drives water
flow. Groundwater moves from areas where the water table is high to
areas where the water table is low. This downward slope of the water
table is called the *hydraulic gradient* (Figure 28.8).

The hydraulic gradient determines the direction of groundwater
flow. And this is where groundwater flow differs from surface water flow.
In a surface stream, water flows through a stream channel. But groundwater does not flow through a channel—it flows through all the open
pore spaces in an aquifer. Groundwater flow depends on the hydraulic

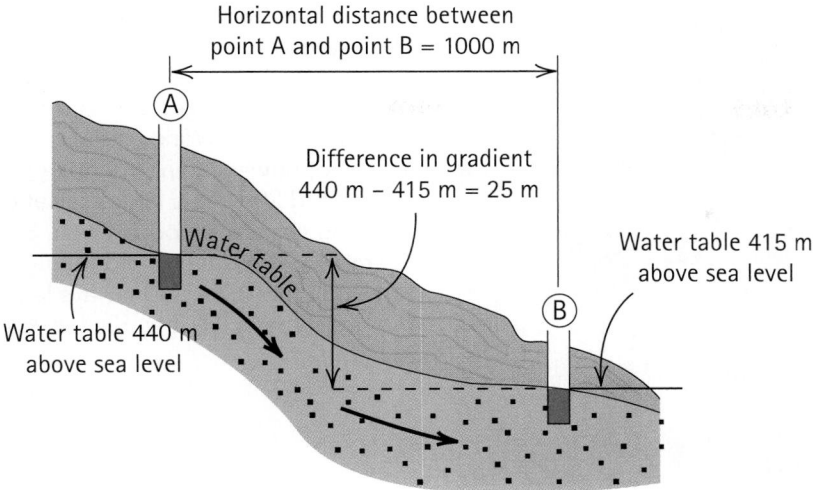

The hydraulic gradient can be expressed like any slope—"rise over run." It is the difference in slope between any two locations divided by the horizontal distance between the locations. In this example, we have (440 m – 415 m)/1000 m.

gradient. And because hydraulic gradient is not the same in all parts of an aquifer, groundwater often flows in curved pathways (Figure 28.9).

✔ So, the movement of groundwater is a result of two factors: permeability—the more permeable the material, the faster the flow—and gravity—water moves down the hydraulic gradient.

The speed of groundwater movement is generally very slow compared to the flow speed in rivers and streams. The more permeable the aquifer, the faster the flow; the greater the hydraulic gradient, the faster the flow. The speed and route of groundwater flow can be measured by introducing dye into a well and noting the time it takes to travel to the next well. In most aquifers, groundwater speed is only a few centimeters per day, enough to keep underground reservoirs full.

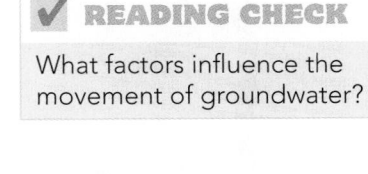

✔ **READING CHECK**

What factors influence the movement of groundwater?

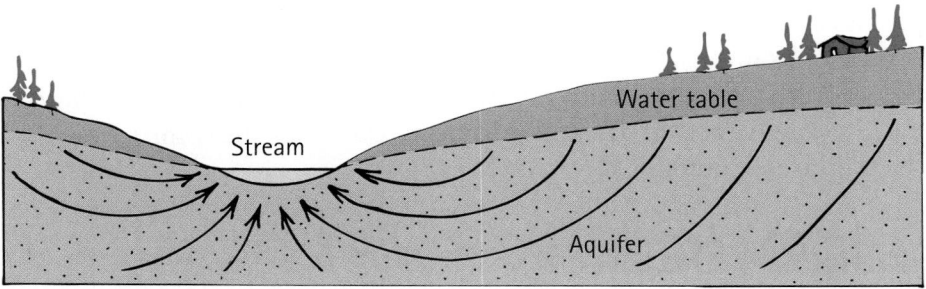

◀ **FIGURE 28.9**
Groundwater flows from an area where the water table is high, such as beneath a hill, to where the water table is low, such as beneath a stream valley. The curved arrows indicate that the stream is fed from below.

28.3 The Work of Groundwater

Flowing groundwater—no matter how slow—can cause large changes in landscapes. These changes can occur because of humans, but more often than not they occur without any human interference.

Pumping Can Cause Land Subsidence

Most wells are drilled so that groundwater can be pumped from the ground. ✔ In areas where groundwater withdrawal has been extreme, the land surface is lowered—it *subsides*. The problem of land subsidence is most noticeable where the ground and underlying

◀ **FIGURE 28.10**

The Leaning Tower of Pisa. Construction began in 1173 but was suspended when builders realized the foundation was inadequate. Work was later resumed, however, and the tower was completed 200 years later. Deviation from the vertical is about 4.6 m. Efforts to stabilize the tower are ongoing. Although the tower will always lean, the goal is to make it stable for years to come.

subsurface is made up of thick layers of loosely packed sediments rather than rock. These thick layers of sediments usually have many layers of easily compressed, water-bearing clays sandwiched between a series of sandy aquifers. Recall that clay has a high porosity and a very low permeability. As water is pumped from the aquifer, water slowly leaks out of the clay layers to replenish the aquifer. As the clays lose water, they compact, causing the land surface to subside.

A famous example of land subsidence is the Leaning Tower of Pisa in Italy. The tilting tower was built on loose sediments from the Arno River. Over the years, as sediments were compacted, the tilt of the tower increased (Figure 28.10). In the United States, large amounts of groundwater have been pumped for irrigation in the San Joaquin Valley of California. This process caused the water table to drop 75 m in 50 years, lowering the land surface by as much as 9 m (Figure 28.11). Because canals now provide water for irrigation, the sandy aquifers are slowly being recharged. Unfortunately, though, most of the land subsidence caused by the compaction of the clay layers cannot be reversed.

CONCEPT CHECK

Why is land subsidence most evident in regions where the underlying geology is a series of clay layers sandwiched between sandy aquifers?

Check Your Answer

Clay layers lose water and compact as water is pumped from the adjacent aquifers. Compaction causes the land to subside.

FIGURE 28.11 ▲

The land surface in California's San Joaquin Valley lowered more than 9 m (30 ft) over a 50-year period because of groundwater withdrawal and the resulting compaction of sediments.

Some Rocks Are Dissolved by Groundwater— Carbonate Dissolution

The vast carbonate deposits (such as limestone) that underlie millions of square kilometers of Earth's surface provide storage areas for groundwater. The effect of groundwater on limestone is unique. It creates some very interesting landforms. Recall from Chapter 25 that limestone is made of the mineral calcite ($CaCO_3$). ✔ On its way to becoming groundwater, rainwater naturally reacts with carbon dioxide in the air and soil to produce carbonic acid. When this slightly acidic water comes in contact with limestone, the carbonic acid partially dissolves the rocks. As groundwater steadily dissolves the limestone, it creates unusual erosional features, like sinkholes and caverns.

Caverns and Caves   The dissolving action of underground water has carved out magnificent caverns and caves (a cavern is simply a large cave). Groundwater flow in limestone aquifers occurs mostly through fractures in the rock, rather than through pores. It is in limestone that we find the only true underground rivers. (In other rocks and soils, underground water is found only in pore spaces, not in large, open channels.)

Rainwater (enriched in carbonic acid) soaking into the limestone flows downward through the fractures toward the water table, dissolving rock as it goes. As groundwater flows toward its outlet, say a stream, the slightly acidic streamwater also dissolves the surrounding limestone. Fractures become larger and eventually underground channels and caves begin to form (Figure 28.12). Water dripping from the cave ceiling (rich in dissolved calcium carbonate) creates icicle-shaped stalactites as water evaporates and the calcium carbonate precipitates. Some of the water solution drips off the end of the stalactites to build corresponding cone-shaped stalagmites on the floor (Figure 28.13).

Sinkholes   Sinkholes are funnel-shaped cavities in the ground that are open to the sky. They are formed in much the same way as caves.

Impressive caves and caverns are found in many places—Carlsbad Caverns in New Mexico, Blanchard Springs in Arkansas, and Mammoth Cave in Kentucky, to name just a few.

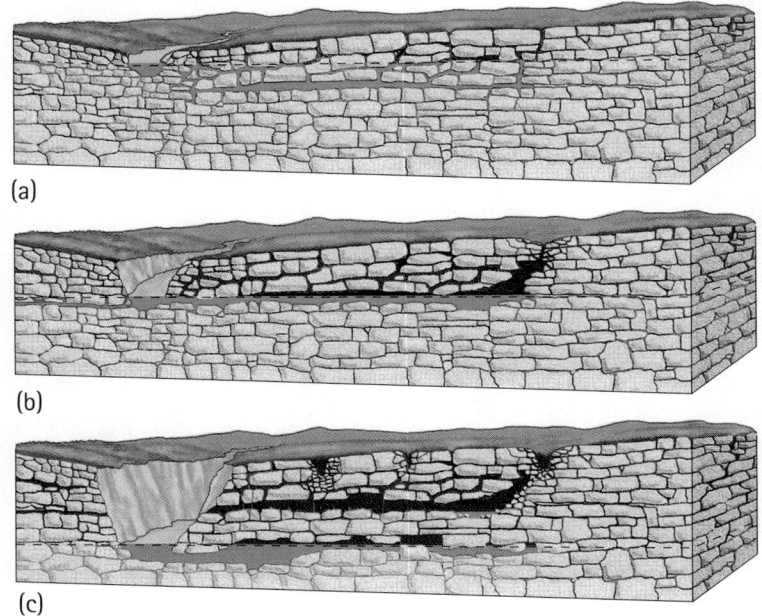

(a)

(b)

(c)

◀ **FIGURE 28.12**
The formation of a cave begins with layers of carbonate rock, mildly acidic groundwater, and an enormous span of time. (a) Groundwater makes its way toward a stream. (b) As the stream channel deepens, the water table is lowered. The carbonate rock is dissolved as acidified water erodes and enlarges the existing fractures into small caves. (c) Further deepening of the stream channel causes the water table to drop even lower; water in the cave seeps downward, leaving an empty cave above a lowered groundwater level.

FIGURE 28.13 ▲
Cave dripstone formations.

FIGURE 28.14 ▲
Karst topography covered by vegetation makes up the rolling hills in south central Kentucky.

✔ **READING CHECK**

How do caves, caverns, and sinkholes form?

Groundwater dissolves limestone and eventually the surface collapses in on itself. Some sinkholes are caves whose roofs have collapsed. Some sinkholes are formed by drought conditions or excessive groundwater pumping.

Karst Regions   When sinkholes, caves, and caverns define the land surface, the terrain is called *karst topography,* named after the Karst region of Yugoslavia. The pattern of streams in this type of landscape is very irregular; streams and rivers disappear into the ground and reappear as springs. Some karst areas appear as soft, rolling hills with large depressions that dot the landscape; the depressions are old sinkholes now covered with vegetation (Figure 28.14). In general, karst areas have sharp, rugged surfaces and thin to nonexistent soils as a result of high runoff and dissolution of surface material (Figure 28.15).

FIGURE 28.15 ▶
The karst landscape of China has been an inspiration to classical Chinese brush artists for centuries.

28.4 Streams and Rivers—Water at Earth's Surface

Streams—by which we mean all flowing surface water, from the Mississippi River to the shallowest woodland creek—are dynamic systems that affect the surface of the land and the people who live on that land. Streams have many benefits to offer: They provide energy, irrigation, and a means of transportation.

Streams also carve out and alter the landscape. The Grand Canyon is testimony to the mighty erosive power of surface water. For millions of years, the Colorado River has been carving out the canyon walls, cutting deeper and deeper into the rock as it makes its way to the ocean. Yet surface water plays another important yet contrasting role as it shapes the landscape—it deposits sediments. In this way, surface water is both a destroyer and creator of sediments and sedimentary rocks.

Stream Flow

Streams come in a variety of forms—straight or curved, fast or slow. At their headwaters (the stream origin), stream channels are narrow and water flows quickly through deeply incised, V-shaped mountain valleys. Farther downstream, channels widen so that water flows into and along broad, low valleys.

✓   There are three variables that influence the speed of water in a stream—*stream gradient, stream discharge,* and *channel geometry.* The **gradient** is the vertical drop in the elevation of the stream channel divided by the horizontal distance for that drop. If we look at a long profile of a stream (Figure 28.16), we see that the gradient is steep near the stream's headwaters and gentler, almost horizontal, near its mouth. Because of gravity, stream speed tends to be greater where the gradient is steep. Downstream, discharge and channel geometry also influence stream speed.

Discharge is the volume of water that passes a given location in a channel in a certain amount of time. It is directly proportional to the cross-sectional area of the channel—the width times the depth—and to the *average* stream speed:

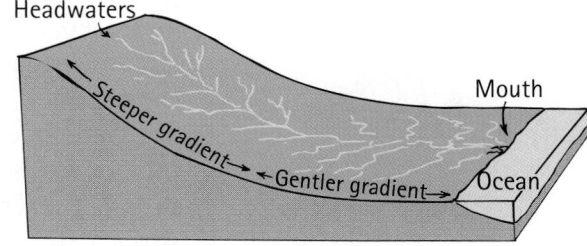

FIGURE 28.16 ▲
The long profile of a stream. At a stream's headwaters, the gradient is high, the channels narrow and shallow, and the stream flow is rapid. As the stream progresses downslope, the gradient decreases, the channel widens, and discharge increases.

$$\textbf{Discharge} = \textbf{cross-sectional area} \times \textbf{average stream speed}$$

Or put another way,

$$\textbf{Average stream speed} = \frac{\textbf{discharge}}{\textbf{cross-sectional area}}$$

Channel geometry—the shape of the channel—greatly influences stream speed. Consider two streams that have the same cross-sectional

READING CHECK

What three variables influence a stream's speed?

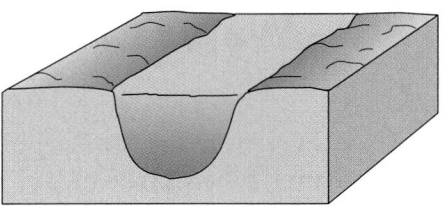 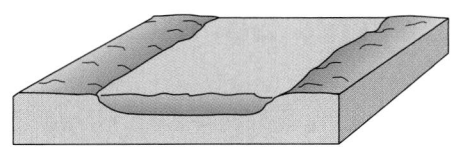

(a) Rounded, deep channel   (b) Wide, shallow channel

FIGURE 28.17 ▲
(a) In a rounded, deep channel, the speed of water flow is relatively high because there is less water in contact with the channel (there is less friction). (b) Wide, shallow channels tend to have slower flows because more water is in contact with the channel (there is more friction).

Squeeze the end of a garden hose and you'll see that water speeds up when the passage for flow becomes narrower. Likewise for the flow of streams.

area. Water flowing in the channel touches the channel bottom and sides. Friction between the water and the channel slows stream speed. The cross-sectional shape of a channel determines the amount of water in contact with the channel. The greater the contact area, the greater the friction (Figure 28.17). If the stream channel is rounded and deep, as opposed to flat-bottomed and relatively shallow, the stream speed will be faster because there is less water in contact with the channel.

Water speed also varies at different locations within the channel. Flow speed is slower along the streambed where water in contact with the channel creates friction; and flow speed is faster near the water's surface. In a large stream flowing in a straight channel, the maximum flow speed is found midchannel (Figure 28.18b). In a stream running through a bending, looping channel, the maximum flow speed is found toward the *outside* of each bend (Figure 28.18a and c).

FIGURE 28.18 ▲
In a stream that bends, (a) and (c), flow speed is greatest toward the outside of each bend and slightly below the surface. In a straight-channel stream, (b), maximum speed is midchannel and near the water's surface. Erosion of the stream channel occurs where stream speed is greatest (cut bank); deposition occurs where stream flow slows (point bar).

CONCEPT CHECK

1. Consider a stream in which discharge doubles downslope, but the channel stays the same size and shape. Now look at the equation for stream speed. What happens to stream speed?
2. Now consider a stream in which discharge doubles and the cross-sectional area of the channel also doubles. Assume that no water enters from side tributaries and friction is insignificant. What happens to stream speed?

Check Your Answers

1. Stream speed doubles.
2. Cross-sectional area and discharge of the stream channel increase by the same percentage, so stream speed does not increase.

28.5 The Work of Surface Water

Landscape evolution—progressive changes to Earth's surface—is driven by surface water flow. The pattern of flow has a large effect on how water alters the landscape. There are two major flow patterns—*turbulent* and *laminar* (Figure 28.19). Water moving erratically downstream, stirring everything it comes in contact with, is **turbulent flow.** Water flowing steadily downstream with no mixing of sediment is **laminar flow.** In general, slow, shallow flows tend to be laminar and faster-moving flows tend to be turbulent. Whether a flow is laminar or turbulent depends on the nature and geometry of the stream channel and the speed of the flow.

Erosion and Transport of Sediment

We learned in Chapter 25 that weathering and erosion create and move sediment. Erosion by water is the most common way sediments are carried away from the places where they formed. ✓ Surface water erodes sediment and rocks, transports them downstream, and eventually deposits them in another place. In this way, surface water reshapes our landscape.

Channel geometry, gradient, discharge, sediment load, and velocity all work together to influence stream flow. A river is a system of interdependence—a change in one variable changes the entire system.

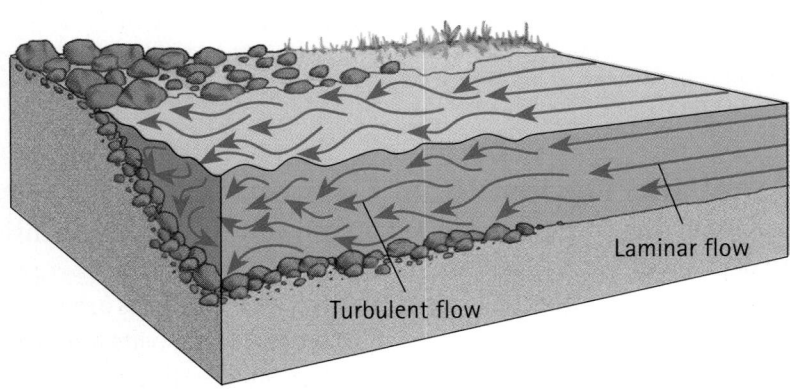

Turbulent flow

Laminar flow

◀ **FIGURE 28.19**
Laminar flow is slow and steady, with no mixing of sediment in the channel. Turbulent flow is fast and jumbled, stirring up everything in the flow.

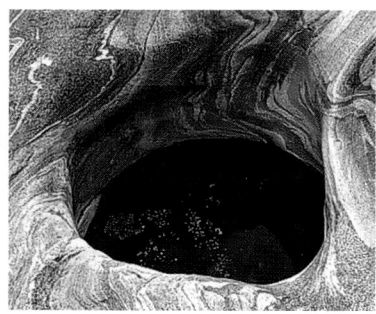

FIGURE 28.20 ▲
When powered by turbulent circular currents of water, rock particles rotate like drill bits and carve out deep potholes.

READING CHECK

How does a flowing stream change the land surface?

Flowing water erodes stream channels in several different ways. First of all, streamwater contains many dissolved substances that *chemically weather* and erode rock material. Another powerful mechanism for erosion is *hydraulic action*—the sheer force of running water. Swiftly flowing streams and streams at flood stage have strong erosive power as they break up and loosen great quantities of sediment and rock. The most powerful type of erosion, however, is *abrasion.* Abrasion occurs when sediments and particles actually scour a channel, much like sandpaper scraping on wood. When powered by turbulent spiraling water, rock particles rotate like drill bits and carve out deep potholes (Figure 28.20). The faster the current, the greater the turbulence and the greater the erosion.

Erosion is only the beginning of the story of how surface water alters Earth's surface. Streams carry more than just water—they transport great amounts of sediment from one location to another. In general, *laminar* flows can lift and carry only the very smallest and lightest particles. A *turbulent* flow, however, depending on its speed, can move and carry a range of particle sizes—from small particles of clay to large pebbles and cobbles. A turbulent current gathers and moves particles downstream mainly by lifting them into the flow or by rolling and sliding them along the channel bottom. The smaller, finer particles are easily lifted into the flow and they remain suspended to make the water murky.

As expected, faster currents can carry larger particles. Also, larger volumes of water can carry greater volumes of sediment. So, streams that have a higher discharge can carry larger volumes of sediment, and streams in which the water is moving fast can carry larger sizes of sediment. The continuous abrasion of sediment in the stream channel breaks up the sediments and thus contributes to an overall decrease in particle size as they are moved downstream. At the river's mouth, only finer particles of sand, silt, and clay remain. As we shall soon see, these tiny particles are deposited to form deltas when a stream loses speed as it enters the sea.

CONCEPT CHECK
Which is more effective in transporting sediment, laminar or turbulent flow?

Check Your Answer
Turbulent flow, because the water's motion is irregular and sediments have a greater tendency to remain in suspension. The energy of the churning water allows the sediments to be carried. In laminar flow, water moves steadily in a straight-line path with no mixing of sediment in the channel.

Erosional and Depositional Environments
Eventually, particles transported by surface water drop out of suspension—they are deposited. ✔ Deposition happens when water loses energy and slows down. As a river gradually loses energy, larger particles are deposited first and then smaller ones. In this way, surface water deposits tend to be well sorted. The most dominant feature of deposited sediments is the way

particles of sediment are laid down, layer upon horizontal layer. These layers are referred to as *beds*. Varying in thickness and area, each bed represents one episode of deposition. For example, flooding in a particular year might produce a layer of sediment next to a river. A flood at any time after that produces an overlying layer.

The deposition and erosion of sediments occur in many different environments, including oceans and shorelines, rivers and streams, deserts, and deltas. Each environment in which erosion, transportation, and deposition occur has its own specific characteristics.

Stream Valleys and Floodplains

As rainfall hits the ground, it loosens soil and washes it away. As more and more rain falls, and the ground continues to lose soil, gullies form. Once water and soil particles funnel into such a gully, a stream channel is created. This erosive action may be extremely rapid, as in the erosion of unconsolidated sediments, or very slow, as in the erosion of solid rock. Water's erosive power enables a stream to widen and deepen its channel, to transport sediment away, and, in time, to create a valley. In high mountain areas, the erosive action of a stream cuts down into the underlying rock to form a narrow V-shaped valley. Because the valley is narrow, the stream channel dominates the whole valley bottom. Fast-moving rapids and beautiful waterfalls are characteristic of V-shaped mountain stream valleys (Figure 28.21).

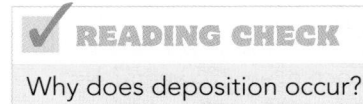

Why does deposition occur?

FIGURE 28.21 ▲
At a stream's headwaters, high gradients contribute to fast-moving rapids. When there is an abrupt increase in gradient, we see beautiful cascading waterfalls.

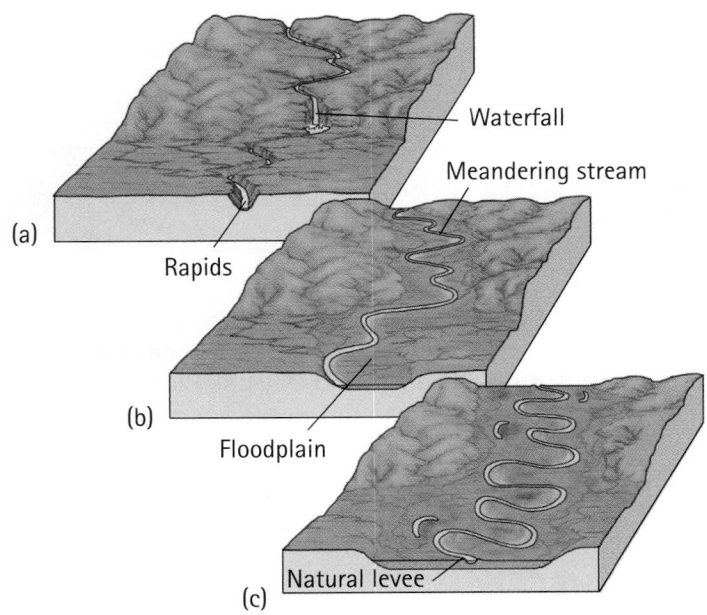

(a)

Waterfall

Rapids

Meandering stream

(b)

Floodplain

(c)

Natural levee

FIGURE 28.22 ▲
The evolution of a stream valley and development of a floodplain. (a) At the headwaters, the V-shaped stream valley is characterized by steep gradients and fast-moving water that cuts down into the stream channel. Features in this area include cascading rapids and waterfalls. (b) Downstream, with reduced gradient, the stream focuses its erosive action in a side-to-side sinuous manner, thereby widening the stream valley. (c) Farther downstream, meandering increases and further widens the stream valley to form a large floodplain.

Does a 100-year flood occur once every 100 years? Not exactly. In fact, a big flood can happen in any year. The term *100-year flood* means that there is a 1-in-100 chance that a "big flood" will occur during any year. Perhaps a better term would be the *1-in-100-chance-flood*.

Stream speed plays an important role in both erosion and deposition. As a stream flows downhill, its gradient becomes gentler and its speed slows. The focus of its energy changes from eroding downward (deepening the channel) to eroding laterally in a side-to-side motion. As a result, the stream develops a more sinuous, *meandering* form (Figure 28.22).

The meandering movement creates a wide belt of almost flat land—a **floodplain** (Figure 28.23). As the name implies, it is this section of the river valley that becomes flooded with water and sediments when a river overflows its banks. As discharge and flow speed increase in a flood, so does the stream's ability to carry sediment. Thus, when a stream overflows its banks, sediment-rich water spills out onto the floodplain. Because the speed of the water quickly decreases as it spreads out over the wide, flat floodplain, a progression of large to small particles is deposited. As expected, larger, coarse-grained sediments are deposited along the edges of the channel and smaller, fine-grained sediments are deposited farther away from the stream channel on the floodplain. The larger particles deposited close to the stream channel form *natural levees* that help to confine future floodwaters (Figure 28.24). The widening of the valley shown in Figure 28.22 occurs because sediment deposited by the stream, especially during floods, progressively fills the valley.

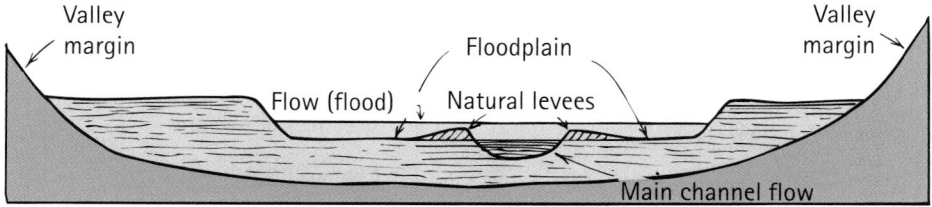

FIGURE 28.23 ▲
Cross section of a river valley. A floodplain is created when a river overflows its banks. Sands and gravels settle out first and act as natural levees to confine the river. Because the finer silt and clay particles are able to flow as a suspended load, they move beyond the levees and settle on the floodplain.

CONCEPT CHECK
Floodplains are often prime agricultural areas. Why would people want to work and live in areas so prone to flooding?

Check Your Answer
People live and work in floodplain areas because such plains are next to rivers that provide easy access to water, food, and a means of transportation. Also, because of periodic flooding, floodplain soils are often extremely fertile and thus serve as prime farmland. Still, a floodplain is always susceptible to flooding.

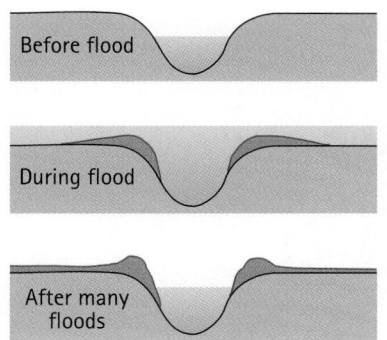

◀ **FIGURE 28.24**
With each successive flood, the height of the levees increases. In time, this can even raise the overall elevation of the channel bed so that it is higher than the surrounding floodplain.

Deltas Are the End of the Line for a River

As a stream flows into a standing body of water, such as a sea, bay, or lake, the moving water gradually loses its forward momentum. With reduced energy, stream speed slows and the stream loses its ability to carry sediment. These changes cause the stream to dump its sediment load. In this way, ✔ the mouth of the stream and the area immediately offshore become filled with sediment. The dumped sediment forms a fan-shaped deposit called a **delta** (Figure 28.25). Sediment is deposited in order of decreasing weight, with heavy, coarse particles deposited first, at and near the shoreline. Light, fine particles are deposited farther offshore. With the continual addition of incoming sediment, the delta progressively builds itself outward as an extension of land into the body of water. Thus delta environments are areas where new land is continuously created (Figure 28.26).

✔ **READING CHECK**

What is a delta?

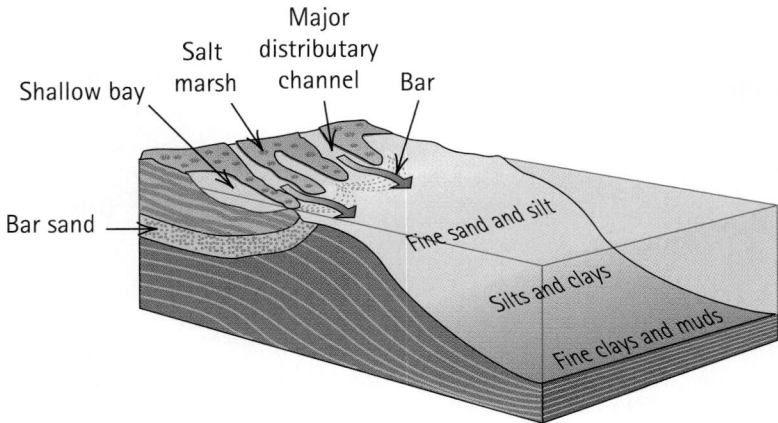

FIGURE 28.25 ▲
Deltas are areas of land generation. As streams flow to the sea, they carry sediment. These sediments are deposited in order of decreasing weight, with heavy, coarse particles settling at or near the shoreline and light, fine particles settling farther offshore.

Millions of years ago, the mouth of the Mississippi River was where Cairo, Illinois, is today. Since that time, the delta has extended 1600 km south to the city of New Orleans. Less than 5000 years ago the site of New Orleans was underwater in the Gulf of Mexico!

FIGURE 28.26 ▲
Satellite image of the Mississippi Delta.

DRAINAGE BASINS

A stream is one small segment of a much larger system called a *drainage basin.* Drainage basins are separated from one another by *divides,* lines tracing out the highest ground between streams. A divide can be either very long, if it separates two enormous drainage basins, or a mere ridge separating two small gullies. The *Continental Divide,* a continuous line running north to south down the length of North America, separates the Pacific basin on the west from the Atlantic basin on the east. Water west of the divide eventually flows to the Pacific Ocean, and water east of it flows to the Atlantic Ocean.

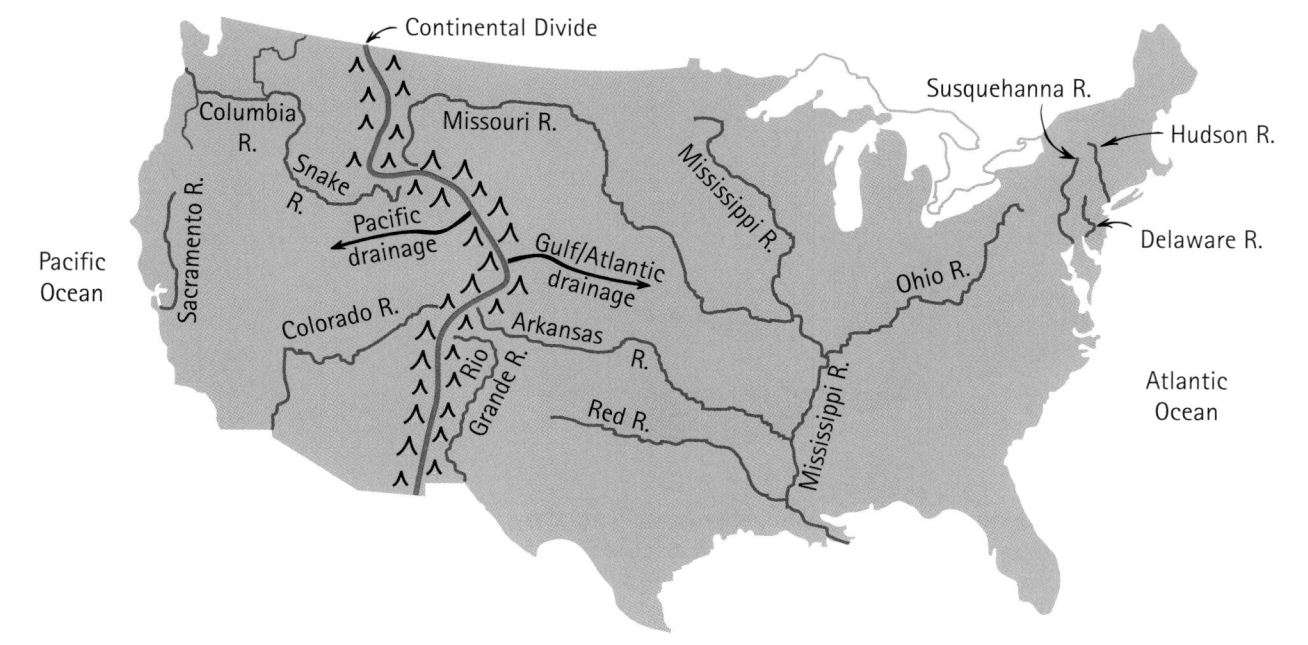

28.6 Glaciers and Glaciation— Earth's Frozen Water

The mightiest rivers on Earth are frozen solid and normally flow a sluggish few centimeters per day. These great icy rivers are called **glaciers.** Glaciers covered significant portions of Earth several times in the distant past. Glaciation is still at work in many regions of the world—small alpine glaciers in mountainous areas, large alpine ice fields, and the huge Arctic and Antarctic continental ice sheets.

Glacier Formation and Movement

The ice of a glacier is formed from recrystallized snow. After snowflakes fall, their accumulation slowly changes the individual flakes to rounded lumps of icy material. As more snow falls, the bottom layers of the icy snow become compacted and recrystallize into glacial ice.

✓ This ice does not become a true glacier, however, until it moves under its own weight. This happens when the thickness of the ice is about 50 m. The pressure exerted by the overlying material causes ice crystals at the base of the glacier to deform *plastically* and flow downslope. This plastic deformation is like what happens to a deck of playing cards that is pushed at one end (Figure 28.27). As the individual cards slide past one another, the entire card deck shifts. Plastic deformation in a glacier is greatest at the base of the ice, where pressure is greatest.

FIGURE 28.27 ▲
When a deck of cards is pushed from one side, the individual playing cards slide past one another, thus shifting the whole deck.

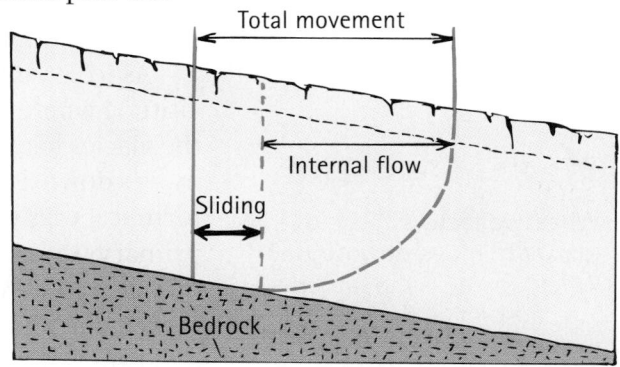

FIGURE 28.28 ▶
Cross section of a glacier. Glaciers move in two ways: internal flow and basal sliding. Movement is slowest at the base because of frictional drag and fastest at the surface. The upper parts of the glacier are carried along in piggyback fashion by plastic flow within the ice.

✓ Plastic flow from slipping ice crystals is not the only way that glaciers move. When melted ice, called *meltwater,* forms at the base of the glacier, a process called *basal sliding* comes into play. The entire glacier slides downslope, with the meltwater acting as a lubricant. The speed of the glacial ice increases from the base up. So the fastest movement is at the glacier's surface (Figure 28.28). Carried along by basal sliding and internal plastic flow, the surface of a glacier behaves like a rigid, brittle mass. Huge, gaping cracks called *crevasses* sometimes develop in this surface ice (Figure 28.29).

FIGURE 28.29 ▶
Crevasses can extend to great depths and can be quite dangerous for people attempting to cross a glacier.

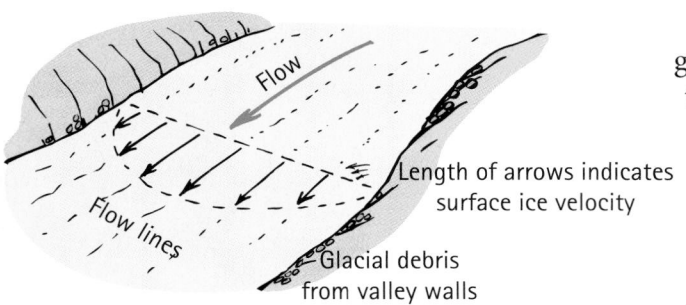

FIGURE 28.30 ▲
Top view of a glacier. Movement is fastest at the center and gradually decreases along the edges because of friction.

Average glacier speed varies from glacier to glacier and can range from only a few centimeters to a few hundred centimeters per day. Such slow speeds are measured by placing a line of markers across the ice and recording their changes in position over a period of time, ranging from days to years. Ice moves fastest in the center and more slowly at the edges because of frictional drag (Figure 28.30). Some glaciers experience surges, or periods of much more rapid movement. These surges are probably caused by periodic melting of the base and sudden redistribution of mass. The flow rate in these relatively brief surges can be 100 times the normal rate. Viewed from above, flow bands of rock debris and ice normally have a parallel pattern, but during a surge the flow bands become intricately folded (Figure 28.31).

Glacial Mass Balance

From season to season, and over longer periods of time, the mass of a glacier changes. Typically, a glacier grows in the winter as snow accumulates on its surface. The amount of snow added and the process of adding snow to a glacier is called **accumulation.**

As ice accumulates and begins to flow downhill, it may move to an altitude where temperatures are warmer. When the ice begins to melt, the glacier loses some of its mass. A glacier may also lose mass as it moves downslope to a shoreline. There ice may break off, or *calve*, to form icebergs that float away to sea. Melting and calving are the two primary ways that a glacier loses mass. Although less noticeable, glaciers may also lose mass as the ice *sublimates* to water vapor. By whatever means, the total amount of ice lost and the process of losing ice is called **ablation** (Figure 28.32).

✔ **READING CHECK**

What two factors cause a glacier to move downslope? What "force" is the underlying cause of all downslope movement?

FIGURE 28.31 ▶
Glacial flows: (a) normal flow; (b) surge flow.

(a)

(b)

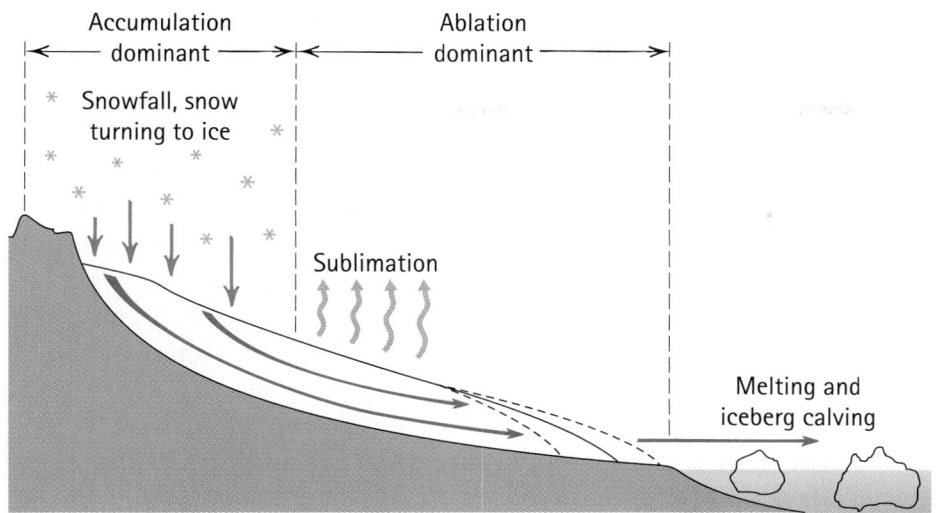

Accumulation on a glacier takes place at high elevations as snow falls on the glacier and turns to ice. Ablation takes place at lower elevations as ice melts or calves into icebergs or is lost through sublimation.

When accumulation equals ablation, the size of the glacier remains constant. For example, in a mountain glacier, accumulation occurs with winter snowfall in the farther-back, higher-elevation parts of the glacier. Ablation occurs at lower elevations, where spring and summer melting is greatest. When accumulation rate and ablation rate are equal, the melting of the lower portion of the glacier is offset by the downslope flow of ice from higher elevations. As a result, the location of the front edge of the glacier does not change. ✓ When accumulation exceeds ablation, the glacier advances—it grows. When ablation exceeds accumulation, the glacier retreats—it shrinks. Naturally, in all these cases, the ice of the glacier always flows downslope.

With global warming, ablation is exceeding accumulation in more and more places. As glaciers calve into the sea, or as ice sheets on land melt, sea level will rise.

CONCEPT CHECK
Under what conditions does the front of a glacier remain at the same location from year to year?

Check Your Answer
The front of a glacier remains at the same location when the rate of growth (accumulation) equals the rate of shrinking (ablation). In the spring, as ice at the glacier's front melts, the glacier retreats upslope. At the same time, the increased mass from the prior winter's accumulation causes the glacier to move forward. When the rate at which this forward movement matches the rate of melting, the location of the front edge doesn't change.

✓ READING CHECK

When does a glacier grow? When does a glacier shrink?

28.7 The Work of Glaciers

Like flowing water in streams, glaciers can erode as well as deposit sediment. Both processes produce characteristic landforms. When we find such characteristics on lands that are glacier-free, we have evidence of glaciers from long ago.

FIGURE 28.33 ▲
Striations mark the presence of a former glacier.

Glacial Erosion and Erosional Landforms

Glaciers are powerful agents of erosion. Glaciation has created the beautiful landscapes of the Himalayas in Asia, the Alps of Switzerland, the fjords of Norway, and Yosemite Valley and the Great Lakes in North America. In many ways, a glacier is like a plow as it scrapes and plucks up rock and sediment. It is also like a sled as it carries its heavy load to distant places. As it moves across Earth's surface, a glacier loosens and lifts up blocks of rock, mixing them into the ice. The large rock fragments carried at the bottom of a glacier scrape the underlying bedrock and leave long, parallel scratches (like sled tracks) aligned in the direction of ice flow (Figure 28.33). These scratches are called *striations*.

The two main types of glaciers, *alpine* and *continental,* have different erosional effects and produce different landforms. Alpine glaciers develop in mountainous areas and are often confined to individual valleys. Alpine glaciers are found in mountain chains all over the world—the Cascades, the Rockies, the Andes, and the

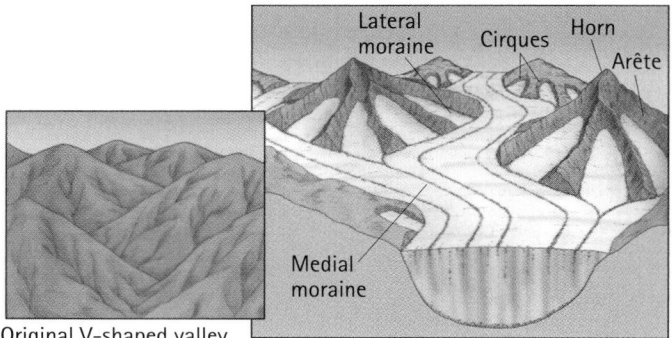

Original V-shaped valley

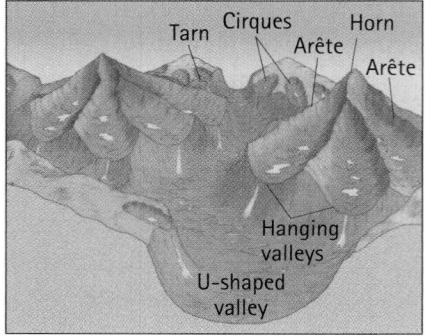

(a)

(b)

◄ **FIGURE 28.34**
The many erosional features of alpine glaciation. (a) The Matterhorn—named for its characteristic "horn" feature. (b) Hanging valleys are a spectacular feature found in areas that have been shaped by alpine glacial erosion. Bridalveil Falls in Yosemite National Park spills out of a hanging valley into the larger valley that was once occupied by the main glacier.

Himalayas. The erosional features of alpine glaciation are depicted in Figure 28.34.

Continental glaciers cover much larger areas as they spread over the land surface, smoothing and rounding the underlying surface. Although striations are produced by both alpine and continental glaciers, they have played a larger role in the study of ancient continental glaciers. ✔ Because a continental glacier scours very large tracts of land, it tends to leave behind few obvious valleys (making it difficult to determine the glacier's direction of flow). By mapping striations on land once covered by continental glaciers, geologists can decipher the flow direction of the ice. The direction of ice flow is also indicated by small, asymmetrical hills called *roches moutonnées* (singular, *roche moutonnée;* Figure 28.35). In the direction of ice flow, the hill's slope is smooth and striated from the abrasion of ice on bedrock. On the downflow side, the slope is rough and steep because the moving ice plucked rock fragments away from cracks in the bedrock.

Glacial Sedimentation and Depositional Landforms

As a glacier advances across the land, it picks up and transports great quantities of debris. When the glacier retreats, this debris is left behind because it is melted out of the ice. Because a glacier abrades and picks up everything in its path, glacial deposits are characteristically made up of unsorted, angular rock fragments in a variety of shapes and sizes. ✔ This wide range of particle sizes is the hallmark that differentiates glacial sediment from the much-better-sorted material deposited by streams and winds. Glacial deposits are collectively called **drift,** a term that dates back to the 19th century, when it was thought that all such debris had been "drifted in" by the great biblical flood.

Drift is deposited in two main ways. When glacial sediment is released into meltwater, it is carried and deposited like any other

▲ Geologist Bob Abrams observes the grandeur of the Juneau Ice Field in Alaska.

> **✔ READING CHECK**
>
> Why are striations so important in continental-type glaciers?

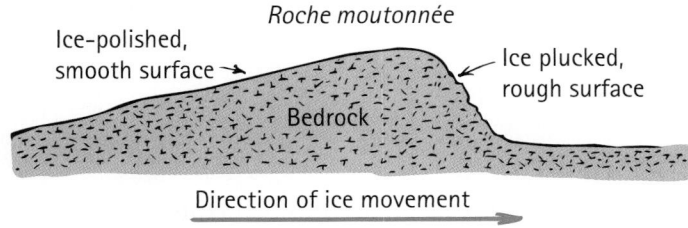

FIGURE 28.35 ▲
Small asymmetrical hills called *roches moutonnées* show the direction of glacial movement. On the side of the hill facing the approaching glacier, the slope is smooth and gentle. On the side facing away from the approaching glacier, the slope is rough and steep with a plucked appearance.

> Remember Alfred Wegener? He used glacial striations to reconstruct formerly joined landmasses.

✔ **READING CHECK**

How are glacial sediments different from stream-deposited sediments?

▲ Geologic features are best viewed from an airplane. Next time you fly in an airplane, request a window seat and enjoy the geology below.

waterborne sediment. Thus, it tends to be well sorted. This type of drift is called *outwash*. Material deposited directly by melting ice—an unsorted mixture of rock debris—is called *till*. Many of the old stone walls and fences of New England are found in areas where the surface material is glacial till. Settlers who tried to farm this land had to remove all the larger boulders before they could plow, and they piled them along the edges of their fields.

The most common landform created by glaciers is the *moraine,* a ridge-shaped landform that marks the boundaries of ice flow. Of all the different types of moraines, probably the most important is the *terminal moraine,* as it marks the farthest point of a glacier's advance (Figure 28.36). Another distinctive landform consisting of glacial sediments is the *drumlin,* an elongated hill shaped like the back of a whale. Formed by continental glaciation and lined up in the direction of ice flow, drumlins have a steep, blunt end in the direction from which the ice came and a tapered gentle slope on the downstream side (Figure 28.37). Perhaps the most famous drumlin in the United States is Bunker Hill in Massachusetts.

Many of the world's lakes, small and large, are the products of glacial action. Glaciers deepened valleys and deposited sediments that acted as dams, blocking stream drainage within some valleys and creating lakes. The Finger Lakes in upstate New York, the "10,000 Lakes" of Minnesota, and the Great Lakes of North America are all products of glacial action.

FIGURE 28.36 ▶
Glacial depositional landforms. Of special importance is the terminal moraine, which marks the farthest point of a glacier's advance.

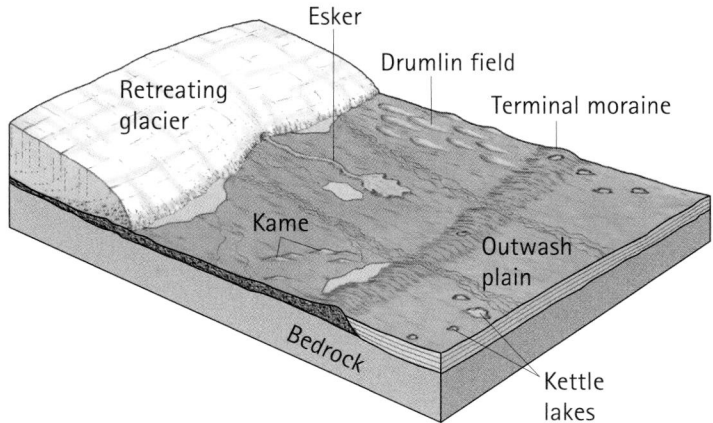

CONCEPT CHECK
What land surface forms can be used to determine the direction of ice flow?

Check Your Answer
The direction of ice flow can be determined from striations (long, parallel scratches aligned in the direction of ice flow), *roches moutonnées* (small, asymmetrical hills), and *drumlins* (elongated hills shaped like the back of a whale).

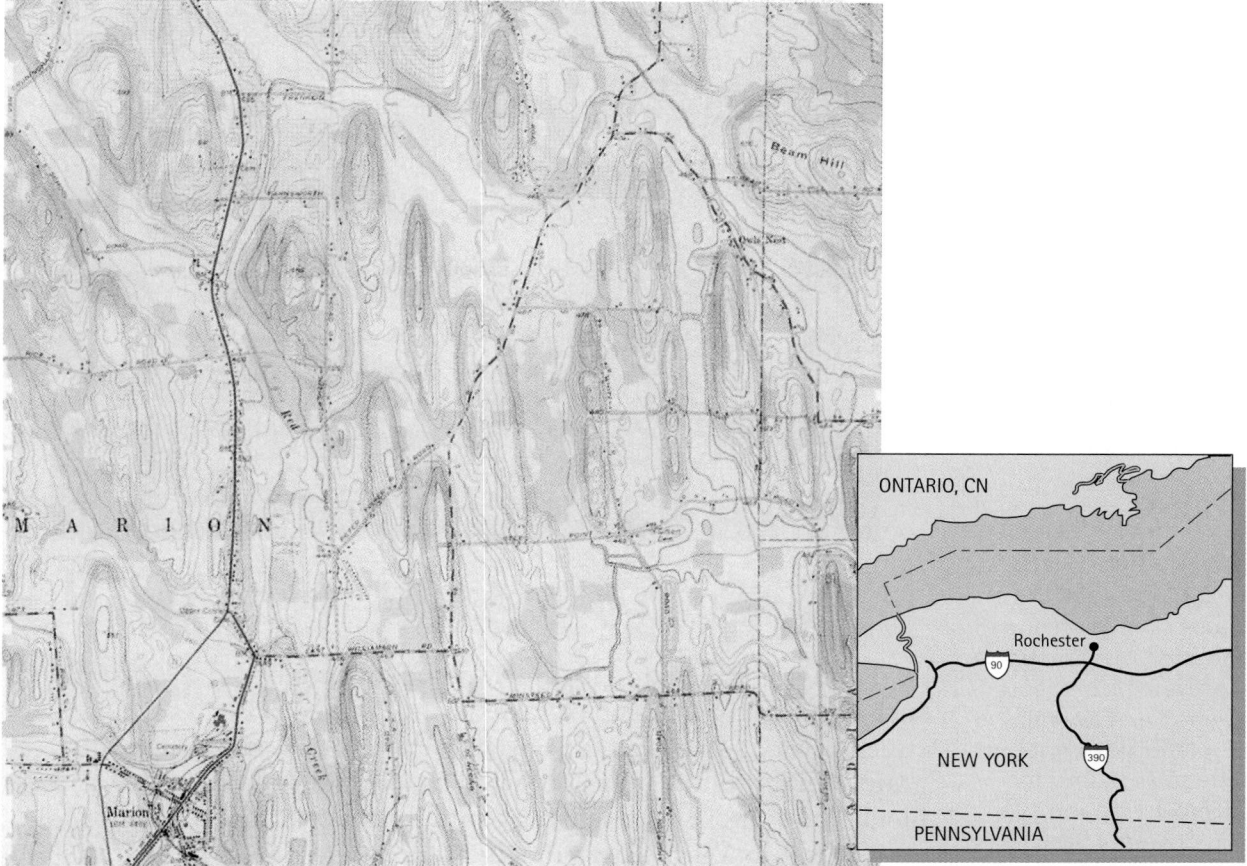

FIGURE 28.37 ▲
Topographic map showing numerous oval-shaped drumlins in upstate New York. Drumlins are steep and blunt on the side that faced the approaching glacier but tapered and gently sloping on the downflow side. Looking at the map, can you tell the direction of continental ice flow?

28.8 The Work of Air

Water is the dominant agent of change altering our natural landscape, but air plays a role too. If you've ever been in a windstorm or at the beach on a windy day, you may have felt the sandblasting effect of the wind. Once in the air, particles of sediment can be carried great distances by the wind. Red dust from the Sahara is found on glaciers in the Swiss Alps and on islands in the Caribbean Ocean. Fine grains of quartz from central Asia blow onto the Hawaiian Islands.

In the desert, winds move over dry sand surfaces. The wind picks up the small, more easily transported particles but leaves the large, harder-to-move particles behind. The small particles bounce across the desert floor. In doing so, they knock more particles into the air. In this way, tiny sand dunes called *ripple marks* begin to form (Figure 28.38). Ripple marks also form as water currents move sand grains. You can see these ripple marks in shallow streams or under the waves at beaches.

FIGURE 28.38 ▶
Generated by blowing winds, ripple marks are narrow ridges of sand separated by wider troughs. They are small, elongated sand dunes. Large sand dunes can be seen in the background of the photograph.

fyi

▪ Even though deserts lack moisture, water is still the main cause of erosion and transportation of sediments. Scarce as water is in the desert, when a heavy rain falls, rainwater does not have time to soak into the ground. No infiltration results in powerful flash floods. These flash floods transport and then deposit great quantities of debris and sediment at the bases of mountain slopes and on the floors of wide valleys and basins.

Sand dunes begin to form where airflow is blocked by an obstacle, such as a rock or a clump of vegetation (Figure 28.39). As the wind sweeps over and around the obstacle, the wind speed slows. Sand grains fall out of the air in the wind shadow (the place where wind is blocked). As more sand falls, a mound forms that blocks the flow of air even more. With more sand and more wind, the mound grows into a dune, which, with continued growth, begins to "move" downwind. The dune moves because grains on the windward slope are blown up and over the crest of the dune, falling on the leeward slope. In this way, wind removes sand from the back of the dune and redeposits it on the front of the dune. Over time, this continuous process moves the entire dune.

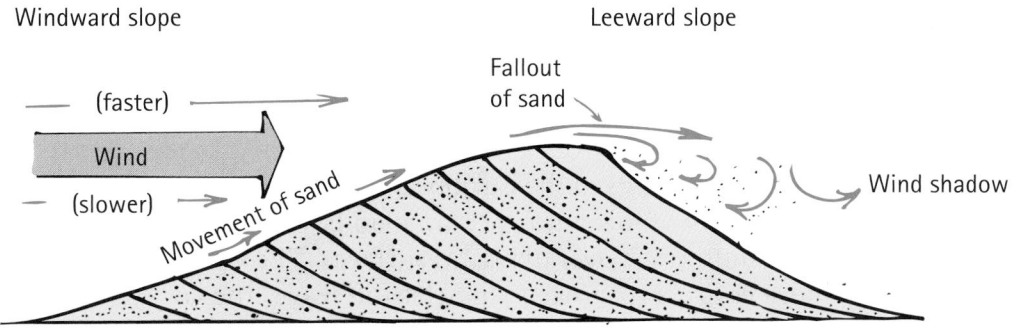

FIGURE 28.39 ▲
Formation of a sand dune. When airflow is obstructed, wind speed drops and, as a result, sand grains settle in the wind shadow. With more wind, more sand settles, and a dune is formed. As the dune grows, sand grains on the windward slope move up and over the crest to fall on the leeward slope, which slowly causes movement of the whole dune downwind.

28 CHAPTER REVIEW

KEY TERMS

Ablation The amount of ice lost, and the process of losing ice, from a glacier.

Accumulation The amount of snow added, and the process of adding snow, to a glacier.

Artesian system A system in which confined groundwater under pressure can rise above the upper boundary of an aquifer.

Channel geometry The shape of a stream channel—the cross-sectional area.

Delta An accumulation of sediments, commonly forming a triangular or fan-shaped plain, deposited where a stream flows into a standing body of water.

Discharge The volume of water that passes a given location in a stream channel in a certain amount of time.

Drift A general term for all glacial deposits.

Floodplain A wide plain of almost flat land on either side of a stream channel. Submerged during flood stage, the plain is built up by sediments discharged during floods.

Glacier A large mass of ice formed by the compaction and recrystallization of snow, which is able to move downslope under its own weight.

Gradient The vertical drop in the elevation of a stream channel divided by the horizontal distance for that drop; the steepness of the slope.

Groundwater Underground water in the saturated zone.

Hydrologic cycle The natural circulation of all states of water from ocean to atmosphere to land, back to ocean.

Laminar flow Water flowing smoothly and fairly slowly in straight lines with no mixing of sediment.

Permeability A measure of the ability of a porous rock or sediment to permit fluid to flow.

Porosity The volume of open pore space in rock or sediment compared to the total volume of solids plus open pore space.

Sand dune A landform created when airflow is blocked by an obstacle, slowing air speed and therefore promoting the deposition of airborne sand.

Turbulent flow Water flowing rapidly and erratically in a jumbled manner, stirring up everything it touches.

Water table The upper boundary of the saturated zone, below which every pore space is completely filled with water.

REVIEW QUESTIONS

The Hydrologic Cycle

1. What is the hydrologic cycle?
2. As water is precipitated onto the land, where does it go?

Groundwater—Water Below the Surface

3. Distinguish between *porosity* and *permeability*.
4. If a hole is dug in the unsaturated zone, does it fill with water? Why or why not?
5. What types of soil allow the greatest amount of rainfall to soak in?

The Work of Groundwater

6. What can happen to the land surface when the pumping of groundwater is greater than the recharge?
7. How does rainwater become acidic? How does this affect limestone?
8. Name three erosional features caused by groundwater in carbonate rocks.
9. What is the difference between a cave and a cavern?

Streams and Rivers—Water at Earth's Surface

10. How does a stream's gradient affect stream speed?
11. What happens to stream speed when the discharge of a stream increases?
12. How does the shape of a stream channel affect flow?

The Work of Surface Water

13. What is the greater transporter of sediment, a laminar flow or a turbulent flow? Why?
14. Name three ways that flowing water erodes a stream channel.
15. What is a delta?

Glaciers and Glaciation—Earth's Frozen Water

16. In what two main ways do glaciers flow?
17. Under what conditions does a glacier front advance? How about retreat?

The Work of Glaciers

18. What is the most common landform created by glaciers?
19. What land features are formed from glacial deposits?

The Work of Air

20. How do sand dunes migrate?

4. In an unconfined aquifer, how high can water rise in a well that is not pumped?
5. What is meant by channel geometry?
6. What happens to stream speed if the discharge in a stream doubles while the channel remains the same size and shape?
7. What three variables influence the speed of stream flow?
8. How does "frictional drag" play a role in the external movement of a glacier? How does this drag affect the internal movement?
9. In the formation of a river delta, why are larger particles deposited first, followed by smaller particles farther out?
10. What is a sinkhole? What factors contribute to its formation?
11. Which of the three agents of transportation—wind, water, or ice—transports the largest boulders? Why?
12. Describe the formation of stalactites.
13. How is a *roche moutonnée* different from a drumlin?
14. What well-known landscapes have been carved by glaciers?
15. Why do crevasses form on the surfaces of glaciers?
16. How are sand dunes formed?
17. What happens to stream speed if discharge doubles and the cross-sectional area of the stream channel also doubles?
18. Which of the three agents of transportation—water, ice, or wind—is limited to transporting small rocks?
19. Is water in the unsaturated zone called groundwater? Why or why not?
20. What is a continental divide?

THINK AND EXPLAIN

1. What percentage of Earth's supply of water is freshwater, and where is most of it located?
2. Where does most rainfall on Earth finally end up before becoming rain again?
3. In a confined aquifer, the water in a well can rise above the top of the aquifer. What is this system called?

THINK AND SOLVE

1. We know that most of Earth's water is in the oceans. The remaining 2.76% is Earth's freshwater supply. Of this freshwater supply, what percentage is found in the polar ice caps? In groundwater? In streams, lakes and rivers? (*Hint:* Calculate the freshwater supplies to be equal to 100%.)

2. A particular stream widens as it progresses down-stream. Using your answers for parts (a) and (b), briefly describe the changes in discharge.
   (a) If the cross-sectional area of the stream is 1 m^2 and the stream speed is 0.5 m/s, what is the stream's discharge?
   (b) If the cross-sectional area of the stream increases to 2 m^2 and the stream speed remains 0.5 m/s, what is the stream's discharge?

3. If the water table at location X is lower than the water table at location Y, does groundwater flow from X to Y or from Y to X?

READINESS ASSURANCE TEST (RAT)

If you have a good handle on this chapter—if you really do—then you should be able to score 7 out of 10 on this RAT. Check your answers with your teacher. If you score less than 7, study further before moving on.

Choose the best answer to each of the following.

1. As a stream moves from a location where gradient is steep to where it is more level, its speed tends to
   (a) become turbulent.
   (b) meander.
   (c) decrease.
   (d) increase.

2. Most of Earth's freshwater is found
   (a) in lakes.
   (b) in ice caps and glaciers.
   (c) in rivers.
   (d) underground.

3. The work of surface water does all of the following except
   (a) erosion.
   (b) deposition.
   (c) land subsidence.
   (d) delta formation.

4. The maximum amount of water a particular soil can hold is determined by its
   (a) porosity.
   (b) permeability.
   (c) degree of saturation.
   (d) amount of recharge.

5. Precipitation that does not soak into the ground or evaporate becomes
   (a) groundwater.
   (b) the water table.
   (c) soil moisture.
   (d) runoff.

6. Sand dunes form as wind
   (a) disperses sand.
   (b) blows sand from the back to the front of the dune.
   (c) blows sand from the front to the back of the dune.
   (d) interrupts the normal sequence of deposition.

7. Deltas form as
   (a) periodic flooding clogs stream channels.
   (b) erosion clogs stream channels.
   (c) stream gradient decreases.
   (d) streams enter a standing body of water.

8. What factors affect stream speed?
   (a) Discharge
   (b) Channel length
   (c) Stream gradient
   (d) Both (b) and (c).
   (e) Both (a) and (c).

9. Snow converts to glacial ice when subjected to
   (a) decreasing temperature.
   (b) pressure.
   (c) rain.
   (d) basal sliding.

10. Underground water in the saturated zone is called
    (a) groundwater.
    (b) soil moisture.
    (c) the water table.
    (d) an artesian system.

EARTH SCIENCE
≋ YUM ≋

29

GEOLOGIC TIME— READING THE ROCK RECORD

THE MAIN IDEA

Rocks record Earth's history. Changes in climate, life forms, and tectonics are all recorded in the rocks. We only need to know how to read the rocks.

Earth is very old—some 4.5 billion years old. This vast span of time, called geologic time, can be difficult to comprehend. But let's try. Imagine that we can compress 4.5 billion years into a single year. Planet Earth would form from matter in the primordial solar system on January 1. The oldest known rocks would appear at the end of February. Simple bacterial life would appear in the sea at the end of March. Plants and animals would emerge in late October or early November. Dinosaurs would rule Earth in mid-December but disappear by December 26. *Homo sapiens* (humans) would appear at 11:50 PM on the evening of December 31. And all of recorded human history would take place in the last minute of New Year's Eve!

Explore!

How Do Rock Layers Show the Order of Events?

1. Think about the letters of the alphabet. When you recite the alphabet, do you say it in the same sequence each time?
2. Put each letter on an index card and stack the cards so that the first letter of the alphabet is at the bottom and the last letter is at the top. Which card comes first? Which card comes last?

Analyze and Conclude

1. **Observing** The letters of the alphabet have an ordered sequence. A always comes first, and Z always comes last.

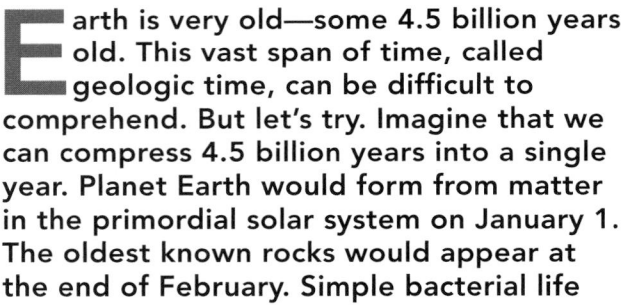

2. **Predicting** Rock layers, like the letters in the alphabet, also have a sequence. So the oldest rocks are at the bottom of the sequence and the youngest layers are at the top of the sequence.
3. **Making Generalizations** Although rock layers have a sequential order, sometimes tectonic disturbances flip the order over. And erosional disturbances can cause some layers to be missing. Because rock layers form over long periods of time, many things can disturb the sequence.

◀ **FIGURE 29.1**
The lowermost layers of the Grand Canyon are older than the uppermost layers—the principle of superposition.

29.1 Relative Dating—The Placement of Rocks in Order

Sedimentary rock layers are deposited layer on top of layer. This sequence of layering provides evidence of relative rock ages. The lower layers were formed before the upper layers, and so they are older than the upper layers. Rock layers in the Grand Canyon of the Colorado River in Arizona give an excellent example of relative dating (Figure 29.1). The many different layers, and the thickness of the layers, chronicle geologic activity over millions of years. The conditions under which the sedimentary layers were deposited varied—changing from season to season, year to year, and millennium to millennium. Some layers reveal climatic cycles that span centuries, other layers indicate times when the land surface was submerged beneath shallow seas, and still other layers show periods of increased rainfall accompanied by the gradual uplift of the entire area. Millions of years after the top layer was deposited, the Colorado River cut through the accumulated layers like a knife cutting into a layer cake. The result of the river's erosive power is the canyon we see today.

In the Grand Canyon, and elsewhere, geologists use several key principles to determine the relative ages of rocks. **Relative dating** is the ordering of rocks in sequence by their comparative ages. Relative dating doesn't tell the actual date when a rock layer formed, but rather its timing relative to other episodes in Earth's past. The relative dating principles include:

1. **Original horizontality** Layers of sediment are deposited as an even horizontal surface. So, layers that are tilted and folded must

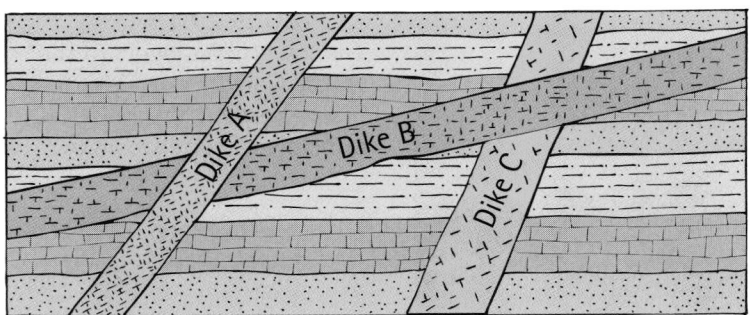

FIGURE 29.2 ▲
Dikes cutting into a rock body are younger than the rock into which they cut. In the diagram, dike A cuts into dike B, and dike B cuts into dike C. From the principle of cross-cutting relationships, A is the youngest dike, B the next youngest, and C the oldest of the three. The horizontal layers, which are cut by all three dikes, are all older than C.

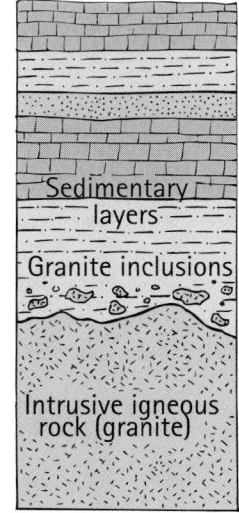

FIGURE 29.3 ▲
The rocks locked in the sedimentary layer existed before the sedimentary layer formed—the principle of inclusion.

have been moved into that position by disturbances, such as earthquakes and mountain building, after deposition.

2. Superposition In an undeformed (horizontal) sequence of sedimentary rocks, each layer was deposited on top of preexisting layers below. Like the newspapers in a recycling bin, older papers are found below newer, more recent papers.

3. Cross-cutting In cross-cutting, an igneous intrusion or fault cuts through preexisting rock and is younger than the rock through which it cuts (Figure 29.2).

4. Inclusion Inclusions are pieces of one rock contained within another. Any inclusion is older than the rock containing it, just as small pieces of rock incorporated into a slab of concrete were formed before the concrete was formed (Figure 29.3).

These four principles are fairly straightforward. They can be used to determine the ages of rock formations relative to one another—which formation was formed first, second, and so on—for a particular area or rock outcrop. But to piece together the relative age of rocks for a larger area, more information is needed. Once again, the Grand Canyon provides the ideal example. The Grand Canyon's horizontal layers, stacked one on top of another, chronicle a great span of time. But because the rock layers stretch continuously for hundreds of miles, they also show the extent of the depositional area. Unfortunately, not all rock layers stretch on like those in the Grand Canyon. Over spans of time, rock layers can be broken by faulting and/or folding or covered up by younger sediments. When this happens, we find geographically separated rock outcrops. In this case, we use the principle of lateral continuity.

5. Lateral continuity Sedimentary layers are deposited in all directions over a large area unless some sort of obstruction or barrier limits their deposition. Faulting, folding, and erosion can separate originally continuous layers into isolated outcrops.

Lateral continuity can be used to match isolated rock outcrops to other rock outcrops over a large area (Figure 29.4). If the various rock layers in isolated outcrops have similar characteristics (such as color, minerals, and fossils), and the vertical sequence of layers is consistent, we can assume that the layers were once continuous. So when combined with other relative dating principles—namely

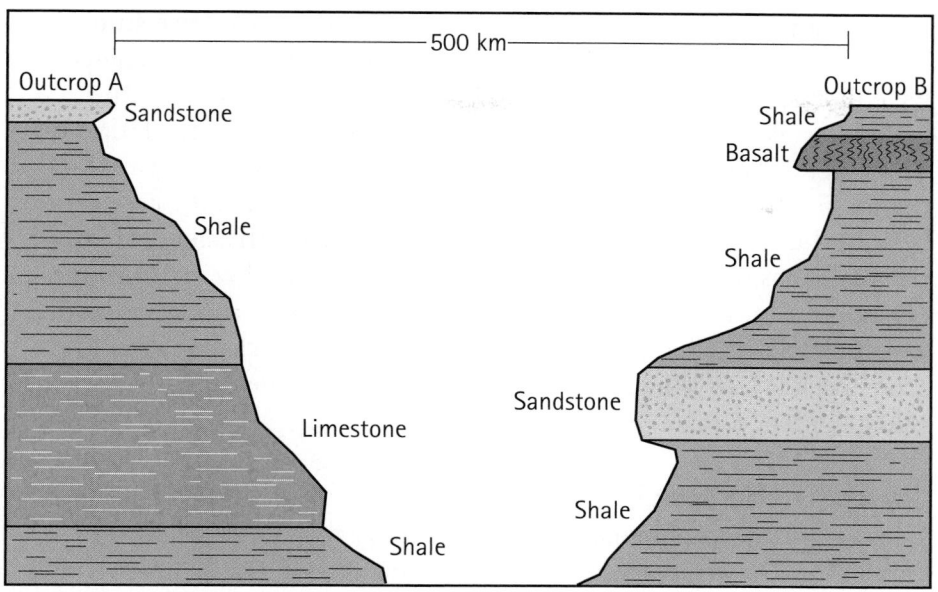

◀ **FIGURE 29.4**
Lateral continuity can be used to determine the relative age of rocks in widely separated areas. Outcrops A and B share similar characteristics, but are separated by 500 km. Is it possible that these outcrops were, at one time, continuous? By looking at the sequence of layers, can you tell which rocks are older and which are younger?

superposition and original horizontality—lateral continuity allows relative age relationships to be applied over a much larger area.

As discussed in Chapter 25, fossils are the remains or impressions of ancient life forms preserved in rock. Today, it is common knowledge that fossils record the evolution of life on Earth, but in the 1700s, fossils were thought of as mere curiosities. Then along came William "Strata" Smith (1769–1839). Working as a surveyor for a canal project, Smith observed that certain rock layers contained different kinds of fossils. He also found that these fossil-bearing layers followed a consistent, predictable sequence. Smith noted the succession of rock types and the fossils within each layer and then used the fossils to correlate (match up) rock layers at various locations. He discovered that fossils could be used to chronologically order the vertical sequence of rock layers. With this knowledge, Smith established the principle of faunal succession.

6. **Faunal succession** Fossil organisms follow one another in a definite, irreversible time sequence. Fossil communities change through time as some species become extinct and new ones appear. Such changes are reflected in the rock record. In this way, fossils provide a key tool for recognizing the relative age of rocks.

Smith's observation that fossils are found in rocks in a definite order not only helped refine relative dating but also helped with the correlation of rock layers on a worldwide scale. Fossils—identified and categorized—could be correlated to specific times in Earth's geologic history. Once scientists established a time period, the fossils in the rock could be used to identify other rocks of the same age in other regions of Earth.

Fossils tell us about past environments. For example, certain present-day corals are found in warm tropical waters. When a similar fossilized coral is found, we can safely assume that a warm and tropical shallow sea once covered the area where it was found. Fossils help unravel Earth's history.

Similar to the way we can tell the age of a tree by its rings, we can tell the age of a canyon by its layers.

In many ways, studying the rock record is like a detective studying a crime scene. Both studies involve looking for clues. ✔ Just as detectives have their methods for solving crimes, Earth scientists rely on the principles of relative dating—age relationships and the correlation of rock layers. They correlate the rocks, the sequences of rock layers, and the fossils within the rock layers. As with a crime scene, the case cannot be solved until sufficient evidence is found. The Earth science detective often has little evidence on which to operate, and sometimes the evidence is completely lost. Earth's internal processes can fold and distort rock layers (Chapter 26) and external processes can weather and erode rock layers (Chapters 25 and 28). Whatever the cause, the result is lost evidence.

Gaps in the Rock Record

The deposition and formation of sedimentary rock layers is always happening. Go to any riverbed or beach and you will see sediment being transported and deposited. Yet a continuous sequence of rock layers from Earth's formation up to the present time is not found anywhere on Earth. ✔ Although deposition is ongoing, so are the processes of weathering and erosion, folding and faulting, and crustal uplift. And these processes remove rock layers or interrupt deposition. This creates time gaps in the rock record (Figure 29.5). These gaps, called **unconformities,** are found by observing the relationships of layers and fossils.

The most easily recognized of all unconformities is an **angular unconformity.** An angular unconformity forms when older, previously horizontal rock layers are uplifted and tilted or folded by movements within Earth (Figure 29.6). During and after the uplift, erosion wears down the tilted layers to create a flat land surface. When the period of erosion is over, new sediment layers are deposited horizontally over the flat surface. The angular unconformity is the "erosional surface" that separates the older, tilted layers from the younger, horizontal layers. It represents the long interval of time during which uplift and erosion took place. The part of the rock record representing this long interval is now missing because of erosion, and the unconformity is the evidence that remains.

When overlying sedimentary rocks are found on an eroded surface of metamorphic or igneous plutonic rocks, the unconformity is called a *nonconformity.* The igneous

FIGURE 29.5 ▲

The age of the Grand Canyon is told by its sequence of rock layers. As in other places, the sequence is not continuous, and there are time gaps. (1) A nonconformity separating older metamorphic rocks from sedimentary layers. (2) An angular unconformity separating older tilted layers from horizontal layers. Time gaps are also represented between horizontal sedimentary layers. The disconformities (3)–(5) are difficult to identify, and they often require both a good eye and knowledge of fossils.

(a) Sediments are deposited layer upon layer beneath the sea.

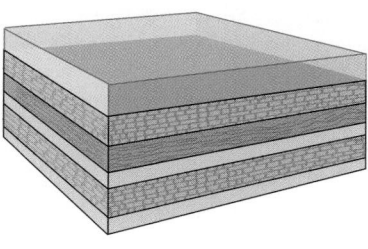

(b) During mountain building, solidified sediment layers become folded and deformed. Erosion begins.

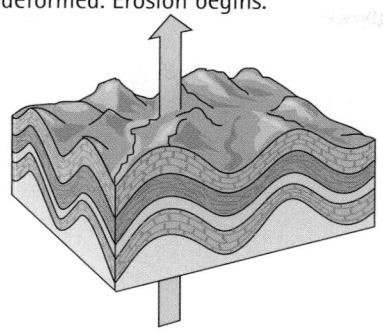

◀ **FIGURE 29.6**
The sequence of events that create an angular unconformity.

(c) As mountain building wanes, the exposed surface is eroded to a more or less even plain.

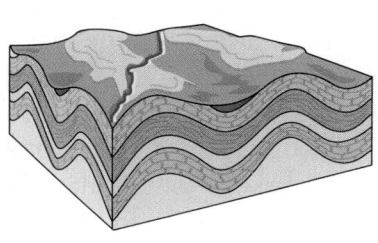

(d) As the land subsides below sea level, younger sediments are deposited on the former erosional surface.

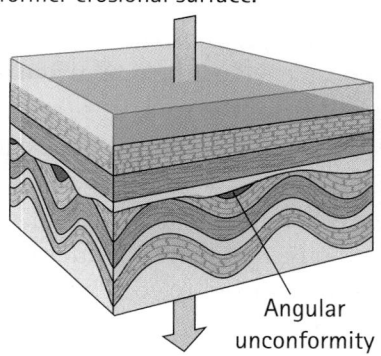

Angular unconformity

or metamorphic rocks formed deep beneath Earth's surface but were present at the surface when the overlying sedimentary rocks were deposited on top of them. Therefore, a nonconformity shows that a great deal of uplift and erosion occurred before the sedimentary layers were deposited, with a large stretch of time "missing" from the rock record. A more subtle type of unconformity, a *disconformity*, is a time gap between parallel layers of sedimentary rock. Because the rocks above and below the disconformity can be very similar, these time gaps can be difficult to identify.

✓ **READING CHECK**

How do "time gaps" occur in the rock record?

CONCEPT CHECK

1. If a granitic intrusion, such as a dike, cuts into or across sedimentary layers, which is older: the granite or the sedimentary layers?
2. Look at outcrops A and B in Figure 29.4, which are separated by 500 km. Is it possible that these rock layers were once continuous? If so, which rocks are older and which are younger?

Check Your Answers

1. The intrusion is new rock in the making. Therefore, the sedimentary layers are older than the intrusions that cut into them.
2. Yes, it is possible. If continuous, the sandstone capping outcrop A matches the sandstone sandwiched between the two shales in outcrop B. So, all rock layers above sandstone B are younger. For outcrop A, original horizontality and superposition tells us that the bottom shale is oldest, followed by limestone, shale, then sandstone. We assume the same sequence in outcrop B even though we cannot see the limestone layer. If fossils were present, we could support our conclusion even more.

29.2 Radiometric Dating Reveals the Actual Time of Rock Formation

Relative dating tells us which parts of Earth's crust are older or younger. But it doesn't tell us the actual age of a rock—the amount of time that has passed since the rock was formed. ✓ The actual age of a rock can be determined by **radiometric dating.** This process measures the ratio between radioactive isotopes and their decay products. The half-life of a radioactive isotope is the amount of time it takes for half of the material to decay to its daughter product (Figure 29.7). Some common radioactive isotopes used for dating and estimates of geologic time are shown in Table 29.1.

To date geologically young objects, and especially for dating organic matter, carbon-14 is the isotope of choice. With its relatively short half-life (5760 years), carbon-14 is useful for dating events within the last 50,000 years or so (see Section 16.6). To date older materials, radioactive elements such as uranium are used. Many common rocks contain trace amounts of uranium, and only a small amount is needed to perform the laboratory tests. Uranium-238 decays to its stable daughter isotope

fyi

- The oldest rocks found are the Acasta gneisses in northwestern Canada. Zircon crystals in the gneiss are dated at 4.03 billion years. The oldest mineral found is a zircon crystal from a sandstone in Australia. This zircon is dated at 4.4 billion years!

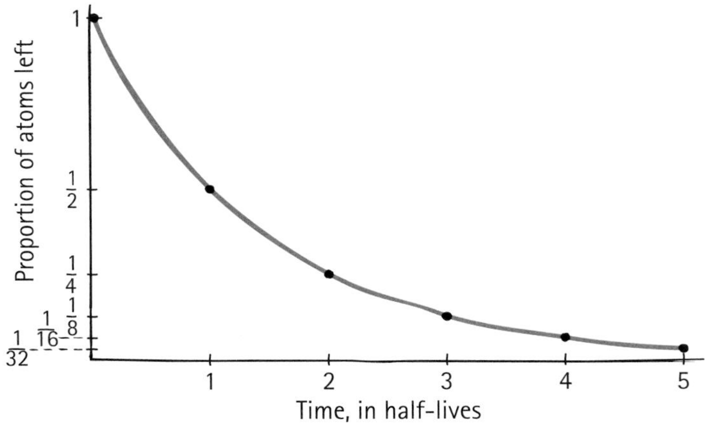

FIGURE 29.7 ▲
The amount of parent material versus the number of half-lives remaining as the radioactive parent decays.

TABLE 29.1 Isotopes Most Commonly Used for Radiometric Dating

Radioactive Parent	Stable Daughter Product	Half-life Value
Uranium-238	Lead-206	4.5 billion years
Uranium-235	Lead-207	704 million years
Potassium-40	Argon-40	1.3 billion years
Carbon-14	Nitrogen-14	5760 years

Radiometric dating tells us the age of rock formation. But for a rock reheated by metamorphism, the "time clock" is reset. So, for metamorphic rocks we get the date of the metamorphism.

lead-206, and uranium-235 decays to the stable isotope lead-207. No other natural sources exist for these two isotopes of lead. Therefore, any lead-206 and lead-207 found in a rock today were at one time uranium. If, for example, a sample contains equal numbers of uranium-235 and lead-207 atoms, the age of the sample is one uranium-235 half-life— 704 million years. If, on the other hand, a sample of uranium contains only a small amount of lead-207, the sample is younger than one half-life (a single half-life has not yet passed).

CONCEPT CHECK

1. Could carbon-14 be used for dating 100-million-year-old rocks?
2. How can we determine the age of sedimentary rock layers?

Check Your Answers

1. No. Carbon-14 has a half-life of 5760 years and can be used to date only relatively younger rocks. Any carbon-14 (from calcite, for example) in rocks this old would have long since been reduced to undetectable amounts.
2. If we know the maximum age (meaning the rock can be no older than the age of the datable minerals within it) of an overlying and an underlying rock layer, we can bracket the age of the sedimentary layer in between by using the principle of superposition.

fyi

- Recall that sedimentary rocks are made from pre-existing rocks. Because of this, we can date only the minerals in the rock, but not the sedimentary rock itself. The rock can be no older than the age of the datable minerals within it (principle of inclusions). To date sedimentary rock layers, we use relative dating combined with radiometric dating.

 **READING CHECK**

How is the actual age of rock layers determined?

29.3 Geologic Time

The time line for the history of Earth is called the *geologic time scale.* The scale is based on the relative ages of rock layers and their fossils. Recall the principle of faunal succession—fossil plants and animals are found in the rock record in a chronological sequence. Because blocks of time are represented by different types of fossils, the geologic time scale subdivides Earth's 4.5-billion year history into time units of different sizes.

Eons are the largest unit of geologic time. The eon we are living in began about 543 million years ago. It is called the Phanerozoic eon, which means "visible life." The Phanerozoic eon is subdivided into three eras: the Paleozoic era (time of ancient life), the Mesozoic era (time of middle life), and the Cenozoic era (time of recent life). Each of the three eras is further divided into periods, which are further divided into epochs. ✔ Periods are the fundamental time interval because each *period* represents a major change in life forms. With radiometric dating, scientists can assign an actual age to the various periods. This gives accuracy to the time scale.

Note that most of Earth's history occurred before the Paleozoic era. This vast span of time, called **Precambrian time,** accounts for about 4 billion years of Earth's history! The Precambrian is divided into three eons: the Hadean, the Archean, and the Proterozoic.

Eon	Era	Period	Subperiod	Epoch	Ma
Phanerozoic	Cenozoic	Quaternary		Holocene	0.01
				Pleistocene	1.8
		Tertiary		Pliocene	5.3
				Miocene	23.8
				Oligocene	33.7
				Eocene	54.8
				Paleocene	65
	Mesozoic	Cretaceous			144
		Jurassic (first bird)			206
		Triassic			248
	Paleozoic	Permian (first reptiles)			290
		Carboniferous	Pennsylvanian		323
			Mississippian		354
		Devonian (first amphibians)			417
		Silurian (first insect fossils)			443
		Ordovician (first vertebrate fossils)			490
		Cambrian (first plant fossils)			543
Precambrian Time		Proterozoic			2500
		Archean			3800
		Hadean			4500

The geologic time scale ▶ divides Earth's history into time units of different size. Units of time on this scale are Ma, which stands for Mega-annum, or "1 million years ago." For example, the Paleozoic era began about 543 Ma, or 543 million years ago.

CONCEPT CHECK

1. Describe the present time in Earth's history in terms of all units of the geologic time scale from eons to epochs.
2. The time units on the geologic scale are Ma. What does Ma stand for?

Check Your Answers

1. We are living in the Phanerozoic eon, in the Cenozoic era, in the Quaternary period, and in the Holocene epoch. Question: Are the conditions of the Holocene still relevant to our time?
2. Ma stands for Mega-annum, which means "1 million years ago."

 **READING CHECK**

Why is the "period" considered to be the fundamental time interval?

29.4 Precambrian Time—A Time of Hidden Life

Precambrian time ranges from about 4.5 billion years ago, when Earth formed, to about 543 million years ago, when abundant macroscopic life appeared. The Precambrian—the time about which we know the least—comprises almost 90% of Earth's history! Most of the rocks that formed in this early part of Earth's history have been eroded away, metamorphosed, or recycled into Earth's interior. Organisms of that time did not have easily fossilized hard body parts, which evolved only later in the history of life. Thus, fossils from this huge span of time are scarce.

The beginning of the Precambrian—the Hadean eon—was a time of considerable volcanic activity and frequent meteorite impact. Large and small chunks of interplanetary debris left over from the formation of the solar system continually smashed into Earth to scar its surface. Earth was an oceanless planet covered with volcanoes erupting gases and steam from its scorching interior. The Hadean atmosphere consisted mostly of gases erupted from the many volcanoes. Carbon dioxide was the dominant gas, with water vapor, and molecular nitrogen, ammonia, sulfur dioxide, and nitric oxide as minor constituents. There was no free oxygen.

In the mid-Precambrian—the Archean eon—Earth's surface began to cool (Figure 29.8). This allowed the formation of the oceans. With cooler temperatures, water vapor in the atmosphere condensed to form clouds. Out of the clouds poured torrents of rain, enough to cover Earth's surface with shallow seas. Earth's interior was still quite hot, and erupting volcanoes dotted Earth's surface to form small islands. These small volcanic islands, carried by convection in Earth's interior, collided with other small islands to form larger islands. These larger islands eventually became the continents. Evidence of these early collisions can be found in certain folded and faulted rocks (for example, the Acasta Gneiss region in northwest Canada), which now form the "cores" of present-day continents.

✓ Simple life emerged during the Archean, and this event touched all other events that followed. Fossils of simple organisms called *stromatolites* are dated at 3.5 billion years old. Stromatolites are algaelike colonies of cyanobacteria interlayered with carbonate sediments (Figure 29.9). The success of these early organisms depended on their ability to survive in the primitive, oxygen-poor environment. These microorganisms evolved a simple version of

fyi

- Very early in Earth's history (between 4.5 and 4.3 billion years ago), a Mars-sized space object struck Earth. The impact created tremendous heat and produced orbiting debris that coalesced to form our Moon. The impact also knocked Earth off its axis of rotation to the present tilt of 23.5°. Without such an impact, Earth's axis of rotation would not be tilted relative to the Sun and we would have no seasons!

FIGURE 29.8 ▶
This artwork of the Archean eon shows characteristic features of the time, including space debris, such as meteorites and comets, crashing into Earth's surface. The bright green material on the edge of the water is primitive bacteria (*archaea*), while cyanobacteria thrive in the darker green, round structures called stromatolites in the water.

◀ **FIGURE 29.9**
Primitive stromatolites found in western Australia are dated as old as 3.5 billion years. They are very similar in structure to the present-day stromatolites pictured here. The first stromatolites evolved in an oxygen-poor environment. They evolved the ability to use sunlight to convert carbon dioxide to food, generating oxygen as a waste product. In this way, stromatolites changed Earth's history as the atmosphere became oxygen-rich.

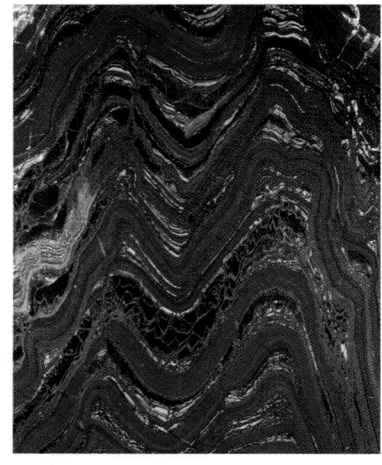

FIGURE 29.10 ▲
Varying oxygen levels in our atmosphere contributed to the formation of alternating layers of iron oxide minerals and iron-free sediment—the banded iron formation. The world's iron and steel industry is based almost exclusively on iron ores formed during the Precambrian.

✔ READING CHECK

What are two *related* and important highlights of Precambrian time?

photosynthesis. Photosynthetic organisms combine carbon dioxide and the Sun's energy to make simple sugars—a biologically usable energy source.

Oxygen is a waste product of the photosynthesis reaction. The expelled oxygen dissolved in seawater and went into the production of certain rocks. The dissolved oxygen reacted with dissolved iron, which was also abundant in the early oceans. The reaction produced solid iron oxides, which chemically precipitated to form *banded iron formations* (Figure 29.10). These ancient rock formations are a testimony to the great amount of oxygen that was expelled during this time.

The end of the Precambrian—the Proterozoic eon—lasted another 2 billion years! Let us highlight a few changes that occurred during this expanse of time. Recall the large islands formed during the Archean—these islands grew to become continents. Carried along by plate tectonics, the continents merged to become one supercontinent, then broke up to form smaller landmass configurations. These changes influenced Earth's climate. Earth was very cold and glaciers covered much of its surface. The reasons for such drastic cooling are unclear, but evidence suggests that many of the continents were near the polar regions.

Life in the Proterozoic flourished in new directions. Evolution produced single-celled organisms and primitive multicellular plants and animals. Proterozoic rocks in southern Australia contain diverse fossils of soft-bodied animals and provide us with the first evidence of an animal community that lived in shallow marine water.

✔ The most important highlight of this eon was the accumulation of free oxygen in the atmosphere. The early Proterozoic atmosphere was still mostly nitrogen, with a little water vapor and carbon dioxide. But free oxygen released by photosynthesizing plants in the oceans finally began to collect in the air. Up until this time, expelled oxygen had been locked up in the great iron deposits around the world. But as oxygen production outpaced its uptake by iron, free oxygen began to enrich the atmosphere. Then, with sufficient free oxygen (O_2) in the atmosphere, a primitive ozone layer (O_3) began to develop above Earth's surface. The ozone layer was very significant—it reduced the amount of harmful ultraviolet (UV) radiation reaching Earth's surface. The accumulation of free oxygen in Earth's atmosphere, and the added protection from the newly formed ozone layer, led to the emergence of new life.

CONCEPT CHECK

1. In what two ways was "free oxygen" essential to the development of new life forms?
2. Where did the carbon dioxide that characterized the early atmosphere go?

Check Your Answers

1. Free oxygen, in the form of O_3, provided protection from harmful UV rays; in the form of O_2, it provided oxygen for respiration.
2. Most of the carbon dioxide dissolved in the oceans combined with dissolved calcium and formed limestone ($CaCO_3$). Much carbon dioxide was consumed by one-celled photosynthesizing organisms, which, when dead, were also incorporated into rocks.

> Rains reduced not only the amount of water vapor in the atmosphere but also the amount of carbon dioxide, which dissolved into water droplets and the newly formed oceans. In this way, the atmosphere started to become more nitrogen enriched.

29.5 The Paleozoic Era—A Time of Life Diversification

More is known about the **Paleozoic era** than the Precambrian, but the Paleozoic is very short in comparison. The Paleozoic era began about 543 million years ago and lasted about 295 million years. During this time, sea levels rose and fell worldwide several times. This allowed shallow seas to partially cover the continents and marine life to flourish. Varying sea levels greatly influenced the progression and diversification of life forms—from marine invertebrates to fishes, amphibians, and reptiles. An important event in the Paleozoic era was the evolution of shelled organisms. In fact, it is because of shelled organisms that we know so much more about the Paleozoic than the Precambrian time. Shelled organisms have hard parts that are more likely to be preserved as fossils. The Paleozoic era is divided into six periods, each characterized by changes in life forms and changes in land configurations.

The Cambrian Period—An Explosion of Life Forms

The Cambrian period marks the beginning of the Paleozoic era. Global temperatures were much warmer than during the icy Proterozoic. In fact, fossil evidence suggests that temperatures were even warmer than they are today. Because the ice sheets and glaciers from the Proterozoic melted, sea level was high. Low-elevation areas of the continents were covered by shallow seas. This expanded the habitats for early Paleozoic marine organisms that evolved to produce a great diversity of life. Hence, this part of Earth's history has become known as "the Cambrian explosion."

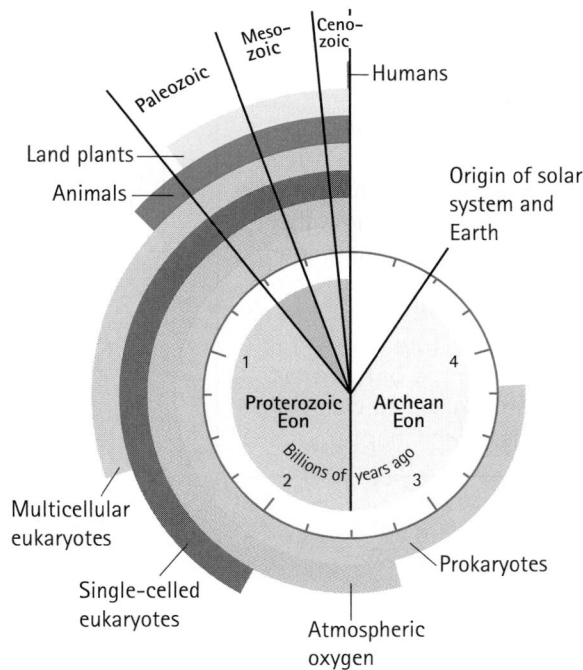

FIGURE 29.11 ▲
This "clock" shows the time when major groups of organisms appeared on Earth. The Paleozoic marked the development of a variety of animal and plant life.

FIGURE 29.12 ▲
Trilobites are the dominant fossils of the Cambrian period.

 **READING CHECK**

Why do we know so much about life in the Paleozoic?

How do warming tempera-tures raise sea levels? Several ways: When ice caps and glaciers on land melt, the water flows into the ocean, raising its level. Also, ocean water expands in volume as its temperature rises—this is an example of thermal expansion, which was discussed in Chapter 9.

FIGURE 29.13 ▶
This map represents the ancestral continents as they may have been positioned during the Silurian period.

✓ The most important event of this time was that many organisms evolved the ability to secrete calcium carbonate and calcium phosphate for the formation of outer skeletons, or shells. Shells helped organisms become less vulnerable to predators, and they provided protection against harmful UV rays. With outer skeletons, many organisms moved to shallower marine habitats. It is because of these hard body parts that the fossil record of Cambrian life is well preserved. A variety of organisms flourished, including the trilobite, the armored "cockroaches" of the Cambrian sea (Figure 29.12).

The Ordovician Period—The Explosion of Life Continues

During the Ordovician, the ancestral continents of South America, Africa, Australia, Antarctica, and India began to merge together to form a new supercontinent called *Gondwanaland.* Gondwanaland was centered over the South Pole, making the latter part of the Ordovician one of the coldest times in Earth history. Ice and massive glaciers covered much of Gondwanaland, draining the shallow seas. As sea level dropped, many shallow-water invertebrates were deprived of their habitat.

Fossil records show that the early to mid-Ordovician period was a time of great diversity and abundant marine life. The Ordovician marks the first arrival of the vertebrates. The end of the Ordovician period is marked by a surge of extinctions, probably the result of the widespread cooling, glaciation, and lowered sea level. The extinctions mainly affected shallow-water marine groups because of habitat loss, whereas deep-water organisms were relatively untouched.

The Silurian Period—Life Begins to Emerge on Land

During the Silurian period, Gondwanaland remained close to the South Pole, but the ancestral continents of North America, Europe, and Siberia were located near the equator (Figure 29.13). The world's climate began to stabilize and warm, which resulted in the melting of many large

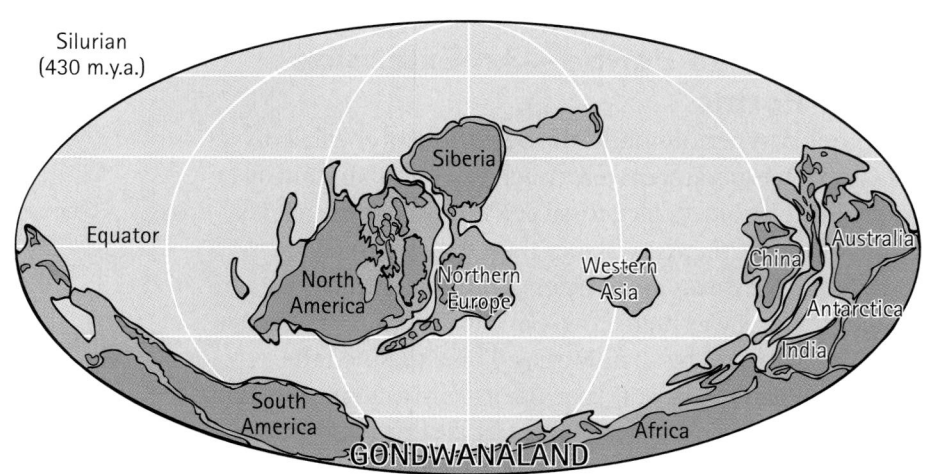

glaciers and a general rise in sea level. Shallow seas moved in to cover many continental interiors. In North America, inland shallow seas were surrounded by coral reefs. These reefs blocked the circulation of water between the inland seas and the open oceans. As water in the inland seas evaporated, deposits of gypsum and other evaporite minerals were left behind. Evaporite beds formed in the Silurian are found today in Ohio, Michigan, and New York.

During the Silurian, terrestrial life, in the form of plants, emerged. Tied to where they originated, these plants inhabited low wetland areas. As plants moved ashore, so did other terrestrial organisms. Air-breathing scorpions and millipedes were common land creatures that evolved during this period in Earth's history.

The Devonian Period—The Age of Fishes

During the Devonian period, the ancestral continents of Europe, North America, and Siberia merged to form another supercontinent known as *Laurasia.* Laurasia was located near the equator. Gondwana-land remained in the Southern Hemisphere.

The Devonian climate was generally warm and moist. Plants and lowland forests of trees and ferns flourished. In the seas, fishes diversified into many new groups, which is why the Devonian is known as the "age of the fishes." Some groups, such as sharks and bony fishes, are still present today. ✔ Among the bony fishes, lobe-finned fishes are of particular interest because they gave rise to land-living, terrestrial vertebrates. Some lobe-finned fishes evolved internal nostrils, which enabled them to breathe air. In addition, the fins of these fishes were lobed and muscular, enabling the animals to support their bodies and "walk." Today, lungfishes and the coelacanth (pronounced "SEE-la-kanth") are the only lobe-finned fishes still in existence. The first land vertebrates, descendants of lobe-finned fishes, evolved during the late Devonian. These vertebrates share many features with the amphibians of today. For example, like today's amphibians, they laid unshelled eggs and could live only in wet environments (Figure 29.14).

The Carboniferous Period—A Time of Great Swampy Forests

The Carboniferous period includes two subperiods—the Mississippian and the Pennsylvanian. During this time, the Paleozoic ocean between Laurasia and Gondwanaland began to close as the two landmasses began merging to form the supercontinent of **Pangaea** (Greek for "all lands"). The collision of these landmasses contributed to great mountain building—the Appalachian Mountains in North America, the Hercynian and Caledonian Mountains in Europe, and the Ural Mountains in Russia.

Pangaea stretched more or less continuously from pole to pole. Most of Gondwanaland was located near the South Pole and the northernmost tip of Laurasia was at the North Pole. Such a stretch of land contributed to alternating periods of ice ages and warming

The coelacanth fish is a "living fossil." This fish was thought to be extinct, but in 1938 one was caught off the coast of East Africa. Since then, other specimens have been discovered in the Madagascar area.

FIGURE 29.14 ▲
This artwork of a late Devonian forest shows an *Acanthostega tetrapod,* an amphibian, climbing over a rock while a dragon-fly flies above. Tetrapods flourished in the Devonian, having evolved from fishes.

◄ **FIGURE 29.15**
Warm, moist climates contributed to the lush vegetation and swampy forests of the Carboniferous period. These forests produced most of the coal deposits around the world.

throughout the Carboniferous. By the late Carboniferous, southern Gondwanaland was completely buried by large sheets of ice. The majority of the Laurasian landmass was still located along the equator, where warm, moist, tropical conditions contributed to lush vegetation, forests, and swamps.

The name Carboniferous or "carbon-bearing" refers to the swampy conditions that produced widespread coal deposits that characterize this part of Earth's history. In fact, dense swamps covered large portions of what are now North America, Europe, China, and Siberia (Figure 29.15). As plants and trees died, their remains settled to the bottoms of the stagnant swamps and decayed to produce coal (see "Fossil Fuels," in Chapter 25). Most of the coal used today is derived from these Carboniferous coal swamps.

Life forms continued to evolve. Insects became very diverse; giant cockroaches and dragonflies with wingspans of 80 centimeters were common. The first amniotes, the vertebrate group that includes today's reptiles and mammals, also evolved at this time. Amniotes have a shelled egg. The amniote egg provides a completely self-contained environment for an embryo—the shell protects the embryo from drying out. This allowed animals to move completely to land. Thanks to the amniote egg, reptiles do not need to lay their eggs in water the way amphibians do.

> ✔ **READING CHECK**
>
> Why are the bony fishes so important?

The Permian Period—The Beginning of the Age of Reptiles

The Permian period marks the end of the Paleozoic era. During this period, all the world's landmasses were completely joined together as Pangaea (Figure 29.16). Like a chain reaction, the formation of Pangaea had many consequences. Continental collisions that started in the Carboniferous affected not only the edges of the continents but also their inner regions. As mountains were uplifted, they blocked

FIGURE 29.16 ►
With the collision of continental landmasses, the supercontinent Pangaea was formed.

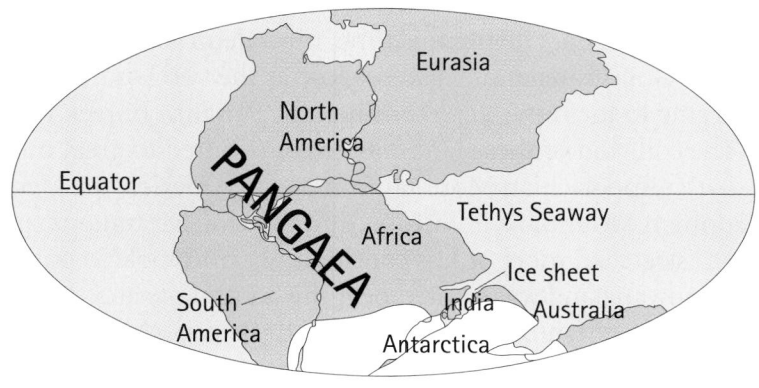

moisture-rich winds from reaching the continental interiors. This caused much of Pangaea's inland areas to become dry and arid. So, changes in land configurations resulted in changes in climate.

At the beginning of the Permian period, glaciers covered much of the Southern Hemisphere and swampy forests covered tropical equatorial regions. By the mid-Permian however (after Pangaea was fully formed), the climate became warmer and milder—glaciers decreased and continental interiors became drier. All landmasses being together also changed the level of the oceans—sea level was lowered. When sea level is lowered, more of the continental shelf (the area between land and the open ocean) is exposed. Because so much life lives in these shallow areas, these organisms began to lose their habitat. This is but one of the factors that may have led to the extinction of many life forms at the end of the Permian.

During the late Permian, one of the greatest extinctions of animal life in Earth's history occurred. Marine life was affected more than land life. About 95% of all marine species and 70% of all land species became extinct. The cause of the extinction is not well understood. ☑ One likely explanation is the domino effect caused by the formation of Pangaea. The redistribution of land and water, the changes in landmass elevations, the change in the world's climate (temperature and precipitation), and the lowering of sea level all come into play.

Terrestrial life, although affected, continued to evolve. As sea levels lowered, new land habitats became available and terrestrial life expanded. The evolution of reptiles continued throughout the Permian. As we will soon see, the reptiles, well suited to their environment, eventually gave rise to the dinosaurs of the Mesozoic era.

CONCEPT CHECK

As continental landmasses merged, coastlines and continental shelf areas were reduced. Look at Figure 29.16 and describe why such a reduction might have occurred.

Check Your Answer

Some of the former coastlines and continental shelves were no longer next to the ocean—they were now in the middle of Pangaea! With all landmasses crunched together, less water was displaced and sea level went down. So some former continental shelf areas became dry inland areas.

READING CHECK

In what way did the formation of Pangaea contribute to the Permian extinction?

29.6 The Mesozoic Era—The Age of Reptiles

Profound changes in fossils separate the Mesozoic era from the Paleozoic era. The **Mesozoic era** consists of three periods—the Triassic, Jurassic, and Cretaceous—and is informally known as "the age of reptiles." Reptiles that survived the Permian extinction evolved to become the

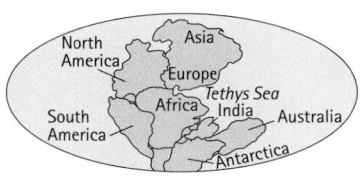

200 million years ago
Mesozoic Era

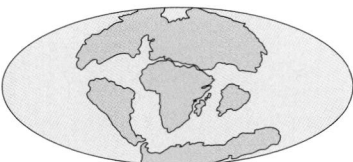

65 million years ago
Cenozoic Era

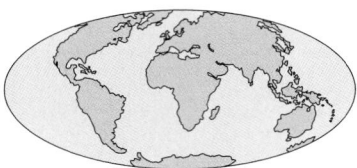

Present

FIGURE 29.17 ▲
Stages during the breakup
of Pangaea.

The Andes Mountains
and the Sierra Nevada have
their beginnings in the great
volcanic activity that rimmed
the eastern Pacific basin
during the Mesozoic.

rulers of the Mesozoic world—the dinosaurs. Mammals also evolved from reptiles early in the Mesozoic, but they were relatively small and insignificant compared to the great dinosaurs.

Land plants greatly diversified. True pines and redwoods appeared and rapidly spread throughout the land. Flowering plants emerged and diversified so quickly that they became the dominant plants of the Mesozoic. Flowering plants also accelerated the evolution and specialization of insects.

The major geological event of the Mesozoic was the breakup of Pangaea. Just as Pangaea was formed in different phases throughout the Paleozoic, Pangaea broke apart phase by phase throughout the Mesozoic and into the Cenozoic (Figure 29.17). The first phase began with the formation of a *rift zone* between what are today's North American and African continents. This was the beginning of the central Atlantic Ocean basin. As the rift zone spread, Europe split from North Africa. In this manner, Laurasia became completely separated from Gondwanaland.

The second phase was the breakup of Gondwanaland. Rifting in the Atlantic Ocean basin spread southward and resulted in the separation of South America from Africa. Similarly, ancestral India/Madagascar rifted away from Antarctica/Australia to open up the Indian Ocean basin. Of all the former continental unions that existed in Paleozoic time, only that of Europe and Asia survive to the present time.

As rifting occurred *within* Pangaea, subduction ensued along Pangaea's margins. Of special interest is the subduction of the "Pacific" oceanic crust beneath the "North American" and "South American" continental crust. This produced widespread deformation, volcanism, and mountain building all along the entire western coast—from present-day Alaska to Chile.

The separation of landmasses had several worldwide consequences. Coastlines were once again more extensive, which increased shallow marine habitat. Land-based life—plants and animals—also changed. As formerly connected habitats became separated, members of the same species also became separated. Now in new, unconnected habitats, these separated organisms began to diverge from one another—they became less and less alike. In this way, new species evolved according to their new habitats.

Climate changed as well. Because most landmasses were no longer centered at the equator, the circulation pattern of the oceans and the atmosphere changed. Warm water from equatorial regions circulated northward to warm northern landmasses. Also, the rifting activity that initiated Pangaea's breakup resulted in a worldwide rise of sea level. So with shallow seas covering the continents, climate conditions became very mild. Overall, the Mesozoic climate was much warmer than conditions today.

The Cretaceous Extinction—An Extraterrestrial Calamity

The end of the Cretaceous period, 65 million years ago, brought another mass extinction that killed more than 60% of Earth's species. Many dinosaurs, flying reptiles, and marine reptiles were wiped out, as were other organisms, both on land and in the seas. The most accepted hypothesis is that the great extinction was caused by the impact of a very large meteorite. The meteorite hit Earth with such force that it wreaked havoc in many ways. The impact caused large, widespread earthquakes and huge shock waves to form. It also released enormous amounts of heat, which resulted in extreme firestorms. In the meteorite impact's aftermath, a gigantic light-blocking cloud of dust enveloped Earth. The dust cloud spread throughout the atmosphere and lasted long enough to stop the process of photosynthesis. So, the cloud not only chilled Earth but also devastated the food supply.

Evidence for this theory is found in the rock record. As the dust settled, a layer of iridium-enriched sediment was deposited. ✓ The element iridium is rare at Earth's surface, but it is abundant in meteorites. So the iridium layer that marks the boundary between the Cretaceous and Tertiary periods (called the K-T boundary) is hypothesized to have been spread worldwide by the extraterrestrial impact event. The Chicxulub crater (dated at 65 million years), located off the Yucatan coast in Mexico, is the hypothesized site of impact (Figure 29.18). Both the K-T boundary layer and the Chicxulub crater coincide with the time of the great Cretaceous extinction—the event that marks the end of the Mesozoic era.

FIGURE 29.18 ▲
Artwork of the Chicxulub crater on the Yucatan Peninsula of Mexico. The crater is about 180 km in diameter. The impact that produced the crater is thought to have caused the mass extinction that ended the reign of the dinosaurs, 65 million years ago at the end of the Cretaceous.

CONCEPT CHECK

1. Could the break up of Pangaea account for the extinction at the end of the Cretaceous?
2. In terms of the number of species destroyed, which was the bigger extinction—the Permian or Cretaceous?

Check Your Answers

1. There is no doubt that the reconfiguring of landmasses when Pangaea broke up put added stress on many areas of life. Climate changed, shallow marine habitats changed, species became separated and diverged from one another, moving landmasses triggered widespread volcanic activity, and so on. But the breakup of Pangaea cannot explain the high concentration of iridium at the K-T boundary.
2. The Permian was the bigger mass extinction, having killed about 95% of all marine species and 70% of all land species. The Cretaceous extinction resulted in the demise of about 60% of Earth's species.

✓ **READING CHECK**

How does iridium at the K-T boundary suggest extraterrestrial activity?

29.7 The Cenozoic Era—The Age of Mammals

The **Cenozoic era,** known as the "age of the mammals," is made up of two periods—the Tertiary and the Quaternary. From oldest to youngest, these two periods are broken into the Paleocene, Eocene, Oligocene, Miocene, and Pliocene epochs (for the Tertiary period) and the Pleistocene and Holocene epochs (for the Quaternary

The shifting of Earth's tectonic plates change not only land configurations but also climate and even life forms! When you change one factor, many other factors change. Hmmm . . . Like Newton's third law, "You cannot touch without being touched."

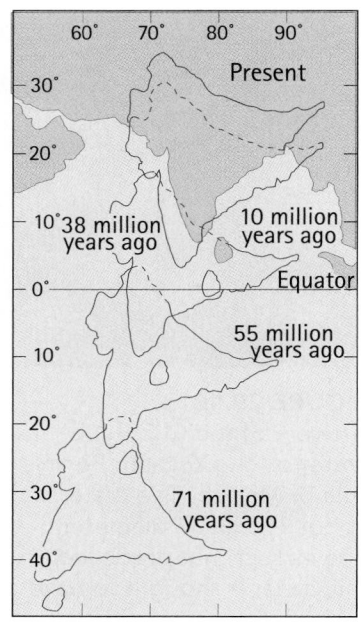

FIGURE 29.19 ▲
The formation of the Himalayas was a result of the collision of India with Asia. Because this was a continent-to-continent collision, the Himalayas have an unusually large thickness of continental crust.

period). Although it is not yet formally established, many scientists claim we are entering a new epoch—the Anthropocene. This new addition to the geologic time scale is attributed to human impact. The impact of humans on Earth—agriculture, industry, deforestation, animal domestication and extinction, and climate change—has changed our planet so much that we are no longer in the "traditional Holocene." So, change is ongoing, even in the making of the geologic time scale.

The third and final phase of Pangaea's breakup took place during the early Cenozoic. North America and Greenland split away from Europe, and Antarctica split from Australia. Earth's surface was very active and landmass collisions were numerous. Considerable mountain building activity occurred when Africa-Arabia collided with Europe to produce the Alps, and India collided with Asia to produce the Himalayas. As shown in Figure 29.19, India moved toward Asia at a rate of 15 to 20 cm per year—a tectonic speed record! The collision produced many folds and reverse and strike-slip faults. Also, India partially wedged itself below Asia, which generated an unusually thick accumulation of continental land. Due to isostasy, this continental land atop continental land provided additional height to the Himalayas.

The late-Mesozoic collision that began the formation of the Sierra Nevada and Andes Mountains continued (and continues) into the Cenozoic. Off the western coastline of North America, a mid-ocean ridge formed in the Pacific Ocean—the Pacific Ridge system. This ridge separated two different oceanic plates—the Pacific Plate on the west and the Farallon Plate on the east (Figure 29.20). As the Farallon Plate subducted below the North American Plate, the Pacific Ridge began to approach the North American continental margin. This not only reactivated mountain building throughout the west but also contributed to widespread deformation. The collision of the Pacific Ridge with the westward-moving North American Plate occurred about 30 million years ago and gave birth to the San Andreas Fault (Figure 29.21). In time, as the fault grew, Baja California was torn away from the Mexican mainland, and the Gulf of California was created. Because these plates are still moving, western California and Baja California will eventually become completely detached from the mainland or will find themselves joined to western Canada.

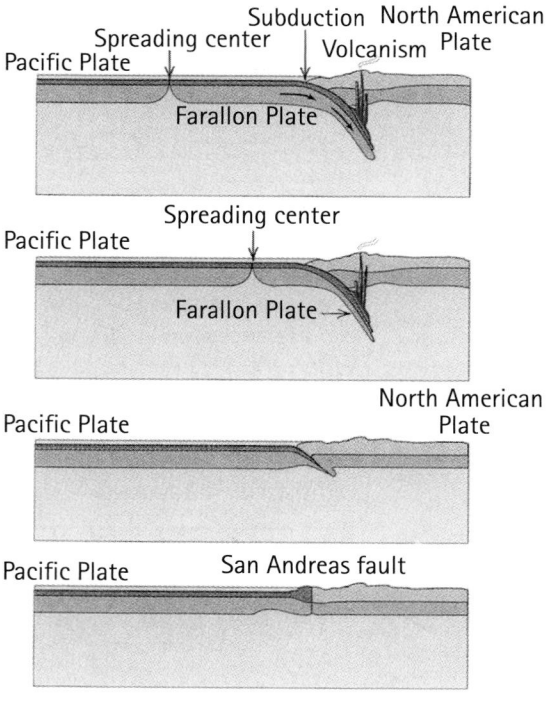

◀ **FIGURE 29.20**
Subduction sequence of the Farallon Plate beneath the North American Plate. As the spreading center moved toward the North American Plate, the San Andreas Fault began to form.

The Hawaiian Island/Emperor Seamount chain (Figure 29.22) provides additional evidence of Cenozoic deformation: the change in direction of the Pacific Plate. The bend in the chain of islands occurred between 30 and 40 million years ago (mid-Tertiary) when the direction of plate motion changed from nearly due north to northwesterly. The change in direction occurred at about the same time as the collision of northern Mexico (the North American Plate) with the Pacific Ridge System.

Climates cooled during much of the Cenozoic era, resulting in the widespread glaciation that characterized the Pleistocene. Although this ice age continues today, there have been many alternations between glacial and interglacial conditions (we are currently in an interglacial period). During the glacial peak, as much as one-third of Earth's surface was covered by ice. On land, great thicknesses of ice formed continental glaciers. The great weight of ice depressed the land and altered the courses of many streams and rivers. The glaciers eroded and scratched the land in some places and deposited huge *moraines*, leaving behind abundant evidence of the extent of their former existence.

Cenozoic Life

After the mass extinctions at the end of the Mesozoic era, many environmental niches—habitats—were left vacant. These openings allowed the rapid evolution of mammals. Bats, some large land mammals, and such marine animals as whales and dolphins evolved to occupy niches left vacant by the extinction of many of the Mesozoic reptiles. Later in the Cenozoic, as global temperatures cooled, the Ice Age began. The cool temperatures had a profound impact on life. Mammoths, rhinos, bison, reindeer, and musk oxen all evolved warm, woolly coats for protection from the frigid cold.

✔ Humans evolved in the Quaternary period during the Pleistocene epoch. Extensive glaciation caused sea level to drop as water became bound up in glaciers and ice caps. Even though the distribution of land was essentially the same as it is today, the lowered sea level resulted in "land bridge" connections between landmasses that are now separated

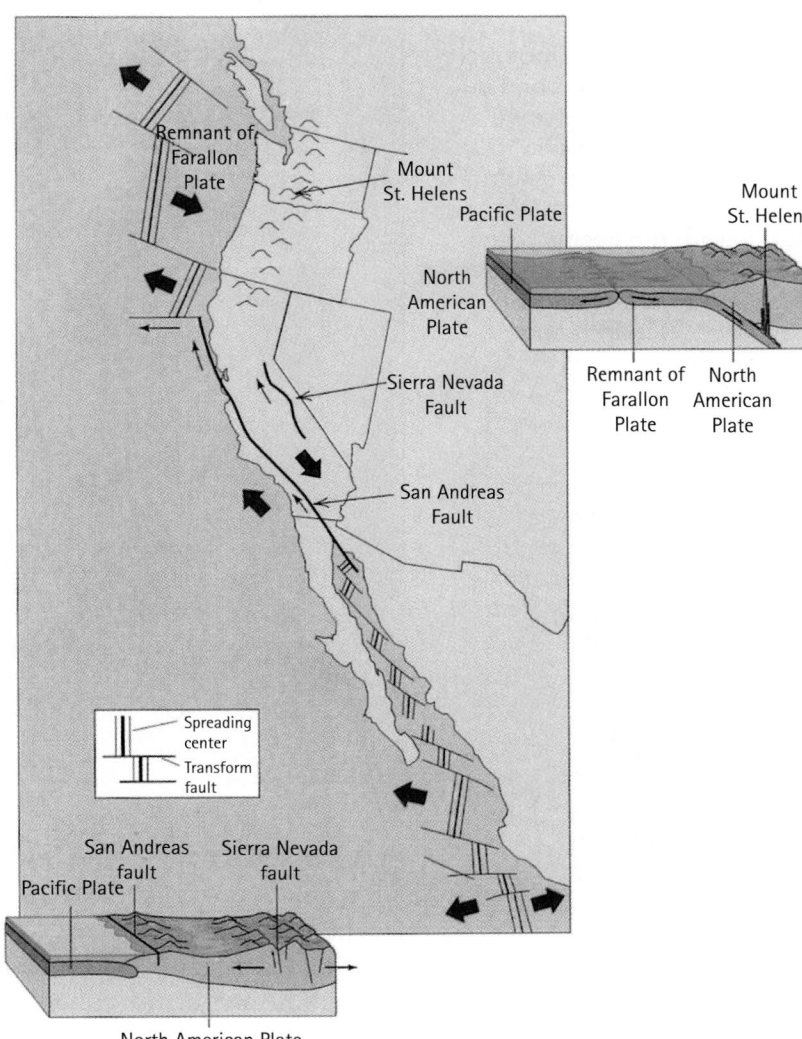

FIGURE 29.21 ▲
The San Andreas Fault is the result of an encounter between the North American Plate and the Pacific Ridge System. As the fault grew longer, the area of Baja California was torn from the Mexican mainland.

FIGURE 29.22 ▶
The Hawaiian Island/Emperor Seamount chain. The bend in the chain shows the change in direction of the Pacific Plate, which is likely a result of the collision of northern Mexico with the Pacific Ridge. The red numbers indicate the age (in millions of years) of the individual islands and seamounts.

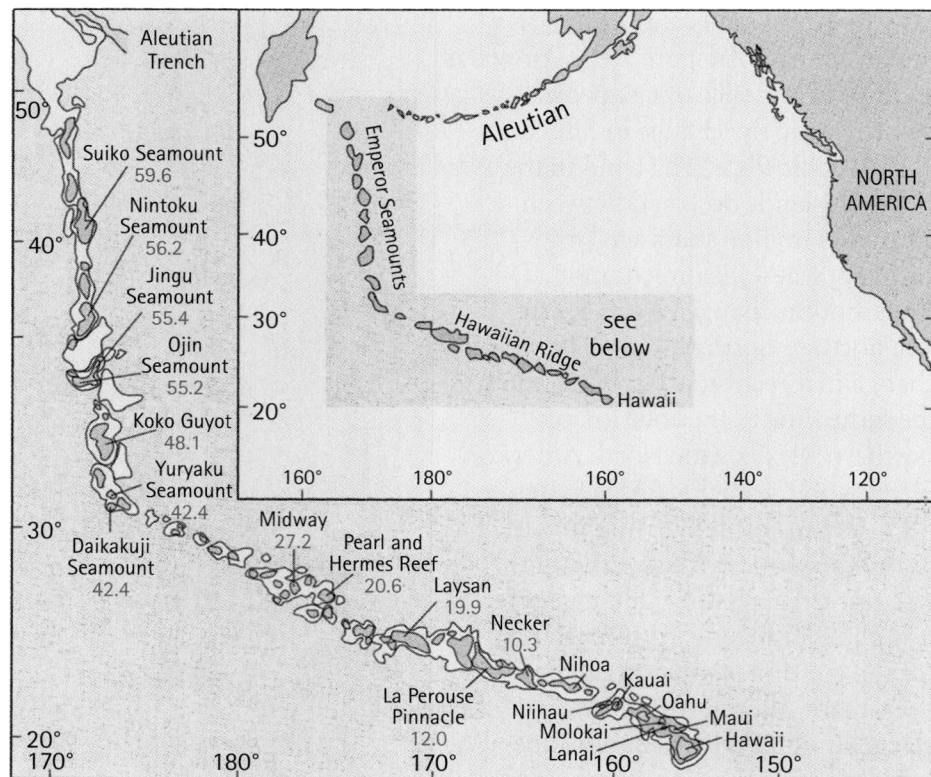

It would not have been possible for large mammals to evolve if dinosaurs had not gone extinct. Such mammals would have been tasty treats for large meat-eating dinosaurs!

by water. One of these land bridges existed across the present-day Bering Strait, and it provided the route for the human migration from Asia to North America.

The expansion of humans, not only into North America but also throughout the world, coincided with a period of extinctions that occurred during the Pleistocene. The extinctions primarily involved large terrestrial mammals; marine animals were for the most part unaffected. In North America, many large mammals became extinct after humans arrived, and in Africa, mammalian extinctions can be related to the appearance of the Stone Age hunters.

The cause of the Pleistocene extinction is a much-debated issue. The extreme climate variation during this time was certainly a factor, as was the hunting and eating of large animals by humans. As with past extinctions, a single cause is unlikely. The Pleistocene extinction resulted from a combination of factors.

Following the Pleistocene is the Holocene epoch. The Holocene accounts for the most recent 10,000 years or so of Earth's history. Although there has been some climatic variation—for example, the "Little Ice Age" (~1300 to 1800)—in general, the Holocene has been a relatively warm interglacial period. The Holocene is sometimes called the "Human Age," but this is somewhat inaccurate because *Homo sapiens* had evolved and occupied many regions of the globe well before the start of the Holocene. Yet the Holocene has witnessed all of humanity's recorded history, including the rise and fall of all

READING CHECK

What life form is the Quaternary period most known for?

HUMAN GEOLOGIC FORCE

Although the "human age" amounts to only a brief 0.002% of geologic time, we are almost certainly the most clever and adaptable organisms to have evolved on the planet. All life forms alter their environments. Humans do it more, as we manipulate our environment to meet our needs. We have but to look at the irrigation systems of Mesopotamia, the cultivation along the Nile, the plowing of the prairies in the Great Plains, the invention of machines to further utilize the land, and the dams and locks on the Mississippi, Missouri, and Colorado rivers to illustrate the human role in geologic changes. Will the human activities that adversely affect our planetary life-support system—the large-scale burning of fossil fuels, the deterioration of the ozone layer, the destruction of the Amazon rain forest, and of particular concern, climate change—ultimately lead to a new period of extinctions? Perhaps our own? Because we have the capacity to cause geologic change, it is imperative that we take care of our terrestrial home. It's the only one we have!

historic civilizations. Humanity has had a great impact upon the Holocene environment. All organisms influence their environments to some degree, but few have ever changed Earth as much, or as fast, as we are doing in modern times. It is because of these changes that many scientists claim we have entered a new epoch—the Anthropocene.

29.8 Earth History in a Capsule

Our Earth has a long and exciting history—4.5 billion year's worth! To unravel this history we look at the rock record. Each event in Earth's long history is recorded in its rock layers. The order of geologic events is deciphered by relative dating, and absolute age is determined by radiometric dating. The geologic time scale provides a chronological listing of different time intervals according to the fossil record. Major changes in life are marked by the beginning of new eras and periods.

Throughout geologic time, life has changed, landforms have changed, and climate has changed. Change in one area influences change in another area. Each change is recorded in the rocks. As mentioned earlier, the Earth scientist is like a crime scene detective. The "crime scene" is Earth, and the rock record provides us with clues to unravel *what* happened and *when* it happened. Earth is like a big puzzle—we can only understand the picture by fitting in one piece at a time.

The Anthropocene epoch is young and dates back to the time of the Industrial Revolution—the 1800s. The changes to our environment because of human population and economic development have been so drastic that post-industrialized Earth is very different from conditions in the Holocene.

29 CHAPTER REVIEW

KEY TERMS

Angular unconformity An unconformity in which older, tilted rock layers are covered by younger, horizontal rock layers.

Cenozoic era The time of recent life, which began 65 million years ago and is still ongoing.

Cross-cutting Where an igneous intrusion or fault cuts through other rocks, the intrusion or fault is younger than the rock it cuts.

Faunal succession Fossil organisms succeed one another in a definite, irreversible, and determinable order.

Inclusion Any inclusion (pieces of one rock type contained within another) is older than the rock containing it.

Lateral continuity Sedimentary layers are deposited in all directions over large areas until some sort of obstruction, or barrier, limits their deposition.

Mesozoic era The time of middle life, from 245 million years ago to about 65 million years ago.

Original horizontality Layers of sediment are deposited evenly, with each new layer laid down nearly horizontally over the older sediment.

Paleozoic era The time of ancient life, from 543 million years ago to 245 million years ago.

Pangaea The late-Paleozoic supercontinent made up of Gondwanaland (ancestral South America, Africa, Australia, Antarctica, and India) and Laurasia (ancestral North America, Europe, and Siberia/Asia).

Precambrian time The time of hidden life, which began about 4.5 billion years ago when Earth formed, lasted until about 543 million years ago (beginning of Paleozoic), and makes up almost 90% of Earth's history.

Radiometric dating A method for calculating the age of geologic materials based on the nuclear decay of naturally occurring radioactive isotopes.

Relative dating The ordering of rocks in sequence by their comparative ages.

Superposition In an undeformed sequence of sedimentary rocks, each bed or layer is older than the one above and younger than the one below.

Unconformity A break or gap in the geologic record, caused by erosion of preexisting rock or by an interruption in the sequence of deposition.

REVIEW QUESTIONS

Relative Dating—The Placement of Rocks in Order

1. What is relative dating?
2. A granitic dike is found across a sandstone layer. What can be said about the relative ages of the dike and the sandstone? What principle applies here?
3. How are fossils used in determining geologic time?
4. In a sequence of sedimentary rock layers, the oldest layer is on the bottom and the youngest layer is at the top. What relative dating principle is this?

Radiometric Dating Reveals the Actual Time of Rock Formation

5. What is radioactive half-life?
6. What are the half-lives of uranium-238, potassium-40, and carbon-14?

Geologic Time

7. Which of the geologic time units spans the greatest length of time?
8. How old is Earth?

Precambrian Time—A Time of Hidden Life

9. What key developments in life occurred during Precambrian time?

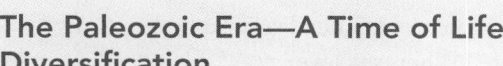

The Paleozoic Era—A Time of Life Diversification

10. The Paleozoic era experienced several fluctuations in sea level. What effect did this have on life forms?

11. Name the periods of the Paleozoic era.

12. Why are the lobe-finned fishes significant?

13. When did Laurasia merge with Gondwanaland to form Pangaea?

14. What animal group evolved from the amphibians with the arrival of the amniote egg?

The Mesozoic Era—The Age of Reptiles

15. What is the most likely cause of the Cretaceous extinction that wiped out the dinosaurs?

16. What effect did the breakup of Pangaea have on sea level?

17. What evidence supports the meteorite impact at the end of the Cretaceous?

The Cenozoic Era—The Age of Mammals

18. Which epochs make up the Tertiary period? The Quaternary period?

19. How did the San Andreas Fault form?

20. What important life forms evolved during the Cenozoic era?

THINK AND EXPLAIN

1. What six principles are used in relative dating?

2. Why don't all rock formations show a continuous sequence from the beginning of time to the present?

3. Explain how fossils of fishes and other marine animals occur at high elevations, such as the Himalayas.

4. What isotope is preferred in dating very old rocks?

5. What isotope is commonly used for dating sediments or organic material from the Pleistocene?

6. What was the most noteworthy evidence of life in the Precambrian?

7. What life form emerged in the Silurian period?

8. Why are internal nostrils in the lobe-finned fishes an important step in the evolution of life on Earth?

9. What time period is associated with coal deposits?

10. What life forms are associated with the Devonian period?

11. By what informal name is the Mesozoic era known?

12. How does the element iridium relate to the time of the extinction of the dinosaurs?

13. Suppose you see a sequence of sedimentary rock layers covered by a basalt flow. A fault cuts through the sedimentary rock layers but does not cut into the basalt. Relate the fault to the ages of the two rock types in the formation.

14. If a sedimentary rock contains inclusions of metamorphic rock, which rock is older?

15. Which isotopes are most appropriate for dating rocks from early Precambrian time? How about the late Pleistocene epoch?

16. Before the discovery of radioactivity, how did geologists estimate the age of rock layers?

17. How did the generation of free oxygen during the late Precambrian affect our planet?

18. In what ways can sea level be lowered? What effect might this have on existing life forms?

19. What can cause a rise in sea level? Is this likely to happen in the future? Why or why not?

20. Why are Paleozoic marine sedimentary rocks such as limestone and dolomite found widely distributed in the continental interiors?

THINK AND SOLVE

Refer to the figure. Using the principles of relative dating, determine the relative ages of the rock bodies and other lettered features. Start with this question: What was there first?

Sequence of events

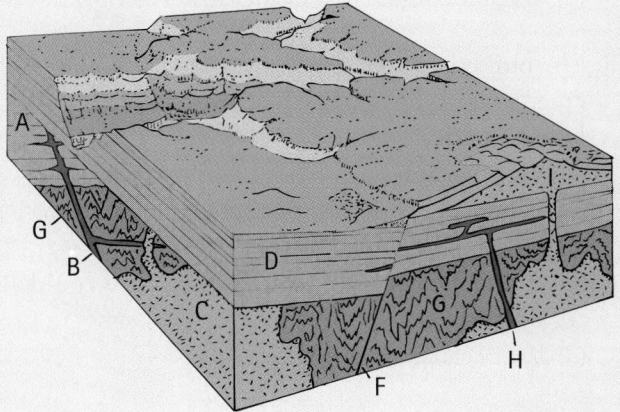

READINESS ASSURANCE TEST (RAT)

If you have a good handle on this chapter—if you really do—then you should be able to score 7 out of 10 on this RAT. Check your answers with your teacher. If you score less than 7, study further before moving on.

Choose the best *answer to the following.*

1. The principle of superposition is that each new
   - (a) sedimentary layer is older than the layer above.
   - (b) sedimentary layer is younger than the layer below.
   - (c) layer of sediment is laid down nearly horizontally.
   - (d) layer of sediment is laid down accordingly.

2. Life forms throughout Earth's past have occurred in a definite order. This is called the principle of
   - (a) fossil assemblage.
   - (b) faunal succession.
   - (c) conformable fossils.
   - (d) fossil determination.

3. The time it takes for 50% of a radioactive substance to decay is known as
   - (a) radiometric dating.
   - (b) carbon-14.
   - (c) the proportion of atoms remaining.
   - (d) the half-life.

4. Development of Earth's oceans was likely due to
   - (a) water-rich meteors bombarding Earth's surface.
   - (b) volcanic outgassing, followed by cooler temperatures in Precambrian time.
   - (c) slow convection in the mantle.
   - (d) volcanic outgassing in the early Paleozoic.

5. The buildup of free oxygen was crucial to the emergence of life on Earth because it led to the formation of
   - (a) O_2 for plants.
   - (b) ozone, O_3, which helped screen Earth from harmful incoming UV radiation.
   - (c) ozone, O_3, which primitive organisms could breathe.
   - (d) the oceans, where life emerged.

6. The Paleozoic experienced several fluctuations in sea level. When sea level rises
   - (a) shallow seas cover the continents.
   - (b) more water is tied up in glaciers, making the climate colder.
   - (c) the climate turns warmer and swamps form.
   - (d) ocean basins become shallow.

7. The most important event during the Cambrian period was the
   - (a) emergence of the fishes.
   - (b) ability of organisms to form an outer skeleton.
   - (c) emergence of the trilobite.
   - (d) ability of organisms to develop lungs.

8. The formation of the supercontinent of Pangaea
   - (a) resulted from the collision of all major landmasses.
   - (b) produced widespread mountain building in the Himalayas.
   - (c) resulted in extensive volcanic activity and flood basalts.
   - (d) All of these.

9. Glaciation during the Cenozoic resulted in
   - (a) lowering of sea level worldwide.
   - (b) carving Earth's surface (for example, the Swiss Alps).
   - (c) land bridge connections between various continents.
   - (d) All of these.

10. The birth of the San Andreas Fault corresponds to
    - (a) the collision of the Pacific Plate and the Hawaiian Islands.
    - (b) the collision of the Pacific Ridge System and North America.
    - (c) the movement of the Pacific Plate over a turbulent zone.
    - (d) the collision of India and Eurasia.

30

THE ATMOSPHERE, THE OCEANS, AND THEIR INTERACTIONS

THE MAIN IDEA

Air moves because of the unequal heating of Earth's surface by the Sun. As air moves over the oceans, the seawater moves.

How did Earth get its atmosphere? Where did the oceans come from? How do the atmosphere and ocean interact? Is it true that without an atmosphere, the ocean would boil away? What causes seasons? Is it true that America is warmer when it is farthest from the Sun? What causes wind? Can ocean water freeze over and become ice? How salty are the oceans? How the atmosphere interacts with the oceans affects all of us. Let's explore and learn why!

Explore!

How Does Wind at Earth's Surface Cause Water in the Ocean to Circulate?

1. Fill a shallow aluminum baking pan with water.
2. With a hole punch, cut out about 10 circles of paper. Float the circles on one side of the pan.
3. Blow across the surface of the pan.

Analyze and Conclude

1. **Observing** Blowing across the right side of the pan causes the paper circles to move in a clockwise direction. When you blow across the left side of the pan, the paper circles move in a counterclockwise direction.

2. **Predicting** Wind on the water creates a surface current. Does the water move in the same direction as the wind? What causes the wind to move?
3. **Making Generalizations** The atmosphere and the oceans circulate. What factors contribute to their circulation pattern? How about the rotation of Earth? Temperature differences? Pressure differences? Does circulation of the atmosphere affect the circulation of the oceans?

Islands have very little temperature differences. The temperature we feel at the island's beach is very similar to the temperature we feel inland.

30.1 Earth's Atmosphere and Oceans

As seen from space, planet Earth is blue with wisps of silver. It is blue because of the oceans. The wispy silver is clouds in the atmosphere. Seventy percent of Earth's surface is covered by water (Figure 30.1). The remaining 30% is land, most of which is located in the Northern Hemisphere (Figure 30.2). Although the many oceans are named for their various locations, they are connected. All of Earth's oceans are actually a single continuous ocean.

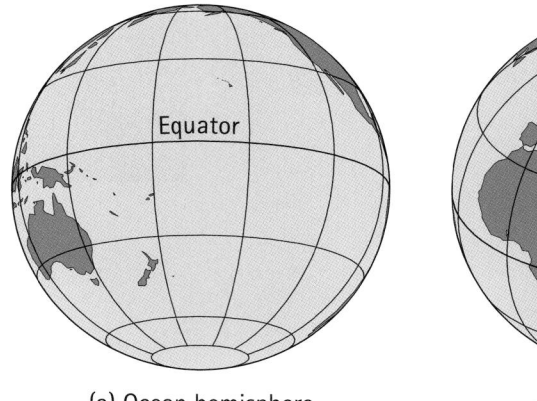

(a) Ocean hemisphere

(b) Land hemisphere

FIGURE 30.1 ▲
Most of Earth's surface is covered by water. We can divide Earth into (a) an ocean-dominated hemisphere and (b) a land-dominated hemisphere.

Earth's Oceans Moderate Land Temperatures

As we learned in Section 28.1, the oceans are the reservoir from which water evaporates into the atmosphere to precipitate later as rain and snow. The oceans play a major role in moderating Earth's temperature and climate. Recall from Chapter 9 that water has a high specific heat capacity—water is slow to heat up and slow to cool down. Water transfers large amounts of heat energy to its surroundings when it cools. Likewise, water absorbs large amounts of heat energy from its surroundings when it warms. Because of water's high specific heat capacity, lands nearer to oceans have moderate temperatures. The moderating influence of the oceans can be seen when we look at seasonal temperature variations for two cities at the same latitude: coastal San Francisco, California, and continental Wichita, Kansas (Figure 30.3). Whereas temperatures in San Francisco tend to have small seasonal variations, temperatures in Wichita show strong seasonal fluctuations—cold winters and hot summers. The oceans do a great job of both making summers cooler and winters warmer.

Evolution of Earth's Atmosphere and Oceans

Earth probably had an atmosphere before the Sun was fully formed. This primitive atmosphere was possibly composed of only hydrogen and helium, the two most abundant gases in the universe, along with trace amounts of ammonia and methane. The early atmosphere contained

FIGURE 30.2 ▲
When a map is centered over Antarctica, the expanse of the world ocean can be seen. In terms of size and volume, the Pacific Ocean accounts for more than half of the world ocean and is thus the largest ocean. In fact, the Atlantic and Indian Oceans combined would easily fit into the space occupied by the Pacific.

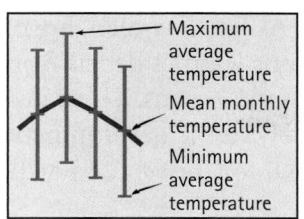

Maximum average temperature
Mean monthly temperature
Minimum average temperature

◀ **FIGURE 30.3**
Comparison of seasonal temperature ranges for coastal San Francisco, California, and continental Wichita, Kansas.

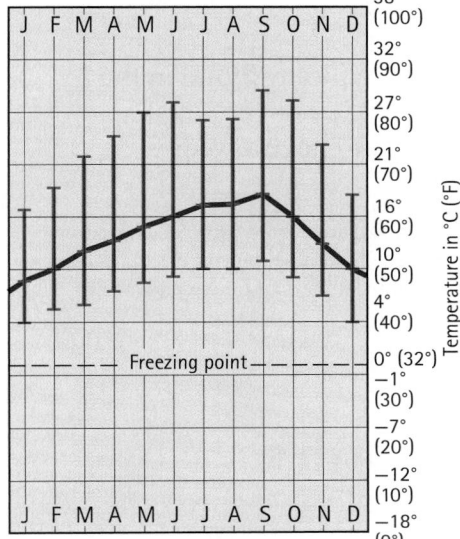

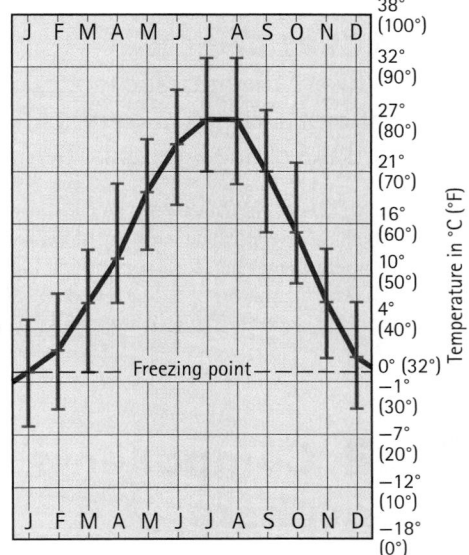

Station: San Francisco, California
Latitude/longtitude: 37°37' N, 122°23' W
Average annual temperature: 14°C (57.2°F)
Total annual precipitation: 47.5 cm (18.7 in.)
Elevation: 5 m (16.4 ft)
Population: 750,000
Annual temperature range: 9°C (16.2°F)

Station: Wichita, Kansas
Latitude/longtitude: 37°39' N, 97°25' W
Average annual temperature: 13.7°C (56.6°F)
Total annual precipitation: 72.2 cm (28.4 in.)
Elevation: 402.6 m (1321 ft)
Population: 350,000
Annual temperature range: 27°C (48.6°F)

no free oxygen. When the Sun was born, the blast from its formation produced a strong outflow of charged particles—an outflow strong enough to sweep Earth of its first atmosphere.

☑ The next atmosphere formed as gases trapped in Earth's hot interior escaped through volcanoes and fissures at the surface. The gases spewed out by these early eruptions were probably much like the gases found in the volcanic eruptions of today—water vapor and carbon dioxide. But in this early atmosphere, carbon dioxide was the dominant gas, with water vapor and other trace elements (nitrogen, ammonia, sulfur dioxide, and nitric oxide) as minor constituents. There was no free oxygen.

As we learned in Chapter 29, free oxygen did not occur in Earth's atmosphere until primitive organisms such as stromatolites appeared. Like the green plants that followed, stromatolites used *photosynthesis* to convert carbon dioxide and water to carbohydrates and free oxygen:

$$CO_2 + H_2O + light \rightarrow CH_2O + O_2$$

With the accumulation of free oxygen, an ozone (O_3) layer formed in the upper atmosphere. The ozone layer acts like a filter, it reduces the amount of ultraviolet radiation that reaches Earth's surface. With less radiation, the surface became able to support life.

Volcanic eruptions release a great deal of carbon dioxide. Yet, there is not a lot of CO_2 in Earth's atmosphere. So where is all the expelled CO_2?

The CO_2 is gobbled up by the ocean, where it dissolves and ends up as calcium carbonate.

✓ As Earth cooled, huge amounts of water vapor condensed to form the oceans. Comet debris from interplanetary space also contributed water to the oceans. These oceans, essential to the evolution of life and ultimately to the development of the present global environment, have remained for the rest of Earth's history.

CONCEPT CHECK

1. Why are the hottest climates on Earth typically found in the interior sections of continents?
2. Did the ozone layer exist before stromatolites evolved?

Check Your Answers

1. The large specific heat capacity of water keeps coastal areas from experiencing extreme temperatures. Therefore, very hot climates are usually some distance from the ocean.
2. No. The formation of ozone, O_3, came after the introduction of free oxygen, which came from photosynthesizing stromatolites.

✓ **READING CHECK**

After the Sun's formation, what contributed to the formation of Earth's early atmosphere? How did the oceans form?

30.2 Components of Earth's Atmosphere

✓ If gravity did not exist, gas molecules in Earth's atmosphere would fly off into outer space. Gases are compressible. This allows the invisible force of gravity to squeeze and hold a great number of gas molecules close to Earth's surface (where gravity is strongest). Thus the density of air molecules is greatest at Earth's surface and gradually decreases with height.

Because air has weight, it exerts pressure on Earth's surface. This pressure is known as *atmospheric pressure* or, simply, *air pressure.* The more weight, the more pressure. Like the atmosphere's density, air pressure also decreases with increasing height above Earth's surface. The higher you go, the lower the air pressure. Interestingly, the weight of air on the ocean's surface keeps the ocean from boiling away. Recall from Chapter 9 that water boils at 0°C when no air pressure acts on it. So fish as well as birds appreciate the existence of the atmosphere.

Table 30.1 shows that Earth's atmosphere is a mixture of various gases—primarily nitrogen and oxygen, with small percentages of water vapor, argon, and carbon dioxide and trace amounts of other elements and compounds.

The condensation of water vapor resulted in torrential rains. These rains not only formed the oceans but also washed lots of carbon dioxide into the ocean. As the carbon dioxide dissolved, it formed into limestone.

PhysicsPlace.com

Tutorial
Vertical Structure of the
Atmosphere

✓ **READING CHECK**

What invisible force holds air molecules near to Earth's surface?

Vertical Structure of the Atmosphere

If you have ever gone mountain climbing, you probably noticed that the air grows cooler and thinner with increasing elevation. At sea level, the air is generally warmer and denser. The greater density near Earth's surface is due to gravity. The density of the air, like the density of a deep pile of feathers, is greatest at the bottom and least at the top. More than half the atmosphere's mass lies below an altitude of 5.6 km, and about 99% lies below an altitude

TABLE 30.1 Composition of the Atmosphere

Permanent Gases			Variable Gases		
Gas	Symbol	Percentage by Volume	Gas	Symbol	Percentage by Volume
Nitrogen	N_2	78%	Water vapor	H_2O	0 to 4
Oxygen	O_2	21%	Carbon dioxide	CO_2	0.038**
Argon	Ar	0.9%	Ozone	O_3	0.000004*
Neon	Ne	0.0018%	Carbon monoxide	CO	0.00002*
Helium	He	0.0005%	Sulfur dioxide	SO_2	0.000001*
Methane	CH_4	0.0001%	Nitrogen dioxide	NO_2	0.000001*
Hydrogen	H_2	0.00005%	Particles (dust, pollen)		0.00001*

* Average value in polluted air.
** The amount of CO_2 in the atmosphere is increasing, as we will see later in this chapter.

of 30 km. Unlike a pile of feathers, however, the atmosphere doesn't have a distinct top. It gradually thins to the near vacuum of outer space.

The atmosphere is divided into layers, each distinct in its characteristics (Figure 30.4). ✓ The lowest layer, the **troposphere,** is where weather occurs—it contains 90% of the atmosphere's mass and almost all of its water vapor and clouds. The troposphere is thin and extends to a height of 16 km over the equatorial region and 8 km over the polar regions. Commercial jets generally fly at the top of the troposphere to minimize the bumpiness caused by weather disturbances. Temperature in the troposphere decreases steadily (at about 6°C per kilometer) with increasing altitude. At the top of the troposphere, temperature averages a freezing −50°C.

Above the troposphere is the **stratosphere,** which reaches a height of 50 km above the ground. Ozone molecules (O_3) form in the stratosphere and absorb ultraviolet (UV) radiation from the Sun. The absorption of UV radiation by the ozone layer causes temperature to rise from about −50°C at the bottom to about 0°C at the top.

Above the stratosphere, the **mesosphere** extends upward to about 80 km. The gases that make up the mesosphere absorb very little of the Sun's radiation. As a result, temperature decreases again from about 0°C at the bottom of the layer to about −90°C at the top.

The density of air molecules is greatest at Earth's surface. Air density gradually decreases with height.

✓ **READING CHECK**

In which atmospheric layer does weather occur?

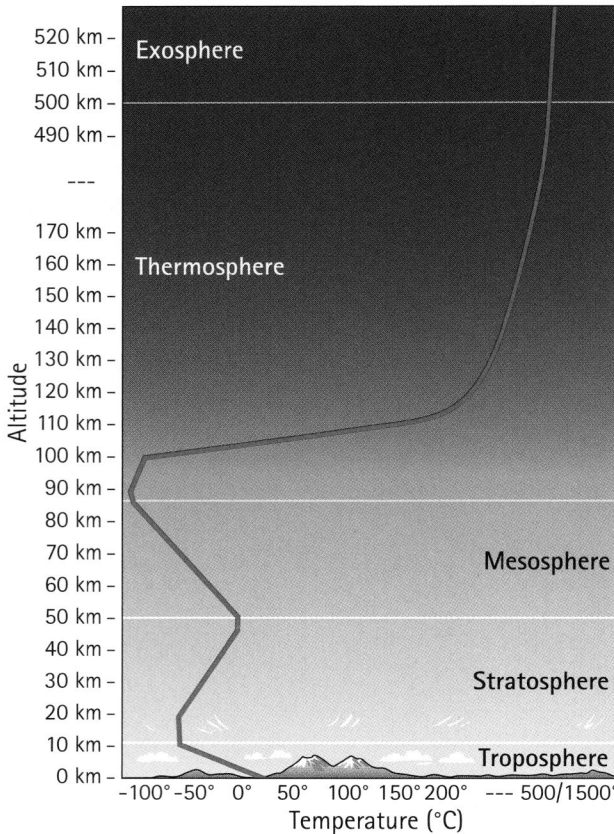

FIGURE 30.4 ▲
The atmospheric layers. The average temperature of the atmosphere varies in a zigzag pattern with altitude.

The situation is just the opposite in the layer above the mesosphere, the **thermosphere.** Here, temperature generally increases with altitude. But because this layer contains very little air, the extreme temperature has little significance. In fact, it would actually be quite chilly if you could visit the thermosphere.

The **ionosphere** is an ion-rich region within the thermosphere and uppermost mesosphere. The ions are produced by the interaction of high-frequency solar radiation with atoms of atmospheric gases. Incoming solar rays strip electrons from nitrogen and oxygen atoms, producing a large concentration of free electrons and positively charged ions in this layer. The degree of ionization depends on air density and on the amount of solar radiation. Ionization is greatest in the upper part of the ionosphere, where air density is low and solar radiation is high.

Ions in the ionosphere cast a faint glow that prevents moonless nights from becoming stark black. Near Earth's magnetic poles, fiery light displays called *auroras* occur as the solar wind (high-speed charged particles ejected by the Sun) strikes and excites molecules of atmospheric gases in the ionosphere (Figure 30.5). These auroral displays are particularly spectacular during times of solar flares—storms or eruptions of hot gases on the Sun.

FIGURE 30.5 ▶
The *aurora borealis* over Alaska is created by solar-charged particles that strike the upper atmosphere and light up the sky (just as similar particles on a smaller scale light up a fluorescent lamp).

the warming of Earth's lower atmosphere by about 1.4°F (0.7°C) since 1880 (Figure 30.9). Concentrations of other greenhouse gases—carbon dioxide (CO_2), methane (CH_4), nitrous oxide (N_2O), ozone (O_3), and chlorofluorocarbons (CFCs)—are also on the rise. And the increase is due to human activity. As the concentrations of these gases increase, the atmosphere's ability to absorb and trap reradiated terrestrial heat energy also increases. The result is *climate change*.

The effects of climate change are not fully known. One consequence that is now occurring is the melting of the polar ice caps on Antarctica and Greenland. If this trend continues, sea level will rise and low-lying coastal lands will be flooded. Warming will also change rainfall patterns, and this will seriously affect agriculture. The grain-growing regions of North America and Asia might shift northward as local climates warm and growing seasons lengthen. On the other hand, deserts in continental interiors might spread to cover much larger areas. We don't know. We do know that Earth has experienced warmer and colder periods in times past and that global-scale climatic changes may have contributed to many of the extinctions discussed in Chapter 29. Ongoing research is becoming more focused on understanding the current impacts of this present-day climate change.

CONCEPT CHECK

1. What does it mean to say the greenhouse effect is like a "one-way valve"?
2. Which gas in the atmosphere is the greatest contributor to the greenhouse effect?
3. What is the primary "natural" contributor to greenhouse gases in Earth's atmosphere?

Check Your Answers

1. The transparent material—atmosphere for Earth and glass for the florist's green-house—passes only incoming short waves and blocks outgoing long waves. In other words, radiation travels only one way.
2. Water vapor.
3. Volcanic eruptions.

30.4 Driving Forces of Air Motion

As warm air rises, it expands and cools. As the air rises, cooler air sinks to occupy the region left vacant by the rising warm air. Such motion constitutes a convection cycle and thermal circulation of the air—in other words, a *convection current*. As convection currents stir the atmosphere, the result is *wind*—defined as air with an average horizontal motion. ✔ Wind is generated in response to pressure differences in the atmosphere, which are largely the result of temperature differences. A difference in pressure between two different locations is called a *pressure gradient,* and the force that causes air to move is the **pressure-gradient force.** The pressure-gradient force drives air from areas of high pressure toward areas of lower pressure.

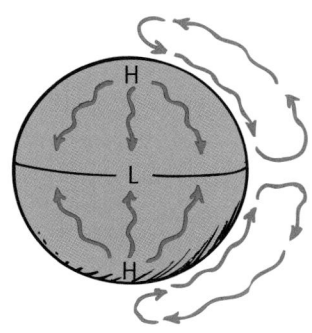

FIGURE 30.10 ▲
If Earth were simply a nonrotating sphere, air circulation would be restricted to a single Northern Hemisphere cell and a single Southern Hemisphere cell. In each cell, heated air would rise at the equator and move toward the polar regions, where it would cool, sink, and be drawn back to the warmer regions of the equator.

PhysicsPlace.com

Tutorial
The Coriolis Effect

The underlying cause of general air circulation is the unequal heating of Earth's surface. On a global level, equatorial regions receive optimum radiant energy from the Sun and as a result have higher average temperatures than other regions. As air heated by the hot ground or ocean at the equator rises, it moves toward the polar regions, cooling gradually in the upper atmosphere. This cooled air then sinks at the poles and is drawn back to the warmer regions of the equator. If we assume Earth to be a nonrotating sphere, the effect is one, simple, single-cell circulation pattern in the Northern Hemisphere and another in the Southern Hemisphere, as shown in Figure 30.10.

But Earth rotates, which greatly affects the path of moving air. Think of Earth as a large merry-go-round rotating in a counterclockwise direction (the same direction Earth spins, as viewed from above the North Pole). Imagine that you and a friend are playing catch on this merry-go-round. When you throw the ball to your friend, the circular movement of the merry-go-round affects the direction in which your friend sees the ball travel. Although the ball travels in a straight-line path, it appears to curve to the right, as shown in Figure 30.11. (The ball travels straight but your friend never catches it, because the movement of the merry-go-round causes her position to change.) This apparent curving is similar to what happens on Earth. As Earth spins, all free-moving objects—air and water, aircraft and ballistic missiles, and even snowballs, to a small extent—appear to deviate from their straight-line paths as Earth rotates under them. This deflection due to the rotation of Earth is called the **Coriolis force.**

Because of the Coriolis force, winds in the Northern Hemisphere deflect to the right and winds in the Southern Hemisphere deflect to the left (Figure 30.12). The Coriolis force varies with wind speed. The

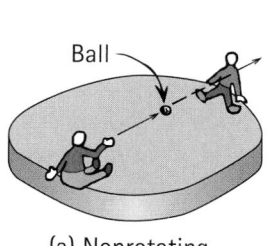

(a) Nonrotating   (b) Rotating

FIGURE 30.11 ▲
(a) On the nonrotating merry-go-round, a thrown ball travels in a straight line. (b) On the counterclockwise-rotating merry-go-round, the ball also moves in a straight line. However, because the merry-go-round is rotating, the ball appears to deflect to the right of its intended path.

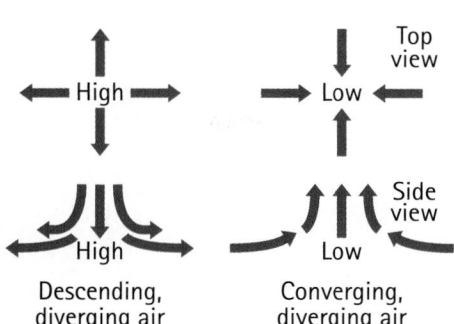

Descending, diverging air

Converging, diverging air

(a) Pressure-gradient force

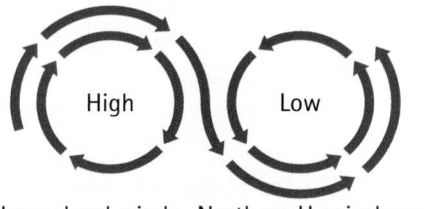

Upper-level winds—Northern Hemisphere

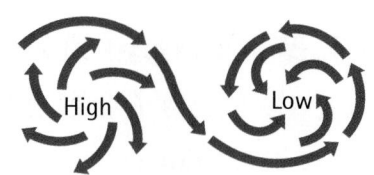

Surface winds—Northern Hemisphere

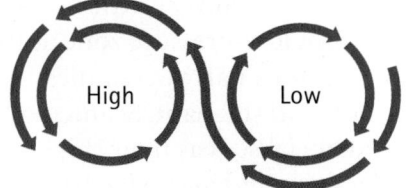

Upper-level winds—Southern Hemisphere

(b) Coriolis effect

Surface winds—Southern Hemisphere

(c) Frictional force at the ground

FIGURE 30.12 ▲
The Coriolis force—the apparent deflection of winds from straight-line paths by Earth's rotation—is a principal force in the production of wind. It is, however, not the only force. (a) Air moves because of pressure differences—the pressure-gradient force. (b) Once the air is moving, it is affected by Earth's rotation—the Coriolis force. (c) As air moves close to the ground, it slows because of frictional force.

faster the wind, the greater the deflection. Latitude also influences the degree of deflection. Deflection is greatest at the poles and decreases to zero at the equator. As Figure 30.13 shows, the Coriolis force has a significant impact on atmospheric motion—and airplanes—in the midlatitudes.

Air moving close to Earth's surface also encounters a *frictional force.* The rougher the surface, the greater the friction, and so the greater the drag on the wind. Because surface friction reduces wind speed, it also reduces the effect of the Coriolis force. This causes winds in the Northern Hemisphere to spiral out clockwise from a high-pressure region and spiral counterclockwise into a low-pressure region (top part of Figure 30.12c). In the Southern Hemisphere, these circulation patterns are reversed (bottom part of Figure 30.12c).

READING CHECK

What starts the wind blowing?

FIGURE 30.13 ▶
Latitude influences the apparent deflection resulting from the Coriolis force. A free-moving object heading east (or west) appears to deviate from its straight-line path as Earth rotates beneath it. Deflection is greatest at the poles and decreases to zero at the equator.

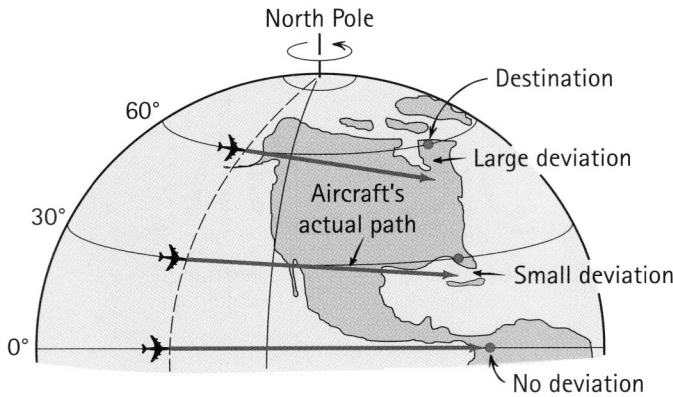

30.5 Global Atmospheric Circulation Patterns

Cell-like circulation patterns are responsible for the redistribution of heat across Earth's surface and for global winds (Figure 30.14). At the equator, warmed air flows straight up with very little horizontal movement, resulting in a vast low-pressure zone at the surface. This rising motion creates a narrow, windless realm of air that is still, hot, and stagnant. Seamen of long ago cursed the equatorial seas as their ships floated listlessly for lack of wind and referred to the area as the *doldrums*. When the moist air from the doldrums rises, it cools and releases torrents of rain. When over land areas, these frequent rains give rise to the tropical rain forests that characterize the equatorial region.

The air of the sweltering doldrums rises to the boundary between the troposphere and stratosphere, where it divides and spreads out to the north and south. (Very little wind crosses the equator into the neighboring hemisphere.) By the time it has reached about 30° N and 30° S latitudes, this air has cooled enough to descend toward the surface. The descending air warms as it is compressed. A resulting high-pressure zone girdles Earth, creating a belt of hot, dry surface air. On land, these high-pressure zones account for the world's great deserts—the Sahara in Africa, the Arabian Desert in the Middle East, the Mojave Desert in the United States, and the Great Victoria Desert in Australia. At sea, the hot, descending air produces very weak winds. According to legend, early sailing ships were frequently stalled at these latitudes, both north and south. As food and water supplies dwindled, horses on board were either eaten or cast overboard to conserve fresh water and reduce the load of the ship. As a result, this region is now known as the *horse latitudes*.

The thermal convection cycle that starts at the equator is completed when air flowing southward from the horse latitudes in the Northern Hemisphere and northward in the Southern Hemisphere is deflected westward by the Coriolis force to produce the *trade winds*.

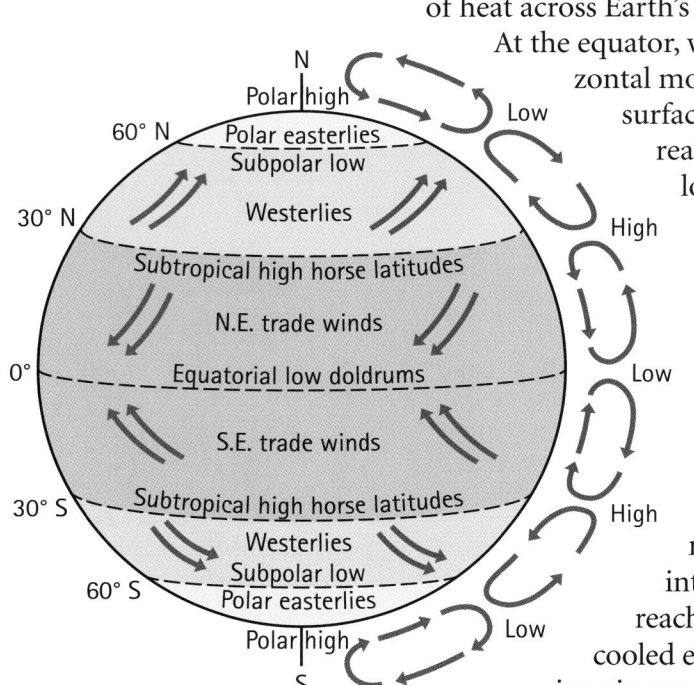

FIGURE 30.14 ▲
Global circulation of the atmosphere results from a combination of two main factors: unequal heating of Earth's surface (which sets up convection cells) and Earth's rotation. The atmosphere has six cell-like circulation patterns; prevailing winds blow in the directions indicated by the arrows. The major prevailing winds are the *westerlies*, the *easterlies*, and the *trade winds*.

Air that flows northward from the horse latitudes in the Northern Hemisphere and southward in the Southern Hemisphere is deflected eastward to produce the prevailing *westerlies.*

Near the poles, frigid air continuously sinks, pushing the surface air outward. The Coriolis force is quite evident in the polar regions, as the wind deflects to the west to create the *polar easterlies* (Figure 30.14). The cool, dry polar air meets the warm, moist air of the westerlies at latitudes 60° N and 60° S. This boundary, called the *polar front,* is a zone of low pressure where contrasting air masses converge, often resulting in storms.

The middle latitudes are noted for their unpredictable weather. Although the winds tend to be from the west, they are often quite changeable, as the temperature and pressure differences between the subtropical and polar air masses at the polar front produce powerful winds. As air moves from regions of high pressure, where air is denser, toward regions of low pressure, the result is a whirlwind effect.

Irregularities in Earth's surface also influence wind behavior. Mountains, valleys, deserts, forests, and great bodies of water all play a part in determining which way the wind blows.

Upper Atmospheric Circulation

In the upper troposphere, "rivers" of rapidly moving air meander around Earth at altitudes of 9–14 km. These high-speed winds are the *jet streams.* With wind speeds averaging between 95 and 190 km/h, the jet streams play an essential role in the global transfer of heat from the equator to the poles.

The two most important jet streams, the *polar jet stream* and the *subtropical jet stream,* form in both the Northern and Southern Hemispheres. ✓ The formation of polar jet streams is a result of a temperature gradient at the polar front—at about 60° N and 60° S latitude—where cool polar air meets warm tropical air. This temperature gradient causes a steep pressure gradient that increases the wind speed. During the winter, the polar jet is strong and migrates to lower latitudes, bringing strong winter storms and blizzards to the United States. In summer, the polar jet stream is weaker and migrates to higher latitudes. The subtropical jet stream is generated by warm air carried from the equator to the poles. This produces a sharp temperature gradient along the subtropical front— about 30° N and 30° S latitude. Once again, a pressure gradient caused by the temperature gradient generates strong winds.

The subtropical jet stream above Southeast Asia, India, and Africa merits special mention (Figure 30.15). The formation of this jet stream is related to the warming of air above the Tibetan highlands. During the summer, the air above the continental highlands is warmer than the air above the ocean to the south. The warmer air rises, drawing in cooler, moist air from over the ocean. Thus, temperature and pressure gradients generate strong onshore winds that contribute to the region's *monsoon* climate. During winter, the winds change direction to produce a dry season. This cycle of winds characterizes the climates of much of Southeast Asia. The predictable rain-bearing summer wind from the sea that moves over the heated land is called the *summer monsoon;* the prevailing wind from land to sea in winter is called the *winter monsoon.*

Air currents are named for the direction the wind is coming from. So the westerlies blow from the west—but air moves to the east.

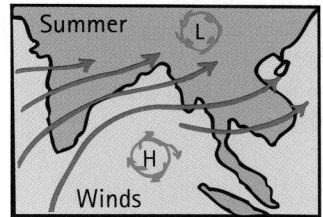

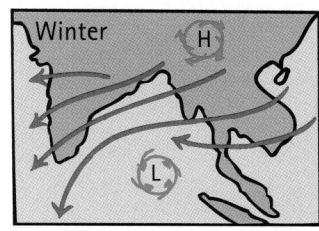

FIGURE 30.15 ▲
Winds over Southeast Asia. (a) During the summer months, air over the oceans is cooler than the air over land. The summer monsoon brings heavy rains as the winds blow from sea to land. (b) During the winter months, air over continents is cooler than air over oceans. The winter monsoon generally has clear skies and winds that blow from land to sea.

730 PART FOUR Earth Science

READING CHECK

What is the underlying
cause of jet stream
formation?

CONCEPT CHECK

1. What are the underlying causes of the trade winds, jet streams,
   and monsoons?
2. In the midlatitudes, airlines schedule shorter flight times for
   planes traveling west to east and longer flight times for planes
   traveling east to west. Why are eastbound planes faster?

Check Your Answers

1. Simply enough, the unequal heating of Earth's surface coupled with Earth's rotation.
2. The upper-level westerly moving winds of the jet stream account for faster-moving
   eastbound aircraft. As the jet stream moves from west to east it carries along everything in its path. To save time and fuel, air pilots seek the jet stream when traveling
   west to east and avoid it when traveling east to west.

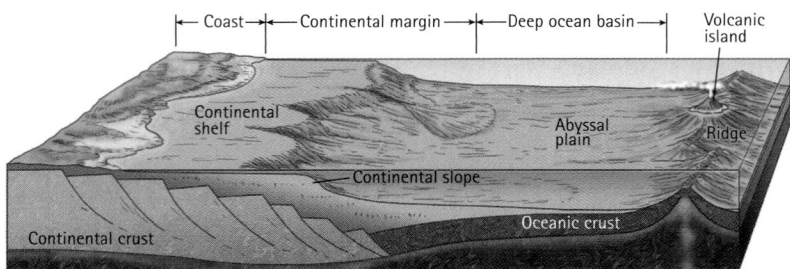

FIGURE 30.16 ▲
Profile of the continental
margin going from land to
the deep-ocean bottom.

30.6 Components of Earth's Oceans

The world ocean is the dominant feature
of our planet. If we could drain the water
from Earth's oceans, we'd find a varied
landscape. First, we'd encounter the
continental margin—the boundary
between the continents and the oceans (Figure 30.16). Next we would
find sediment-covered deep ocean basins, mid-ocean ridges that encircle
the globe, and deep seafloor trenches near some of the continental borders
(Figure 30.17).

The ocean is an immense body of salty water. Its salty nature
contributes to its variable density, which contributes to the movement
of ocean currents. Waves move through the ocean in response to atmospheric movements, but the ocean also responds to extraterrestrial influences—the Moon and the Sun, which create tides. With so much of Earth's
surface covered by oceans, the oceans hold a wealth of scientific treasures.
And because it is so vast, we still have much to discover—we know more
about the surface of the Moon than we know about the ocean floor!

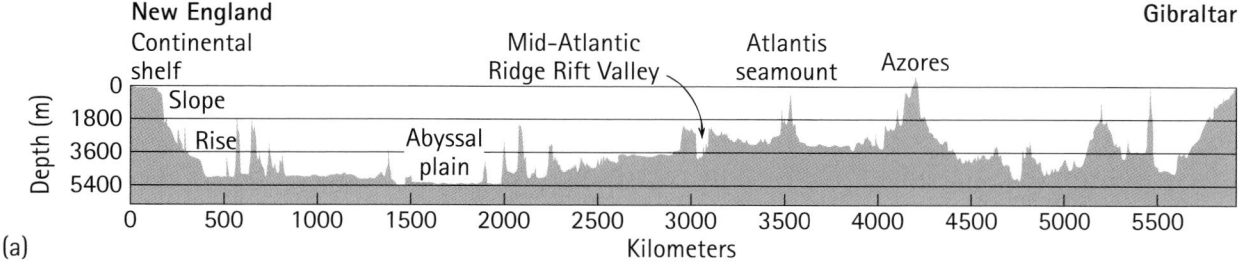

FIGURE 30.17 ▲
Map of the ocean floor showing variation in topography. (a) Atlantic profile.
(b) Pacific profile.

TABLE 30.2 Principal Elements in Sea Salts

Element	Chemical Symbol	Percentage by Weight
Chloride	Cl^-	55.07
Sodium	Na^+	30.62
Sulfate	SO_4^-	7.72
Magnesium	Mg^{2+}	3.68
Calcium	Ca^{2+}	1.17
Potassium	K^+	1.10
Total		99.36

Seawater

Have you ever wondered why seawater is salty? Seawater is a complex solution of dissolved minerals, dissolved gases, and decomposing biological material. Just about every natural compound can be found in the ocean at some concentration. But overall, the composition of seawater is surprisingly simple—only a few elements and compounds are present in abundance. Chloride, sodium, sulfate, magnesium, calcium, and potassium make up more than 99% of the salts in the sea (Table 30.2). ✓ Hmmm, think about the combination of chloride and sodium—NaCl. Ahh! Common table salt (halite), that's why seawater tastes salty!

The amount of dissolved salts in seawater is measured as **salinity**—the mass of salts dissolved in 1000 grams (g) of seawater. Salinity varies from one part of the ocean to another, but the overall composition of seawater is the same from place to place—a mixture of 96.5% water and 3.5% salt. Salinity variation is influenced by factors that increase or decrease supplies of freshwater (Figure 30.18). Freshwater enters the ocean in three ways: runoff from streams and rivers, precipitation, and melting of glacial and sea ice. Freshwater leaves the ocean in two ways: evaporation and the formation of sea ice. Evaporation increases salinity because only pure water vapor leaves the seawater solution—the salts are left behind. And when sea ice forms, only the water molecules freeze. Salts are once again left behind in solution. Overall balance is maintained when evaporation is offset by precipitation and runoff and ice formation is offset by ice melting.

✓ READING CHECK

Why is seawater salty?

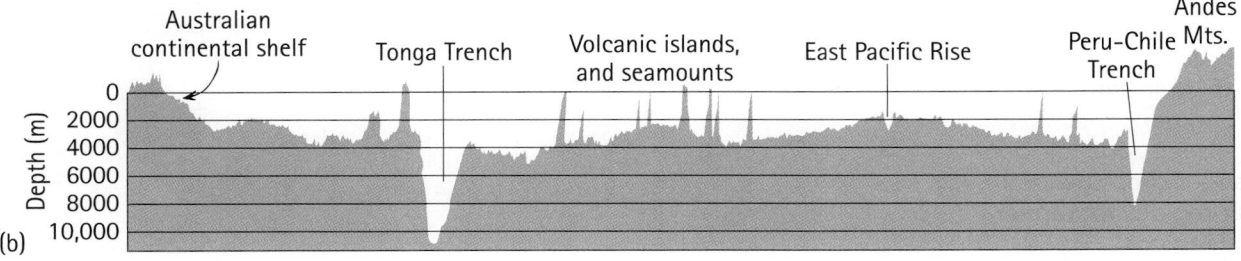

(b)

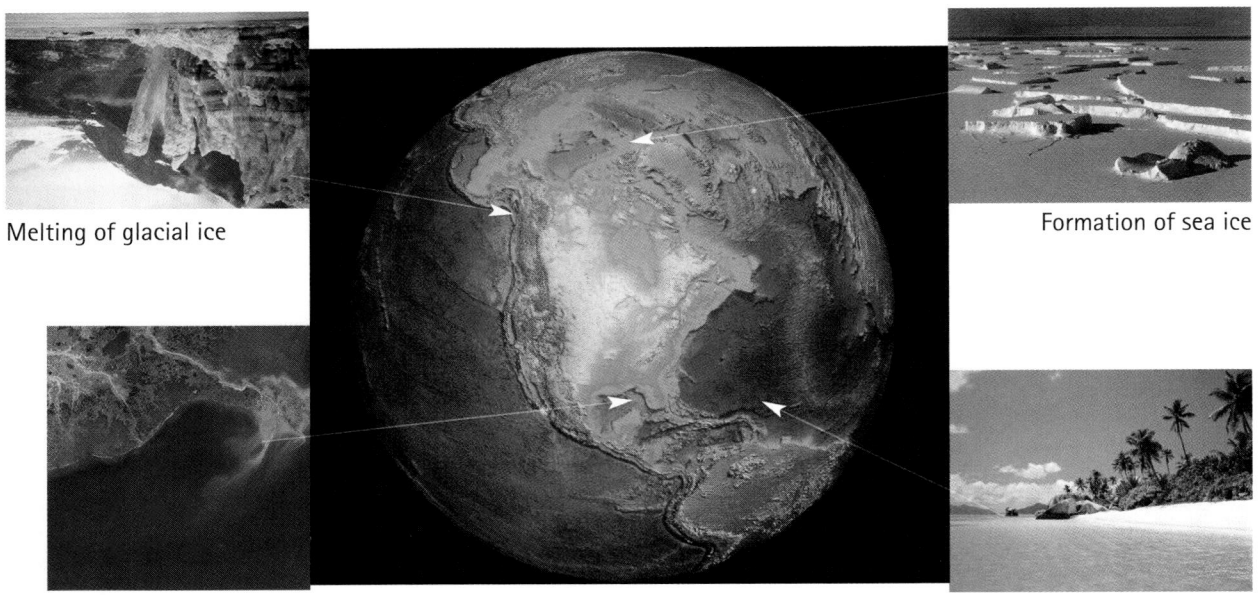

Melting of glacial ice

Formation of sea ice

Runoff

Evaporation

FIGURE 30.18 ▲
Salinity increases as the supply of freshwater decreases. Factors that increase salinity include formation of sea ice and evaporation. Salinity decreases as the supply of freshwater increases. Factors that decrease salinity include the runoff from streams and rivers, precipitation, and the melting of glacial ice, sea ice, and icebergs.

Vertical Structure of Ocean Water

Like the atmosphere, the ocean can be divided into several vertical layers—the surface zone, a transition zone, and the deep zone (Figure 30.19). Scuba divers notice an increase in water pressure when swimming to lower depths. The deeper you descend, the greater the water pressure. The pressure is simply the weight of the water above pushing down on you. Another factor that changes as you descend is temperature. Deeper waters are cooler. So, in addition to variations in salinity, seawater also varies in temperature and pressure. Because cold water is denser than warm water, cold seawater sinks below warmer seawater. Salinity also affects density: The greater the salinity, the greater the density (Figure 30.19).

FIGURE 30.19 ▶
The ocean's vertical structure. In the surface zone, water moves in response to temperature and density changes, and in response to wind. Water in the transitition zone moves along density surfaces and in the deep zone it is density driven—it circulates from cold polar regions to warmer equatorial regions.

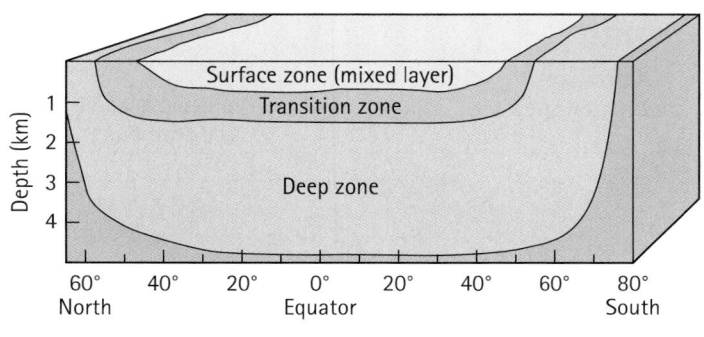

to move westward into the Gulf of Mexico and once again become part of the Gulf Stream.

Oceanic circulation in the North Pacific is similar to that in the North Atlantic. The Pacific counterpart of the Gulf Stream is the warm, northward-flowing current known as the *Kuroshio*. In the Southern Hemisphere, surface oceanic circulation (with the exception of the Antarctica Circumpolar Current) is similar except that the gyres move counterclockwise.

Nature, a never-ending series of cycles!

Deep-Water Currents

✔ Surface waters are driven by winds, but deeper waters are driven by gravity. In essence, deep water flows because dense water sinks. Although deep water flows more slowly than surface water, the volume of deep-water flow is like a large global conveyor belt (Figure 30.24).

In the high latitudes, where seawater in the deep zone interacts with seawater in the surface zone, a very slow, worldwide, north–south circulation pattern develops. To understand how this pattern comes about, we need to look at what happens when seawater begins to freeze.

First of all, seawater does not freeze easily. When it does, however, only the water freezes, and the salt is left behind. Thus, the seawater that does not freeze experiences an increase in salinity, which in turn causes an increase in density. The cold, denser, saltier seawater sinks, which sets up a pattern of vertical movement. There is also horizontal movement as the dense water that sinks in the polar regions flows along the bottom to the deeper parts of the ocean floor.

Thus conveyor-belt circulation begins in the North Atlantic as dense, cold, salty seawater around Greenland and Iceland sinks and flows along the ocean bottom toward the equator then onto the Antarctic Ocean. Once near Antarctica, the water flows eastward around the continent, then northward into the Pacific and Indian Oceans. Thus deep-water currents flow in an overall north–south circulation pattern.

✔ **READING CHECK**

How is the movement of deeper waters different from movement of surface waters?

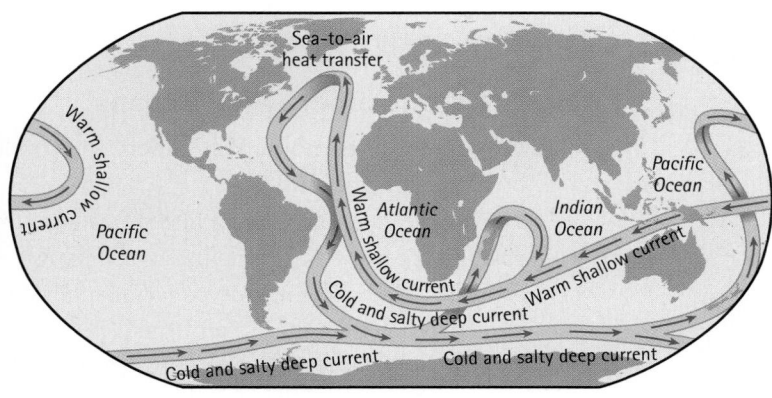

◀ **FIGURE 30.24**
Deep-ocean currents act like a conveyor belt, transporting cold water from the North Atlantic past the equator and on to the Antarctic. From the Antarctic, water flows eastward and then northward into the Pacific Ocean and the Indian Ocean.

KEY TERMS

Continental margin The boundary between the continents and the oceans.

Coriolis force The apparent deflection from a straight-line path observed in any body moving near Earth's surface, caused by Earth's rotation.

Exosphere The fifth atmospheric layer above Earth's surface, extending from the thermosphere upward and out into interplanetary space.

Greenhouse effect Warming caused by short-wavelength radiant energy from the Sun that easily enters the atmosphere and is absorbed by Earth. This energy is then reradiated at longer wavelengths that cannot easily escape Earth's atmosphere.

Gyre A circular or spiral whirl pattern, usually referring to very large current systems in the open ocean.

Ionosphere An electrified region within the thermosphere and uppermost mesosphere where fairly large concentrations of ions and free electrons exist.

Mesosphere The third atmospheric layer above Earth's surface, extending from the top of the stratosphere to 80 km.

Neap tide A tide that occurs when the Moon is midway between new and full, in either direction. The pulls of the Moon and Sun are perpendicular to one another, so the solar and lunar tides do not overlap. This makes high tides not as high and low tides not as low.

Pressure-gradient force The force that moves air from a region of high-pressure air to an adjacent region of low-pressure air.

Salinity The mass of salts dissolved in 1000 g of seawater.

Spring tide A high or low tide that occurs when the Sun, Earth, and Moon are aligned so that the tides due to the Sun and Moon coincide, making the tides higher or lower than average. Occurs during the full Moon or new Moon.

Stratosphere The second atmospheric layer above Earth's surface, extending from the top of the troposphere up to 50 km. This is where stratospheric ozone forms.

Thermosphere The fourth atmospheric layer above Earth's surface, extending from the top of the mesosphere to 500 km.

Troposphere The atmospheric layer closest to Earth's surface. This layer contains 90% of the atmosphere's mass and essentially all of its water vapor and clouds. Weather occurs in the troposphere.

REVIEW QUESTIONS

Earth's Atmosphere and Oceans

1. Why do we have moderate temperatures on lands bordering the oceans?
2. Earth's present atmosphere likely developed from gases that escaped from its interior during volcanic eruptions. What two principal atmospheric gases were produced by these eruptions?

Components of Earth's Atmosphere

3. What elements make up today's atmosphere?
4. Why doesn't gravity flatten the atmosphere against Earth's surface?

5. In which atmospheric layer does all our weather occur?

6. Does temperature increase or decrease as one moves upward in the troposphere? As one moves upward in the stratosphere?

Solar Energy

7. What does the angle at which sunlight strikes Earth have to do with the temperate and polar regions?

8. What does Earth's tilt have to do with the change of seasons?

9. How does radiation emitted from Earth differ from that emitted by the Sun?

10. How is the atmosphere near Earth's surface heated from below?

Driving Forces of Air Motion

11. What is the underlying cause of air motion?

12. What causes pressure differences to arise, and hence causes the wind to blow?

13. In what direction does Earth spin—west to east or east to west?

14. What does the Coriolis force do to winds? To ocean currents?

Global Atmospheric Circulation Patterns

15. Why are most of the world's deserts found in the area known as the horse latitudes?

16. Why are eastbound aircraft flights usually faster than westbound ones?

Components of Earth's Oceans

17. Salinity of the ocean varies from one place to another. What two factors lead to an increase in salinity? What two factors lead to a decrease in salinity?

Oceanic Circulation

18. Why do waves become taller as they enter shallow water?

19. Why are all tides greatest at the time of a full or new Moon?

20. When would the highest high tides occur: during a spring tide or during a neap tide?

THINK AND EXPLAIN

1. Earth is closest to the Sun in January, but January is cold in the Northern Hemisphere. Why?

2. If the composition of the upper atmosphere were changed so that it permitted a greater amount of terrestrial radiation to escape, what effect would this have on Earth's average temperature?

3. If the composition of the upper atmosphere were changed so that less terrestrial radiation could escape, what effect would this have on Earth's average temperature?

4. Why is it important that mountain climbers wear sunglasses and use sunblock even when the temperature is below freezing?

5. Relate the jet stream to upper-air circulation. How does this circulation pattern relate to airline schedules from New York to San Francisco and the return trip to New York?

6. What are the jet streams and how do they form?

7. How do the total number of hours of sunlight in a year compare for equatorial regions and polar regions of Earth? Why are polar regions so much colder?

8. Why are temperature fluctuations greater over land than over water? Explain.

9. Because seawater does not freeze easily, sea ice never gets very thick. So from where do large icebergs originate?

10. The Mediterranean Sea is very salty. Does evaporation exceed precipitation, or does precipitation exceed evaporation over the Mediterranean?

11. How does the ocean influence weather on land?

12. What effect does the formation of sea ice in polar regions have on the density of seawater? Explain.

13. Why is there more concern about the melting of polar ice caps than there is about melting icebergs?

14. As ocean waves approach shallow water, those with longer wavelengths slow down before those with shorter wavelengths do. Why?

15. With respect to spring and neap ocean tides, when are the lowest tides? That is, when is it best for digging clams?

16. If Earth were not spinning, in what direction would the surface winds blow where you live? In what direction does it actually blow on Earth at 15° S latitude and why?

17. Why is the thermosphere so much hotter than the mesosphere?

18. Explain the circulation pattern of the Gulf Stream.
19. If it is winter and January in Chicago, what are the corresponding season and month in Sydney, Australia?
20. What causes the fiery displays of light called the auroras?

READINESS ASSURANCE TEST (RAT)

If you have a good handle on this chapter—if you really do—then you should be able to score 7 out of 10 on this RAT. Check your answers with your teacher. If you score less than 7, study further before moving on.

Choose the best answer to each of the following.

1. Earth's lower atmosphere is kept warm by
   (a) solar radiation.
   (b) terrestrial radiation.
   (c) short wave radiation.
   (d) the ozone layer.

2. When compared to lands far from the oceans, lands that border the oceans tend to have
   (a) extreme seasonal variations.
   (b) small seasonal variations.
   (c) wet and cold winters.
   (d) dry, hot summers.

3. Which pulls with the greater *force* on Earth's oceans?
   (a) The Sun.
   (b) The Moon.
   (c) Both the Sun and the Moon; they have equal force.
   (d) There is no force.

4. Air motion is greatly influenced by
   (a) pressure differences.
   (b) temperature differences.
   (c) the Coriolis force.
   (d) All of the above.

5. Ocean tides are caused by the differences in the
   (a) gravitational pull of the Sun on opposite sides of Earth.
   (b) force of the Moon.
   (c) gravitational pull of the Moon on opposite sides of Earth.
   (d) distance of the Sun from the Moon.

6. A factor that increases ocean salinity is
   (a) runoff from streams and rivers.
   (b) formation of sea ice.
   (c) precipitation.
   (d) glacial melting.

7. The wind blows in response to
   (a) frictional drag.
   (b) Earth's rotation.
   (c) pressure differences.
   (d) moisture differences.

8. Earth experiences changes of the seasons because of
   (a) incoming solar radiation.
   (b) Earth's rotation around the Sun.
   (c) the Coriolis force.
   (d) the tilt of Earth's axis.

9. The most significant result of the Coriolis force is
   (a) the deflection of air and water currents.
   (b) the creation of jet stream transport.
   (c) the reflection of air currents and global air circulation.
   (d) the creation of the jet streams.

10. The ultimate cause of ocean surface currents is
    (a) divergence in equatorial regions.
    (b) the gradient between the doldrums and the horse latitudes.
    (c) density contrasts.
    (d) frictional drag by prevailing winds.

31

WEATHER

THE MAIN IDEA

The weather depends on several factors—temperature, pressure, density, and moisture content. The factors are interrelated. Change in one factor impacts change in another factor.

We've all seen weather reports on the news. But how do meteorologists figure out whether it will rain or not? What causes air to warm up? What makes air cool down? Why do deserts occur on one side of a tall mountain range? Which side is the desert on? And what do most people talk about in idle conversation? That's right, what this chapter is about—the weather!

Explore!

What Happens When Water Vapor in a Can Suddenly Condenses?

1. Put a small amount of water in an aluminum soft-drink can. Heat it on a stove or heater until steam comes out of the opening.
2. With a pair of tongs, invert the can into a pan of water.
3. Be prepared to see atmospheric pressure suddenly crush the can.

Analyze and Conclude

1. **Observing** Steam coming from the opening is evidence that air has been driven from the can and only water-vapor molecules remain inside.

2. **Predicting** When the can with vapor is inverted into the pan of water, the water vapor inside encounters the surface of water in the pan. Do water molecules bounce off the water, or do they stick to it?
3. **Making Generalizations** When vapor sticks to a liquid, the process is called *condensation*. The condensation of molecules in the can leave behind a vacuum. The pressure of the atmosphere, with nothing inside pushing back, crushes the can.

31.1 Water in the Atmosphere

Water is vital to life on Earth. But think for a moment about water's role in the physical processes of Earth—water shapes Earth's surface and governs its weather. Did you know that no matter how "dry" the air may feel at times, it is never completely dry? There is always some water vapor in the air. **Humidity** is a measure of the water vapor in the air. Specifically, humidity is the mass of water per volume of air.

When you hear TV weather forecasters describing humidity, they are actually talking about **relative humidity.** Relative humidity depends on temperature—it describes the amount of water vapor currently in the air compared to the maximum amount of water vapor the air can hold at that temperature. The maximum amount varies depending on the temperature. For example, a relative humidity of 50% means that the water-vapor content is half the amount the air can hold at that temperature.

When air contains as much water vapor as it can possibly hold, the air is *saturated.* Saturation occurs when the air temperature drops and water-vapor molecules in the air begin to condense to form droplets. Because slower-moving molecules characterize lower air temperatures, saturation and condensation are more likely to occur in cool air than in warm air (Figure 31.1). As such, warm air can hold more water vapor than cold air.

✓ As air rises, it cools and expands. The expansion occurs because air moves to a region of lower air pressure. As the air cools, water molecules move slower and condensation occurs. When microscopic particles of dust, smoke, and salt are in the air, they can act as *cloud condensation nuclei.* When water vapor condenses on these particles, a cloud forms. As the size of the cloud droplets grow, they fall to Earth and we have rain.

When the relative humidity is high, hot weather feels hotter, and cold weather feels colder.

fyi

■ In rainy weather, when your car windshield fogs up, make sure the air conditioner is on when in defrost mode, even when blowing heated air. What causes window fogging is the humidity in the car caused by rain, by wet clothes, and by the breath of the passengers. Since the air from the air conditioner is very dry, it clears a foggy windshield very nicely in a very short time.

Is fog a low-altitude cloud, or is a cloud a high-altitude fog?

Fast-moving H$_2$O molecules rebound upon collision

Slow-moving H$_2$O molecules condense upon collision

FIGURE 31.1 ▲
Condensation of water molecules.

CALCULATION CORNER: HUMIDITY

Humidity is the mass of water vapor per volume of air. Relative humidity is the ratio of the air's water vapor content compared with the air's water vapor capacity at a certain temperature.

For these three problems, consider a small, experimental air mass at 30°C that weighs 90 N.

Sample Problem 1

If the air density is 1.25 kg/m^3, what is the volume of the air mass?

Solution:

Newton's second law tells us that

$$a = \frac{F}{m}$$

Rearrange the equation and get

$$\text{Mass} = \frac{\text{force}}{\text{acceleration due to gravity}}$$
$$= \frac{90 \text{ N}}{10 \text{m/s}^2} = 9 \text{ kg}$$

The volume of 9 kg can be found in this way:

$$9 \text{ kg} \times \frac{1 \text{ m}^3}{1.25 \text{ kg}} = 7.2 \text{ m}^3$$

Sample Problem 2

If there is 0.13 kg of water vapor in the air mass, what is the humidity of the air mass?

Solution:

$$\text{Humidity} = \frac{\text{mass of water}}{\text{volume of air}}$$
$$= \frac{0.13 \text{ kg}}{7.2 \text{ m}^3} = 0.018 \text{ kg/m}^3$$

Sample Problem 3

At 30°C, the maximum amount of water vapor in the air mass is 30 g/m^3 of water vapor. What is the relative humidity of the air mass?

Solution:

First, we must convert the units:

$$\frac{30 \text{ g}}{\text{m}^3} \times \frac{1 \text{ kg}}{1000 \text{ g}} = 0.03 \text{ kg/m}^3$$

Then,

$$\text{Relative humidity} = \frac{\text{amount of water vapor in air}}{\text{maximum amount of water vapor in air at 30°C}}$$
$$= \frac{0.018 \text{kg/m}^3}{0.03 \text{kg/m}^3} \times 100\% = 60\%$$

Rain is one form of precipitation. Other familiar forms of precipitation are mist, hail, snow, and sleet. Precipitation comes from water vapor in the air that condenses to make clouds, which then falls as liquid water or ice.

Water vapor does not need to be high in a cloud to form precipitation; condensation can also occur in air close to the ground. When condensation occurs at or near Earth's surface, we call it *dew, fog,* or *frost.* On cool, clear nights, objects near the ground cool down more rapidly than the surrounding air. As air cools below a certain temperature, called the *dew point,* the air becomes saturated—relative humidity is 100%—and condensation occurs. In this case, cloud condensation nuclei are not needed. Water from the now-saturated air condenses on any available surface. This may be a twig, a blade of grass, or the windshield of a car. We often call this type of condensation *early morning dew,* because it occurs when the daily temperatures are the

FIGURE 31.2 ▶
San Francisco is well known for its summer fog.

coldest, just before sunrise. When the dew point is at or below water's freezing point, we have frost. When a large mass of air cools and reaches its dew point, we get a cloud near the ground—fog.

CONCEPT CHECK
What is the major difference between fog and a cloud?

Check Your Answer
Altitude.

31.2 Weather Variables

✓ Air pressure, temperature, and density are three key variables that control how air behaves. As such, they control the weather. To understand and predict the weather, we must understand all three. First, consider air pressure. Air is a mixture of molecules that move randomly and collide with one another like billiard balls. When a molecule bumps into something, it exerts a small push on whatever it hits. Such pushes, by countless molecules, produce *air pressure.*

The faster the air molecules move, the greater their kinetic energy. The greater the kinetic energy, the greater the impact of molecular collisions—and the greater the air pressure. All else being equal, air composed of fast-moving molecules—warm air—exerts more air pressure on its surroundings than cooler air. So warm air has higher air pressure than cold air.

Another factor that affects air pressure is density. The denser the air, the more molecules present and hence the greater the number of molecular collisions. And more collisions mean greater air pressure. Air becomes denser when it is compressed, and it becomes less dense when it expands. So changes in air density occur when the volume of a given air mass is made smaller by compression or larger by expansion.

Adiabatic Processes in Air

PhysicsPlace.com

Video
Adiabatic Process

The concept of *heat exchange* shows us that air pressure, temperature, and density are interrelated. When heat is added to an air mass, air temperature increases, air pressure increases, or both increase. Heat can be added to air by solar radiation, by moisture condensation, or by contact with warm ground. When heat is subtracted from an air mass, the temperature or the pressure of the air falls. Heat can be subtracted from air by radiation to space, by the evaporation of rain falling through dry air, or by contact with cold surfaces.

But air can change temperature without the loss or gain of heat. When heat transfer is zero, or nearly so, and there is a change in air temperature, we have an **adiabatic** process. Adiabatic processes occur when air is expanded or compressed. For example, when air is suddenly compressed or allowed to expand, its temperature changes as its pressure changes. So the air warms or cools even though no heat exchange takes place. Large bodies of air in the atmosphere often follow adiabatic processes. To illustrate how bodies of air behave, imagine a body of air enclosed in a very thin plastic garment bag—an *air parcel*. Like a free-floating balloon, the parcel can expand and contract freely without heat transfer to or from air outside the parcel.

Recall from Chapter 30 that air pressure decreases with increasing height. For example, as an air parcel flows up the side of a mountain, the air pressure within the parcel decreases. So the air parcel expands and cools without any heat exchange. In general, temperature goes down when pressure goes down—and air pressure goes down when air expands. Conversely, temperature goes up when pressure goes up—and air pressure goes up when air is compressed.

With adiabatic expansion, the temperature of a dry air parcel decreases about 10°C for each kilometer it rises (Figure 31.3). This rate of cooling for dry air is called the *dry adiabatic lapse rate.* Air flowing up and over tall mountains or rising in thunderstorms may change elevation by several kilometers. When a dry air parcel at ground level with a comfortable temperature of 25°C rises 6 kilometers, its temperature drops to a frigid −35°C! On the other hand, if air at 6 kilometers elevation, at a typical temperature of −20°C, descends to the ground, its temperature rises to a whopping 40°C!

A dramatic example of this type of adiabatic warming is the Chinook—a dry wind that blows down from the Rocky Mountains across the Great Plains (Figure 31.4). Cold air moving down a mountain slope is compressed as it moves to lower elevations (where air pressure is greater than at higher elevations), and it becomes much warmer. The effect of expansion or compression of gases is quite impressive.

Adiabatic processes are not restricted to dry air. As rising air cools, its ability to hold water vapor decreases so the relative humidity of the rising air increases. If the air cools to its dew point, the relative humidity climbs to 100%, water vapor condenses, and clouds form. Because condensation releases heat, the surrounding air is warmed. This added

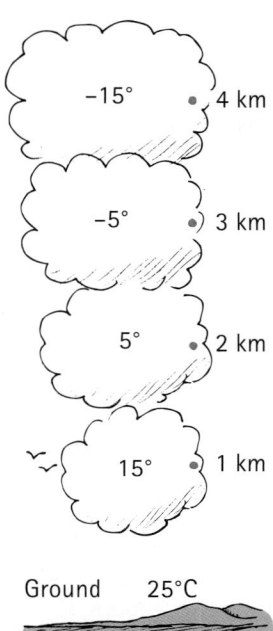

FIGURE 31.3 ▲
The temperature of a parcel of dry air that expands adiabatically changes by about 10°C for each kilometer of elevation.

The temperature of an air parcel decreases when lifted, and air temperature increases when an air parcel descends.

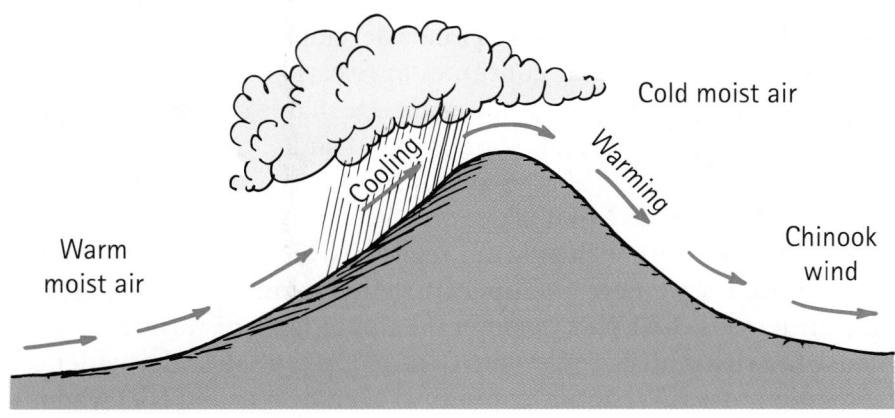

◀ **FIGURE 31.4**
Chinooks—which are warm, dry winds—occur when high-altitude air descends and is adiabatically warmed.

heat offsets the cooling due to expansion, making the air cool at a lesser rate—a *moist adiabatic lapse rate.* On average, a saturated air parcel cools by about 6°C for every kilometer it rises.

CONCEPT CHECK

1. If a parcel of dry air initially at 0°C expands adiabatically while flowing upward alongside a mountain, what is its temperature when it has risen 2 km? 5 km?
2. What happens to the air temperature in a valley when dry, cold air blowing across the mountains descends into the valley?
3. Imagine a gigantic dry-cleaner's garment bag full of dry, −10°C air floating 6 km above the ground like a balloon with a string hanging from it. If you yank it suddenly to the ground, what will its temperature be?

Check Your Answer

1. The air cools at the dry adiabatic rate of 10°C for each kilometer it rises. When the parcel rises to an elevation of 2 km, its temperature is −20°C. At an elevation of 5 km, its temperature is −50°C.
2. The air is adiabatically compressed, and so its temperature increases. Residents of some valley towns in the Rocky Mountains, such as Salida, Colorado, benefit from this adiabatic compression and enjoy "banana belt" weather in midwinter.
3. If the bag of air is pulled down quickly and heat conduction is negligible, the atmosphere adiabatically compresses the air and its temperature rises to a piping-hot 50°C.

Wind is like the air, only pushier.

Atmospheric Stability

Stable air is air that resists upward movement. When a parcel of rising air is cooler than its surroundings, it is also denser than its surroundings. The denser air tends to sink. The two effects—rising and sinking—balance each other and the air is stable. Stable air that is forced to rise spreads out horizontally. When clouds develop in stable air, they too spread out into thin horizontal layers having flat tops and bottoms.

On the other hand, when a rising air parcel is warmer than its surroundings, it is less dense and continues to move upward until its temperature equals the temperature of its surroundings. In this case, the air is unstable and favors upward movement. Rising dry air cools at the dry adiabatic rate while air at the surface warms up. When the rising air is moist, billowy and towering clouds develop.

✓ A rising parcel of air continues to rise as long as it is warmer and less dense than the surrounding air. Air that is cooler and denser than its surroundings does the opposite—it sinks. Under some conditions, large parcels of cold air sink and remain at low elevations. This results in air above that is warmer. When upper regions of the atmosphere are warmer than lower regions, which is opposite of what normally occurs, we have a **temperature inversion.** In such cases, most rising air parcels can't pass through the upper layer of warmer air, because the rising air is cooler and denser. On a small scale, evidence of a temperature inversion is commonly seen over a cold lake when

Stable air resists upward movement. Unstable air favors upward movement.

HANDS-ON-EXPLORATION: ADIABATIC EXPANSION

To see that expanding air cools, blow on your hand with your mouth wide open. Can you feel that your breath is warm? Now do this again, but this time pucker your lips so that your mouth opening is very small. This time your breath expands as it leaves your mouth. Is your breath appreciably cooler? Adiabatic expansion, hooray!

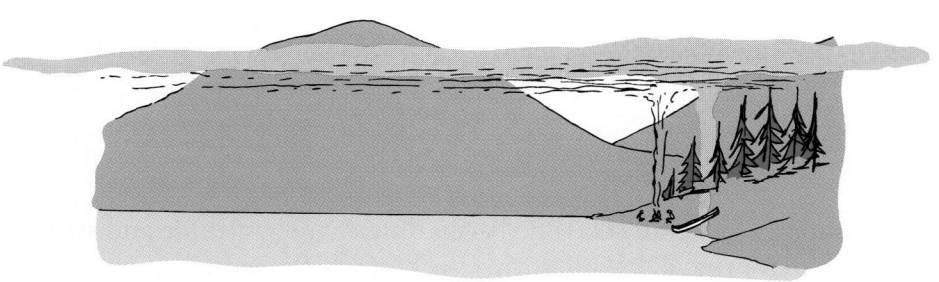

◀ **FIGURE 31.5**
The layer of campfire smoke over the lake indicates a temperature inversion. The air above the smoke is warmer than the smoke, and the air below is cooler.

visible gases and small particles such as smoke spread out in a flat layer above the lake rather than rising and dissipating higher in the atmosphere (Figure 31.5).

The smog of Los Angeles is trapped by such an inversion, caused by cold air from the ocean being capped by a layer of hot air moving westward over the mountains from the hot Mojave Desert. The west-facing side of the mountains helps to confine the trapped air (Figure 31.6). The Rocky Mountains on the western edge of Denver play a similar role in trapping smog beneath a temperature inversion.

✓ **READING CHECK**

What causes the air to rise? What causes the air to sink?

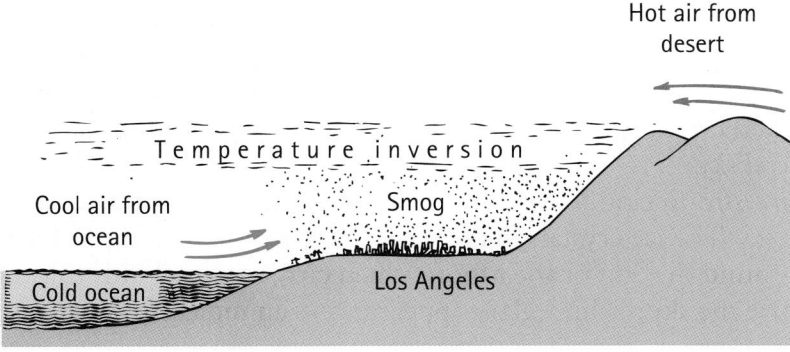

◀ **FIGURE 31.6**
Smog in Los Angeles is trapped by the mountains and a temperature inversion caused by warm air from the Mojave Desert overlying cool air from the Pacific Ocean.

31.3 There Are Many Different Clouds

Clouds are a mixture of suspended water droplets and ice crystals. They are formed from rising, warm, moist air. As the warm air rises, it cools and therefore becomes less able to hold water vapor. The water vapor condenses into tiny droplets, and clouds are formed.

TABLE 31.1 The Four Major Cloud Groups

1. High clouds (above 6000 m)	3. Low clouds (below 2000 m)
Cirrus	Stratus
Cirrostratus	Stratocumulus
Cirrocumulus	Nimbostratus
2. Middle clouds (2000–6000 m)	4. Clouds having vertical development
Altostratus	Cumulus
Altocumulus	Cumulonimbus

The altitude range of the major cloud groups varies somewhat with season and latitude. Also, some clouds extend vertically in more than one altitude range.

Clouds are generally classified according to their altitude and shape. There are ten principal cloud forms, each of which belongs to one of four major groups (Table 31.1).

High Clouds

High clouds are clouds that form at altitudes above 6000 m. High clouds (other than cirrus clouds) are denoted by the prefix *cirro-*. Air at this elevation is quite cold and dry, and clouds this high up are made up almost entirely of ice crystals.

The most common high clouds are thin, wispy *cirrus* clouds. Cirrus clouds are blown by high winds into their well-known wispy shapes, such as the classic "mare's tail" or "artist's brush." Cirrus clouds usually indicate fair weather, but they may also indicate approaching rain.

Cirrocumulus clouds are the familiar rounded white puffs. They are found in patches, and they seldom cover more than a small portion of the sky. Small ripples and a wavy appearance make the cirrocumulus clouds look like the fish-scale markings on the body of a mackerel. Hence, cirrocumulus clouds are often said to make up a *mackerel sky.*

Cirrostratus clouds are thin and sheetlike, and they often cover the entire sky. The ice crystals in these clouds refract light and produce a halo around the Sun or the Moon. When cirrostratus clouds thicken, they give the sky a white, glary appearance—an indication of coming rain or snow.

Middle Clouds

Middle clouds form at elevations between 2000 and 6000 m. Middle clouds are denoted by the prefix *alto-*. These clouds are made up of water droplets and, when cold enough, ice crystals.

Altostratus clouds are gray to blue-gray, and often cover the sky for hundreds of square kilometers. Altostratus clouds are often so thick that

FIGURE 31.7 ▲
The four cloud groups. (a) High clouds: cirrus, cirrostratus, cirrocumulus. (b) Middle clouds: altostratus, altocumulus. (c) Low clouds: stratus, stratocumulus, nimbostratus. (d) Clouds with vertical development: cumulus, cumulonimbus.

they diffuse incoming sunlight to the extent that objects on the ground don't produce shadows. Altostratus clouds often form before a storm. Look on the ground next time you're planning a picnic. If you don't see your shadow, you may want to cancel!

Altocumulus clouds appear as gray, puffy masses in parallel waves or bands. The individual puffs are much larger than those found in cirrocumulus clouds, and the color is also much darker. The appearance of altocumulus clouds on a warm, humid summer morning often indicates thunderstorms by late afternoon.

Low Clouds

Low clouds ranging from the surface up to 2000 m are called *stratus* clouds. They are almost always made up of water droplets. But in cold weather, low clouds may also contain ice crystals and snow. *Stratus* clouds are uniformly gray and often cover the whole sky. They are very common in winter and give the sky a hazy, gray look. They resemble a high fog that doesn't touch the ground. Although stratus clouds are not directly associated with falling precipitation, they sometimes generate a light drizzle or mist.

Stratocumulus clouds either form a low, lumpy layer that grows in horizontal rows or patches or, with weak rising motion, appear as rounded masses. The color is generally light to dark gray. To tell the difference between altocumulus clouds and stratocumulus clouds, hold your hand at arm's length and point toward the cloud in question. An altocumulus cloud commonly appears to be the size of your thumbnail; a stratocumulus cloud appears to be about the size of your fist. Rain and snow do not usually fall from stratocumulus clouds.

Nimbostratus clouds are dark and foreboding. They are a wet-looking cloud layer associated with light to moderate rain or snow.

Clouds That Have Vertical Development

Cumulus clouds are the most familiar of the many cloud types. Cumulus clouds resemble pieces of floating cotton, with sharp outlines and flat bases. They are white to light gray, and they generally occur about 1000 m above the ground. The tops of cumulus clouds are often in the form of rising towers, showing the upward limit of the rising air. These are the clouds of childhood daydreams. When you were younger, did you ever see castles or the shapes of animals in the clouds?

When cumulus clouds turn dark and are accompanied by precipitation, they are referred to as *cumulonimbus* clouds. In this case, they indicate a coming storm. As we shall see, cumulonimbus clouds often become *thunderheads*.

Precipitation Formation

Several things have to happen for precipitation to form. Each step toward precipitation is part of the *collision-coalescence process.* The first requirement is the presence of dust—the condensation nuclei discussed earlier in the chapter.

Water vapor is less dense than air. But once cloud droplets form, the droplets themselves are considerably denser than the air. The gravitational force pulling the droplets downward is enough to make them fall. So why don't all the water droplets in a cloud fall to the ground? The answer is *updrafts*—the upward movement of air. A typical cumulus cloud has an updraft speed of at least 1 m/s, which is faster than the droplets can fall. The droplets are "floated up" by the upward-rising air. Without updrafts, the droplets drift so slowly out of the bottom of the cloud and evaporate so quickly that they have no chance of reaching the ground.

Clouds don't float! They are buoyed up by an invisible conveyor belt of rising air. Clouds are always moving up.

Since clouds are denser than the surrounding air, why don't they fall from the sky? Ah, they do! They fall as fast as the air below rises. So without updrafts, we'd have no clouds.

In the collision-coalescence process, tiny droplets join together to form larger droplets. At first, the updrafts are stronger than the downward motion of the droplets so rain does not fall. But as the droplets grow, they eventually fall at the same rate they are pushed skyward by the updraft. At this stage the droplets are more or less stationary. But these droplets are repeatedly bombarded from below by smaller droplets rising with the updraft. And this is when significant droplet growth occurs. Eventually, the stationary drops of water grow larger and become huge compared to the typical cloud droplets—they become raindrops. Because raindrops fall faster than the updraft can push them upward, precipitation forms.

✓ In order for precipitation to occur there needs to be sufficient vertical development of the cloud. In thicker clouds, the cloud droplets have more time and space to coalesce into heavy-enough-to-fall drops. So, thicker clouds have a higher chance of producing rain.

> Raindrops form because the condensation rate exceeds the evaporation rate.

CONCEPT CHECK
1. If rain falls, why are updrafts so important?
2. What is meant by condensation nuclei?

Check Your Answer
1. The continual lift of updrafts gives the cloud droplets time to coalesce and grow into raindrops.
2. Condensation nuclei are any type of air particle—dust, smoke, salt—where slow-moving water vapor molecules can condense.

✓ **READING CHECK**

What is needed for cloud droplets to grow big enough to fall as rain?

31.4 Air Masses, Fronts, and Storms

An *air mass* is a volume of air much larger than the parcels of air we've discussed so far. Various distinct air masses cover large portions of Earth's surface. Each has its own characteristics. An air mass formed over water in the tropics is different from one formed over land in the polar regions. Air masses are divided into six general categories according to the type of land or water they form over and the latitude where formation occurs (Table 31.2 and Figure 31.8). The type of surface an air mass forms over is designated by a lowercase letter (m for maritime, c for continental). The source region where an air mass forms is designated by an uppercase letter (A for arctic, P for polar, T for tropical.).

Continental polar (cP) and continental arctic (cA) air masses generally produce very cold, dry weather in winter and cool, pleasant weather in summer. Maritime polar (mP) and maritime arctic (mA) air masses, picking up moisture as they travel across the oceans, generally bring cool, moist weather to a region. Continental tropical (cT) air masses are generally responsible for the hot, dry weather of summer, and warm, humid conditions are due to maritime tropical (mT) air masses.

TABLE 31.2 Classification of Air Masses and Their Characteristics

Typical Source Region	Classification	Symbol	Characteristics
Arctic	maritime arctic	mA	cool, moist, unstable
Greenland	continental arctic	cA	cold, dry, stable
North Atlantic and Pacific Oceans	maritime polar	mP	cool, moist, unstable
Canada, Siberia	continental polar	cP	cold, dry, stable
Caribbean Sea, Gulf of Mexico	maritime tropical	mT	warm, moist; usually unstable
Mexico, southwestern United States	continental tropical	cT	hot, dry, stable aloft; unstable at surface

So we see that different types of air masses have their own characteristics. When two different air masses meet, there are a variety of different weather conditions that can develop.

Atmospheric Lifting Creates Clouds

Clouds are great indicators of weather. For clouds to form, air must be lifted. The three principal lifting mechanisms in the atmosphere are convectional lifting, orographic lifting, and frontal lifting.

Convectional Lifting   Earth's surface is heated unequally. Some areas are better absorbers of solar radiation than others and so they heat up more quickly. The air that touches these surface "hot spots" becomes warmer than the surrounding air. This warmed air rises, expands, and cools. The rising air is accompanied by the sinking of cooler air from above. This circulatory motion produces **convectional lifting.**

If cooling occurs close to the air's saturation temperature, the condensing moisture forms a cumulus cloud. Air within the cumulus cloud

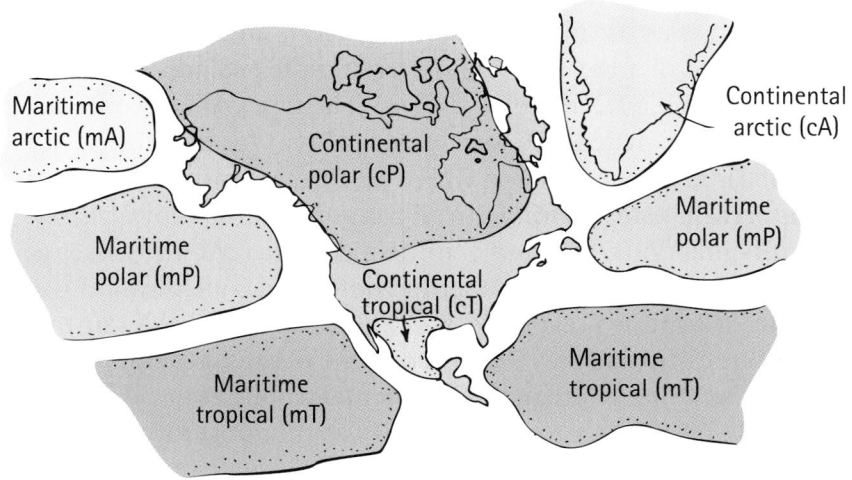

FIGURE 31.8 ▶
Typical source regions of air masses for North America.

◀ **FIGURE 31.9**
Cumulus clouds are often found as individual towering white clouds separated from each other by expanses of blue sky.

moves in a cycle: Warm air rises, cool air descends. Because descending cool air inhibits the expansion of warm air beneath it, small cumulus clouds usually have a great deal of blue sky between them (Figure 31.9).

Cumulus clouds often remain in the same place that they formed, dissipating and reforming many times. As they grow, they shade the ground from the Sun. This slows surface heating and inhibits the upward convection of warm air. Without a continuous supply of rising air, the cloud begins to dissipate. Once the cloud is gone, the ground reheats, allowing the air above it to warm and rise. Convectional lifting begins again, and another cumulus cloud begins to form in the same location.

Orographic Lifting   An air mass that is pushed upward over an obstacle such as a mountain range undergoes **orographic lifting**. The rising air cools. If the air is humid, clouds form. The types of clouds that form depend on the air's stability and moisture content. If the air is stable, a layer of stratus clouds may form. If the air is unstable, cumulus clouds may form. As the air mass moves down the other side of the mountain (the leeward slope), it warms adiabatically. This descending air is dry because most of its moisture was removed in the form of clouds and precipitation on the windward (upslope) side of the mountain. Because the dry leeward (downslope) sides of mountain ranges are sheltered from rain and moisture, they are often referred to as regions of *rain shadow* (Figure 31.10).

Frontal Lifting   In weather reports we often hear about *fronts*. A **front** is the contact zone between two different air masses. ✔ When two air masses make contact, differences in temperature, moisture, and pressure can cause one air mass to ride over the other. When this occurs, we have **frontal lifting.** If a cold air mass moves into an area occupied by a nonmoving warm air mass, the contact zone between them is called a *cold front*, and if warm air moves into an area occupied by a nonmoving mass of cold air, the zone of contact is called a *warm front.* An *occluded*

No way can *all* air rise. Some has to come back down. Where air rises we see clouds; where it descends we see blue sky between the clouds.

PhysicsPlace.com
Tutorial
Rain Shadow Activity

Aha, now I know why mountains are usually wet and green on one side and dry and brown on the other side!

FIGURE 31.10 ▶

A mountain range may produce a rain shadow on its leeward slope. As warm, moist air rises on the windward slope, the air cools and precipitation develops. By the time it reaches the leeward slope, the air is depleted of moisture, so that the leeward side is dry. It lies in a rain shadow.

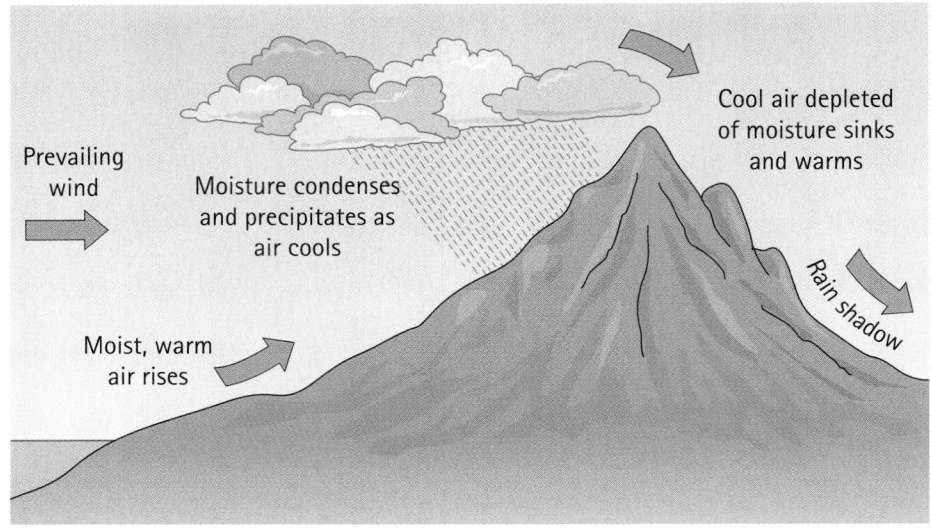

Prevailing wind

Moisture condenses and precipitates as air cools

Cool air depleted of moisture sinks and warms

Rain shadow

Moist, warm air rises

PhysicsPlace.com

Tutorial
Cold Front Activity; Warm Front Activity

front forms when a cold front overtakes a warm front, or vice versa. If neither of the air masses is moving, the contact zone is called a *stationary front.* Fronts are usually accompanied by wind, clouds, rain, and storms.

Meteorologists and other observers of the sky can often tell when a cold front is approaching by observing high cirrus clouds, a shift in wind direction, a drop in temperature, and a drop in air pressure. As cold air moves into a warm air mass, forming a cold front, the warm air is forced upward (Figure 31.11). As it rises, it cools, and water vapor condenses into a series of cumulonimbus or nimbostratus clouds. The advancing wall of clouds at the front develops into thunderstorms with heavy showers and gusty winds. After the front passes, the air cools and sinks, pressure rises, and rain ceases. Except for a few fair-weather cumulus clouds, the skies clear and we have the calm after the storm.

When warm air moves into a cold air mass, forming a warm front, the less-dense warmer air gradually rides up and over the colder, denser air (Figure 31.12). The approach of a warm front, although less obvious and more gradual than the approach of a cold front, is also indicated by cirrus clouds. Ahead of the front, the cirrus clouds descend and thicken

FIGURE 31.11 ▶

INTERACTIVE FIGURE

A cold front occurs when a cold air mass moves into a warm air mass. The cold air forces the warm air upward, where it condenses to form clouds. If the warmer air is moist and unstable, heavy rainfall and gusty winds develop.

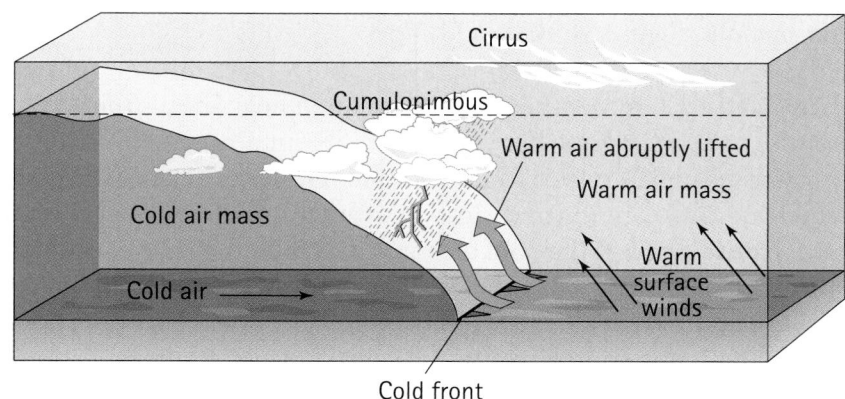

Cirrus

Cumulonimbus

Warm air abruptly lifted

Warm air mass

Cold air mass

Warm surface winds

Cold air

Cold front

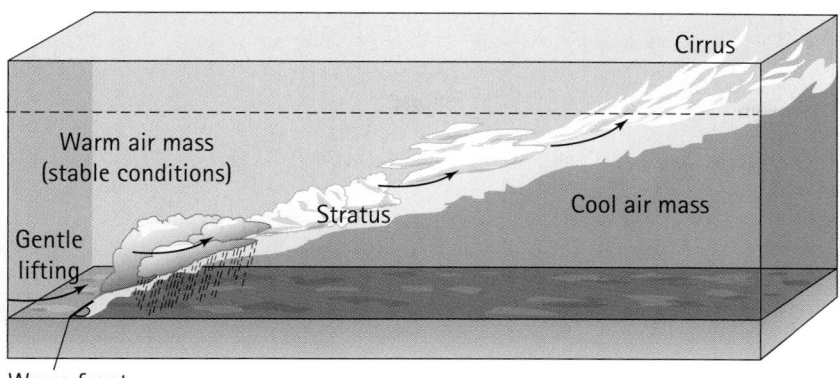

Cirrus

Warm air mass
(stable conditions)

Stratus

Cool air mass

Gentle
lifting

Warm front

FIGURE 31.12 ▲
A warm front occurs when a warm air mass moves into a cold air mass. The less-dense warmer air rides up and over the colder, denser air, resulting in widespread cloudiness and light to moderate precipitation that can cover great areas.

into altocumulus and altostratus clouds that turn the sky an overcast gray. Moving still closer to the front, light to moderate rain or snow develops, and winds become brisk. At the front, air gradually warms, and the rain or snow turns to drizzle. Behind the front, the air is warm and the clouds scatter.

When a cold front and a warm front merge, the result is an occluded front. There are several steps that lead to the formation of an occluded front (Figure 31.13). In the first step, the two fronts are joined at the center of a low-pressure area—they do not overlap. At this stage, warm air is in contact with the ground, between the two fronts (Figure 31.13a). When the cold front first overtakes the warm front, the two fronts meet at the ground, such that the warm air above each front no longer touches the ground (Figure 31.13b). As the cold front continues to invade the warm front, the warm front itself no longer touches the ground (Figure 31.13c). The advancing cold front wedges itself under the warm front so that the intersection between the two fronts is above ground. As you would imagine, a wide area of rainy weather accompanies occluded fronts.

When two different air masses are not strong enough to overtake the other, the boundary between them becomes a *stationary front*. A stationary front is like a stalemate between fronts and as such can remain over an area for several days. Eventually, the stalemate ends and the front either dissipates or, depending on conditions aloft, changes into a cold or warm front.

CONCEPT CHECK
The discussion of advancing and retreating fronts makes the weather sound like a military operation. Why is this?

Check Your Answer
The word *front* is a military term used to describe the boundary line between two different armies. During World War I, Norwegian meteorologists started using "front" as a way to describe the boundary line between two "warring" air masses.

 **READING CHECK**

What happens when two different air masses make contact?

FIGURE 31.13 ▶
Steps in the formation of an occluded front for the case when a cold front overtakes a warm front. The right side shows representative vertical cross sections for the map views to the left. The location of each cross section is shown on its respective map.

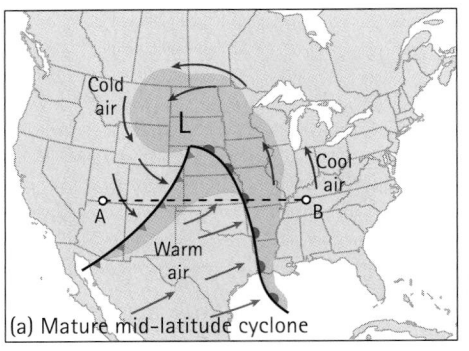

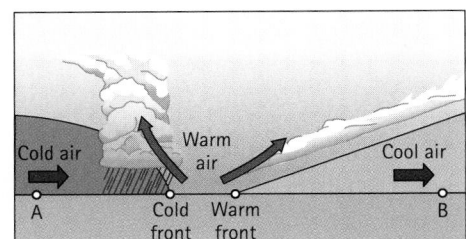

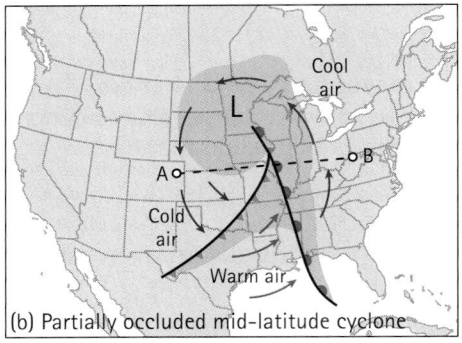

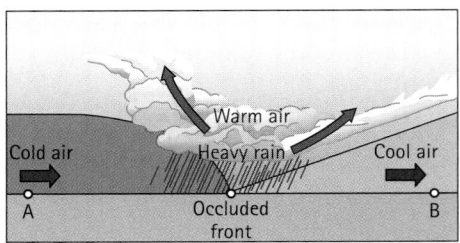

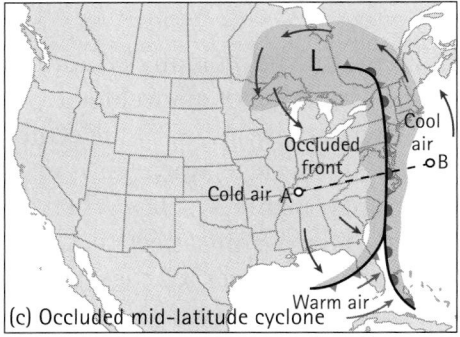

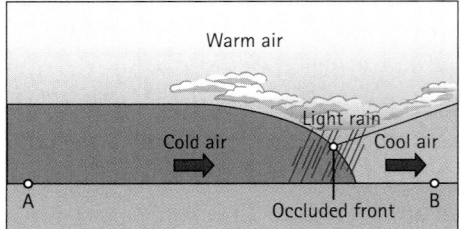

31.5 Weather Can Be Violent

The three types of lifting just discussed bring about many different weather conditions. Weather resulting from air masses in contact depends on the conditions of their source regions. Weather changes can occur slowly or very quickly. The most rapid changes, and the most violent ones, occur with three major types of storms: thunderstorms, tornadoes, and hurricanes.

Thunderstorms

A thunderstorm begins with humid, unstable air that rises, cools, and condenses into a single cumulus cloud. This cloud builds and grows upward as long as it is fed by an updraft of rising warm air from below. Cloud droplets grow larger and heavier within the cloud until they eventually begin to fall as rain. The falling rain drags some of the cool air along with it, creating a downdraft—the chilled air is colder and denser than the air around it. Together, the rising warm updraft and the sinking, chilled downdraft turn the cloud into a storm cell. This is the mature stage, where the thunderstorm cloud appears as a lonely giant—dark and brooding in the sky. It typically has a base several kilometers in diameter, and it can tower to altitudes up to 12 km. At such high altitudes, horizontal winds and lower temperatures flatten and stretch the thunderhead crown into a characteristic anvil shape (Figure 31.14). After the thunderstorm dissipates, it leaves behind the cirrus anvil as a reminder of its once mighty presence.

At any given time, there are about 1800 thunderstorms in progress in Earth's atmosphere. Wherever thunderstorms occur, there is lightning and thunder. As water droplets in the cloud bump into and rub against one another, the cloud becomes electrically charged. Rather than being uniformly distributed throughout the cloud, the electrical charge separates—it is usually positively charged where there are ice crystals (at the colder cloud top) and negatively charged where the cloud is warmer (at the cloud bottom). As electrical stresses between the oppositely charged regions build up, the charge becomes great enough that electrical energy is released and passed to other points of opposite charge, which quite often means the ground. The electrical energy flowing from cloud to ground is lightning (Figure 31.15). As lightning heats up the air, the air expands and we hear lightning's noisy companion—thunder. Lightning strikes Earth roughly 100 times every second, with some bolts having an electric potential of as much as 100 million volts. Lightning claims about 100 human victims per year in the United States alone.

FIGURE 31.14 ▲
The mature stage of a thunderstorm cloud appears as a towering cumulonimbus cloud that reaches up to about 12 km. Strong horizontal winds and icy temperatures flatten and distend the cloud's crown into a characteristic anvil shape.

◀ **FIGURE 31.15**
Time exposure of cloud-to-ground lightning during an intense thunderstorm.

Tornadoes

A revolving object, such as a whirling ball on a string, speeds up when pulled toward its axis of revolution. Similarly, winds slowly rotating over a large area speed up when the radius of rotation decreases. This increase in speed can produce a *tornado,* which is a funnel-shaped cloud that extends downward from a large cumulonimbus cloud. Produced as an extension of a powerful thunderstorm, the funnel cloud is called a tornado only after it touches the ground. The winds of a tornado travel in a counterclockwise direction (clockwise in the Southern Hemisphere) at wind speeds as low as 65 kilometers per hour but up to 450 kilometers per hour depending on the tornado's strength.

As a tornado moves across the land, at speeds from 45 to 95 km/h, it follows a path controlled by its parent thundercloud. The tornado can bounce and skip, as it rises briefly from the ground and then touches back down again. A tornado acts like a gigantic vacuum cleaner, picking up everything in its path. It wreaks havoc not only by suction but also by the battering power of its whirling winds. In its wake, a trail of flying dirt and debris is left behind (Figure 31.16).

Tornadoes occur in many parts of the world. In the flat central plains of the United States, a tornado zone extends from northern Texas through Oklahoma, Kansas, and Missouri. In this area, more than 300 tornadoes touch down each year. Hence the name for this area: Tornado Alley. Tornadoes are so frequent in this part of the country that many homes are built with underground storm shelters. The power of a tornado is terrifying and devastating.

FIGURE 31.16 ▶
Like a gigantic vacuum cleaner, the strong wind of a tornado can pick up and obliterate everything in its path.

Hurricanes

✓ In the steamy tropics, where the Sun warms the oceans, heat transfer to the atmosphere by evaporation and conduction is so thorough that air and water temperatures are about equal. The high humidity in this part of the world favors the development of cumulus clouds and afternoon thunderstorms. Most of the individual storms are not severe. However, as the moisture content and the temperature of the air increase and surface winds collide, a strong vertical wind shear can cause the rising warm, moist air to be diverted horizontally, forming a spiral. If this spiraling storm isn't broken apart by upper-level winds, it can develop into a *tropical depression* (wind speed less than 60 km/h), so called because of a central area of low pressure. If the storm intensifies, it progresses into a *tropical storm,* with increased wind speeds above 60 km/h. If favorable conditions continue, a more violent storm develops—a hurricane—with wind speeds above 120 km/h and up to nearly 300 km/h hour.

Hurricanes gain energy from the heat released by the condensation of water (Chapter 9). The condensation produces the vast amounts of rain that are typical of such storms. The heat warms the surrounding, upper-level air, causing it to rise. As the upper-level air rises, surface air is sucked upward, intensifying the low-pressure center—the eye of the hurricane. Horizontal airflow spirals counterclockwise (in the Northern Hemisphere) around the eye. The spiral bands of cumulonimbus clouds give the hurricane its familiar appearance (Figure 31.17).

The airflow that sets up in the eye forms a positive feedback loop. Condensation releases heat, which draws moist air upward from the ocean surface. The moist air cools and more condensation occurs, which releases more heat, which draws more warm, moist air upward. This cycle—essentially a natural heat engine—continues unless strong,

fyi

■ Hurricane Katrina formed on August 23, 2005 near the Bahamas. It struck the Gulf Coast of the United States 6 days later, wreaking havoc. New Orleans, Louisiana, and coastal Mississippi were hit particularly hard. Hurricane Katrina, however, was such a large storm that it devastated areas of the Gulf Coast up to 160 km from the eye of storm. At least 1836 people were killed as a result of the hurricane and the floods that followed. It was the deadliest U.S. hurricane since the Okeechobee Hurricane of 1928. It was a costly storm, too. Over $100 billion was needed to repair storm damage and fund cleanup operations. Some areas still haven't recovered. It was the costliest natural disaster in U.S. history.

◀ **FIGURE 31.17**
This satellite image of Hurricane Katrina shows the characteristic appearance of a hurricane. Bands of cumulonimbus clouds spiral around the low-pressure eye of the storm.

READING CHECK

Why do hurricanes form in tropical waters?

upper-level winds from outside the storm disrupt the upward flow pattern or the hurricane moves over land. Once over land, the hurricane is deprived of its energy source.

CONCEPT CHECK

Would storms occur if all parts of Earth's surface were heated evenly?

Check Your Answer

No. The principal factor in the formation of storms is contact between warm air and cool air.

31.6 The Weather—Number-One Topic of Conversation

Meteorologists have the important job of forecasting hurricanes and other storms. Weather forecasting is, in part, a matter of determining air mass characteristics, predicting how and why the characteristics might change, and in what direction air masses might move. In the case of hurricanes and tornadoes, such predictions are lifesaving. Meteorologists have a long and remarkable record of saving human lives and reducing property loss.

There are several methods of weather forecasting. Some forecasts are based on the *continuity* of a weather pattern, such as rain today likely means rain tomorrow. Or, because surface weather systems tend to move in the same direction and at the same speed, a forecast is based on the *trend* of the weather pattern. For example, if a cold front is moving eastward at an average speed of 20 km/h, it can be expected to affect the weather 80 km away in 4 hours. We also hear about the *probability* of a weather condition—for example, the probability of rain is 70%. This is an expression of chance, meaning that there is a 70% chance that rain will fall somewhere in the forecast area. So you should probably carry an umbrella. Another forecast we often hear about is the *extended forecast*. This forecast is based on weather types that develop in certain areas. Recall the classification of air masses and their characteristics; if a continental polar air mass is approaching, we can expect cold dry weather, whereas if a maritime polar air mass is approaching we can expect cold moist weather. All these methods of weather prediction are based on the statistical analysis of weather information.

Weather forecasting involves great quantities of data from all over the world. Meteorologists use numerical models and computers not only to plot and analyze data but also to help predict the weather. The computer draws maps of projected weather conditions, which the weather forecaster uses as a guide for predicting weather. Even so, the many variables involved are not exactly predictable and it may unexpectedly rain on your parade!

fyi

■ Swirling storms and disturbances caused by solar activity produce "space weather" between the Sun and Earth. Solar flares, coronal mass ejections, and magnetic storms affect not only Earth satellites but Earth's surface environment as well. Communications system failures, power blackouts, and brownouts are often attributed to space weather. As our use of space grows, so must our ability to predict its weather.

WEATHER MAPS

The weather forecaster's primary tool is the weather map or chart. A weather map is essentially a representation of the frontal systems and the high-pressure and low-pressure systems that overlie the areas outlined in the map. Symbols on such a map are a shorthand notation to represent data gathered from various observation stations. These symbols are called weather codes.

This shorthand notation compiles 18 categories of data into a very small area called a *station model*. The circle at the center describes the overall appearance of the sky. Jutting from the circle is a wind arrow, its tail in the direction from which the wind comes and its feathers indicating wind speed. The other 15 weather elements are in standard position around the circle.

A weather map is covered with lines—isobars—that connect points of equal pressure. As air moves from a high-pressure region to a low-pressure region, it rises and cools and the moisture in it condenses into clouds. In the vicinity of the low (L on map), we see an extensive cloud cover. In the vicinity of the high (H on map), we see clear skies. In a high-pressure region, air sinks and warms adiabatically. Because sinking air does not produce clouds, we find clear skies and fair weather. The heavy lines on a weather map represent fronts. Because fronts generally mean a change in the weather, they are of great importance on weather maps.

Weather Symbols

Total Sky Cover

- No clouds
- Less than one–tenth or one–tenth
- Two–tenths or three–tenths
- Four–tenths
- Five–tenths
- Six–tenths
- Seven–tenths or eight–tenths
- Nine–tenths or overcast with openings
- Completely overcast
- Sky obscured

Pressure Tendency

- Rising, then falling
- Rising, then steady; or rising, then rising more slowly
- Rising steadily, or unsteadily
- Falling or steady, then rising; or rising, then rising more quickly

  Barometer no higher than 3 hours ago

- Steady, same as 3 hours ago
- Falling, then rising, same or lower than 3 hours ago
- Falling, then steady; or falling, then falling more slowly

  Barometer no lower than 3 hours ago

- Falling steadily, or unsteadily
- Steady or rising, then falling; or falling, then falling more quickly

Wind Entries

	Miles (Statute) per hour	Knots	Kilometers per hour
(calm)	Calm	Calm	Calm
	1–2	1–2	1–3
	3–8	3–7	4–13
	9–14	8–12	14–19
	15–20	13–17	20–32
	21–25	18–22	33–40
	26–31	23–27	41–50
	32–37	28–32	51–60
	38–43	33–37	61–69
	44–49	38–42	70–79
	50–54	43–47	80–87
	55–60	48–52	88–96
	61–66	53–57	97–106
	67–71	58–62	107–114
	72–77	63–67	115–124
	78–83	68–72	125–143
	84–89	73–77	135–143
	119–123	103–107	144–198

Common Weather Symbols

- Light rain
- Moderate rain
- Heavy rain
- Light snow
- Moderate snow
- Heavy snow
- Light drizzle
- Ice pellets (sleet)
- Freezing rain
- Freezing drizzle

- Rain shower
- Snow shower
- Showers of hail
- Drifting or blowing snow
- Dust storm
- Fog
- Haze
- Smoke
- Thunderstorm
- Hurricane

Front Symbols

- Cold front (surface)
- Warm front (surface)
- Occluded front (surface)
- Stationary front (surface)
- Warm front (aloft)
- Cold front (aloft)
- Squall line

Station Model

- Wind speed
- Wind direction
- Type of middle cloud
- Temperature
- Present weather
- Visibility
- Dew point
- Type of low cloud
- Base height of low clouds
- Amount of low clouds
- Barometric pressure reduced to sea level
- Pressure higher or lower than 3 hours ago
- Amount of barometric change in last 3 hours
- Barometric tendency in last 3 hours
- Time precipitation began or ended
- The weather during past 6 hours
- Amount of precipitation during past 6 hours

250

31
24
30
+28
−6
2
4
.45

KEY TERMS

Adiabatic A term that describes temperature change in the absence of heat transfer—expanding air cools and compressing air warms.

Clouds The condensation of water droplets above Earth's surface.

Convectional lifting An air-circulation pattern in which air warmed by the ground rises while cooler air aloft sinks.

Front The contact zone between two different air masses.

Frontal lifting The lifting of one air mass by another as two air masses converge.

Humidity A measure of the concentration or amount of water vapor in the air—the mass of water vapor per volume of air.

Orographic lifting The lifting of an air mass over a topographic barrier such as a mountain.

Relative humidity The amount of water vapor in the air at a given temperature expressed as a percentage of the maximum amount of water vapor the air can hold at that temperature.

Temperature inversion A condition in which the upper regions of the troposphere are warmer than the lower regions.

REVIEW QUESTIONS

Water in the Atmosphere

1. What is the difference between humidity and relative humidity?
2. Why does relative humidity increase at night?
3. As air temperature decreases, does relative humidity increase, decrease, or stay the same?
4. What does saturation point have to do with dew point?

5. What happens to the water vapor in saturated air as the air cools?

Weather Variables

6. Explain why warm air rises and cools as it expands.
7. When a parcel of air rises, does it become warmer, become cooler, or remain at the same temperature?
8. When does an adiabatic process happen in the atmosphere?
9. What is a temperature inversion? Give one location where these inversions may occur.
10. What happens to the air pressure and temperature of an air parcel as it flows up the side of a mountain?

There Are Many Different Clouds

11. How do clouds form?
12. Rain or snow is most likely to be produced by which of the following cloud forms? (a) cirrostratus, (b) nimbostratus, (c) altocumulus, (d) stratocumulus
13. Are clouds that have vertical development characteristic of stable air, stationary air, unstable air, or dry air?
14. Which type of clouds can become thunderheads?

Air Masses, Fronts, and Storms

15. Explain how convectional lifting plays a role in the formation of cumulus clouds.
16. Does a rain shadow occur on the windward side of a mountain range or on the leeward side? Explain.
17. What are the three main atmospheric lifting mechanisms?

Weather Can Be Violent

18. How do downdrafts form in thunderstorms?
19. Briefly describe how thunder and lightning develop.

The Weather—Number-One Topic of Conversation

20. What information is needed to predict the weather?

THINK AND EXPLAIN

1. Does condensation occur more readily at high temperatures or low temperatures? Explain.
2. When water vapor condenses to liquid water, is heat absorbed or released?
3. Why do clouds tend to form above mountain peaks?
4. Why does warm, moist air blowing over cold water result in fog?
5. Why does dew form on the ground during clear, calm summer nights?
6. In simplest terms, what is an occluded front?
7. Why are saturation and condensation more likely to occur on a cold day than on a warm day?
8. Why does surface temperature increase on a clear, calm night as a low cloud cover moves overhead?
9. During a summer visit to Cancun, Mexico, you stay in an air-conditioned room. Getting ready to leave your room for the beach, you put on your sunglasses. The minute you step outside your sunglasses fog up. Why?
10. Why must an air mass rise in order to produce precipitation?
11. As an air mass moves first upslope and then downslope over a mountain, what happens to the air's temperature and moisture content?
12. What steps need to happen before precipitation occurs?
13. What accounts for the large spaces of blue sky between cumulus clouds?
14. Why don't cumulus clouds form over cool water?
15. What is the difference between rainfall that accompanies the passage of a warm front and rainfall that accompanies the passage of a cold front?
16. How do fronts cause clouds and precipitation?

17. What part of the United States has the highest frequency of tornadoes?
18. In which atmospheric layer does all our weather occur?
19. Why are clouds that form over water more efficient in producing precipitation than clouds that form over land?
20. What is the source of the enormous amount of energy released by a hurricane?

READINESS ASSURANCE TEST (RAT)

If you have a good handle on this chapter—if you really do—then you should be able to score 7 out of 10 on this RAT. Check your answers with your teacher. If you score less than 7, study further before moving on.

Choose the best *answer to each of the following.*

1. Air that contains the maximum amount of water vapor for the temperature of the air mass is considered to
   (a) have a relative humidity of 100%.
   (b) be saturated.
   (c) have an evaporation rate equal to its condensation rate
   (d) All of these.

2. Fast-moving air molecules have
   (a) greater kinetic energy.
   (b) greater molecular collisions.
   (c) warmer temperatures.
   (d) All of these.

3. An adiabatic process has occurred when
   (a) air is warmed by solar radiation.
   (b) heat is subtracted by evaporation.
   (c) air expands and cools.
   (d) air expands and warms.

4. When air sinks,
   (a) it compresses and warms.
   (b) it reaches its equilibrium level and then begins to sink.
   (c) it expands and cools.
   (d) it forms into clouds.

5. When upper regions of the atmosphere are warmer than lower regions we have
   (a) convective lifting.
   (b) a temperature inversion.
   (c) absolute instability.
   (d) an adiabatic process.

6. A key factor needed for precipitation to occur is
   (a) updraft motion in relatively thick clouds.
   (b) a condensation rate that exceeds the evaporation rate.
   (c) condensation nuclei.
   (d) All of these.

7. For clouds to form, air must be lifted. The principal lifting mechanisms are
   (a) convectional, orographic, and frontal lifting.
   (b) continental, orogeny, and occluded lifting.
   (c) conversational, orthodontic, and face lifting.
   (d) stationary, occluded, and contact lifting.

8. As air temperature decreases, relative humidity
   (a) increases.
   (b) decreases.
   (c) stays the same.
   (d) None of these.

9. As air flows up the side of a mountain, the air pressure
   (a) increases.
   (b) decreases.
   (c) gets warmer.
   (d) gets colder.

10. When air is saturated,
    (a) the condensation rate is greater than the evaporation rate.
    (b) the condensation rate is less than the evaporation rate.
    (c) the condensation rate equals the evaporation rate.
    (d) the condensation rate depends on temperature.

32

THE SOLAR SYSTEM

 **THE MAIN IDEA**

Many different types of objects orbit our Sun including planets, asteroids, comets, and dwarf planets.

Five billion years ago, there was no Sun. Instead, the region of the galaxy that would become our solar system was a gently swirling, diffuse cloud of gas and dust. How then did this cloud evolve into our present solar system? How are the planets similar, and how are they different?

How did our Moon form and how does it go through phases? Why do we see only one side of the Moon? What are meteors, asteroids, and comets? How frequently do they collide with our planet, and why does a comet's tail always point away from the Sun?

Explore!

How Large Is the Solar System?

Use a tennis ball to represent the Sun and place it in a corner of the room. To scale, the four inner planets would each be about as large as a grain of sand. Place a single grain of sand representing Mercury about 3 meters away from the tennis ball. For Venus, place a grain about 5 meters away. The grain representing Earth should be placed 7 meters away and Mars about 11 meters away.

Analyze and Conclude

1. **Observing** Is the solar system made mostly of planets or of empty space?
2. **Predicting** If the Sun were the size of a tennis ball, the largest planet in our solar system, which

is Jupiter, would be about as large as a green pea. How far should this pea be placed away from the tennis ball? (*Hint:* Jupiter is 5 times farther away from the Sun than Earth.)

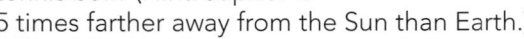

3. **Making Generalizations** Our most distant planet, Neptune, is about 30 times farther from the Sun than Earth. If the Sun were the size of a tennis ball, would all the planets fit within your school grounds? How about the outermost reaches of our solar system, which are about 50,000 times as far from the Sun?

32.1 The Solar System Is Mostly Empty Space

Our solar system is a collection of objects gravitationally bound to the Sun. Along with the Sun itself, the solar system contains eight **planets.** The planets all lie roughly in the same plane. ✓ This plane, called the **ecliptic,** is defined as the plane of Earth's orbit. Within the solar system are also numerous moons (objects orbiting planets), asteroids (small, rocky bodies), comets (small, icy bodies), and a collection of miniature planets known as dwarf planets, or *plutoids*, that orbit on the outer edges of the solar system. The most well-known plutoid is Pluto, which was downgraded from planet status in 2006. The planets and all these other objects are quite small compared to the Sun.

The vast distances between the Sun and the objects orbiting it can be grasped by imagining the Sun reduced to the size of a tennis ball. The four innermost planets would each be about as large as a grain of sand, all within 11 meters of this tennis-ball-sized Sun.

These first four **inner planets**—Mercury, Venus, Earth, and Mars— are small, solid rocky planets. The **outer planets** are larger gaseous planets located much farther away. The first outer planet is Jupiter, which on the scale mentioned would be the size of a pea more than 25 m away. The second outer planet, Saturn, famous for its extensive ring system, would be the size of a smaller pea more than 65 m away. Planets Uranus and Neptune would both be about the size of grains of rice located 135 and 210 m away, respectively. We see that solar system objects are mere specks in the vastness of the space about the Sun.

Because of these vast interplanetary distances, astronomers use the *astronomical unit* to measure them. One **astronomical unit** (AU) is the distance from Earth to the Sun, which is about 150,000,000 km. Table 32.1 gives the distances of planets from the Sun in AU. The data in Table 32.1 also show the division of the planets into two groups with similar properties. The inner planets—Mercury, Venus, Earth, and Mars—are solid and relatively small and dense. They are called the

The stars appear fixed in their patterns in the sky, but the planets wander from night to night. The term *planet* is derived from Greek for "wandering star."

FIGURE 32.1 ▶

This illustration shows the order and relative sizes of planets. Moving away from the Sun, we have in order: Mercury, Venus, Earth, Mars, Jupiter, Saturn, Uranus, and Neptune. The planets range greatly in size, but the Sun dwarfs them all—containing more than 99% of the mass in the solar system. (*Note:* distances are not to scale in this illustration.)

TABLE 32.1 Planetary Data

	Mean Distance from Sun (Earth distances, AU)	Orbital Period (years)	Diameter		Mass		Density (g/cm³)	Inclination to Ecliptic
			(km)	(Earth = 1)	(km)	(Earth = 1)		
Sun			1,392,000	109.1	1.99×10^{30}	3.3×10^5	1.41	
Terrestrial								
Mercury	0.39	0.24	4,880	0.38	3.3×10^{23}	0.06	5.4	7.0°
Venus	0.72	0.62	12,100	0.95	4.9×10^{24}	0.81	5.2	3.4°
Earth	1.00	1.00	12,760	1.00	6.0×10^{24}	1.00	5.5	0.0°
Mars	1.52	1.88	6,800	0.53	6.4×10^{23}	0.11	3.9	1.9°
Jovian								
Jupiter	5.20	11.86	142,800	11.19	1.90×10^{27}	317.73	1.3	1.3°
Saturn	9.54	29.46	120,700	9.44	5.7×10^{26}	95.15	0.7	2.5°
Uranus	19.18	84.0	50,800	3.98	8.7×10^{25}	14.65	1.3	0.8°
Neptune	30.06	164.79	49,600	3.81	1.0×10^{26}	17.23	1.7	1.8°
Plutoid								
Pluto	39.44	247.70	2,300	0.18	1.3×10^{22}	0.002	1.9	17°
Eris	67.67	557	2,400	0.19	1.6×10^{22}	0.002	1.9	44°

terrestrial planets. The outer planets are large, have many rings and satellites, and are composed primarily of hydrogen and helium gas. These are called the *jovian* (Latin for Jupiter) planets because their large sizes and gaseous compositions resemble Jupiter.

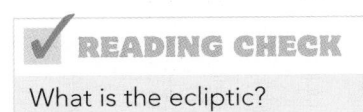

✔ **READING CHECK**

What is the ecliptic?

32.2 Solar Systems Form from Nebula

Any theory of solar system formation must be able to explain (1) *the orderly motions among the bodies of our solar system* and (2) *the division of planets into two main types—terrestrial and* jovian. The modern scientific theory that meets these requirements is called the *nebular theory.* The **nebular theory** proposes that a solar system forms from a cloud of gas and dust. The word *nebula* is Latin for "cloud." With telescopes, we can see many distant nebula, all at different stages of solar system formation. An example is shown in Figure 32.2.

FIGURE 32.2 ▶
This photograph, provided by the Hubble Telescope, shows the Orion Nebula. The Orion Nebula, like the nebula from which our solar system formed, is an interstellar cloud of gas and dust and the birthplace of stars.

According to the nebular theory, our own solar system began to condense from a cloud of gas and dust about 5 billion years ago. The cloud would have been very diffuse and large, with a diameter thousands of times larger than Neptune's orbit.

Within this nebula, the gravitational pull on particles was strong enough to cause the nebula to collapse. This collapse began slowly but then gradually accelerated as the force of gravity became stronger. Recall from Chapter 7 that the universal law of gravity is an inverse-square law: The strength of the gravitational force increases dramatically as the particles become closer.

The nebula maintained a constant mass as it shrank, gravitational forces grew ever stronger, and the nebula took on a spherical shape. This nebula must also have had a slight rotation, possibly due to the rotation of the galaxy itself (Figure 32.3a).

✓ As the nebula shrank under the influence of gravity, much heat was released upon the collision of particles. To see for yourself how collisions release heat, try hitting a penny with a hammer. After a few hits you'll find that the penny becomes quite warm.

Also, as the nebula shrank, it spun faster and faster. As any spinning object contracts, the speed of its spin increases such that angular momentum is conserved. A familiar example is an ice skater whose spin rate increases when her extended arms are pulled inward. A nebula does the same (Figure 32.3b).

What happens to the shape of a sphere as it spins faster and faster? The answer is, it flattens. A familiar example is the chef who turns a ball of pizza dough into a disk by spinning it on his hands. Even planet Earth is a slightly "flattened" sphere because of its daily spin. So the initially spherical nebula progressed to a spinning disk, the center of which became the *protosun* (Figure 32.3c).

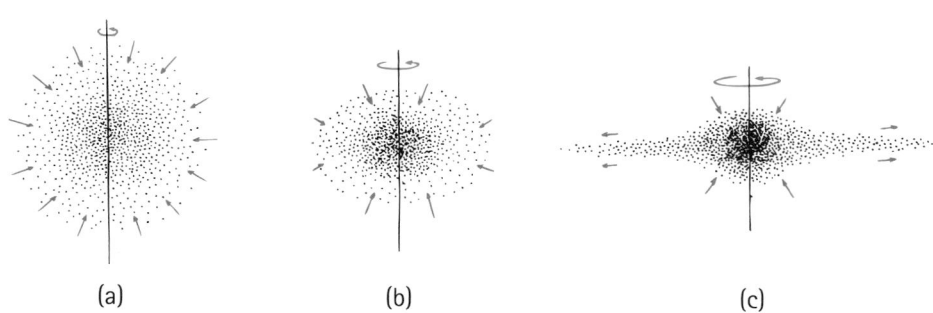

(a)   (b)   (c)

FIGURE 32.3 ▲
(a) The nebula from which the solar system formed was originally a large, diffuse cloud that rotated quite slowly. The cloud began to collapse under the influence of gravity. (b) As the cloud collapsed, it heated up as gravitational potential energy converted to heat. It spun faster by the conservation of angular momentum. (c) The cloud flattened into a disk as a result of its fast rotation. A spinning, flattened disk was produced with mass concentrated at its hot center.

The formation of the spinning disk explains the orderly motions of our solar system today. All planets orbit the Sun in nearly the same plane because they formed from that same flat nebular disk. The direction in which the disk was spinning became the direction of the Sun's rotation and the orbits of planets.

The central portion of the condensed nebula became so hot that thermonuclear fusion ignited within it, creating our Sun. The surrounding disk was the source of material that became the planets, which grew as the material steadily clumped together over time. Once the Sun began to radiate energy, the nebular disk warmed. The inner portions reached higher temperatures than the outer portions. As a result, the inner and outer planets developed differently. The inner planets formed from materials that remained solid at high temperatures; hence, the inner planets are rocky. The outer planets, by contrast, consist mainly of hydrogen and helium gas that condensed in the cold regions of the solar system far from the Sun.

What was released as the nebula shrank?

CONCEPT CHECK
According to the nebular theory, what force causes a nebula to contract into a solar system?

Check Your Answer
The gravitational force causes a nebula to contract. Nebular theory also states that as the nebula contracts it spins faster. As the contracting nebula spins faster it flattens out into a solar system.

32.3 The Sun Is Our Prime Source of Energy

The Sun produces energy from the thermonuclear fusion of hydrogen to helium. Each second, approximately 657 million tons of hydrogen is fused to 653 million tons of helium. The 4 million tons of mass lost is discharged as radiant energy. This conversion of hydrogen to helium in the Sun has been going on since it formed nearly 5 billion years ago, and it is expected to continue at this rate for another 5 billion years.

✓ Solar energy is generated deep within the core of the Sun. The solar core constitutes about 10% of the Sun's total volume. It is very hot—more than 15,000,000 K. The core is also very dense, with more than 12 times the density of solid lead. Pressure in the core is 340 billion times Earth's atmospheric pressure! Because of these intense conditions, the hydrogen nuclei move fast enough to undergo nuclear fusion, as discussed in Section 16.9. The energy released from this nuclear fusion rises to the surface, where it causes gases to emit a broad spectrum of electromagnetic radiation, from infrared to X-rays.

From where does the Sun's energy originate?

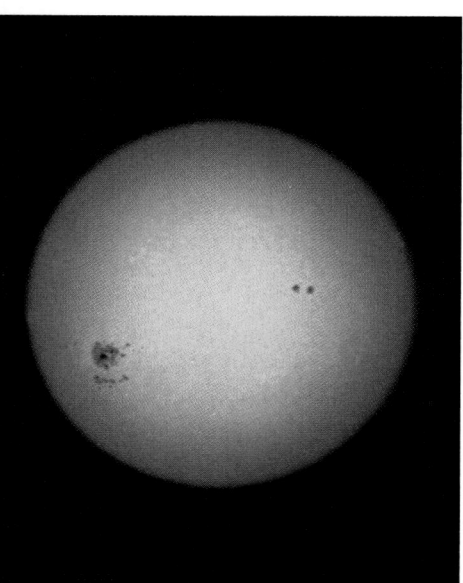

FIGURE 32.5 ▲
Sunspots on the solar surface are relatively cool regions. We say relatively cool because they are hotter than 4000 K. They look dark only in contrast with their 5800 K surroundings.

FIGURE 32.4 ▶
Never directly look at the Sun! Instead, you can get a nice view of the Sun by focusing the image of the Sun from a pair of binoculars onto a white surface. If the Sun is eclipsed by the Moon, which is a rare event, the Sun will be seen as a crescent. More commonly, the Sun's image may reveal sunspots.

The Sun's surface is 5800 K, which is much cooler than the Sun's core but hot enough to generate lots of light. This surface, called the *photosphere* (sphere of light), is about 500 km deep. Within the photosphere are relatively cool regions that appear as *sunspots* when viewed from Earth. **Sunspots** are cooler and darker than the rest of the photosphere and are caused by magnetic fields that impede hot gases from rising to the surface. As shown in Figure 32.4, sunspots can be seen by focusing the image of the Sun from a telescope or pair of binoculars onto a flat white surface. Sunspots are typically twice the size of Earth and move around because of the Sun's rotation, lasting about a week or so. Often, they cluster in groups, as shown in Figure 32.5.

The Sun spins slowly on its axis. Because the Sun is a fluid rather than a solid, different latitudes of the Sun spin at different rates. Equatorial regions spin once in 25 days, but higher latitudes take up to 36 days to make a complete rotation. This differential spin means the surface near the equator pulls ahead of the surface farther north or south. The Sun's differential spin wraps and distorts the solar magnetic field, which bursts out to form the sunspots mentioned earlier. A reversal of magnetic poles occurs every 11 years, and the number of sunspots also reaches a maximum every 11 years.

Above the Sun's photosphere is a transparent 10,000-km-thick shell called the *chromosphere* (sphere of color), seen during an eclipse as a pinkish glow surrounding the eclipsed Sun (Figure 32.6). The chromosphere is hotter than the photosphere, reaching temperatures of about 10,000 K.

Beyond the chromosphere are streamers and filaments of outward-moving, high-temperature plasmas curved by the Sun's magnetic field. This outermost region of the Sun's atmosphere is the *corona*, which

◀ **FIGURE 32.6**
The pink chromosphere becomes visible when the Moon blocks most of the light from the photosphere during a solar eclipse.

FIGURE 32.7 ▶
The pearly-white solar corona is visible only during a solar eclipse. Notice how this exceptional photo of the corona also captures some of the pink of the chromosphere as well as the face of the new Moon, which is faintly illuminated by light reflected from the full Earth.

extends out several million kilometers (Figure 32.7). The temperature of the corona is amazingly high—on the order of 1 million K—and it is where most of the Sun's powerful X-rays are generated. Because the corona is not very dense, its brightness is not as intense as the Sun's surface, which makes the corona safe to observe during (and only during) a total solar eclipse when the Moon completely covers the Sun. High-speed protons and electrons are cast outward from the corona to generate the solar wind, which powers the aurora on Earth, as was discussed in Section 30.4. The solar wind also produces the tails of comets, which we discuss later in this chapter.

CONCEPT CHECK

1. Was the Sun more massive 1000 years ago than it is today?
2. Of the photosphere, chromosphere, and corona, which is thinnest? Which is hottest? Which is between the other two?

Answers

1. Yes, although only slightly compared to its great mass. The Sun loses mass as hydrogen nuclei combine to make helium nuclei.
2. The photosphere is the thinnest. The corona is the hottest. The chromosphere is the pinkish layer above the photosphere and below the vast corona.

32.4 The Inner Planets Are Rocky

Compared to the outer planets, the four planets nearest the Sun are close together. These are Mercury, Venus, Earth, and Mars. These rocky planets each have a mineral-containing solid crust.

Mercury

Mercury (Figure 32.8) is somewhat larger than Earth's Moon and similar in appearance. It is the closest planet to the Sun. Because of this closeness it is the fastest planet, circling the Sun in only 88 Earth days—which thus equals one Mercury "year." Mercury spins about its axis only 3 times for each 2 revolutions about the Sun. This makes its daytime very long and very hot, with temperatures as high as 430°C.

Because of Mercury's small size and weak gravitational field, it holds very little atmosphere. Mercury's atmosphere is only about a trillionth as dense as Earth's atmosphere—a better vacuum than laboratories on

FIGURE 32.8 ▶
NASA's *Messenger* spacecraft passed by Mercury in 2008 during which this high-resolution color image was taken. The *Messenger* spacecraft goes into orbit with Mercury in 2011, when it will begin detailed mapping of the planet's surface.

fyi
- We have seven days in a week because ancient Europeans decided to name days after the seven wandering celestial objects they could observe. The English day names were derived from the language of the Teutonic tribes who lived in the region that is now Germany. In Teutonic, the Sun is *Sun* (Sunday), the Moon is *Moon* (Monday), Mars is *Tiw* (Tuesday), Mercury is *Woden* (Wednesday), Jupiter is *Thor* (Thursday), Venus is *Fria* (Friday), and Saturn is *Saturn* (Saturday).

- Ancient American cultures ran their lives using three calendars. Their secular calendar, which told them when to plant seeds and so forth, followed the 365-day orbit of Earth. Their religious calendar centered on the roughly 260-day orbit of Venus, which gave them a calendar with 20-day weeks and 13-day months. Both these calendars were cyclical and couldn't account for succeeding years. For that purpose they developed the "long-count" calendar, which, interestingly enough, employed the concept of zero. They did this centuries before the concept of zero was recognized by accountants in India.

Earth can produce. So without a blanket of atmosphere, and because there are no winds to transfer heat from one region to another, nighttime on Mercury is very cold, about −170°C. Mercury is a fairly bright object in the nighttime sky and is best seen as an evening "star" during March and April or as a morning star during September and October. When seen, it is always seen near the Sun either just before sunup or just after sunset.

Venus

Venus, the second planet from the Sun, is frequently the first starlike object to appear after the Sun sets, so it is often called the evening "star," as illustrated in Figure 32.9. Compared with other planets, Venus most closely resembles Earth in size, density, and distance from the Sun. However, as shown in Figure 32.10, Venus has a very dense atmosphere and opaque cloud cover that generate high surface temperatures (470°C)—too hot for oceans. The atmosphere of Venus is about 96% CO_2. Remember from Chapter 30 that carbon dioxide is a "greenhouse gas." By this we mean that CO_2 blocks the escape of infrared radiation from a planet's surface and contributes to global warming. The thick blanket of CO_2 surrounding Venus effectively traps heat near the Venusian surface. This and Venus's proximity to the Sun make Venus the hottest planet in the solar system.

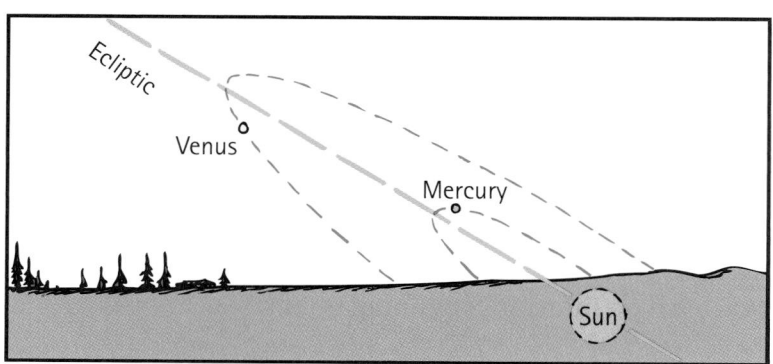

FIGURE 32.9 ▲
Because the orbits of Mercury and Venus lie inside the orbit of Earth, they are always near the Sun in our sky. Near sunset (or sunrise) they are visible as "evening stars" or "morning stars."

FIGURE 32.10 ▲
This photo of Venus, taken by NASA's Pioneer Venus Orbiter with cameras sensitive to ultraviolet light, reveals many of the planet's cloud features.

Another difference between Venus and Earth is in how the two planets spin about their axes. Venus takes 243 Earth days to make one full spin and only 225 Earth days to make one revolution around the Sun. Venus spins in a direction opposite to the direction of Earth's spin. The slow spin of Venus means that the atmosphere is not disturbed by the Coriolis effect, which was described in Chapter 30. As a result, there is very little wind and weather on the surface of Venus. Instead, the stifling hot dense air sits still through its long days and nights. In recent years, many probes have either landed on the surface of Venus or flown by on their way to other destinations.

CONCEPT CHECK
Why are Mercury and Venus never seen at night straight-up, high in the sky?

Check Your Answer
Mercury and Venus are closer to the Sun than Earth, which means they always appear close to the Sun. So, Mercury and Venus never appear high up in the sky at night for pretty much the same reason that the Sun doesn't either.

Earth

Our home planet Earth resides within the Sun's *habitable zone,* which is a region not too close and not too far from the Sun, so water can exist predominantly in the liquid phase, as shown in Figure 32.11. Earth has an abundant supply of liquid water covering about 70% of its surface, which makes our planet the blue planet (Figure 32.12). ✓ Earth's oceans support the *carbon dioxide cycle,* which acts as a thermostat to keep global temperatures from reaching harsh extremes. For example, if Earth froze over completely, carbon dioxide released by volcanoes would no longer be absorbed by the oceans. Instead, the carbon dioxide would build up in the atmosphere, which would warm the atmosphere and hence melt the frozen Earth. Conversely, if Earth became hotter and hotter, more water would evaporate. This would lead to more precipitation, which would remove carbon dioxide from the atmosphere. With less carbon dioxide in the atmosphere, the greenhouse effect would be minimized and Earth would cool.

So not only are we a nice distance from the Sun, but our atmosphere contains just enough water vapor and carbon dioxide to keep temperatures favorable for life. Furthermore, our relatively high daily spin rate allows only a brief and small lowering of temperature on the nighttime side of Earth. So temperature extremes of day and night are also kept moderate.

CONCEPT CHECK
Does Venus have a carbon dioxide cycle like Earth?

Check Your Answer
No, the carbon dioxide cycle requires oceans of water, which Venus does not have.

Mars

Mars captures our fancy as a world with life because it resides on the outer fringes of the habitable zone. Mars has a mass about one-ninth that of Earth; and it has a core, a mantle, a crust, and a thin, nearly cloudless atmosphere. It has polar ice caps and seasons that are nearly twice as long as Earth's because Mars takes nearly 2 Earth years to orbit the Sun. Mars and Earth spin on their axes at about the same rate, which

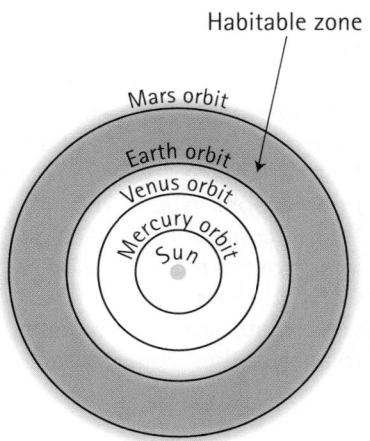

Habitable zone

Solar System

FIGURE 32.11 ▲
Earth resides on the inner side of the Sun's habitable zone, which is where conditions are favorable for life as we know it.

FIGURE 32.12 ▲
Earth, the blue planet. This famous photo was taken by Apollo 17 astronauts as they returned from the last manned mission to the Moon in 1972. Can you see that it was taken in the summer months of the Southern Hemisphere?

 **READING CHECK**

What cycle acts as Earth's thermostat?

FIGURE 32.13 ▶

(a) NASA's Mars Exploration Rover *Spirit,* with cameras mounted on the white mast, took panoramic photos of the Martian surface. (b) *Spirit* took photographs in June 2004 for this composite, true-color image of the region named Columbia Hills on Mars. The vehicle later traveled to the hills to analyze their composition.

(a)

(b)

means the lengths of their days are about the same. When Mars is closest to Earth, a situation that occurs once every 15 to 17 years, its bright, red color outshines the brightest stars.

The Martian atmosphere is about 95% carbon dioxide, with only about 0.15% oxygen. Yet because the Martian atmosphere is relatively thin, it doesn't trap heat via the greenhouse effect as much as Earth's and Venus's atmospheres do. So the temperatures on Mars are generally colder than on Earth, ranging from about 30°C in the day at the equator to a frigid −130°C at night. If you visit Mars, never mind your raincoat, for there is far too little water vapor in the atmosphere for rain. Even the ice at the planet's poles consists primarily of frozen carbon dioxide.

Conditions on Mars indicate that water was once abundant in its distant past. Channels on the Martian surface that appear to have been carved by water are seen by orbiting spacecraft. Landings on Mars, however, show it now to be a very dry and windy place.

In 2004, spacecraft orbiting Mars detected signs of the organic compound methane, CH_4, within the atmosphere. This is unusual because methane decomposes fairly rapidly with the conditions on Mars, which tells us that this compound is currently being produced. The likely source is ongoing volcanic activity, which could potentially melt underground ice into liquid water. Indeed, scientists have found evidence of the leakage of underground liquid water onto the surface occurring since we started surveying Mars from space. Once on the surface, this water would evaporate or freeze. Underground pools of volcanically warmed liquid water may harbor microscopic life forms.

Mars has two small moons—Phobos, the inner one, and Deimos, the outer. Both are potato-shaped and have cratered surfaces. They are likely captured asteroids.

Why are planets round? All parts of a forming planet pull close together by mutual gravitation. No "corners" form because they're simply pulled in. So gravity is the cause of the spherical shapes of planets and other large celestial bodies.

32.5 The Outer Planets Are Gaseous

The outer planets—Jupiter, Saturn, Uranus, and Neptune—are gigantic, gaseous, low-density worlds. They each formed from rocky and metallic cores that were much more massive than the terrestrial planets. ✓ The gravitational forces of these cores were strong enough to sweep up gases of the early planetary nebula, primarily hydrogen and helium.

The cores continued to collect gases until the Sun ignited and the solar wind blew away all remaining interplanetary gases. The core of Jupiter was the first to develop, and hence it had the longest time to collect gas before solar ignition. This is why Jupiter is the largest of the outer planets. Another commonality is that they all have ring systems, Saturn's being the most prominent. We will explore the outer planets in the order of their distance from the Sun.

✓ **READING CHECK**

What were the cores of the outer planets powerful enough to do in the early solar system?

Jupiter

Jupiter is the largest of all the planets in our solar system. Its yellow light in our night sky is brighter than that of any star (Figure 32.14). Jupiter spins rapidly about its axis in about 10 hours, a speed that flattens it so that its equatorial diameter is about 6% greater than its polar diameter. Jupiter's atmosphere is about 82% hydrogen, 17% helium, and 1% methane, ammonia, and other gaseous molecules.

Jupiter's mass is greater than the combined masses of all the other planets. More than half of Jupiter's volume is an ocean of liquid hydrogen. Beneath the hydrogen ocean lies an inner layer of hydrogen compressed into a liquid metallic state. In it are abundant free electrons that flow to produce Jupiter's enormous magnetic field.

If you're planning to visit Jupiter, choose one of its moons instead. It has over 60 of them, in addition to a faint ring. The four largest moons were discovered by Galileo in 1610; Io and Europa are about the size of our Moon, and Ganymede and Callisto are about as large as Mercury (Figure 32.15). Jupiter's moon Io has more volcanic activity than any other body in the solar system. Perhaps most intriguing of all, however, is Europa, whose surface is made of frozen water. As

FIGURE 32.14 ▲
Jupiter, with its moons Io (orange dot in front of planet) and Europa (white dot to right of planet), as seen from the *Voyager I* spacecraft in February 1979. The great red spot (lower left), larger than Earth, is a cyclonic weather pattern of high winds and turbulence.

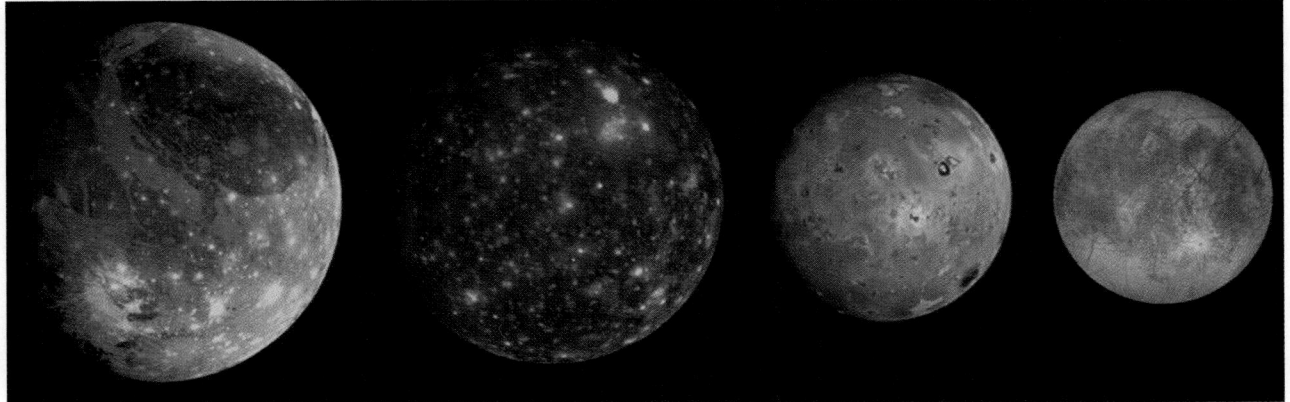

FIGURE 32.15 ▲
Shown here are Jupiter's four largest moons (from left to right): Ganymede, Callisto, Io, and Europa. They were discovered in 1610 by Galileo, who was the first to point the recently invented telescope toward the heavens. He noted the changing positions of these moons and concluded that they were orbiting Jupiter, which was a violation of the then widely held belief that all heavenly objects orbited the Earth. His discovery was revolutionary. In his honor, these four moons are known as the Galilean moons.

FIGURE 32.16 ▶
A model of the interior of Europa with a zoomed-in view of its ice-capped ocean, which, according to magnetic measurements, likely covered the entire sphere at one time.

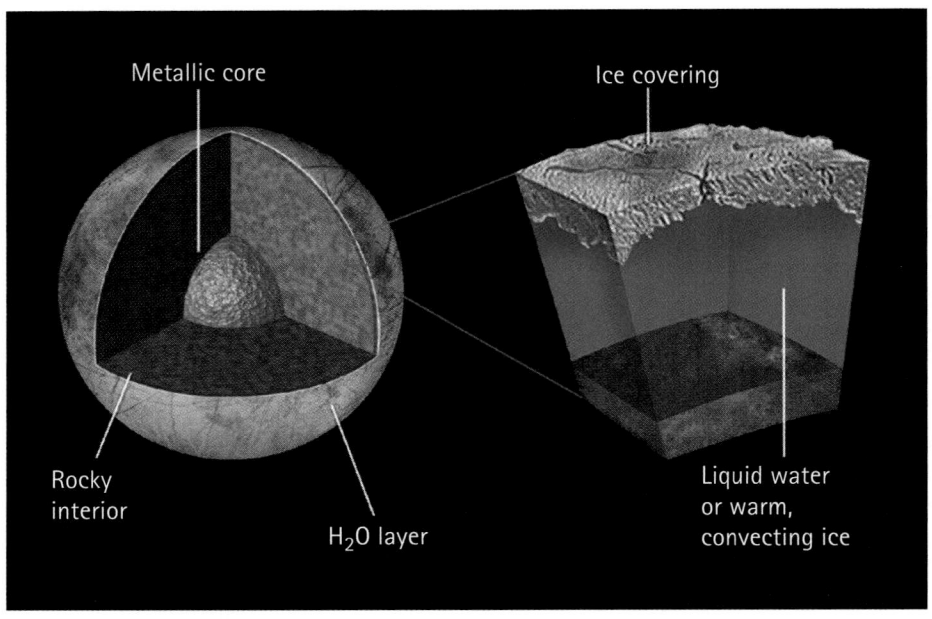

shown in Figure 32.16, deep beneath this ice is likely an ocean of water kept warm by the strong tidal forces from nearby Jupiter. If life were to be found anywhere in this solar system besides Earth, it would likely be on the floor of Europa's ocean adjacent to volcanic thermal vents. Such extraterrestrial life forms may be similar to the bizarre forms of life recently discovered adjacent to deep thermal vents on Earth's ocean floor. Alternatively, they may be single-cell organisms, such as bacteria. Then again, there may be nothing. So is life rare or common in this universe? The answer may be waiting for us right here in our own solar system.

Saturn

Saturn is one of the most remarkable objects in the sky, with its rings clearly visible through a small telescope. It is quite bright—brighter than all but two stars—and it is second only to Jupiter in mass and size. Saturn is twice as far from Earth as is Jupiter. It is composed primarily of hydrogen and helium, and it has the lowest density of any planet, only 0.7 times the density of water. These characteristics mean that Saturn would easily float in a bathtub, if the bathtub were large enough. Its low density and its 10.2-hour rapid spin produce more polar flattening than can be seen in the other planets. Notice its oblong nature in Figure 32.17.

Saturn's rings have been known for many years. They are only a few kilometers thick and are composed of chunks of frozen water and rocks,

Like Jupiter, Saturn radiates about twice as much heat energy as it receives from the Sun.

◀ **FIGURE 32.17**
Saturn surrounded by its famous rings, which are composed of rocks and ice.

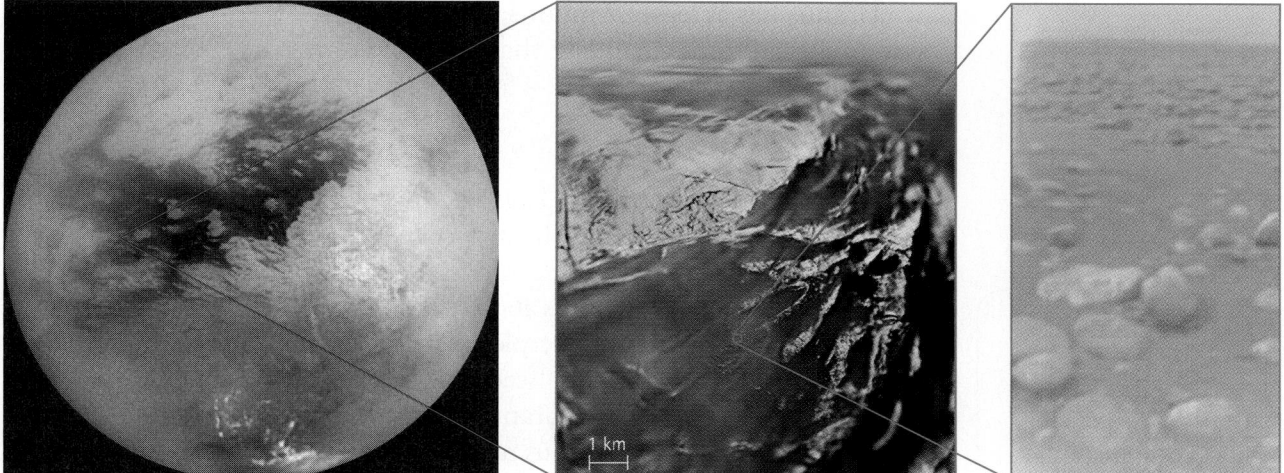

FIGURE 32.18 ▲
Images from Saturn's largest moon, Titan, taken by the *Cassini* spacecraft and its space probe, the *Huygens*, which successfully descended to the surface.

believed to be the material of a moon that never formed or the remnants of a moon torn apart by tidal forces.

Saturn has some 47 moons beyond its rings. The largest is Titan, 1.6 times as large as our Moon and even larger than Mercury. It spins once every 16 days and has a methane atmosphere (not likely from any biological source) with atmospheric pressure that is greater than Earth's. Its surface temperature is cold—roughly −170°C. A space probe sent by NASA and the European Space Agency landed on Titan in 2005. Remarkably, photos revealed a landscape similar to Earth's despite the fact that the materials are completely different (Figure 32.18). Lakes and streams are filled not with water but with liquid methane. Rocks are made of ice. Instead of lava, Titan has a flowing slush of ice and liquid ammonia. No life is expected to be found on this moon because of the intensely cold temperatures. Titan, however, holds an intriguing soup of organic molecules whose chemistry may provide a clue to what Earth was like during the time before life arose here.

Uranus

Uranus (pronounced "YUR-uh-nus," accent on the first syllable) is twice as far from Earth as Saturn is, and it can barely be seen with the naked eye. Uranus was unknown to ancient astronomers and not discovered as a planet until 1781. The *Voyager 2* spacecraft first visited this planet in 1986. It has a diameter 4 times that of Earth and a density slightly greater than that of water. So if you could place Uranus in a giant bathtub, it would sink. The most unusual feature of Uranus is its tilt. Its axis is tilted 98° to the perpendicular of its orbital plane, so it lies on its side (Figure 32.19). Unlike Jupiter and Saturn, it appears to have no appreciable internal source of heat. Uranus is a cold place.

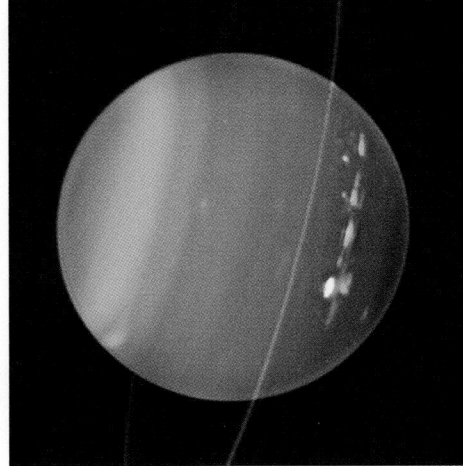

FIGURE 32.19 ▲
This infrared image of Uranus was gathered by the 10-m Keck Telescope in Hawaii on July 11–12, 2004. Astronomers believe that Uranus tilts on its rotational axis as a result of a collision it had with a large body early in the solar system's history. The bright white and blue spots in the southern hemisphere of Uranus are clouds. Methane in the upper atmosphere absorbs red light, giving Uranus its blue-green color.

Uranus has at least 27 moons, in addition to a complicated faint ring system. Recall from Chapter 7 that disturbances in the orbit of Uranus led to the discovery in 1846 of a farther planet, Neptune.

Neptune

Neptune has a diameter about 3.9 times that of Earth and a mass 17 times as great. Its atmosphere is mainly hydrogen and helium, with some methane and ammonia, which makes Neptune bluer than Uranus (Figure 32.20). Like Jupiter and Saturn, it emits about 2.5 times as much heat energy as it receives from the Sun. This is perplexing because calculations show that, like Uranus, Neptune should have already lost all of its original heat. One possible explanation is that, unlike Uranus, Neptune is still contracting. As discussed in Section 32.2, as gaseous material contracts it heats up.

The *Voyager 2* spacecraft flew by Neptune in 1989. It showed that Neptune has at least 13 moons in addition to a ring system. The largest moon, Triton, orbits Neptune in 5.9 days in a direction opposite to the planet's eastward spin. This suggests that Triton is a captured object. Triton's diameter is three-quarters the size of our Moon's diameter, and yet Triton is twice as massive as Earth's Moon. It has bright polar caps and geysers of liquid nitrogen.

Recent studies of Galileo's notebooks show that Galileo saw Neptune in December 1612 and again in January 1613. He was interested in Jupiter at the time, and so he merely plotted Neptune as a background star.

FIGURE 32.20 ▲
Cyclonic disturbances on Neptune in 1989 produced a great dark spot, which was larger than Earth and similar to Jupiter's Great Red Spot. The spot has now disappeared. The gray horizon in the foreground of this computer-generated montage is a close-up of Neptune's moon Triton, which has a composition and size similar to that of Pluto.

32.6 Earth's Moon

Earth's Moon is puzzling. It is close to the size of Mercury, which is a planet and *not* a moon. The composition of Earth's Moon is nearly the same as Earth's mantle. Furthermore, the Moon possesses a rather small iron core. To explain these and many other facts about the Moon, scientists have pieced together the following probable scenario for its origin.

During the early history of the solar system, the young Earth had a Mercury-sized companion that formed within an orbit close to that of Earth. Eventually, a random event, such as the passing of an asteroid or comet, caused our companion to fall toward and collide with Earth. The collision would have been massively spectacular, spewing debris everywhere while turning Earth fully molten. The impact sent Earth into a wild spin, rotating once every 5 hours. The debris soon collected as a ring around Earth, and then, within about 1000 years, the ring accreted (clumped together) into the Moon, as shown in Figure 32.21. This scenario is known as the *giant impact theory* of the origin of

A Mercury-sized planetesimal crashes into the young Earth, shattering both the planetesimal and our planet.

Hours later, our planet is completely molten and rotating very rapidly. Debris splashed out from Earth's outer layers is now in Earth orbit. Some debris rains back down on Earth, while some will gradually accrete to become the Moon.

Less than a thousand years later, the Moon's accretion is rapidly nearing its end, and relatively little debris still remains in Earth orbit.

FIGURE 32.21 ▲
Three steps to the formation of Earth's Moon. A Mercury-sized object collides with Earth, which turns molten. Debris collects in a ring that accretes into the Moon, which is quite close to the rapidly rotating Earth. Over the next billion years, tidal forces slow the rate of Earth's rotation while also causing the Moon to move farther away.

FIGURE 32.22 ▲
Earth and the Moon as photographed in 1977 from the *Voyager I* spacecraft on its way to Jupiter and Saturn.

Earth's Moon. It explains why the Moon is so large (we started out as twin planets) and why its composition is similar to Earth's (it formed from our mantle and our mantle formed from it).

From a distance, Earth and the Moon still resemble a twin planet system, as you can see in Figure 32.22. Compared to Earth, however, the Moon is relatively small, with a diameter of about the distance from San Francisco to New York City. The Moon is too small with too little gravitational pull to have an atmosphere, and so, without weather, the only eroding agents have been meteoroid impacts. The Moon's gravity, however, is strong enough to hold down an astronaut, as shown in Figure 32.23. The astronaut's weight is only about one-sixth that experienced on Earth.

The Phases of the Moon

✓ The Moon is a sphere and sunshine always lights up half of the Moon's spherical surface. The Moon shows us different amounts of its sunlit half as it circles Earth each month. These changes are the **Moon phases** (Figure 32.24). The Moon cycle begins with the **new Moon.** In this phase, its dark side faces us and we see darkness. This occurs when the Moon is between Earth and the Sun (position 1 in Figure 32.25).

During the next 7 days, we see more and more of the Moon's sunlit side (position 2 in Figure 32.25). The Moon is going though its waxing crescent phase (waxing means "increasing"). At the first quarter, the

FIGURE 32.23 ▲
Edwin E. Aldrin, Jr., one of the three Apollo 11 astronauts, stands on the dusty lunar surface. To date, 12 people have stood on the Moon.

| Waxing crescent | First quarter | Waxing gibbous | Full moon | Waning gibbous | Last quarter | Waning crescent |

FIGURE 32.24 ▲
The Moon in its various phases.

angle between the Sun, the Moon, and Earth is 90°. At this time, we see half the sunlit part of the Moon (position 3 in Figure 32.25).

   During the next week, we see more and more of the sunlit part. The Moon is going through its waxing gibbous phase (position 4 in Figure 32.25). (*Gibbous* means "more than half.") We see a **full Moon** when the sunlit side of the Moon faces us squarely (position 5 in Figure 32.25). At this time, the Sun, Earth, and the Moon are lined up, with Earth in between. To view this full Moon you need to be on the nighttime side of Earth, at sunset when the full Moon rises from the east, or at sunrise when it sets in the west.

   The cycle reverses during the following two weeks, as we see less and less of the sunlit side while the Moon continues in its orbit. This movement produces the waning gibbous, last quarter, and waning crescent phases. (*Waning* means "shrinking.")

✔ **READING CHECK**

How much of the Moon is always lit up by the Sun?

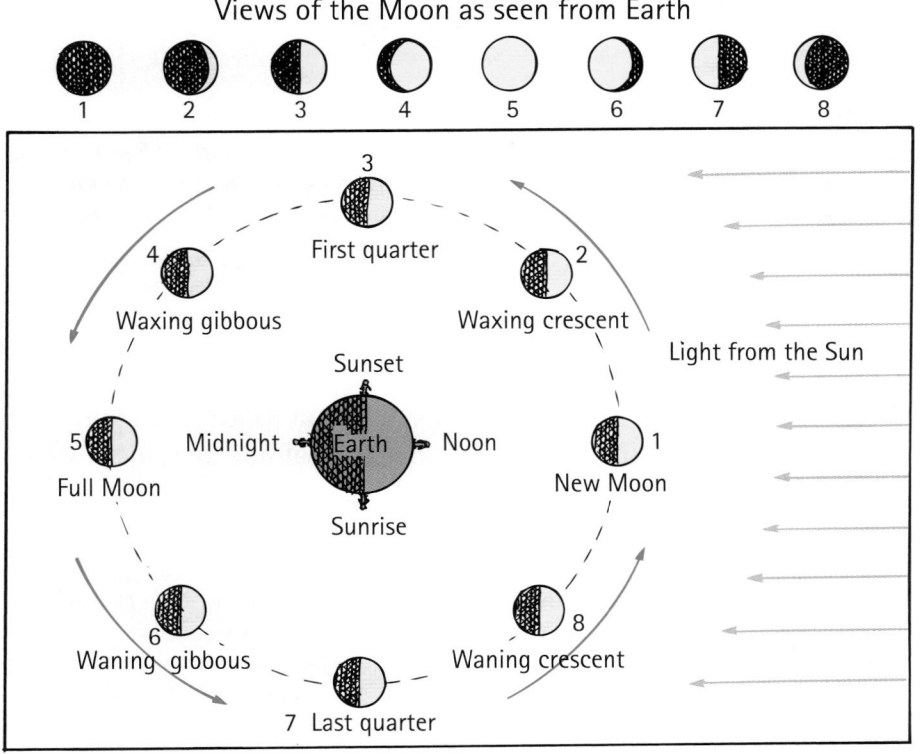

Views of the Moon as seen from Earth

◀ **FIGURE 32.25**
Sunlight always illuminates half of the Moon. As the Moon orbits Earth, we see varying amounts of its sunlit side. One lunar phase cycle takes 29.5 days.

CONCEPT CHECK

Can a full Moon be seen at noon? Can a new Moon be seen at midnight?

Check Your Answer

Inspection of Figure 32.25 shows that, at noontime, you would be on the wrong side of Earth to view the full Moon. Likewise, at midnight, the new Moon would be absent. The new Moon is in the sky in the daytime, not at night.

The first human witnesses of the Moon's backside were Apollo 8 astronauts, who orbited the Moon in 1968.

Why One Side Always Faces Us

From Earth, we see only a single side of the Moon. The familiar facial features of the "man in the Moon" are always turned toward us on Earth. Does this mean that the Moon doesn't spin about its axis as Earth does daily? No, relative to the stars, the Moon in fact does spin, although quite slowly—about once a month. This explains why the same side of the Moon always faces Earth (Figure 32.26). After you answer the following Concept Check, we'll explore why.

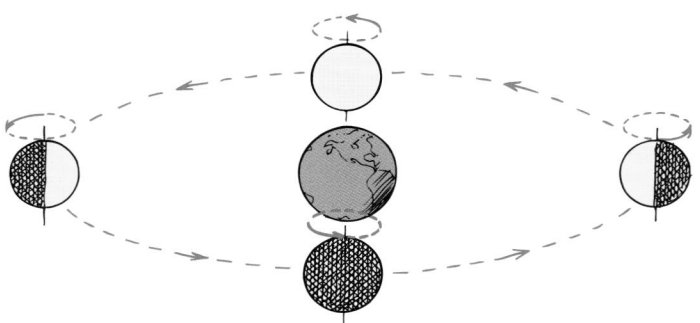

FIGURE 32.26 ▲
The Moon spins about its own polar axis just as often as it circles Earth. So as the Moon circles Earth, it spins so that the same side (shown here in yellow) always faces Earth. In each of the four successive positions shown here, the Moon has spun one-quarter of a turn.

CONCEPT CHECK

A friend says that the Moon does not spin about its axis, and the evidence for a nonspinning Moon is the fact that its same side always faces Earth. What do you say?

Check Your Answer

Place a quarter and a penny on a table. Pretend the quarter is Earth and the penny the Moon. Keeping the quarter fixed, revolve the penny around it in such a way that Lincoln's head is always pointed to the center of the quarter. Ask your friend to count how many rotations the penny makes in one revolution (orbit) around the quarter. He or she will see that it rotates once with each revolution. The key concept is that the Moon takes the same amount of time to complete one rotation as it does to revolve around the Earth.

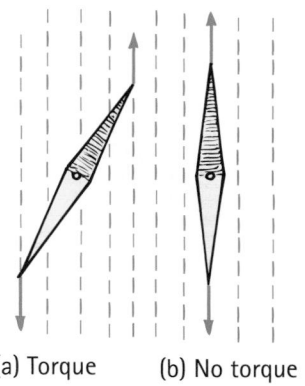

FIGURE 32.27 ▶

(a) When the compass needle is not aligned with the magnetic field (dashed lines), the forces represented by the blue arrows at either end produce a pair of torques that rotate the needle. (b) When the needle is aligned with the magnetic field, the forces no longer produce torques.

(a) Torque   (b) No torque

Think of a compass needle that lines up with a magnetic field. This lineup is caused by the pull of the magnetic field, as shown in Figure 32.27. The needle rotates counterclockwise until it aligns with the magnetic field. In a similar manner, the Moon aligns with Earth's gravitational field.

We know from the law of universal gravitation that gravity weakens with the inverse square of distance, so the side of the Moon nearer to Earth is gravitationally pulled more than the farther side. This stretches the Moon out to a football shape. (The Moon does the same to Earth and gives us tides.) If its long axis doesn't line up with Earth's gravitational field, the forces act as shown in Figure 32.28. Like a compass in a magnetic field, it turns into alignment. So the Moon lines up with Earth in its monthly orbit. One hemisphere always faces us. We say the Moon is "tidally locked."

Eclipses

Although the Sun is 400 times larger in diameter than the Moon, it is also about 400 times as far away. So from Earth, both the Sun and Moon measure the same angle (0.5°) and appear to be the same size in the sky. This coincidence allows us to see solar eclipses.

Both Earth and the Moon cast shadows when sunlight shines on them. When the path of either of these bodies crosses into the shadow cast by the other, an eclipse occurs. A **solar eclipse** occurs when the Moon's shadow falls on Earth. Because of the large size of the Sun, the rays taper to provide an umbra and a surrounding penumbra, as shown in Figures 32.29 and 32.30.

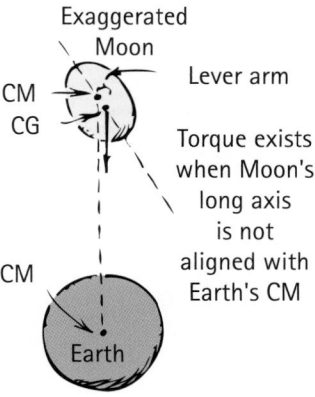

FIGURE 32.28 ▲

When the long axis of the Moon is not aligned with Earth's gravitational field, Earth exerts a torque that rotates the Moon into alignment (CM: center of mass; CG: center of gravity).

FIGURE 32.29 ▲

A solar eclipse occurs when the Moon passes in front of the Sun as seen from Earth. The Moon's shadow has two portions: a dark, central umbra surrounded by the lighter penumbra. A total eclipse is seen from within the umbra and may last several minutes.

During the time of the dinosaurs, a day on Earth was only about 19 hours. Today our days have slowed down to about 24 hours. Billions of years from now, a day will last about 47 days. Why? Because tidal forces from the Moon are slowing us down.

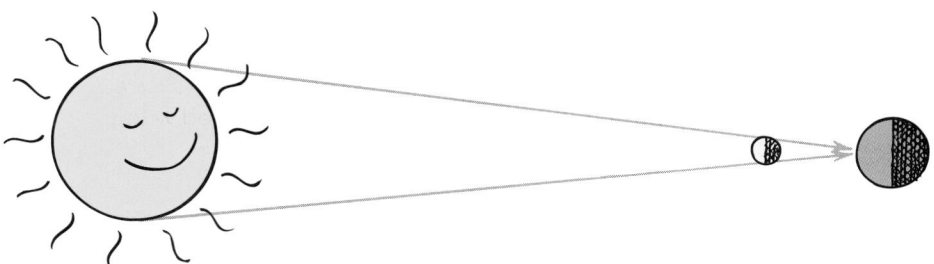

FIGURE 32.30 ▲

Geometry of a solar eclipse. During a solar eclipse, the Moon is directly between the Sun and Earth and the Moon's shadow is cast on Earth. Because of the small size of the Moon and tapering of the solar rays, a solar eclipse occurs only on a small area of Earth.

An observer in the umbra part of the shadow experiences darkness during the day—a total eclipse, totality. Totality begins when the Sun disappears behind the Moon and ends when the Sun reappears on the other edge of the Moon. The average time of totality at any location is only about 2 or 3 minutes. The eclipse time in any location is brief because of the Moon's motion. During totality, what appears in the sky is an eerie black disk surrounded by the pearly white streams of the corona, as was shown in Figure 32.7. It is a rare experience one can never forget. Many solar eclipse enthusiasts travel the world to view this inspiring natural phenomenon.

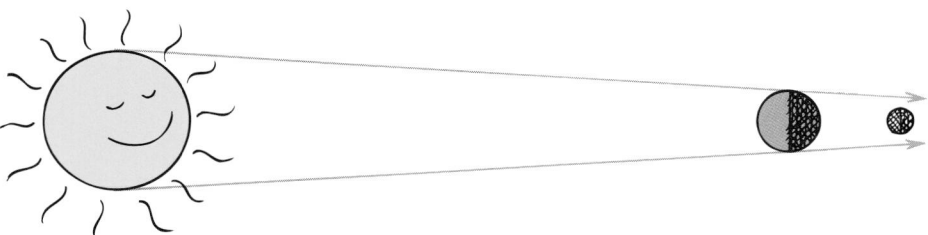

FIGURE 32.31 ▲

A lunar eclipse occurs when Earth is directly between the Moon and the Sun and Earth's shadow is cast on the Moon.

FIGURE 32.32 ▲

A fully eclipsed Moon is not completely dark in the shadow of Earth but is quite visible. This is because Earth's atmosphere acts as a lens and refracts light into the shadow region—sufficient light to faintly illuminate the Moon.

The alignment of Earth, the Moon, and the Sun also produces a **lunar eclipse** when the Moon passes into the shadow of Earth, as shown in Figure 32.31. Just as all solar eclipses involve a new Moon, all lunar eclipses involve a full Moon. They may be partial or total. All observers on the dark side of Earth see a lunar eclipse at the same time. Interestingly enough, when the Moon is fully eclipsed, it is still visible, as is shown and discussed in Figure 32.32.

Why are eclipses relatively rare events? This has to do with the different orbital planes of Earth and the Moon. Earth revolves around the Sun in a flat planar orbit. The Moon similarly revolves about Earth in a flat planar orbit. But the planes are slightly tipped with respect to each

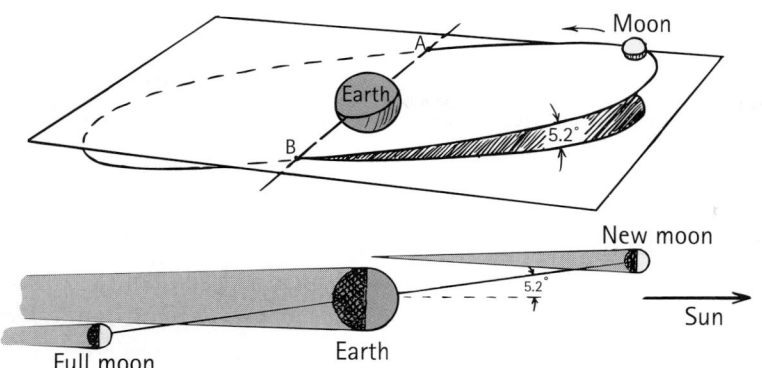

◀ **FIGURE 32.33**
The Moon orbits Earth in a plane tipped 5.2° relative to Earth's orbit around the Sun. A solar or lunar eclipse occurs only when the Moon intersects the Earth–Sun plane (points A and B) at the precise time of a three-body alignment.

other—a 5.2° tilt, as shown in Figure 32.33. If the planes weren't tipped, eclipses would occur monthly. Because of the tip, eclipses occur only when the Moon intersects the Earth–Sun plane at the time of a three-body alignment (Figure 32.34). This occurs about twice per year, which is why there are at least two solar eclipses per year (visible only from certain locations on Earth).

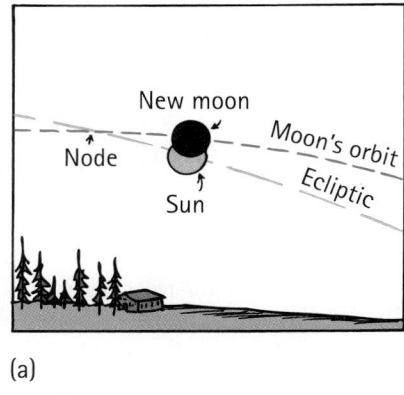

(a)

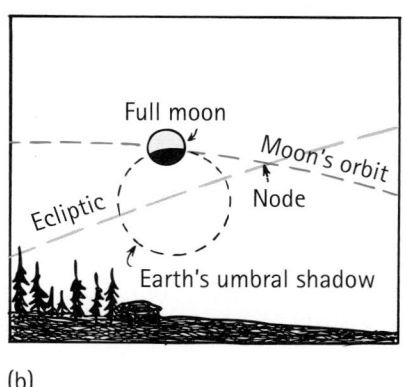

(b)

FIGURE 32.34 ▲
A total eclipse can occur only when the Moon's orbit intersects with the plane of Earth's orbit, which is the ecliptic. A solar eclipse occurs during the day as the new Moon passes in front of the Sun. A lunar eclipse happens only at night when the full Moon passes through Earth's shadow.

CONCEPT CHECK

1. Does a solar eclipse occur at the time of a full Moon or a new Moon?

2. Does a lunar eclipse occur at the time of a full Moon or a new Moon?

Check Your Answers

1. A solar eclipse occurs at the time of a new Moon, when the Moon is directly in front of the Sun. Then the shadow of the Moon falls on part of Earth.

2. A lunar eclipse occurs at the time of a full Moon, when the Moon and Sun are on opposite sides of Earth. Then the shadow of Earth falls on the full Moon.

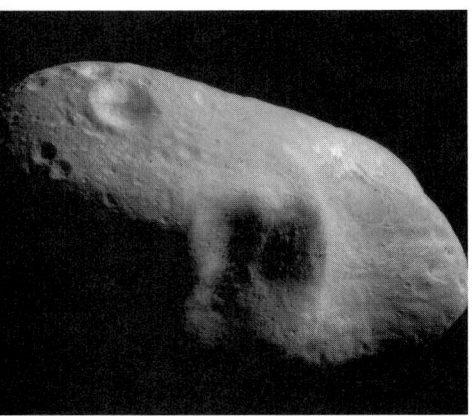

FIGURE 32.35 ▲
The asteroid Eros is about 40 km long, and, like other small objects in the solar system, is not spherical.

FIGURE 32.36 ▲
A meteor is produced when a meteoroid enters Earth's atmosphere. Most are sand-sized grains, which are seen as "falling" or "shooting stars."

FIGURE 32.37 ▲
The Barringer Crater in Arizona, made 25,000 years ago by an iron meteorite with a diameter of about 50 m. The crater extends 1.2 km across and reaches 200 m deep.

32.7 Failed Planet Formation

In three regions of our solar system, we find the remains of material that failed to collect into planets. These regions are the asteroid belt, the Kuiper belt, and the Oort Cloud.

The Asteroid Belt and Meteors

The **asteroid belt** is a collection of rocks located between the orbits of Mars and Jupiter. More than 150,000 asteroids have been cataloged so far, but many more no doubt have yet to be discovered. They come in all shapes and sizes, but the largest asteroid, Ceres, is just under a thousand kilometers in diameter. Although Ceres is large enough to be fairly round, most asteroids are shaped more like a potato, as shown in Figure 32.35. If all the asteroids were scrunched together, they would make a sphere less than half the size of our Moon.

Asteroid fragments known as **meteoroids** frequently find their way to Earth, where they are heated white-hot by friction with the atmosphere. As they descend with a fiery glow, they are called **meteors** (Figure 32.36). If the meteoroid is large enough, it may survive to reach the surface, where it is called a **meteorite**. Most meteoroids, meteors, and meteorites are from asteroids, but many are also from comets, as we discuss later.

Fortunately, smaller meteorites hit us more frequently than larger ones do. About 200 tons of small meteorites strike Earth every day. Every 10,000 years or so we are hit with a meteorite big enough to create a large crater, such as the one shown in Figure 32.37. Every 100 million years or so we are hit with one big enough—about 10 km in diameter—to cause a mass extinction, as occurred 65 million years ago at the end of the Cretaceous period, discussed in Chapter 29.

Meteorites fall all over our planet but the easiest place to find them is on the icy white surfaces found in polar regions. Are you looking to collect your own meteorites? Head south to Antarctica!

The Kuiper Belt and Plutoids

Beyond Neptune at a distance from about 30 to 50 AU is a region known as the **Kuiper belt** (pronounced "KI-pur," "KI" as in *kite*, rhymes with *hyper*). The Kuiper belt is occupied by many rocky, ice-covered objects. The most well-known Kuiper belt object is Pluto, which until recently was classified as a planet. Since its discovery in 1930, however, astronomers knew that Pluto was quite different from all the other known planets. For example, Pluto orbits at an angle to the plane of the solar system. Also, Pluto is quite small, being only one-seventh as massive as our Moon.

Starting in the 1990s, astronomers began discovering many more Kuiper belt objects as large as or larger than Pluto. So in 2008 these

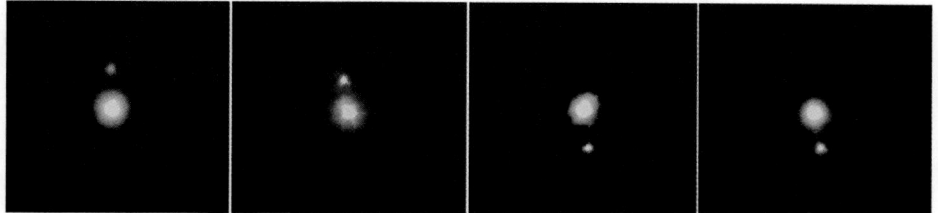

◀ **FIGURE 32.38**
These infrared images of Pluto being orbited by its moon Charon are fuzzy because of the small size of these bodies and their great distance from Earth.

Pluto-sized Kuiper belt objects were officially classified as **plutoids.** The main reason they do not meet full planet status is that they have yet to collect all the material in their orbital paths. In the outer edges of our solar system, however, matter is simply too sparse for that to happen. Interestingly, if the Kuiper belt were more dense with material, then these dwarf planets could have served as cores for additional jovian planets. But that never happened, and so the Kuiper belt is another zone of failed planet formation.

Space probes have yet to visit any of the plutoids of the Kuiper belt. However, Pluto and its moon Charon, shown in Figure 32.38, are due to be visited by the *New Horizons* spacecraft in 2015. We may have already had a preview, however, when the *Voyager 2* spacecraft took pictures of Neptune's moon Triton. Astronomers now suspect that Triton is a Kuiper belt plutoid pulled off course and captured into orbit around Neptune.

The larger Kuiper belt objects, such as Pluto, have a fair amount of inertia and so are not so easily thrown off course. Lighter Kuiper belt objects, however, are thrown off course quite frequently. Sometimes they are thrown toward the Sun, where the added heat and solar wind cause the ice and other volatile materials to be ejected, always in a direction away from the Sun, as illustrated in Figure 32.39. We see these objects as **comets,** which are characterized by their long and sometimes

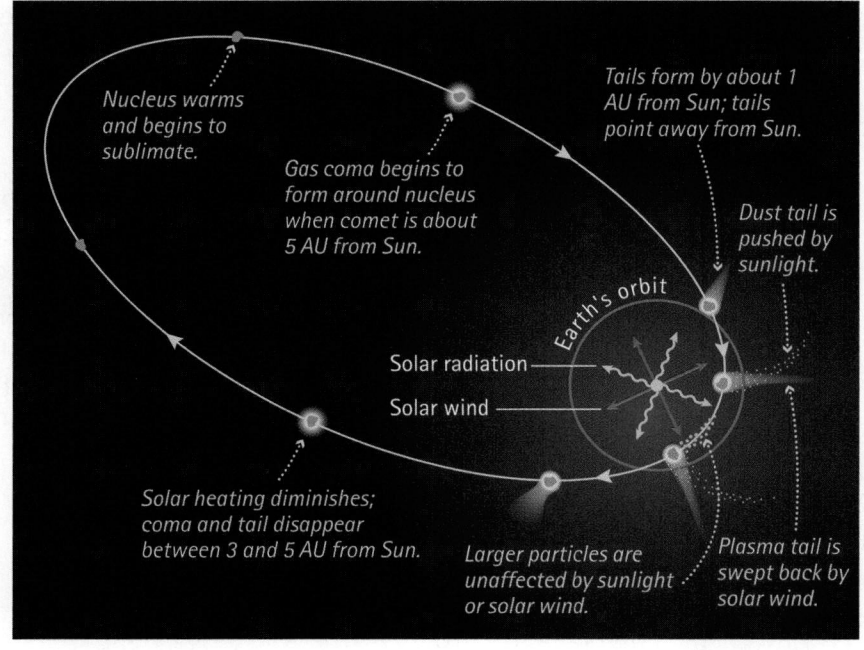

◀ **FIGURE 32.39**
The comet warms as it gets closer to the Sun and initially develops a coma, which is a halo of gases surrounding the comet nucleus. From this coma arises the tail, which is blown outward by solar wind. Note how the tail always extends away from the Sun. Most comets never make this journey and instead remain perpetually frozen within the outer reaches of our solar system.

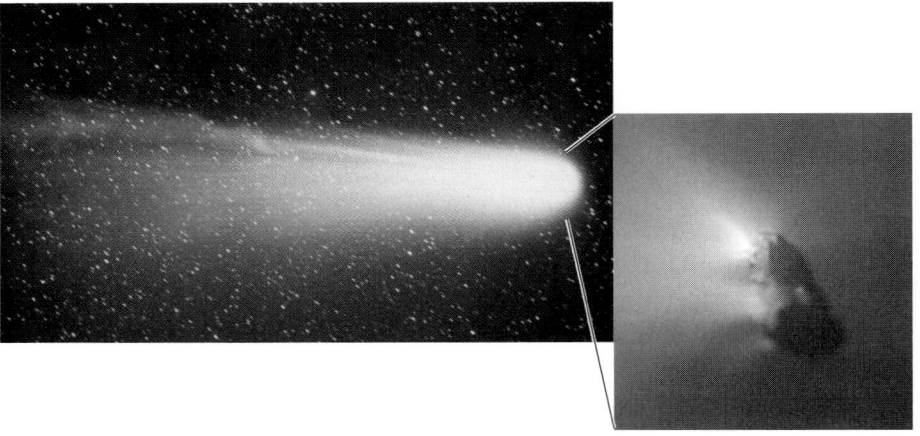

FIGURE 32.40 ▲
Observations of Comet Halley have been recorded for thousands of years. Although it usually provides a brilliant display, its last visit in 1986 was not so spectacular when viewed from Earth. We were ready, though, with space probes that flew close enough to Halley to capture dramatic images of its nucleus.

READING CHECK

Why are large Kuiper belt objects, such as Pluto, not classified as planets?

quite brilliant tails. Comets that come from the Kuiper belt tend to have orbital periods of less than 200 years. An example is Comet Halley, which returns to the inner solar system every 76 years—once in an average lifetime (Figure 32.40). Its next scheduled return is in 2061.

The Oort Cloud and Comets

As the jovian planets grew, their gravitational pulls became stronger, which made them even more effective at pulling in additional interplanetary debris. Not all debris, however, was pulled fully into the jovian planets. In many cases, a chunk of rock or ice just missing a planet was instead whipped around the planet and then flung violently outward in any direction. Over billions of years this created a sphere of far-out objects just barely held to our solar system. We refer to this collection of far-out objects as the **Oort cloud** (*Oort* rhymes with *court*).

Evidence suggests that the Oort cloud consists of a trillion objects extending as far out as 50,000 AU, which brings the cloud about a quarter of the way to the nearest star. A few of these objects periodically fall toward and then around the Sun, where they appear as comets. The orbital periods of comets originating from the Oort cloud is on the order of thousands or even millions of years. They come from nearly any angle. Whether the comet comes from the Kuiper belt or the Oort cloud, it still has the potential for

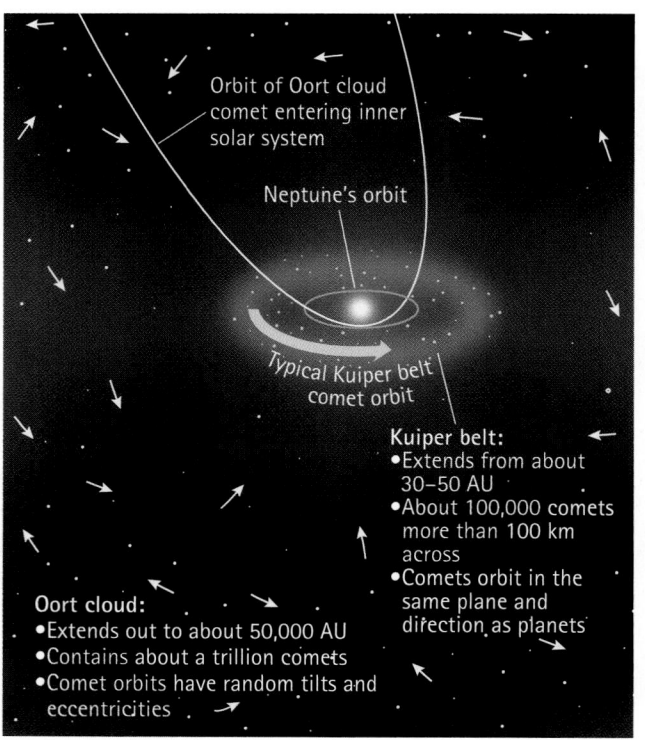

Orbit of Oort cloud comet entering inner solar system

Neptune's orbit

Typical Kuiper belt comet orbit

Kuiper belt:
•Extends from about 30–50 AU
•About 100,000 comets more than 100 km across
•Comets orbit in the same plane and direction as planets

Oort cloud:
•Extends out to about 50,000 AU
•Contains about a trillion comets
•Comet orbits have random tilts and eccentricities

◀ **FIGURE 32.41**
There are two major reservoirs for comets: the Kuiper belt and the Oort cloud.

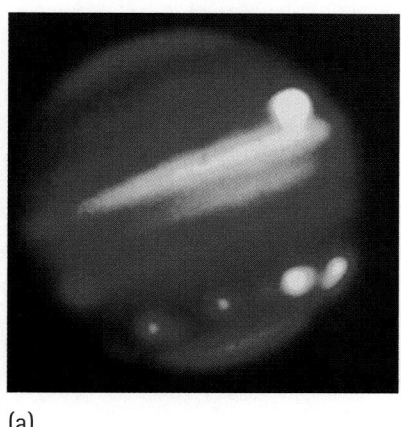
(a)

(b)

◀ **FIGURE 32.42**
Comet Shoemaker-Levy was already broken up into a string of objects just before it collided with Jupiter in 1994. The image on the left (a) shows an infrared view of the collision, which produced much heat as well as visible scars (the black dots), as shown in the photograph on the right (b).

colliding with a planet. In 1994, Comet Shoemaker-Levy collided spectacularly with Jupiter, as shown in Figure 32.42. The large meteorite that collided with Earth 65 million years ago causing the mass extinction of the dinosaurs may have been a comet.

Comet tails leave behind a shower of particles. Each year Earth passes through the remnants of comet tails that create annual meteor showers, as indicated in Table 32.2. Meteor showers are beautiful to watch. Just go outside, look up at the sky, and with clear moonless skies every minute or so you will see a shooting star. Each streak is a tiny chip of a comet, once so very far away, that has fallen into Earth's neighborhood (Figure 32.43).

One of NASA's goals is to detect up to 90% of all large, near-Earth objects. If we can detect a dangerous space fragment early enough, we can take actions to alter its orbital path sufficiently to avoid impending disaster.

TABLE 32.2 Meteor Shower Data

Shower Name	Radiant*	Dates	Peak Dates	Meteors per Hour
Quadrantids	Pegasus	Jan. 1–6	Jan. 3	60
Eta Aquarids	Aquarius	May 1–10	May 6	35
Perseids	Perseus	Jul. 23–Aug. 20	Aug. 12	75
Orionids	Orion	Oct. 16–27	Oct. 22	25
Germids	Germini	Dec. 7–15	Dec. 13	75

* Meteors appear to radiate from a certain region of the sky, appropriately called a *radiant*. Radiants refer to constellations. See Chapter 27 for more on where the various constellations are located in the night sky.

CONCEPT CHECK

Of the asteroid belt, the Kuiper belts, and the Oort cloud: (1) Which is closest to the Sun? (2) Which generate comets? (3) Which gives us the most meteorites? (4) Which gives us the brightest meteor showers? (5) Which consists of fragments that never formed into planets?

Check Your Answers

(1) The asteroid belt. (2) The Kuiper belt and Oort cloud. (3) The asteroid belt. (4) The Kuiper belt and Oort cloud. (5) All of them.

FIGURE 32.43 ▲
When Earth crosses the orbit of a comet, we see a meteor shower.

32 CHAPTER REVIEW

KEY TERMS

Asteroid belt A region between the orbits of Mars and Jupiter containing small, rocky, planetlike fragments that orbit the Sun. These fragments are called *asteroids*, which in Latin means "small star."

Astronomical unit (AU) The average distance between Earth and the Sun, which is about 1.5×10^8 km (about 9.3×10^7 mi).

Comet A body composed of ice and dust that orbits the Sun, usually in a very eccentric orbit, and that casts a luminous tail produced by solar radiation pressure when close to the Sun.

Ecliptic The plane of Earth's orbit around the Sun. All major objects of the solar system orbit roughly within this same plane.

Full Moon The phase of the Moon when its sunlit side faces Earth.

Inner planets The four planets orbiting within 2 AU of the Sun, including Mercury, Venus, Earth, and Mars. These planets are all rocky planets, known as the *terrestrial* planets.

Kuiper belt (pronounced KI-pur) The disk-shaped region of the sky beyond Neptune populated by many icy bodies and a source of short-period comets.

Lunar eclipse The phenomenon whereby the shadow of Earth falls upon the Moon, producing the relative darkness of the full Moon.

Meteor The streak of light produced by a meteoroid burning in Earth's atmosphere; a "shooting star."

Meteorite A meteoroid, or a part of a meteoroid, that has survived passage through Earth's atmosphere to reach the ground.

Meteoroid A small rock in interplanetary space, which can include a fragment of an asteroid or comet.

Moon phases The cycles of change of the "face" of the Moon, changing from *new*, to *waxing*, to *full*, to *waning*, and back to *new*.

Nebular theory The idea that the Sun and planets formed together from a cloud of gas and dust, a *nebula*.

New Moon The phase of the Moon when darkness covers the side facing Earth.

Oort cloud The region beyond the Kuiper belt populated by trillions of icy bodies and a source of long-period comets.

Outer planets The four planets orbiting beyond 2 AU of the Sun including Jupiter, Saturn, Uranus, and Neptune—all gaseous and known as the *jovian* planets.

Planets The major bodies orbiting the Sun that are massive enough for their gravity to make them spherical and small enough to avoid having nuclear fusion in their cores. They also have successfully cleared all debris from their orbital paths.

Plutoid A relatively large icy body, such as Pluto, originating within the Kuiper belt.

Solar eclipse The phenomenon whereby the shadow of the Moon falls upon Earth, producing a region of darkness in the daytime.

Sunspots Temporary, relatively cool and dark regions on the Sun's surface.

REVIEW QUESTIONS

The Solar System Is Mostly Empty Space

1. How many known planets are in our solar system?
2. What plutoid was downgraded from planetary status in 2006?

Solar Systems Form from Nebula

3. How are the outer planets different from the inner planets aside from their location?
4. Why does a nebula spin faster as it contracts?
5. According to the nebular theory, did the planets start forming before or after the Sun ignited?

The Sun Is Our Prime Source of Energy

6. What happens to the amount of the Sun's mass as it "burns"?
7. What are sunspots?
8. What is the age of the Sun?

The Inner Planets Are Rocky

9. Why are days on Mercury very hot and the nights very cold?
10. What two planets are evening or morning "stars"?
11. What gas makes up most of the Martian atmosphere?

The Outer Planets Are Gaseous

12. What main feature do the outer planets have in common?
13. How tilted is Uranus's axis?
14. Why is Neptune bluer than Uranus?

Earth's Moon

15. Why does the Moon have no atmosphere?
16. Where are the Sun and the Moon located at the time of a new moon?
17. Why don't eclipses occur monthly, or nearly monthly?

Failed Planet Formation

18. Between the orbits of what two planets is the asteroid belt located?
19. What is the difference between a meteor and a meteorite?
20. What is a falling star?

THINK AND DO

1. Find yourself a Ping-Pong ball on the next clear day when the Moon is out. Hold the Ping-Pong ball with your arm stretched out toward where the Moon is so that the ball overlaps the Moon. Look carefully at how the Ping-Pong ball is lit by the Sun. Notice that this is the same way that the Moon is also lit by the Sun! For an example, see Question 9 of Think and Explain. To see the different phases that the Moon would have if it were elsewhere in the sky, move your Ping-Pong ball around.

2. Simulate the lunar phases. Insert a pencil into a Styrofoam ball. This will be the Moon. Position a lamp (representing the Sun) in another room near the doorway. Hold the ball in front and slightly above yourself. Slowly turn yourself around keeping the ball in front of you as you move. Observe the patterns of light and shadow on the ball. Relate this to the phases of the Moon.

3. When viewed from the North Pole, Earth spins counterclockwise. This means that the stars appear to move in the opposite direction, which is toward the west. Just as Earth spins counterclockwise, the Moon revolves around us also counterclockwise, which is eastward. Look where the Moon is located one night, say 11:00. Look for the Moon the next night at the same time and you'll see that it has moved eastward (a counterclockwise direction) from where it was the previous night.

4. The crescent moon always points to the Sun. You can use this fact to estimate your latitude. Down by the equator (0° latitude) the setting crescent moon lies flat to the horizon, while up close to the North Pole (90° latitude) the crescent moon stands on its end. Next time you see the crescent moon close to the horizon, look carefully at its angle and use that to estimate your latitude. This method is most accurate when the Moon is passing through the ecliptic, that is, when the Moon falls in line with the planets.

THINK AND COMPARE

1. Rank the following in order of increasing mass: Mercury, Venus, Earth, Mars.
2. Rank the following in order of increasing distance from the Sun: the outer planets, the Kuiper belt, the Oort cloud.
3. Rank the following in order of increasing density: Earth, Saturn, Pluto.
4. Rank the following in order of increasing volume: Jupiter, Earth, the Sun.
5. Rank the following in order of increasing density: the core of Jupiter, the core of Earth, the core of the Sun.

THINK AND EXPLAIN

1. According to the nebular theory, what happens to a nebula as it contracts under the force of gravity?
2. What happens to the shape of a nebula as it contracts and spins faster?
3. What energy processes make the Sun shine? In what sense can it be said that gravity is the prime source of solar energy?
4. The greenhouse effect is very pronounced on Venus but doesn't exist on Mercury. Why?
5. What does Jupiter have in common with the Sun that the terrestrial planets don't? What differentiates Jupiter from a star?
6. Why are the seasons on Uranus different from the seasons on any other planet?

Uranus

7. Why are many craters evident on the surface of the Moon but not on the surface of Earth?
8. Is the fact that we see only one side of the Moon evidence that the Moon spins or that it doesn't? Defend your answer.
9. Photograph (a) shows the Moon partially lit by the Sun. Photograph (b) shows a Ping-Pong ball in sunlight. Compare the positions of the Sun in the sky when each photograph was taken.

(a)   (b)

10. If we never see the back side of the Moon, would an observer on the back side of the Moon ever see Earth?
11. In what alignment of Sun, Moon, and Earth does a solar eclipse occur?
12. In what alignment of Sun, Moon, and Earth does a lunar eclipse occur?
13. What does the Moon have in common with a compass needle?
14. If you were on the Moon and you looked up and saw a full Earth, would it be nighttime or daytime on the Moon?
15. If you were on the Moon and you looked up and saw a new Earth, would it be nighttime or daytime on the Moon?
16. Do astronomers make stellar observations during the full Moon part of the month or during the new Moon part of the month? Does it make a difference?
17. Nearly everybody has witnessed a lunar eclipse, but relatively few people have seen a solar eclipse. Why?
18. Why are meteorites so much more easily found on Antarctica than on other continents?
19. What would be the consequence of a comet's tail sweeping across Earth?
20. In terms of the conservation of mass, describe why comets eventually burn out.

THINK AND SOLVE

1. Knowing that the speed of light is 300,000 km/s, show that it takes about 8 minutes for sunlight to reach Earth.
2. How many days does sunlight take to travel the 50,000 AU from the Sun to the outer reaches of the Oort cloud?
3. The light-year is a standard unit of distance used by astronomers. It is the distance light travels in 1 Earth year. In units of light-years, what is the approximate diameter of our solar system including the outer reaches of the Oort cloud? (Assume that one light year equals 63,000 AU.)

READINESS ASSURANCE TEST (RAT)

If you have a good handle on this chapter—if you really do—then you should be able to score 7 out of 10 on this RAT. Check your answers with your teacher. If you score less than 7, study further before moving on.

Choose the best answer to the following.

1. The Sun contains what percentage of the solar system's mass?
   (a) About 35%.
   (b) 85%.
   (c) The percentage varies over time.
   (d) Over 99%.

2. The solar system is like an atom in that both
   (a) are governed principally through the electric force.
   (b) consist of a central body surrounded by objects moving in elliptical paths.
   (c) are composed of plasma.
   (d) are mainly empty space.

3. The nebular theory is based upon the observation that the solar system
   (a) is highly ordered, indicating it formed in a stepwise manner from physical processes.
   (b) has a structure much like an atom.
   (c) is highly complex and appears to have been built by chaotic processes.
   (d) appears to be very old.

4. Where is the Sun located when you view a full Moon?
   (a) Directly in back of you.
   (b) Directly in front of you.
   (c) To your right.
   (d) The Sun could be anywhere.

5. Each second, the burning Sun's mass
   (a) increases.
   (b) remains unchanged.
   (c) decreases.

6. What evidence tells us that Mars was at one time wetter than it presently is?
   (a) The frozen polar ice caps.
   (b) Recorded history that Mars was once blue.
   (c) Dried-up ocean beds and river channels.
   (d) A halo of water vapor in orbit around the planet.

7. When the Moon assumes its characteristic thin crescent shape, the position of the Sun is
   (a) almost directly in back of the Moon.
   (b) almost directly behind Earth, so that Earth is between the Sun and the Moon.
   (c) at right angles to the line between the Moon and Earth.

8. When the Sun passes between the Moon and Earth, we have
   (a) a lunar eclipse.
   (b) a solar eclipse.
   (c) met our end.

9. Asteroids orbit the
   (a) Moon.　　(b) Earth.
   (c) Sun.　　(d) All of the above.
   (e) None of the above.

10. With each pass of a comet about the Sun, the comet's mass
    (a) remains virtually unchanged.
    (b) actually increases.
    (c) is appreciably reduced.

ASTRONOMY IS FAR OUT STUFF

33

STARS

THE MAIN IDEA

Stars are large thermonuclear explosions contained by gravity.

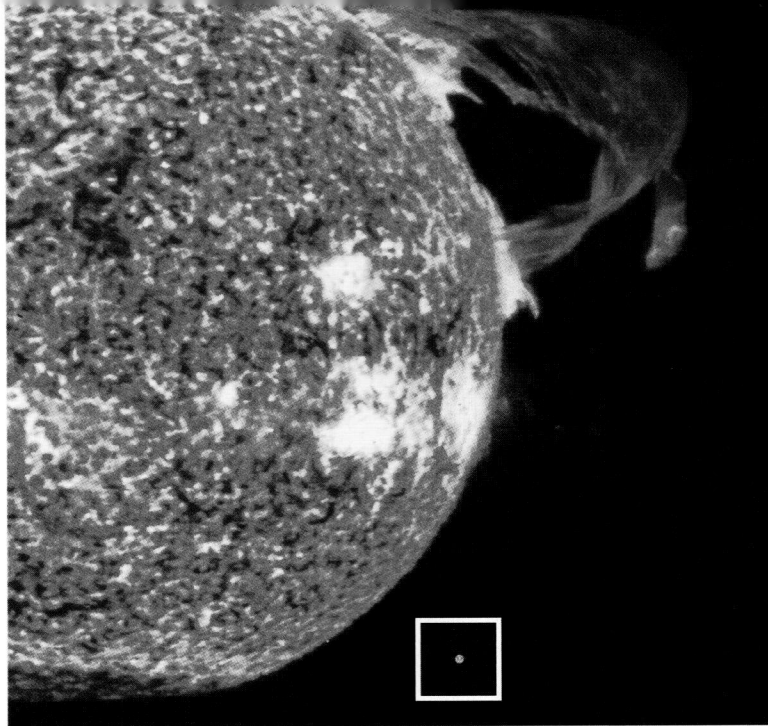

Our Sun is a star. The warmth you feel from this star is not so much because it is hot. Indeed, the Sun's surface temperature of about 6000°C is no hotter than the flame of some welding torches. The primary reason you are warmed by the Sun is because it is so big. Look at the lower right corner of the photograph above, and you'll see a small blue dot. This dot (painted on the photograph) is the approximate relative size of Earth. Are there stars even larger than our Sun? Why are some stars brighter than others? Why do stars have different colors? When are stars born? How do they die? How do some transform into black holes?

Explore!

Why Do We See Different Constellations at Different Times of the Year?

Cut three sheets of 8.5 × 11 paper into quarters to make 12 panels. Draw many random dots on each panel. Better yet, dip an old toothbrush into black paint and use your thumb to spray flecks of the paint randomly onto each panel. Connect some dots to create a constellation on each panel. Tape the short edges of all 12 panels together to make a loop that can stand up as shown in Figure 33.2. Place a tennis ball in the center of the loop to represent the Sun. Use a Ping-Pong ball to represent the Earth, which can revolve around the Sun.

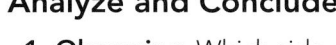

Analyze and Conclude

1. **Observing** Which side of the Ping-Pong ball is the day side? Which is the night side?
2. **Predicting** Will the constellations one sees at midnight be the same constellations one sees 6 months later also at midnight? How about a year later?
3. **Making Generalizations** Why do we only see one star in the daytime? If the Sun at midday were suddenly blacked out, say during a solar eclipse, we would be able to see many constellations of stars. What constellations would we be seeing?

33.1 Observing the Night Sky

The roots of astronomy reach back to prehistoric times when humans began to observe star patterns in the night sky. Though ancient people developed ways to measure the position and periodic movement of the stars, they knew nothing about the stars themselves. Today we know that Earth orbits a star—our Sun. We know that all the stars we see in the nighttime sky are much farther away from us than our fellow planets. On a moonless night we may guess we see many thousands or even millions of stars. But the unaided eye sees at most about 3000 stars, horizon to horizon. Many more stars become visible with a telescope, especially when the telescope is pointed toward a cloudlike band of light that stretches north to south. The ancient Greeks called this diffuse band of light the Milky Way.

Today we know the Milky Way to be a vast conglomeration of more than 100 billion stars. Our Sun and all the stars in our nighttime sky are located at the outer edges of this conglomeration.

When viewed from afar, as we discuss in the next chapter, all of these stars appear as a great swirl of stars known as a *galaxy*. The band of light identified by the ancient Greeks is the edge-on view of this galaxy. When modern telescopes peer away from the Milky Way, other galaxies come into view. How many? Again, the answer is more than 100 billion. So there are more than 100 billion galaxies, each containing, on average, more than 100 billion stars. ✓ The total number of stars in our observable universe, therefore, is more than 100 billion × 100 billion, which is much greater than the total number of grains of sands on all the beaches found on Earth. Stars are numerous.

Early astronomers divided the night sky into groups of stars, called constellations, such as the group of seven stars we now call the Big Dipper. The names of the constellations today carry over mainly from the names assigned to them by early Greek, Babylonian, and

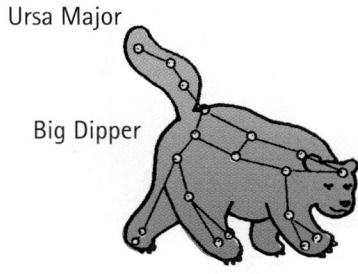

FIGURE 33.1 ▲
The constellation Ursa Major, the Great Bear. The seven stars in the tail and back of Ursa Major form the Big Dipper.

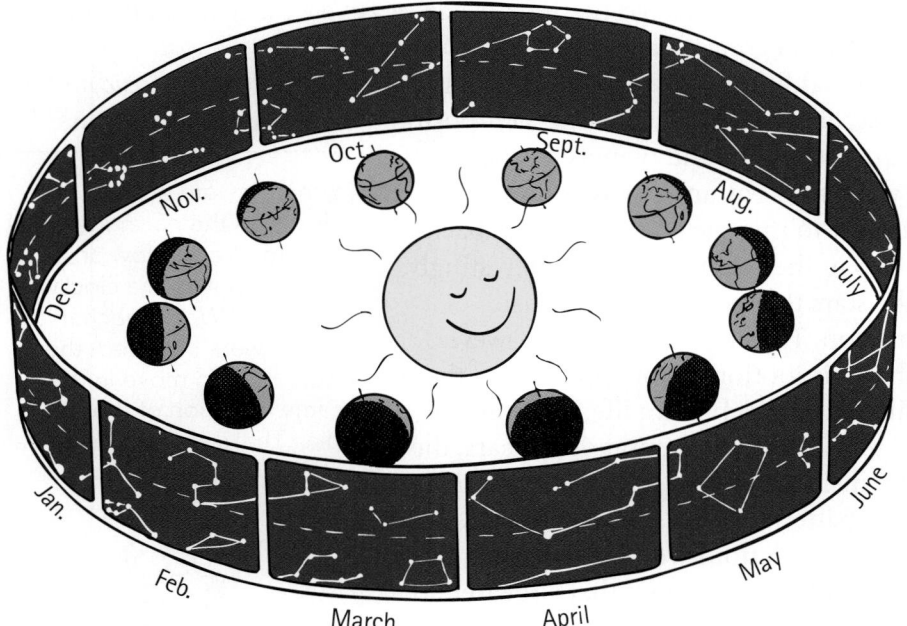

◄ **FIGURE 33.2**
The night side of Earth always faces away from the Sun. As Earth circles the Sun, different parts of the universe are seen in the nighttime sky. Here the circle, representing 1 year, is divided into 12 parts—the monthly constellations. The stars in the nighttime sky change in a yearly cycle.

The grouping of stars into constellations tells us about the thinking of astronomers of earlier times, but it tells us nothing about the stars themselves.

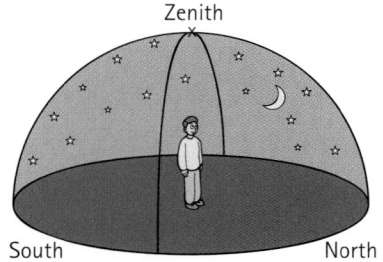

FIGURE 33.3 ▲
The celestial sphere is an imaginary sphere to which the stars are attached. We see no more than half of the celestial sphere at any given time. The point directly over our heads at any time is called the zenith.

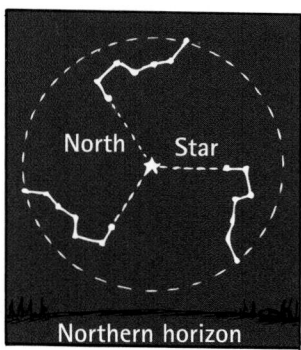

FIGURE 33.4 ▲
The pair of stars in the end of the Big Dipper's bowl point to the North Star. Earth rotates about its axis and therefore about the North Star, so over a 24-hour period the Big Dipper (and other surrounding star groups) makes a complete revolution.

Egyptian astronomers. The Greeks, for example, included the stars of the Big Dipper in a larger group of stars that outlined a bear. The large constellation Ursa Major (the Great Bear) is illustrated in Figure 33.1. The groupings of stars and the significance given to them has varied from culture to culture. In some cultures, the constellations stimulated storytelling and the making of great myths; in other cultures, the constellations honored great heroes, such as Hercules and Orion; in yet others, they served as navigational aids for travelers and sailors. For many cultures, including the African Bushmen and Masai, the constellations provided a guide for planting and harvesting crops because they were seen to move in the sky along with the seasons. Charts of the periodic movement of stars became some of the first calendars. We can see in Figure 33.2 why the background of stars varies throughout the year.

The stars are at different distances from Earth. However, because all the stars are so far away, they appear equally remote. This illusion led the ancient Greeks and others to conceive of the stars as being attached to a gigantic sphere surrounding Earth, called the **celestial sphere.** Though we know it is imaginary, the celestial sphere is still a useful construction for visualizing the motions of the stars (Figure 33.3).

The stars appear to turn around an imaginary north–south axis once every 24 hours. This is the *diurnal motion* of the stars. Diurnal motion is easy to visualize as a rotation of the celestial sphere from east to west. This motion is a consequence of the daily west to east (counterclockwise) rotation of Earth on its axis. Figure 33.4 shows the diurnal motion of the stars making up the Big Dipper. Time-exposure photographs, like the one seen on the cover of this textbook and also in Figure 33.5, show how stars appear to move in circles around the North Star. The North Star appears stationary as the celestial sphere rotates because it lies very close to the projection of Earth's rotational axis. Interestingly, the North Star has been the North Star for only the past couple hundred years. The Earth's axis slowly wobbles—about once every 26,000 years. About 14,000 years ago the "north star" was Vega, which will be the "north star" once again in about 12,000 years.

In addition to the diurnal motion of the sky, there is *intrinsic motion* of certain bodies that change their positions with respect to the stars. The Sun, the Moon, and planets—called "wanderers" by ancient astronomers—appear to migrate across the fixed backdrop of the celestial sphere. Interestingly, the stars themselves have intrinsic motion. They are so far away, however, that this motion is not apparent on the time scale of a human life. As shown in Figure 33.6, over thousands of years, the intrinsic movement of stars results in new patterns of stars. In other words, the constellations we see today are quite different from the ones that appeared to our earliest ancestors.

Sundials were our first clocks. In the northern hemisphere, the shadow on a sundial moves in a clockwise direction. When mechanical clocks were invented, they were built to move in the same direction of the sundial. That's why clocks move clockwise.

FIGURE 33.5 ▲
A time exposure of the northern night sky from an observatory in California.

✓ **READING CHECK**

What is the total number of stars in our observable universe?

CONCEPT CHECK
1. Which celestial bodies appear fixed relative to one another, and which celestial bodies appear to move relative to the others?
2. What are two types of observed motions of the stars in the sky?

Check Your Answers
1. The stars appear fixed as they move across the sky. The Sun, the Moon, and planets move relative to one another as they move across the backdrop of the stars.
2. One type of motion of the stars is their nightly rotation as if they were painted on a rotating celestial sphere; this is due to Earth's rotation on its own axis. Stars also appear to undergo a yearly cycle around the Sun because of Earth's revolution about the Sun.

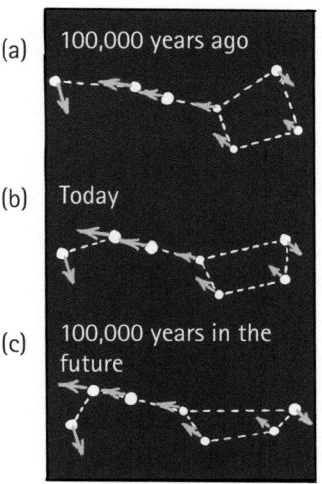

(a) 100,000 years ago

(b) Today

(c) 100,000 years in the future

FIGURE 33.6 ▲
The present pattern of the Big Dipper is temporary. Here we can see its pattern (a) 100,000 years ago; (b) as it appears at present; and (c) as it will appear in the future, about 100,000 years from now.

Astronomers measure the vast distances between Earth and the stars in units of *light-years.* One **light-year** is the distance that light travels in 1 year, nearly 10 trillion km. For perspective, the diameter of Neptune's orbit is about 0.001 light-year. The distance from the Sun to the outer edges of the Oort cloud (the full radius of our solar system) is about 0.8 light-year. The star closest to our Sun, Proxima Centauri, is about 4.2 light-years away. The diameter of the Milky Way Galaxy is about 100,000 light-years. The next closest major galaxy, the Andromeda Galaxy, is about 2.3 million light-years distant. Figure 33.7 shows the distances to the seven stars making up the Big Dipper in light-years.

The speed of light (as we know from Chapter 13) is 3×10^8 m/s. Although this is very fast, it nevertheless takes light significant time to travel large distances. And so when you see a very distant object, you are actually seeing the light it emitted long ago—you are looking back in time. Consider the example of Supernova 1987A (a supernova is the

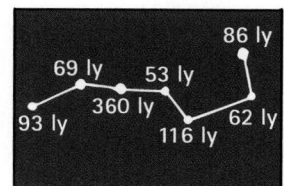

FIGURE 33.7 ▲
Interestingly, the seven stars of the Big Dipper are at varying distances from Earth. Note their varying distances in light-years. Note also that they are our immediate neighbors in a much larger Milky Way Galaxy.

explosion of a star, as you will learn more about in Section 33.4). This supernova occurred in a galaxy 190,000 light-years from Earth. Although we witnessed the supernova in 1987, the light from this explosion took 190,000 years to reach our planet, so the explosion actually occurred 190,000 years earlier. "News" of the supernova took 190,000 years to reach Earth!

33.2 Stars Have Different Brightness and Color

All stars have much in common with the Sun. All are born from clouds of interstellar dust with roughly the same chemical composition as the Sun, as was discussed in Chapter 32. About three-fourths of the interstellar material from which any star forms is hydrogen; one-fourth is helium; and no more than 2% of the material from which a star forms consists of heavier chemical elements. Stars shine brilliantly for billions of years because of the nuclear fusion reactions that occur in their cores, as was discussed in Chapter 16. And all stars, the Sun included, ultimately exhaust their nuclear fuel and die.

Yet not all stars are the same. ✔ If you look into the night sky, you will see that stars differ in two very visible ways: brightness and color. Brightness depends upon the star's energy output and also upon how far away it is from Earth. Recall from earlier chapters the *inverse-square law:* The intensity of light diminishes as the reciprocal of the square of the distance from the source. For example, the stars Betelgeuse and Procyon appear equally bright even though Betelgeuse emits about 5000 times as much light as Procyon. The reason? Procyon is much closer to Earth than is Betelgeuse.

Astronomers clearly distinguish between apparent brightness and the more important property, *luminosity.* Apparent brightness is the brightness of a star as it appears to our eyes. **Luminosity,** on the other hand, is the total amount of light energy that a star emits into space. Luminosity is usually expressed relative to the Sun's luminosity, which is noted L_{Sun}. For example, the luminosity of Betelgeuse is 38,000 L_{Sun}. This indicates that Betelgeuse is a very luminous star emitting about 38,000 times as much light into space as the Sun. On the other hand, Proxima Centauri is quite dim, with a luminosity of 0.00006 L_{Sun}. Astronomers have measured the luminosity of many stars and found that stars vary greatly in this respect. The Sun is somewhere in the middle of the luminosity range.

Besides apparent brightness, star color is another property that varies widely. As shown in Figure 33.8, stars come in every color of the rainbow. A star's color directly tells you about its surface temperature—for example, a blue star is hotter than a yellow star, and a yellow star is hotter than a red star. In fact, astronomers use color to measure the temperatures of stars. Why is it that a star's color corresponds to its temperature?

As you learned in Chapters 9 and 13, all objects with a temperature above absolute zero emit energy in the form of electromagnetic radiation.

READING CHECK

In what two visible ways do stars differ?

◀ **FIGURE 34.4**
The giant elliptical galaxy M87, one of the most luminous galaxies in the sky, is located near the center of the Virgo cluster, some 50 million light-years from Earth. It is about 120,000 light-years across and about 40 times as massive as our own galaxy, the Milky Way.

Spiral galaxies, such as the Andromeda Galaxy, shown in Figure 34.1, are perhaps the most beautiful arrangements of stars. Some spirals, such as the Sombrero Galaxy of Figure 34.5, have a round central hub. Others, as shown in Figure 34.6, have a hub shaped like a bar. The Milky Way Galaxy is thought to look much like the NGC 6744 spiral galaxy, which is an intermediate between a barred and unbarred spiral (Figure 34.7).

fyi

- The Andromeda Galaxy is our closest spiral neighbor, being only some 2.2 million light-years away. It contains many more stars than the Milky Way, which makes it more luminescent. Also, its diameter is about 220,000 light-years, compared to the Milky Way's 100,000 light-years. Thus, our view of the Andromeda is likely more spectacular than the Andromeda's view of us.

FIGURE 34.5 ▶
The Sombrero Galaxy, cataloged as M104, is about 80,000 light-years in diameter and about 32 million light-years from Earth. At its center is one of the most supermassive black holes measured in any nearby galaxy.

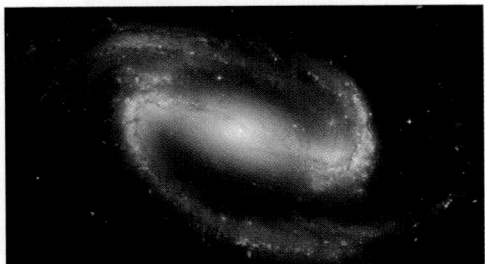

◀ **FIGURE 34.6**
The beautiful barred spiral galaxy NGC 1300 is about 100,000 light-years across and some 70 million light-years away.

FIGURE 34.7 ▲
The NGC 6744 galaxy is an intermediate between a barred and unbarred spiral galaxy. Studies of the Milky Way suggest that it too is an intermediate spiral. In other words, this is what we may look like from afar.

Irregular galaxies are normally small and faint and are difficult to detect. They tend to contain large clouds of gas and dust mixed with both young (blue) and old (yellow) stars. The irregular galaxy first described by the navigator on Magellan's voyage around the world in 1521 is our nearest neighboring galaxy—the Magellanic Clouds. This galaxy consists of two "clouds," called the Large Magellanic Cloud (LMC) and the Small Magellanic Cloud (SMC), both of which are slowly being pulled into the

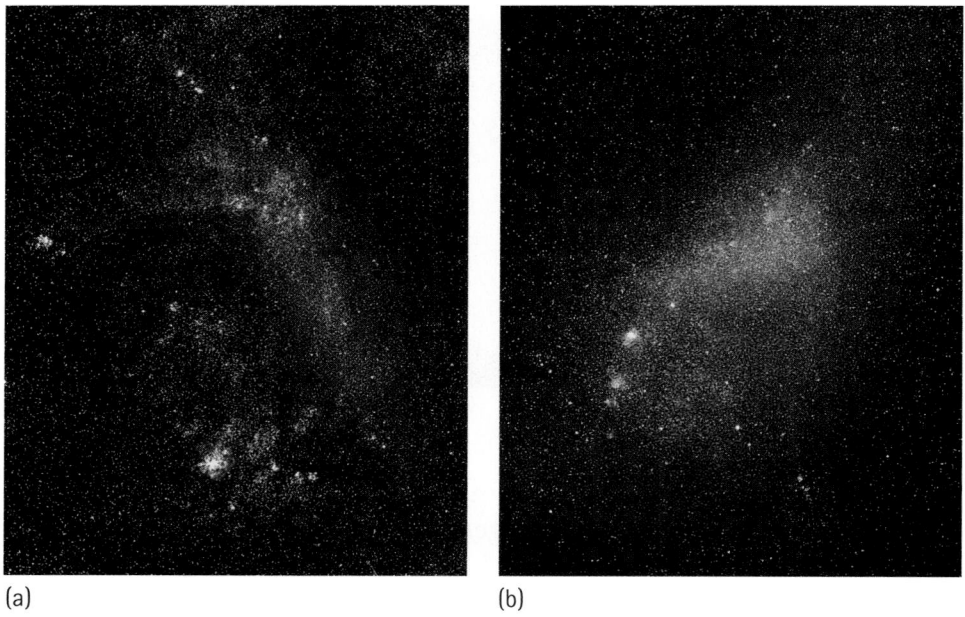

(a)   (b)

FIGURE 34.8 ▲

(a) The Large Magellanic Cloud and (b) the neighboring Small Magellanic Cloud are a pair of irregular galaxies. The Magellanic Clouds are our closest galactic neighbors, about 150,000 light-years distant.

Milky Way. The LMC is dotted with hot young stars with a combined mass of some 20 billion solar masses, and the SMC contains stars with a combined mass of about 2 billion solar masses (Figure 34.8). Some irregular galaxies, such as NGC 4038, shown in Figure 34.9, are the aftermaths of galactic collisions.

Galaxies often collide. But because the stars within a galaxy are so far apart from each other, the chances of collisions among the stars from each galaxy is quite small. Instead, gravitational forces merely cause each of the colliding galaxies to become distorted in shape.

FIGURE 34.9 ▲

Shown in black and white is the ground-telescope view of an irregular galaxy resulting from the collision of two galaxies. Note the remnant arms that suggest two former spiral galaxies. The inset shows a close-up color view taken by the Hubble Space Telescope. Evident is the rapid formation of new stars (blue) occurring as the mass of the two galaxies combined.

CONCEPT CHECK
Is it possible for one type of galaxy to turn into another?

Check Your Answer
Yes, and this occurs as two symmetrically shaped galaxies collide to form an asymmetrically shaped irregular galaxy.

READING CHECK
What are the three main classes of galaxies?

34.3 Active Galaxies Emit Huge Amounts of Energy

By galactic standards, our Milky Way Galaxy is a pretty mellow place. Peering away from our galactic home, however, astronomers find galaxies, called *active galaxies,* that are brimming with energy. One type of active galaxy is the **starburst galaxy,** in which stars are forming at an unusually high rate. A starburst galaxy can produce more than 100 new stars per year. By comparison, our Milky Way produces an average of about 1 new star per year. ✓ A starburst's high rate of star formation is often the result of some violent disturbance, such as a collision between two galaxies. The irregular galaxy shown in Figure 34.9 is an example of a starburst galaxy. Another example is the Cigar Galaxy, M82, which is being deformed by the tidal forces from its much larger neighbor, M81 (Figure 34.10).

A starburst tends to die down once the disturbance is removed or after the starburst galaxy consumes of all its interstellar fuel. Many elliptical galaxies are thought to be former starburst galaxies because of their low abundance of interstellar dust and gases.

Other active galaxies are active by virtue of their galactic core, which hosts a black hole more massive than millions or even billions of Suns. The event horizons of these black holes are about as large as our solar

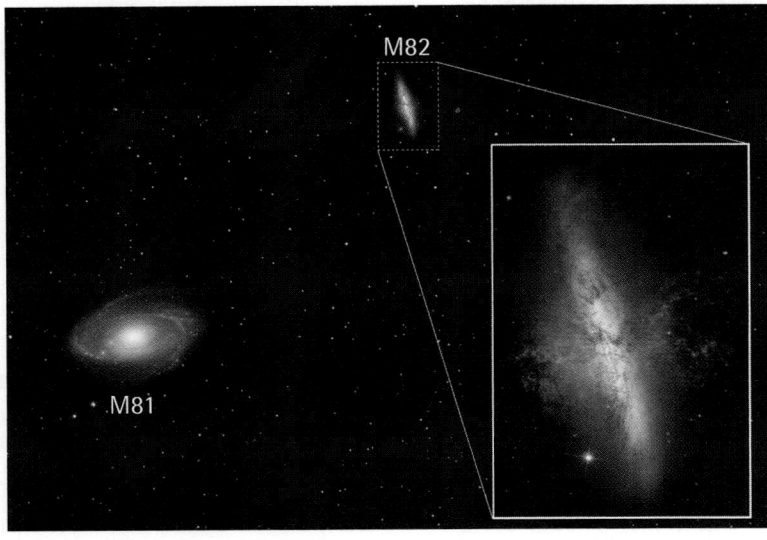

◀ **FIGURE 34.10**
The Cigar Galaxy, M82, is a spiral galaxy tilted away from us so that we see it from an edge-on view. Tidal forces from the nearby M81 galaxy disturb the distribution of matter within M82, which clumps, allowing for the formation of many new stars, as evidenced by M82's remarkable blue color. The red gases above and below the galactic plane are primarily hydrogen being pushed out by abundant stellar wind.

FIGURE 34.11 ▲
Material falling into the super-massive black hole in the center of M87 generates powerful jets that shoot out at near light speed.

system! Most large galaxies, including the Milky Way, have such black holes in their centers. What sets the active galaxies apart is the large amount of matter that continually falls into their supermassive black holes. Before falling into the black hole, the doomed mass forms a rapidly spinning disk, called an *accretion disk,* around the equator of the black hole. Charged particles in this hyperspinning disk create a narrow yet ultrastrong magnetic field that aligns with the black hole's poles. Rather than falling into the black hole, some of the charged particles, such as electrons, are accelerated outward through these magnetic fields to nearly the speed of light. This results in two extremely long streams of particles, called *jets,* extending thousands of light-years away from the galactic center, which is called an **active galactic nucleus** (AGN).

A relatively close AGN is found within the large elliptical galaxy M87, which was shown in Figure 34.4. High-resolution images of this galaxy, as shown in Figure 34.11, reveal a jet of material streaming away from the center of this galaxy to a distance of about 7000 light-years. Interestingly, the jet is angled toward us. This plus the great speed of the jet (99.5% of the speed of light) helps make the jet appear more luminous. The opposite "counterjet" receding away from us at such great speeds is invisible for reasons having to do with Einstein's special theory of relativity, mentioned at the end of Chapter 33.

READING CHECK

What is a probable cause of a starburst galaxy's high rate of star formation?

CONCEPT CHECK
Is the Milky Way a starburst galaxy?

Check Your Answer
No. Starburst galaxies produce hundreds of new stars per year. The Milky Way Galaxy produces only about 1 new star per year.

34.4 Galaxies Form Clusters and Superclusters

Galaxies are not the largest structures in the universe. They tend to cluster into distinguishable groups. Our Milky Way Galaxy, for example, is part of a cluster of local galaxies that include two other major spiral galaxies, namely, the Andromeda Galaxy and the Triangulum Galaxy. Also included are more than a dozen smaller elliptical galaxies, such as the Leo I galaxy shown in Figure 34.3, and a few irregular galaxies, such as the Large Magellanic Cloud. Altogether this cluster of galaxies is called the **Local Group.** Their approximate distributions are shown in Figure 34.12. To scale, Andromeda is only about 20 Milky Way diameters away from the Milky Way. The Triangulum Galaxy, so named because it completes a triangle between the spirals, is even closer to Andromeda, but farther away from us.

Our Local Group cluster is also under the gravitational influence of neighboring galactic clusters. Our cluster plus all these other clusters

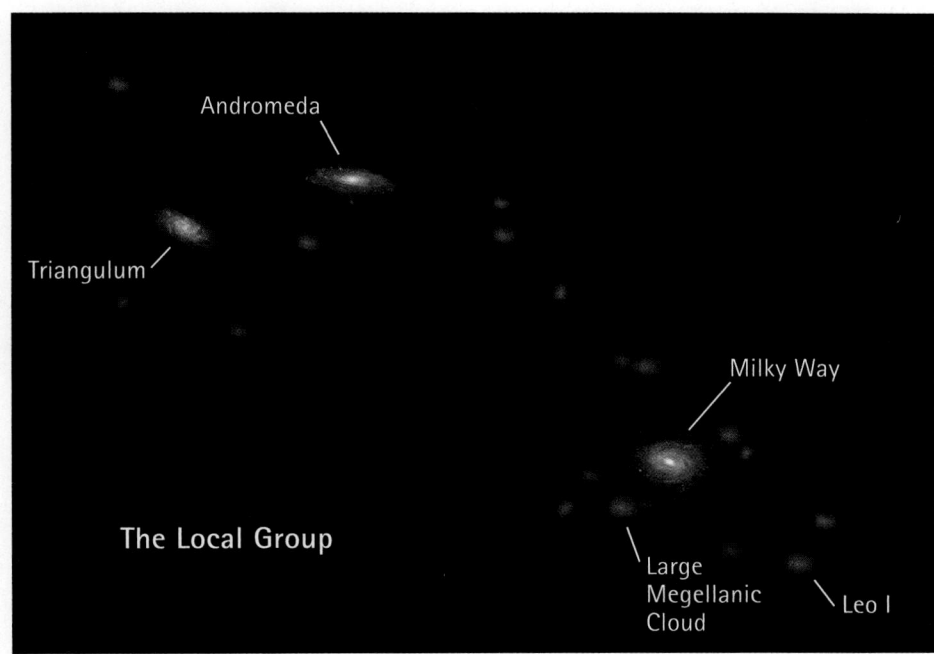

This two-dimensional compo-sition shows the approximate relative distances between the members of our Local Group of galaxies. These galaxies are all moving toward each other and will one day collide into a larger supergalaxy.

makes for what is called a *supercluster,* which is a cluster of galactic clusters. Our Local Group is a rather minor component of our **Local Supercluster,** as is illustrated in Figure 34.13.

Our Local Supercluster is tied in with an elaborate network of many other superclusters, as shown in Figure 34.14. Together, these superclus-ters appear as though they reside on the surface of foam inside which are large *voids* of empty space. Zooming out farther we find that the network of superclusters extends to the edges of the *observable universe,*

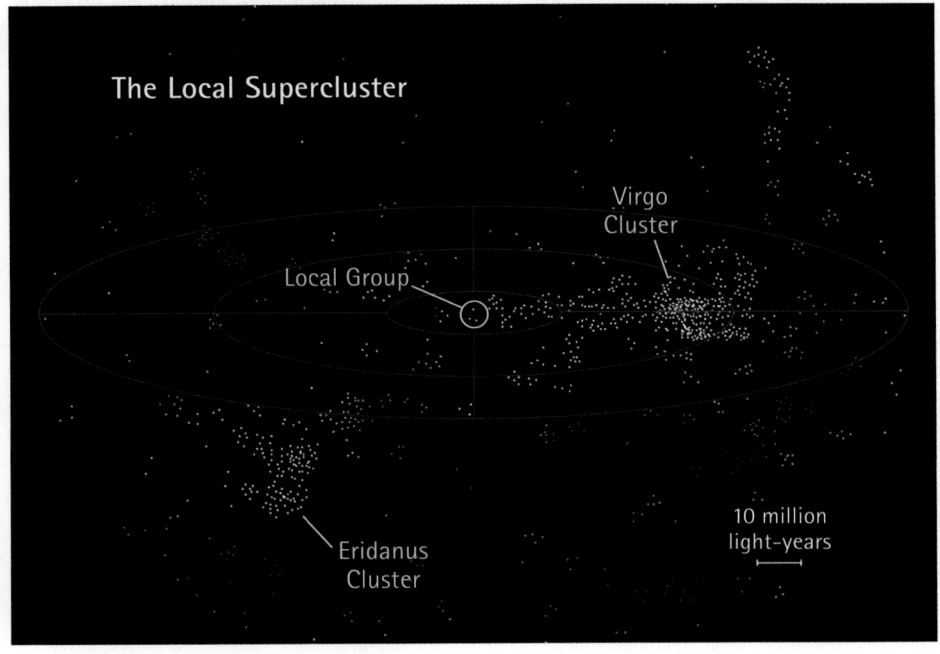

◀ **FIGURE 34.13**
A supercluster is a cluster of galactic clusters. Each dot represents a galaxy. Note that our Local Group is found midway between two much larger clusters, the Virgo and Eridanus clusters.

FIGURE 34.14 ▶
Each cloud represents a supercluster. Note that the superclusters are strung together as though on the surface of foam.

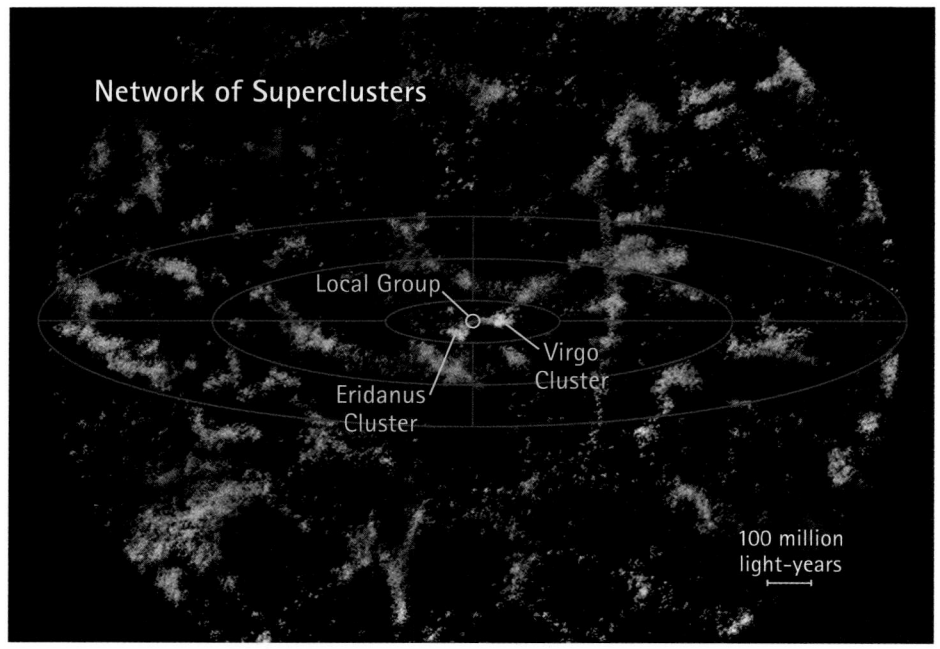

Network of Superclusters

Local Group

Virgo Cluster

Eridanus Cluster

100 million light-years

Think of the foam that you get when blowing bubbles into an ice cream soda. That's the structure of our universe! On the surface of the foam are superclusters, which are separated from each other by huge amounts of empty space, called *voids*.

as illustrated in Figure 34.15. ✔ By **observable universe** we mean all that we are able to see given the fact that the universe is only about 14 billion years old. The light coming from any object farther than 14 billion light-years has not had sufficient time to reach us.

So our observable universe is huge. Utterly huge. How large, then, might be the entire universe—the whole shebang? We don't know, and maybe never will.

But that doesn't stop cosmologists and mathematicians from developing models proposing possible answers. One such model suggests that

FIGURE 34.15 ▶
The network of superclusters extends to the edges of the observable universe, which is no farther than 14 billion light-years away. This illustration, however, shows a hypothetical bird's-eye view of this observable universe fully matured to the present moment, which, because of cosmic expansion, would place those most distant objects now some 42 billion light-years away.

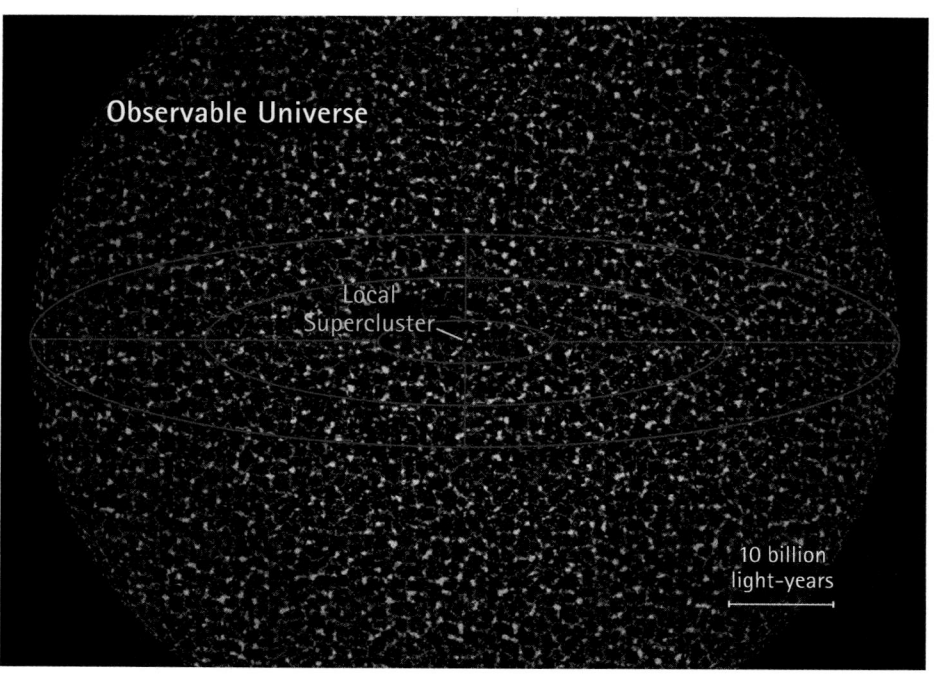

Observable Universe

Local Supercluster

10 billion light-years

if the observable universe were the size of a proton, the entire universe would be about as large as planet Earth. Consider how many protons fit within the dimensions of Earth. This would be the number of observable universes within the entire universe. The number is so enormous that if you could travel $10^{10^{118}}$ m in any direction, you would have a large probability of coming across another observable universe that looks very much like the one you left. Continuing this speculation, you'd look for somebody like yourself who is reading a book exactly like the one you are reading now. What if this person is actually a future you? Besides living very, very far away, the only measurable difference is that he or she has already finished reading this paragraph. Scientists call this the "many worlds" model, in which each observable universe is static, representing one possible arrangement of matter. We don't move through time. Rather, we jump from one observable universe to the next, which gives the appearance of moving through time. Welcome. You have just jumped to a new observable universe. The old one you left six sentences ago is now ever so distant.

CONCEPT CHECK

Which is greater, the number of stars in our galaxy or the number of galaxies in the universe?

Check Your Answer

There are far more galaxies within the entire universe than there are stars in our galaxy. Recall from the beginning of this chapter that astronomers estimate that there are about 100 billion stars in our galaxy and about 100 billion galaxies in our observable universe. If true, that means there are about 10^{22} stars in our observable universe, which is about the same number of water molecules in a drop of water. As large as the observable universe is large, it is as small as the fundamental building blocks of our body are small. As humans we are nicely situated between these two extremes.

34.5 Galaxies Are Moving Away from One Another

Hubble's discovery of galaxies beyond our own was a remarkable achievement. But Hubble took his research a step further and discovered something even more amazing—he discovered that the galaxies were all moving away from one another at fantastic speeds.

What was the evidence for this discovery? Hubble knew that the color of light emitted by a star or galaxy receding away from us shifts to the red because of the Doppler effect (Section 12.10). The degree of redshift could be measured quantitatively by focusing on the light coming from heated hydrogen atoms. The greater the shift in hydrogen's spectrum, the faster the receding speed. His research team measured both the distances and redshifts of numerous galaxies and discovered that the farther the galaxy, the greater the redshift. This meant that the galaxies

FIGURE 34.16 ▲
Every ant on the expanding balloon sees all other ants moving farther away. Each ant may therefore think that it is at the center of the expansion. Not so! There is no center to the surface of the balloon, just as there are no edges.

✔ **READING CHECK**

Is the universe in space, or is space in the universe?

If two galaxies are close together, then gravity causes those two galaxies to move towards each other. Such is true of the Milky Way and Andromeda galaxies, which are on course to collide in about 5 billion years.

were not static islands. Rather, they were moving away from us in every direction, which meant that the universe itself was expanding.

If distant galaxies were all moving away from one another, that could only mean that they were once much closer together. Running the cosmic movie backward would inevitably lead to a moment when all the galaxies were together within a single point. The universe as we know it, therefore, had a beginning. This moment has come to be known as the **Big Bang**. Modern-day measurements of the motions of galaxies tell us that the Big Bang gave rise to the universe about 13.7 billion years ago.

An important implication of Hubble's discovery is as follows:

✔ The universe is not contained within a region of space. Rather, space is *in* the universe and this space is rapidly expanding. This is peculiar because you may first think that the Big Bang occurred within an already existing infinite space and that matter and energy flew outward from this Big Bang to occupy this space. But this is not the case. Instead, when we talk about the expansion of the universe, we are referring to an expansion of the very structure of space itself. A useful analogy is a group of ants on an expanding balloon, as shown in Figure 34.16. As the balloon is inflated, every ant sees every other ant moving farther away. Likewise, in an expanding universe, any observer sees all distant galaxies moving away. So the Big Bang marked not only the beginning of time, but the beginning of space.

Aside from studies of galaxies, there are two other lines of evidence that strongly support the notion of a Big Bang. This includes *cosmic background radiation* and the abundance of hydrogen and helium in the universe. We explore these two lines of evidence next.

34.6 Further Evidence for the Big Bang

In 1964, scientists Arno Penzias and Robert W. Wilson, working at Bell Labs in New Jersey, used a simple radio receiver to survey the heavens for radio signals (Figure 34.17). No matter which way they directed their receiver, they detected microwaves with a wavelength of 7.35 cm coming toward Earth. Penzias and Wilson were puzzled. With no identifiable source of the radiation, where were the microwaves coming from?

Remember from Chapter 13 that any object above absolute zero emits energy in the form of electromagnetic radiation. The frequency of this radiation is proportional to the absolute temperature of the emitter. Theorists at Princeton University, working around the same time as Penzias and Wilson, showed that if the universe began in an explosion as described by the Big Bang, it would still be cooling off. Further, they showed that the temperature of the early universe would have cooled by today to an average temperature of 2.73 K. A universe of this temperature would be expected to emit microwave radiation of just the frequency observed by Penzias and Wilson. Thus the influx of microwave radiation that initially puzzled Penzias and Wilson was found to be

◀ **FIGURE 34.17**
Arno Penzias and Robert Wilson in front of the microwave receiver they used to detect the afterglow of the Big Bang.

emitted by the cooling universe itself. ☑ This faint microwave radiation is now referred to as **cosmic background radiation** and is taken as strong evidence of the Big Bang (Figure 34.18).

The Big Bang answers another cosmic mystery involving the element helium. Measurements show that matter in the universe is about 75% hydrogen and 25% helium. Hydrogen is the simplest of all elements, consisting of a single proton nucleus. It makes sense that hydrogen was the original element. Helium, however, is a more complex element containing a nucleus of two protons and two neutrons. We know helium is produced from the fusion of hydrogen in stars. But the number of stars is insufficient to account for all the observed helium—not more

If you want to point to the location of the Big Bang, just point your finger to the tip of your nose, or anywhere for that matter. You can't miss.

◀ **FIGURE 34.18**
This all-sky map of the cosmic background radiation taken by the Wilkinson Microwave Anisotropy Probe (WMAP) satellite reveals an average temperature of about 2.73 K everywhere. This is the cooled-off remnants of the Big Bang. The color shows minor temperature variation on the order of 0.0001 K.

READING CHECK

What is considered to be the "afterglow" of the Big Bang itself?

than 10% of the observed helium could have originated in stars. Most of the helium observed in the universe must have been created elsewhere. The Big Bang model predicts that the early universe would have been favorable to the formation of helium, but not the formation of other elements. A more detailed analysis shows that the amount of helium created just after the Big Bang should be about that which we observe in the universe today.

In summary, three major lines of evidence strongly support the Big Bang hypothesis. The first is the current expansion of space, which causes galaxies to recede from one another. The second is the discovery of cosmic background radiation, which is the Big Bang's afterglow. The third is the Big Bang's ability to explain the observed proportions of elements. Because of these and other similar lines of evidence, the Big Bang has come to be widely accepted by the scientific community.

34.7 Dark Matter Is Invisible

Evidence now suggests that the Big Bang generated matter in at least two different forms—one we can see and another that we can't. The visible form of matter is the "ordinary matter" made of subatomic particles such as protons, neutrons, and electrons. As you learned in earlier chapters, these particles combine to make the atoms of the periodic table. You, the planets, and the stars are made of this form of matter, which we will from here on refer to as **ordinary matter.**

The second form of matter generated from the Big Bang is quite unlike ordinary matter. This form of matter does not interact with the strong nuclear force, which means it cannot clump to form atomic nuclei. Neither does this second form of matter recognize the electromagnetic force, which makes it invisible to light as well as our sense of touch. As was explained in Chapter 15, the electromagnetic force is responsible for the repulsion between electrons. The reason you can't walk through a wall is because of the repulsions between the electrons in your body and the electrons in the wall. If the wall were made of this invisible matter, you would be able to walk right through it. Of course, you wouldn't be able to see the wall either. This invisible form of matter that we cannot see or touch is known as **dark matter.**

So if dark matter is invisible, how do we know it's there? This ghostly form of matter gives itself away by its gravitational effects. One of the first clues came to us as we were mapping the speeds at which stars orbit our galactic center. According to the laws of gravity, orbital speed is a function of the force of gravity between the orbiting object and the object being orbited—the greater the force of gravity, the greater the orbital speed. The inner planets of our solar system, for example, orbit the Sun much faster than the outer planets because they are closer to the Sun and experience greater gravitational forces. Relative to our galaxy, we might expect the same trend—stars closest to the galactic center should have faster orbital speeds than stars farther out. Interestingly, that's not what we observe! Instead, stars closer to the galactic center and those farther out orbit with about the same speed. How can this be?

For a solar system, planets orbit as they do because most of the solar system's mass is concentrated within the central Sun. For a galaxy such as the Milky Way or the Andromeda, it sure looks as though most of the mass is concentrated within the central bulge. ✔ The measured orbital speeds of stars, however, tell us that the largest bulk of the galaxy's mass lies outside the galaxy itself within a diffuse (spread out) yet massive invisible halo many times the diameter of the visible galaxy, as shown in Figure 34.19. We know it's invisible because all of our telescopes see right through it! Either our equations are wrong or something must be there affecting stellar orbital speeds. Because our equations are so well tested, most scientists tend to favor the presence of dark matter.

Another bit of evidence for dark matter comes from measuring the speeds of galaxies as they orbit one another within clusters. The measured speeds tell us that the masses of these galaxies are many times greater than the total mass of all their stars.

Lastly, we know that the path of light is bent by gravity much as it is bent by an optical lens. A cluster of galaxies, therefore, can bend the light from an even farther cluster lying directly behind it—we say the foreground cluster behaves as a *gravitational lens.* The degree to which the light from the distant cluster bends depends upon the mass of the foreground cluster. Once again, the degree of light bending tells us that the mass of the closer cluster far exceeds that which we would expect based solely on the cluster's luminosity. So by carefully studying the bending of light from distant galaxies, we can build a map of the dark matter's distribution. Such a study using the Hubble Space Telescope is illustrated in Figure 34.20.

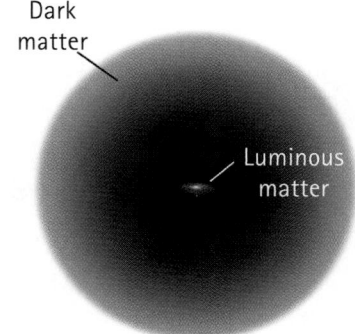

Dark matter

Luminous matter

FIGURE 34.19 ▲
As large as a galaxy is, its diffuse halo of dark matter is much larger. This halo may measure up to 10 times the diameter of the luminous galaxy and be about 6 times as massive.

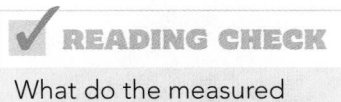

READING CHECK

What do the measured orbital speeds of stars strongly indicate to us?

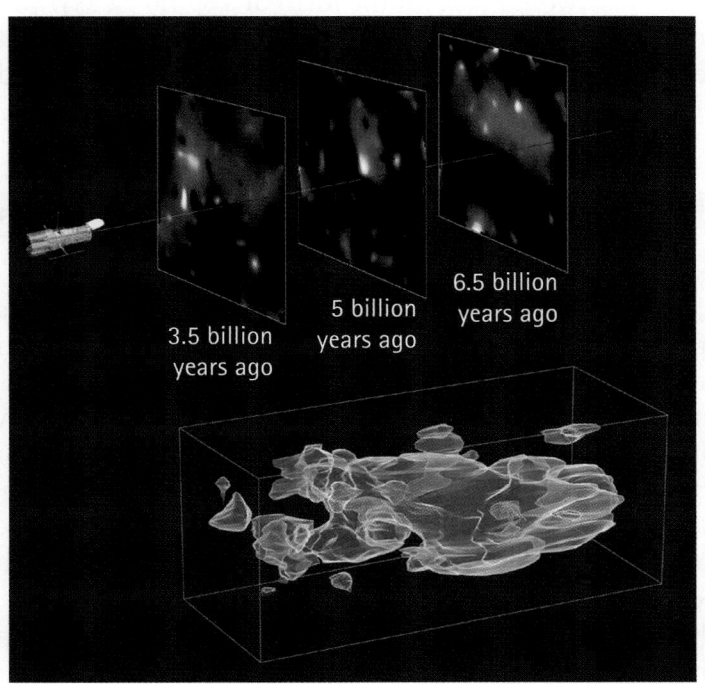

3.5 billion years ago

5 billion years ago

6.5 billion years ago

◀ **FIGURE 34.20**
Images of dark matter, shown in blue, were created through the Hubble Space Telescope's Cosmic Evolution Survey. Of course, the dark matter is not blue. This graphic, however, shows dark matter's distribution over a narrow region of the sky back to about 6.5 billion years ago.

- An alternative theory to dark matter is Modified Newtonian Dynamics (MOND), proposed by the physicist Mordehai Milgrom in the early 1980s. According to MOND, Newton's equation $a = F/m$ fails when the force is exceedingly weak. A modified version of this equation was thus created to account for the observed orbital velocities of stars within galaxies. The theory is controversial but has yet to be fully refuted.

FIGURE 34.21 ▲
Ordinary matter condensed out of a mixture of dark and ordinary matter.

Although *dark energy* and *dark matter* both begin with the word *dark*, they are uniquely different. One is a form of matter, the other is a form of energy. Both are still mysterious. By the next edition of this textbook, our understandings may be quite different!

So the evidence for dark matter is strong. Our current problem, however, is trying to figure out exactly what dark matter is made of. It's clear that dark matter is not simply ordinary matter, such as expired stars, that have gotten so cold that they emit no light. Although numerous theories abound about the fundamental nature of dark matter, no dark matter particles have been detected. Until that happens, we are left with yet another fascinating mystery of our universe.

Dark Matter and Galaxy Formation

We can speculate that when the universe formed, ordinary matter plus an even greater amount of dark matter were produced. Held together by gravity, the ordinary matter and dark matter would have been strewn outward in a clumpy fashion. Within a clump, ordinary and dark matter may initially have been uniformly mixed together.

These two forms of matter differ significantly in that when ordinary matter collides with ordinary matter, energy is released as heat. With this loss of energy, the ordinary matter loses orbital speed and thus falls closer to the center of the clump. Over time, while all dark matter stayed distributed throughout the clump, the ordinary matter became concentrated at the center (Figure 34.21). This concentration of ordinary matter at the center of the clump allowed for the formation of stars.

Also, as ordinary matter congregated toward the center, the rate of rotation would increase. If the original clump of ordinary and dark matter was just barely spinning, then the stars forming at the center would take on the form of an elliptical galaxy. If the original clump was spinning a bit faster, then the new stars would be spinning fast enough to flatten the galaxy, much like a rapidly spinning ball of pizza dough. The resulting disk would take on the form of a spiral galaxy.

So from a clump of ordinary and dark matter, ordinary matter condensed to form a central galaxy. The dark matter remained diffuse, forming an invisible halo surrounding the newly formed galaxy.

34.8 Dark Energy Opposes Gravity

In the years just before Hubble's discovery of the expansion of the universe, Einstein was struggling to understand why gravity wasn't causing the universe to collapse in a Big Crunch. He was thinking of the universe as static, neither collapsing nor expanding. But in order for the universe to remain static against the inward pull of gravity, there would need to be another fundamental outward push counteracting gravity. In other words, if gravity is the "pull inward," there should be a phenomenon that creates a "push outward." To allow for such a balance, he introduced into his equations the idea of what he called the *cosmological constant*. He had no proof for the existence of such a phenomenon. Rather, he just made it up to make his equations appear more natural.

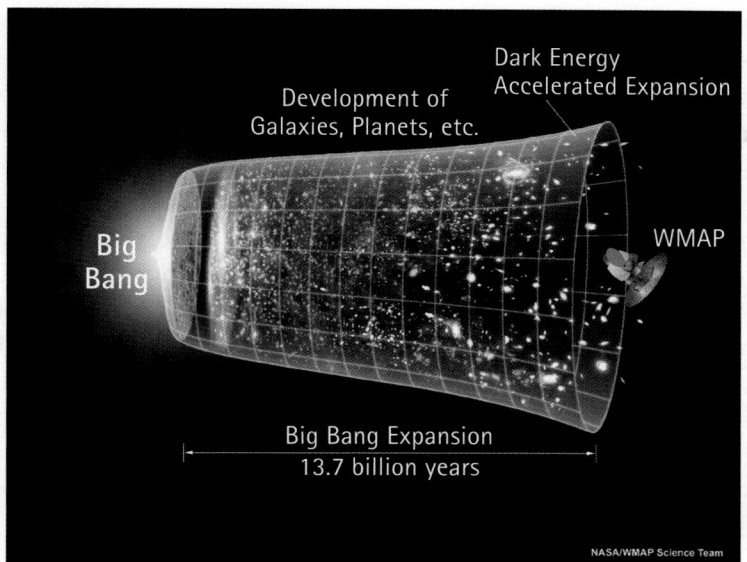

◀ **FIGURE 34.22**
The expansion of space started to accelerate about 7.5 billion years ago, which is shown on this diagram as a gradual widening just after the development of galaxies.

Then in the 1990s, some 40 years after Einstein's death, two teams of astronomers made a startling discovery. ✔ High-resolution data from very distant galaxies showed that space, beginning about 7.5 billion years ago, started to accelerate in its expansion. Galaxies are not simply coasting away from each other. Rather, some unknown form of energy is causing an increase in the rate at which galaxies are receding, as illustrated in Figure 34.22. Here was a phenomenon that acted as the opposite of gravity, causing an outward push. Here was evidence of Einstein's proposed cosmological constant!

At this writing, less than 20 years since the discovery of this phenomenon, there is much speculation about how these findings affect the fate of the universe. This form of unknown energy is generally described as **dark energy** (Figure 34.23). Current astronomical data suggest that dark

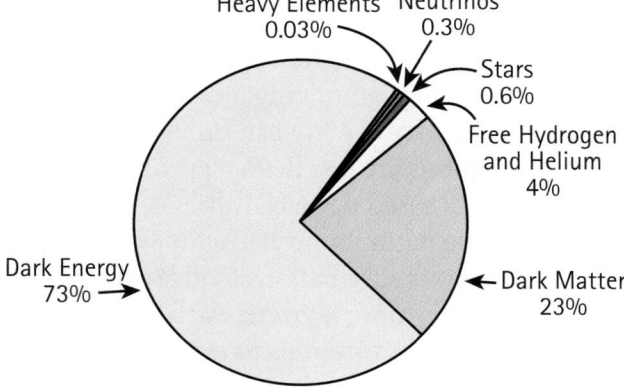

FIGURE 32.23 ▲
Ordinary matter, the stuff from which we and the galaxies we live in are made, comprises not more than 4% of the composition of the universe. The remainder is primarily dark matter (23%) and dark energy (73%), about both of which we know very little.

fyi

- The space through which our planets orbit contains about 100 hydrogen atoms per liter. The space between the stars of our galaxy contains about 2 hydrogen atoms per liter. If you want really empty space, you'll need to travel to the vast voids that separate the superclusters. It's only from these regions that dark energy appears to be taking hold. The space within and even between our local galaxies is too dense!

READING CHECK

What happened about 7.5 billion years ago that strongly suggests the existence of dark energy?

energy is, in fact, Einstein's famed cosmological constant. Other possibilities, however, cannot yet be ruled out.

Some current models suggest that dark energy finds its source within the emptiness of space. Matter has the effect of pulling things together. The classic example is what happens upon the formation of a black hole. In the absence of matter—within a perfect vacuum—empty space is brimming with an energy that creates an opposite effect. As more empty space is created, the dark energy becomes more predominant, which accelerates the formation of even more empty space. This may explain the large voids found between superclusters, as was shown in Figure 34.14. Distant galaxies are thus seen to be accelerating away from each other at ever increasing speeds. It is as though gravity and dark energy are opposites of each other. When gravity gains full control, the result is infinite density—a black hole. When dark energy gains full control, the result may be a perfect vacuum—an empty hole.

Epilogue

What, then, of the fate of our universe? It was until recently thought that the gravity within the universe might be strong enough to pull all objects back together in a *Big Crunch*. With the discovery of dark energy, this possibility has since been eliminated. Now we are at the point of wondering what might happen after eternal expansion. Might the acceleration of the universe continue to the point of a *Big Rip,* where even subatomic particles are torn apart and hurled to infinite distances? Might a Big Rip lead to another Big Bang?

It is interesting how the focus of our speculations has been narrowing down over our history. In the beginning, it seemed anything was possible. But as we opened our eyes and minds to the natural universe, we learned that some speculations were more worthy than others. We once viewed the constellations as heavenly gods. Of course, Earth was the center of the universe. Then we realized that Earth was just a planet orbiting the Sun, which itself was a medium-sized star among many. Our universe was the galaxy until Hubble proved otherwise. And now we find that galaxies are not just moving away from each other, they are *accelerating* away from each other. We can only speculate as to why.

Our current speculations are just that—speculations. But they are sophisticated speculations based upon centuries of collected evidence. As we continue to look to the natural universe with ever more powerful telescopes, we can expect our speculations to give way to deeper understandings. At the same time, more mysteries will likely be revealed. Our quest for new knowledge will prompt us to even further speculations. This is the mind-opening art of science, which seeks to learn the nature of the universe for what it is—not for what we might wish it to be. Stay tuned.

34 CHAPTER REVIEW

KEY TERMS

Active galactic nucleus A supermassive black hole at the center of a galaxy into which matter is falling at a high rate, thereby releasing astronomical amounts of energy.

Big Bang The primordial explosion of space at the beginning of time.

Cosmic background radiation The faint microwave radiation emanating from all directions that is the remnant heat of the Big Bang.

Dark energy An unknown form of energy that appears to be causing an acceleration of the expansion of space. It is thought to be associated with the energy exuded by a perfect vacuum.

Dark matter Matter that responds only to weak nuclear and gravitational forces. This form of matter is invisible but reveals itself to us by its gravitational effects.

Galaxy A large assemblage of stars, interstellar gas, and dust held together by gravity, usually categorized by its shape: elliptical, spiral, or irregular.

Local Group Our immediate cluster of galaxies, including the Milky Way, Andromeda, and Triangulum spiral galaxies plus a few dozen smaller elliptical and irregular galaxies.

Local Supercluster A cluster of galactic clusters in which our Local Group resides.

Observable universe All that we are able to see given the fact that the universe is only about 14 billion years old. The observable universe is one small sector of the universe as a whole.

Ordinary matter Matter that responds to the strong nuclear, weak nuclear, electromagnetic, and gravitational forces. This is matter made of protons, neutrons, and electrons, which includes the atoms and molecules that make us and our immediate environment.

Starburst galaxy A galaxy in which stars are forming at an unusually high rate.

REVIEW QUESTIONS

A Galaxy Is an Island of Stars

1. Why did people once think there was blackness between the stars?
2. What is a Cepheid?

Elliptical, Spiral, and Irregular Galaxies

3. What types of galaxies tend to be easy to see through?
4. What type of galaxy is the Milky Way?
5. Why are the Magellanic Cloud galaxies distorted?

Active Galaxies Emit Huge Amounts of Energy

6. What is a starburst galaxy?
7. What is thought to be at the center of most large galaxies?

Galaxies Form Clusters and Superclusters

8. How many spiral galaxies are within the Local Group?
9. Is the Local Group a relatively small or large cluster of galaxies?
10. What are three galactic clusters found within our Local Supercluster?

Galaxies Are Moving Away from One Another

11. What did Hubble observe in the light coming from glowing hydrogen atoms in distant galaxies?
12. Is the universe in space or is space in the universe?

Further Evidence for the Big Bang

13. According to the cosmic background radiation, what is the average temperature of the universe today?

14. The universe is about 25% helium. From where did most of the helium come?

Dark Matter Is Invisible

15. What type of matter is visible?

16. If we can't see dark matter, how do we know it is there?

17. Is dark matter found mostly within or just outside a galaxy?

Dark Energy Opposes Gravity

18. Did Einstein first believe that the universe was static or dynamic?

19. What was Einstein's cosmological constant?

20. What is likely the major component of our universe?

THINK AND DO

1. Find a quiet and comfortable place to sit where you won't be disturbed. Place your hands on your legs and straighten your back. Imagine a string attached to the top of your head lifting you upwards. Gaze to the floor a couple feet ahead of you. Relax your jaw so that your lips remain slightly open. Once you are comfortable, begin focusing on your out-breath. Simply notice you're out-breath as it leaves your body and disperses into your surroundings. When you find yourself carried away by a thought, label the thought as "thinking" and let the thought drift away. New thoughts will keep popping in your mind, but return your attention to your out-breath. Try doing this for at least 10 minutes. Using this technique, you may come to the point of being able to distinguish your bare presence from your random thoughts. This is your inner universe.

2. After contemplating your inner universe, try contemplating your outer universe. Envision yourself in your chair, then mentally zoom out to see yourself in the room, then in the building, then on the planet. Zoom out stepwise so that you can experience each stage. Zoom out to the solar system circling within a galaxy that orbits neighboring galaxies within a supercluster that is one supercluster of billions upon billions of superclusters. Once you have zoomed fully out, focus on that sensation of bigness. This is your outer universe.

THINK AND COMPARE

1. Rank the following in order of increasing size: cluster, supercluster, galaxy, solar system.

2. Rank the following galaxy types in order of increasing transparency: elliptical, spiral, irregular.

3. Rank the following in order of increasing energy output: spiral galaxy, supernova, active galactic nucleus.

4. Rank the following in order of increasing distance from Earth: the Andromeda Galaxy, the Virgo cluster, the nearest void.

5. Rank the following in order of increasing abundance: dark energy, ordinary matter, dark matter.

THINK AND EXPLAIN

1. Are there galaxies other than the Milky Way that can be seen with the unaided eye? Discuss.

2. If there are so many stars and galaxies, why do we see so much darkness in the clear night sky?

3. If the universe were unchanging and there were an infinite number of stars, what effect might this have on the darkness of a clear night sky?

4. Will the Milky Way Galaxy ever turn into an irregular galaxy? If so, when?

5. Does the collision of two galaxies involve the collisions of many, many stars?

6. Does the Milky Way Galaxy contain an active galactic nucleus?

7. There are instances where the jets of an active galactic nucleus point right towards our planet. What we see looks like a very powerful entity called a *quasar*. Might there be any quasars within the Milky Way Galaxy?

8. What is the difference between the universe and the observable universe?

9. If we were to define a *structure* as a group of objects held together, what would be the largest structure known?

10. If we are made of stardust, what are stars made of?

11. Why should anyone care about how the universe is organized? After all, how could galaxies and the like possibly have an impact on our daily living?

12. Are astronomers able to point their telescopes in the direction of where the Big Bang occurred?

13. If the initial universe remained hotter for a longer period of time, would there likely be more or less helium?

14. No galaxy found so far is made of less than 25% helium. If not from the stars, where did this helium come from?

15. A helium balloon here on Earth pops, releasing direct remnants of the Big Bang. True or false? Explain.

16. Early astronomers such as Kepler and Newton developed the laws of gravity based upon the motion of the planets around the Sun. How might these laws have been different if our solar system had been surrounded by a thick halo of dark matter?

17. If dark matter is affected by gravity, might there be lots of it surrounding us here on the surface of the Earth?

18. A police officer pulls you over for speeding. He tells you that his radar tracked you moving at a rate of 45 mph away from his parked police car. Were you really speeding away from him or was the space between the two of you simply expanding?

19. Is it dark matter or dark energy that keeps a galaxy from collapsing in on itself?

20. If we can't even predict the weather, how can we ever expect to predict the fate of the universe?

THINK AND SOLVE

1. If you were to travel straight up from the core of our galaxy and then look back, you would have a grand view of the Milky Way's spiral shape. If the distance from the core to the outer edges was 50,000 light-years, how much surface area are you looking at? Assume the galaxy to be a circle whose area can be found by the equation area $= \pi r^2$.

2. Assume the Milky Way contains 100 billion stars evenly distributed with none concentrated towards the center. What would be the surface area density of stars? Use the equation: surface area density $=$ number/surface area.

3. Using your answer to the previous question, would it be possible for two galaxies with stars evenly distributed to pass right through each other?

READINESS ASSURANCE TEST (RAT)

If you have a good handle on this chapter—if you really do—then you should be able to score 7 out of 10 on this RAT. Check your answers with your teacher. If you score less than 7, study further before moving on.

Choose the best answer to the following.

1. Scientists estimate the age of the universe to be about
   (a) 5000 years old.
   (b) 1 billion years old.
   (c) 14 billion years old.
   (d) 42 billion years old.

2. Starburst galaxies
   (a) tend to be spiral.
   (b) produce hundreds of stars per year.
   (c) have jets coming from an active galactic nucleus.
   (d) are noted for their supermassive black holes.

3. When two galaxies collide
   (a) the result is a massive explosion.
   (b) they become greatly distorted in shape.
   (c) the smaller one is destroyed.
   (d) Galaxies have not been observed colliding.

4. Which of the following is not accepted evidence for the Big Bang?
   (a) Cosmic background radiation
   (b) The nearly uniform temperature of the universe
   (c) The abundance of helium
   (d) Dark energy

5. Edwin Hubble is famous for
   (a) building the first telescope.
   (b) discovering the nature of Cepheids.
   (c) discovering Cepheids in the Andromeda nebula.
   (d) determining the mass of the Milk Way Galaxy.

6. The average temperature of the universe right now is about 2.73 K. Over the next billion years this temperature will likely
   (a) go down because of expansion.
   (b) go down because space is very cold.
   (c) remain constant because matter and energy cannot be destroyed.
   (d) increase because of the gravitational attractions among superclusters.

7. Elements heavier than helium come from
   (a) the Big Bang.
   (b) the Big Crunch.
   (c) nuclear fusion within supermassive stars.
   (d) volcanic activity of rocky planets, such as Earth.

8. Dark matter is
   (a) ordinary matter that is no longer emitting light.
   (b) altering the orbits of our planets.
   (c) attracted to ordinary matter.
   (d) repelled by ordinary matter.

9. In a huge cloud ordinary and dark matter are uniformly mixed together. Over time, the ordinary matter becomes concentrated toward the center of this cloud because ordinary matter
   (a) slows down as ordinary matter particles collide.
   (b) experiences a strong gravitational pull.
   (c) is lighter than dark matter.
   (d) is not affected by dark energy.

10. Dark energy is thought to be found primarily within the
   (a) cores of supermassive stars.
   (b) deepest regions of space where matter is absent.
   (c) event horizon of a black hole.
   (d) the sunspots of our Sun.

Two major systems of measurement prevail in the world today: the United States Customary System (USCS, formerly called the British system of units), used in the United States of America, Burma/Myanmar, and Liberia, and the Système International (SI) (known also as the international system and as the metric system), used everywhere else. Each system has its own standards of length, mass, and time. The units of length, mass, and time are sometimes called the fundamental units because, once they are selected, other quantities can be measured in terms of them.

United States Customary System

Based on the British Imperial System, the USCS is familiar to everyone in the United States. It uses the foot as the unit of length, the pound as the unit of weight or force, and the second as the unit of time. The USCS is presently being replaced by the international system—rapidly in science and technology (all Department of Defense contracts since 1988) and some sports (track and swimming), but so slowly in other areas and in some specialties it seems the change may never come. For example, we will continue to buy seats on the 50-yard line.

For measuring time, there is no difference between the two systems except that in pure SI the only unit is the second (s, not sec) with prefixes; but in general, minute, hour, day, year, and so on, with two or more lettered abbreviations (h, not hr), are accepted in the USCS.

Système International

PhysicsPlace.com
Tutorial
Metric System

During the 1960 International Conference on Weights and Measures held in Paris, the SI units were defined and given status. Table A.1 shows SI units and their symbols. SI is based on the metric system, originated by French scientists after the French Revolution in 1791. The orderliness of this system makes it useful for scientific work, and it is used by scientists all over the world. The metric system branches into two systems of units. In one of these the unit of length is the meter, the unit of mass is the kilogram, and the unit of time is the second. This is called the *meter-kilogram-second* (mks) system and is preferred in physics. The other branch is the *centimeter-gram-second* (cgs) system, which, because of its smaller values, is favored in chemistry. The cgs and mks units are related to each other as follows: 100 centimeters equal 1 meter; 1000 grams equal 1 kilogram. Table A.2 shows how several units of length are related to each other.

TABLE A.1 SI Units

Quantity	Unit	Symbol
Length	meter	m
Mass	kilogram	kg
Time	second	s
Force	newton	N
Energy	joule	J
Current	ampere	A
Temperature	kelvin	K

TABLE A.2 Conversions Between Different Units of Length

Unit of length	Kilometer	Meter	Centimeter	Inch	Foot	Mile
1 kilometer =	1	1000	100,000	39,370	3280.84	0.62140
1 meter =	0.00100	1	100	39,370	3.28084	6.21×10^{24}
1 centimeter =	1.0×10^{25}	0.0100	1	0.39370	0.032808	6.21×10^{26}
1 inch =	2.54×10^{25}	0.02540	2.5400	1	0.08333	1.58×1025
1 foot =	3.05×10^{24}	0.30480	30.480	12	1	1.89×1024
1 mile =	1.60934	1609.34	160,934	63,360	5280	1

Prefix	Definition
micro-	One-millionth: a microsecond is one-millionth of a second
milli-	One-thousandth: a milligram is one-thousandth of a gram
centi-	One-hundredth: a centimeter is one-hundredth of a meter
kilo-	One thousand: a kilogram is 1000 grams
mega-	One million: a megahertz is 1 million hertz

One major advantage of the metric system is that it uses the decimal system, where all units are related to smaller or larger units by dividing or multiplying by 10. The prefixes shown in Table A.3 are commonly used to show the relationship among units.

Meter The standard of length of the metric system originally was defined in terms of the distance from the North Pole to the equator. This distance was thought at the time to be close to 10,000 kilometers. One ten-millionth of this, the meter, was carefully determined and marked off by means of scratches on a bar of platinum-iridium alloy. This bar is kept at the International Bureau of Weights and Measures in France. The standard meter in France has since been calibrated in terms of the wavelength of light—it is 1,650,763.73 times the wavelength of orange light emitted by the atoms of the gas krypton-86. The meter is now defined as being the length of the path traveled by light in a vacuum during a time interval of 1/299,792,458 of a second.

Kilogram The standard unit of mass, the kilogram, is a block of platinum, also preserved at the International Bureau of Weights and Measures located in France (Figure A.1). The kilogram equals 1000 grams. A gram is the mass of 1 cubic centimeter (cc) of water at a temperature of 4°C. (The standard pound is defined in terms of the standard kilogram; the mass of an object that weighs 1 pound is equal to 0.4536 kilogram.)

Second The official unit of time for both the USCS and the SI is the second. Until 1956, it was defined in terms of the mean solar day, which was divided into 24 hours. Each hour was divided into 60 minutes and each minute into 60 seconds. Thus, there were 86,400 seconds per day, and the second was defined as 1/86,400 of the mean solar day. This proved unsatisfactory because the rate of rotation of Earth is gradually becoming slower. In 1956, the mean solar day of the year 1900 was chosen as the standard on which to base the second. In 1964, the second was officially defined as the time taken by a cesium-133 atom to make 9,192,631,770 vibrations.

FIGURE A.1 The standard kilogram.

Newton One newton is the force required to accelerate 1 kilogram at 1 meter per second per second. This unit is named after Sir Isaac Newton.

Joule One joule is equal to the amount of work done by a force of 1 newton acting over a distance of 1 meter. In 1948, the joule was adopted as the unit of energy by the International Conference on Weights and Measures. Therefore, the specific heat of water at 15°C is now given as 4185.5 joules per kilogram Celsius degree. This figure is always associated with the mechanical equivalent of heat— 4.1855 joules per calorie.

Ampere The ampere is defined as the intensity of the constant electric current that, when maintained in two parallel conductors of infinite length and negligible cross section and placed 1 meter apart in a vacuum, would produce between them a force equal to 2×10^{27} newton per meter length. In our treatment of electric current in this text, we have used the not-so-official but easier-to-comprehend definition of the ampere as being the rate of flow of 1 coulomb of charge per second, where 1 coulomb is the charge of 6.25×10^{18} electrons.

Kelvin The fundamental unit of temperature is named after the scientist William Thomson, Lord Kelvin. The kelvin is defined to be 1/273.15 the thermodynamic temperature of the triple point of water (the fixed point at which ice, liquid water, and water vapor coexist in equilibrium). This definition was adopted in 1968 when it was decided to change the name *degree Kelvin* (°K) to *kelvin* (K). The temperature of melting ice at atmospheric pressure is 273.15 K. The temperature at which the vapor pressure of pure water is equal to standard atmospheric pressure is 373.15 K (the temperature of boiling water at standard atmospheric pressure).

Area The unit of area is a square that has a standard unit of length as a side. In the USCS, it is a square with sides that are each 1 foot in length, called 1 square foot and written 1 ft². In the international system, it is a square with sides that are 1 meter in length, which makes a unit of area of 1 m². In the cgs system it is 1 cm². The area of a given surface is specified by the number of square feet, square meters, or square centimeters that would fit into it. The area of a rectangle equals the base times the height. The area of a circle is equal to πr^2, where $\pi \approx 3.14$ and r is the radius of the circle. Formulas for the surface areas of other objects can be found in geometry textbooks.

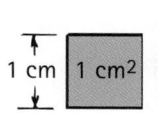
1 cm   1 cm²

Volume The volume of an object refers to the space it occupies. The unit of volume is the space taken up by a cube that has a standard unit of length for its edge. In the USCS, one unit of volume is the space occupied by a cube 1 foot on an edge and is called 1 cubic foot, written 1 ft³. In the metric system it is the space occupied by a cube with sides of 1 meter (SI) or 1 centimeter (cgs). It is written 1 m³ or 1 cm³ (or cc). The volume of a given space is specified by the number of cubic feet, cubic meters, or cubic centimeters that will fill it.

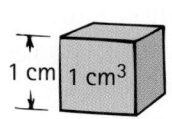
1 cm   1 cm³

In the USCS, volumes can also be measured in quarts, gallons, and cubic inches as well as in cubic feet. There are 1728 (12 × 12 × 12) cubic inches in 1 ft³. A U.S. gallon is a volume of 231 in³. Four quarts equal 1 gallon. In the SI volumes are also measured in liters. A liter is equal to 1000 cm³.

Unit Conversion

Often in science, and especially in a laboratory setting, it is necessary to convert from one unit to another. To do so, you need only multiply the given quantity by the appropriate *conversion factor*.

All conversion factors can be written as ratios in which the numerator and denominator represent the equivalent quantity expressed in different units. Because any quantity divided by itself is equal to 1, all conversion factors are equal to 1. For example, the following two conversion factors are both derived from the relationship 100 centimeters = 1 meter:

$$\frac{\textbf{100 centimeters}}{\textbf{1 meter}} = 1 \qquad \frac{\textbf{1 meters}}{\textbf{100 centimeters}} = 1$$

Because all conversion factors are equal to 1, multiplying a quantity by a conversion factor does not change the value of the quantity. What does change are the units. Suppose you measured an item to be 60 centimeters in length. You can convert this measurement to meters by multiplying it by the conversion factor that allows you to cancel centimeters.

CONCEPT CHECK
Convert 60 centimeters to meters.

CHECK YOUR ANSWER

(60 centimeters)   (1 meter) (100 centimeters) = 0.6 meter

↓   ↓   ↓

quantity in centimeters   conversion factor   quantity in meters

To derive a conversion factor, consult a table that presents unit equalities, such as Table A.2. Then multiply the given quantity by the conversion factor, and voilà, the units are converted. Always be careful to write down your units. They are your ultimate guide, telling you what numbers go where and whether you are setting up the equation properly.

Advanced Concepts of Motion

PhysicsPlace.com
Tutorial
Rotational Motion

To take the concepts of motion developed in the physics part of this text to the next level of complexity, we must be mindful of certain subtleties. In some instances, the idea of the *reference frame* is important.

When we describe the motion of something, we say how it moves relative to something else (Chapter 2). That "something" we measure motion with respect to is a *reference frame*. A reference frame can be defined mathematically by an origin and axes; in everyday terms, a reference frame is like a "background" we measure motion against. We are free to choose this frame's location and to have it moving relative to another frame. When our frame of motion has zero acceleration, it is called an *inertial frame*. In an inertial frame, force causes an object to accelerate in accord with Newton's Second Law. But when our frame of reference is accelerating, we observe fictitious forces and motions. Observations from a carousel, for example, are different when it is rotating and when it is at rest. Our description of motion and force depends on our "point of view."

We distinguish between speed and velocity (Chapter 2). Speed is how fast something moves, or the time rate of change of position (excluding direction): a scalar quantity. Velocity includes direction of motion: a vector quantity whose magnitude is speed. Objects moving at constant velocity move the same distance in the same time in the same direction.

But here's the subtlety: There is a distinction between speed and velocity, which has to do with the difference between distance and net distance, or *displacement*. Speed is distance per duration while velocity is displacement per duration. Displacement differs from distance. For example, a commuter who travels 10 kilometers to work and back travels 20 kilometers, but has "gone" nowhere. The distance traveled is 20 kilometers and the displacement is zero. Although the instantaneous speed and instantaneous velocity have the same value at the same instant, the average speed and average velocity can be very different. The average speed of this commuter's round-trip is 20 kilometers divided by the total commute time—a value greater than zero. But the average velocity is zero. In science, displacement is often more important than distance. (To avoid information overload, we have not treated this distinction in the text.) As you learned in Chapter 3, acceleration is the rate at which velocity changes. This can be a change in speed only, a change in direction only, or both. Negative acceleration is often called *deceleration*.

Finally, in Newtonian space and time, space has three dimensions—length, width, and height—each with two directions. We can go, stop, and return in any of them. Time has one dimension, with two directions—past and future. We cannot stop or return, only go. In Einsteinian space-time, these four dimensions merge (but this *very* advanced topic awaits you in a follow-up course.)

Computing Velocity and Distance Traveled on an Inclined Plane

A staple of any physics course is the study of motion on an inclined plane. We develop this topic here to help you sharpen your analytical concepts of motion. Recall from Chapter 2 Galileo's experiments with inclined planes. We considered a plane tilted such that the speed of a rolling ball increases at the rate of 2 meters per second each second—an acceleration of 2 m/s^2. So at the instant it starts moving its velocity is zero, and 1 second later it is rolling at 2 m/s, at the end of the next second 4 m/s, the end of the next second 6 m/s, and so on. The velocity of the ball at any instant is simply velocity = acceleration × time. Or, in shorthand notation $v = at$. (It is customary to omit the multiplication sign, ×, when expressing relationships in mathematical form. When two symbols are written together, such as the *at* in this case, it is understood that they are multiplied.)

How fast the ball rolls is one thing; how far it rolls is another. To understand the relationship between acceleration and distance traveled, we must first investigate the relationship between *instantaneous* velocity (velocity at a particular

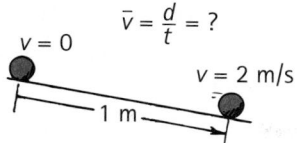

FIGURE B.1 The ball rolls 1 meter down the incline in 1 s and reaches a speed of 2 m/s. Its average speed, however, is 1 m/s. Do you see why?

point in time) and *average* velocity (which is computed over some extended time interval.) If the ball shown in Figure B.1 starts from rest, it will roll a distance of 1 meter in the first second. What will be its average speed? The answer is 1 m/s (it covered 1 meter in the interval of 1 second). But we have seen that the instantaneous velocity at the end of the first second is 2 m/s. Since the acceleration is uniform, the average in any time interval is found the same way we usually find the average of any two numbers: add them and divide by 2. (Be careful not to do this when acceleration is not uniform!) So if we add the initial speed (zero in this case) and the final speed of 2 m/s and then divide by 2, we get 1 m/s for the average velocity.

In each succeeding second we see the ball roll a longer distance down the same slope in Figure B.2. Note the distance covered in the second time interval is 3 meters. This is because the average speed of the ball in this interval is 3 m/s. In the next 1-second interval the average speed is 5 m/s, so the distance covered is 5 meters. It is interesting to see that successive increments of distance increase as a sequence of odd numbers. Nature clearly follows mathematical rules!

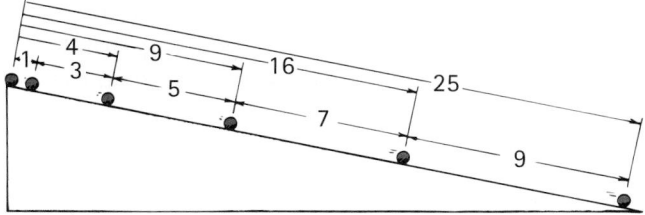

FIGURE B.2 If the ball covers 1 m during the first second, then it will cover the odd-numbered sequence of 3, 5, 7, 9 m, in each successive second. Note that distance increases as the square of time.

Investigate Figure B.2 carefully and note the total distance covered as the ball accelerates down the plane. The distances go from zero to 1 meter in 1 second, zero to 4 meters in 2 seconds, zero to

When Chelcie Liu releases both balls simultaneously, he asks, "Which will reach the end of the equal-length tracks first?" (*Hint*: On which track is the average speed of the ball greater?)

9 meters in 3 seconds, zero to 16 meters in 4 seconds, and so on in succeeding seconds. The sequence for total distances covered is of the squares of the time. We'll investigate the relationship between distance traveled and the square of the time for constant acceleration more closely in the case of free fall.

Computing Distance When Acceleration Is Constant   How far will an object released from rest fall in a given time? To answer this question, let us consider the case in which it falls freely for 3 seconds, starting at rest. Neglecting air resistance, the object will have a constant acceleration of about 10 meters per second each second (actually more like 9.8 m/s^2, but we want to make the numbers easier to follow).

$$\text{velocity at the beginning} = 0 \text{ m/s}$$
$$\text{velocity at the end of 3 seconds} = (10 \times 3) \text{ m/s}$$
$$\text{average velocity} = \frac{1}{2} \text{ the sum of these two speeds}$$
$$= \frac{1}{2} \times (0 + 10 \times 3) \text{ m/s}$$
$$= \frac{1}{2} \times 10 \times 3 = 15 \text{ m/s}$$
$$\text{distance traveled} = \text{average velocity} \times \text{time}$$
$$= \left(\frac{1}{2} \times 10 \times 3\right) \times 3$$
$$= \frac{1}{2} \times 10 \times 3^2 = 45 \text{ m}$$

We can see form the meanings of these numbers that

$$\text{distance traveled} = \frac{1}{2} \times \text{acceleration} \\ \times \text{square of time}$$

This equation is true for an object falling not only for 3 seconds but for any length of time, as long as the acceleration is constant. If we let d stand for the distance traveled, a for the acceleration, and t for the time, the rule may be written, in shorthand notation,

$$d = \frac{1}{2}\, at^2$$

This relationship was first deduced by Galileo. He reasoned that if an object falls for, say, twice the time, it will fall with twice the average speed. Since it falls for twice the time at twice the average speed, it will fall four times as far. Similarly, if an object falls for three times the time, it will have an average speed three times as great and will fall nine times as far. Galileo reasoned that the total distance fallen should be proportional to the square of the time.

In the case of objects in free fall, it is customary to use the letter g to represent the acceleration instead of the letter a (g because acceleration is due to gravity). Although the value of g varies slightly in different parts of the world, it is approximately equal to 9.8 m/s² (32 ft/s²). If we use g for the acceleration of a freely falling object (negligible air resistance), the equations for falling objects starting from a rest position become

$$v = gt$$
$$d = \frac{1}{2}\, gt^2$$

Much of the difficulty in learning physics, like learning any discipline, has to do with learning the language—the many terms and definitions. Speed is somewhat different from velocity, and acceleration is vastly different from speed or velocity.

CONCEPT CHECK

1. An auto starting from rest has a constant acceleration of 4 m/s². How far will it go in 5 s?
2. How far will an object released from rest fall in 1 s? In this case the acceleration is $g = 9.8$ m/s².

3. If it takes 4 s for an object to freely fall to the water when released from the Golden Gate Bridge, how high is the bridge?

CHECK YOUR ANSWERS

1. distance $= \frac{1}{2} \times 4 \times 5^2 = 50\,\text{m}$

2. distance $= \frac{1}{2} \times 9.8 \times 1^2 = 4.9\,\text{m}$

3. distance $= \frac{1}{2} \times 9.8 \times 4^2 = 78.4\,\text{m}$

Notice that the units of measurement when multiplied give the proper unit of meters for distance:

$$d = \frac{1}{2} \times 9.8 \times 16 = 78.4\,\text{m}$$

Circular Motion

Linear speed is what we have been calling simply *speed*—the distance traveled in meters or kilometers per unit of time. A point on the perimeter of a merry-go-round or turntable moves a greater distance in one complete rotation than a point nearer the center. Moving a greater distance in the same time means a greater speed. The speed of something moving along a circular path is **tangential speed,** because the direction of motion is tangent to the circle.

Rotational speed (sometimes called angular speed) refers to the number of rotations or revolutions per unit of time. All parts of the rigid merry-go-round turn about the axis of rotation *in the same amount of time.* All parts share the same rate of rotation, or *number of rotations or revolutions per unit of time.* It is common to express rotational rates in revolutions per minute (rpm).* Phonograph records that were common some years ago rotate at 33 1/3 rpm. A ladybug sitting anywhere on the surface of the record revolves at 33 1/3 rpm (Figure B.3).

Tangential speed is *directly proportional* to rotational speed (at a fixed radial distance). Unlike rotational speed, tangential speed depends on the distance from the axis (Figure B.4). Something at the center of a rotating platform has no tangential

---

* Physics types usually describe rotational speed in terms of the number of "radians" turned in a unit of time, for which they use the symbol ω (the Greek letter omega). There's a little more than 6 radians in a full rotation (2π radians, to be exact).

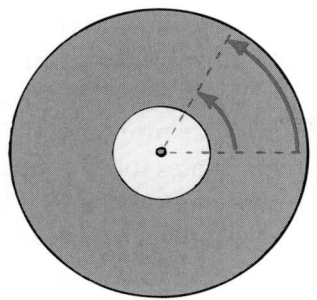

FIGURE B.3 When a phonograph record turns, a ladybug farther from the center travels a longer path in the same time and has a greater tangential speed.

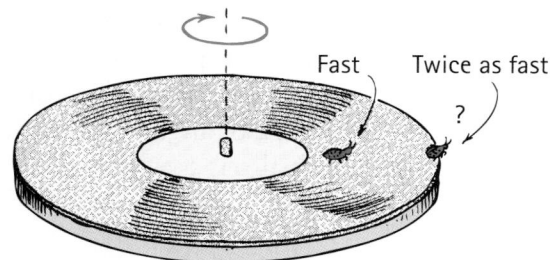

Fast   Twice as fast
?

FIGURE B.4 The entire disk rotates at the same rotational speed, but ladybugs at different distances from the center travel at different tangential speeds. A ladybug twice as far from the center moves twice as fast.

speed at all and merely rotates. But, approaching the edge of the platform, tangential speed increases. Tangential speed is directly proportional to the distance from the axis (for a given rotational speed). Twice as far from the rotational axis, the speed is twice as great. Three times as far from the rotational axis, there is three times as much tangential speed. When a line of people locked arm in arm at the skating rink makes a turn, the motion of "tail-end Charlie" is evidence of this greater speed. So tangential speed is directly proportional both to rotational speed and to radial distance.*

Torque

Whereas force causes changes in speed, *torque* causes changes in rotation. To understand torque (rhymes with *dork*), hold the end of a meter stick horizontally with your hand (Figure B.5). If you dangle a weight from the meter stick near your

* When customary units are used for tangential speed v, rotational speed ω, and radial distance r, the direct proportion of v to both r and ω becomes the exact equation $v = r\omega$. So the tangential speed will be directly proportional to r when all parts of a system simultaneously have the same ω, as for a wheel, disk, or rigid wand. (The direct proportionality of v to r is not valid for the planets because planets don't all have the same ω.)

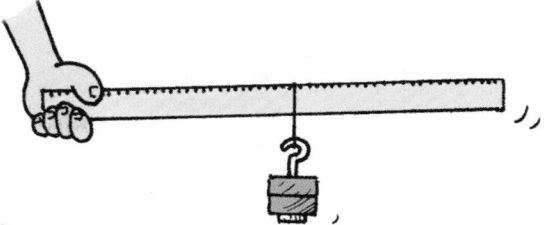

FIGURE B.5 If you move the weight away from your hand, you will feel the difference between force and torque.

hand, you can feel the meter stick twist. Now if you slide the weight farther from your hand, the twist you feel is greater, although the weight is the same. The force acting on your hand is the same. What's different is the torque.

torque = lever arm × force

Lever arm is the distance between the point of application of the force and the axis of rotation—the axis about which the body turns around. The lever arm is the shortest distance between the applied force and the rotational axis. Torques are intuitively familiar to youngsters playing on a seesaw. Kids can balance a seesaw even when their weights are unequal. Weight alone doesn't produce rotation. Torque does, and children soon learn that the distance they sit from the pivot point is every bit as important as weight (Figure B.6). When the torques are equal, making the net torque zero, no rotation is produced.

Recall the equilibrium rule in Chapter 2—that the sum of the forces acting on a body or any system must equal zero for mechanical equilibrium. That is, $\sum F = 0$. We now see an additional condition. The net torque on a body or on a system must also be zero for mechanical equilibrium. Anything in mechanical equilibrium doesn't accelerate—neither linearly nor rotationally.

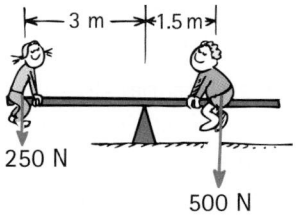

3 m   1.5 m

250 N

500 N

FIGURE B.6 No rotation is produced when the torques balance each other.

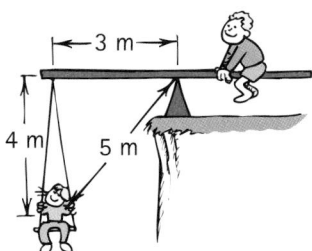

FIGURE B.7 The lever arm is still 3 m.

Suppose that the seesaw is arranged so that the half-as-heavy girl is suspended from a 4-meter rope hanging from her end of the seesaw (Figure B.7). She is now 5 meters from the fulcrum, and the seesaw is still balanced. We see that the lever-arm distance is 3 meters, not 5 meters. The lever arm about any axis of rotation is the perpendicular distance from the axis to the line along which the force acts. This will always be the shortest distance between the axis of rotation and the line along which the force acts.

This is why the stubborn bolt shown in Figure B.8 is turned more easily when the applied force is perpendicular to the handle, rather than at an oblique angle, as shown in the first figure. In the first figure, the lever arm is shown by the dashed line and is less than the length of the wrench handle. In the second figure, the lever arm is equal to the length of the wrench handle. In the third figure, the lever arm is extended with a pipe to provide more leverage and a greater torque.

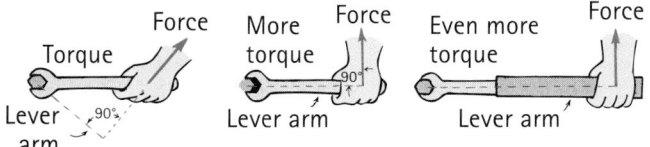

FIGURE B.8 Although the magnitudes of the force in each case are the same, the torques are different.

Angular Momentum

Things that rotate—whether a cylinder rolling down an incline, an acrobat doing a somersault, or the cloud of gas and dust that would become the solar system (Chapter 27)—keep on rotating until something stops them. A rotating object has an "inertia of rotation." Recall, from Chapter 32, that all moving objects have "inertia of motion" or

momentum—the product of mass and velocity. This kind of momentum is **linear momentum.** Similarly, the "inertia of rotation" of rotating objects is called **angular momentum.**

Any object that rotates turns about its *axis of rotation*. For the case of an object that is small compared with the radial distance to its axis of rotation, like a tetherball swinging from a long string or a planet orbiting around the Sun, the angular momentum can be expressed as the magnitude of its linear momentum, mv, multiplied by the radial distance, r (Figure B.9).* In shorthand notation, angular momentum = mvr. Like linear momentum, angular momentum is a vector quantity and has direction as well as magnitude. In this appendix, we won't treat the vector nature of angular momentum (or even of torque, which also is a vector).

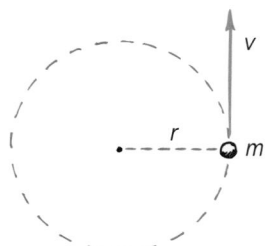

FIGURE B.9 A small object of mass m whirling in a circular path of radius r with a speed v has angular momentum mvr.

Just as an external net force is required to change the linear momentum of an object, an external net torque is required to change the angular momentum of an object. We can state a rotational version of Newton's First Law (the Law of Inertia):

An object or system of objects will maintain its angular momentum unless acted upon by an unbalanced external torque.

We see application of this rule when we look at a spinning top. If friction is low and torque also is low, the top tends to remain spinning. Earth and the planets spin in torque-free regions, and once they are spinning, they remain so.

---

* For rotating bodies that are large compared with radial distance—for example, a planet rotating about its own axis—the concept of rotational inertia must be introduced. Then angular momentum is rotational inertia × rotational speed. See any of Paul Hewitt's *Conceptual Physics* textbooks for more information.

FIGURE B.10 Angular momentum keeps the wheel axle nearly horizontal when a torque supplied by Earth's gravity acts on it. Instead of causing the wheel to topple, the torque causes the wheel's axis to turn slowly around the circle of students. This is called precession.

Angular momentum is a vector. However, in this book we will not treat the vector nature of momentum except to acknowledge the amazing action of the gyroscope. The rotating bicycle wheel in Figure B.10 shows what happens when a torque supplied by Earth's gravity acts to change its angular momentum (which is along the wheel's axle.) The pull of gravity that normally acts to topple the wheel over and change its rotational axis causes it instead to *precess* about a vertical axis, like a gyroscope. The best way to appreciate this is to hold a spinning bicycle wheel while standing on a platform that is free to rotate—the wheel turns you around. Precession is another advanced motion topic you may learn more about in a follow-up physics course.

Conservation of Angular Momentum

Just as the linear momentum of any system is conserved if no net forces are acting on the system, angular momentum is conserved if no net torque acts on the system. In the absence of an unbalanced external torque, the angular momentum of that system is constant. This means that its angular momentum at any one time will be the same as at any other time.

Conservation of angular momentum is shown in Figure B.11. The man stands on a low-friction turntable with weights extended. To simplify, consider only the weights in his hands. When he is slowly turning with his arms extended, much

FIGURE B.11 Conservation of angular momentum. When the man pulls his arms and the whirling weights inward, he decreases the radial distance between the weights and the axis of rotation, and the rotational speed increases correspondingly.

of the angular momentum is due to the distance between the weights and the rotational axis. When he pulls the weights inward, the distance is considerably reduced. What is the result? His rotational speed increases!* This example is best appreciated by the turning person, who feels changes in rotational speed that seem to be mysterious. But it's straight physics! This procedure is used by a figure skater who starts to whirl with her arms and perhaps a leg extended and then draws her arms and leg in to obtain a greater rotational speed. Whenever a rotating body contracts, its rotational speed increases.

The Law of Angular Momentum Conservation is seen in the motions of the planets and the shape of the galaxies. When a slowly rotating ball of gas in space gravitationally contracts, the result is an increase in its rate of rotation. The conservation of angular momentum is far-reaching.

---

* When a direction is assigned to rotational speed, we call it rotational velocity (often called *angular velocity*). By convention, the rotational velocity vector and the angular momentum vector have the same direction and lie along the axis of rotation.

APPENDIX C WORKING WITH VECTOR COMPONENTS

A vector quantity is a directed quantity—one that must be specified not only by magnitude (size) but by direction as well. Recall from Chapter 2 that velocity is a vector quantity. Other examples are force, acceleration, and momentum. In contrast, a scalar quantity can be specified by magnitude alone. Some examples of scalar quantities are speed, time, temperature, and energy.

Vector quantities may be represented by arrows (Figure C.1). The length of the arrow tells you the magnitude of the vector quantity, and the arrowhead tells you the direction of the vector quantity. Such an arrow drawn to scale and pointing appropriately is called a vector.

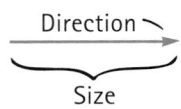

FIGURE C.1

Adding Vectors

PhysicsPlace.com
Tutorial
Vectors

Vectors that add together are called **component vectors**. The sum of component vectors is called a **resultant**. To add two vectors, make a parallelogram with two component vectors acting as two of the adjacent sides (Figure C.2). (Here our parallelogram is a rectangle.) Then draw a diagonal from the origin of the vector pair; this is the resultant (Figure C.3).

FIGURE C.2

Caution: Do not try to mix vectors! We cannot add apples and oranges, so velocity vectors combine

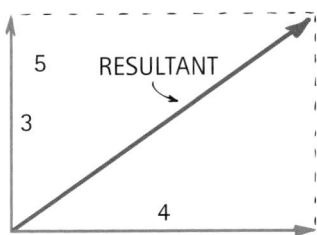

FIGURE C.3

only with velocity vectors, force vectors combine only with force vectors, and acceleration vectors combine only with acceleration vectors—each on its own vector diagram. If you ever show different kinds of vectors on the same diagram, use different colors or some other method of distinguishing the different kinds of vectors.

Finding Components of Vectors

Recall from Chapter 2 that to find a pair of perpendicular components for a vector, first draw a dashed line through the tail of the vector (in the direction of one of the desired components as in Figure C.4). Second, draw another dashed line through the tail end of the vector at right angles to the first dashed line (Figure C.5). Third, make a rectangle whose diagonal is the given vector (Figure C.6). Draw in the two components. Here we let **F** stand for "total force," **U** stand for "upward force," and **S** stand for "sideways force."

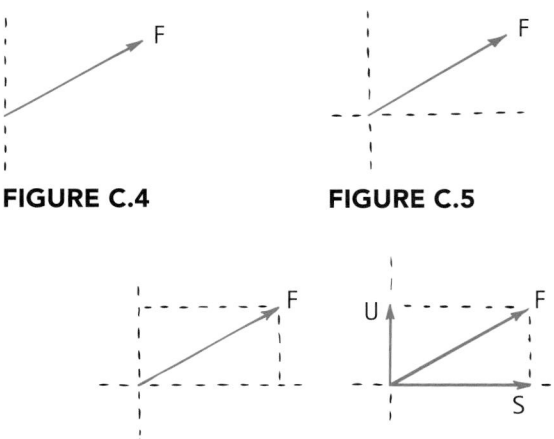

FIGURE C.4   **FIGURE C.5**

FIGURE C.6

846

Examples

1. Ernie Brown pushes a lawnmower and applies a force that pushes it forward and also against the ground. In Figure C.7, **F** represents the force applied by the man. We can separate this force into two components. The vector **D** represents the downward component, and **S** is the sideways component, the force that moves the lawnmower forward. If we know the magnitude and direction of the vector **F**, we can estimate the magnitude of the components from the vector diagram.

FIGURE C.7

2. Would it be easier to push or pull a wheelbarrow over a step? Figure C.8 shows a vector diagram for each case. When you push a wheelbarrow, part of the force is directed downward, which makes it harder to get over the step. When you pull, however, part of the pulling force is directed upward, which helps to lift the wheel over the step. Note that the vector diagram suggests that pushing the wheelbarrow may not get it over the step at all. Do you see that the height of the step, the radius of the wheel, and the angle of the applied force determine whether the wheelbarrow can be pushed over the step? We see how vectors help us analyze a situation so that we can see just what the problem is!

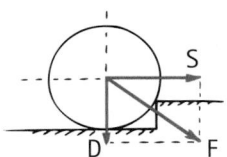 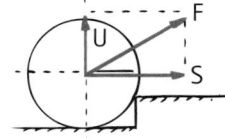

FIGURE C.8

The Polarization of Light—An Application of Vector Components

Recall from Chapter 14 that light travels as a transverse electromagnetic wave. You can create a transverse wave in a rope by shaking it. If you shake the rope vertically up and down, the wave vibrates in a vertical plane. If you shake the rope horizontally side-to-side, the wave vibrates in a plane that is horizontal, as the wave moves forward (Figure C.9). The vibrations in each case are to and fro in one direction. For this reason we say the wave is **polarized**—it vibrates in a single plane.

FIGURE C.9 A vertically plane-polarized plane wave and a horizontally plane-polarized plane wave.

A light wave is made up of an oscillating electric field vector and an oscillating magnetic field vector (Figure C.10). It is the orientation of the electric field vector that defines the direction of the polarization of light. A single vibrating electron emits an electromagnetic wave that, like the rope shaken in one direction, is polarized. Its electric field vibrates in a single plane.

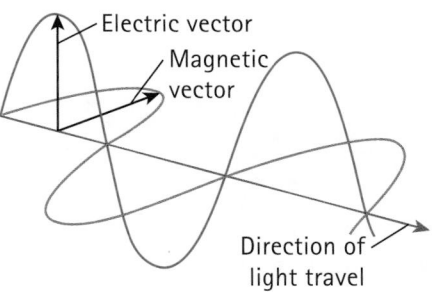

FIGURE C.10

However, the electric field vectors in waves of light from the Sun or from a lamp vibrate in all conceivable directions as they move. That is, they are *unpolarized* because the light waves are emitted from so many electrons vibrating in so many directions.

Now, light can be polarized, or filtered so that it consists only of electromagnetic waves with electric vectors vibrating all in the same plane. One type of polarizing filter is the material known as Polaroid, invented in the 1930s and now so widely used in sunglasses. Regular light incident upon a polarizing filter, such as the lenses in Polaroid sunglasses, emerges as polarized light. Think of a beam of unpolarized light coming straight toward you. Consider the electric field vector in that beam. Some of the possible directions of the vibrations are shown in Figure C.11. There are as many vectors in the horizontal direction as there are in the vertical direction since the light is unpolarized. The center sketch shows the light falling on a polarizing filter with its polarization axis vertically oriented. Only vertical components of the light pass through the filter, and the light that emerges is vertically polarized, as shown on the right.

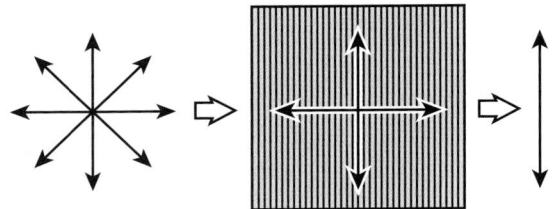

FIGURE C.11

Figure C.12 shows that no light can pass through a pair of Polaroid sunglasses when their axes are at right angles to one another, but some light does pass through when their axes are at a nonright angle. This fact can be understood with vectors and vector components.

Recall from Chapter 3 that any vector can be thought of as the sum of two components at right angles to each other. The two components are often chosen to be in the horizontal and vertical directions, but they can be in any two perpendicular directions. In fact, the number of sets of perpendicular components possible for any vector is infinite. A few of them are shown for the vector **V** in Figure C.13. In every case, components A and B make up the sides of a rectangle that has **V** as its diagonal.

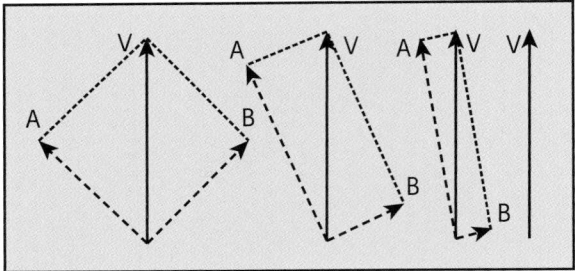

FIGURE C.13

You can see this somewhat differently by thinking of component **A** as always being vertical and **B** as being horizontal and picturing vector **V** as rotating instead (Figure C.14). This time the different orientations of **V** are superimposed on a polarizing filter with its polarization axis vertical. In the first sketch on the left, all of **V** gets through. As **V** rotates, only the vertical component **A** passes through, and it gets shorter and shorter until it is zero when **V** is completely horizontal.

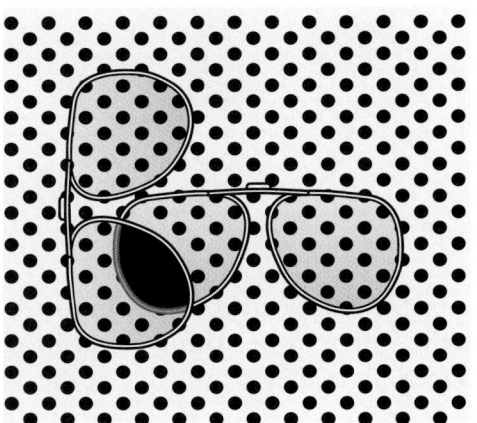
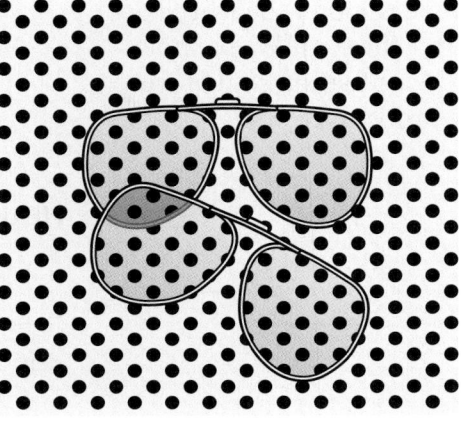

FIGURE C.12

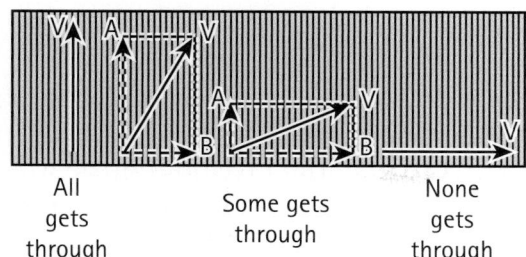

All gets through

Some gets through

None gets through

FIGURE C.14

Can you now understand how light gets through the second pair of sunglasses in Figure C.12? Look at Figure C.15, where for clarity, the two crossed lenses of Figure C.12, which are one atop the other, are

instead shown side by side. The vector **V** that emerges from the first lens is vertical. However, it has a component **A** in the direction of the polarization axis of the second lens. Component **A** passes through the second lens, while component **B** is absorbed.

To really appreciate this, you must play around with a couple of polarizing filters, which you can do in a lab exercise. Rotate one above the other and see how you can regulate the amount of light that gets through. Can you think of practical uses for such a system?

CONCEPT CHECK

As shown in Figure C.16, light is transmitted when the axes of the polarizing filters are aligned (left), but absorbed when they are at right angles to each other (center). Interestingly enough, when a third polarizing filter is sandwiched between the crossed filters (right), light is transmitted. Why?

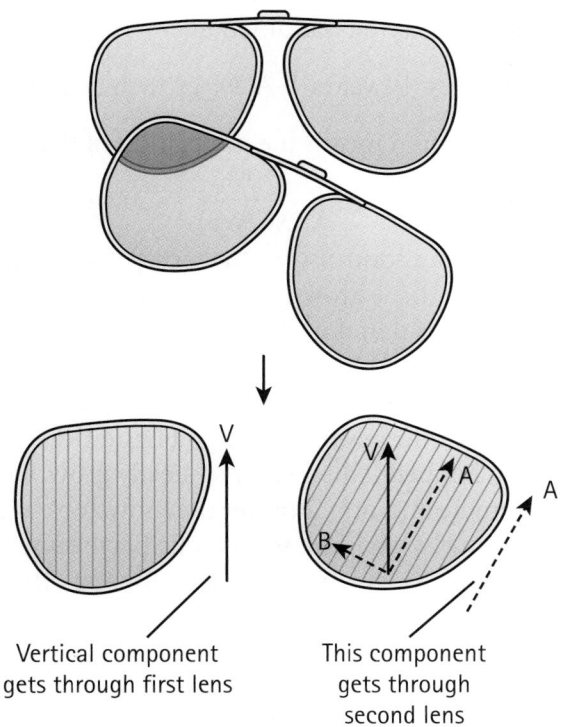

Vertical component gets through first lens

This component gets through second lens

FIGURE C.15

FIGURE C.16 Two polarizing filters at right angles transmit light when a third filter with a polarization axis at an intermediate angle is sandwiched between them. Why?

CHECK YOUR ANSWER

The explanation is shown in the following diagram.

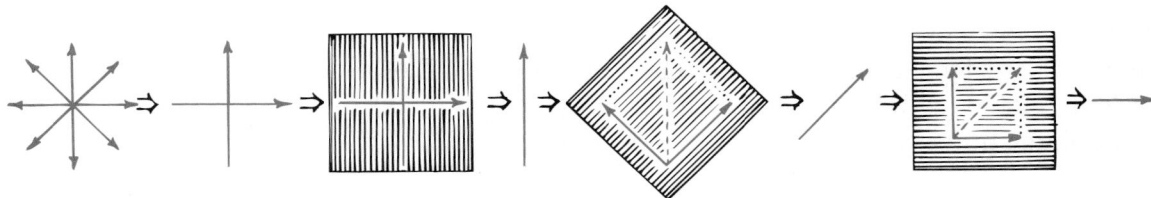

Try to fold a piece of paper in half and then fold it again upon itself successively 9 times. You won't be able to do it. It gets too thick for folding. Even if you could fold a fine piece of tissue paper upon itself 50 times, it would be more than 20 kilometers thick! The continual doubling of a quantity builds up exponentially. Double one penny 30 times, so that you can begin with one penny, then have two pennies, then four, and so on, and you'll accumulate a total of $10,737,418.23! One of the most important things we seem unable to perceive is the process of exponential growth. If we could, we could "see" why compound interest works the way it does, prices of goods rise the way they do, and populations and pollution proliferate out of control.

When a quantity such as money in the bank, population, or the rate of consumption of a resource steadily grows at a fixed percent per year, we say the growth is *exponential*. Money in the bank may grow at 4% per year; electric power generating capacity in the United States grew at about 7% per year for the first three-quarters of the twentieth century. The important thing about exponential growth is that the time required for the growing quantity to double in size (increase by 100%) is also constant. For example, if the population of a growing city takes 12 years to double from 10,000 to 20,000 inhabitants and its growth remains steady, in the next 12 years the population will double to 40,000 and in the next 10 years to 80,000 and so on.

When a quantity is *decreasing* at a rate that is proportional to its value, that quantity is undergoing *exponential decay*. As discussed in Chapters 10 and 26, radioactive elements are subject to radioactive decay, meaning that "parent" isotopes are transformed by nuclear processes to "daughter" isotopes (see Figure 10.15). The time required for an exponentially decaying quantity to be reduced to half of its initial value is called its half-life. Just as doubling time is constant for an exponentially

growing quantity, half-life is constant for an exponentially decaying quantity.

There is an important relationship between the percent growth rate and its *doubling time*, the time it takes to double a quantity:

$$\text{doubling time} = \frac{69.3}{\text{percent growth per unit time}} \approx \frac{\sim 70}{\%}$$

So to estimate the doubling time for a steadily growing quantity, we simply divide the number 70 by the percentage growth rate. For example, the 7% growth rate of electric power generating capacity in the United States means that in the past the capacity has doubled every 10 years $\left(\frac{70\%}{70\%/\text{year}} = 10 \text{ years}\right)$. A 2% growth rate for world population means the population of the world doubles every 35 years $\left(\frac{70\%}{2\%/\text{year}} = 35 \text{ years}\right)$.

A city planning commission that accepts what seems like a modest 3.5% growth rate may not realize that this means that doubling will occur in 70/3.5 or 20 years; that's double capacity for such things as water supply, sewage-treatment plants, and other municipal services every 20 years (Figure D.1).

What happens when you put steady growth in a finite environment? Consider the growth of bacteria that grow by division, so that one bacterium becomes two, the two divide to become four, the four divide to become eight, and so on. Suppose

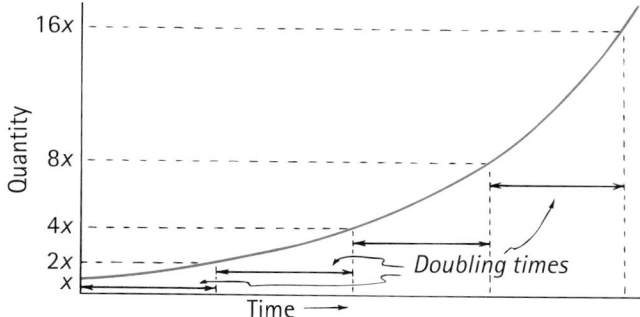

FIGURE D.1 An exponential curve. Notice that each of the successive equal time intervals noted on the horizontal scale corresponds to a doubling of the quantity indicated on the vertical scale. Such an interval is called the *doubling time*.

---

* This appendix is drawn from material by University of Colorado physics professor Albert A. Bartlett, who strongly asserts, "The greatest shortcoming of the human race is man's inability to understand the exponential function."

the division time for a certain strain of bacteria is 1 minute. This is then steady growth—the number of bacteria grows exponentially with a doubling time of 1 minute. Further, suppose that one bacterium is put in a bottle at 11:00 A.M. and that growth continues steadily until the bottle becomes full of bacteria at 12 noon. Consider seriously the following question.

CONCEPT CHECK
When was the bottle half full?

CHECK YOUR ANSWER
11:59 A.M.; the bacteria will double in number every minute!

It is startling to note that at 2 minutes before noon the bottle was only $\frac{1}{4}$ full. Table D.1 summarizes the amount of space left in the bottle in the last few minutes before noon. If you were an average bacterium in the bottle, at which time would you first realize that you were running out of space? For example, would you sense there was a serious problem at 11:55 A.M., when the bottle was only 3% filled, $\left(\frac{1}{32}\right)$ and had 97% of open space (just yearning for development)? The point

here is that there isn't much time between the moment that the effects of growth become noticeable and the time when they become overwhelming.

Suppose that at 11:58 A.M. some farsighted bacteria see that they are running out of space and launch a full-scale search for new bottles. Luckily, at 11.59 A.M. they discover three new empty bottles, three times as much space as they had ever known. This quadruples the total resource space ever known to the bacteria, for they now have a total of four bottles, whereas before the discovery they had only one. Further suppose that, thanks to their technological proficiency, they are able to migrate to their new habitats without difficulty. Surely, it seems to most of the bacteria that their problem is solved—and just in time.

CONCEPT CHECK
If the bacteria growth continues at the unchanged rate, what time will it be when the three bottles are filled to capacity?

CHECK YOUR ANSWER
12:02 P.M.!

We see from Table D.2 that quadrupling the resource extends the life of the resource by only two doubling times. In our example the resource is space—but it could as well be coal, oil, uranium, or any nonrenewable resource.

Continued growth and continued doubling lead to enormous numbers. In two doubling times, a quantity will double twice ($2^2 = 4$; quadruple) in size; in three doubling times, its size will increase eightfold ($2^3 = 8$); in four doubling times, it will increase sixteenfold ($2^4 = 16$); and so on. This is

TABLE D.1 The Last Minutes in the Bottle

Time Empty	Part Full (%)	Part
11:54 A.M.	$\frac{1}{64}$ (1.5%)	$\frac{63}{64}$
11:55 A.M.	$\frac{1}{32}$ (3%)	$\frac{31}{32}$
11:56 A.M.	$\frac{1}{16}$ (6%)	$\frac{15}{16}$
11:57 A.M.	$\frac{1}{8}$ (12%)	$\frac{7}{8}$
11:58 A.M.	$\frac{1}{4}$ (25%)	$\frac{3}{4}$
11:59 A.M.	$\frac{1}{2}$ (50%)	$\frac{1}{2}$
12:00 noon	full (100%)	none

TABLE D.2 Effects of the Discovery of Three New Bottles

Time	Effect
11:58 A.M.	Bottle 1 is $\frac{1}{4}$ full
11:59 A.M.	Bottle 1 is $\frac{1}{2}$ full
12:00 noon	Bottle 1 is full
12:01 P.M.	Bottles 1 and 2 are both full
12:02 P.M.	Bottles 1, 2, 3, and 4 are all full

best illustrated by the story of the court mathematician in India who years ago invented the game of chess for his king. The king was so pleased with the game that he offered to repay the mathematician, whose request seemed modest enough. The mathematician requested a single grain of wheat on the first square of the chessboard, two grains on the second square, four on the third square, and so on, doubling the number of grains on each succeeding square until all squares had been used (Figure D.2). At this rate there would be 2^{63} grains of wheat on the 64th square. The king soon saw that he could not fill this "modest" request, which amounted to more wheat than had been harvested in the entire history of Earth!

FIGURE D.2 A single grain of wheat placed on the first square of the chessboard is doubled on the second square, this number is doubled on the third, and so on, presumably for all 64 squares. Note that each square contains one more grain than all the preceding squares combined. Does enough wheat exist in the world to fill all 64 squares in this manner?

It is interesting and important to note that the number of grains on any square is 1 grain more than the total of all grains on the preceding squares. This is true anywhere on the board. Note from Table D.3 that when 8 grains are placed on the fourth square, the eight is 1 more than the total of 7 grains that were already on the board. Or the 32 grains placed on the sixth square is one more than the total of 31 grains that were already on the board. We see that in one doubling time we use more than all that had been used in all the preceding growth!

So if we speak of doubling energy consumption in the next however many years, bear in mind that this means in these years we will consume more energy than has heretofore been consumed during the entire preceding period of steady growth. And if power generation continues to use predominantly

TABLE D.3 Filling the Squares on the Chessboard

Square Grains Number	Total Grains on Square	Thus Far
1	1	1
2	2	3
3	4	7
4	8	15
5	16	31
6	32	63
7	64	127
.	.	.
.	.	.
.	.	.
64	2^{63}	$2^{64}-1$

fossil fuels, then except for some improvements in efficiency, we would burn up in the next doubling time a greater amount of coal, oil, and natural gas than has already been consumed by previous power generation, and except for improvements in pollution control, we can expect to discharge even more toxic wastes into the environment than the millions upon millions of tons already discharged over all the previous years of industrial civilization. We would also expect more human-made calories of heat to be absorbed by Earth's ecosystem than have been absorbed in the entire past! At the previous 7% annual growth rate in energy production, all this would occur in one doubling time of a single decade. If over the coming years the annual growth rate remains at half this value, 3.5%, then all this would take place in a doubling time of two decades. Clearly this cannot continue!

The consumption of a nonrenewable resource cannot grow exponentially for an indefinite period, because the resource is finite and its supply finally expires. The most drastic way this could happen is shown in Figure D.3a, where the rate of consumption, such as barrels of oil per year, is plotted against time, say in years. In such a graph the area under the curve represents the supply of the resource. We see that when the supply is exhausted, the consumption ceases altogether. This sudden change is rarely the case, for the rate of extracting the supply falls as it becomes more scarce. This is shown in Figure D.3b. Note that the area under the curve is equal to the area under the curve in (a). Why? Because the total supply is the same in both cases. The principal difference is the time taken to finally extinguish the

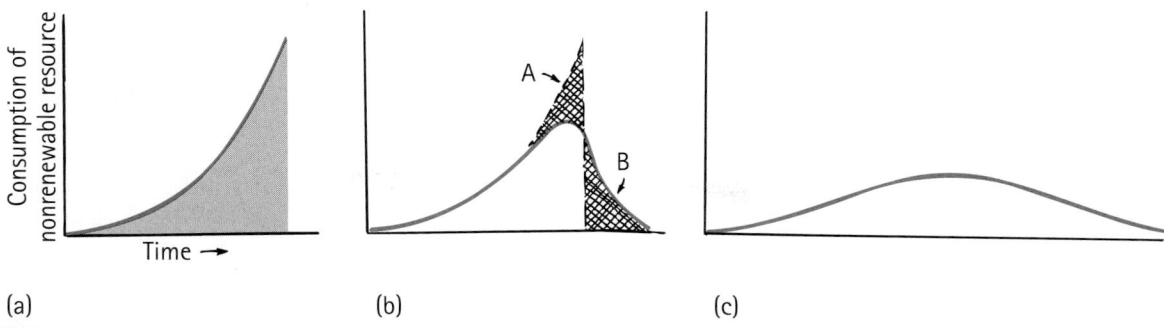

(a)   (b)   (c)

FIGURE D.3 If the exponential rate of consumption for a nonrenewable resource continues until it is depleted, consumption falls abruptly to zero. The shaded area under this curve represents the total supply of the resource. (b) In practice, the rate of consumption levels off and then falls less abruptly to zero. Note that the crosshatched area A is equal to the crosshatched area B. Why? (c) At lower consumption rates, the same resource lasts a longer time.

supply. History shows that the rate of production of a nonrenewable resource rises and falls in a nearly symmetric manner, as shown in Figure D.3c. The time during which production rates rise is approximately equal to the time during which these rates fall to zero or near zero.

Production rates for all nonrenewable resources decrease sooner or later. Only production rates for renewable resources, such as agriculture or forest products, can be maintained at steady levels for long periods of time (Figure D.4), provided such production does not depend on waning nonrenewable resources such as petroleum. Much of today's agriculture is so petroleum-dependent that it can be said that modern agriculture is simply the process whereby land is used to convert petroleum into food. The implications of petroleum scarcity go far beyond rationing of gasoline for cars or fuel oil for home heating.

The consequences of unchecked exponential growth are staggering. It is important to ask: Is growth really good? In answering this question, bear in mind that human growth is an early phase of life that continues normally through adolescence. Physical growth stops when physical maturity is reached. What do we say of growth that continues in the period of physical maturity? We say that such growth is obesity—or worse, cancer.

PROBLEMS

1. According to a French riddle, a lily pond starts with a single leaf. Each day the number of leaves doubles, until the pond is completely covered by leaves on the 30th day. On what day was the pond half-covered? One-quarter covered?

2. In an economy that has a steady inflation rate of 7% per year, in how many years does a dollar lose half its value?

3. At a steady inflation rate of 7%, what will be the price every 10 years for the next 50 years for a theater ticket that now costs $30? For a coat that now costs $300? For a car that now costs $30,000? For a home that now costs $300,000?

4. If the sewage treatment plant of a city is just adequate for the city's current population, how many sewage treatment plants will be necessary 42 years later if the city grows steadily at 5% annually?

5. If world population doubles in 40 years and world food production also doubles in 40 years, how many people then will be starving each year compared to now?

6. Suppose you get a prospective employer to agree to hire your services for wages of a single penny for the first day, 2 pennies for the second day, and double each day thereafter, providing the employer keeps to the agreement for a month. What will be your total wages for the month?

In the preceding exercise, how will your wages for only the 30th day compare to your total wages for the previous 29 days?

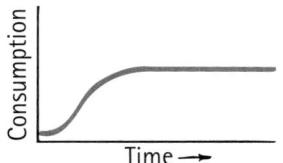

FIGURE D.4 A curve showing the rate of consumption of a renewable resource such as agricultural or forest products, where a steady rate of production and consumption can be maintained for a long period, provided this production is not dependent upon the use of a nonrenewable resource that is waning in supply.

Hopefully, you will have the opportunity to perform many of the explorations and other activities presented in this textbook. In performing these, you should always keep the safety of yourself and others in mind. Most safety rules that must be followed involve common sense. For example, if you are ever unsure of a procedure or chemical, ask your teacher or parent who should always be there to help you.

To guide you to safe practices, in this textbook you will find several types of icons posted by selected activities. Here are the icons and what they indicate:

 Wear approved safety goggles. Wear goggles when working with a chemical, solution, or when heating substances.

 Wear gloves. Wear gloves when working with chemicals.

Flame/heat. Keep combustible items, such as paper towels, away from any open flame. Handle hot objects with tongs, oven mitts, pot holders, or the like. Do not put your hands or face over any boiling liquid. Use only heat-proof glass, and never point a heated test tube or other container at anyone. Turn off the heat source when you are finished with it.

Here are some specific safety rules that should be practiced at all times:

1. Do not eat, drink, or smoke in the laboratory.
2. Maintain a clean and orderly work space. Clean up spills at once or ask for assistance in doing so.
3. Do not perform unauthorized experiments. First obtain permission from your teacher or parent. It is most important that others know what you are doing and when you are doing it. Think of it this way: If you are doing an experiment alone and something goes terribly wrong, there will be no one there to help you. That's not using your common sense.
4. Do not taste any chemicals or directly breathe any chemical vapors.
5. Check all chemical labels for both name and concentration.
6. Do not grasp recently heated glassware, clamps, or other heated equipment because they remain hot for quite a while.
7. Discard all excess reagents or products in the proper waste containers.
8. If your skin comes in contact with a chemical, rinse under cold water for at least 15 minutes.
9. Do not work with flammable solvents near an open flame.
10. Assume any chemical is hazardous if you are unsure.

Chemical equation A representation of a chemical reaction in which reactants are drawn before an arrow that points to the products.

Chemical formula A notation used to indicate the composition of a compound, consisting of the atomic symbols for the different elements of the compound and numerical subscripts indicating the ratio in which the atoms combine.

Chemical property A property that characterizes the ability of a substance to undergo a change that transforms it into a different substance.

Chemical reaction Synonymous with chemical change.

Chemical sediments Sediments that form by the precipitation of minerals from water on Earth's surface.

Chemotherapy The use of drugs to destroy pathogens without destroying the animal host.

Cleavage The tendency of a mineral to break along planes of weakness.

Clouds The condensation of water droplets above Earth's surface.

Combustion An exothermic oxidation–reduction reaction between a nonmetallic material and molecular oxygen.

Comet A body composed of ice and dust that orbits the Sun, usually in a very eccentric orbit, and that casts a luminous tail produced by solar radiation pressure when close to the Sun.

Complementary colors Any two colors that when added together produce white light.

Compound A material in which atoms of different elements are bonded to one another.

Concentration A quantitative measure of the amount of solute in a solution.

Condensation A transformation from a gas to a liquid.

Condensation polymer A polymer formed by the joining together of monomer units accompanied by the loss of small molecules, such as water.

Conduction The transfer of internal energy by molecular and electronic collisions within a substance (especially a solid).

Configuration A term used to describe how the atoms within a molecule are connected. For example, two structural isomers will consist of the same number and same kinds of atoms, but in different configurations.

Conformation One of a wide range of possible spatial orientations of a particular configuration.

Conservation of energy Energy cannot be created or destroyed; it may be transformed from one form into another, but the total amount of energy never changes. In an ideal machine, where no energy is transformed into heat, $work_{input}$ = $work_{output}$ and $(Fd)_{input} = (Fd)_{output}$.

Continental drift A hypothesis by Alfred Wegener that the world's continents are mobile and have moved to their present positions as the ancient supercontinent Pangaea broke apart.

Continental margin The boundary between the continents and the oceans.

Convection The transfer of internal energy in a gas or liquid by means of currents in the heated fluid. The fluid flows, carrying energy with it.

Convectional lifting An air-circulation pattern in which air warmed by the ground rises while cooler air aloft sinks.

Convergent plate boundary A plate boundary where tectonic plates move toward one another; an area of compressive stress where lithosphere is recycled into the mantle, or shortened by folding and faulting.

Converging lens A lens that is thicker in the middle than at the edges and refracts parallel rays passing through it to a focus.

Core The central layer of Earth's interior, divided into an outer liquid core and an inner solid core.

Coriolis force The apparent deflection from a straight-line path observed in any body moving near Earth's surface, caused by Earth's rotation.

Corrosion The deterioration of a metal, typically caused by atmospheric oxygen.

Cosmic background radiation The faint microwave radiation emanating from all directions that is the remnant heat of the Big Bang.

Coulomb The unit of electrical charge. It is equal in magnitude to the total charge of 6.25×10^{18} electrons.

Coulomb's law The electrical force between two charged bodies is directly proportional to the product of the charges and inversely proportional to the square of the distance between them:

$$F = k \frac{q_1 q_2}{d^2}$$

Covalent bond A chemical bond in which atoms are held together by their mutual attraction for two or more electrons they share.

Covalent compound An element or chemical compound in which atoms are held together by covalent bonds.

Critical mass The minimum mass of fissionable material in a reactor or nuclear bomb that will sustain a chain reaction.

Cross-cutting Where an igneous intrusion or fault cuts through other rocks, the intrusion or fault is younger than the rock it cuts.

Crust Earth's outermost layer.

Crystal form The outward expression of the orderly internal arrangement of atoms in a crystal.

Crystallization The growth of a solid from liquid or gas whose atoms come together in specific chemical proportions and crystalline arrangements.

Dark energy An unknown form of energy that appears to be causing an acceleration of the expansion of space. It is thought to be associated with the energy exuded by a perfect vacuum.

Dark matter Matter that responds only to weak nuclear and gravitational forces. This form of matter is invisible but reveals itself to us by its gravitational effects.

Delta An accumulation of sediments, commonly forming a triangular or fan-shaped plain, deposited where a stream flows into a standing body of water.

Density The amount of matter per unit volume:

$$\text{Density} = \frac{\text{mass}}{\text{volume}}$$

Weight density is expressed as weight per unit volume.

Deoxyribonucleic acid A nucleic acid containing a deoxygenated ribose sugar, having a double helical structure, and carrying genetic code in the nucleotide sequence.

Deposition The stage of sedimentary rock formation in which eroded particles come to rest.

Diffraction The bending of light as it passes around an obstacle or through a narrow slit, causing the light to spread and to produce light and dark fringes.

Diffuse reflection Reflection in irregular directions from an irregular surface.

Dipole A separation of charge that occurs in a chemical bond because of differences in the electronegativities of the bonded atoms.

Direct current (dc) Electrically charged particles flowing in one direction only.

Discharge The volume of water that passes a given location in a stream channel in a certain amount of time.

Dissolving The process of mixing a solute in a solvent to produce a homogeneous mixture.

Distillation A purifying process in which a vaporized substance is collected by exposing it to cooler temperatures over a receiving flask, which collects the condensed purified liquid.

Divergent plate boundary A plate boundary where lithospheric plates move away from one another (a spreading center); an area of tensional stress where new lithospheric crust is formed.

Diverging lens A lens that is thinner in the middle than at the edges, causing parallel rays passing through it to diverge as if from a point.

Doppler effect The change in frequency of wave motion resulting from motion of the wave source or receiver.

Drift A general term for all glacial deposits.

Earthquake The shaking or trembling of the ground that happens when rock under Earth's surface breaks.

Ecliptic The plane of Earth's orbit around the Sun. All major objects of the solar system orbit roughly within this same plane.

Efficiency The percent of the work put into a machine that is converted into useful work output.

Elastic collision A collision in which colliding objects rebound without lasting deformation or the generation of heat.

Electric current The flow of electric charge that transports energy from one place to another. Measured in amperes.

Electric power The rate of energy transfer, or rate of doing work. Measured by

$$\text{Power} = \text{current} \times \text{voltage}$$

Measured in watts (or kilowatts), where $1\,\text{A} \times 1\,\text{V} = 1\,\text{W}$.

Electrical resistance The property of a material that resists the flow of charged particles through it. Measured in ohms (Ω).

Electrically polarized Term applied to an atom or molecule in which the charges are aligned so that one side has a slight excess of positive charge and the other side a slight excess of negative charge.

Electrochemistry The branch of chemistry concerned with the relationship between electrical energy and chemical change.

Electrolysis The use of electrical energy to produce chemical change.

Electromagnet A magnet whose field is produced by an electric current. Electromagnets are usually in the form of a wire coil with a piece of iron inside the coil.

Electromagnetic induction A magnetic field is induced in any region of space in which an electric field is changing with time. The magnitude of the induced magnetic field is proportional to the rate at which the electric field changes.

Electromagnetic spectrum The range of electromagnetic waves extending in frequency from radio waves to gamma rays.

Electromagnetic wave A wave emitted by vibrating electrical charges (often electrons) and composed of vibrating electric and magnetic fields that regenerate one another.

Electron An extremely small, negatively charged subatomic particle found outside the atomic nucleus.

Electron-dot structure A shorthand notation of the shell model of the atom, in which valence electrons are shown around an atomic symbol.

Electronegativity The ability of an atom to attract a bonding pair of electrons to itself when bonded to another atom.

Element Any material that is made up of only one type of atom.

Elemental formula A notation that uses the atomic symbol and (sometimes) a numerical subscript to denote how atoms of the element are bonded together.

Ellipse The oval path followed by a satellite. The sum of the distances from any point on the path to two points called foci is a constant. When the foci are together at one point, the ellipse is a circle. As the foci get farther apart, the path gets more "eccentric."

Endothermic A term that describes a chemical reaction in which there is a net absorption of energy.

Energy The property of a system that enables it to do work.

Equilibrium Rule $\Sigma F = 0$.

Erosion The wearing away of rocks and the processes by which rock particles are transported by water, wind, or ice.

Escape speed The speed that a projectile, space probe, or similar object must reach to escape the gravitational influence of the Earth or celestial body to which it is attracted.

Ester An organic molecule containing a carbonyl group, the carbon of which is bonded to one carbon atom and one oxygen atom bonded to another carbon atom.

Ether An organic molecule containing an oxygen atom bonded to two carbon atoms.

Evaporation A transformation from a liquid to a gas.

Event horizon The boundary region of a black hole from which no radiation may escape. Any events within the event horizon are invisible to distant observers.

Exosphere The fifth atmospheric layer above Earth's surface, extending from the thermosphere upward and out into interplanetary space.

Exothermic A term that describes a chemical reaction in which there is a net release of energy.

Fact A phenomenon about which competent observers can agree.

Faraday's law The induction of voltage when a magnetic field changes with time. If the magnetic field within a closed loop changes in any way, a voltage is induced in the loop:

$$\text{Voltage induced} \sim \text{number of loops} \times \frac{\text{magnetic field change}}{\text{time}}$$

This is a statement of Faraday's law. The induction of voltage is the result of a more fundamental phenomenon: the induction of an electric *field*.

Fat A biomolecule that packs a lot of energy per gram and consists of a glycerol unit attached to three fatty acid molecules.

Fault A fracture along which movement of rock on one side relative to rock on the other has occurred.

Faunal succession Fossil organisms succeed one another in a definite, irreversible, and determinable order.

Floodplain A wide plain of almost flat land on either side of a stream channel. Submerged during flood stage, the plain is built up by sediments discharged during floods.

Force A push or a pull.

Force pair The action and reaction pair of forces that occur in an interaction.

Forced vibration The setting up of vibrations in an object by a vibrating force.

Fracture A break that does not occur along a plane of weakness.

Free fall Motion under the influence of gravitational pull only.

Freezing A transformation from a liquid to a solid.

Frequency For a vibrating body or medium, the number of vibrations per unit time. For a wave, the number of crests that pass a particular point per unit time.

Friction The resistive force that opposes the motion or attempted motion of an object past another with which it is in contact, or through a fluid.

Front The contact zone between two different air masses.

Frontal lifting The lifting of one air mass by another as two air masses converge.

Full Moon The phase of the Moon when its sunlit side faces Earth.

Functional group A specific combination of atoms that behaves as a unit in an organic molecule.

Galaxy A large assemblage of stars, interstellar gas, and dust held together by gravity, usually categorized by its shape: elliptical, spiral, or irregular.

Gamma rays High-frequency electromagnetic radiation emitted by the nuclei of radioactive atoms.

Gas Matter that has neither a definite volume nor a definite shape, always filling any space available to it.

Generator An electromagnetic induction device that produces electric current by rotating a coil within a stationary magnetic field. A generator converts mechanical energy to electrical energy.

Giant stars Cool giant stars above main-sequence stars on the H–R diagram.

Glacier A large mass of ice formed by the compaction and recrystallization of snow, which is able to move downslope under its own weight.

Gradient The vertical drop in the elevation of a stream channel divided by the horizontal distance for that drop; the steepness of the slope.

Greenhouse effect Warming caused by shortwavelength radiant energy from the Sun that easily enters the atmosphere and is absorbed by Earth. This energy is then reradiated at longer wavelengths that cannot easily escape Earth's atmosphere.

Groundwater Underground water in the saturated zone.

Group A vertical column in the periodic table, also known as a family of elements.

Gyre A circular or spiral whirl pattern, usually referring to very large current systems in the open ocean.

H–R diagram (Hertzsprung–Russell diagram) A plot of intrinsic brightness versus surface temperature for stars. When so plotted, stars' positions take the form of a main sequence for average stars, with exotic stars above or below the main sequence.

Half-life The time required for half the atoms in a sample of a radioactive isotope to decay.

Heat The thermal energy that flows from a substance of higher temperature to a substance of lower temperature, commonly measured in calories or joules.

Heat of fusion The amount of energy needed to change any substance from solid to liquid, which is the same amount of energy released when the substance transforms from a liquid to a solid.

Heat of vaporization The amount of energy required to change any substance from liquid to gas, which is the same amount of energy released when the substance transforms from a gas to a liquid.

Hertz The SI unit of frequency. It equals one vibration per second.

Heteroatom Any atom other than carbon or hydrogen in an organic molecule.

Heterogeneous mixture A mixture in which the various components can be seen as individual substances.

Homogeneous mixture A mixture in which the components are so finely mixed that the composition is the same throughout.

Humidity A measure of the concentration or amount of water vapor in the air—the mass of water vapor per volume of air.

Hydrocarbon A chemical compound containing only carbon and hydrogen atoms.

Hydrogen bond A strong dipole–dipole attraction between a slightly positive hydrogen atom on one molecule and a pair of nonbonding electrons on another molecule.

Hydrologic cycle The natural circulation of all states of water from ocean to atmosphere to land, back to ocean.

Hydronium ion A water molecule after accepting a hydrogen ion.

Hydroxide ion A water molecule after losing a hydrogen ion.

Hypothesis An educated guess; a reasonable explanation that is not fully accepted as factual until tested over and over again by experiment.

Igneous rocks Rocks formed by the cooling and crystallization of hot, molten rock material called magma (or lava).

Impulse The product of the force acting on an object and the time during which it acts. In an interaction, impulses are equal and opposite.

Impulse-momentum relationship Impulse is equal to the change in the momentum of the object that the impulse acts on. In symbol notation, $Ft = \Delta mv$.

Impure In chemistry, this term refers to a material that is a mixture of more than one element or compound.

Inclusion Any inclusion (pieces of one rock type contained within another) is older than the rock containing it.

Induced dipole A dipole temporarily created in an otherwise nonpolar molecule, induced by a neighboring charge.

Inelastic collision A collision in which the colliding objects become distorted, generate heat, and possibly join together.

Inertia The property of things remaining at rest if at rest, and in motion if in motion.

Inner planets The four planets orbiting within 2 AU of the Sun, including Mercury, Venus, Earth, and Mars. These planets are all rocky planets, known as the *terrestrial* planets.

Insoluble Not capable of dissolving to any appreciable extent in a given solvent.

Interaction Mutual action between objects where each object exerts an equal and opposite force on the other.

Interference The pattern formed by superposition of different sets of waves that produces mutual reinforcement in some places and cancellation in others.

Inverse-square law A law relating the intensity of an effect to the inverse square of the distance from the cause:

$$\text{Intensity} \sim \frac{1}{\text{distance}^2}$$

Inversely When two values change in opposite directions, so that if one increases and the other decreases by the same amount, they are said to be inversely proportional to each other.

Ion An electrically charged particle created when an atom either loses or gains one or more electrons.

Ionic bond A chemical bond in which an attractive electric force holds ions of opposite charge together.

Ionic compound Any chemical compound containing ions.

Ionosphere An electrified region within the thermosphere and uppermost mesosphere where fairly large concentrations of ions and free electrons exist.

Isotopes Any member of a set of atoms of the same element whose nuclei contain the same number of protons but different numbers of neutrons.

Ketone An organic molecule containing a carbonyl group, the carbon of which is bonded to two carbon atoms.

Kilogram The fundamental SI unit of mass.

Kinetic energy Energy of motion, described by the relationship kinetic energy $= 1/2\ mv^2$.

Kuiper belt (Pronounced KI-pur) The disk-shaped region of the sky beyond Neptune populated by many icy bodies and a source of short-period comets.

Laminar flow Water flowing smoothly and fairly slowly in straight lines with no mixing of sediment.

Lateral continuity Sedimentary layers are deposited in all directions over large areas until some sort of obstruction, or barrier, limits their deposition.

Lava Molten magma that moves upward from inside Earth and flows onto the surface. The term *lava* refers both to the molten rock and to the solid rocks that form from it.

Law A general hypothesis or statement about the relationship of natural quantities that has been tested over and over again and has not been contradicted. Also known as a *principle*.

Law of conservation of momentum When no external net force acts on an object or a system of objects, no change of momentum takes place. Hence, the momentum before an event involving only internal forces is equal to the momentum after the event: $mv_{(before\ event)} = mv_{(after\ event)}$.

Law of mass conservation Matter is neither created nor destroyed during a chemical reaction—atoms merely rearrange, without any apparent loss or gain of mass, to form new molecules.

Law of reflection The angle of a reflection equals the angle of incidence. The reflected and incident rays lie in a plane that is normal to the reflecting surface.

Law of universal gravitation Every body in the universe attracts every other body with a mutually attracting force. For two bodies, this force is directly proportional to the product of their masses and inversely proportional to the square of the distance separating them:

$$F \sim \frac{m_1 m_2}{d^2}$$

Light-year The distance light travels in 1 year.

Lipid A broad class of biomolecules that are not soluble in water.

Liquid Matter that has a definite volume but no definite shape, assuming the shape of its container.

Lithosphere The entire crust plus the rigid portion of the mantle that is situated above the asthenosphere.

Local Group Our immediate cluster of galaxies, including the Milky Way, Andromeda, and Triangulum spiral galaxies plus a few dozen smaller elliptical and irregular galaxies.

Local Supercluster A cluster of galactic clusters in which our Local Group resides.

Lock-and-key model A model that explains how drugs interact with receptor sites.

Longitudinal wave A wave in which the medium vibrates in a direction parallel (longitudinal) to the direction in which the wave travels. Sound is an example.

Love waves Surface waves that move in a side-to-side, whiplike motion.

Lunar eclipse The phenomenon whereby the shadow of Earth falls upon the Moon, producing the relative darkness of the full Moon.

Luster The appearance of a mineral's surface when it reflects light.

Machine A device such as a lever or pulley that increases (or decreases) a force or simply changes the direction of a force.

Magma Molten rock in Earth's interior.

Magnetic domains Clustered regions of aligned magnetic atoms. When these regions are aligned with one another, the substance containing them is a magnet.

Magnetic field The region of magnetic influence around either a magnetic pole or a moving charged particle.

Magnetic force (1) Between magnets, it is the attraction of unlike magnetic poles for each other and the repulsion between like magnetic poles. (2) Between a magnetic field and a moving charged particle, it is a deflecting force due to the motion of the particle. It is perpendicular to the velocity of the particle and perpendicular to the magnetic field lines. Also, it is greatest when the particle moves perpendicular to the field lines and zero when the particle moves parallel to the field lines.

Main sequence The diagonal band of stars on an H–R diagram; such stars generate energy by fusing hydrogen to helium.

Mantle The middle layer in Earth's interior, between the crust and the core.

Mass The quantity of matter in an object. More specifically, it is the measure of the inertia or sluggishness that an object exhibits in response to any effort made to start it, stop it, deflect it, or change in any way its state of motion.

Mass number The total number of protons and neutrons within an isotope.

Maxwell's counterpart to Faraday's law An electric field is induced in any region of space in which a magnetic field is changing with time. The magnitude of the induced electric field is proportional to the rate at which the magnetic field changes.

Melting A transformation from a solid to a liquid.

Mesosphere The third atmospheric layer above Earth's surface, extending from the top of the stratosphere to 80 km.

Mesozoic The time of middle life, from 245 million years ago to about 65 million years ago.

Metabolism The general term describing all chemical reactions in the body.

Metallic bond A chemical bond in which positively charged metal ions are held together within a "fluid."

Metamorphic rocks Rocks formed from preexisting rocks that have been changed or transformed by high temperature, high pressure, or both.

Metamorphism The changes in rock that happen as physical and chemical conditions change.

Meteor The streak of light produced by a meteoroid burning in Earth's atmosphere; a "shooting star."

Meteorite A meteoroid, or a part of a meteoroid, that has survived passage through Earth's atmosphere to reach the ground.

Meteoroid A small rock in interplanetary space, which can include a fragment of an asteroid or comet.

Mineral A naturally formed crystalline solid, composed of an ordered arrangement of atoms with a specific chemical composition.

Mixture A combination of two or more substances in which each substance retains its properties.

Mohorovičić discontinuity (Moho) The crust–mantle boundary. This marks one of the depths where the speed of P-waves traveling through Earth increases.

Mohs scale of hardness A ranking of a mineral's hardness, which is its resistance to scratching.

Molarity A unit of concentration equal to the number of moles of a solute per liter of solution.

Mole The amount of any pure substance that contains as many atoms, molecules, ions, or other elementary units as the number of atoms in 12 grams of carbon-12. This is equal to 6.02×10^{23} particles.

Molecule A group of atoms held tightly together by covalent bonds.

Momentum The product of the mass of an object and its velocity.

Monomers The small molecular units from which a polymer is formed.

Moon phases The cycles of change of the "face" of the Moon, changing from *new,* to *waxing,* to *full,* to *waning,* and back to *new.*

Motor A device employing a current-carrying coil that is forced to rotate in a magnetic field. A motor converts electrical energy to mechanical energy.

Natural frequency A frequency at which an elastic object naturally tends to vibrate, so that minimum energy is required to produce a forced vibration or to continue vibrating at that frequency.

Neap tide A tide that occurs when the Moon is midway between new and full, in either direction. The pulls of the Moon and Sun are perpendicular to one another, so the solar and lunar tides do not overlap. This makes high tides not as high and low tides not as low.

Nebular theory The idea that the Sun and planets formed together from a cloud of gas and dust, a *nebula.*

Net force The combination of all forces that act on an object.

Neuron A specialized cell capable of receiving and sending electrical impulses.

Neurotransmitter An organic compound capable of activating receptor sites on proteins embedded in the membrane of a neuron.

Neutral solution A solution in which the hydronium-ion concentration is equal to the hydroxide-ion concentration.

Neutralization A reaction in which an acid and base combine to form a salt.

Neutron An electrically neutral subatomic particle of the atomic nucleus.

Neutron star A small, extremely dense star composed of tightly packed neutrons formed by the welding of protons and electrons.

New moon The phase of the moon when darkness covers the side facing Earth.

Newton The scientific unit of force.

Newton's First Law of Motion Every object continues in a state of rest, or in a state of motion in a straight line at constant speed, unless a net force acts on it.

Newton's second law The acceleration produced by a net force on an object is directly proportional to the net force, is in the same direction as the net force, and is inversely proportional to the mass of the object.

Newton's third law Whenever one object exerts a force on a second object, the second object exerts an equal and opposite force on the first. Or put another way, "To every action there is always an opposed equal reaction."

Nonbonding pairs Two paired valence electrons that tend not to participate in a chemical bond.

Nonpolar Said of a chemical bond that has no dipole.

Nonsilicate A mineral that does not contain silica (silicon + oxygen).

Nova An event wherein a white dwarf suddenly brightens and appears as a "new" star.

Nuclear fusion The combining of nuclei of light atoms to form heavier nuclei, with the release of much energy.

Nucleic acid A long polymeric chain of nucleotide monomers.

Nucleon Any subatomic particle found in the atomic nucleus; another name for either proton or neutron.

Nucleotide A nucleic acid monomer consisting of three parts: a nitrogenous base, a ribose sugar, and an ionic phosphate group.

Observable universe All that we are able to see given the fact that the universe is only about 14 billion years old. The observable universe is one small sector of the universe as a whole.

Ohm's law The statement that the current in a circuit varies in direct proportion to the voltage across the circuit and inversely with the circuit's resistance:

$$\text{Current} = \frac{\text{voltage}}{\text{resistance}}$$

Oort cloud The region beyond the Kuiper belt populated by trillions of icy bodies and a source of long-period comets.

Opaque The term applied to materials through which light cannot pass.

Ordinary matter Matter that responds to the strong nuclear, weak nuclear, electromagnetic, and gravitational forces. This is matter made of protons, neutrons, and electrons, which includes the atoms and molecules that make us and our immediate environment.

Ore A geologic deposit containing relatively high concentrations of one or more metal-containing compounds.

Organic chemistry The study of carbon-containing compounds.

Original horizontality Layers of sediment are deposited evenly, with each new layer laid down nearly horizontally over the older sediment.

Orographic lifting The lifting of an air mass over a topographic barrier such as a mountain.

Outer planets The four planets orbiting beyond 2 AU of the Sun including Jupiter, Saturn, Uranus, and Neptune—all gaseous and known as the *jovian* planets.

Oxidation The process whereby a reactant loses one or more electrons.

pH A measure of the acidity of a solution, equal to the negative of the base-10 logarithm of the hydronium-ion concentration.

Paleomagnetism The natural, ancient magnetization in a rock that can be used to determine the polarity of Earth's magnetic field and the rock's location of formation.

Paleozoic The time of ancient life, from 543 million years ago to 245 million years ago.

Pangaea A single large landmass—a supercontinent—that existed in the geologic past and was composed of all the present-day continents.

Parabola The curved path followed by a projectile near the Earth under the influence of gravity only.

Parallel circuit An electric circuit in which electrical devices are connected so that the same voltage acts across each one and any single one completes the circuit independently of all the others.

Pascal's principle A change in pressure at any point in an enclosed fluid at rest is transmitted undiminished to all points in the fluid.

Period The time required for a vibration or a wave to make a complete cycle; equal to 1/frequency.

Period A horizontal row in the periodic table.

Periodic table A chart in which all known elements are listed in order of atomic number.

Permeability A measure of the ability of a porous rock or sediment to permit fluid to flow.

Photoelectric effect The emission of electrons from a metal surface when light shines on it.

Photon A particle of light, or the basic packet of electromagnetic radiation.

Physical change A change in which a substance changes one or more of its physical properties without transforming it into a new substance.

Physical dependence A dependence characterized by the need to continue taking a drug to avoid withdrawal symptoms.

Physical property Any physical attribute of a substance, such as color, density, or hardness.

Pitch The "highness" or "lowness" of a tone, as on a musical scale, which is mainly governed by frequency.

Planetary nebula An expanding shell of gas ejected from a low-mass star during the latter stages of its evolution.

Planets The major bodies orbiting the Sun that are massive enough for their gravity to make them spherical and small enough to avoid having nuclear fusion in their cores. They also have successfully cleared all debris from their orbital paths.

Plate tectonics The theory that Earth's lithosphere is broken into pieces (plates) that move over the asthenosphere; boundaries between plates are where most earthquakes and volcanoes occur and where lithosphere is created and recycled.

Plutoid A relatively large icy body, such as Pluto, originating within the Kuiper belt.

Plutonic rock Intrusive igneous rock formed from magma that cools beneath Earth's surface. Granite is a plutonic rock.

Polar Said of a chemical bond that has a dipole.

Polarization The alignment of the transverse electric vectors that make up electromagnetic radiation.

Polymer A long organic molecule made of many repeating units.

Porosity The volume of open pore space in rock or sediment compared to the total volume of solids plus open pore space.

Potential energy The stored energy that a body possesses because of its position.

Power The time rate of doing work: Power = work/time.

Precambrian time The time of hidden life, which began about 4.5 billion years ago when Earth formed, lasted until about 543 million years ago (beginning of Paleozoic), and makes up almost 90% of Earth's history.

Precipitate A solute that has come out of solution.

Pressure The ratio of force to the area over which that force is distributed:

$$\text{Pressure} = \frac{\text{force}}{\text{area}}$$

Liquid pressure = weight density × depth

Pressure-gradient force The force that moves air from a region of high-pressure air to an adjacent region of low-pressure air.

Primary wave (P-wave) A longitudinal body wave that compresses and expands the material through which it moves; it travels through solids, liquids, and gases and is the fastest seismic wave.

Principle of flotation A floating object displaces a weight of fluid equal to its own weight.

Products The new materials formed in a chemical reaction.

Projectile Any object that moves through the air or through space under the influence of gravity.

Protein A polymer of amino acids, also known as a polypeptide.

Proton A positively charged subatomic particle of the atomic nucleus.

Pseudoscience Fake science that has no tests for its validity.

Psychoactive Said of a drug that affects the mind or behavior.

Psychological dependence A deep-rooted craving for a drug.

Pure Having a uniform composition, or being without impurities. In chemistry, the term is used to denote a material that consists of a single element or compound.

Radiation The transfer of energy by means of electromagnetic waves.

Radioactivity The process whereby unstable atomic nuclei break down and emit radiation.

Radiometric dating A method for calculating the age of geologic materials based on the nuclear decay of naturally occurring radioactive isotopes.

Rayleigh waves Waves that move in an up-and-down motion along the Earth's surface.

Reactants The reacting substances in a chemical reaction.

Reaction rate A measure of how quickly the concentration of products in a chemical reaction increases or the concentration of reactants decreases.

Real image An image formed by light rays that converge at the location of the image—which can be displayed on a screen.

Reduction The process whereby a reactant gains one or more electrons.

Reflection The return of light into the medium from which it came.

Refraction The bending of a wave through either a nonuniform medium or from one medium to another, caused by differences in wave speed.

Relative dating The ordering of rocks in sequence by their comparative ages.

Relative humidity The amount of water vapor in the air at a given temperature expressed as a percentage of the maximum amount of water vapor the air can hold at that temperature.

Resonance The response of a body when a forcing frequency matches its natural frequency.

Ribonucleic acid A nucleic acid containing a fully oxygenated ribose sugar.

Rift (rift valley) A long, narrow gap that forms as a result of two plates diverging.

Rock An aggregate of minerals. Some rocks are aggregates of fossil shell fragments, solid organic matter, or any combination of these components.

Rock cycle A sequence of events involving the formation, destruction, alteration, and re-formation of rocks as a result of the generation and movement of magma; the weathering, erosion, transportation, and deposition of sediment; and the metamorphism of preexisting rocks.

Saccharide Another term for carbohydrate. The prefixes *mono-*, *di-*, and *poly-* are used before this term to indicate the length of the carbohydrate.

Salinity The mass of salts dissolved in 1000 g of seawater.

Salt An ionic compound formed from the reaction between an acid and a base.

Sand dune A landform created when airflow is blocked by an obstacle, slowing air speed and therefore promoting the deposition of airborne sand.

Satellite A projectile or small body that orbits a larger body.

Saturated hydrocarbon A hydrocarbon containing no multiple covalent bonds, with each carbon atom bonded to four other atoms.

Saturated solution A solution containing the maximum amount of solute that will dissolve in its solvent.

Science Organized common sense. Also, the collective findings of humans about nature and a process of gathering and organizing knowledge about nature.

Scientific method An orderly method for gaining, organizing, and applying new knowledge.

Seafloor spreading The moving apart of two oceanic plates at a rift in the seafloor.

Secondary wave (S-wave) A transverse body wave that vibrates the material through which it moves side to side or up and down; an S-wave cannot travel through liquids and so does not travel through Earth's outer core.

Sedimentary rocks Rocks formed from the accumulation of weathered material (sediments) that has been eroded by water, wind, or ice.

Sedimentation The stage of sedimentary rock formation in which deposited sediments accumulate and change into sedimentary rock through the processes of compaction and, usually, cementation.

Series circuit An electric circuit in which electrical devices are connected so that the same electric current exists in all of them.

Shell A region of space around the atomic nucleus in which electrons may reside.

Shock wave The cone-shaped wave created by an object moving at supersonic speed through a fluid.

Silicate A mineral that contains both silicon and oxygen and (usually) other elements in its chemical composition; silicates are the largest and most common rock-forming mineral group.

Solar eclipse The phenomenon whereby the shadow of the Moon falls upon Earth, producing a region of darkness in the daytime.

Solid Matter that has a definite volume and a definite shape.

Solubility The ability of a solute to dissolve in a given solvent.

Soluble Capable of dissolving to an appreciable extent in a given solvent.

Solute Any component in a solution that is not the solvent.

Solution A homogeneous mixture in which all components are in the same phase.

Solvent The component in a solution that is present in the largest amount.

Sonic boom The loud sound resulting from the incidence of a shock wave.

Sound wave A longitudinal vibratory disturbance that travels in a medium, which a young ear can hear in the approximate frequency range 20–20,000 Hz.

Specific heat capacity The quantity of heat required to raise the temperature per unit mass of a substance by 1 degree Celsius.

Speed The distance traveled per time.

Spring tide A high or low tide that occurs when the Sun, Earth, and Moon are aligned so that the tides due to the Sun and Moon coincide, making the tides higher or lower than average. Occurs during the full Moon or new Moon.

Standing wave A stationary wave pattern formed in a medium when two sets of identical waves pass through the medium in opposite directions.

Starburst galaxy A galaxy in which stars are forming at an unusually high rate.

Stratosphere The second atmospheric layer above Earth's surface, extending from the top of the troposphere up to 50 km. This is where stratospheric ozone forms.

Streak The name given to the color of a mineral in its powdered form.

Structural isomers Molecules that have the same molecular formula but different chemical structures.

Subduction The process in which one tectonic plate bends and descends beneath another plate at a convergent boundary.

Submicroscopic Refers to the realm of atoms and molecules, which is a realm so small that we are unable to observe it directly with optical microscopes.

Subtractive primary colors The three colors of absorbing pigments—magenta, yellow, and cyan—that when mixed in certain proportions reflect any other color in the visible-light part of the electromagnetic spectrum.

Sunspots Temporary, relatively cool and dark regions on the Sun's surface.

Supernova The explosion of a massive star caused by gravitational collapse with the emission of enormous quantities of matter.

Superposition In an undeformed sequence of sedimentary rocks, each bed or layer is older than the one above and younger than the one below.

Support Force The force that supports an object against gravity.

Surface wave A type of seismic wave that travels along Earth's surface.

Suspension A homogeneous mixture in which the various components are in different phases.

Synaptic cleft A narrow gap across which neurotransmitters pass either from one neuron to the next or from a neuron to a muscle or gland.

Syncline A down-fold in rock with relatively young rocks at the fold core; rock age increases with horizontal distance from the fold core.

Synergistic effect One drug enhancing the effect of another.

Tangential velocity Velocity that is parallel (tangent) to a curved path.

Technology Method and means of solving practical problems by applying the findings of science.

Tectonic plates Sections into which Earth's crust is broken up; they move in response to heat flow and convection in Earth's interior.

Temperature A measure of the hotness or coldness of substances, related to the average kinetic energy per molecule in a substance, measured in degrees Celsius, or in degrees Fahrenheit, or in kelvins.

Temperature inversion A condition in which the upper regions of the troposphere are warmer than the lower regions.

Terminal speed The speed at which the acceleration of a falling object terminates because air resistance balances its weight.

Terminal velocity Terminal speed with direction of motion (down for falling objects).

Terrestrial radiation The radiant energy emitted by Earth.

Theory A synthesis of a large body of information that encompasses well-tested hypotheses about certain aspects of the natural world.

Thermal energy The total energy (kinetic plus potential) of the submicroscopic particles that make up a substance.

Thermal expansion The expansion of a substance due to increased molecular motion in that substance.

Thermonuclear fusion Nuclear fusion produced by high temperature.

Thermosphere The fourth atmospheric layer above Earth's surface, extending from the top of the mesosphere to 500 km.

Transform plate boundary A plate boundary where two plates are sliding horizontally past each other, without appreciable vertical movement.

Transmutation The conversion of an atomic nucleus of one element into an atomic nucleus of another element through a loss or gain in the number of protons.

Transparent The term applied to materials through which light can pass in straight lines.

Transverse wave A wave in which the medium vibrates in a direction perpendicular (transverse) to the direction in which the wave travels. Light is an example.

Troposphere The atmospheric layer closest to Earth's surface. This layer contains 90% of the atmosphere's mass and essentially all of its water vapor and clouds. Weather occurs in the troposphere.

Turbulent flow Water flowing rapidly and erratically in a jumbled manner, stirring up everything it touches.

Unconformity A break or gap in the geologic record, caused by erosion of preexisting rock or by an interruption in the sequence of deposition.

Universal gravitational constant, G The proportionality constant in Newton's law of universal gravitation.

Unsaturated hydrocarbon A hydrocarbon containing at least one multiple covalent bond.

Unsaturated solution A solution that is capable of dissolving additional solute.

Valence electron An electron located in the outermost occupied shell in an atom that can participate in chemical bonding.

Vector quantity A quantity that specifies direction as well as magnitude.

Velocity The speed of an object and specification of its direction of motion.

Virtual image An image formed by light rays that do not converge at the location on the image. Cannot be displayed on a screen.

Vitamin Organic chemicals that assist in various biochemical reactions in the body and can be obtained only from food.

Volcanic rocks Extrusive igneous rocks formed by the eruption of molten rock at Earth's surface. Basalt is a volcanic rock.

Volcano A central vent through which lava, gases, and ash erupt and flow.

Voltage A form of "electrical pressure":

$$\text{Voltage} = \frac{\text{potential energy}}{\text{charge}}$$

Volume The quantity of space an object occupies.

Water table The upper boundary of the saturated zone, below which every pore space is completely filled with water.

Watt The unit of power, the joule per second.

Wave A disturbance or vibration propagated from point to point in a medium or in space.

Wave speed The speed with which waves pass a particular point:

$$\text{Wave speed} = \text{frequency} \times \text{wavelength.}$$

Wavelength The distance between successive crests, troughs, or identical parts of a wave.

Weathering Disintegration and/or decomposition of rock at or near Earth's surface.

Weight The force due to gravity on an object.

White dwarf A dying star that has collapsed to the size of Earth and is slowly cooling off; located at the lower left on the H–R diagram.

Work The product of the force and the distance through which the force moves: $W = Fd$.

Work–energy theorem The work done on an object is equal to the energy gained by the object. Work $= \Delta E$. The energy change can be PE or KE.

PHOTO CREDITS

1: NASA Goddard Laboratory for Atmospheres
5, clockwise from top left: Lennard Lessin, Peter Arnold Inc.; Ray Nelson/Phototake; Paul G. Hewitt; Francois Gohier/Photo Researchers; David Parker/Photo Researchers
15: Paul G. Hewitt III
17: Paul G. Hewitt
18: Photos.com
20: The Granger Collection, NY
21, top: Rick Lucas, Collection of Paul Hewitt
21, bottom: Peter Austin/iStock Photo
23: Photos.com
26: Paul G. Hewitt
37: Leslie A. Hewitt
50, top: John Dalton/Photo Researchers
50, bottom: Fundamental Photographs
57: Paul G. Hewitt
68: Henry R. Fox/Animals Animals/Earth Scenes
69: John Suchocki
71: Paul G. Hewitt
74: R. Llewellyn/Superstock
75: Craig Aurness/Corbis
76: Palm Press
77: Glen Allison/Stone/Getty
78: Paul G. Hewitt
79, top right: Paul G. Hewitt
79, bottom left: Paul G. Hewitt
84: Paul G. Hewitt
90: Peter Bowater/Alamy
90: NASA/John F. Kennedy Space Center
94: Paul G. Hewitt
96: Paul G. Hewitt
97, left and right: Paul G. Hewitt
110: Paul G. Hewitt
112: NASA
128: Paul G. Hewitt
137: Paul G. Hewitt
140: Paul G. Hewitt
141: Paul G. Hewitt
145, left: Paul G. Hewitt
145, right: Paul G. Hewitt
148: The Granger Collection
151: Paul G. Hewitt
154: Construction Photography.com
157: Paul G. Hewitt
165: Paul G. Hewitt
167: Tracy Suchocki
168, top left: Paul G. Hewitt
168, top center: AKG Images
169: Paul G. Hewitt
176: Wide World Photos
177, top left: LU Engineers
177, top right: Meidor Hu/Collection of Paul G. Hewitt
178: Nuridsany & Perennou/Photo Researchers
181, middle right: Paul G. Hewitt
181, bottom: Don Hynek
183: Paul G. Hewitt
185: David J. Green/Alamy
186: Paul G. Hewitt
192: Steve Hunt/The Image Bank
198: Pearson Science
199: Zig Leszczynski/Animals Animals/Earth Sciences
200: Lightfoot, Collection of Paul Hewitt
204: Pearson Science
206: Pearson Science
207: Pearson Science
216: Bob Abrams
218: Richard Megna/Fundamental Photographs
219, left and right: Richard Megna/Fundamental Photographs
221: Paul G. Hewitt

222, top left: Richard Megna/Fundamental Photographs
222, middle left: Richard Megna/Fundamental Photographs
222, bottom left: Richard Megna/Fundamental Photographs
222, bottom right: John Suchocki
223: AP Wide World Photos
225, left: Pearson Science
225, right: Pearson Science
235: Paul G. Hewitt
238: Brenda Tharp/Jupiter Images
240: Paul G. Hewitt
242: Newscom
243, middle right: Leslie A. Hewitt
243, bottom: Eniko Balogh/iStock Photo
246, left: AP Wide World Photos
246, middle: AP Wide World Photos
246, right: AP Wide World Photos
247, left: Educational Development Center
247, right: Richard Megna/Fundamental Photographs
248: Norman Synnestvedt/Paul G. Hewitt
253: Collection of Paul G. Hewitt
257: Paul G. Hewitt
260: Fred R. Myers Jr.
264: Paul G. Hewitt
265: Paul G. Hewitt
267: Paul G. Hewitt
269: David Nunuk/Photo Researchers
270: Institute of Paper
272: Ted Mahieu
273: Robert Greenler
274: Meidor Hu/Collection of Paul G. Hewitt
277: Paul G. Hewitt
278: John Suchocki
279: Paul G. Hewitt
280, top left: Meidor Hu/Collection of Paul G. Hewitt
280, bottom right: Stephen Mcsweeny/Shutterstock
282, middle left: Takahiro Miyamoto/Getty Images
282, bottom: Image Bank/Getty Images Inc.
285, top right: Paul G. Hewitt
285, middle right: Paul G. Hewitt
285, bottom right: Paul G. Hewitt
287: Paul G. Hewitt
290: Paul G. Hewitt
293: Maksym Gorpenyuk/iStock Photo
296: Collection of Paul G. Hewitt
299, top right: Education Development Center
299, middle right (two photos): Education Development Center
299, bottom right: Ken Kay/Fundamental Photographs, NYC
304: Paul G. Hewitt
307: Diane Schiumo/Fundamental Photographs, NYC
308, left: Paul G. Hewitt
308, middle: Paul G. Hewitt
308, right: Paul G. Hewitt
311: Albert Rose
315: European Communities, 1995–2006
318, left: Peter Arnold
318, middle: IBM Corporate Archives
318, right: IBM Corporate Archives
319: Photo Researchers
321, left: Rachel Epstein/SKA/Stuart Kenter Associates
321, middle: Rachel Epstein/PhotoEdit Inc.
321, right: Tony Freeman/PhotoEdit Inc.
328: Bettmann/Corbis
331: Paul G. Hewitt
334: Joe Sohm/Chromoshom/The Stock Connection
336: Larry Mulvehill/Photo Researchers
338, top: International Atomic Energy Agency
338, bottom: Richard Megna/Fundamental Photographs
342, left: Jerry Nulk and Sarah Joshua Baker
342, middle: Paul G. Hewitt

INDEX